The Primordial Language

Confirmation of the Divine Creator

by
Anita Meyer

Presented before you is the one book that combines and engulfs Religion / Spirituality / Science, which gives the reader the ultimate evidence that G-d exists. A feat that almost all have failed to prove beyond any spiritual perspective. It is the holy grail of all grails!!! There is nobody on the face of the Earth that has the full encompassing knowledge that this book contains. The reader might be thinking that this is a similar story that they've heard many times before. Nevertheless, this one IS DIFFERENT, I can assure you!

The Primordial Language

Confirmation of the Divine Creator

by
Anita Meyer

Published By
In Search Of The Universal Truth Publisher
Denver, Colorado

The Primordial Language - Confirmation of the Divine Creator
by Anita Meyer

Cover Graphic Design by Mike Salituro

First Edition
First Printing December 2009

ISBN 13: 978-1-61500-002-9
ISBN 10: 1-61500-002-X

Library Congress Control Number: 2009939397

Publisher info:
In Search Of The Universal Truth Publisher
8824 E. Florida Ave. Unit G-13
Denver, CO. 80247 USA

www.insearchoftheuniversaltruthpublisher.com

Published in
Denver, Colorado, USA

Contents:

My special dedication and thanks:

I would like to give special thanks to my loving parents, Judy and Stewart, though unsuspectingly, you both were necessary components to instill within me the beliefs that I carry today, starting from a young age to set into motion the forthcoming of this book.

To my charismatic lifelong sweetheart and husband Curt, in the making of this book - five years and 500+ pages later - you are the one (the flowing fountain) who has elevated and challenged me in my thinking. Some might say you were the ghostwriter, but you are really the ghost rider!

To Joshua my alter ego son, you were the conduit by which I was first impelled and propelled to start writing this book. Although you are a slow work in progress, I know you are going to eventually do great things - G-d has His hand on you! In addition, to Rachael my fun-loving and charming daughter, beauty, smartness and talent are within you… there has never been a step you have walked alone - nor will there ever be.

I would also like to give special thanks to Michael J. Kiser (and staff) the book publisher of "In Search Of The Universal Truth" (ISOTUT) who made the publishing of this book possible. Thank You!

Everyone mentioned were necessary threads woven into the fabric of my life, and for that I give G-d great praise! Knowing that great works are happening (according to His will), not a day goes by that I can live without His all consuming passion inside me!

INTRODUCTION...

This book is an interlocking mesh between (1) religion, (2) spirituality and (3) science. Whereupon reading this book one may be taken aback to learn and gain the knowledge that the three are very much interrelated.

Not only are the three interrelated, but this book goes beyond this unity to present beyond-belief evidence that confirms as well as illustrates a creator in action. Some may not make the connection to various things I talk about in this book, and therefore simply loose interest. While others might find it as ground breaking evidence that leads to an elevated or higher-minded aspect of life. But I hope that for many it will be an uplifting and unstoppable book from start to finish and and worth keeping or passing onto another person. It is for those who are on a quest for ageless answers - but surely I tell you, it is for ALL walks of life, young and old, Jew, Christian, Muslim to name a few, even the unbeliever (the atheist), as well as the scientifically minded person - to find meaning in life and to strive for something much more divine and significant.

In the same breath, it is also an inward journey in pursuit of the source of ones being.

I anticipate that there will of course be some who will oppose my religious approach as they read. However, ultimately (in the end), some will appreciate it as heralding the most important central (and highest) secret knowledge that directly links us to G-d and the universe BY WAY OF A "PRIMORDIAL LANGUAGE" that has intentionally guised itself and eluded us so that the material aspect of all life should not have its existence nullified. In truth, it is the mechanism that is essential to the whole of creation. In revealing this, I will let the evidence speak for itself as the scientific point of view is melded with the religious aspect.

This knowledge is not for the dim at heart. Whichever it may be for you, I ask you to please read the entirety of this book before you can make a worthy assessment of it. Contained within, one may obtain a knowledge that will miraculously transform the soul in ways that might not at first be perceived, but will later discover it as a gift that helps us navigate this thing called life. It is a story that has never been told in such a way before that its impact may be a life altering experience! It is a story that's filled with such wisdom (information, facts and data) that goes beyond any/all texts books, leaving no question answered inadequately, insufficiently or unsatisfactorily. Its physical truth is so magnificent, awe-inspiring and overwhelming, that soon all the words will fade, the parchment will crumble (figuratively), and this book will drop until all that's left is the vaporous remains of spiritual completeness.

I have also written this book with the critics in mind, as well as those that are impassioned by twisting and spinning the words (interpretations and terminology) of others. In spite of this, these people all happen to be my best allies. Additionally as anticipated and intended, there are those that will blaspheme this book with any bad exposure or publicity they can muster up. Either way, ANY PUBLICITY IS GOOD PUBLICITY as there is no such thing as bad publicity. As the old saying goes: Beware the trap of publicity for publicity's sake! After all, even a bad review gives a bit of write-up in which readers might become intrigued and curious about. In effect, its alluring gravity will invite humankinds innate curiosity, which ultimately in due course, leads virtually to the undeniable reality of G-d.

Additionally, I am not writing this book to quote facts, or to cite scientific references. Though I may use some in this book, they are used merely to express my opinion. You, the reader can of course have your own opinions as you continue to read. I of course fall short of being any kind of great zircon in the diadems of teaching or literature aside from the great Bible itself. It was also not my intention to write this book in a glitzy or showy way using fancy words befuddled in an eye-candy of words. Neither is this book written with any perfect skillfulness such as in proper sentence structuring, or even correct usage of grammar - as that it not its true purpose. Additionally, I am not trying to win an argument here either. In the end, all I truly want is to have finally expressed all the life force and knowledge that is within me to be passed unto someone else. I have no other mission in life as important as this.

Yes, you have picked up a book to read about "religion", but not just any ordinary book, but one that integrates religion, science and spirituality and exposes the raw truth. A subject that few really hit home base on.

I invite and challenge ALL to embark upon this unequivocal journey of truth.

1 - MEDIATOR AND MODERATOR...

To begin with, I must make known that I didn't officially write this book, but that of an energy working through me did - guiding me along step by step. Whether you wish to call it G-d, source, spirit or Holy Spirit, this energy writes all the books, delivers all the speeches, paints all the pictures, and miraculously it also has the ability to correct (set straight) all things in its path (as you will find out). It is this source that I'd like to identify with as the Holy Spirit (denoted as the third person of the Trinity in Christianity) that allows the words to flow through me, through my mind and into this book. In effect when any of us are in thought, or inspired by something, we are essentially in spirit transcending all limitations of our consciousness. Following this thought, I soon realized that there was something greater (of a higher purpose) going on here. I was not only in the act of creating this great book, but fulfilling the divine purpose.

Just as well, my whole purpose in writing this book is to reveal this divine source. I have been compelled to write it, and have been made a mouthpiece for G-d. Exodus 4:12 - Now therefore go, and I will be with thy mouth, and teach thee what thou shalt say. However let me clarify that I am in no way a prophet, I am only adding to the understanding of the Bible and to bring to light very strong evidence of G-d's existence for all those whose lights are dimmed or badly lit. Let me also clarify that there are no more apostles or prophets. The last book to be written was the book of Revelation written by the apostle John. John was the last when he said: Revelation 22:18 - I warn everyone who hears the words of the prophecy of this book: If anyone adds anything to them, G-d will add to him the plagues described in this book. There are no more authentic books to be written and included in the Bible simply because everything that G-d has to say has already been said and recorded. There is no more revelation, and thus no more sign miracles either happening anywhere in the world to confirm any revelation because it's all FINISHED! The main goal of the prophets and apostles has been achieved. This also includes the great task that Jesus the Messiah came here to do has also been accomplished. We are now just waiting for prophesy to completely fulfill itself.

And to know that whatever else this story has done, (after you have finished reading it) that there is a deeper matter as well, a higher purpose that I, the writer vis-à-vis a divine source almost certainly intended. This deeper matter is that I have been given an in-depth understanding that illuminates G-d. And this must be shared for it is not for me alone. If I do not share it, I am the selfish looser. Psalm 89:1 - I will sing of the Lord's great love forever; with my mouth I will make your faithfulness known through all generations. To be truly effective, the objective is to question all that I write and then go on the search for answers yourself - feeling a powerful pull of truth afterwards (though I make that search effortless for you). You will begin to ponder more and more on such thoughts until the truth finally sets in.

In effect, everything you are subjected to in life (including this book) may all of a sudden start making sense to you. "Coincidence" my friends is not a kosher word - meaning that everything has a purpose in the makers works.

Before we continue reading, I would also like to point out that contained within the pages of this book you will come across many captioned segments highlighted in red (like observed above). These are either quotes from famous people, or Biblical references straight out of the Bible, and others are simply remarks of explanation.

You will notice that the Bible speaks in parables. Some people find them cloaked and hard to follow, but the parables of the Bible are very important, as you will come to find out. They are not merely just short blind alley statements. They point out the relationship between two or more allegories, which convey a higher spiritual truth. These truths are signified as comparisons between two subjects, issues or topics and are used as an illustrative device (a tool employed by G-d). Thus, a parable may contain one main word, which is used to mirror (reflect) another singular concept. An example of a parable comparison can be found in the "Old Testament" with its allegory backed up in the "New Testament". The one you are about to read happens to be of a prophetic nature, which as alleged conveys a higher spiritual truth. Found in the "Old Testament" book of Psalms an expression and quality of G-d is noted. Psalm 78:2 - I will open my mouth in parables, I will utter hidden things, things from of old. The Parable comparison to Psalms 78:2 is found in the "New Testament" spoken by Jesus and recorded by the disciple Matthew. Matthew 13:35 - **That it might be fulfilled** which was spoken by the prophet, saying, I will open my mouth in parables; I will utter things which have been kept secret from the foundation of the world.

The spiritual truth of this parable comparison is that Jesus is matching Himself up with G-d. Keep reading to find out if this comparison (as well as others) rings true. Proverbs 1:5 - Let the wise listen and add to their learning, and let the discerning (discriminating) get guidance.

Additionally, I would like all who do not believe in G-d or Jesus (atheists) including those who believe in another religion, deity, evolution or science to open their minds to what I am putting forth. Most of what I write about is based off the Old Hebrew Testament and the New Testament. The reason for this is because (as you will come to find out) the Old Hebrew Testament, which was based off the "ten commandments", happened to be written by the FINGER OF G-D HIMSELF (a supernatural being) and given to Moses on Mount Sinai. Exodus 31:18 - When the Lord finished speaking to Moses on Mount Sinai, he gave him the two tablets of the Testimony, the tablets of stone **inscribed by the finger of G-d.** Accompanying this, starting from the book of Genesis and finishing off at Deuteronomy is the oldest (subsequent to the flood) documented evidence that was written by Moses when taking dictation from G-d. In addition, the New Testament, which happens to be based off the Old Hebrew Testament as we briefly witnessed with the parable comparison above. It will be demonstrated that the Bible is not simply just meaningless words. Its original text, which is written in Hebrew, bares evidence of the supernatural... and this is what you will come to find out as you continue to read on. If you are in unbelief to what you are reading at the present moment, or if you are one of those people that need immediate proof especially before you continue to read any further, then by all means you may skip directly to chapter 8 (precisely pages 155-161) of this book to view how these Hebrew letters bare witness to G-d's testimony along with all of His creation, handiwork and design which is seen in ALL of the natural world, but be aware of the factor that there is a fruition to fully understanding that needs to come beforehand.

Additionally as you continue to read a phenomenon will also reveal itself that the Bible, both Old and New Testament bares witness of prophetic (predictive) prophesy. These statements consisting of what happened in the past and statements that will happen/transpire in the future.

Moreover, (I cannot stress enough) as you continue to read, do not be overconfident of your own intellect (or understanding) as to ridicule or mock the nature of religion or G-d. I think Albert Einstein said it best: We should take care not to make the intellect our G-d; it has, of course, powerful muscles, but no personality - Albert Einstein.

I guess you might also be wondering who I am. I am someone that believes in the literal word of the Bible (which is of G-d), both Old and New Testament. Within this book, I will attempt to demonstrate beyond a reasonable doubt that not one single word of the Bible is in error. I will also make it obvious that the Bible is unquestionably divinely inspired. Additionally, I cannot count how many times I have heard the argument that there is no scientific proof that G-d exists. And it is because of statements like this that have also compelled me to write this book. I am therefore on the rebound with such a posing question as this - that there is indeed very strong scientific evidence of G-d's existence available to all of us, and we shall examine a great deal of them. I will start out by telling you, one of these proofs is evidence of a divine event that was witnessed firsthand by 3 million Jews when they heard and felt G-d personally speaking to them at Mount Sinai in the desert. 3 million people - this is an entire nation! Such an event like this is called a "National Revelation". G-d did not just appear to Moses in a private rendezvous, He appeared to EVERYONE, some 3 million people. This claim is mentioned many times in the Torah. (Moses told the Israelites): Deuteronomy 4:9-13 - Only beware for yourself and greatly beware for your soul, lest you forget the things that your eyes have beheld. Do not remove this memory from your heart all the days of your life. Teach your children and your children's children about the day that you stood before the Lord your G-d at Horev (Mount Sinai). G-d spoke to you from the midst of the fire, you were hearing the sound of words, but you were not seeing a form, only a sound. He told you of His covenant, instructing you to keep the Ten Commandments, and He inscribed them on two stone tablets. Deuteronomy 4:32-36 - You have been shown in order to know that G-d, He is the Supreme Being. There is none besides Him. From Heaven, He let you hear His voice in order to teach you, on earth, He showed you His great fire, and you heard His words amid the fire. Deuteronomy 5:1-4 - Moses called all of Israel and said to them: Hear, O Israel, the decrees and the ordinances that I speak in your ears today - learn them, and be careful to perform them. The Lord your G-d sealed a covenant with us at Horev (Mount Sinai). Not with our forefathers did G-d seal this covenant, but with us - we who are here, all of us alive today. Face to face did G-d speak with you on the mountain from amid the fire.

How do we actually know that it was indeed approximately 3 million people that witnessed G-d speaking? Well as it happens, this valuable piece of information is scrupulously recorded in the Bible as well. Exodus 12:37 - The children of Israel journeyed from Rameses to Succoth, **about six hundred thousand on foot**, the men, besides the young children. Numbers 1:1-3 - The Lord spoke to Moses in the Sinai Desert, in the Tent of Meeting on the first day of the second month, in the second year after the exodus from the land of Egypt, saying. Take the sum of all the congregation of the children of Israel, by families following their fathers' houses; a head count of every male according to the number of their names. From twenty years old and upward, all who are fit to go out to the army in Israel, you shall count them by their legions you and Aaron. Numbers 1:46-47 - The sum of all those who were counted: **six hundred and three thousand, five hundred and fifty.** But the Levites, according to their father's tribe were not numbered among them. Numbers 3:43 (regarding the Levites) - And all the firstborn males by the number of names, from a month old and upward, of those that were numbered of them, were **twenty, two thousand two hundred and threescore, and thirteen.** Now if you take the number 600,000 (or more accurately 603,550) and add Levites (which in NUMBER 3, is placed at about 22,000) and add women and children - a conservative estimate of one child per family, would net about 2 million people. In reality, it was most probably over 3 million.

So, we have the Torah that claims that the ENTIRE JEWISH NATION heard and felt G-d speak at Sinai, an assertion that has been accepted as part of their nation's history for over 3,000 years. Moreover, this central defining event makes Judaism different from every other religion in the world. Though some people continue to refute this by saying that this event never actually happened. They claim it was only a mythical saga written down and recorded some time afterwards by the writers of the Bible as a ploy to keep believers. However, in this book I give you a precious and indispensable piece of information that not only connects this main event, but proves beyond a reasonable doubt that this incident did actually happen. The proof for this (as you will come to find out in a future chapter) has to do with the actual letters of the Hebrew alphabet that were written with the finger of G-d onto the stone tablets (that are mentioned in the Bible). Deuteronomy 9:10 - The Lord gave me two stone tablets inscribed by the finger of G-d. On them were all the commandments the Lord proclaimed to you on the mountain out of the fire, on the day of the assembly. These tablets with the Hebrew letters were then given to Moses and thus passed down to the Hebrew people in what is now the conventional Hebrew Torah. These were the same Hebrew letters that the Jews had carried down for generation upon generation without even slightly altering any letters or leaving even the smallest letter out (this was a prerequisite). If they had altered any letter in any way, or left one out, it would have spelled destruction for the entire Universe since the Torah is a blueprint of creation itself. (When you delve into the science of the Bible Code which will be addressed in a future chapter, you will then begin to realize why). It is these stone tablets that included the Hebrew alphabet (which we have come to know today) that offers us a phenomenal piece of evidence (opening up at center stage) because it allows for verification and confirmation (both religiously and scientifically). And even though we were not ALL there to witness this great event of G-d speaking to us, the proof is still there and continues to remain open for all to view. Please keep reading to fully understand this, and to witness (view) the marvelous display of these divinely written Hebrew letters. All must know!

The next question you might be asking is why do I hyphenate the word G-d? The reason for this can be found In Genesis 1:3 were we read that G-d said, let there be light and there was light. Genesis 1:3 - And G-d said, Let there be light: and there was light. From this passage alone, we learn that there is great power in "words". Then at Mount Sinai when G-d is giving the 10 Commandments to MOSES in EXODUS 20:7, He gave The Third Commandments in Exodus 20:7, G-d gave the third commandment - you shall not take the name of the Lord your G-d in vain. Exodus 20:7 - Thou shalt not take the name of the Lord thy G-d in vain; for the Lord will not hold him, guiltless that taketh his name in vain. For that reason, by leaving the "o" out of the word G-d it makes His name unpronounceable. Not only unpronounceable, but we should constantly be reminded of the power and holiness of G-d's name (after all He is the author of life, which you will shortly come to find out). Thus, by placing a hyphen in the middle of G-d's name shows a certain amount of respect for Him. I am merely just trying to remind myself and those who read the things I write, to focus on the proper use of G-d's name, and to give Him all the glory due His name.

2 - RELIGION VERSES EVOLUTION...

This chapter is a book in itself.

I've got a lot of work to do in this book (starting here), since we live in a world that has programmed us to think that the Bible is old and outdated and not a valid source, when in fact (as it happens) the Bible remains fresh off the press as today's newspaper (implying that things that are happening now in the world seem to foreshadow earlier period prophesy). To prove this will require substantial support of evidence as well as a whole lot of reprogramming. For now, let me start with explaining why reading the Bible is of immense importance. It is not merely for religious reasons, but for common everyday knowledge that pertains to the modern world. Even some of the best-read people are illiterate concerning the Bible. In fact, I'm surprised that students aren't required to read portion of the Bible in high school and college, the way they are required to read Shakespeare or the U.S. Constitution. However, it is in my opinion that everyone should read it, whether religious or not. Let me explain why. One cannot get through a single chapter of the Bible without encountering a common saying, phrase, a name, a character, or an idea that has come down to us approximately 3,000 years later. From the coining of phrases and metaphors such as "the writing on the wall" (from the book of Daniel - Daniel 5:1-30), to the famous words of Martin Luther King's "I have a Dream" speech (from the books of Amos and Isaiah - Amos 5:24 and Isaiah 40:4). Many of these types of things have left a lasting imprint on our world. The Bible is also the leading source of everything from military plots to the very issues of morality that to this day shape our politics, law and justice system. So if you want to know where/how such things came about, you can be sure to find the answers in the Bible.

Moving forth... As you know, this chapter is about religion verses evolution. If you go to the library or bookstore, you might find numerous books claiming to provide the answers. Many people even spend hours researching on the Internet seeking-out and madly hunting such information. It is often found though that one book contradicts the other, while others seem valid and convincing at the moment but soon become outdated and are either revised or replaced. I have spent many hours (actually years) researching and compiling a multitude of information myself, and I put before you hopefully the one book that contains ALL the answers, leaving no stone unturned. However I am in no way claiming to know-it-all, nor am I boasting. None of us should be engrossed by the illusion of "finality" as there is plenty of ongoing research in this arena. One must continue to read if they truly and sincerely desire a well-rounded view. Additionally, one must study the atheists view, the skeptic and doubter views, the scientific view along with the religious views, even the bizarre, strange and nutty, not just to merely have balance, but to find truth. The more one reads, the more educated and knowledgeable they become. Reading is to the mind what exercise is to the body - Richard Steele

For the first phase in this book, I'm going to talk about that one subject that everyone seems to label as the opium of the mass - religion! In addition, to let you know, by far this is not the only subject that I'm going to be focusing on, but it is the strongest of them all. And it is specifically intended and designed for the atheist at heart who suppresses all G-dly intent and inspiration, and additionally for those who already have faith but need more strength in believing. If I have written this chapter correctly, hopefully at the end of this book everyone will be able to confidently make this discriminate determination for themselves (strongly ruling in favor of a supernatural being). Now let me just quickly say that many people are under the impression that religion is the broad and easy path to take (by just believing, or having faith in G-d). Rather it is the other way around. Religion is by far not the easiest route to take, for the Bible itself tells us that narrow is the door to Heaven. And all fall short of the glory of G-d. Matthew 7:13 - Enter through the narrow gate. For wide is the gate and broad is the road that leads to destruction, and many enter through it. Matthew 7:14 - But small is the gate and narrow the road that leads to life, and only a few find it. Luke 13:24 - He said to them, Make every effort to enter through the narrow door, because many, I tell you, will try to enter and will not be able to. Romans 3:23 - For all have sinned and fall short of the glory of G-d. Now, those who frown on the subject of religion or the Bible along with those who may have prejudice views against it are truthfully not being intellectually honest or fair as to completely or even partially rule it out. If you are one of these people, you may not have completely read the entirety of the Bible or may have simply interpreted its words wrongly, or you most likely just cannot relate to it, and for what ever reasons just flat-out don't understand it. You may even be a person who feels that they don't need religion or feel the need for a G-d (this may be because religion teaches the cure often without the greater understanding of the disease). You may also be a person who for what ever reasons just views life as material (matter) and not anything of the supernatural - seeing the world in a way that suggests that life is a purely natural phenomenon which is material and dependent upon the chemical workings of the brain (or the natural world). Therefore one may come to the conclusion that there is no good reason to even bother believing. Such people simply do not believe that life is much more than just the flesh and matter we see around us. For these reasons, I realize that I must touch on all bases and be as completely thorough as I can. This is not going to be an easy task, but I will attempt to give it my all. But remember it is not my will alone that is acting here. Something is working inside of me which is much more bigger than me.

So here, we go... breathing in and letting everything out.

I'd like to first point out religions worst enemy in this era of science, which is the "Evolution Theory".

This theory purposes that the Earth is billions of years old (approximately 4.6 billion years old). This large span of time involves the development of life on the Earth and is a process by which organisms pass on traits from generation to generation (in the attempts) of explaining the origin of new species which are related to each other through a common ancestor.

This is the evolutionists purposed timeline:

3.8 billion years of simple cells
3 billion years of photosynthesis
2 billion years of complex cells
1 billion years of multi-cellular life
600 million years of simple animals
570 million years of arthropods (insects)
550 million years of complex animals
500 million years of fish and amphibians
475 million years of land plants
400 million years of insects and seeds
360 million years of amphibians
300 million years of reptiles
200 million years of mammals
150 million years of birds
130 million years of flowers
65 million years dinosaurs died out
2.5 million years the appearance of the modern human
200,000 humans started looking like they do today
25,000 Neanderthals died out

Since this relatively recent theory came forward, it seems that religion is slipping ever so slyly away, or rather should I say being questioned as technology and science progresses. However the statistics tell us that religious convictions are still strongly prevalent in the world with Christianity ranking as the number 1 most prominent and preferred religion in the United States and steadily increasing in leaps and bounds with an overall outreach across the entire world. But the war is waging in the scientific community and the new generations of children as well as a large percentage of elders that are now being directed and taught in schools that "evolution" is a correct methodology and as a means to how we came about verses G-d creating us. But please bear in mind that the evolution theory is only a relatively recent belief. The misconception about evolution is that it is a science, but in actuality, it is not a science at all - it is only a theory and a philosophy. This is because (to date) there is no real substantial evidence of any kind WHATSOEVER of evolution ever happening in the past, or currently transpiring in the present! In addition, even though it is only a theory and philosophy, it has somehow managed to work its way into the minds of many people causing them to completely abandon the concept of G-d. It has also left many believers making exceptions or compromising with the idea that perhaps G-d only played a small part by seeding us, and from here, evolution drove us to what we are today. Nevertheless, let me give you the lowdown: This whole "evolutionary view" is in downright OPPOSITION to science as well as a creator.

So, what has so strongly persuaded some scientists along with the board of education (around the world) to take up the "evolutionary" idea and concept? Well for starters, scientists cannot seem to wrap their minds around the concept of a supernatural entity that instantaneously created life. They question where is the observable proof for this? Since they don't currently see, this actual proof they believe that life created itself in the duration of extremely long periods of time and did not need a supernatural creator. Yet, simultaneously, scientists undeniably and invariably know (without fail) that living things cannot come about from non-living things. Life coming about from non-living things would require non-living things to naturally turn into living organisms all by themselves - which is scientifically impossible! Yet, they stubbornly look for another explanation so that they will not have to except this self-evident fact and can rule out the supernatural - in effect, the "evolution theory" was born. As a result, they are producing only theories, which are of no scientific value because they do not provide any ideas of how new genetic information is produced. But it doesn't seem to matter that scientists cannot come up with how new genetics are produced... all that seems to matter (even though there is no demonstrational evidence) is that evolution is the only "other explanation" aside from supernatural creation (that's stated in the Bible). Therefore, it is their conclusion that the only possible scientific "alternative" aside from supernatural creation for how living things came about is by evolutionary means. Again, the evolutionary view is completely in disagreement with what the Bibles tell us (and we will delve into just why).

The scientific evolutionists tell us that evolution is widely accepted as the "only explanation" for how living things came about, and is now a fact. However, if indeed this scientific estimation is correct and based on facts, where are the rigorous repeated, demonstrated and observable(ed) tests which science is based on? I bet you really believe there are such tests. In certainty there are no tests to determine or prove that evolution took place (or is taking place at present) because it was/is the unseen, and science cannot test what is not seen by our human means, (it leans on assumptions). Not even in the most sophisticated laboratories today (where test-after-tests are being performed) are we witnessing any sort of evolution taking place (however, blindingly or rather imaginably in the eyes of the evolutionists they believe we are). But it is for these "unobservable" reasons that the evolution theory lies in the realm of the mysterious and therefore it can only be believed, and under those grounds is based on faith (the substance of things not seen, but believed). Rather, what we do find evidence for, (even though it too is/was the unseen) happens to be groundbreaking evidence that actually supports G-dly creation - where science actually backs up what the Bible says. In fact, you will find as you read along that TRUE SCIENCE WILL ALWAYS SUPPORT G-D'S WORD! Moreover, the origin of most science, theories and ideas (if not all) about the nature of things have originated from the Bible that was written thousands of years ago, and continues to do so even in what we consider this advanced age. (Keep reading to find out what these proofs are). Additionally, a whole chapter will be devoted to these advanced scientific proofs.

Sounds a little far fetched so far? Therefore, to back this rational thinking up I have identified some contradictions and inconsistencies in the theory taught on Evolution. And by far in my analysis all data strongly nominates (and continues to do so) that the "Evolution Theory" is INACCURATE.

First, one of the best evidences we have to determine the truthfulness of creation over evolution is the fossil record of earths history. For instance, a fish of the water growing / developing walking legs to walk onto the land as they assumed and wrongly associated about the prehistoric coelacanth fish. Come to find out this alluded rare fish, was recently rediscovered in 1938 off the east coast of Africa to still exist today. It was determined the proper use of its fins was to swim along the sea bottom in a motion that looks similar to walking. It was afterward concluded that this fish WAS NOT evidence of evolution taking place as it's evident that this creature still exists. In addition, this same view can be applied to other living things that have been suggested to evolve over time. Such as the "Tictolic", a creature that looked like a crossbreed between a alligator and a fish having short arms and legs, and an arm/hand that resembled the bone structure of a human hand suggesting that fish grew legs and arms to become land animals. Another related transitional form was thought to be the "Archaeopteryx" which is a bird like creature (between a dinosaurs and a bird) and is proclaimed the earliest and most primitive bird known/found. However again, the evolutionist imaginably visualize in their minds eye that indeed these are evidences of transitional forms. However, they are nothing more than false interpretations. (Keep reading to find out why they are only imaginary).

Now I've only just begun to scratch the surface concerning such things... Don't be farsighted to think you've heard it all already. There is more, lots more! And it's going to get a lot more intense.

Moving on here… The Giraffe is another prime example that they claim evolved. They assert that early Giraffes had necks of various lengths starting out small and then due to "natural selection or mutation", they had to develop longer necks to reach higher tree branches for food and eventually only long necked Giraffes survived. You can find many pictures of over imaginative drawings in science books of Giraffes starting out with small necks. Nevertheless, let me put forth the real evidence about Giraffes. NO fossil evidence has ever been unearthed showing Giraffes with shorter necks. Amazingly, Giraffes also have a miraculous mechanism in their brains that allow the flow of blood to pass through their neck on the way to the brain so they don't pass out when they bend their long neck down to the ground for water (this is like a regulating valve). If evolution is true, how did something like this evolve when we find no evidence of Giraffes with shorter necks? This is contrary to what the evolutionary theory tells us - which is that this valve mechanism was put into place by a series of natural selections and survival of the fittest.

Additionally today with all these fossilized prehistoric bones evolutionists have come to the resolute conclusion that dinosaurs lived millions of years before man. However, little does the commoner know that there is NO REAL WAY we could date those bones simply by looking or measuring them, not even using carbon 14 dating or radioisotope testing. You cannot tell how old the bone is, but what you CAN tell is how old the rock was that you dug the bone out of, but you cannot accurately measure the date and age of that rock either apart from the catalog fossils that are in it. This is because living creatures that have become petrified rock contain no carbon. So how do they date these bones? They date these bones by what's called a "geologic time table" of the earths "ASSUMED" history, which is based upon approximately 100 years of leading professionals with the "evolutionist understanding" that living things have evolved for billions of years. In effect when they find dinosaur bones, they identify that those bones must be at least million of years before humans were here with the indoctrination in mind that it took a long stretch of time for reptiles to evolve into mammals, then primates, and then into humans. So their whole format of dating things depends on their whole premise of evolutionism, NOT CARBON OR RADIOISOTOBE DATING! (I will further explain this in greater detail as you read along in this chapter).

Now before we continue, keep in mind that everything we are being told, taught and led to believe, as "fact" is simply not fact at all. It is only the "current theory" and "thought of the day", which historically is/has often shown not to be true. And as such, there are no such things as facts, everything is simply an opinion based on the information we have at the current time. It was the Greek philosopher Democritus who said: Nothing exists except atoms and empty space, everything else is opinion. Additionally, it should be noted that we are lead to believe that if something is labeled as "scientific fact" that it is true, and this is not the case either. A belief may be larger than a fact - Vannevar Bush. Even Albert Einstein himself (one of the greatest minds in science and philosophy) broke the jurisdictions regarding "facts" when he quoted: If the facts don't fit the theory, change the facts - Albert Einstein.

However, what I'm truly trying to get at here is that the evolution theory is only a recent belief and based off the "current theory of the day" which is also subject to change. In addition, I'd like to make known that science does not really tell us the "origin" of ANYTHING!

Again, I'd like to put forth as I've already previously stated in the opening chapter, that I am not purposely writing this book to quote facts, or to cite scientific references. Though I may use some in this book, they are used merely to express my opinion. You, the reader may of course have your own opinions as you continue to read. You may also find that this chapter is very choppy in sentence structuring - going from one extreme to the other. But this should be of little concern to those that truly desire to grab hold of the full complexity of the message.

Now for the reasons why carbon dating and radioisotope dating is an imprecise science. Firstly, as I've already mentioned if we continue to go by the assumptions of the evolutionists we will continue to derive a million/billion years scenario. But if we do the correct version of radioisotope dating it reveals to us a much younger age for the Earth of only six or seven thousand years. Firstly, lets examine what Radioisotope dating really is.

Radioisotope dating is the process of estimating the age of rocks from the decay of their radioactive elements. These are rocks that form when molten hot (supposedly at the time of creation). During the molten state, it is assumed that the intense heat will force any gaseous elements to escape. Once the rock cools it is assumed that no more atoms can escape and any radioactive element that's found in a rock will be the result of radioactive decay. Therefore, the radioisotope dating begins at the point when a rock cools and traps these radioactive elements. The dating process then requires knowing the decay rate of the elements.

Now the scientists say that the rate of radioactivity has been constant in such rocks therefore determining the age of the Earth. But this formative conclusion remains in question? This is because we may have failed to consider one BIG FLAW in the whole dating process. And that flaw is a "starting date". Remember, we did not witness such a starting date. Therefore, determining how the environment might have affected a rock also falls under witnessable evidence. For that reason, and in all truth, we cannot directly measure the age of something because its starting date (coming into existence) was not directly observed. Therefore, since we were not eyewitnesses and did not observe the initial conditions of the early earth, we must make "assumptions" which can affect our time (age) calculations.

If scientists fail to consider this critical factor, then radioisotope dating can give incorrect ages. Let me go on to explain… This can be likened to the newly formed Toutle river (canyon) at Mount St. Helens, which formed in 1980 after an eruption. If we did not actually witness this eruption and creation of a newly formed huge canyon that was created in a single day (and incidentally is now in the process of creating peat that forms coal), and we had arrived on the scene some thousand years later, we would most likely estimate and calculate this canyon to be perhaps millions of years old (like we do with the Grand Canyon that is also "assumed" to be millions of years old).

Additionally, they have taken samples of many of these NEWLY FORMED rocks from Mount St. Helens and radioisotope them, and they estimate in the millions of years. So we know this cannot be so since these rocks were recently formed during the eruption in 1980! Its not just the rocks from the Mount St. Helens eruption, there are also numerous false readings from other known newly formed rocks such as from Mount Ngauruhoe (an active volcano) located in New Zealand. These rocks are known to have been newly formed from the eruptions starting in 1949 and they have been dated to be millions of years old. Therefore, it's clear to see that something is evidently wrong with our "starting assumptions". Additionally, the erosion features at Mount St. Helens are not unique, but are similar to those observed elsewhere. Other examples of rapid erosion are**:** Lituya Bay, Alaska created in 1958... The volcanic island Surtsey south of Iceland created in 1964... Lake Peigneur, Louisiana 1980... A rainstorm in southern Brazil 1974... The Waiho River of New Zealand 1965... Providence Canyon State Park near Lumpkin, Georgia 1820... The Imperial Valley of California by the Colorado River from 1905 to 1907 (the Salton Sea).

Additionally, we have the evolutionary geologists disclaiming that such samples collected from the Mount Saint Helens eruption were by dishonest people seeking to discredit real science. My conclusion in all this is that, not all the samples can be wrong! However, for those rocks consistently analyzing in the millions/billions we must additionally find out why they are dating so old. In addition, we will examine the reason why shortly.

There are also other glitches that scientist have run into that the Earth is not really millions of years old… One of the most telling and startling glitches is through radiocarbon dating of a variety of diamonds using highly sensitive C14 (Carbon 14) detectors in which it was discovered that there is ten times the detection limit of C14 in diamonds. Now, diamonds which are said to be billions of years old SHOULD NOT have any C14 in them do to the short half life of C14 (a half life is equal to roughly 5,700 years). Logically this tells us that the C14 found in the diamonds should have been all decayed by now. Furthermore, carbon 14 has also been found in coal that is also said to be billions of years old. Doesn't this have any of us questioning all the conditioning of the evolution theory? And of course, there are those that try to refute it and find another explanation (like contamination) for why C14 is found in diamonds and coal. But by far the evidence is prevailing against the "Evolution Theory" as well as the proper age of the earth.

The evolutionists persist… They say, the Mount Saint Helens canyon exists only on the side of a mountain, as the result of a massive slide. And exert that a "canyon" produced by "subsidence" (the sinking down of land level) has very different properties from a canyon produced by "water erosion over level ground", such as the Grand Canyon, which pose entirely different geologic processes at work.

Firstly, we don't entirely know this?! Yes, I realize there are differences between subsidence and water erosion. But again, since we were not eyewitnesses to the formation of the Grand Canyon we cannot verify what acts of creation actually went into its "initial" forming. We know at Mount Saint Helens that the canyon was created by subsidence from the act of the volcano's eruption, but shortly afterwards we also witnessed this canyon filling with water and eventually erosion became manifest, and now even peat forming at the bottom (peat that forms coal). Furthermore, we find the same fossilized tree trunks ripped from their roots and found in seams - just like at the Grand Canyon. (The peat and the seams will be examined again shortly).

Therefore, since we are faced with that whole eyewitness thing again, it's all hearsay. Actually, its not all hearsay when it is evident that we have the fundamental supporting conditions for coal to form rapidly, (which again will be examined and explained in further detail shortly).

There is also another surprising factor that comes into play regarding radioisotope and carbon dating, that gives an ample answer to why some of these things are analyzed and found to be millions or billions of years old, and this has to do with the Great Flood (Noah's flood). Ironically, proof of a worldwide flood has captured the eye of geologists who are now fully discovering that the ENTIRE earth was once completely covered by "marine flood sedimentation" (salt water from the sea) consisting of marine fossils, which are found in all rock strata. This being the case, rocks that are exposed to water for a length of time (such as the great flood) can cause the elements to leak out do to solubility. Such as salts of uranium and other radioactive elements which are quite capable of dissolving in water, making all the age measurements flat-out USELESS! The proof of how water can effect radioactive dating can be found when the Hawaiian volcano "Kilauea" was recorded to have erupted less than 200 years ago and the lava from this eruption was submerged under water. It was later dated to be 22 million years old. Other samples come from "Hualalai" which erupted in 1800 were dated to be 300 million years old.

However, perhaps the most telling proof of how old the Earth really is lies within "granite rock". Granite is the only rock that cannot be reproduced. Diamonds and other gemstones can be reproduced in a laboratory, but granite cannot. This is because granite has what's called "polonium" in it (a radioactive element). The interesting thing about this radioactive element is that it only has a half-life of ABOUT 3 MINUTES. This logically tells us that polonium must have been encapsulated in granite within at least 30 minutes of it coming into existence. The Earths upper crust and continents consist of granite. Therefore, this is an indicator that tells us that the Earth must have been created in less than an hour. Some question the evidence contradicting it by claiming that polonium can be produced from uranium. This is true, however, when polonium is found in granite alone without the uranium, this then becomes a bit of an unsolved mystery. We then have to question, how is it that the polonium is there without whatever produced it? Finding only polonium suggests that it was formed with the rock itself. The only other explanation is that there must be some other source that allows the polonium in. Some suggest that polonium could have entered the rock when it was molten hot because heat in itself cannot destroy the actual polonium remnants. However, you'll have stripped off the electrons from fervent heat, creating positively charged plasma, but they say that shouldn't affect the nucleus or the rate of decay.

I'm sure that there could be many other explanations brought up concerning where the polonium came from, but at hand does anyone believe that it could have been G-d when He originally created the Earth that these radioactive elements were already in existence. However, how can this be when the evolutionists claim Granite is millions of years old? The evolutionists continue to dredge up several good reasons why the polonium found in granite is mistaken. Suggesting that some gasses in the radioactive decay sequence might migrate through the crystal lattices, as well and the simple mistake of not properly measuring the halos. Additionally, there are also decay sequences with nearly identical decay energies and it may be exceedingly difficult to distinguish between them. The evolutionists propound - why polonium? Is there no other possible explanation more satisfactory than saying G-d created a short lived radioactive isotope as some obscure sort of signature in granite as a sign of a miraculous creation as evidence of a young Earth? They proclaim, surely G-d could have done much better than that!

Firstly, I think these "might" be supportive alternative factors. But the key word here is "might" (which is a very elusive word). Yes, the polonium might migrate through crystal lattices. And yes, there could be identical decay energies at play here. However, I tend to think that G-d did indeed use polonium to prove miraculous creation. In addition, that He did create a short-lived radioactive element to prove beyond a reasonable doubt that the earth is young - just as His word claims.

Other proponents remain a mystery in granite rocks as well. This expounds on the amount of helium found in granite rocks. Granite also contains within it tiny zircon crystals, which contain radioactive elements. During this process, atoms migrate out of the zircons and granite rapidly. Within the zircon crystals, any helium atoms generated by nuclear decay in the distant past should have long ago dissipated away escaping from the tiny zircon crystals present in the granite rock. However, this is not what is found... what is found is large amounts of helium that are still presently trapped inside the zircons. This should not be the case! Yet there it is.

Additionally, what becomes more apparent that the polonium trapped in granite is genuinely aged correctly is that the helium that's also found in granite rocks is one of the byproducts of uranium decay. Uranium is a heavy element that breaks down into lead and it produces helium at the same time. The uranium is called the "parent element" and the lead and helium are the "daughter elements". Radioisotope dating of lead is analyzed to be millions/billions of years old, but miraculously when we measure the helium it gives us a completely different date for the age of the Earth of ONLY 6,000 YEARS (which fits perfectly into the Biblical time span). We have been measuring the wrong clock! All this time radioisotope dating has been analyzing the "lead" in granite, which estimates in the millions of years, however the "helium" gives us a reading of only 6,000 years.

Therefore, as you can see there are many good evidences that support the young age of the Earth. Evolutionists try to undermine creation with millions/billions of years calculations... However, it appears that the Bible was again one step ahead of us in this argument.
1 Timothy 1:3-4 - Teach no other doctrine, neither give heed to fables and ENDLESS GENEALOGIES. (Plainly put, endless genealogies imply the study of the history of the human race (and even animals) and their lines of descent from their ancestors that can be traced directly back from an earlier relative or form).

The conclusion (as well as the demolition) in all of this radioactive dating is that there was not a human eyewitness to give us a starting time. Of course, all of these kinds of arguments (including scientific testing) involve "assumptions" about the past, which quickly fall away becoming completely futile (fruitless and pointless). That is why, ultimately, the only way to know about the past is to have a reliable historic documentation (a record) written by an eyewitness. IN ADDITION, THIS IS EXACTLY WHAT WE HAVE IN THE BIBLE! However the evolutionist try and downplay the Bible by proclaiming that it was written by man and has no real G-d-link, and assert that it‘s all of mythical origins. I think you will find out otherwise as you continue to read.

Well if all these little tidbits are not enough proof automatically, I don't know what is. But this is only the beginning; we have so much more to expand on. So while keeping these bits of information in the back of your mind, lets reverse and examine some more details on just what the "evolution model" has to tell us.

Now here is where the dilemma starts... The evolution model predicts that we should have many "intermediates" (transitional forms), creatures that are slowly evolving (either by natural selection or mutating) their way up to some other creature all the way up through the fossil record, and that we should find many/thousands of these transitions (leftovers or mistakes). However, on the contrary, what we see is a "sudden" appearance of complex creatures from the earliest of recorded earth history created after their "same kind" with no transitions and with no observable intermediates. So where are all the fossils showing evolution? Even the father of the Evolution Theory (Darwin) said he couldn't find them. So where is the evidence? Where are the missing links? How did these creatures do this "gradually"(or even rapidly) without leaving a single trace of how it was done? Though some believe that we simply haven't found any of the evidence yet, but if we keep digging, it will eventually reveal itself in time. In the meantime scientists have found that fossil beds are loaded with dead end "one hit wonders" in which (as theory goes) some genes were purposed and the environment was the ultimate decider (though we find no evidence for this). The evolutionist concludes that fossilization is a very rare event, and indeed, it is! It needs certain conditions and circumstances for it to occur such as being buried instantly by water, mud or icy conditions. Additionally, soft tissues do not fossilize well. And to make an even stronger point for a catastrophic flood (Noah's Great Flood), coincidentally most fossilized creatures seem to have been fossilized from floodwaters. These factors are all very true about fossilization. To make an even stronger point regarding the nonexistence of transitional forms in the fossil record... if indeed evolution occurred slowly creating small changes in living creatures, then logically there would have to be thousands of more "incomplete" transitional forms found in fossil beds than just the complete forms that we find which all have fully functional parts. A true transitional (intermediate) form would have non-functioning (malfunctioning) faulty parts or appendages, such as the nub (root) of a leg, arm or wing. But features like an extra wing, fin, arm or leg being produced don't count since it isn't a new feature (a part that we know already exists and is consistent with that creature). These extra parts are only gene duplicates. We should at least be finding some fossils that show a deformed fin in the process of halfway changing into an arm, leg or wing. Logically if they were the weak link and unable to swim, walk, fly or feed themselves they would have died. There must be at least a few that fell into water, mud or covered instantly with ice that were rapidly buried. Let me rectify this, we shouldn't be finding just a few of these incomplete transitional forms, we should be finding abundantly more, since they would logically outnumber fully-formed creatures if transitions occurred from one kind into another new kind. Yet, we find none! However, the evolutionists continue to persist that they exist, not realizing that their imagination about this resembles the wings of an Ostrich.

But we must also keep in mind that for evolution to have occurred it is based on "genetic change" and there is no way of studying the DNA of fossils - so we can never have any way of relating fossil organisms to one another or to modern ones genetically. Additionally, just because two fossils look the same does not mean they are the same species. So these are factors that fossils will never tell us. Now, please keep in mind that NOT ONE transitional or intermediate creature has been found in all these alleged millions of years. However, as I've expressed above some people have a misconception of what is actually a transitional form. They believe that we do have proof of transitional forms when we can "visibly see the resemblance" between monkeys, apes, and humans as well as other living creatures (including fossilized ones). They also seem to believe that DNA reveals that we share common genes with other living things such as monkeys, apes and other small creatures, and believe that these mutations took place over time. The truth in the matter (fly in the ointment) is that we actually do share similarities in genes with all living things (these are called "tool kit" genes), but to properly understand these "tool kit" genes, scientist cannot find a single mutation that ever took place throughout time. However, the scientific evolutionists claim they have located where the chromosome fusion occurred between man and monkey. However finding the fusion point still does not in any way indicate it was not originally created this way by G-d. The similarities, mixture, or fusion points in genes only means that G-d had used a method of sharing similarity (using the same ingredients) just like an artist who uses the same primary colors and mixes them into different shades. To visualize this more in depth for why genes are identical… Lets look at several manmade objects. If you take a "principle or standard" design you can build a car, an airplane, even an air condition system using little bits and pieces (components) of the same design mechanisms that has been developed by technology over the years. Continually these components have been tried and proven to work repeatedly. Some of these parts may go into the structure of a bridge, a building, an elevator, a tractor, and some may work in an airplane or a car. The same gears that are found in a clock also work in an engine. Therefore, as one can see there are many things that are manmade that share the same design. Likewise, G-d said to Himself, I can use this design in the bird, or that would also function good in the mouse or the fish. And while I'm at it (creating man), I think I'll use the same DNA in the Human Genome.

However, in the scheme of things the evolutionist cries that "similarities in the genome" point directly to evolution. (Where living things progressed taking steps forward throughout time). They reply why would G-d try to trick us with these similarities. But the truth is in the pudding, that throughout the immensity of time the evolutionists CANNOT find a single genuine (accurate) intermediate transforming, or a single beneficial mutation! (A beneficial mutation will be talked about shortly). For the evolutionists there remains NO EVIDENCE for how evolution occurred, only irrational and unfounded theories that don't add up.

Consequently, there is another reason for why G-d used shared similarities in all living things? The answer to this is that G-d wants us to look deeper and deeper into His creation (possibly to learn something about our connection with Him). That is why He made us intelligent. Eventually, we will arrive at the "real reason" why G-d has allowed (and chosen to use) so many similarities. Remember true science will always support G-d's word. In addition, we share these similarities with other creatures that help us in fighting some of our most debilitating human diseases that affect us today. One such animal happens to be the Cat. The Cat for some odd reason shares the same X and Y sex chromosomes of humans, which enable scientists to better, understand male infertility. Cats may also be a valuable surrogate for studying such shared hereditary diseases that manifest in humans, such as diabetes, Tay Sachs disease, hemophilia, Feline leukemia and the Feline Immunodeficiency Virus (FIV) (which is very similar to HIV)(also known as AIDS). Eventually we will come to realize that we need to stop animal extinction and preserve all of G-d's creatures as was done with Noah's Ark. Genesis 6:19 - You are to bring into the ark two of all living creatures, male and female, to keep them alive with you. This is because we were created in a balanced ecosystem where everything has a purpose in the creators workings (why else would G-d request that Noah save the animals). Today science knows that everything in nature is linked to everything else, that when animal waste and dead plants rot, other plants grow. Animals eat those plants and create waste, which rots and feeds new plants - an ecosystem in a nutshell. And it is my theory that science and new technological discoveries about how things work and their purposes for working will eventually reveal their true intentions and conduct us through the whole range of nature until we arrive, at length, at G-d. But keep in mind that the reason we share a "tool kit" of genes with other creatures does not in any way reveal that mutations took place throughout time.

The evolutionists persist in their theories… suggesting that all living things evolved from a simple one celled organism and became more complex. They claim that nature (all on its own) has created life's countless varieties. However, the biggest gap in this theory is that we can find these one-celled creatures, but the next simple creatures that we find contain thousands of cells. Why is nobody questioning where the two-celled transitions or the four, eight, sixteen, thirty-two or sixty four-celled links are? This is because they are not there - they are missing! And what gets me is that evolutionists (with their indoctrinated evolutionary precepts of an "assumed" earth history) determine through radiometry / radioisotope dating that simple creatures are older than creatures that are more complex. They go on to surmise that a directed force in nature did this through what's called "natural selection", thus allowing some living things a better chance at survival. The living things that survived had sharper eyes or quicker arms and legs to run faster to escape a predator. These creatures are than more likely to live long enough to produce offspring and pass on their strong genetic characteristics to the next generation. The evolutionists go on to profess that these small differences can than grow into large differences leading to new species. Additionally as their theory goes, the environment plays a large role in this, as well as sex (female choice). But we don't find the evidences for this in the fossil record either. In addition, something else is not realized.

That natural selection is actually a loss of information. In fact, it's the "opposite" of evolution since living things are being segregated and then isolated moving to different parts of the earth. What's actually happening here are that these creatures are loosing information in their DNA (as you split up these populations and some die). You see over a period of time natural selection results in loss of information, specialization (adaptation or condition in response to environmental conditions), eventually getting to the stage were they can't interbreed anymore.

Taking another approach, the scientists go onto elate that the very first cell and its fundamental characteristics originated from the early rich and creamy seas of our planet. They go on to compare this by referring to tiny teeming little creatures of individual marine bacteria that "come together" in synchrony to light up and illuminate. This "coming together" can be explained as a type of chemical exchanging communication system (a language). The purpose of this coming together and lighting up is to attract fish to ingest/eat them - thus becoming a host to feed off of their body - growing and maturing and then detaching and starting the whole process again. The evolutionists believe that this very first communication system of how complex cells come together, work in the same ways that the evolved/advanced cells in our human body's now do, only to accomplish other tasks - but through the same process of chemical communication/language. The evolutionists put forward that this is the way that groups of cells work together to do things cooperatively. And they go on to assert that such cooperation among cells is what creates and maintains the organs that run our advanced and elaborate bodies today. The mechanisms that bacteria use to do this chemical communication are very comparable to the method and stratagems used by the different cells in our body to carry out particular tasks and work together for the whole. This analogy is again very deceptive… yes, our body does work in this way, and yes, bacteria do too, but do not loose sight here, the creationist should remember just because there are similarities in how bacteria communicate in comparison to the cells of our human body does not in any way prove evolution of this kind took place. What it truly signifies (again) is that G-d had used a method of sharing similarity (using the same ingredients) just like an artist who uses the same primary colors and mixes them into different shades.

Then comes the gap between creatures like invertebrates and vertebrates. Invertebrates have no backbone, while vertebrates are creatures with backbones. Therefore, the invertebrates have a shell like a lobster or turtle, and the vertebrates have a spine like an alligator or human. The question is how could soft tissue evolve into a lobster or turtle shell with no transitional phases on the way. And how about the spiny vertebrates, did their shell suddenly fold up and begin working as a complicated spine with no in-between stages? Houdini himself couldn't pull a stunt of this nature off. In addition, the fossil record does not show evidence of this happening.

How about Whales… They give birth tail-first and upside-down. And it usually takes a long time. If they gave birth headfirst the baby would die of lack of oxygen. Now how did evolution know how to do that? Did it wait until all the baby whales drowned and then realized that they needed to give breach birth?

Oh yes, and it only gets better. The evolutions push their theories to the extremes sounding even more convincing… They surmise that random mutations in nature can happen when one set of genes is copied to another, the enzymes that are responsible for constantly replicating DNA do not do it with 100% accuracy causing errors to be made. They say this is a built in mechanism that allows change to happen that causes one creature to evolve into another - thus evolution. However, this is not correct! As it turns out nature is programmed to protect genes from changes and to correct errors that occur. What we are witnessing here are only "varieties" of species (such as varieties of fruits, vegetables, insects, animals, as well as humans) that have been concentrated, but there are no new genes.

They even conduct such experiments in the lab… They purpose that life may have advanced from the water to the land, or a single organism to a more complex creature because of its "heart". Our human heart and all mammals have four chambers, while amphibians, snakes and frogs have three chambers, fish two chambers, and small creatures only one chamber. Theory is, that a creature that has more chambers to its heart can increase blood and oxygen flow, allowing the tissues to produce more energy, thus it becomes a higher evolutionary creature from the water to the land. There's only one problem here… for this theory to hold true it has to happen naturally in nature, and has thus far only been achieved in the lab by human intervention of adding a single gene into an egg of a one-hearted creature. Additionally, the fossil record does not show evidence of this happening.

They also discovered something else about this “tool kit” genome in the lab. Specific genes that tell other genes where to put the body parts such as head, arms and legs. Now that Genetic Scientist have learned which genes do what, they can now manipulate these genes. However, to no avail, such manipulations only prove to be hazardous to the creature, which usually results in death. To date, no “good” mutations have ever been observed (to note, we cannot really declare such changes as “mutations“, but rather just a reorganizing or taking away of the genes that are already present - this is because no new or “additional” information has been added.

Additionally, there has been numerous genetic experiments done that have tried to induce/cause mutations to flies, insects, rats, mice and birds and other organisms in an effort to prove evolution. Yet, none of these experiments have been successful in producing a single permanent beneficial morphological change in any organism. However, the evolutionists say that this is not true! They talk about such things as chromosomal crossover, which they say, can cause morphological changes in the produced offspring. They also say that another good example of evolution. at work is the nylon-eating bug, which can live on the waste products of nylon. They say in order for this to occur it requires the evolution of new information by either random mutations or natural selection.

However, these types of things have already been debunked. This is because the nylon-eating bug (bacterium) contains within in what's called plasmids (DNA that can duplicate itself and often contain genes that can produce antibodies that are resistant to many types of things). These plasmids can become rapidly adaptive (by some designed mechanism still not clearly understood) making it capable of surviving in new situations as well as acclimatizing to new food sources and even toxins. However, nothing has truly changed within its chromosomes, nor has it changed into a different type of bacterium. In other words, if this bacterium didn't die from poisons, it was already resistant, and if it died it had a decrease (or reduction) of information. Actually, these reductions (mutations) are due to the curse the fall of man since a copying error has occurred. I find it rather ironic that the evolutionists directly claim morphological or mutational change, because in my observation this actually shows signs of intelligent design!

Another scanty attempt the evolutionist throws into the salad is something called “reversing evolution“. And this is where a few genetic experiments have been done on chicks where “atavistic features” (genetic reappearances) have been turned back on during development, forming things such as teeth (which no modern bird has, but reptilian bird ancestors did) including long bony tails with scales (which no modern bird has either). This is very deceptive and can almost buy the mind of creationists if one is not careful. Firstly, many animals and insects alive today have atavistic features… some of the commonly known ones are dew claws in dogs, wings in earwigs which are normally wingless, extra toes in guinea pigs and salamanders, and even various atavisms in humans including Siamese twins, cleft palates, harelips and human tail spine in which many evolutionists view this appendage as tail-like enough to be interpreted as evidence of man's primitive evolutionary ancestry from ape/monkey. However, let me reassure you that these types of things are not any sorts of primitive reclusive genes that suggest evolution of any type. As for the atavistic characteristics of the chicks, these are things that “natural selection” originally selected as advantages at one time and later choose to disregard them when they became disadvantaged. If this were the case that animals have within them the genes to regress into their primitive forms, why then doesn't a bird have the DNA that changes them back into a lizard, snake or a rodent type creature? (Where are those genes?) Again, there is no new information here of any kind going on, nor can it truly be categorized as any old information either. Some of these other things are just physical abnormalities since they reveal "asymmetry" like producing more toes, fingers, and fins. An appendage like the human tailbone is not a primate tail either, but rather a structural deformity and variant of development. In other words, it is a birth abnormality, which occurred during embryologic development. We know this because this appendage is not connected to the vertebral column, as are the tails of other vertebrates. Additionally, the appendage is not even in line with the vertebral column. The appendage contains no bony structures, as do the tails of all other vertebrates. It is not a "true tail", it is a deformity. The most likely diagnosis for what appears as a human tail is a skin remnant of the outer layer of the caudal region. Additionally some people even say that the Coccyx (tailbone) is another left over appendage that we truly do not need. However, this is not so! It indeed does have an important function. The Coccyx is there because there are 5 muscles that attach to it that keep the lower body organs from falling through. Additionally, we can also say that the Coccyx is there for the muscles to keep them attached to something. It has also been thought that the Tonsils and Thymus were just appendages that we truly don't need either, however it has recently come to light that they are necessary for the immune system.

Also as a side note to birds, the evolutionist claim that feathers evolved from reptile scales. However, let me tell you that this is the farthest thing from the truth. This is because if one looks under a microscope (with their own eyes) they would soon see that there are vast differences between feathers and scales. For instance, scales are folded in skin (and in pockets) in which the whole skin comes off when shed. While feathers like hairs grow out of follicles. But the evolutionists argue that scales became frayed into hair like texture. But when we look under the microscope there is no mechanism that is seen that could even slightly change a scale into a feather. No matter how hard the evolutionists try to pull this hare/rabbit out of the hat, the evidence remains that scales and feathers are BIOLOGICALLY different. And no matter how hard ones imagination tries to make scales into feathers - it just cannot happen! Since the mechanism to do so is simply not there. Nor has any creature been discovered that shows the transition from scales to feathers, or anything in-between.

Scientists also theorize that we took the leap from apes to human because of a commonly known disease that is built into every human today (that apes/monkeys don't have). This disease is Muscular Dystrophy (MD). The scientists are calling this a "mutation". This disease has turned off the chemical enzymes that help build muscle tissue in humans. They claim that this mutation allows the skull and jaw bone of the human head to allow our brains to grow bigger with less muscle (because of muscular dystrophy). It also allows our skull to stay flexible for approximately 30 years and allow for further brain growth. However, it is NOT this disease that makes us human. It just allows for continued brain growth. Nobody can rightfully answer if this is actually a mutation or the ORIGINAL DESIGN. What truly makes Muscular Dystrophy are the genetic variants. Although there are many different types of Muscular Dystrophy, they all involve the degeneration and weakening of skeletal/muscle tissue. This may also be manifested in body organs like the heart, gastrointestinal, nervous system, glands, skin and eyes. The variants and severities of MD may be brought on by the onset of age, diet, smoking and occupation. However, they all include a defect in muscle proteins.

All of these theories about evolution will be revisited again and put to rest below. However, quickly before we move on here, the question must be asked: What does all this genetic business adds up to? After reading all this, I guess the forefront question remains for why we should accept the evolution model when the evidence cannot be produced? All of what you've just read from the many paragraphs above logically dictates that the Evolution Theory has no substantial evidence of having actually happened or is currently occurring in this day and age. Can we therefore safely say that the evolution theory is not actually based on science, but rather on a misguided faith since it consists mainly on a theory that has not been seen occurring or witnessed in nature or in any laboratory in the entire world. In addition, we are allowing a theory that's based on misguided faith to be the guiding mechanism of science. It fatefully concerns me that throughout this scientific century we have been so imbued/instilled with a strong sense of evolutionist thinking that we cant reason and weigh up the truth for ourselves. And even though the evolution theory may seem rational, logical, and even conceivable, it in turn is just invalid assumptions and excuses for ruling out the truth of a creator. What's more, the Bible appears to be the only book that gives any kind of logical answer for how life came about.

Now for those of you who do not believe, or are leery about believing in anything let alone the Bible or taking it for its literal word, (a bunch of hokum you might think to yourself). I challenge you to be intellectually honest and fair, to put aside any prejudice views you may have about it for now, and to postulate and give reasonable doubt to a written work of literature (the buzz killers say is conjecture) translated from a higher source (G-d) and documented by a man named Moses which are included in what's called the Old Testament (The first 5 books of Moses - Genesis, Exodus, Leviticus, Numbers and Deuteronomy) otherwise known as the Torah or the Pentateuch WRITTEN FIRST IN HEBREW, then later translated into Greek and then eventually into almost all languages of the world today which now consist of Old Testament (Moses and the prophets) and New Testament (Jesus and the disciples). And to which all religions of the Earth were established and comprised of, in one form or another, the same "basic/similar" biblical foundational accounts (even the aborigines of Australia have a account similar to the Genesis account which talks about a tree of good and evil). Moreover, I bet you're thinking that there were other forms of writing that came before Hebrew like Phoenician, cuneiform, Sanskrit, Paleo writings, or better yet, Petroglyphic cave drawings? You might be surprised to find out that they are all forms of adjacent (broken up) Hebrew. But in order to properly understand this, you will have to wait until you get to the chapter's about the "Primordial language" and "Language similarities", so please keep reading while keeping this in mind for the following chapters. However, I will clue you in a little and tell you that I find it fascinating how linguists have largely (and wrongly) suspected that language was derived from genetic evolutionary means. For the most part, it can be shown geographically how one language was derived from another. The evolutionists wrongly think that if the Tower of Babel event mentioned in the Bible would have resulted in a random and unrelated distribution of tongues, it would not be a distribution that can be so closely tied to linguistic evolution. In other words, the evolutionists think that there WOULD NOT be any relationships at all between languages. However, this assumption is quite wrong. We should not listen to the highfalutin language scholars who tell us that language started from simple grunts and cues in the process from ape to man. Because when one reads the Bible properly, it tells us that G-d "confused" languages... Genesis 11:9 - That is why it was called Babel - because there the Lord **confused** the language of the whole world. From there, the Lord scattered them over the face of the whole earth. The word "confused" is very different from changing completely. G-d had probably done so, so that the people who could understand and relate to each other would group off together. What the evidence leaves behind is actually what we find today (a commonality in all languages), which is an amazing revelation as to the evidence that we spoke a "mother language"... and this mother language was Hebrew, G-D'S LANGUAGE! Therefore, we can blow the evolutionist theory of linguistic evolution (from ape to man) back into the sewer from where it originally came from. In addition, one might ask, what was the purpose of G-d confusing languages? Well, the main purpose of G-d doing this was to ensure the survival of the human race. And He had to do this because the Bible tells us that all the earths human inhabitants after the flood settled in (one particular place) the land of Shinar. Theory is, if G-d had not done this, Satan could have come down in one full swoop and destroyed all of humankind. Therefore, "diversity" is G-d's tool to insure survival. Incidentally, G-d employed this exact system into nature… G-d insured the system in such a way that if one creature were to be completely killed (extinct) it would not harm the total system - one species could compensate for the other. And it was for the very same exact reason why G-d destroyed Babel and confused languages... separating them by languages and sending them out to different regions of the earth BECAUSE THERE IS STRENGTH IN DIVERSITY (this is also why we look different racially today). By the way this is also why we find more of a variety of unknown creatures in the fossil record that aren't alive today because of G-d's great adversary in nature and not because of evolution. There were more things alive back then. Now creatures are going extinct today caught-up in the downward spiral of entropy along with us (because of sin). Entropy and sin will also be addressed shortly as you continue to read.

Now getting back on track here… according to the Old Testament dictated and recorded down by Moses directly from G-d, the Bible tells us a different history of how all life came about. Telling us in Genesis 1:1 that in the beginning G-d created the Heaven and the Earth. It goes on to tell us in Genesis 1:25 that G-d made the beast of the earth AFTER HIS KIND, and the cattle AFTER THEIR KIND, and everything that creepeth upon the earth AFTER HIS KIND, and G-d saw that it was good. The Bible clearly states that G-d created all life AFTER ITS KIND. Nowhere in the Bible does it state that we evolved out of the water from a single celled creature into a more advanced life form. Lets dig further for the answers to prove that ultimately the Bible is a valid source of the truth that supports instantaneous creation. For instance, In our tiny corner of the world science knows that for life to be possible it needs to have the following consistency of atoms, which turn into molecules (amino acids and proteins). Upon which there have been scientific experiments that set out to try to recreate the building blocks of life (amino acids) in a laboratory by simulating earths natural condition. These experiments claim that they acquired amino acids to create life. And yes, they did! But if we analyze just what kind of amino acids they really created, we would later learn that they were NOT the right type of amino acids for active living life! There are over 2,000 types of amino acids and only 20 are used in life. Amino acids come in two shapes, "right (D) and left (L)", and it should be known that every single amino acid and protein that is found in "ALL" living things is "LEFT". Nobody seems to know why (or can answer) why there are no "right" amino acids used in proteins anywhere in life? In addition, what they failed to tell us in these experiments is that they CANNOT create life consisting of all "left" amino acids. Consistently, test after test, they end up with a mixture of left and right amino acids. Now when the left amino acids (in a living thing) start dying something surprising begins to happen, all the left amino acids start reverting to a mixture of left and right. In other words, the natural tendency of amino acids is to revert away from life and never progress toward it.

So this leaves us with the mystery of how life got started on earth, or how life for that matter could get started anywhere else in the Universe? In addition, of course there are always those scientists that say that we just haven't finagled around enough to figure out how to get all "left" amino acids. However, let me repeat, that to date, they CANNOT create life consisting of ALL "left" amino acids. Furthermore, there is no known natural process that makes only left amino acids. However, they will continue trying to achieve them with only theories. Some scientist hypothesize that there had to be some natural process that separated and concentrated the left and right amino acids from each other forming chain like molecules of pure left amino acids. They hypothesize that they arrived on meteorites, and other theories suggest that it doesn't matter whether the amino acids rained down on the earth from meteorites or not, because once they arrived and mixed with the environment, the commonplace chemical reactions would have erased the left amino acid signature. AND REMEMBER, THAT RIGHT (D) AMINO ACIDS NEVER "NATURALLY" PROGRESS TOWARD LIFE.

Here's another huge contradictory misconception... The evolutionist claim that the beginning stages of earths atmosphere did not contain oxygen because we know from observable repeatable experiments that in the presents of oxygen, amino acids necessary for life will NOT bond together. This is because oxygen is like a corrosive and pulls these bonds apart. But the geologists know that there WAS oxygen in the early atmosphere (much higher levels) because they find this evidence in rocks. Therefore, this brings us to a very unsettling predicament that life cannot start "with or without" oxygen in the atmosphere. So the question remains, how did it start then? This remains unanswered for the scientists as well as the evolutionist, but the creationist knows otherwise.

Then science came up with the hypothesis that we evolved (crawled) out of the water and onto the land. But neither can this be a correct synopsis when chemistry demonstrates and teaches us about something called HYDROLYSIS, (this is the action of water DECOMPOSING molecules) specifically amino acids. We all know that water is vitally necessary for life, but it is also DISADVANTAGEOUS (unfavorable and damaging) to the ORIGINS of life. And we know this because test after test has proved it so. For the doubters at heart this information can be found in a "basic" chemistry book. To deny this evidence is the surest sign that a conspiracy is working. So let not our thoughts on this matter be only 6 inches long and 3 pounds oppressed, rather that they should be leaping across mountains.

Furthermore, new evidence now suggests that water came from space in meteorites, but if this were so, how did it barrel through the earths atmosphere without burning up? Well it is possible that both water as well as amino acids can survive in space and travel on asteroids (since we find the evidence for this in fallen asteroids). But the conditions for this require that an asteroid be of considerable size so that it will not burn away as it enters the Earths atmosphere. So the bigger the asteroid, the better the chance of it not completely heating up and destroying the amino acids on its way down to Earth. Additionally, the speed at which the asteroid falls to Earth also matters. The slower the trajectory and landing, the more chance for the amino acids to survive. Furthermore, we must also take into account cosmic rays, ultraviolet light and radiation that might kill the amino acids or organisms along the way. The problem that pertains to the Earth itself with no oxygen present at the beginning would also mean no ozone (which is oxygen with an extra atom per molecule). Therefore with no ozone there would be nothing to filter the Earth from harmful radiation from the Sun.

This includes dangerous ultra-violet rays, gamma rays, infrared light, and others. However, focusing specifically on the UV (ultra-violet rays), as these rays are fatal to forming life. This is because UV light also destroys amino acids. So there would have to have been oxygen at the very beginning of Earths existence, which logically fits accurately into the biblical Genesis account of being created.

However, whether or not these amino acids (or even organisms) can survive and multiply depends, of course, on conditions of the Earth, and they don't stand a very high percentage of survival when they mix with the environment as previously mentioned - the commonplace chemical reactions would have erased the left amino acid signature. Additionally, it should also be mentioned that all such experiments to create amino acids in a lab isolated the products (both oxygen and water) in order to avoid this destructive reaction.

Along with amino acids, it has also been observed that "microorganisms" can additionally survive long exposure to the space environment. They can tolerate extremely high radiation doses. Even in the vacuum of space, some bacteria quickly dehydrate and enter a state of suspended animation from which they are readily revived by contact with water and nutrients. However, this does not mean that they can survive millions or billions of years necessary for interplanetary travel.

The thing that irks me in all this is that NASA has awarded astrobiology millions of dollars in grants to study the origins, evolution, distribution, and future of life in the universe when we might be missing one vital flaw. Why do we always seem to look outwards and not inwards, (which is obviously due to the indoctrination of evolutionary precepts), But how do we know that these asteroids (that contain amino acids) that are floating around in our solar system (for said billions of years) and then falling to Earth DID NOT ORIGINALLY COME FROM THE EARTH ITSELF? Rather they were spewed into space from the trajectory that the Great Flood caused.

This is the big question whether as bizarre or ridiculous as it sounds. I know the NASA evolutionists would have a field day over this theory. However, despite this discrimination, and the notion that our Earth was seeded by asteroids from another planet, I strongly believe that the origin of asteroids, meteoroids and comets in what is now the asteroid belt of the solar system resulted from the explosive events at the beginning of Noah's Great Flood. There is reason to believe this based on all the space littered debris that falls to Earth - in that it contains mostly earthly ingredients. Yes, you heard me right - earthy ingredients! In effect, Earth is the mother body of all comets, asteroids and meteorites, most of which came from the "pillars" (at the time of the Great Flood) specified in the Bible. I don't readily believe that the Earth was created to be bombed by comets or killer asteroids or meteoroids, and what's more, specifically aimed at earth. If this were so, their presence at the end of the Genesis creation week would NOT have been "very good". Genesis 1:31 - **G-d saw all that He had made, and it was "very good".** In addition, there was evening, and there was morning—the sixth day.

To better understand the mechanism of the "flood and its pillars" we need to examine what triggered the Great Flood - and how the flood could have produced (and projected into space) comets, asteroids and meteoroids? The Bible gives some intriguing clues about this starting in Genesis. Genesis 1:9 - And G-d said, "Let the water under the sky be gathered to one place, and let dry ground appear". And it was so. The Bible also says the earth was founded on pillars. Psalm 75:3 - When the earth and all its people quake, it is I who hold its pillars firm. Selah.

In fact, the Bible is very steadfast (and unfailing) about the facts concerning the Great Flood, and puts additional effort into wielding more scripture for its support.

The "foundation" of the Earth is noted as pedestals and sockets, which held pillars… Job 38:4-6 - Where wast thou when I laid the foundations of the earth? Declare, if thou hast understanding. Who hath laid the measures thereof, if thou knowest? Or who hath stretched the line upon it? Whereupon are the foundations thereof fastened? Or who laid the corner stone thereof.

In Job, G-d speaks of a confined sea of water that burst forth and exploding the sea... Job 38:8-11 - Or who shut up the sea with doors, when it brake forth, as if it had issued out of the womb? When I made the cloud the garment thereof, and thick darkness a swaddling band for it, And brake up for it my decreed place, and set bars and doors, And said, Hitherto shalt thou come, but no further: and here shall thy proud waves be stayed?

Additionally, Genesis tells us… Genesis 7:11 - In the six hundredth year of Noah's life, in the second month, the seventeenth day of the month, the same day **were all the fountains of the great deep broken up, and the windows of heaven were opened.**

It also specifically states that G-d founded it! Psalm 24:2 - For he hath founded it upon the seas, and established it upon the floods.

Psalm 33:7 - He gathereth the waters of the sea together as an heap: he layeth up the depth in storehouses.

This is where the Pillars were formed... Psalm 104:3 - Who layeth the beams of his chambers in the waters: who maketh the clouds his chariot: who walketh upon the wings of the wind.

Psalm 136:6 - To him that stretched out the earth above the waters: for his mercy endureth forever.

2 Peter 3:5 - But they deliberately forget that long ago by G-d's word the heavens existed and the earth was formed out of water and by water.

In effect, it was these waters (underground subterranean waters) that burst forth bringing the flood into fruition… Genesis 7:11-12 - In the six hundredth year of Noah's life, on the seventeenth day of the second month—on that day all the springs of the great deep burst forth, and the floodgates of the heavens were opened. And rain fell on the earth forty days and forty nights.

Psalm 18:15 - The valleys of the sea were exposed and the foundations of the Earth laid bare at your rebuke, O Lord, at the blast of breath from your nostrils.

Proverbs 3:20 - By his knowledge the deeps were divided, and the clouds let drop the dew.

Genesis 7:12 - And rain fell on the earth forty days and forty nights.

Genesis 7:18-19 - The waters rose and increased greatly on the Earth, and the ark floated on the surface of the water. They rose greatly on the earth, and all the high mountains under the entire Heavens were covered.
Genesis 7:24 - The waters flooded the earth for a hundred and fifty days.

Mountains dramatically formed as the floodwaters receded... Psalm 104:6-9 - You covered it with the deep as with a garment; the waters stood above the mountains. But at your rebuke the waters fled, at the sound of your thunder they took to flight; they flowed over the mountains, they went down into the valleys, to the place, you assigned for them. You set a boundary they cannot cross; never again will they cover the earth.

Proverbs 8:27-29 - I was there when he set the heavens in place, when he marked out the horizon on the face of the deep, when he established the clouds above and fixed securely the fountains of the deep, when he gave the sea its boundary so the waters would not overstep his command, and when he marked out the foundations of the earth.

Additionally, all of this might very well explain why our moon has so many craters. And specifically why the Moon has significantly more craters on the side facing the Earth, than on the backside (picture shown below). In fact, it is always the same side of the Moon that faces the Earth. We did not see the backside of the Moon until NASA landed there and photographed it (now we also have telescopes). Astronomers had always thought that the backside should be similar to the front with lots of meteorite craters and lava seas. However, the photographs show a much different landscape with the backside of the Moon being very bumpy and hilly with mostly small craters and mountain ranges with only a few lava seas. Astronomers and Scientists are stumped by this... they are unable to explain the differences. Theoretically, the probability of being hit by meteorites should be the same for both sides of the Moon. Why is there a difference? There might be a logical answer to this question. Maybe it wasn't from "incoming" meteorites from outer space that hit the Moon; maybe it was from "outgoing" debris from the Earth itself? And it would explain why we find "amino acids" (both a mixture of left and right) in asteroids - simply because they actually originated from the Earth itself and not some other fabled distant alien planet. Additionally, this also leads to an upcoming chapter where we will also come to find out that there is no other life anywhere else in the entire Universe except for on the Earth. We are the pinnacle (and center) of G-d's eye! So much so, that He sent His only son to die for us here. I will eventually expose the evidence that will make this clear as day to you.

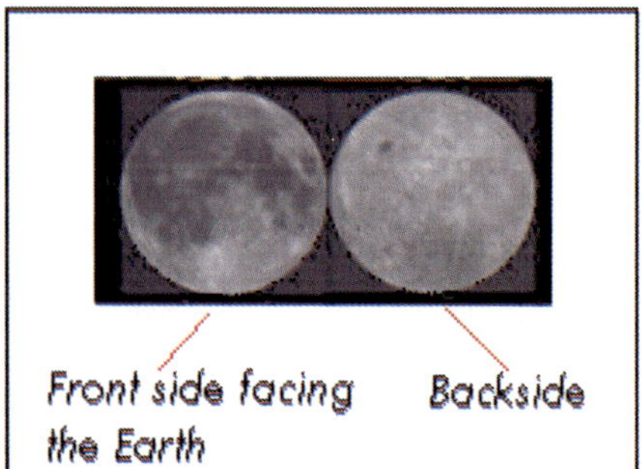

Now to support this idea that Earth is the mother planet from which all comets, asteroids and meteorites come from... there are many other clues found in meteorites that are remarkably similar to those elements found on earth which is highly indicative that Earth is their origin... Meteorites containing chondrules, salt crystals, limestone, water, cellulose, left-handed amino acids, sugars, living bacteria, terrestrial-like brines (salt solution), earthlike minerals, isotopes, including a nuclear substance of excess heavy hydrogen called (deuterium) which became concentrated. This came from pressurized subterranean chambers under the ground and concentrated upon explosion like hydrogen bombs - which was the powerful launcher. However, the evolutionists would tell you that these materials such as ice, frozen carbon dioxide, methane, and organic material came from the Kuiper Belt beyond Neptune (where Pluto lies). And they tell you that Deuterium (heavy hydrogen) comes from an off-worldly place.

Contradictory to this, living bacteria has also been found in meteorites (and the evolutionists cry - contamination!) However since we know that bacteria need fluid water to grow, live, and replicate, and we know that liquid water does not exist inside meteoroids because the temperatures in outer space are freezing. Logically this tells us that the bacteria must have been living in the presence of liquid water before being launched into space. When they arrived in space they than quickly froze and became dormant. The question remains that if bacteria had originated in freezing cold space, what would they have eaten?

Additionally, the evolutionists say that another explanation needs to be given for the higher ratio of heavy hydrogen (deuterium) that is found in meteorites because they say that deuterium is an origin off worldly. However, this is not so. Our oceans contain a small percent of deuterium. What the evolutionists fail to realize is that deuterium could have been "concentrated" upon nuclear reaction. To understand this lets firstly look at our oceans again. Salt deposits in deep oceanic areas are considered deposits from hot salt solutions originating at great depths in the Earth during tectonic movements. Now, salt is actually a result of "degasification" of the earth's interior and consequential in thermo dynamical conditions. In addition, the subterranean chambers absorbed both "chemical and nuclear" reactions as the flood began. Additionally these nuclear reactions are the result of the earths "radioactivity" caused by powerful pressure from the trembling that the crust generated, via the piezoelectric effect (electrical voltages within rock passing through conductive salt water). So briefly, it's also quite possible that the resulting electrical surges could account for today's (high ratio) radioisotope and carbon dating readings (and all the exaggeratedly aged daughter products) that are analyzed to be in the millions and billions of years old.

So how did this energy generate? Most likely tidal draining (sucking in and drawing out) which squeezed together the pillars, which caused them to heat up along with the water. This whole process involved lots of quartz melting… hot, salty and therefore, electrically conducting - effectively containing staggering amount of nuclear energy. This most likely also included burning, combustion and fire within the subterranean chambers thus releasing oxygen united with methane which releases even more oxygen in water combined with carbon. All of these factors are equivalent to hydrogen bombs! The hydrogen bomb uses electricity to break apart water molecules into hydrogen and oxygen. Then all it needs is a spark of electricity to explosively recombine the gases into high-pressure steam, which has the power to propel a stream of water high into the air.

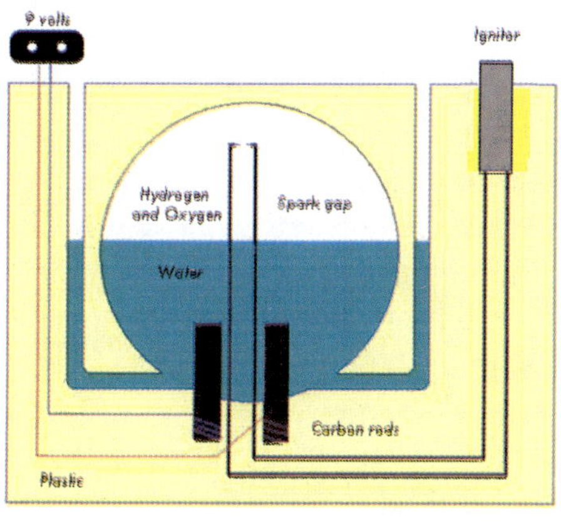

One can experiment with this on their own making a miniature replica hydrogen bomb. All they need is a 9-volt battery that is connected to carbon rods, which are under water in a chamber. The water breaks up into hydrogen and oxygen, which is than trapped in the upper half of the chamber. Than when the piezoelectric ignite from an electronic lighter is pushed, which causes a spark to jump the spark gap, which will ignites the hydrogen and oxygen? The resulting high pressure steam forces the water out through the exit tubes and high into the air.

It's important to realize that most of the energy from the erupting subterranean chambers of the earth did not come from chemical energy alone, but from "nuclear energy!" Atomic nuclei that quickly decayed and released their binding energy. During the flood, that huge energy was released as the plates sank and the subterranean waters violently escaped upward. In this whole process, the new super heavy radioisotopes (deuterium) rapidly formed when gigantic amounts of heat were released and spewed into space, (the asteroid belt we see today). Therefore, the concentrated deuterium we find today in asteroids and comets was due to Earth's nuclear engine! Therefore, the other logical explanation for the high ratios of radioisotope and carbon dating is most likely due to the Great Flood, which increased and concentrated nuclear products. Thus giving the impression today that all of Earth's products of radioactive decay are millions and billions of years old, when indeed they are not. So as you can see there are at least a few other explanations other than just taking the stanch word of current science.

Some might pose the question, how do we know these subterranean chambers actually existed? Well for starters, the Genesis account tells us that much of the water came from inside the Earth. Today we have large underground reservoirs of water and an amazing amount of underground streams that exist throughout the earth (but in the pre-flood world these underground systems were even more common). We also find many geologic formations that appear to have been made under water. In fact, most of our planet's surface is covered with sedimentary rocks - those that are often deposited under water or associated with water. Therefore, as subterranean water escaped upward to flood the earth - mountains and continents had to sink. Therefore, the sinking continents were a consequence of the rising water.

Again the Bible tells us that the "the fountains of the great deep were broken up". We find this information cited very clearly in Genesis 7:11. Genesis 7:11 - In the six hundredth year of Noah's life, in the second month, the seventeenth day of the month, the same day **were all the fountains of the great deep broken up, and the windows of heaven were opened.** We can postulate that there was a massive upheaval and eruptions, which caused tremendous movements of land and continents (shifting of the Earth's plates) causing the ocean floors to rise and break forth the subterranean waters.

So the next logical question would be… is it actually possible that rocky bodies launched from earth (that travel in elliptical orbits) COULD become concentrated in circular orbits between Mars and Jupiter?

Just how it could do this relies on gases, such as water vapors which were most likely abundant in the inner solar system for many years after the flood. The closer an asteroid is to the Sun, the greater the outward thrust. This is because when hot gas strikes an asteroid's (especially on its hot side) it creates great thrust. This would have continued to slowly push asteroids away from the Earth and the Sun until no more available gas remained in the inner solar system to provide any more thrust - thus arriving between the orbits of Mars and Jupiter. Incidentally, NASA has devised this same approach using lasers to deflect approaching asteroids and comets. (Do the research for yourself).

The next question is, could the Earth actually expel all the asteroids that we see in the asteroid belt.

The answer to this is YES! It has been estimated that the total mass of the asteroid belt which lies between Mars and Jupiter may total less than 1/1000th of the mass of the Earth. Indeed, if all asteroids boulders were combined together, the resulting object would measure less than 1,300 to 1,500 km (810 to 930 miles) across, which is less than "one third to one half the diameter" of the Earth's Moon. Therefore, the main asteroid belt is only a small remnant of material that resides in the region between Mars and Jupiter. So, as you can see, its quite possible that ALL OF IT came from the Earth.

Additionally we need to examine the Earths atmosphere before the flood. The Bible says it was quite different with a separating water canopy. This may have been likened onto a pressurized water canopy - similar to a hyperbaric chamber. Genesis 1:6-8 - And G-d said, Let there be a firmament in the midst of the waters, **and let it divide the waters from the waters.** And G-d made the firmament, and divided the waters, which were under the firmament from the waters, which were above the firmament: and it was so. And G-d called the firmament Heaven. These early earth conditions may have also been similar to a giant terrarium in which no rain is needed. Genesis 2:5-6 - And every plant of the field before it was in the earth and every herb of the field before it grew: for **the Lord G-d had not caused it to rain upon the earth,** and there was not a man to till the ground. But **there went up a mist from the earth, and watered the whole face of the ground.** A mist that came up from the face of the ground is the exact effect expected if the Earth was surrounded by a vapor canopy. The everyday environment (temperature) was like spring 24/7.

Now a hyperbaric chamber is a pressurized vessel used to administer oxygen at elevated pressure (at least 2 ½ times normal atmospheric pressure), which involves the breathing of pure oxygen while sealed in a chamber that has been pressurized at several times the normal atmospheric pressure. The medical field uses hyperbaric therapy for many things. It seems that when one is subjected to higher levels of oxygen they heal twice as fast. Not only do they heal faster, but tissue also seems to be preserved from the long-term effect of aging. Research has shown that hyperbaric therapy has a healing effect on bone damage, carbon monoxide poisoning, decompression sickness, gangrene, brain abscess, air bubbles in the blood and injuries in which tissues are not getting enough oxygen. In fact, higher levels of oxygen in our blood would allow us to run twice the distance without getting tired. Amazingly it also has the ability to destroy disease-causing microorganisms, cancer, allergy symptoms, help AIDS patients, arthritis, multiple sclerosis, autism, stroke, cerebral palsy, senility, ulcers, anemia due to blood loss, poor wound healing and is helpful in treating burns and fast-spreading infections of the skin and underlying tissues. It may also have the ability to deter cancer, which cannot survive in an oxygen rich environment.

Such a hyperbaric chamber theory for the early earth is laughable to scientific evolutionists because they automatically assume that if such a change (depressurizing) occurred it would have blown things up to smithereens. They conveniently dismiss even reading the biblical texts to determine this. The Bible specifies that the flood lasted for 40 days and 40 nights. That sounds like an adequate time to depressurize. Additionally, because the ark floated on top of the water resting 20 feet above, this may of had some effect on depressurizing. Genesis 7:17 - For forty days the flood kept coming on the earth, and as the waters increased they lifted the ark high above the earth. The waters rose and increased greatly on the earth, and the ark floated on the surface of the water. They rose greatly on the earth, and all the high mountains under the entire heavens were covered. The waters rose and covered the mountains to a depth of more than twenty feet. Furthermore, the early Earth atmosphere could have been like heat coming out of a kettle from the earth itself. Water vapor is visibly clear to the eye… if you watch steam coming out of the spout of a tea kettle you will notice that the very lip/tip (or mouth/spout) has a clear layer that you can see through before it gets foggy. It's likened unto driving through the hills of Tennessee in the early morning and witnessing the foggy clouds on the ground rising to dissipate as the sun comes up. This could have been the condition on the early earth prior to the flood. Science also know that the early Earth had lots more trees and plants with an atmosphere that contain twice as much oxygen.

Secondly, we find clues of "atmospheric change" in the Bible when it distinctively tells us that the animals journeyed, or somewhat instinctively migrated to Noah's ark. Genesis 6:19-21 - You are to bring into the ark two of all living creatures, male and female, to keep them alive with you. Two of every kind of bird, of every kind of animal and of every kind of creature that moves along the ground **will come to you to be kept alive.** You are to take every kind of food that is to be eaten and store it away as food for you and for them. Genesis 7:15 - Pairs of all creatures that have the breath of life in them **came to Noah and entered the ark.** Maybe sensing the pressurized conditions like the elephants and animals that took higher ground which was witnessed just before the 2005 tsunami of Indonesia hit. Furthermore, Noah was commanded by G-d to take food for him and the animals, but the Bible says nothing about Noah actually feeding the animals. However if he did, he could have easily done it with a pulley system to feed all the animals at one specific time, or another possible (more likely) scenario is that innate in all animals (especially after traveling long periods of time and eating food on the way) animals will then go into a state of hibernation and sleep (also called "aestivation") in which "ALL" animals under extreme conditions are capable of doing and living off of reserves of body fat). Most likely Noah didn't have to worry about feeding them; since they were in a state of sleep conserving energy from the food they ate beforehand on their way to higher ground. Another explanation for the food that G-d told Noah to bring for the animals may have been to feed the animals who were hurried on their long walk to the Ark and didn't have time to eat on the way… and upon their arrival Noah fed them, which then put them into a long hibernated sleep where their bodily functions were reduced to a minimum.

Additionally, looking at the Earths early atmosphere we find buried and fossilized in coal "lycopods". These are scale trees, the types of which have not existed anywhere on our earth for a long period of time. In this case, as the evolutionist say - millions of years. However, this is the thing, just because they "say" that these plants have not existed for millions of years does not mean it is so. These numbers are so garbled its unbelievable! They could be gravely mistaken for one obvious reason. And this reason is due to the Great Flood (Noah's flood). It's just a guess that Lycopods are trees that are in the fern family. In addition, other lycophytes formed extensive moss carpets in the under story of wet tropical forests. Therefore, what becomes clear to me is that these plants fed off a moist humid environment - one that was said to exist in the Bible prior to the Great Flood (with a water vapor canopy circling the Earth causing a giant greenhouse effect). When the environment changed after the flood, these types of plants went extinct because they could no longer thrive in the new environment. This is obviously logical. However, for plants like the dry desert redwood bristle cone pine, the new environment allowed it to thrive. A greenhouse effect would also be expected because of the heat generated by the sun-warmed canopy. We know that this warn canopy circled the entire earth because we find palm tree fossils in Alaska and broad leaf ferns in the Arctic which are warm climate plants. Additionally, scientists have found tropical forests and coal deposits in Antarctica along with plum trees found frozen in the New Siberian Islands. This is because they grew there before the great flood when the earth was a pole-to-pole warm greenhouse. Furthermore, it has been observed that the greenhouse effect along with the hyperbaric chamber can create some interesting effects. In fact experiments with growing plants such as baby cherry tomatoes (which are tiny) grown in a green house, which filtered out UV light (radiation) and pressurized CO2 to the plants stems to imitate the hyperbaric effect. Just after two years, just one plant was measured at 16 feet tall and produced almost 900-1000 tomatoes. After 10 years, the plant was enormously 32 feet tall and produced 10,000+ tomatoes. What's even more remarkable is that these tiny cherry tomatoes were now the size of baseballs.

Additionally other experiments were performed which resulted in some amazing effects… Fruit Flies were put in a hyperbaric chamber and their life span increased to 500%. Fruit flies normally live for approximately only 10-18 days. Additionally, a molecular change in the venom of snakes rendered them to a non-toxic state. And experiments with fish have succeeded in accelerated growth (doubled the normal size). However, most miraculously, an open wound healed almost overnight.

What might also hugely explain some of the highly exaggerated carbon dates for millions and billions of years may again actually have something to do with the earth early atmosphere (likened onto a hyperbaric chamber in combination of a terrarium and greenhouse effect). With more plants and trees, there was obviously more oxygen which allowed things to grow much bigger. Plants, insects and animals. This high concentration most likely contained enormous amounts of carbon. Additionally, when insects eat trees and plants even more carbon is released into the air and eventually the ground.

This might also explain how reptiles survived. Today we see that they function best as cold-blooded animals in warm temperature climates. But in the past the different atmospheric condition kept them growing in an efficient high-pressure atmosphere with plenty of warmth and unlimited supplies of flourishing vegetation to eat. Dinosaurs were never cold-blooded creatures this is because scientists are beginning to realize that there is not enough air pressure to enable a 300,000-pound cold-blooded body to circulate properly and exist in our present environment. This would also explain how giant flying creatures with 50-foot long wingspans such as the pterosaurs (pterodactyls and pteranodons) could have flown. They could not fly in our present atmosphere; they needed a heavier pressurized atmospheric condition to get enough air for lift. After the great flood, and the lighter atmosphere change, which also resulted in cooler temperatures, ultimately inhibited these reptiles to live long lives or grow so enormous again. Interestingly scientists are now discovering new remarkable evidences that all creatures were vegetarians. They have found that Tyrannosaurus rex indeed did eat plants and used its long sharp teeth to strip leaves from plants. They assess this from the Tyrannosauruses three to five inch long teeth whose roots are too short to support tearing and crunching meat or bone. Additionally, green chlorophyll has been found between their teeth in fossils. It will also be talked about in chapters to come that dinosaurs lived during (and alongside) the time of man. There are pictographs all around the world including India and China that foretell of dinosaurs (dragons and leviathans). In fact, Alexander the great 2300 years ago mentioned that they were afraid of the dragons that lived there in India. He brought back a report of seeing a great hissing dragon that lived in a cave. Some 2000 years ago Pliny the elder wrote that elephants are constantly at war with dragons. This also included Apollonius of Tyana whom in 1900 mentioned that dragons are killers of the elephants. 750 years ago (1271) Marco Polo went to China and wrote about dragons that they used to pull chariots. There are also records mentioning a dragon being slain in Nerluc, France. Amidst this, there are many cave drawings of dinosaurs ranging from 500-2000 years old, some of which can be found at the Grand Canyon, another at Ta Prohm Cambodian temple. Inasmuch as these claims may seem bizarre, the patriarch "Job" recants this for us in the Bible.

Job 41:1 - (Leviathan)... Job 41:1 - Canst thou draw out leviathan with a hook? Or his tongue with a cord which thou lettest down?
Job 41:10 - (none so fierce)... Job 41:10 - None is so fierce that dare stir him up: who then is able to stand before me?

Job 41:21 - (a flame goeth out of his mouth)... Job 41:21 - His breath kindleth coals, and a flame goeth out of his mouth.

As mythical as a fire breathing dragon may sound in this day and age, its not out of the realm of reasonableness that such a creature may have been equipped with a mechanism that used chemical combustion components similar to the workings of the bombardier beetle. These beetles are most notable for the defense mechanism that gives them their name... When threatened, this beetle ejects a noxious chemical spray (gas) in a rapid burst of pulses (accompanied by a popping sound). These chemicals are produced from special glands in its abdomen which stores two reactant chemical compounds, (hydroquinone and hydrogen-peroxide) in separate compartments in the rear-end of its abdomen. When threatened, it contracts the muscles that force the two reactants through regulating tubes into a mixing chamber. When combined, the reactants undergo a violent outside-heating (exothermic) chemical reaction raising the temperature almost to the boiling point of water. The beetle also has a mechanism that seals and blocks these chemicals to protect the internal organs.

Finally and ultimately sealing up the hyperbaric chamber theory, when we start to realize things about the earths early atmosphere, and take into account the Great Flood, things come into greater perspective and we soon begin to realize that the long exaggerated ages of carbon and radioisotope dating could have been effected by it. Solar shielding by the water canopy above the atmosphere would also affect carbon dating techniques. Another result of the water vapor above the firmament (specified in the Bible) would be for the defensive shielding effects from cosmic radiation. Scientists have studied the amounts of solar radiation that is filtered by water and found that if indeed our atmosphere consisted of a higher water vapor (which shields solar radiation) people could live to be very old. This is because scientists have determined that one of the primary aging factors is found to be from solar radiation and polluted air. And it has been estimated that by filtering out the harmful radiation (as a water canopy would do) humans might be able to live close to 1,000 years old. This factor becomes very fascinating when we consider that the Bible speaks about the old ages of the patriarchs such as Noah who lived to 950 years, and the conditions that he lived in prior to the Great Flood which was pressurized with lots more abundant oxygen like a hyperbaric chamber... and since the medical field knows that when these proponents are exposed to the human body, it has some miraculous effects on healing as well as aging, we can then begin to realize that the long ages of the patriarchs was indeed very possible. Not only that, but there may also be a very subtle clue in the Bible that confesses to this high oxygenated pressurized system and its new effects when Noah experienced the difference between the pre-flood system (heavy atmospheric pressure) and the later (post-flood lighter atmospheric pressure) causing the rapid fermentation of alcohol which was quite possibly the reason for Noah's drunkenness. This can be found in Genesis 9:20-21 - Noah, a man of the soil, proceeded to plant a vineyard. When he drank some of its wine, he became drunk and lay uncovered inside his tent. The Bible clearly tells us that Noah was a man of the soil and he was much practiced in the knowledge of planting vineyards. Surely, he would have known the effects of wine prior to drinking any. He obviously did not realize its alcohol potency, which resulted in a stronger reaction (the result of rapid fermentation caused by a lighter atmospheric pressure). This may also explain why Noah was so mad that he had cursed Canaan for unwisely, irresponsibly and unthinkably not knowing what to do with his father when he became drunk. This is because drunkenness was not regarded as something pleasant, but possibly a forbidding and frightening thing. Its also quite possible that nobody knew what getting drunk was prior to the flood. Genesis 9:22-27 - Ham, the father of Canaan, saw his father's nakedness and told his two brothers outside. But Shem and Japheth took a garment and laid it across their shoulders; then they walked in backward and covered their father's nakedness. Their faces were turned the other way so that they would not see their father's nakedness. When Noah awoke from his wine and found out what his youngest son had done to him, he said, Cursed be Canaan! The lowest of slaves will he be to his brothers. He also said, Blessed be the Lord, the G-d of Shem! May Canaan be the slave of Shem? May G-d extend the territory of Japheth; may Japheth live in the tents of Shem, and may Canaan be his slave.

But there is still another huge criterion which must be satisfied here, and we aren't just talking about such things like radioisotope and carbon dating, or the earths early atmosphere, but a unavoidable contingent property that combats and succumbs us throughout our lifespan. This is ENTROPY (disorder) acknowledged as one of the basic laws of physics. Entropy is also in correlation with the Second Law of Thermodynamics. Entropy substantiates that under natural conditions all materialism (whether living or not) eventually wears out and deteriorates in time. This law cannot be avoided; it is an undeniable fact of all life whether living or nonliving. It is something that all of us observe and experience through the act of living and dying - as in the development from embryo to adult which eventually increases from a ordered state of birth to the deterioration of overall disorder and death. In addition, it has been observed that our ENTIRE expanding universe is unavoidably proceeding towards a more expanding disordered state. Jumping ahead of myself here, there is another reason for why we have the law of entropy - whose end result is death. That very reason can be found within the Bible as the faulty thing that caused the laws of entropy to kick-in in the first place - which was the initial act of sin itself! (This will also be addressed in this chapter as well as future chapters).

Now, the mechanism offered by evolution totally contradicts the second law of thermodynamics. It argues that disordered atoms and molecules impulsively came together over time and formed all by themselves under natural condition to assemble complex molecules such as proteins and DNA, whereby millions of different living species with more organized configuration gradually emerged and are (at present) continually in the process of evolving into higher levels of order. But this is conflicting with the law of thermodynamics and entropy which clearly reveals to us that systems should DECAY through time thus giving less, not more order, because in the process of evolution (things changing), it would dissipate the available energy in the development of that life form. For instance, in the process of embryo to adult or seed to a tree, the DNA information only has time to follow its given instruction. It does not have the planned (preprogrammed) energy within its given information to make new DNA; it can only work with existing information. No new energy is available to generate new DNA (information) since time equals available energy. Moreover, specific mechanisms are needed to make the energy functional, and as you can see, that INFORMATION (instructions) already had to exist PRIOR to the fact of the existing thing emerging or growing into the full course of its lifespan (example, seed into a tree). In addition, to note that energy by "itself" is incapable of making amino acids form the more complex molecules of proteins. Therefore, inescapably, without known exception there has to be a coded pre-planned program from whence these codes came from. Disorder will NEVER randomly become order. Something must direct the impending physical energy before it can "KNOW" how to organize the components. This NEED (by itself) is of little whim to the mindless interplaying atomic particles of the elements that they were to form, and then latter get together and program the code of the replication process themselves. In the example of a plant, the necessary program for its growth has already been written within its seed, which directs the assimilation of chemicals into the resulting plant. In crystals, each compound by its chemical and elemental property is directed into a geometric pattern of its pre-existing structure. It seems that a code always requires an intelligent order as well an intelligent planner. So, as one can see the much-needed ingredient to all of this is "INFORMATION". It deserves independence of its own. In that, INFORMATION is something comparatively different from matter. It can never be deduced to just a material aspect. For instance, I would like you to think of this book you are now reading. It consists of paper, ink, and the information within it. The paper and ink are material substances composed of matter. The paper is made of fiber, and the ink of different chemicals. However, the "INFORMATION" in this book is NONMATERIAL, and it came from my conscious mind that is the writing source behind the material aspect of it. I put it into material form. I translated the information in my mind into letters, using a keyboard and computer, which stored the data that my laser printer later printed out by my push of the keyboard button command onto the paper, and taking the form of a book. As a product and outcome, this "INFORMATION" is passed onto you. In conclusion, you can understand that mind, matter and design constructed this book from a source beyond just the material aspects of it. Similarly, the source of information in nature cannot be matter itself either, but another influence at work. A superior source beyond matter and existing PRIOR to matter. The administrator of this information is what we can interpret as G-d who brought everything into existence, formed it, and so eloquently organized it all.

Additionally, information is a vital clue to the mystery of life itself. We cannot produce a living thing by merely putting together all its parts. An example of this is a dead animal that has been driven over by a car and smashed instantly. In one moment, it was once alive, and in the next moment now it's dead. This tells us that life is not the sum total of its physical parts. Nonsensically nobody really knows what the difference is between a living animal and a dead animal other than once it was consciously moving and alive and now its unconscious, motionless, lifeless and dead. But the question remains, "WHAT" did it loose? It obviously lost its life (its spirit), but where did it go? This is what nobody seems to know - the mystery of life has eluded us. It's a mysterious organizing, coordinating and orchestrating principle, and only G-d alone holds the answer to solving the mystery of life. The theory of evolution reduces life down to a mere physical chemical thing, and its not! The Bible tells us that G-d breathed His breath into Adam to make him alive (a living spirit and soul). For a moment, we need to stop and consider just what the living breath of life is. Genesis 2:7 - The Lord G-d formed the man Adam from the dust of the ground and breathed into his nostrils the breath of life, and the man became a living being.

Additionally science continually observes in nature the apparent fact that something cannot come about from nothing (in other words, living things cannot come from non living things). We witness this with fruit fly's and bacteria, which cannot come about by themselves (microbes will not grow in water without microbes). The fruit fly eggs were laid in the fruit prior to mature fruit fly's appearing, and bacteria and microbes cannot come about in water or elsewhere without being transmitted there by another source.

These things simply cannot just appear out of thin air by themselves. That is why according to the creation story in Genesis; G-d had created everything already fully made (with seed inside itself). Such as fruit trees so that the man would have something to eat when he was placed in the Garden. If Adam had to wait for the trees to grow, he would have starved. Genesis 1:11 - And G-d said, Let the earth bring forth grass, the herb yielding seed, and the fruit tree yielding fruit after his kind, whose seed is in itself, upon the earth: and it was so. In addition, another baffling mystery surrounds this. If man had written the Bible without the divine intervention of G-d telling us (through Moses in Genesis) how earthly things came about, how did man alone contrive this piece of information? Lets not forget that the Bible is thousands of years old and just recently in this modern day of science we are finding these very same facts out under microscopes as well as through general DNA testing that we cannot create any new creature that isn't already in existence. We cannot even create a new color in the rainbow!

Getting back on track with the evolutionists view… They additionally hypothesize as mentioned above that living things MUTATED over time. But if one really understands what mutation is, it is when one of the DNA genes either goes missing, is replaced by a duplicate, or ADDED with other already existing information. In addition, in the case of humans evolving from apes, let me confirm to date that not one DNA gene in humans that we have available on record has ever been found to have been "ADDED" by mutation. Now if this were true that we evolved by mutation thereby stepping a notch up in intelligence, then we would have an "ADDED" gene somewhere down the line in our human DNA, which we DO NOT. Neither by nuclear radiation has it ever been found to have "added" DNA, only mangle and confuse the existing DNA. Therefore, mutation is not evolution. And it is not a promotion of any sort of life; it is a demotion so to speak.

What we see as evolution is something called NATURAL SELECTION. Now to clarify "natural selection". Natural selection "cannot" cause one type of creature to evolve into another, because natural selection ONLY works with "existing information" (DNA). It never adds anything new to the genes! Because in order to have evolution you must "add" new information to the genes, and let me remind you that not one mutation has been found to have added information to the genes. To explain natural selection in better terms… When the environment changes, lakes dry up or new ones emerge, weather patters change, a new vegetation grows or one dies, any organism or living thing that possess characteristics that are suited for the environment will have a better chance at surviving. For instance, birds that have poor eyesight will go hungry and birds that have short beaks that need to feed on a certain narrow elongated flower will not be able to reach into the flower to retrieve the necessary food without long beaks, so eventually if they don't adapt they will go hungry and die out. As a result longer beaks and better eyesight are necessary to survive, and those that don't have these accommodating features will die off or become extinct. In effect and to the contrary, what transpires is something marvelous! The birds with the short beaks begin to grow longer beaks so that they could get into the narrow elongated flowers for food. Perhaps, they have to sharpen them on wood or rock or perhaps after all the ware and tear of pushing pressing or maneuvering to get into the narrow elongated flowers for food, there skull along with their beaks gradually changes form to accommodate them. Little by little, each following generation will become accustomed to the food dilemma and have the new trait needed for survival. And now this creature will LOOK DIFFERENT, but nothing new has been added to the DNA it is still the same, and it is STILL a bird! Likewise, a dog is still a dog, a cat is still a cat, a bear is a bear, even a flower is still a flower no matter how many different breeds or variations (big or small) in appearance they have! NOTHING can change that fact. Additionally evolutionists associate a mutation of the DNA to be "changed" genes and categorize them as tandem repeats, duplications, copies and sequencing (such as in dogs when cross breading). The tandem repeats are small sections of DNA that replicate over and over again changing the protein made by the cell, thus leading to changes in cell function and ultimately physical appearance. All living things have tandem repeats in their genes but the dog has significantly more than most, that is why the dog is so malleable in appearance. And again this is NOT the mutation that accomplishes "steps forward" in evolution, they have confused the word to make the dog look like something other than what it is - another species! Again, there is no ADDED gene, everything that was originally there is still EXSISTING. And they go on perplexing about mutations, incorporating that every once in awhile, something goes wrong with the DNA, the instructions get botched, and they call these random mistakes mutations. They go on to acknowledge the fact that mutations are invariably harmful, but they say that's not always the case because many mutations have little or no effect, and some mutations are lifesavers. They go on to relate an example in human DNA when certain people carry a mutated gene that resists the HIV virus that causes AIDS, or another example of what they call a mutation is when a bacteria becomes resistant to antibiotics (but there is still no new information that happens here either). But it should be noted that in most (if not all cases) that occur by tandem repeats, duplications, copies and sequencing such as in dogs, they have been associated with disease (such as in hip displasia (dysphasia) in dogs, white fur (such as in Dalmatians causing deafness), epilepsy, and dogs that were bread to be herding dogs have been found to have neurological disorders if they are house pets and don't herd. These cases are not considered beneficial. Leviticus 19:19 - Keep my decrees. Do not mate different kinds of animals. Do not plant your field with two kinds of seed. Do not wear clothing woven of two kinds of material.

Additionally, evolutionists claim that it is these random mistakes in the genes that had propelled evolution into action. The geneticists say that the genes are designed to allow for these random mistakes and that this is a good thing, which points directly to the evolution theory being correct. However, this is not correct! As (was also mentioned earlier) it turns out that nature is programmed to protect genes from changes and to correct errors that occur. What we are witnessing here are only "varieties" of species (such as varieties of fruits, vegetables, insects, animals, as well as humans) that have been concentrated, but there are no new genes. Again, don't be confused readers; all of this is still "Natural Selection"! (And again, natural selection is not evolution). No genes have been ADDED. Only ADDING genes can be considered a higher evolutionary step. Just as well, the laws of

"THERMODYNAMICS" do not allow for mutation by adding new DNA to the genes. Remember, the genes have no available energy to endure and undergo adding onto them, they conserve what they have by using only existing information. Thermodynamics does not permit the genes to add additional DNA. It only has enough energy and time to create, to live and to die. Ending here, "Natural Selection" may explain the "survival" of the fittest, but it doesn't explain the "arrival" of the fittest.

One final thing pertaining to "natural selection"... The evolutionists are very slick in the way they portray "evolution in action". They use the example of a lizard called a "skink" to prove that evolution is occurring right-smack in the face of science. These unique lizards can loose their fingers and even go completely legless just like the alleged evolutionist theory that transpired with snakes. In addition, they can rapidly change in an exceedingly fast short period of time depending upon their environment and if they are subjected to spending most of their years swimming though water, sand or soil. Since the limbs are unnecessary for this environment (and a hindrance) they become obsolete, therefore their body looses the fingers and the limbs and becomes snake-like. In addition, once a skink has lost its limbs, there is no evidence of reversals for growing back.

However, is this really "evolution in action"? NO, IT IS NOT! Moreover, it is again entirely in the evolutionists' imagination. But the sharp-witted (on the ball) creationist knows that this indeed is driven by natural selection, but it remains to be seen as any "permanent change" (manifesting in all skinks), and remember that "natural selection" is not any sort of real evolution. On the other hand, if one defines "change" to mean the ability of an organisms to generate new genetic information (which let me remind you again has NEVER been observed) and evolve into an entirely new kind of creature, this would then be considered "evolution". But this is not what's happening with the skink. We are not seeing any new genetic information added to the skink lizard causing it to grow extra fingers or legs or evolving into a true snake. This would fall into the same category as a fish evolving into a vertebrate with four limbs and legs, or a small mammal resembling a mouse changing into a monkey/ape. So as you can see, skinks evolving into snakes in no way supports evolution. In other words, the skinks are losing genetic information, not gaining it - and that is the complete opposite of what evolution requires. In fact, creationism backs us on this point since no new fingers or limbs can be added back in the skinks. This is permanent damage and not evolution.

Here is something else that needs serious thought... Scientists have found that certain living cells come with a built-in self-destruct mechanism. For instance as a caterpillar turns into a butterfly, or a tadpole into a frog. When these types of things morph (metamorphosis) they no longer need their former apparatuses (tails, arms, legs or gills). This happens when a special morphing gene gives the order at the appropriate programmed time for the tail cells of a tadpole to begin to die, or the caterpillar to grow butterfly wings. What is actually happening here is that programmed within these specific living cells G-d had programmed a certain gene that signals the death of those cells at the appointed time of His choosing. Perhaps G-d created metamorphosis in some creatures to help pollinate, provide food for predators, and help plants grow through many mechanisms. However, the million dollar question remains, if this is not G-d's doing, then why would evolution develop genes that order their own death, because such a gene would not aid survival? I'll tell you what else is amazing about the metamorphosis of frogs and butterfly (including other things that morph like beetles)… and this has to do with continuing reproduction. If one evolved where was the other to mate with?

I know I am erratically skipping around here from one subject to the next like a poet using “free verse” which is a method of writing poetry, which does not essentially follow any structure or style. However, even in poetry the reader is sure to evoke meaning. With this in mind, I can move onto the next subject.

Now I must back up here a bit, making an addendum, and confer on an implication that the evolutionist throws forth about the law of “entropy and thermodynamics”. The evolutionists say this law is quickly coming to an end because we witness examples of natural order from disorder constantly occurring in every time it snows, whereupon billions upon billions of beautiful examples of “reverse entropy“ spring into existence and are blown about by the wind which are highly structured crystalline patterns of simple molecules and each one unique. Additionally, salt crystals and deoxyribonucleic acid (DNA) is another molecular form of crystal even more complex than that of water crystallizing into ice. The evolutionists even make the case of water running upstream, metabolism, growth, digestion, condensation, and using the example of a seed and an embryo as reverse entropy. In addition, they use the example that energy is being transferred from one location to another and conclude that overall, entropy MUST increase and this energy transfer CAN reverse “local entropy“ without any violation of thermodynamic principles whatsoever. However, these types of things such as “growth” are not an example of improvement or defeating entropy. This is because everything that’s needed for the entire life of an organism is present in the genes. Growth is not an improvement of any sort since it requires outside fuel, thus the provider undergoes entropy. Additionally, the evolutionists view of growth is lapsed because as something grows it needs more and bigger organs to accomplish less than the original single-celled organism. Overall, growth still does not explain how one kind of creature can change into another, which is what evolution requires.

Therefore, lets get this straight! What the laws of entropy and thermodynamics prove is that evolution is absurdly impossible because it claims that we have evolved progressively. We are not evolving, we are actually de-evolving! The very notion opposites what evolution says? Evolution is in the position of assuming that random mutations somehow reverse the tendency of destruction inherent in the 2nd law as well as the mutations themselves, but they do not! What on earth can make anybody think that we can become another kind of creature?! Something just does not make sense here… how can evolutionists believe that the very same processes that have been observed only to cause decay, death, and destruction caused NOTHING to create something, and then that something to make itself come alive and then to evolve into something more complex as man?

Evolutionists like using all sorts of defensible examples to try to break the law of entropy. They throw forth examples like: We could not build houses, roads, computers, bridges, and so on. Because all of these things (they say) represent reversals of entropy where more energy is expended then what is required to build such structures, and all of them are subject to decay, which is why they require maintenance. The fact that entropy must increase explains why everything eventually breaks, but it in no way precludes things from being fixed or built up - otherwise plants and people could not grow. Additionally they note that the ultimate source of energy for “reversing entropy” comes from the sun and solar energy.

Again, let me strongly stress here that nothing “NEW“ is really happening here either (I will further explain why). Conversely, how chaos is turned into order (like snowflakes) is not a mystery, it clearly shows intelligent design! These same geometric patterns of creation (like in snowflakes and crystals) are continually being discovered, but often without the greater context of the spiritual understanding... Job 38:22 - Hast thou entered into the treasures of the snow? Or hast thou seen the treasures of the hail. Additionally “free energy“ within a “closed loop“ system such as the Universe - as a whole is experiencing decreased entropy, and will inevitably die out as the sun uses up all its free energy. However there are those who envelope the idea that the Universe is not a closed loop system but rather an open looped system. But the scientific evidence is present that the overall amount of gas in the entire Universe is dissipating. Therefore, we can validate that it truly is a closed looped system in which entropy will eventually win in the end and the solar system (indeed, the whole universe) will die a death at maximum entropy. Life on earth is a temporary blip in the process of universal down winding. It is an ILLUSION that things in life are “reversing“ and can temporarily produce small areas of organization. In reference to a “closed looped system“. Ecclesiastes 1:9 - What has been will be again, what has been done will be done again; there is nothing new under the sun.

I've heard it all, even such things like the claim about greenhouse gases violating the second law of thermodynamics. The claim that the air can become hotter than the ground it receives heat from, and then heats the ground. The claim that CO2 becomes hotter by taking in (absorbing) infrared radiation. However, nothing new is happening here either since the absorption of specific wavelengths of light by an atom/molecule changed the energy state of the atom’s electrons rather than causing an increase in motion (or heat). Moreover, before an atom could absorb additional radiation of that wavelength it had to first “emit“ radiation of that wavelength. So you see again, there is an even exchange of energy transpiring here.

But eventually, what goes up must come down. Do you think that DNA has the “time + energy“ to change direction from its given instructions? It is a copy of the same DNA information from the one before it (such as in a plant seed). But herein lies the answer to this question... has anything new “ever“ came out of anything living that we have observed? And I'm not talking snowflakes or uphill water streams, I'm talking about a rat giving birth to a cat (almost like the fairy tale of the Frog and the Prince). Even though evolutionists admit this cannot happen because they are two different species, I find it hilarious that evolutionists really cannot tell which came first between the chicken and the egg! Moreover, experiments have even been done with radiation on living things to try to get them to mutate, and the result from this was a colossal failure. There was no increase in any genetic material, but even more devastating was that there was actually a decrease in the genetic information resulting in deformity and death.

Moreover, in DNA, each cell reads and interprets the code and follows the instructions that it is given. Each cell knows which part of the instructions apply to it. Its likened unto a team of construction workers building a house in which each worker recognizes and understands the part of the architectural schematics that apply to his/her specific project. One worker has the given knowledge on how to interpret the instructions for installing windows and doors, while another worker installs the electrical wiring and connections, and another the plumbing. In the end everything fits together flawlessly according to the master instructions - likened onto the instruction that DNA has in all living things. Stem cells are magnificent in that they can form to become any cell in the body. However, even in this respect the role and amount of stem cells at the given time of conception was determined by its mother hosts DNA, and there are also cells that instruct the direction or the places that the stem cells are needed in the body.

However, the question remains, where did these detailed instructions come from? What directs each cell into knowing what portion of the architectural schematics to perform? The window installer could care less about how to install plumbing or the electrical. That is why the information in DNA and how it is used specifically and at the precise times necessary cannot be explained by the laws of thermodynamics alone. We need another law before we can even hope to understand the origin of information like this. This law lies in the foundation that all life is the SPOKEN AND WRITTEN WORD OF G-D!

The conclusion in all of this is that no matter what any evolutionary proposal may suggest, spontaneous generation never happens, not even in any advanced scientific laboratory and no matter what oxygen-free atmosphere is tried. In fact, it has been tested time-and-time again that living amino acid proteins necessary for life have never been observed to form outside of a living cell in any conditions that exist in nature. The answers to why this never happens are because life only comes from life! (Again, the spoken and written word of G-d!) This is a basic principle of science that cannot be evaded under any circumstance. Remember, true science always backs up what the Bible says and will continue to do so until the result of a creator is finally acknowledged.

What is also interesting to think about is that the evolutionists keep saying that we came from a systematic process. They INSIST that life started (with amino acids) in hot water vents in the deep ocean. But we just discussed earlier that life could not have come out of the water since water pulls amino acid bonds apart. Let me remind you that (to date) there is NO SCIENTIFIC PROOF that living things evolved from water. Yet, evolutionists continue to imagine (like teddy bears eating cookies) that it somehow found a way, which let me remind you has not been discovered yet, nor will it ever be. No Laboratory in the world is anywhere near creating life consisting of 100% left handed amino acids (which are completely evident in all forms of living things). These deep hot water ocean vents are not creating anything new. They are only uses existing materials that were all ready present. And simulating this environment in a laboratory is just using and manipulating the same ingredients that were already previously/present on earth. But like the deep hot water vents of the ocean purposed by the evolutionists to have started the process of life on earth, it remains a mystery when we observe the intricate mechanisms of living things as to how their internal organs came about to work in synchronization. To examine this scenario all we need to do is to look at our own human body. Such as how the human body amid its complete system works simultaneously. For instance, like the scenario of the chicken and the egg (which one came first?) We can ask the same question of the workings of the human body. Which one came first in the human body, "the nerves, or the organs" since the nerves are controlling the organs, and if we don't have the nerves around we cant digest food or process information in our brain. We can also ask the same question of the heart and lungs since the heart itself is dependant on oxygenated blood. What's more, the blood is dependant of the nervous, muscular and digestive system, but ever has it come to the mind of the evolutionist to conceive of just how we (or other creatures) survived during these stages and changes of development? Since blood is the main transporter of life, is it safe to say that blood contained all its mechanisms such as the white and red corpuscles as well as the platelets and plasma? So as to see that they had to come about SIMULTANEOUSLY (at the same time in order to function efficiently) and not by way of a step-by-step process. To obtain a better understanding of this lets use the example of hydrochloric acid which is made in the human stomach for digestion? This is a gastric chemical that is highly acidic, it can dissolve many metals. However, the stomach itself is protected from the strong acid by the secretion of a thick mucus layer and a neutralizing component called sodium bicarbonate, which is found in the small intestines. The question is, if this is an evolutionary scenario, did we evolve those acid-producing cells first, or the container to store them first? This "interdependence" strongly argues against the possibility of things evolving. Which begs the question of how amino acids in hot water vents of the deep ocean could have evolved into complex synchronizing living mechanisms? The only way for this to have occurred is for the DNA instructions to have already been immediately there. As well as all of the systems in the first human, animal or creature must have been fully functional in the first moments of life. Because of this "mutually dependent elements factor" it clearly indicates that evolution could not have occurred, (and the fossil record backs this up!)

So how did life get started?! Answer: IT WAS CREATED!

And there is so many other questions that we should be asking, things that we take for granted, such as where did all the 90-or so "elements" come from like: Silver, nickel, calcium, iron, etc… and what factor determined how many bonds each element would have for combining with other elements? Additionally, we should also be asking where did the thousands of "compounds" we find in the world come from: Such as chlorophyll, sucrose, carbon dioxide, propane, benzene, hydrochloric acid, boric acid etc…

Additionally, what would cause millions of species such as fish, birds and insects as well as humans to all have symmetrical features where one side is a mirror image of the other like two arms, two legs, two eyes…? The question is, would random (hit and miss) processes (in the theory of evolution) create balanced proportion. Moreover, we should be questioning the odds that in all the millions of species that exists why, is it that only both a male and female exists. Genesis 1:27 - So G-d created man in his own image, in the image of G-d created he him; male and female created he them. Even Jesus restated this fact in Matthew 19:4 - Haven't you read, He replied, that at the beginning the Creator made them male and female. There are only two sexes, which is a clear indication of a planned framework. This would also reflect that each creature (male and female) had to develop at the same time (as well as location) so that the species could reproduce. Furthermore, one should be asking how did the first mating pair get to the mating point. This is impossible since a species would have to skip from a primitive form to a fully developed male or female, and each with the capability and instinct to mate. The Bible clearly states that each creature was created after it own kind. Genesis 1:11 - And G-d said, Let the earth bring forth grass, the herb yielding seed, and the fruit tree yielding fruit after his kind, whose seed is in itself, upon the earth: and it was so. Genesis 1:12 - And the earth brought forth grass, and herb yielding seed after his kind, and the tree yielding fruit, whose seed was in itself, after his kind: and G-d saw that it was good. Genesis 1:21 - And G-d created great whales, and every living creature that moveth, which the waters brought forth abundantly, after their kind, and every winged fowl after his kind: and G-d saw that it was good. Genesis 1:24 - And G-d said, Let the earth bring forth the living creature after his kind, cattle, and creeping thing, and beast of the earth after his kind: and it was so.

What is marvelous to witness with today's technology is peering into the womb during the developmental stages of the human fetus. We witness the neural tube and the brain forming first, and then the organs and the rest of the body. But this takes place only during development and not by a step-by-step evolutionary process OUTSIDE the womb. This happens only inside the womb (or egg) simply because (as just mentioned above) it is impossible since a species would have to skip from a primitive form to a fully developed form with the capability and instinct to mate. However, there is one marvelous defining detail in this, which is the umbilical cord that attaches, connects, and facilitates the nutrients and oxygen supply between mother and baby. We are no differently dependant upon G-d since we were created by Him (in His image). Genesis 1:27 - So G-d created man in his own image, in the image of G-d he created him; male and female he created them.

How about such things as the law of gravity and where it comes from? Its only logically to deduce that when matter was created, the law of gravity was also established at the same time to stabilize matter (a state of equilibrium). Yet, why don't the evolutionists see from this vantage point?

Moreover, it is utterly impossible to overlook the supernatural! How else can one explain the origin of matter, the origin of energy, and the origin of life? Secular (non-religious) minds need to address this issue.

Hop, skipping and jumping around here… lets take a better look at "natural selection" again… In defense of "natural selection" (not evolution) its also interesting to note that environmental pressures have produced more noticeable differences among peoples, particularly in body shapes. Among the tallest and long limbed people in the world who live in the hot, dry areas of the world, and at the opposite extreme in body shape are the Eskimos who live in a cold climate, have smaller bodies and relatively short arms and legs. The reasons for this have to do with bodily heat loss. If you have a large body, you desperately need to get rid of heat, while small Eskimos have opposite problems of heat balance. Because of the cold climate they live in, they desperately need to conserve heat. Thus, natural selection molded their body shapes oppositely, based on their contrasting climates.

This also applies to skin color and hair type. Some people even facilitate their racist beliefs that the origin of afro-like kinky hair (tightly coiled) was the next step up from chimpanzee hair to gradually straight human hair. However, this is the most absurd idea from the truth! The reason why some people have straight hair rather than afro-like kinky hair is due to light exposure and absorption from the sun (vitamin D). This is because the roots of straight human hair may act as fiber optic tubes that allow UV ultraviolet light to pass into the skin. However, "kinks" in fiber optic tubes are known to PREVENT UV from passing through. In addition, this is also linked and has everything to do with the gradual "natural selection" of dark skin. We find the evidence for this in those who left the African continent and migrated to the northern regions of the world. While in Africa upon exposure to "UV-B" light, (the ultraviolet wavelength that stimulates our bodies to produce vitamin D) their dark skin and kinky hair had developed the natural ability to manufacture vitamin D (which is essential for bone development). In effect because the degrees of UV-B where much higher in this region, their skin darkened with skin cells called melanocytes that produce melanin so as to block the overabundance of UV-B so that the skin could guard against allowing to much vitamin D absorption into their body. But when they moved to the northern regions of the world the UV-B light was (a lower degree) and thus was not strong enough to penetrate the dark pigmented skin and therefore left them with a vitamin D deficiency in which malformed their bones. Hence, lighter skin had to gradually evolve in order to allow UV-B light into the skin. And it was during this transition period from dark to light skin, that the need for vitamin D grew so intense that the hair of these northerners simultaneously straightened so that UV-B light could pass into the body. Additionally, we also find that persons with blond hair have lighter complexions and those with dark brown or black hair tend to have darker complexions. This is because blonde hair reflects sunlight (but allows more UV-B in), while black hair absorbs it (but allows less UV-B in). Incidentally, the need for vitamin D absorption may also serve as the reason why infants of all races (including African infants) are born with fairer skin, looking like white infants so that they can absorb more vitamin D (which they desperately need) from the sun for development.

Another example of "natural selection" (not evolution) is the Hobbits of the small island Flores who were cut off from the mainland and began to shrink in order to adapt to the limited resources of their environment which included less food, whereby their smaller bodies could get by on less food. The smallest elephant of the world also comes from Flores and was determined by "natural selection" to be "island dwarfing". Another thing that is baffling about the Hobbits is that they had a very small body and thus a small head that contained a smaller than typical sized brain. Yet we know they were intelligent because we find remnants of stone tools and the use of fire. This alone should suggest something very important to evolutionists - that big brains are not always smarter or better. We find these today in dwarfs - a condition called "microcephaly" which is an abnormal smallness of the head. Additionally, other miniature living humans can be viewed in the world Guinness book of records. One such person was Lucia Zarate born 1864 in San Carlo, Mexico and died in 1890 of hypothermia. She was 26 years old (fully grown) and weighed only 4.7 pounds and measured at only 21.5 inches tall. She was found to be healthy and intelligent and could speak two languages (Spanish and English). THESE PEOPLE ARE STILL HUMAN AND AS INTELLIGENT AS ANYONE ELSE IS! However, the evolutionists still try and smash the long-cherished religious belief that our human species has had the earth to ourselves for approximately 6,000 years when they examined three little bones from the most complete Hobbit's left wrist (there has been 12 hobbits discovered). They determine that this one Hobbit's three left wrist bone pieces is "indistinguishable" from an African ape as well as a modern-like human wrist. It does not resemble modern humans or ape/monkey. Therefore, they classify the Hobbit as representing a different species of human - thus an evolutionary step for humankind. Again, for the creationists, be extremely careful here! This was determined from only 1 Hobbits left wrist. It is neither ape nor humans, which quite possibly offers the logical explanation that it was not found truly complete. Additionally, the dispute has not been entirely resolved since hundreds of genetic diseases could affect the size and shape of human bones.

So let it be known that humans whether big or small, black, tan, yellow or white, (best suited for their environment with varying traits) are still of the human species and NOT THAT OF THE MONKEY FAMILY! To no accord, the evolutionists put forward that we evolved through various stages of development (from monkey to caveman to human) becoming increasingly erect in our body posture until finally we became bipedal and big braided animals living in caves and using primitive tools. However, make no mistake about any of this, some of the greatest people who have ever lived were cavemen… "Lot" was the nephew of the patriarch Abraham, who ended up living in a cave with his two daughters after the destruction of Sodom and Gomorrah. Additionally, King David and his army hid in caves, as well as the Essenes who wrote the Dead Sea scrolls also lived in caves. In addition, they all used primitive unrefined tools at that time. Even Jesus Himself was born in a cave (animal manger). Additionally, walking upright is NOT convincing evidence that an ape is becoming human, since there are species in the monkey and ape family living today that walk upright, like orangutans, chimpanzees and spider monkeys?

The truth is that there were never any ape-men or women, only humans. In fact "Job" in the Old Testament tells us about cavemen, (Job 30:5-6) - They were driven out from among men. They shouted at them as at a thief. They had to live in the clefts of the valleys, in caves of the earth and the rocks. (These were not ape-men, but descendants of those who were scattered from Babel). Hebrews 11:37-38 - They were stoned, they were sawn asunder, were tempted, were slain with the sword: they wandered about in sheepskins and goatskins; being destitute, afflicted, tormented; (Of whom the world was not worthy:) they wandered in deserts, and in mountains, and in dens and caves of the earth. Even Jesus Himself was homeless without a place to sleep. Matthew 8:20 - Jesus replied, Foxes have holes and birds of the air have nests, but the Son of Man has no place to lay his head. Regrettably in the end times instead of evolving into more civilized beings, G-d is going to push us right back down into the ground again because of our pride. Isaiah 2:17-22 - And the loftiness of man shall be bowed down, and the haughtiness of men shall be made low: and the Lord alone shall be exalted in that day. And the idols he shall utterly abolish. And they shall go into the holes of the rocks, and into the caves of the earth, for fear of the Lord, and for the glory of his majesty, when he ariseth to shake terribly the earth. In that day a man shall cast his idols of silver, and his idols of gold, which they made each one for himself to worship, to the moles and to the bats; To go into the clefts of the rocks, and into the tops of the ragged rocks, for fear of the Lord, and for the glory of his majesty, when he ariseth to shake terribly the earth. Cease ye from man, whose breath is in his nostrils: for wherein is he to be accounted of? Revelation 6:15-17 - And the kings of the earth, and the great men, and the rich men, and the chief captains, and the mighty men, and every bondman, and every free man, hid themselves in the dens and in the rocks of the mountains; And said to the mountains and rocks, Fall on us, and hide us from the face of him that sitteth on the throne, and from the wrath of the Lamb. For the great day of his wrath, is come; and who shall be able to stand?

What the evolutionists are doing is twisting things to fit within a frame of reference that is contradictory to established evidence. In addition, ALL of the "few" discovered missing links between ape and man ever to be discovered are still those of monkey and ape species. (Please readers, do the research for yourself). However, I will give a quick breakdown… All of these discovered missing links (mentioned as transitional forms) (hominids), from monkey/ape to human are Nebraska man, Piltdown man, Swanscombe Cro-Magnon, Neanderthal, Pithecanthropus, Homo erectus, Lucy, Ardipithecus, and Australopithecines. The history of hominid paleontology is a history of shifting theories and many uncertainties. Simply put, it doesn't fall neatly together into a nice evolutionary package. Many scientists spend their careers jousting over the proper interpretation of ancient teeth, bone chips and dust. In fact, all the hominid bones in the entire world would'nt fill one large coffin, and the very best skeletons are only less than half complete. Therefore, we're really working with a lack of evidence.

Starting at Nebraska man… this whole reconstruction was based on one tooth that turned out to be the tooth of an extinct pig.

Piltdown Man… The skull was a human skull, a combination of a human and ape jaw, eight teeth were filed, and the bone was treated with iron salts to give the appearance of age.

Cro-Magnon Man, Neanderthal man and Swanscombe Man… Now considered Homo sapiens - modern humans. Their brain capacity was actually larger than todays modern man. Their skeletal features are virtually indistinguishable from modern man. If you took any of these races of men, gave them a shave, dressed them up in a suit of clothes and walked them down Hollywood, no one would take notice. (In other words, they were human).

Now, one of the earliest of those transitional types was known as Pithecanthropus, or Homo erectus, also known as Java man, Heidelberg man and Peking man. The archeologist who has discovered these bones (actually only a ape skull fragment) have admitted that they were found within a few feet away from that of human skulls and bones and confided that the bones could have been mixed.

Following this is Ramapithecus… Consisting of only several teeth, jaw fragments and canine teeth. These are simply just apes.

Then we have Ardipithecus, which means ground ape. This creature even though it has a upright walking pelvis like the orangutan today, still has the ape-like skull no matter how human they try to make it look. Additionally, what does not make logical sense here is that this creature had very prominent divergent big toes, which are known to be associated with grasping. Therefore, why would a creature that's fully adapted to support its weight on its forelimbs in the trees elect to walk bipedal on the ground? Thus, it is my estimation that this is just another species of ape. Let me also add that when the archeologists found these bones they were so brittle that they literally crumbled when touched (like chalk). The bones were also scattered into many fragments. The bones were so distorted they needed digital reconstruction. I heard someone remark, how much would you be sure about a "Rosetta stone" that was trampled and disfigured to the point of illegibility?

Australopithecines Africanus also known as Eoanthropus, Anamensis, Homo Habilus or more popularly known as Lucy (said to have walked upright). However, they determined this from just little bits and pieces of bones. Lucy's pelvis bone was altered to make it fit, and the knee bone was not even found with Lucy (it was found a mile away).

I've been told by some evolutionists as well as highly educated and intelligent people (salt of the earth) that words matter, and in a debate, even the littlest inaccuracy of words can have a bearing. I try to explain to them that everything said and perceived lies in a "point of view". I think the Greek philosopher Democritus said it best: Nothing exists except atoms and empty space, everything else is opinion. Most scientist and Evolutionist tend to mock people who are unknowledgeable on scientific words as if they don't understand all the schooling terms involved. When they say words matter, I fall back on what words actually are (the words of man that is), which are just labels as well as idiosyncrasies and technicalities. These are all word games that are full of hot air and only serve to pull apart the argument and the true reality of the discussion and debate to minuscule pieces of irrelevant garbage. They always try to cage me into the isolated palindromes of labeled science! Again, these are all POINTS OF OPINION whether cited from scientists, specialists, original sources or not. The squabbling can go on and on only establishing points of view and accomplishing nothing!

To understand just what I'm trying to get at here… The evolutionists continues to argue over the authenticity of Lucy the supposed hominid. The argument revolves around just her pelvis. The contention is that the scientists on this project that put the pieces of Lucy's pelvis together made a plaster mold of what pieces they had, and then in order to make the pieces fit they had to be sanded, sawed and filed to fit together. When the finished product was complete, it had appeared to look somewhat human suggesting that this was a transitional stage between ape and man. My opinion in the whole process of reconstructing Lucy's pelvis is that it is a DISTORTION AND DEVIATION OF THE NORMAL SHAPE (considering what little bone fragments they had to originally work with)! And this is not even to mention that Lucy may have been just another species of ape/monkey that walked upright. We should scrutinize and observe that just because Lucy had a somewhat humanlike pelvis and may have walked upright, it is not an indication that an ape/monkey is becoming human, since there are species in the monkey and ape family living today that walk upright, like orangutans, chimpanzees and spider monkeys.

In response to my point of view, that Lucy's pelvis was a distortion; the scientists and the evolutionists say that my judgment is quite an accusation as well as a poor argument that cannot be back up. So, lets have a better look at what was involved in this whole process to determine whether it can be decided if indeed there were any distortions or deviation of Lucy's pelvis?

For starters, Lucy's left hip joint socket had been misshaped and broken into about forty pieces while it was in the ground. The pelvis was then X-rayed and discovered that the back of it where the sacrum connects with the hip joint had been smashed (by unknown means) thus shattering and twisting the ilium. The ilium is the wide flat upper portion of the pelvis that is connected to the base of the vertebral column. It is a separate bone at birth but later becomes fused with two other bones to form the hipbone.

Months were spent carefully identifying each piece of the ilium then casting it into plaster and sanding out the edges, and then reassembling them like pieces of a puzzle. Every piece had to be sanded to line up and connect with other pieces of bone to cover up any distortion that occurred after the bone was "damaged". Now after the left side of the pelvis was restored, a sculptured duplicate image of the right side of the pelvis was sculpted in plaster which placed Lucy's sacrum in the middle to make Lucy's pelvis have a concaved (curved in look) like a human pelvis, but not quite as protruding. However, remember Lucy's left-half pelvis isn't one piece. It is about forty pieces of plastered and sculpted (sawed, sanded and filed bone replicates) that have been carefully reshaped to remove the "distortion" that occurred during burial. Their strong argument is that this was done by "professionals" who have received a certified education in knowing how to do these types of things. In addition, this very well might be the case. However, "DISTORTION" is by definition, a deviation from the normal shape. But since this is the only pelvis of its type that was found (actually it is just a half-pelvis), how do they know what it is actually supposed to look like? In other words, they had to reshape it to look the way someone thought it should look.

The discerning part in all this is that they take these sculpted models to schools to show kids how Lucy was a "transition" between an ape and human. This is not something that is definite. Like a religion in itself, some of these scientists believe so heavily (almost madly) on the theory of evolution that it blinds them into believing so strongly that they almost seem to meld what they find into some kind of imaginary Golem (in Jewish legend, a creature made of clay and brought to life by magical incantations).

This type of philosophy is actually a whole other religion in itself… that without even realizing it - it too is based solely on faith for the mere reason alone that NOT ONE physically tangible (touchable and concrete) transitional form in the whole history of archeology has ever been found. Therefore, evolution is a preconceived notion (involving vague and general ideas) that living things changed and transpired through time without a single shred of evidence. So why do I say that the theory of evolution is also a religion? Because it is something that has to be believed (through faith) without visible evidence. Hebrews 11:1 - Now faith is the substance of things hoped for, the evidence of things not seen.

All the evidence is clearly expected (not one authentic transitional form has ever been found), man has always been man!

Logically thinking, if man had been evolving from ape-like creatures though MILLIONS of years, do you know how many "missing links and transitional forms" we would be discovering? The answer to this is the amount of people in the world today. It is just about the number you would expect if eight people (Noah and his family) walked off of an ark only 4500 years ago. (I encourage readers to investigate the math of human population growth to discover this for themselves). Briefly, we have a current population of approximately 7 billion people on the planet today. Now we've been taking population censuses ever since biblical times and it has been determined that the worlds population has historically doubled every 150 years. Upon which the 7 billion population we have today is equivalent to 8 people walking off an ark and 150 years later there are only 12 people (give or take a few) on the earth and 300 years later 24 people (give or take a few) on the earth. You find out you only have to double 30 times to get the current population of the world 30x150 years means they walked off that ark after the Great Flood approximately 4500 years ago. Its not a perfect method or estimation, but it sure does agree with G-d's word. Genesis 7:21 - And all flesh died that moved upon the earth, both of fowl, and of cattle, and of beast, and of every creeping thing that creepeth upon the earth, and every man. Job 22:15-17 - Hast thou marked the old way, which wicked men have trodden? Which were cut down out of time, whose foundation was overflow with a flood: Which said unto G-d, Depart from us: and what can the Almighty do for them? 2 Peter 3:6 - Whereby the world that then was, being overflowed with water, perished.

So how do we know the earth is approximately 6,000 years old? Because documented in the Bible (aside from the Great Flood account) we also have a reliable historic record of "genealogies" thereby allowing us to estimate an approximate age of 6,000 years old for the earth in several ways… beginning with Adam and ending with Jesus.

Adam - 930
Lamech - 874-1651
Noah - 1056-2006
Terah - 1879-2084
Abraham - 2009-2184

Flood - 1650

Additionally we know the timeline today from Jesus was 2,000 years, from Jesus to Abraham was 2,000 years, and from Abraham back to Adam is about 2,000 years, which gives us an approximate age of 6,000 years old. Which may also pose the question that the 7,000th year may be the golden jubilee year since 7 is a number throughout the Bible that encompasses completion. Genesis 2:2-3 - By the seventh day G-d had finished the work he had been doing; so on the seventh day he rested from all his work. And G-d blessed the seventh day and made it holy, because on it he rested from all the work of creating that he had done.

The gathering of the first fruits… Leviticus 25:3-7 - For six years sow your fields, and for six years prune your vineyards and gather their crops. 4 But in the seventh year the land is to have a Sabbath of rest, a Sabbath to the Lord. Do not sow your fields or prune your vineyards. The year of Jubilee… Leviticus 25:8-13 - Count off seven Sabbaths of years—seven times seven years—so that the seven Sabbaths of years amount to a period of forty-nine years. Then have the trumpet sounded everywhere on the tenth day of the seventh month; on the Day of Atonement sound the trumpet throughout your land. Consecrate the fiftieth year and proclaim liberty throughout the land to all its inhabitants. It shall be a jubilee for you; each one of you is to return to his family property and each to his own clan. The fiftieth year shall be a jubilee for you; do not sow and do not reap what grows of itself or harvest the untended vines. For it is a jubilee and is to be holy for you; eat only what is taken directly from the fields. In this Year of Jubilee, everyone is to return to his own property. There are far more reaching implication here since it may also signify the day of the Lords coming… Returning… The year of the Lords favor… The Day of Atonement… The holiday of Rosh Hashanah. Jesus' returns to gather the first fruits (the believers). The rapture… For every commandment there is a far reaching reason and purpose behind it in the fulfilling and prophetic culmination of G-d great plan.

In addition, this is the point where things get choppy in the book. There are also other things of importance I'd like to throw into the mix here (little tidbits) that should not be left out of the equation of the evolutionary theory. In addition, these being that there are evidences in the plant world that defy evolution. One such plant is called the Stylidium orchid (a trigger plant) that somehow defies evolution by mimicking a certain species of female wasp. The aroma that the orchid puts out smells just likes the female wasp so that the male wasp is attracted to it. When the male wasp lands on it, it closes on it quickly using its trigger mechanism and thus whacking the pollen onto it for pollination. This is the only way this orchid pollinates itself. The timing is critical… for two weeks, and only once a year when the orchid is ready to pollinate, the male wasp is out flying around looking for the female wasp, but the female wasp isn't there yet, SHE DOESN'T MATURE UNTIL TWO WEEKS LATER. The question is, how could something like this evolve. Simple, IT CAN'T! It had to be created for that very purpose. Additionally, there is also the Lampsilis mussel (in the clam family). This mussel mimics a tiny little fish like a minnow. It puts its orifice tissue on top of its shell and it sits under the water jigging its orifice like a baitfish. In order for it to maintain its future existence as a species, it requires a large mouth bass to swim down and eat the orifice (bait fish). It rejects all other fish except for the bass. It somehow knows the difference and the split second a bass opens its mouth to take the bait, the Lampsilis muscle shoots all its eggs up into the mouth of the bass that then attach to the gills and drink its blood until they are fully grown and then drop off to repeat the cycle again. The question is, how could something like this evolve. In addition, how does the L-muscle know the difference between fish?

Moving on here… If the evolution theory, mutation, natural selection or monkey skeletons has not won the belief of most vulnerable minded people yet… Now they came up with something called "Emergent Complexity" which is a new agent in aiding the theory of evolution of how life began. It seems that they focus on things in nature such as "groups" of animals or clusters of insects, for instance a flock of birds or a school of fish… they focus on how they move as one, as if there was "one" brain behind them all, some sort of mechanism that controls them as "one" living organism, suggesting that the behavior of the whole is more than the sum of the parts. Which is partly true in a sense, but next they go on to suggest that where there is order, there is a leader, so they postulate who is in charge of the school of fish or the heard of deer when swimming or running? They suggest THAT THERE IS NONE (a sort of random chaos) and that they are following their own set of rules (as a group) such as staying aligned in the same direction and maintaining distance by not getting to close to each other but not to far away either, and staying out of harms way of a predator. This grouping of movement is also seen in the growth of molds and is seen in crowds of people where movement is happening at a subconscious level. Even the neuron network of our human brain, they have begun to associate that "consciousness" is one of the "emergent" phenomena that's taking place… that something so complex spontaneously emerged from individual parts following simple rules. To follow, they have also associated the human neuron network with the same laws of a computer that can be programmed by "rules" and they call this "emergent". They declare this is going to be the study of the 21st century proving the evolution theory! They also go on to suggest that complex parts as small as molecules under the right conditions at the creation of the heated earth will form increasingly more complex structures and those complex structures will form even more complex structures until finally you get life - from simplicity to complexity. But let me remind you (as previously brought to your attention) that reconstructing these conditions in a lab has still not brought about all "left" amino acids that are necessary for life. And we cannot say that a flock of birds, pods of fish and people walking in a crowd is "emergent" when we all know that a LEADER (someone or something) had to make the first reflex in movement thus sending a signal to the others whether subconsciously, consciously or even physically to point in the direction necessary for movement. However, little does "emergent complexity" theory know that all these movements are mathematically inclined by an even superior derivative that you will read about in the following chapters? Something even greater is creating these "groups" and instructing their movements. It's an invisible force that's called the "one law" otherwise known as the "Primordial Language" - the creator in action! (keep reading to find out what this is?)

In the meantime, lets get back on track with the evolution theory again. Evolution theory also advocates to us that the Earth is billions of years old… While the Bible clearly tells us a different story, that it was created in a literal six days. It is interesting to note that the word DAY in Hebrew, which is YOM, distinctively indicates a twenty four (24) hour period. It is used exactly 2301 times throughout the entire Old Testament Bible (5 books of Moses) and in every instance; it is always used in a singular way. However, its also interesting to note that in the 6 days of creation G-d reiterated it six times. The reason it was reiterated repeatedly was that G-d purposefully and firmly wanted to get His point across. In foresight, He knew the evolutionist would try and harm the reputation of it. Now notice when we read Genesis that it repeats and reaffirms it repeatedly in Case we did not grasp it the first time, it makes it known again. Genesis 1:5, 8,13,23,31 - And G-d called the light Day, and the darkness he called Night. **In addition, the evening and the morning were the first day.** 8 - And G-d called the firmament Heaven. **And the evening and the morning were the second day.** 13 - **And the evening and the morning were the third day.** 23 - **And the evening and the morning were the fifth day.** 31 - And G-d saw every thing that he had made, and, behold, it was very good. **In addition, the evening and the morning were the sixth day.** Genesis 2:2-3 - **And on the seventh day G-d ended his work which he had made; and he rested on the seventh day** from all his work which he had made. 3 - **And G-d blessed the seventh day, and sanctified it:** because that in it he had rested from all his work which G-d created and made. Additionally, it is also noticed that each one of the creation days has a numerical number attached to it, namely day 1, the second day, 3rd, 4th, 5th and 6th, and lastly 7th day. And every time a "day" is associated to in Genesis it means a 24 hour period that is noted by "evening and morning", which is one cycle of the Earths rotation on its access in reference to a fixed astronomical light (our sun). All of this is to show us that G-d was not just talking about a prolonged (long-drawn-out) time, but about one cycle of the Earths rotation on its access in reference to a fixed astronomical light (our sun).

As an added message, something else entirely astonishing reveals itself when we add the total amount of YOM'S (2301 to be exact) that are used throughout the entire Old Testament Bible… if one adds 2+3+0+1, it equals 6 days… G-d is going out of His way to emphasize that each day is a normal, single (24 hours) day. Something to think about for those who study Torah - it has been noticed that the associated coincidences of words/letters that equate to numbers (same numbers), seem to have a correlation with a significant meaning. This subject is called Gematria. (I will discuss Gematria more in depth in one of the following chapters).

Even more significant to a single day… G-d goes one step further in reaffirming this for us when He conglomerates "one full week" as a period of "seven days".

Those that are familiar with Hebrew know that the literal word for "week" (a period of seven days) is the word "shabuwa" or pronounced "shabuway".

Spelled in Hebrew: שביעי

This very word is found in Genesis 2 as well as Leviticus 12:5 and virtually the whole Old Testament, which refers to the word "week".

So again, not only does G-d tell us that one day is equal to 24 hours as evening and morning which are fixed references that denote planetary revolvement of the Sun and Moon. But He also reaffirms this for us when He refers to one full week as a period of seven days. So, it's not merely that the word "Yom" is used six different times in Genesis, but it is also the word "SHABUWA" that defines a seven day period.

So as we can see these are literal days! Exodus 20:11 - For in six days the Lord made heaven and earth, the sea, and all that in them is, and rested the seventh day: wherefore the Lord blessed the sabbath day, and hallowed it. So important was the confirmation of this message that G-d included it in the first set of laws (the 10 commandments) that He gave to

Moses on Mount Sinai - to remember and keep the 7th day. Exodus 20:8 - Remember the Sabbath day by keeping it holy. Deuteronomy 5:12 - Keep the Sabbath day to sanctify it, as the Lord, thy G-d hath commanded thee.

We also have to question… If a day isn't a literal 24-hour day, then what's a year? We all know that it's one cycle of the Earth orbiting around the Sun (365 days). The message here is, G-d is not exaggerating about the chronological time limit. He laid this message on thick and is quite serious about it! Genesis 1:14 - And G-d said, Let there be lights in the firmament of the heaven to divide the day from the night; and let them be for signs, and for seasons, and for days, and years.

But how can we really-truly be sure these are literal days aside from the 6 days that are mentioned? The answer to this tough question is that G-d protects His sovereign word. When G-d speaks, His word is instantaneous! (As was creation). Testimony of this can be found in the New Testament when Jesus rebuked the storm on the sea of Galilee it was "instantly" gone! Which incidentally proved to the disciples beyond a doubt that they were in the presents of the living G-d and Creator?

Also to mention, G-d deliberately ordered the creation of the fruit trees on the 3rd day (without the sun) for a purpose. Because G-d knew that "evolutionism" could not negotiate with His creation order… and incidentally all the other things G-d ordered also defies the categorized order of the geologic time table of evolutionism which will be discussed in a later chapter. Today science directly KNOWS that there is no known natural process that can create something that was not already there first, (microbes will not grow in water without microbes). Genesis 1:11-12 - Then G-d said, Let the land produce vegetation: seed-bearing plants and trees on the land that bear fruit with seed in it, according to their various kinds. In addition, it was so. The land produced vegetation: **plants bearing seed according to their kinds and trees bearing fruit with seed in it according to their kinds.** And G-d saw that it was good. G-d needed fruit trees already fully grown so that He could feed the man (Adam). The day will NEVER come in which we will catch evolution in the act of happening because it simply didn't happen that way.

Many commentators say that there is a fatal flaw in Genesis… How could the Sun be created on the 4th day and still have light 3 days before it when there wasn't any Sun? Genesis 1: -5 - In the beginning G-d created the heaven and the earth. In addition, the earth was without form, and void; and darkness was upon the face of the deep. In addition, the Spirit of G-d moved upon the face of the waters. **And G-d said, let there be light: and there was light.** And G-d saw the light, that it was good: and G-d divided the light from the darkness. And G-d called the light Day, and the darkness he called Night. And the evening and the morning were the **first day.** Genesis 1:16-19 - **And G-d made two great lights; the greater light to rule the day, and the lesser light to rule the night:** he made the stars also. And G-d set them in the firmament of the heaven to give light upon the earth, and to rule over the day and over the night, and to divide the light from the darkness: and G-d saw that it was good. And the evening and the morning were the **fourth day.**

So as we read light was already established on the first 24 hour day, but the actual Sun was not created until the fourth day. What is the purpose and meaning for two lights? G-d knew you would ask that question! His message here is so spiritually important that an "extra light" existed prior to the actual creation of the Sun itself because G-d is telling us that it is Him alone and His light that sustains us and not actually the material Sun itself. The creation of the actual Sun by G-d was only a supplemental afterthought. It's not fixed, necessary or essential, only G-d is essential.

Something to this effect is mentioned in the New Testament in Revelation 21 and 22 that there is no more need of the Sun or Moon but G-d Himself will provide the illumination for the future New (and eternal) Jerusalem all by Himself. Revelation 21:11 - Having the glory of G-d, and her light was like unto a stone most precious, even like a jasper stone, clear as crystal. Revelation 22:23-24 - **And the city had no need of the sun**, neither of the moon, to shine in it: **for the glory of G-d did lighten it**, and the Lamb is the light thereof. And the nations of them which are saved shall walk in the light of it: and the kings of the earth do bring their glory and honour into it. Revelation 22:5 - And there shall be no night there; and they need no candle, **neither light of the sun; for the Lord G-d giveth them light:** and they shall reign for ever and ever. What's even more interesting is that this was prophesized about before-hand in the Old Testament by the prophet Isaiah. Isaiah 60:19 - The sun shall be no more thy light by day; neither for brightness shall the moon give light unto thee: but **the Lord shall be unto thee an everlasting light**, and thy G-d thy glory.

This clearly indicates to us that since there will be no need for the Sun as a light source in the future, it is certainly possible that there was no need for the Sun as a light source in the beginning.

Furthermore, worshiping the sun as the Egyptians did (or anyone else) is blasphemous! Job 31:26-28 - If I have regarded the sun in its radiance or the moon moving in splendor, so that my heart was secretly enticed and my hand offered them a kiss of homage, then these also would be sins to be judged, for I would have been unfaithful to G-d on high.

However, is it scientifically possible that light can come from another source other than the material Sun? Yes, surprisingly it can! Light does not automatically require the sun since it can also be produced by friction, *sonoluminescence* (created by intensive sound waves), chemical reactions, atomic fission and fusion and by fire.

Light in itself is actually the ONLY thing we do see, since physics says light behaves as a particle and wave, and EVERYTHING is made up of particles that move in waves. Nothing could be closer to the truth, especially when we realize that "G-d is light" Himself. 1 John 1:5 - This then is the message, which we have heard of him, and declare unto you, **that G-d is light**, and in him is no darkness at all.

It is elevating to see the scientific evidence supporting what scripture says… This might very well be one of the most powerful weapons that exist in the whole arsenal of scripture and science - the connection between G-d and light.

More importantly, there is something very VERY special about the wave path that these atomic light particles take (since they move in waves and spiral). And it is this spiraling design that forms the Hebrew letters. In a nutshell, IT IS THE PRIMORDIAL LANGUAGE! To sum it up, (though I don't completely expect you to understand it yet) these Hebrew letters that are reminiscent of light (since they are formed by atomic particles that move in spiraling wave patterns) are the embodiment of G-d through which He forms all things and works His ways (some may visualize this as "String Theory"). Please keep reading, I assure you, you will eventually come to understand this magnificent piece of knowledge when you get to the actual chapter on the "Primordial language". I know the temptation is there to quickly thumb through this book to look at the pictures and skip directly to that chapter... I suppose you could if you really wanted to since comprehension comes to each of us in different ways. But it may grossly impair the full aspect of understanding.

In the meanwhile lets move onto another subject... We briefly talked about this earlier, but now I'm going to expand on it again just a little more... Geologists have the overconfidence to claim that the Grand Canyon is millions of years old... However, there is also substantial evidence there that shows us that it formed rather quickly (research this for yourself). In fact there are many pieces of evidence at the Grand Canyon that tell us that it is NOT millions of years old, but rather only thousands, and it may even corroborate and commemorate the biblical story of Noah's flood. There is a demonstrated reenactment (that might prove otherwise) of the possible creation and age of the Grand Canyon, which can be found at Mount St. Helens. It was in March of 1982 that Mount St. Helens clued us in to how fast a canyon could be created when the north slope of the mountain slid into the Toutle valley below, damming up the Toutle River. The waters breached the dam and it disastrously drained by a massive water and mudflow that formed several complete canyons in just a matter of hours. Over a TWO-YEAR period a huge lake formed (approximately 1000 feet wide, 160 feet deep and 2 miles long). However, the evolutionists say that structures like this form slowly over millions of years. In that case if we hadn't seen or heard the Mount St. Helens eruption or witnessed the canyons creation and arrived on the scene some time later, perhaps we would have estimated the newly created canyon to that of the exaggerated age of the Grand Canyon. Get my point? In addition, additionally, something else phenomenal has occurred here in this newly formed Toutle canyon. COAL IS IN THE PROCESS OF FORMING!

Lets focus our next attention on coal... The evolutionist theory is that coal takes millions of year to form. From what we understand about the formation of coal, it starts out with swamp vegetation that sits for long periods of time gradually decaying and mixing with other minerals and elements to slowly turn into "peat". Then it takes that "peat" long periods of time to slowly, harden and turn into coal. This process according to the evolutionists takes a very VERY long period of time. But there's one flaw in this whole scenario of coal formation. Embedded in coal beds "ALL" over the earth we find polystrate fossils. These are tree fossils spanning many layers, ridges and stratum (seams) of coal. Now here is the problem... if that coal took many hundreds of thousands of years to form, none of those tree "fossils" would be there - they would have all decayed in a matter of years. We also find that the root structures of these trees are not there either, meaning that they did not grow in the coal. So the question is, how did they get there? One of the things that would be required for these trees to be uprooted and moved would be by a great flood. A great world catastrophe - because we find coal beds "ALL" over the Earth! Which (I might add) is testimony of Noah's great flood? Its now been about 20 years since the eruption that formed the Toutle river/lake at Mount Saint Helens... there is now forming at the bottom of this river bed a thick layer of peat. This is observable evidence that coal forms rapidly.

One should also consider that with the advancement of human technology, we can make oil in one day, coal can be made in one week and petrified wood can be petrified in one week (one can get the chemical ingredients for tuning wood into petrified rock which is nothing more than silicon and aluminum compounds). The chemical components used to artificially petrify wood can also be found in natural conditions such as around volcanoes and sedimentary rock and its possible for natural petrification to occur rapidly by these processes. Also stalactites which are claimed to be thousands of years old have been observed to grow as fast as one inch a year, and we can even create cubic zirconia diamonds in one day in a lab which are almost as high-quality as real diamonds (other synthetic gems can also be created except for granite). Now, if man can create what science tells us takes millions of years for the earth to create, imagine how much more potential G-d possess to have created everything within 6 days. Logic tells us it can be done!

Besides just examining what the earth exhibits from its crust, there are also observations of the earths magnetic field. Experiments have been done that measure its strength from the surface of the earth. These test indicate that over the last several hundred years or so the magnetic field of the earth has been "decreasing" in strength, suggesting that the earths magnetic field cannot be any older than 10,000 years old. But its also been estimated that there were many magnetic field reversals and intervals throughout the history of the earth since sparse magnetic readings from around the world suggest this. The bottom line in all this is that we have not physically witnessed (or have recorded in science) a reversal in earths history even though sparse readings suggests this in rocks, lava flows or basalt throughout the world. This again involves unverifiable proof since there was not an eyewitness to record such events. There may indeed be other reasons for such sparse magnetic readings around the world, such as continental drifts. This may also be indicative of a worldwide flood - Noah's flood. Since we find frozen Woolly mammoths in the arctic with warm weather tropical foods still frozen inside their stomachs - meaning, that a catastrophic event occurred since the environment that they were living in was once originally warm weather and not the cold arctic that we are now familiar with today.

The Bible happens to be very descriptive about the flood! AND IT BACKS UP WHAT SCIENCE KNOWS ABOUT HYDROLOGY, as well as the Earths topography and plate tectonics! It tells us that Noah's flood caused the mountains to RISE and the valleys to SINK down with water. Eventually the water was above all the mountain. Why do you think they find so many fossils that are high up in layers of rock and mountains? This is because they were pushed up there by the rising water and died there, (remember it takes the right properties to be fossilized). Biblical scripture backs this geological process up: Psalm 104:6 - You covered it with the deep as with a garment; the waters stood above the mountains. Psalm 104:8 - They flowed over the mountains, they went down into the valleys, to the place, you assigned for them. Psalm 104:10 - He makes springs pour water into the ravines; it flows between the mountains. Psalm 104:13 - He waters the mountains from his upper chambers; the earth is satisfied by the fruit of his work. Clearly, what the Bible is telling us is that G-d altered the earth's topography by causing the flood waters to create new continental landmasses, mountains and deep ocean basins. Which is exactly what we find today?

But herein lies the clue that the Grand Canyon cannot really-truly be millions of year old. If by all the rising and sinking, and rivers pouring, surging and depositing more layers on top of layers until finally creating the number of superimposed layers that we see today… why were there no river "deltas" formed. A delta is a landform where the mouth of a river flows into an ocean, sea, lake, or another river. Point in fact, we cannot have rivers "flowing" without forming deltas… simply put, no rivers in the world can do that! Even if they could, you could not have all the years of rising and sinking of rivers without buckling, bending, sloping, twisting or eroding and wearing down. So where are the deltas? Even Hydraulic engineers (who study the flow of fluids) know that the depths of a deposit is not a factor of "time", but of the enormity of hydrodynamic forces involved. Therefore, a cataclysmic event such as the great flood had to occur, which rapidly deposited the Grand Canyon that we see today. Nevertheless, the evolutionists have a field day with this one... Oh yes, they assert that there is deltas at the Grand Canyon (not in the actual canyon itself) but found at the Colorado River (a distance away). They tell us that rivers, not canyons, have deltas. In addition, that deltas form when water-carrying sediments slows down and deposits the sediment. They claim this is not going to happen in the same place a canyon is being carved! Secondly, as I said they submit that the Colorado River has a delta, and a rather large one. However, because of dams, diversions, and other water management modifications of the Colorado River, very little water still makes its way to the delta at the Gulf of Mexico, and the delta is mostly dry now. However, it is still a mystery as to how the whole system of surrounding land mass, canyons and rivers formed. Today they are still just guesses and assumptions. But what we do know is that Canyons without deltas are formed rapidly, and this rapid pouring of water also fits the precise qualifications for fossilization, which is found at the Grand Canyon. And while trying to figure out how the whole system of the Grand Canyon and other canyons in the region, along with the Colorado River may have worked together in some way, we might want to envision again that the Great Flood broke apart the waters of the deep and the subterranean chambers (exploding from the ground upwards). In effect, the Grand Canyon may have been one of the great remnants of this testimony. Additionally, attesting to this there is unquestionably a vast system of interconnected tunnels, tubes (tunnels through rock and holes), chambers, caverns and caves beneath the Grand Canyon.

However there is still a lingering question here concerning how the canyon layers actually formed? And this observable and quite obvious piece of evidence is happening right before our very eyes today when we stand at the sea shore and observe the sand patterns that the waves make as they go in and out. We notice that it separates itself into gradients and sediments where the larger pieces of sand sink to the bottom and the finer pieces rest at the top surface. However, many of these same patterns are made under water as well from currents. Now when we apply this to strata, layers and banks (such as at the Grand Canyon) we cannot see from just looking at the canyon layers that they were actually formed SIDEWAYS UNDERWATER. And that when the water receded, the sand sediment (rock) was exposed to the air thus drying it out, and in the process BREAKING AND CRACKING into huge fissures. Additionally, we wrongly assume that the bottom layers are older than the layers above it. But the truth in the matter is that strata shows no indication of age, but are rather caused by changes in water currents in which the living creatures that died in the strata cannot be dated either because they are now petrified rock containing no carbon. Therefore, a logical explanation for the position of fossils in the layers is due to the distribution of creatures that lived at different depths.

A picture of how strata forms SIDEWAYS UNDERWATER:

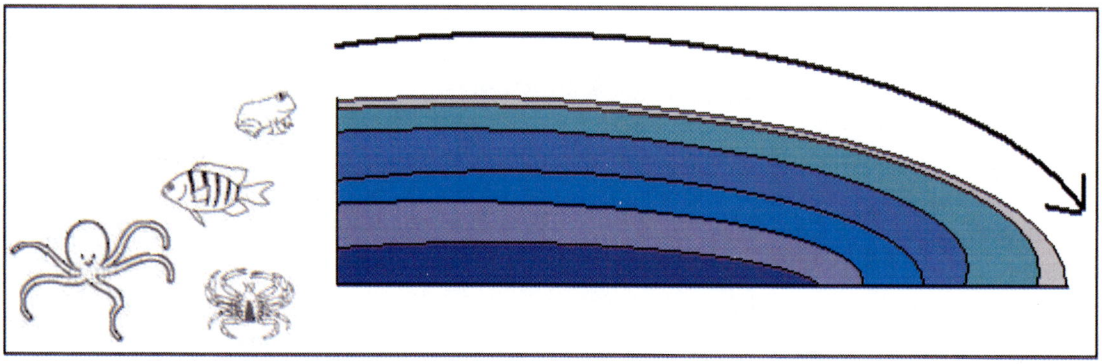

A better understandable view might look like this picture below:

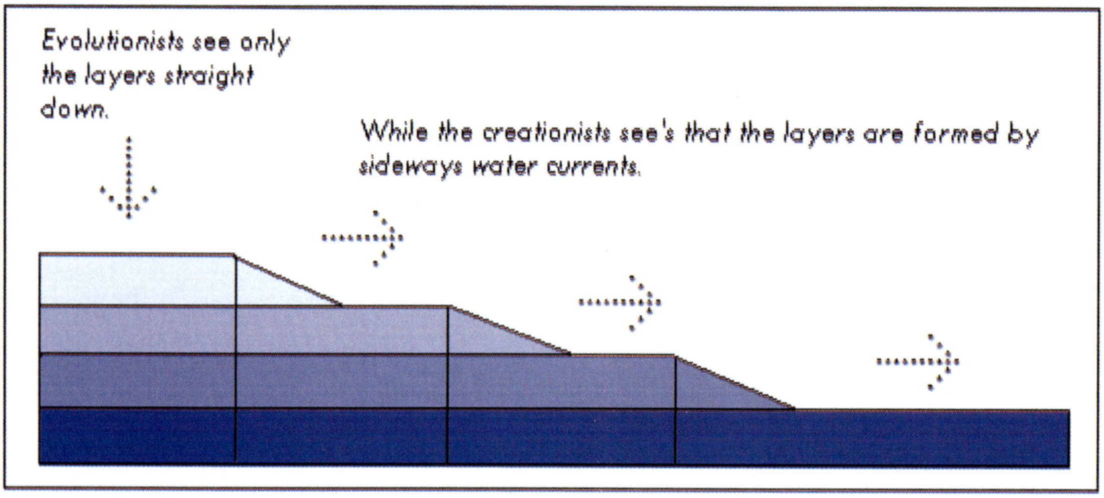

Moving on here to another area still under discussion… Lets examine the properties of light again… Its seems to be that not only does the evolutionists estimate the earth to be billions of years old, but they also join these forces with the properties of light from outer space that additionally fool astronomers into believing that the Universe is also billions of years old. Astronomers relate to it as a "clock in the sky" along with trails of light that are recorded on photographic plates.

This is an illusion. Simply put, we cannot rely on the time dilation that astronomy tells us. We should be asking questions like, DO WE KNOW THAT LIGHT HAS ALWAYS PROPAGATED AT TODAY'S SPEED. Secondly, light is a strange oddity and it does not tell us the whole story, or the "current" whole story. Did you know that it takes approximately 8 minutes just for the Suns rays to reach us here on earth? That would mean that light coming from extremely far distances in space would take many MANY more years just to reach the eyes of our telescopes here on earth. Therefore, what we are seeing is actually past tense already. That is why to me, all of astronomy is still a non-observable science - can that which cannot be observed be considered a science? I mean were still bickering about tangible evidence concerning transient forms like Lucy the speculated hominid - here we have touchable evidence in our own hands and we still cant prove otherwise. Therefore, we should try to stay more focused on what our immediate galaxy tells us. For instance, our own spiral galaxy slowly rotates, but the inner regions of the spiral rotate faster than the outer regions. This means that our spiral galaxy is constantly becoming increasingly twisted up as the spiral becomes tighter. If this were so after a few hundred million years, the galaxy would be wound so tightly that the spiral structure would no longer be recognizable. However, some scientist theorizes through computer simulations that the arms reform themselves and that these types of things constantly occur. This again is conjecture, since nobody has been around long enough to have witnessed or documented any such twisting off and reforming of spiral arms (besides this WOULD take millions if not billions of years). Computer simulation is again hearsay based on prior postulation of un-witnessed history! However remember, we do have a recorded eyewitness - the Bible.

Additionally, the time and distance of light is relative. Which typically means that it is comparative, measured or considered in comparison with something else. It also changes with circumstances and is not permanently fixed, but having a meaning or rate that can only be established in relation to something else and will change according to circumstances. “Big, small, long and short” are relative terms. Now many people assume that time flows at the same rate in all conditions. At first, this seems like a very reasonable assumption, but in fact, this assumption is false. And there are a few different ways in which the nature of time could allow distant starlight to reach earth within the biblical timescale. After all, Albert Einstein discovered that the rate at which time passes is affected by motion and by gravity. For example, when an object moves very fast, close to the speed of light, its time is slowed down. This is called “time-dilation.” Therefore, if we were able to accelerate a clock to nearly the speed of light, that clock would tick very slowly. If we could somehow reach the speed of light, the clock would stop completely. Its not the actual clock, but because time itself is slowed. Likewise, gravity slows the passage of time. A clock at ground level ticks slower than one on a mountain, since the clock at ground level is closer to the source of gravity. We see these same effects when we are traveling in a vehicle. Time appears to flow at the same rate as when we are standing still. This indicates to us that since time can flow at different rates from different points of view, events that would take a long time as measured by one person will take very little time as measured by another person. If you still don’t grasp the concept, try putting your hand on a hot stove for a minute and it seems like an hour, or sit with an attractive guy or girl for an hour and it seems like a minute. That’s relativity. This same concept also applies to distant starlight. Light that would take billions of years to reach earth (as measured by clocks in deep space) could reach earth in only thousands of years as measured by clocks on earth. It may also be relative to us gazing from the earth at distant stars in the sky. So again, time is relative to the observer and it may be relative throughout the entire Universe. Because in “relative” terms (in the colossal grand scheme of things), how are we to really determine what is “big” and what is “small”, or what is “long” and what is “short“, or what is “young” and what is “old” when we compare it to different (conscious) points of view?

There is more… We also know that the Universe is expanding. The Bible indicates in several places that the Universe was stretched out (or expanded). Isaiah 40:22 - It is he that sitteth upon the circle of the earth, and the inhabitants thereof are as grasshoppers; that stretcheth out the heavens as a curtain, and spreadeth them out as a tent to dwell in. This clearly suggest that the Universe has actually increased in size since its creation since astronomers have discovered that virtually all clusters of galaxies appear to be moving away from all other clusters - so indeed it is expanding! This in itself may suggest that light did not always travel at the same speed and may have been billions of times faster at the beginning, and eventually slowed down in transit after creation. In addition, that could well explain why starlight that is not billions of years old is visible on the earth. G-d spread the light out from where the stars were "placed" or He moved the stars from close to the earth to far away. This may also help explain the radiometric dates since most nuclear processes are mathematically related to the speed of light, a faster "c" might well mean a faster rate of radioactive decay, thus explaining much of the evidence used to substantiate the billions of years of geological dating.

Furthermore, when talking about the Universes' age… You hear the constant dispute thrown back and forth, about what created the Universe? Some say it was the Big Bang and others even suggest that our Universe was spawned from a Universe before that. The argument is never ending! And it always seems to cancel out G-d. It almost seems pointless because no one seems to know the real answer to this. But to further assimilate such thinking, it's not about the Big Bang, or even where the notion of a Universe before the one we currently know came from. It's about ENERGY! And there is an answer to this, and its found in the law of entropy (thermodynamics) again. The laws of thermodynamics tells us concerning the Universes beginning that if we go back in time the available energy will increase until it equals the entire (total) energy. We cannot go back any further because this is the point at which time and energy must have come into existence (at its highest quantity) beyond which it could have existed or it would have either obliterated or disintegrated, and since energy cannot create itself, both logic and science drives us to the conclusion that in the beginning it must have been G-d who created the one and only Universe. Therefore, according to the law of thermodynamics it could not have gotten its available energy by chance through time. It had to be designed. But I often question… science is just now discovering this when we already had an answer as to the Universes' true beginning written in a book 4,000 years ago by an eyewitness named G-d who testifies that He alone created it in a literal 6 days. Again, TRUE SCIENCE WILL ALWAYS SUPPORT G-D'S WORD!

As creation scientists research possible solutions to the distant starlight problem along with the available energy in the Universe, we should also remember the body of evidence that is consistent with the young age of the Earth and Universe. Again, a small recapping here of the chapter reveals that the fossil record is null and void on transitional or intermediate life forms… Carbon dating and radioisotope dating is not accurate… The carbon 14 found in diamonds and coal (as well as the fossils found in coal), and the atomic particles found in rocks such as granite that reveal a young age… Such things like canyons and lakes, coal, oil, petrified wood, diamonds, and other synthetic gems can be made in several days to a week and don't need long periods of billions or millions of years to do so… In addition, don't forget the ingredients for the precise amino acids necessary for life. The mystery of how life got started when oxygen pulls apart amino acid bonds, and water decomposes them… The effect that entropy (the law of thermodynamics) has on all things (both living and non-living things) which completely eradicates all evolutionary precepts… The evolutionist dating of using a (assumed) "geologic time table" that living things evolved over billion of years… Rotating spiral galaxies that cannot last billions of years… Decaying magnetic fields that cannot last billions of years, and most profoundly the real nature of light itself. Of course, "ALL" such arguments always involve assumptions about the past. That is why ultimately, the only way to know about the past is to have a reliable historic record written by an eyewitness. That is exactly what we have in the Bible.

There are even things closer that we should pay attention to... Such things like the Moons distance away from the Earth... We know that the moon moves about a half an inch away from the earth every year. This tells us that the moon would have been closer to the Earth in the past. If things have been consistent, then 6,000 years ago the Moon would have been approximately about 800 feet closer to the Earth. This being the case, if the Moon were billions of years old we would have big problems because the Moon would have been so close that it would have been actually touching the earth billions of years ago. Of course, many scientists hypothesize otherwise with other theories. For example, they might assume that the rate at which the moon was receding was actually smaller/slower in the past (for whatever reason). However, this is an extra assumption needed to make their billions-of-years model work. Again, we don't know and should not make assumptions based on computer simulations because we don't know the "starting point" or the rate of past recession. Psalms 19:1 - The heavens declare the glory of G-d; and the firmament sheweth his handywork.

However we do know that the recession of the moon is a big problem for a belief in billions of years if its current recession rate had remained approximately close to what it is today which is currently receding away from the earth at about half an inch every year. If so, it is consistent with a young age.

Additionally, the Earth is literally perfect for life to exist on it for several reasons (to name a few)... It just so happens to be the perfect size and distance from the Sun. How about we stretch our limits to the right type of galaxy for the earth to exists and thrive in and the right location within the galaxy. Incidentally, our solar system is in a part of the Milky Way that scientists call the "galactic habitable zone" because it has just the right concentrations of the chemical elements needed to support life. Farther out, those elements are too scarce, farther in; our galactic neighborhood is too hazardous because of the greater abundance of potentially lethal radiation and other factors. The Earth also spins at the precise speed of rotation (as well as tilt) that it needs to in order to sustain life as well as having the perfect number of moons needed. In addition, the Moon being just the perfect size as well and additionally in the precise location "today" in order to control the waves of the ocean, which in turn creates just the right amount of oxygen that we need to survive. (The ocean produces 90% of the arths oxygen from microscopic plants that depends on the oceans motion). Additionally, the Earth has water for life and a water cycle that keeps it clean and fresh. The Earth also has just the right temperatures maintained by the Sun in order to sustain life. If it were off by just a few degrees (such as the witnessing of the melting of the ice caps) it could wipe out entire species in the warm waters of the oceans as well as on land. It's also imperative to know that the earth orbit's the Sun within only "one-ninth" of an inch every 18 miles. If it were off by one-eighth or one-tenth, we would either heat up or freeze beyond tolerable limits. Few people neglect to realize just how precise everything has to be in order for humans as well as the vast diversity of life on the planet to continue to exist.

However, the evolutionists persist here also. They surmise that if the Earth were a little further from the Sun, it would be better for us. And they go on to persist about the size of the earth saying that Earth could be larger and denser and still have the same surface gravity, or it could be smaller and less dense. Also, it could be larger and less dense, or smaller and more dense, and life would be a bit different from what it is now - persisting that it would not be non-existent. They say, life on Earth has adapted to a Moon, but if it were pulled away, life would get by. A few organisms might go extinct, but over time new ones would evolve to take their place. Additionally, the evolutionists want to know why if the earth were so perfect for life, why is it that most of all the species that ever lived are gone? My answer to all of this can be found in the Bible, and the Bible tells us it was because of SIN! The disobedience (as well as curse) of Adam and Eve, which brought on all of today's conditions. The reason why such creatures are gone is due to "entropy" (the effects of the 2nd law of thermodynamics which took effect when Adam and Eve sinned). I Corinthians 15:22 - All "in Adam" die. From the curse on down throughout history, things die and go extinct (and Genesis 3:14-24 fully explains this). Genesis 3:19 - For dust thou art, and unto dust shalt thou return (implying a physical death from which there was not any before). Nothing is mutating or evolving... NOTHING, only dying! In fact, everything is in "reverse" and the Bible clearly reveals this to us. I will review it quickly and briefly because it will also be mentioned again in a coming chapter. It goes like this: First (in the order of events) there was G-d. Then G-d made man who was in submission (served) to Him. Then G-d made the woman who was in submission to man. Lastly the animal kingdom that was in submission to mankind (which G-d gave man dominion over). Now after the curse/fall, everything is in reverse order. It's now the animal kingdom (the serpent), and then the woman who submitted to the animal (Satan), and the man who now submits to his wife, and the one who is lastly submitted to is G-d. From this point on, everything of G-d's original plan was in reversed order and went from a sustained order into downward disorder (AKA entropy). Now some people might be thinking that this reversal could not have actually caused physical changes in living things, but the Bible clearly states that Adam and Eve felt shame and pain (in nakedness, fear and coldness) immediately after they ate of the "tree of good and evil". Genesis 3:7 - Then the eyes of both of them were opened, and **they realized they were naked**, so they sewed fig leaves together and made coverings for themselves. Genesis 3:10 - He answered, I heard you in the garden, and I **was afraid** because I was naked, so I hid. G-d had warned them before hand that if they ate of this tree they would die. And dying implies a physical death (from which there was none before), maybe not at that exact moment, but an eventual death from the reverse order of things (entropy). Genesis 2:17 - But you must not eat from the tree of the knowledge of good and evil, for when you eat of it **you will surely die.**

Here we can distinguish the actual effects that took place. Genesis 3:14-19 - So the Lord G-d said to the serpent, because you have done this, **Cursed are you above all the livestock and all the wild animals! You will crawl on your belly and you will eat dust all the days of your life.** In addition, I will put enmity between you and the woman, between your offspring and hers; he will crush your head, and you will strike his heel. To the woman he said **I would greatly increase your pains in childbearing; with pain, you will give birth to children.** Your desire will be for your husband, and he will rule over you. To Adam he said, Because you listened to your wife and ate from the tree about which I commanded you, You must not eat of it, **Cursed is the ground** because of you; **through painful toil you will eat of it** all the days of your life. **It will produce thorns and thistles** for you, and you will eat the plants of the field. By the sweat of your brow, **you will eat your food until you return to the ground, since from it you were taken; for dust you are and to dust you will return.** Therefore, in a sense (both mentally, physically and inevitably) they did die and this might have been due to G-d changing something in their nervous system that allowed them to feel shame, pain and coldness (which they hadn't before), which ultimately led to downward gradation to inescapable death itself (the physical body dying). Some say that this biblical account is just a fictional story, but it may actually be based on scientific precepts. If you look at it in a scientific conventional tense, such things like pain can be blocked by inhibiting the spine and its linkages to the brain such with the advent of the epidural (used for alleviating pain during birth). Perhaps left behind from the original human design, some people today are actually born with a defect that inhibits pain. They actually feel no pain. Although today this can be a hindrance since illness or injury could set in and they wouldn't know it. In fact, the mechanism for pain was actually set into motion by G-d for the very reason of learning from mistakes, so that we would revert away from sin/shame/pain. Logically speaking if we hit our hand hard with a hammer and it hurt, chances are we would be much more careful the next time around. This is ultimately G-d's message and lesson to us concerning sinning.

Back on the subject to all these views of evolution, light distance, the time issue, and the placement of our Sun and Moon are truly all narrow minded... and to truly emphasize the importance of the Earth Moon and Sun connection which serves to shed light on the amount of "daylight" that we have in a single 24 hour day (what some call circadian rhythm). What the evolutionists fail to realize is that the "moon" is not only a necessity for oxygen, waves and wind... it is also extremely necessary for producing the hormone "melatonin" in the human brain that brings on sleep. And the human eyes are made in such a way that they register at the sight of dusk to also stimulate this sleep hormone (melatonin). People are not meant to live in darkness without the moon and the sun. Sure they can adapt, but with noted health risks attached. In fact, few people realize the importance of the hormone melatonin. It is linked between the brain, the pituitary gland and the gonads (testes and ovary). The gonads produce a hormone called "gonadotropin" which is the key hormone priming the body for sex. Simply put, if we didn't/don't sleep we simply could not have reproduce offspring, survived or thrived and increased on the planet. Thus, "sleep" is vital for the REPRODUCTION of all living things. Therefore, the combination of the "Moon and Sun" proves to be vital for survival of all living things. Incidentally, the full moons also have an effect on gravity and electricity (electromagnetism) causing the stimulation of plant seeds to grow. So don't have the wool pulled over your eyes by any evolutionists who tells you the Moon and the Suns positions, distance, size and weight do not matter because as was just pointed out, the moon and its precise location as well as its functioning purpose is VITAL for all life. And life for just the way G-d had intended it. And not life brought on by millions/billions of years of mutation or evolution. But the creationists should be deeply seeded to know that the bottom line in all this is that there was NO MILLIONS OF YEARS AGO. The length of a day (given extremely minute differences) has always been the same for roughly 365 days in a year for 6,000 years even though the evolutionists claims that we have proof for millions/billions of years in tidal sedimentations. But neglecting to realize that these layers may be the result of something else such as the catastrophic conditions brought on by the Great Flood. Furthermore, we wont even be around for another millions of years scenario, as we'll come to find out in a future chapter that this is not G-d's plan in the scheme of "prophesy".

Additionally, lets turn our eye to another planet in our galaxy... Saturn has a moon that is still hot which indicates it is still young, and when space telescopes revealed pictures of Saturn's rings they noticed billion of particles in violent turbulence and instability. If these rings existed for billion of years these particles should be in extremely stable orbits, and they are not, which is a strong indicator that Saturn and its rings are roughly only several thousand years old.

Another interesting thing that I found was that the Bible remains consistent with what science/astronomy is discovering today (especially through the eyes of a telescope). We can observe the uniqueness of stars. What we observe is that every star, planet and moon is different, and that no two stars are the same. But the Bible had told us this thousands of years ago already. 1 Corinthians 15:41 - There is one glory of the sun, and another glory of the moon, and another glory of the stars: **for one star differeth from another star in glory.** G-d has even numbered and named them all. He determines the number of the stars and calls them each by name. Psalm 147:4 - He determines the number of the stars and calls them each by name. Isaiah 40:26 - Lift your eyes and look to the heavens: Who created all these? He who brings out the starry host one by one, and calls them each by name. Because of his great power and mighty strength, not one of them is missing. We can also assess here that the stars are divided into constellations by what the book of Isaiah tells us concerning the stars and their light. Isaiah 13:10 - For the stars of heaven and the constellations thereof shall not give their light: the sun shall be darkened in his going forth, and the moon shall not cause her light to shine. The stars also have an order or procession. Song of Solomon 6:10 - Who is this that appears like the dawn, fair as the moon, bright as the sun, majestic as the stars in procession?

The constellations are also referred to as signs. Genesis 1:14-15 - And G-d said, Let there be lights in the expanse of the sky to separate the day from the night, and let them serve as **signs** to mark seasons and days and years, and let them be lights in the expanse of the sky to give light on the earth. And it was so. These are the 12 major constellations (or Mazzaroth as the patriarch Job calls them). Job 38:32 - Can you bring forth the constellations (Mazzaroth) in their seasons or lead out the Bear (Arcturus) with its cubs? It is these constellations, which mark a continuous circle around the earth today, which most of us are familiar with called the Zodiac. Today and even in the recent past these constellations have been twisted to tell a different story. Many of them being used by soothsayers, fortune-tellers and astrologers to mean many mythological things in opposition to G-d. But surprisingly as we are about to find out, the constellations bare no other meaning or purpose other than bringing the promises (prophesies) concerning the coming messiah (Jesus the deliverer) into fulfillment. It should also be noted that the Bible patriarch "Job" was not the first to tell of these constellations (which most people think he was). It was Enoch (before the deluge of Noah's Great Flood) mentioned in the Old Testament (which you'll read more about in a future chapter) was actually the revealer that pointed out and exposed the constellations (given to him by the appointed angel of G-d). Most people don't know about this because the "book of Enoch" was left out of the Bible. And it was this knowledge of the constellations given to Enoch that was later passed down to Abraham, and Abraham who was latter commanded to teach the Egyptians. In addition, the Greeks are said to have received their knowledge of the constellations from the Egyptians. And it is the Egyptians from, which we get what's known as the modern constellations. Again, all of these constellations literally testify and bare record of the coming messiah - Jesus.

This was Enoch's assigned mission from G-d! Again, what all these constellations truly represent was symbolic of the angels message of the coming deliverer (the abolisher of sin). The ancient constellations are all picture prophecies of Jesus, and are in accordance WITH THE WRITTEN WORD OF G-D. In fact, many individual star names are recorded in the Bible as characteristic attributes and aspects of Jesus. Just to get us started, here are a few:

Lamb, Revelation 5:6 - Then I saw a **Lamb,** looking as if it had been slain, standing in the center before the throne, encircled by the four living creatures and the elders. The **Lamb** had seven horns and seven eyes, which are the seven spirits of G-d sent out into all the earth. Revelation 5:8 - And when he had taken it, the four living creatures and the twenty-four elders fell down before the **Lamb**. Each one had a harp and they were holding golden bowls full of incense, which are the prayers of G-d's people. Revelation 6:16 - They called to the mountains and the rocks, Fall on us and hide us from the face of him who sits on the throne and from the wrath of the **Lamb!** Revelation 7:9 - After this I looked, and there before me was a great multitude that no one could count, from every nation, tribe, people and language, standing before the throne and in front of the **Lamb**. They were wearing white robes and were holding palm branches in their hands. Revelation 7:10 - And they cried out in a loud voice: Salvation belongs to our G-d, who sits on the throne, and to the **Lamb**. Revelation 7:17 - For the **Lamb** at the center before the throne will be their shepherd; he will lead them to springs of living water. And G-d will wipe away every tear from their eyes.

Shepherd, Mark 14:27 - And Jesus saith unto them, all ye shall be offended because of me this night: for it is written, **I will smite the shepherd**, and the sheep shall be scattered.

Arm of the Lord, Isaiah 51:9 - Awake, awake, put on strength, O **arm of the Lord**; awake, as in the ancient days, in the generations of old. Art thou not it that hath cut Rahab, and wounded the dragon?

Chief Shepherd, 1 Peter 5:4 - And when the **chief Shepherd** shall appear, ye shall receive a crown of glory that fadeth not away.

Great shepherd of the sheep, Hebrews 13:20 - Now the G-d of peace, that brought again from the dead our Lord Jesus, **that great shepherd of the sheep**, through the blood of the everlasting covenant.

Lion of the tribe of Judah, Revelation 5:5 - And one of the elders saith unto me, weep not: behold, **the Lion of the tribe of Judah,** the Root of David, hath prevailed to open the book, and to loose the seven seals thereof.

King, Matthew 21:5 - Say to Daughter Zion, See, **your king comes to you**, gentle and riding on a donkey, and on a colt, the foal of a donkey.

Prince and Saviour, Acts 5:31 - Him hath G-d exalted with his right hand to be **a Prince and a Saviour**, for to give repentance to Israel, and forgiveness of sins.

Redeemer, Isaiah 59:20 - And the **Redeemer** shall come to Zion, and unto them that turn from transgression in Jacob, saith the Lord.

Governor, Matthew 2:6 - And thou Bethlehem, in the land of Juda, art not the least among the princes of Juda: for out of thee shall come a **Governor** that shall rule my people Israel.

Scepter, Numbers 24:17 - I shall see him, but not now: I shall behold him, but not nigh: there shall come a Star out of Jacob, and **a Sceptre shall rise out of Israel**, and shall smite the corners of Moab, and destroy all the children of Sheth.

Desired of all nations, Haggai 2:7 - And I will shake all nations, and the **desire of all nations** shall come: and I will fill this house with glory, saith the Lord of hosts.

Light, John 8:12 - Then spake Jesus again unto them, saying, **I am the light of the world**: he that followeth me shall not walk in darkness, but shall have the light of life.

Before I show you how the 12 constellation altogether tie in with Jesus, I'd like to make known again that the constellations do not apply to anything other than G-d's message to us of the coming, living, dying, resurrected and reigning messiah. And that they do not pertain to any other persons, or are meant to predict any other futures. The constellations are purely picture components, which tell the complete story of Jesus and not merely for some other hokey twisted, made up fortune telling. All such things about fortune telling are condemned in the Bible!

The Bible is very clear about this!

Deuteronomy 4:19 - And lest thou lift up thine eyes unto heaven, and when thou seest the sun, and the moon, and the stars, even all the host of heaven, shouldest be driven to worship them, and serve them, which the Lord thy G-d hath divided unto all nations under the whole heaven.

Deuteronomy 17:2-3 - If there be found among you, within any of thy gates which the Lord thy G-d giveth thee, man or woman, that hath wrought wickedness in the sight of the Lord thy G-d, in transgressing his covenant, And hath gone and served other gods, and worshipped them, either the sun, or moon, or any of the host of heaven, which I have not commanded.

Deuteronomy 18:10-12 - There shall not be found among you any one that maketh his son or his daughter to pass through the fire, or that useth divination, or an observer of times, or an enchanter, or a witch. Or a charmer or a consulter with familiar spirits, or a wizard, or a necromancer. For all that do these things are an abomination unto the Lord: and because of these abominations the Lord, thy G-d doth drive them out from before thee.

Leviticus 19:31 - Regard not them that have familiar spirits, neither seek after wizards, to be defiled by them: I am the Lord your G-d.

Leviticus 20:6 - And the soul that turneth after such as have familiar spirits, and after wizards, to go a whoring after them, I will even set my face against that soul, and will cut him off from among his people.

Deuteronomy 13:1-4 - If there arise among you a prophet, or a dreamer of dreams, and giveth thee a sign or a wonder, And the sign or the wonder come to pass, whereof he spake unto thee, saying, Let us go after other gods, which thou hast not known, and let us serve them; Thou shalt not hearken unto the words of that prophet, or that dreamer of dreams: for the Lord your G-d proveth you, to know whether ye love the Lord your G-d with all your heart and with all your soul. Ye shall walk after the Lord your G-d, and fear him, and keep his commandments, and obey his voice, and ye shall serve him, and cleave unto him.

2 Kings 21:6 - And he made his son pass through the fire, and observed times, and used enchantments, and dealt with familiar spirits and wizards: he wrought much wickedness in the sight of the Lord, to provoke him to anger.

2 Chronicles 33:6 - And he caused his children to pass through the fire in the valley of the son of Hinnom: also he observed times, and used enchantments, and used witchcraft, and dealt with a familiar spirit, and with wizards: he wrought much evil in the sight of the Lord, to provoke him to anger.

Isaiah 8:19-20 - And when they shall say unto you, Seek unto them that have familiar spirits, and unto wizards that peep, and that mutter: should not a people seek unto their G-d? For the living to the dead? To the law and to the testimony: if they speak not according to this word, it is because there is no light in them.

Isaiah 44:25 - That frustrateth the tokens of the liars, and maketh diviners mad; that turneth wise men backward, and maketh their knowledge foolish.

Isaiah 47:10-15 - For thou hast trusted in thy wickedness: thou hast said, none seeth me. Thy wisdom and thy knowledge, it hath perverted thee; and thou hast said in thine heart, I am, and none else beside me. Therefore shall evil come upon thee; thou shalt not know from whence it riseth: and mischief shall fall upon thee; thou shalt not be able to put it off: and desolation shall come upon thee suddenly, which thou shalt not know. Stand now with thine enchantments, and with the multitude of thy sorceries, wherein thou hast laboured from thy youth; if so be thou shalt be able to profit, if so be thou mayest prevail. Thou art wearied in the multitude of thy counsels. Let now the astrologers, the stargazers, the monthly prognosticators, stand up, and save thee from these things that shall come upon thee. Behold, they shall be as stubble; the fire shall burn them; they shall not deliver themselves from the power of the flame: there shall not be a coal to warm at, nor fire to sit before it. Thus shall they be unto thee with whom thou hast laboured, even thy merchants, from thy youth: they shall wander every one to his quarter; none shall save thee.

Jeremiah 27:9 - Therefore hearken not ye to your prophets, nor to your diviners, nor to your dreamers, nor to your enchanters, nor to your sorcerers, which speak unto you, saying, Ye shall not serve the king of Babylon.

Jeremiah 29:8 - For thus saith the Lord of hosts, the G-d of Israel; Let not your prophets and your diviners, that be in the midst of you, deceive you, neither hearken to your dreams which ye cause to be dreamed.

Jeremiah 10:23 - O Lord, I know that the way of man is not in himself: it is not in man that walketh to direct his steps.

Proverbs 19:21 - There are many devices in a man's heart; nevertheless the counsel of the Lord, that shall stand.

Lamentations 3:37 - Who is he that saith, and it cometh to pass, when the Lord commandeth it not?

Ecclesiastes 3:1 - To every thing there is a season, and a time to every purpose under the heaven.

John 3:27 - John answered and said, A man can receive nothing, except it be given him from heaven.

John 15:5 - I am the vine, ye are the branches: He that abideth in me, and I in him, the same bringeth forth much fruit: for without me ye can do nothing.

Proverbs 3:5-6 - Trust in the Lord with all thine heart; and lean not unto thine own understanding. In all thy ways acknowledge him, and he shall direct thy paths.

Now here are the 12 constellations:

[Capricorn] [December - January] [The Sea Goat] The Sacrifice… Romans 3:25 - Whom G-d hath set forth to be a propitiation through faith in his blood, to declare his righteousness for the remission of sins that are past, through the forbearance of G-d.

[Aquarius] [January - February] [The Water Bearer] The Living Water… John 7:38 - **He that believeth on me, as the scripture hath said, out of his belly shall flow rivers of living water.**

[Pisces] [February - March] [The Fishes] The Liberator… Genesis 48:19 - **And his father refused, and said, I know it, my son, I know it: he also shall become a people, and he shall be great: but truly, his younger brother shall be greater than he shall, and his seed shall become a multitude of nations.** Mark 1:17 - **And Jesus said unto them, Come ye after me, and I will make you to become fishers of men.**

[Aries] [March - April] [The Ram/Lamb] The Crowned Lamb… Revelation 5:6 - **And I beheld, and, lo, in the midst of the throne and of the four beasts, and in the midst of the elders, stood a Lamb as it had been slain, having seven horns and seven eyes, which are the seven Spirits of G-d sent forth into all the earth.** (Incidentally, this is also when the Jewish holiday of Passover occurs in April. When the sun rises in Aries the lamb/ram is sacrificed).

[Taurus] [April - May] [The Bull] The Judge… John 5:22 - **For the Father judgeth no man, but hath committed all judgment unto the Son.**

[Gemini] [May-June] [The Twins] The King… Romans 1:3 - **Concerning his Son Jesus Christ our Lord, which was made of the seed of David according to the flesh.**

[Cancer] [June - July] [The Crab] The Protector… Isaiah 4:5-6 -**And the Lord will create upon every dwelling place of mount Zion, and upon her assemblies, a cloud and smoke by day, and the shining of a flaming fire by night: for upon all the glory shall be a defense. In addition, there shall be a tabernacle for a shadow in the daytime from the heat, and for a place of refuge, and for a covert from storm and from rain.**

[Leo] [July - August] [The Lion] The Victor… Revelation 5:5 - **And one of the elders saith unto me, weep not: behold, the Lion of the tribe of Judah, the Root of David, hath prevailed to open the book, and to loose the seven seals thereof.** (Incidentally, the sun rising in Leo is the resurrection - the time of the second coming of Jesus).

[Virgo] [August - September] [The Virgin] The Incarnate Son... Isaiah 7:14 - **Therefore the Lord himself shall give you a sign; Behold, a virgin shall conceive, and bear a son, and shall call his name Immanuel.**

[Libra] [September - October] [The Balance] The Redeemer… 1 Corinthians 6:20 - **For ye are bought with a price: therefore glorify G-d in your body, and in your spirit, which are G-d's.**

[Scorpio] [October - November] [The Scorpion] The Sufferer…1 Corinthians 15:55-56 - **O death, where is thy sting? O grave, where is thy victory? The sting of death is sin; and the strength of sin is the law.**

[Sagittarius] [November - December] [The Archer] The Conqueror… Revelation 6:2 - **And I saw, and behold a white horse: and he that sat on him had a bow; and a crown was given unto him: and he went forth conquering, and to conquer.**

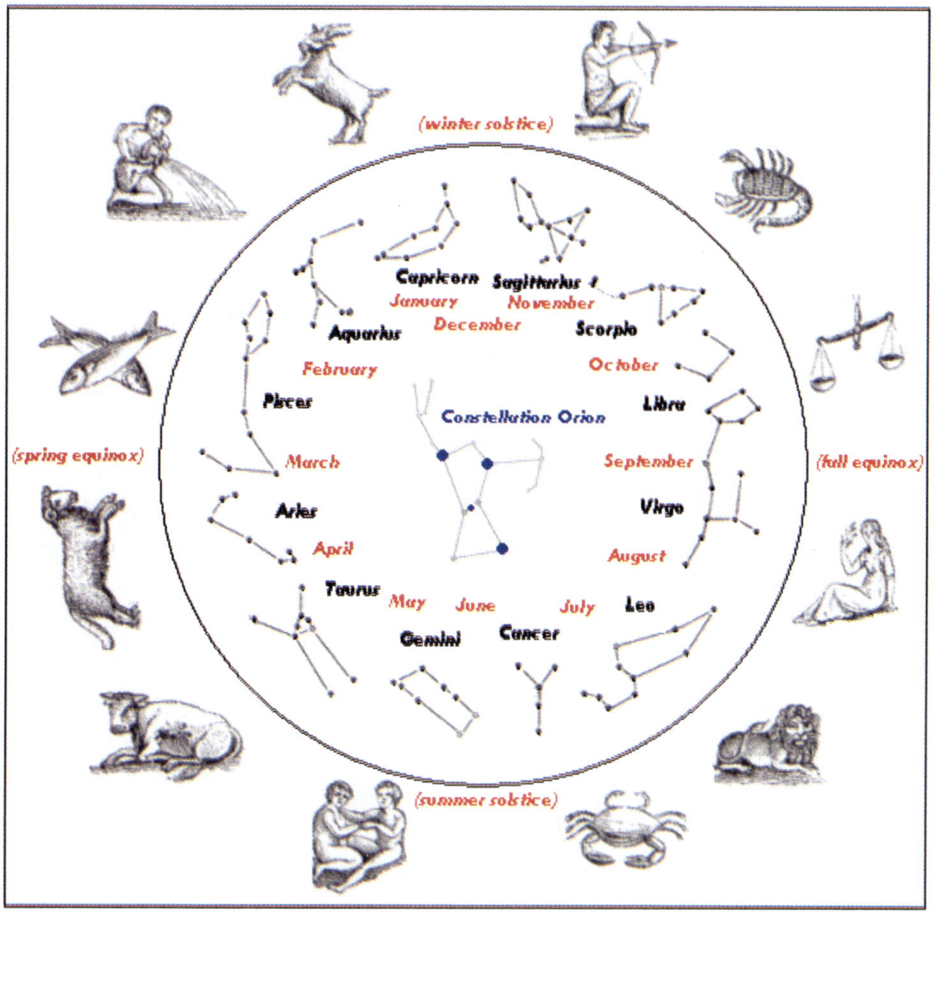

Now there is a grand purpose to why there is so much emphasis given on the constellation "Orion" in the center of the picture above. Orion depicts a man. However, most notoriously seen in the sky as three bright stars in a row. Orion is today mainly known as the "Hunter". But it has also been known as the destroyer and denoted in the Hebrew scripture by the patriarch Job as the Hebrew word "keciyl" (pronounced k'sil) that translates to mean the word "fool". But be careful here in deciphering between fools.

Perhaps the word keciyl/fool was used by Job as a description of the constellation Orion as something that received (or will receive) scorned and severe punishment. So, it's not merely that Job was calling the constellation Orion (a portrayal of Jesus) as a literal fool. It's more likely that he was perhaps describing how Jesus WOULD BE scrutinized, mocked and scorned (as in the future tense).

And at Jesus' crucifixion, He was certainly ridiculed and looked upon by all the people who condemned Him to be a scorned fool (who could not save Himself) with a crown of thorns put on His head labeling Him as King of the Jews. Matthew 27:27-31 - Then the governor's soldiers took Jesus into the Praetorium and gathered the whole company of soldiers around him. They stripped him and put a scarlet robe on him, and then twisted together a crown of thorns and set it on his head. They put a staff in his right hand, knelt in front of him, and mocked him. "Hail, king of the Jews!" they said. They spit on him, and took the staff and struck him on the head repeatedly. After they had mocked him, they took off the robe and put his own clothes on him. Then they led him away to crucify him. Matthew 27:37 - Above his head, they placed the written charge against him: THIS IS JESUS, THE KING OF THE JEWS.

However, aside from the Hebrew word keciyl, the word "Orion" literally means "the coming light" - in light of Jesus being scorned in order to bring judgment on the world. It is also derived from the Sumerian name uru-anna (Urion), meaning "light of heaven". John 8:12 - Then spake Jesus again unto them, saying, I am the light of the world: he that followeth me shall not walk in darkness, but shall have the light of life. John 1:4-5 - In him was life; and the life was the light of men. And the light shineth in darkness; and the darkness comprehended it not. Isaiah 9:2 - The people that walked in darkness have seen a great light: they that dwell in the land of the shadow of death, upon them hath the light shined. What Enoch had in mind (bestowed of G-d) was for "Orion" to be a representation of Jesus the Messiah (the deliverer), and to show the way of His coming. Interestingly the constellation "Orion" also has a special relationship to the Great Pyramid of Giza in Egypt. This celestial knowledge was lost and will be revisited again in a coming chapter.

Now lets thoroughly examine the stars in the constellation Orion… The bright blue star in Orion's right shoulder is the star Betelgeuse and signifies "the coming branch". Isaiah 11:1 - And there shall come forth a rod out of the stem of Jesse, and a Branch shall grow out of his roots. Jeremiah 23:5 - Behold, the days come, saith the Lord, that I will raise unto David a righteous Branch, and a King shall reign and prosper, and shall execute judgment and justice in the earth.

The blue star in Orion's left shoulder arm is the star Bellatrix and signifies "swiftly coming or suddenly destroying". Revelation 11:18 - The nations were angry; and your wrath has come. The time has come for judging the dead and for rewarding your servants the prophets and your saints and those who reverence your name, both small and great— and for destroying those who destroy the earth. Revelation 22:7 - Behold, I come quickly: blessed is he that keepeth the sayings of the prophecy of this book.

Now, the blue star in Orion's elevated foot is the star Rigel, which signifies "the foot that tramples, bruises and crushes". Psalm 91:13-15 - Thou shalt tread upon the lion and adder: the young lion and the dragon shalt thou trample under feet. Because he hath set his love upon me, therefore will I deliver him: I will set him on high, because he hath known my name. He shall call upon me, and I will answer him: I will be with him in trouble; I will deliver him, and honour him. Romans 16:20 - And the G-d of peace shall bruise Satan under your feet shortly. The grace of our Lord Jesus Christ is with you. Amen.

Lastly, the blue star in the middle of Orion's belt (Alnilam) means "the wounded one". Although it is commonly deciphered along with the two other stars of Alnitak and Mintaka as meaning "a belt of pearls". Isaiah 53:5 - But he was wounded for our transgressions, he was bruised for our iniquities: the chastisement of our peace was upon him; and with his stripes, we are healed.

(Again, this subject about the constellations and Orion will be revisited in a future chapter).

Although I'm not quite finished yet, just when you thought you heard it all there is still even more… I'd like to add a few more things to the list of evolutionary qualms before I end this chapter.

According to the evolutionary theory, comets are supposed to be the same age as the solar system, (estimated to be over 5 billion years old) yet the average estimated age of found comments are only 10,000 years.

Even more evidence suggests that there is not enough dirt and sediment on the sea floor at the rate of the earths erosion. If the earths oceans were indeed millions of years old then the oceans would be massively saturated and stifled with sediment dozens of meters deep.

Another revealing clue is that there is not enough sodium (salt) in the sea with a continual source from rivers and other sources that dump millions of tons of sodium into the ocean. If the oceans were indeed millions of years old, no life could exist in the salinity today.

There are other things as well that own up to a young earth, such as coral reef growth and stalactite growth. For example, a coral reef can grow 3.5 meters high, 37 meters wide and 18 kilometers long in just a few hours. This can occur when a growing reef traps sediments as they are moved along by currents, thus adding to its thickness. Storms can dramatically add to the thickness of a reef by bringing in coral from other areas. Many of these storms are due to cyclones or water turbulence, even a great flood (Noah's flood). Such reef growth has been reported as high as 414 millimeters per year. At that rate, a coral reef could be formed in less than 3,500 years. Additionally, coral growth can be doubled by increasing the temperature only five degrees, to which many coral reefs have been found to sit on extinct volcano's. Stalactites can grow very quickly too. Some have been observed to grow more than a half an inch per year. Therefore, it's not an unrealistic estimation that stalactites could form in only a few thousand years.

What's more, real-world evidence exists of the oldest surviving plants living today. One of the oldest is a tree called Methuselah (named after the longest lived of all the biblical patriarchs). A bristle cone pine tree grows on a remote hillside near Las Vegas Nevada. This tree is the oldest known "living" thing on earth. Nearly 5,000 years old. It was seeding when the pyramids were built. Unlike words tree rings never die. However, some times a tree might produce two rings in a year. You can generally tell because the rings are narrower than normal. It is also possible a tree might not produce a ring in a given year because of a succession of droughts, fire, or even a bug infestation can influence tree rings. There have even been some rare finds where a tree can grow continuously and not produce any rings. This can occur if the environments temperature (and moisture) remain consistent. These types of environmental conditions may have certainly been prevalent before the Great Flood.

However, the most telling evidence of all is that there are simply not enough skeletal remains at the current rate of the human population today. There should be millions of skeletons but we can't find them – thus the Earth cannot be millions of years old.

In addition, these things are only the tip of the iceberg.

So as one can see there are quite a few surmounting inconsistencies that the conditions for evolution to have taken place seems to be suggesting otherwise, that the evolution theory is in fallible trouble. It is clear to assess that evolution is solely a THEORY that describes a process. As I said earlier at the beginning of this chapter, there is more evidence for "G-dly creation" then there is for the evolution theory. The evolution theory CANNOT explain what started that process. Therefore, it is only reasonable that we concluded and can now put forth the viewpoint that EVOLUTION IS A RELIGION BASED ON FAITH! This is because the evolutionist cannot base their theory solely on any proof or facts any stronger than that of the creationist because it was the unseen. However, as the evidence was supplied above, the creationists have more substantial evidence than the evolutionists do. And this fact alone is that WE DO has a reliable historic record written by an eyewitness. This eyewitness is G-d, and His gift to us - the Bible. Additionally, true science is continually illuminating G-d's word.

So how do we really-truly know the earth is approximately 6,000 years old aside from all the evidences provided in this chapter? Because we have a reliable historic record written by an EYEWITNESS - G-D HIMSELF! And He tells us in His historically written record (the Bible) that the earth is approximately 6,000 years old… (Again as listed above), we can figure this out by the Bibles genealogy of the patriarchs. Additionally we know (according to the Bible) that the Biblical patriarchs lived longer because the earths atmosphere before the flood was different containing lots more oxygen then we presently have today. And its scientifically known that the more oxygen a living things gets the bigger and longer it can live, and if it gets sick it can regenerate much faster. Therefore long life is not something over-exaggerated when its plainly stated in the Bible. (Additionally in a future chapter, there will also be detailed data supplied that makes obvious the downward diminishing ages of the patriarchs after the flood).

Adam - 930
Lamech - 874-1651
Noah - 1056-2006
Terah - 1879-2084
Abraham - 2009-2184

Flood - 1650

The timeline today from Jesus was 2,000 years, from Jesus to Abraham was 2,000 years, and from Abraham back to Adam is about 2,000 years, which gives us an approximate age of 6,000 years old.

There is much more evidence to be provided, and when it is weighed at the end of this book, IT WILL ULTIMATELY BE DETERMINED THAT CREATIONISM PREVAILS. The Biblical record gives the only logical scientific and satisfying account of the origin of all things.

Nevertheless, on an even higher note, before I end this chapter, there is still one more translucent detail that needs to be properly addressed. And this deals with the "evolutionist thinking". It is in my opinion that the evolutionist thinking is dangerous. Such scientific theories that suggest that (starting from a single cell) life evolved in a series of events. This opinion is based on the way in which I see (and have come to learn) the manipulative ability of evil working its ways. Not only is the evolutionist thinking dangerous, there is simply no actual scientific proof for any of its premises yet (nor will there ever be). To get at the heart of why the evolutionists thinking is dangerous and why we should be fearful of it is because it clearly wants plants, animals, humans and all manner of living things to have come about by chance. Yet, we fail to comprehend how "morality, ethics and principles" came about. Isn't it ironic how we as a human society are inspired by moral laws? One aspect of moral law that is so significant is that we have always buried our dead out of respect with the idea and philosophy of a superior realm (Heaven). Yet, how is it that we can have an ultimate meaning in life, such as life after death, while being tainted by the evolutionist mentality? BECAUSE IF EVOLUTIONARY BIOLOGY IS CORRECT, THEN ALL THESE LOFTY DESIRES ARE HOPELESS. And we begin to see that when the moral aspect of things are perverted the evolutionist thinking becomes the root of all evil... leaving the door wide open for atheism, racism, communism, Nazism, animalistic behavior, crimes, theft, stealing, killing, drugs, economic takeover, anarchism, coveting, adultery, lying, dishonesty, injustice, and all manner of anti-moral beliefs and practices. These of course are the extremes, in which one gives birth to the other. Professor Randy Thornhill wrote an article in 2001 entitled "Rape and Evolution": Rape is evolutionary, biological, and natural. Rape itself is an adaptation, a product of direct selection for rape in the past. Our male ancestors became ancestors in part because they conditionally used rape. Now if we continue with this thinking we can begin to believe that not only is rape acceptable, but this thinking leaps itself forward into possibly believing that killing one of our siblings (the week link to thin out the family because food resources are short) is an expectable alternative, and is a perfectly natural behavior. We see it begins to tear our morality down and succumbs us to nothing more than animals, when we are not animals! G-d made it clear that mankind was distinct from the animal kingdom as well as being valued in a fashion above that of the animals. Genesis 1:26-27 - And G-d said, Let us make man in our image, after our likeness: and let them have dominion over the fish of the sea, and over the fowl of the air, and over the cattle, and over all the earth, and over every creeping thing that creepeth upon the earth. So G-d created man in his own image, in the image of G-d created he him; male and female created he them.

Lets face it, there's no morality without G-d. Humanism simply will not cut it.

Though others see evolutionism in a benign middle ground, seeing it only as science and not as defying a source of morality. They somehow hold two separate standards that separate morality from science, but decline to merge these precepts with the demise of morality. So as we can see, the evolutionists thinking is a very dangerous path to partake down, which can sneakily lead to corruption and the downfall of society. And eventually the complete downfall of humanity. I would caution just what "FACTS" we think are so, since if someone can get people to believe something that is not true, and that belief leads to those people taking actions to change the world (in terms of scientific views) in a manner that the claimant (Satan) wishes, this could be more devastating to an open society than a nuclear bomb! Darwinism and evolutionism may be that very theory through which evil strives to take the "morality" away from humankind. This is one strong challenging reason why G-d would never have created a system that functions off evolutionary precepts. What should clearly be seen is the objectionable face of the tempter who illustrates false impression and destruction. This is the way "Satan" would have it look (evolution/evil-ution). A good example of believing something that is not true (such as evolution), and taking it to the extreme can tear the fabric of morality to shreds. Such as the indoctrination of evil, that Adolph Hitler used to gain power… He was quoted as saying: Let me control the textbooks and I will control the state - Adolph Hitler. Adolph Hitler - during a speech he delivered on November 6, 1939: When an opponent declares I will not come over to your side, I calmly say, your child belongs to us already… what are you? You will pass on. Your descendants, however, now stand in a new camp. In a short time, they will know nothing else but this new community.

Still the evolutionists try to get the last word in on this argument as well, asserting that one can still have morals and at the same time not be religious or believe in G-d (this can also include being totally devoid of spirituality). This however is not true, and I will lay emphasis on just why. Sure, one can still have MORALS (self laws of quality and regulation), including guidelines and rules based on governmental laws. But what such people fail to realize is that morality still triumphs on a higher order of things, notably FAITH! (The substance of things hoped for, but not seen). Everything is of a "spiritual basis"… if our eyes are open and we are aware of it, one cannot deny what is central to our makeup as human beings.

Thus either way, there is no escaping G-d.

One last thing before we move on to the next chapter… This is for all the doubting and unbelieving people who continually beg for proof of G-d. They are the ones who after every religious conversation will say, "show me proof of G-d". Nobody talks so constantly about G-d as those who insist that there is no G-d - Heywood Broun. I love to challenge this question more than any other particular one, to show beyond a reasonable doubt that indeed there is proof of G-d's existence. Firstly, it is true that G-d has not revealed Himself in the same way that He did with Moses during the Exodus, which is of course until Jesus. But actually seeing or talking with G-d (face to face) as well as sign miracles from Him in the Bible are actually very rare events and we find that G-d only intervenes when it involves His plan or program of prophesy, which involves a time-table to which He's not going to cancel or change to accommodate our personal problems, and He's especially not going to show Himself to prove His existence. This is because the Bible tells us that He has already done this through Moses and the prophets. Luke 16:29- Abraham replied, they have Moses and the Prophets, let them listen to them. Luke 16:28-30 - He said to him, if they do not listen to Moses and the Prophets, they would not be convinced even if someone rises from the dead.

But aside from G-d revealing Himself to us face to face, His relations and interactions can still be found and accounted for. And "one" of these leading and foremost proofs can be found in ARCHEOLOGICAL EVIDENCE. In addition, I'll get to the significance of just why.

Many people fail to see the significance of archeological evidence! Not only archeological evidence but in point of fact a piece of historical evidence. And it is this fact that continues to offer a growing mass of surmounting evidence from archaeology as well as history that the Bible accounts deal with real people who lived in real places. Some of these places including many of the patriarchs and people named in the Bible, which have been identified from archeological finds. Just to mention a few of these finds... The city of Jericho has been identified along with Solomon's Temple and the city of AI. There have even been inscriptions found and identified that specify about the Canaanites and the Jews. Even inscriptions indicating the "House of David". Additionally, the store houses of Ramses and Pithim have also been identified which is where the Jews were said to be slaves, as well as the two cities of Sodom and Gomorrah where Lot Abraham's Nephew lived. These two cities (Sodom and Gomorrah) were destroyed because of sin from the likes of brimstone (sulfur - balls of fire) raining down from the sky and burning the city and its inhabitants. Amazingly, archeologists have located these cities and among evidences found they also found crystallized balls of "sulfur" that may have either rained down or came out of the earth (some say from a volcanic eruption). Genesis 19:24 - Then the Lord sent down burning sulfur. It came down like rain on Sodom and Gomorrah. It came from the Lord out of the sky. Luke 17:29 - But on the day Lot left Sodom, fire and sulfur rained down from heaven. In addition, all the people were destroyed. King Hezekiah's tunnel has also been positively identified from an ancient inscription that was found within it carved in the rock at the end of the tunnel. This inscription described how the tunnel was build by two teams working from opposite ends (some 600 yards apart) that somehow managed to meet in the middle. When translated it reads: "The tunneling was completed". The inscription is now in the Istanbul Museum. This rock inscription was also carbon 14 dated to the precise era of King Hezekiah (7th century B.C.).

We also have extensive evidence of the "Cave of Mechpelah" (also known as the "Cave of the Patriarchs"). This cave is located in southern Israel in the city of Hebron. Numbers 13:22 - And they ascended by the south, and came unto Hebron; where Ahriman, Sheshai, and Talmai, the children of Anak, were. (Now Hebron was built seven years before Zoan in Egypt.). Around 1720 BCE (some 3700 years ago) it was purchased by the patriarch Abraham from Ephron the Hittite for 400 silver shekels (Genesis 23) as a family tomb. As recorded in Genesis, the Patriarchs Abraham, Isaac, and Jacob, and the Matriarchs Sarah, Rebecca and Leah, are buried there. The only one who is missing is Rachel, who was buried near Bethlehem where she died in childbirth.

Additionally it has also been revealed to archeology that the existence of the four rivers of the garden of Eden (mentioned in the Bible) showed up on satellite by NASA. The Bible says Eden was located east of Israel where four rivers meet (the Tigris, Euphrates, Pison and Gihon). The later two have dried up and have long been considered mythical, however recent satellite photography from a distance above the earth reveals visible trace features that these rivers did in fact exist in Iraq. Genesis 2:8-14 - Now the Lord **G-d had planted a garden in the east, in Eden**; and there he put the man he had formed. And the Lord G-d made all kinds of trees grow out of the ground—trees that were pleasing to the eye and good for food. In the middle of the garden were the tree of life and the tree of the knowledge of good and evil. A river watering the garden flowed from Eden; from there, **it was separated into four headwaters.** The name of **the first is the Pishon**; it winds through the entire land of Havilah, where there is gold. (The gold of that land is good; aromatic resin and onyx are also there.) The name of **the second river is the Gihon;** it winds through the entire land of Cush. The name of **the third river is the Tigris**; it runs along the east side of Asshur. In addition, **the fourth river is the Euphrates.**

Moreover recent excavations at Tell Mardikh, which is now known to be the site of Ebla. Here they have discovered thousands of tablets, which have been, translated which tell of Sodom and Gomorrah. Other archaeological verifications include testimony that there was actually a ruler named Belshazzar. Even more finds reveal to us that the Hittites actually existed. Other evidence found that King Sargon mentioned in the book of Acts also ruled.

Many of the cities of the Bible have also been identified such as "Nazareth" where Jesus grew up. Additionally, the Bible repeatedly mentions "Capernaum" and this is where Peter's house and synagogue was found. Also "Tiberias"... this is where Jesus calmed the stormy sea and walked on the water. We also have such things in our possession like the "temple warning notice" that was enclosed by a decorative railing and posted at the entrances of the Temple; one of them is now in a museum in Istanbul. Other archeological evidence has been found such as the pool of "Bethesda". John 5:2 - Now there is in Jerusalem near the Sheep Gate a pool, which in Aramaic is called Bethesda and which is surrounded by five covered colonnades. Additionally, there is also the pool of Siloam where Jesus cured a blind man. Nehemiah 3:15 - The Fountain Gate was repaired by Shallun son of Col-Hozeh, ruler of the district of Mizpah. He rebuilt it, roofing it over and putting its doors and bolts and bars in place. He also repaired the wall of the Pool of Siloam, by the King's Garden, as far as the steps going down from the City of David. John 9:7- Go, he told him, wash in the Pool of Siloam. So the man went and washed, and came home seeing. In addition to these places, Caesarea has been identified. Matthew 16:13 - When Jesus came to the region of Caesarea Philippi, he asked his disciples, who do people say the Son of Man is. Mark 8:27 - Jesus and his disciples went on to the villages around Caesarea Philippi. On the way, he asked them, who do people say I am. The city of Derbe. Acts 16:1 - He came to Derbe and then to Lystra, where a disciple named Timothy lived, whose mother was a Jewess and a believer, but whose father was a Greek. Acts 14:8 - In Lystra there sat a man crippled in his feet, who was lame from birth and had never walked. Archeology has even uncovered what has been noted by the Bible to be "Satan's Seat" which German archeologist found in "Pergamos" and took it back to Germany where it can now be found in the Berlin museum. Revelation 2:12-13 - And to the angel of the church in Pergamos write; These things saith he which hath the sharp sword with two edges, I know thy works, and where thou dwellest, even where Satan's seat is: and thou holdest fast my name, and hast not denied my faith, even in those days wherein Antipas was my faithful martyr, who was slain among you, where Satan dwelleth.

Moreover, there is proof of the holy temples remains such as the golden menorah (Lampstand), trumpets, and the table of showbread that was mentioned in the Bible. Exodus 25:31- Make a lampstand of pure gold and hammer it out, base and shaft; its flowerlike cups, buds and blossoms shall be of one piece with it. Exodus 25:30-32 - Six branches are to extend from the sides of the **lampstand**—three on one side and three on the other. Exodus 26:35 - Place the table outside the curtain on the north side of the tabernacle and put the lampstand opposite it on the south side. Exodus 26:34-36 - Command the Israelites to bring you clear oil of pressed olives for the light so that the lamps may be kept burning. Numbers 10:1-3 - The Lord said to Moses: Make two **trumpets of hammered silver**, and use them for calling the community together and for having the camps set out. When both are sounded, the whole community is to assemble before you at the entrance to the Tent of Meeting. Exodus 25:23-30 - Make a **table** of acacia wood—two cubits long, a cubit wide and a cubit and a half high. Overlay it with pure gold and make a gold molding around it. Also, make around it a rim a handbreadth wide and put a gold molding on the rim. Make four gold rings for the table and fasten them to the four corners, where the four legs are. The rings are to be close to the rim to hold the poles used in carrying the table. Make the poles of acacia wood, overlay them with gold and carry the table with them. And make its plates and dishes of pure gold, as well as its pitchers and bowls for the pouring out of offerings. **Put the bread of the Presence on this table to be before me at all times.**

These remains were mentioned in the Bible (and they were the instructions that were given to Moses straight from G-d that the Jews built during the Exodus) to be within the possession of the 2nd temple that was destroyed in 70 A.D. (approximately 40 years later after Jesus' prediction in Luke 21:20) when the Roman army under Titus fulfilled the prediction and they came in and destroyed the Temple and stole the golden valuables. This event was recorded by a Jewish historian named Josephus. It was also recorded by the Romans (in Rome) that they witnessed the golden menorah being carried through the streets of Rome surrounded by Roman soldiers who later used its riches to build the Roman coliseum (picture shown below). The emperor Vespasian received his general share of the vast quantity of gold treasure seized by the Romans and with these riches; he ordered a new amphitheatre (the Roman coliseum we know today) to be erected. There was also an "original inscription" that was carved on a stone stele (ancient stone slab), which was placed above the entrance to the coliseum that alleged that the theater was funded, and build from the riches taken from the 2nd temple in Jerusalem. Archeologists know this from the older Greek letter indentations found under the current inscription.

This stone stele depiction found within the Arch of Titus in Rome provides the proof of what happened to the Jewish Temple treasures. It shows proof that these articles actually existed, and just as the Bible describes them, and in close sight, exactly the same way as the Romans viewed them. This stone depiction clearly shows the Menorah, the trumpets, as well as what might be the Table of Showbread.

Shown here is the Arch of Titus, which represents the triumph commemorating the captured treasure of Jerusalem in 70 A.D. It is located on the Via Sacra just to the southeast of the Forum in Rome.

Continuing with the archeological evidence here… Even something as simple as a mountains name that we take for granted. Such as the mountain name (Mount Ararat) in which Noah's ark is mentioned to have landed is in the Bible and found to be a real mountain still in existence. Genesis 8:4 - And on the seventeenth day of the seventh month the Ark came to rest on the **mountains of Ararat.** This also goes for many other Mountains mentioned in the Bible. They appear to all be there (and noted today by their same names) just as the Bible calls them. There is even evidence of the apostle Paul's journeys traveling in the eastern Roman Empire. The record of Paul's travels are found in the book of Acts and are found to be astonishingly accurate with Roman records even down to the nautical evidence mentioned in Acts 27:12 - 28:1.

Among other archeological evidences in our possession, we also have the "Dead Sea Scrolls" which have been found in the caves at Qumran on the northwest shore of the Dead Sea. Consisting of roughly 900 documents and are written in Hebrew, Aramaic and Greek. They back-up with amazing accuracy the same accounts that are written in the Bible. The significant thing about the Dead Sea Scrolls is that they demonstrate unequivocally the fact that the Jews were faithful in their transcription of biblical manuscripts. Recently there are many more texts of other origins that are coming forth and revealing the standing credibility of the Bible.

There is further archaeology evidence, which supports other Biblical accounts, such as: Abraham's name, which appears in Babylonia. Additionally, the field of Abram in Hebron is mentioned by the Pharaoh Shishak of Egypt inscribed on the walls of his temple at Karnak, proving that Abraham was known in Palestine - the land the Bible places him in. In addition, we also have the Beni Hasan Tomb, which depicts persons coming to Egypt during a famine, corresponding with the Biblical account of the plight of the sons of Abraham, Isaac and Jacob. Archeology has also found such things as the "doors of Sodom" (Tell Beit Mirsim). These are heavy doors needed for security, the same doors which we find in Genesis 19:9. Genesis 19:9 - Get out of our way, they replied. And they said, this fellow came here as an alien, and now he wants to play the judge! We'll treat you worse than them. **They kept bringing pressure on Lot and moved forward to break down the door.** We also know from Biblical accounts that Joseph was sold by his brothers as a slave to Egypt for a price of 20 shekels, which according to trade tablets from that period, is the correct price. Genesis 37:28 - So when the Midianite merchants came by, his brothers pulled Joseph up out of the cistern **and sold him for twenty shekels** of silver to the Ishmaelites, who took him to Egypt. Moreover, Joseph's Tomb has almost certainly been identified in Shechem where a mummy has been found with a distinguished Egyptian sword next to it. Joshua 24:32 - And Joseph's bones, which the Israelites had brought up from Egypt, were **buried at Shechem** in the tract of land that Jacob bought for a hundred pieces of silver from the sons of Hamor, the father of Shechem. This became the inheritance of Joseph's descendants.

The conclusion in all of this archeological evidence is that recent discoveries have tended to confirm the accuracy of many background details in the Bible narratives. These discoveries clearly reveal to us that they are certainly there endorsing the true status of the Bible. They support historical reliability and are based on the testimony of eyewitnesses. This being the case, we can surely say that the Bible is supportive of historical evidences which therefore suggests that IT HIGHLY ELEVATES THE ACTUALITY OF A DIVINE ENTITY

INTERVENING ON ITS BEHALF (and it is this divine author G-d alone, that the Bible is dedicated to).

Still, even after all the evidence I provided smacking one clean in the face, the deniers will still parrot the phrase: Show me proof of G-d! However, if such people want to continue to play this game, I would like them to show me evidence that their tongue is connected to their brain. All kidding aside, even though I have just presented these Biblical proofs, the resolute blind-sited evolutionists continually ignore and block out these proofs. They still cling to evolutionary concepts, and continue to see the evolution of imaginable transitional things happening within their minds. Such a condition is worse than blindness, and that's seeing something that isn't there.

Another thing I would like to bring up is the nature of mean-spirited evolutionists who are continually retorting in opposition to creationists. I had found myself in many arguments with such people. However, in the end their defeat over me is like being savaged by dead sheep. Not to underestimate their deviousness though, evolutionists are masters of speculation. They continually assert that the evolution theory is a fact. They use the typical stratagem of making their statement strong enough and repeating it often enough, that they are able to convince themselves and others that it may be true. Despite this, I would like to remind the evolutionists that there are many educated and informed people who do not believe that the evidence supports the theory of evolution.

Oh, I've been through all of their rigmarole and around the bush many times! They retort back with statements like: Christians, in general, are not renowned for they're critical thinking skills. Additionally, these mean-spirited evolutionists (in their ignorance) blurt out statements like: What a joke, stupid, ignorant and liar! Even name calling, like canard and creobot. They are also notorious for picking apart words like "complex" assigning several meanings to the labeled word (in technical terms) as a sort of scapegoat to dodge the conversation or to trap and redirect the uneducated into the palindromes of their encapsulated thinking realms. They continually tell us to open a book (a real book!) and we'll realize that there are answers to all these questions about evolution occurring. However, little do they realize that all they are doing is leading the league in nostril hair!

One of the best ones they told me (to date) was that I have no imagination! I responded back saying: Oh, I have imagination all right, but it stops when I start dreaming. Additionally they keep repeatedly saying that creationists don't understand what evolution is all about. Meaning that we don't understand all the technical terms that it fully consists of. Something like: Mutations do not cause structural change in an organism (morphological change)… and science does not CLAIM that mutations cause morphological change in organisms. Morphological changes occur in species. It's not individuals that evolve. It is the genetic makeup of populations of individuals (the gene pool) that changes over time and evolves. But the creationists still innately knows that this is nothing but "natural selection" (which is not any sort of evolution as was discussed earlier). The bottom line in all of this is that there still isn't any EVIDENCE of it happening in nature or the laboratory. These kinds of rebuttals make my hair hurt!

They even profess a worthless view of themselves when they make comments like: I could care less if I was created or evolved as a human being. They also believe that there is no need of a G-d or any other magical entity in order for the universe to exist. Nevertheless, to some extent some can acknowledge that they were at least created by something. But then they go on to assume that this something must have been created by something else. They continually want to know that if G-d created the Universe, then who or what created Him. It goes on and on - there is no stop to this madness! They play spinning mind games such as claiming that if creationists can accept the notion that G-d has no beginning and no end, and He always was, why can't creationists also accept the notion that the universe has no beginning and no end and it always was.

But surprisingly the Bible does have an answer to all these types of questions if only one would read and UNDERSTAND. In fact, the Bible directs us to acknowledge just what G-d is. It tells us in so many words that we can only partially comprehend the notion of G-d's existence. To do so, we must use human concepts to speak of G-d. Such concepts described in the Bible say that G-d is "without beginning or end", "eternal" and "infinite" and from "old and everlasting". The Bible tells us that G-d has always existed, even from everlasting to everlasting, thou art G-d. Psalm 90:2 - Before the mountains were brought forth, or ever thou hadst formed the earth and the world, even from everlasting to everlasting, thou art G-d. Psalm 93:2 - Thy throne is established of old: thou art from everlasting. Psalm 41:13 - Praise be to the Lord, the G-d of Israel, from everlasting to everlasting. Amen and Amen. Quite simply, G-d has no beginning and no end. So the questions remain, where did G-d come from? The answer is, He didn't. He always was. To us humans the notion of time is linear so we cannot comprehend what it would be like without a beginning or end, eternal or infinite. For us time continually flows forward. But G-d is BEYOND any (and all) time. He has no beginning or end. He simply has always been and this is because He is outside of time. In addition, it is extremely hard for us to understand and fully grasp this with our human concepts simply because we are governed by "time as well as matter". And even though "time and matter" is physical to us, G-d is not confined by either. It was G-d who had created both "time and matter". So, "time and matter" began when G-d created the universe. Before that, G-d simply existed and time had no meaning (except conceptually), no relation to Him. Therefore, to ask where G-d came from is to ask a question that cannot really be applied to G-d in the first place. Simply put, this is because time has no meaning with G-d in relation to who He is.

Conclusion of this chapter… there is strong proof that life had a designer who created it in a literal 6 days as the dependable, unfailing and consistent Bible claims, verses the speculated evolution theory that (to date) has mustered up no credible proof whatsoever. The Bible not only contains some of the strongest moral attributes that serves to save society (as well as humanity) from downfall, but there is additionally surmounting evidences in the Bible that are not only scientifically, historically, biologically, geologically, geographically, anthropologically, linguistically and archeologically accurate, but also astrologically and astronomically applicable. It will also be demonstrated in a future chapter that the Bible is "prophetically" accurate as well. (More on this in coming chapters).

All human discoveries seem to be made only for the purpose of confirming increasingly the Truths contained in the Sacred Scriptures - Sir William Herschel (1738-1822).

3 - QUANTUM MECHANICS...

Though we cannot see invisible waves of energy we know they exist, such as in how a clicker remote to a television transports an invisible wave to change channels, or a microwave oven heats food without a fire. Additionally, the new era of cell phones, GPS and satellite technology. How about some spooky invisible interactions such as when two identical twins (or even loved ones, friends or family) can communicate with one another at distances apart. It seems that "subconsciously" one knows that the other is in trouble or in pain. And how is it that cells in a test tube, are reacting (somehow still connected) while far distances apart from their mother host? This experiment was done in 1993 under the direction of the United States Army Intelligence and Security Command. It was conducted to demonstrate that our cells, even while outside of our body, would still respond to our emotions as though they were still connected to the body itself. The test involved scraping the mouth (cheek cells) of a volunteer to obtain white blood cells (leukocytes) which were then placed in a centrifuge (which is a rotating device that separates substances) and then placed into a test tube, which was then watched and observed by a probe that was inserted into the test tube while the donor was seated in a separate room down the hall and being shown a television program with many violently intense scenes. When the volunteer watched scenes of fighting and killing, the probe detected extreme stimulations and responses in the mouth cells. The experiment was repeated with donor and cells separated up to fifty miles apart and were found to still be non-locally connected. Two days afterwards, it was still showing the same results.

The same results ensue in the perception of plants. Numerous researchers have conducted experiments (using polygraph equipment) with plants to measure their reaction to outside events. Plants have been found to have what's called "primary perception", which is a sensory system existing in cell life. It works like a telepathic communication system that possesses something closely akin to feelings or emotions (but has no brain). This "primary perception" operates regardless of distance. The plants in these experiments were found to register an awareness, comprehension and understanding to events that took place around them. In one experiment, a burning match was held to a plant leaf, and at the instant of this incident, there was a dramatic change in the tracing pattern on the polygraph, suggesting that this pattern was triggered by the mere "thought" of burning the leaf. To remove any human interference, another experiment was performed using animals and insects. They were put in close proximity to the plants (even put in separate rooms). When harm came to any of the animals or the insects, a polygraph reading showed that (several seconds later) the plants registered a burst of activity. When all the results were analyzed it was determined that plants appreciate being watered, even pleased to be looked at. They worry when an animal or insect comes near as well as sympathizing when harm comes to any of the animals and insects in proximity to them. They even wilt (droop) when violence threatens their own well-being. These experiment prove beyond a doubt that "consciousness" is a fundamental property of nature. There is also another example of consciousness brought forth in another experiment in chapter 17 that can be done as a home-based experiment in which anyone can perform to witness the effects of human consciousness.

There are other things of strange behavior and goings-on that are puzzling, confusing and mysterious… For instance, how is it that some people claim they have been reincarnated and remember a past life (though the Bible renounces reincarnation)? Hebrews 9:27 - Just as man is destined to die once, and after that to face judgment. Could there be another possible explanation for such claims of reincarnation… perhaps something of a different nature such as the transmission of electrical energy (spiritual in nature)? It has even been compellingly proven through computer analysis that we can predict 3 seconds into the future, a sixth sense like that of a gut feeling. This was discovered from experiments that were recently done by using a modified lie detector (hooked up to a computer) which measures an electrical current across the surface of the skin, the computer detects any current changes when a person reacts to a particular picture of a violent, soothing or sexual picture when viewed, the recorded reaction (3 seconds beforehand) is equivalent to a wince. In addition, it was soon revealed that people began reacting to pictures BEFORE they actually saw them. What's truly going on here? Is this just good educated guessing summed up by wisdom and experience, or is it something of a spiritual 6th sense nature? Rather the ultimate question here is, where is this future information coming from.

It "altogether" falls into the field and science of what's called "Quantum Mechanics". Emboldened by the word "quanta" which is the word used to describe the smallest quantity of energy (or smallest unit of something). Quantum Mechanics simply put, is the study of matter and radiation (energy) at an atomic level. It is also a calculating and statistical machine for determining or estimating the actions and behaviors of microscopic particles. This of course is contingent upon (in part) the measuring instruments we use to examine those behaviors. However, Quantum Mechanics should not be confused with "Quantum Evolution", which suggests that genetic mutation can be adaptive, or directed through quantum effects.

The theory for Quantum Mechanics came from physicist Max Planck when he was experimenting with the color spectrum and found that energy existed in individual units. From this theory resulted yet another theory that emerged from the discovery of quanta. And this very existence of quanta applied a completely new fundamental understanding of the laws of nature (general principle based on observation) in relation to strange phenomena. Touching on to these strange phenomena, many people are familiar with the double slit experiment**:**

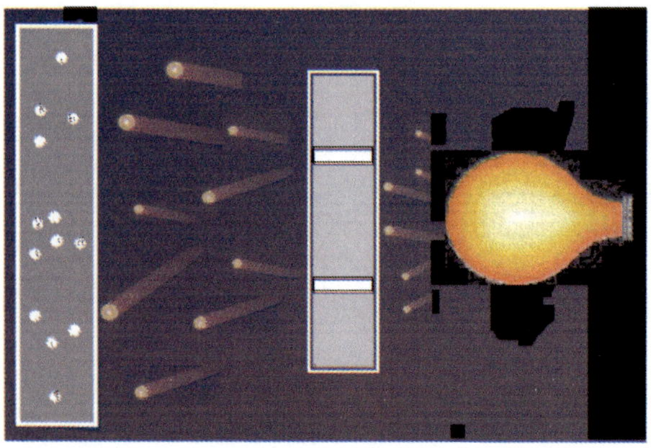

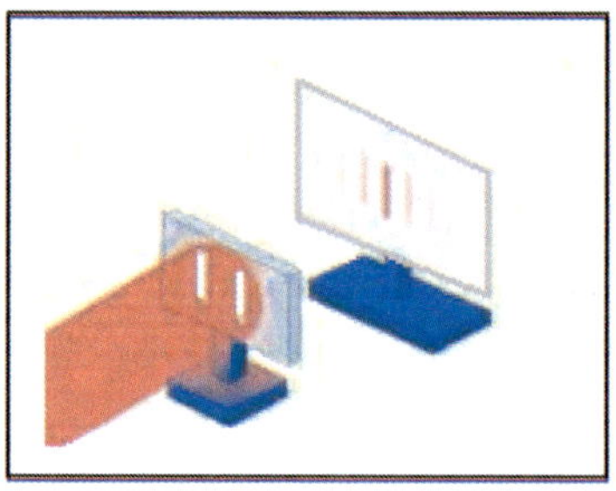

It implies that two particles are in two or more of the same places simultaneously (at the same time).

Scientist originally thought that atoms were bouncing off each other and creating an interference pattern through the two slits. Until they decided to test this theory by firing (and releasing) only "one" particle at a time to make sure that they could not interfere with each other. However, when they did this the same interference pattern continued to appear. The conclusion was astounding! It sets off as a single particle, becomes a wave of potential probabilities, goes through both of the two slits and creates an interference pattern at the other end. This prompted scientist to investigate further, they wanted to catch sight of which slit the one particle actually went through. Therefore, they put a measuring device behind the slits to capture this class act. But when they "observed the particle", it went back to it typical behavior of being a particle producing a pattern of just two bands and did not generate an interference pattern of any kind. The scientists were baffled by this! Nevertheless, when all the data was analyzed, it was soon realized that this "one particular particle" coincided with all of the slits in a state of superposition. In addition, most peculiarly it was realized that the very act of measuring or observing which slit the particle went through actually collapsed the wave function. Thus, the particle decided to act differently as though it was aware it was being watched.

Quantum Mechanics not only represents the double slit experiment, it engulfs everything including "Anthropic Principle" which constitutes that "observers" are necessary to bring the Universe into being. Psalm 8:4 said it ever so eloquently, "What is man that thou are mindful of him". String Theory and Particle Entanglement are also akin and inferred from Quantum Mechanics in the respect that "String Theory" involves multiple dimensions. Additionally Quantum Mechanics encapsulates what's called "Particle Entanglement" in such ways that two identical particles that are ranges of distances apart are still connected and act as though they were still joined in terms of identical response and reaction - make a change to one particle, and that change is instantly reflected in the other(s) - however far apart they may be. This action is amazing, and rigorous tests have virtually proven it true! However, Quantum Mechanics does not and cannot pinpoint the exact values for the position or momentum of a certain particle in a given space in a finite time, but, rather, it only provides a range of "probabilities" of where that particle might be. To get a better understanding of this… when you split an atom you find that it has a nucleus of protons and neutrons along with tiny electrons bustling/fleeting around. Now these protons and neutrons also contain quarks, which are like shifty/dodgy puffy semi-transparent clouds that constantly blink in and out of existence blurring the answers to when and where. Therefore, it becomes necessary to describe these states as uncertain, unknown and unable to be measured. It can also be likened to pixels (tiny dots of light) making little quantum leaps from one pixel to the next.

To touch on a further note… as discussed earlier regarding the double slit experiment, where it was discovered that the very act of "just looking" at these very particles in magnification, "observing" and measuring them caused the atoms to react differently and change course. Oddly enough, it was also discovered that the person performing the experiment actually influences the outcome. It seems the atomic particles have become AWARE! Not only does this apply to observed particles taking different paths through slits, but this strange phenomenon also occurs in the course of our everyday life, activities and choices. This transpires in the routines of everyday life when we look at impending situations in a certain way it becomes one path and if we turn away (avert) it takes another path. An additional way to perceive of it is to look at it as a ceaselessly flowing quantum soup of waves. However, whenever one turns around and tries to see this soupy wave, ones glance instantly congeals and solidifies this wave and turns it into actual reality. It seems the atomic particles are not only AWARE; they are also swayed by the consciousness of the observer.

So lets forget the term Quantum Mechanics, lets call it "Consciousness Mechanics". It's certainly a strange notion to consider that our own minds are creating multiple possibilities in the subconscious. These superposition's of possibilities are in our subconscious and when we become more aware of them they will eventually collapse into either one or the other.

Whether we look at it from the atomic aspect or the superposition of human consciousness. Nobody seems to understand what's going on here, and that's why most scientists don't like that idea. They call it spookiness at a distance because these sorts of behaviors require some kind of "signal" passing between the two particles at a speed faster than light. The scientist tries to explain it off as - it's moving to fast for the given instruments to capture and we just don't know yet what is actually taking place. However, I articulate that maybe it doesn't have to do with the instruments timing at all. Maybe it has everything to do with "conscious observation" and anonymous spiritual matters. Perhaps they haven't truly comprehended the totality of what Quantum Theory really is and what it means to "predicting physical systems using human thought" in which the basic laws of simply looking and observing apply. It is the very first time science has encountered "free will" in which consciousness is "limitless and free". This is simply because there is NO mathematical or measuring description of the subject. Remember, exact values for the particles position cannot be pinpointed, rather only a range of where the particle might be. Therefore, consciousness has become the ultimate voter! Nothing exists except atoms and empty space; everything else is opinion - Democritus (Greek philosopher).

To focus on something involves using our consciousness, but nobody has yet determined what consciousness is, or what impact, intent, weight or sway it has in the field of Quantum Mechanics? But some things are unwavering from a observable physical standpoint of our human viewpoint such as, when we are "in the process of thought" particles are undetermined waves of possibility, and when we are "looking" they are particles of occurrence that give rise to the atoms that (in a literal sense) combine and create molecules that form and grow into cells - the very cells (that make up matter) that are composed of all living things in our world. Therefore, by free will, belief and faith alone, Quantum Mechanics creates choices. Choices that eventually give rise to true reality or constructing it thereof - hence true reality is really WITHIN our brains where physical matter is being perceived through electrical signals. We can also understand this view from another angle of perception that I brought up earlier such as the psychic phenomenon coexisting between identical twins over a distance, or the cells in a test tube that are somehow still connected while they are far distances apart from their mother host, and the many reincarnating claims of past life memories possibly by the transmission of electrical energy (spiritual in nature). As well as the information coming from the future 3 seconds beforehand. Did we ever stop to wonder that this is why cell phones and satellite technology works since these are the things we can call non-local, which means that it does not just relate to a particular area, but a widespread area. It has no particular place yet it is everywhere. If we can understand that this is biophysically possible and then we understand that consciousness is part of this, the constraints of space and distance are no longer relevant.

Surprising enough, there have been rigorous scientific experiments that have been done to show beyond a reasonable doubt that "conscious intention" DOES have an effect over matter" starting at the atomic level of the atom and emanating itself into the physical/material world. One of the early experiments were conducted using the flip of coins that suggested that

INTENTION directed them to their fallen places. Now such test consists of more subtle things as black boxes with LED lights, laser lights and fiber optics. It may still be arguable, debatable and unsettling, but the consistent results speak for themselves. I urge all readers to investigate these experiments and matters for themselves.

And one may ask the really big question of how atomic particles, big and small objects or even future events in the physical world can be affected and ultimately altered by something so fleeting as a though? This may sound outlandish, but recent experiments suggest and demonstrate that all matter in the universe exists in a mesh of interrelationship and continuous influence. For instance, they had discovered that the building units of matter (atoms) have a far and wider instantaneous connection (aforementioned earlier), and that atomic particles linger in a state of potential or probable position where they remain open to outside influences. It has even been observed that all existing things emit a constant tiny current of photons (which are tiny particles of light). Amazing as this all sounds, directed thought which scientist call intention, I call choice (free will) - hence faith, has a powerful constituent over this tiny current of light directing it to create the very reality that we all see. So in theory, yes phenomenally by some unknown means, conscious thought guided by intention, choice, free will and faith does create!

However we are baffled by the fact that in life we may never truly know what is true reality when we observe nature, for what we observe is not nature itself, but nature exposed to "our" human method of questioning. What we are seeing with our human eyes it not really material substance or matter, but only invisible rays of electrical signals (energy) that is transmitted by neurons and deciphered by a tiny part in the back of the brain. Thus, the brain (as we observe and perceive it) is also a material thing in itself. We are not in the world, but the world is within us. It is amusing how we can dream in sleep, and within our brain still have the sensations to see with our eyes and touch with our hands, yet we have neither eyes nor hands. What is left? Only an incorporeal/spiritual matter - hence faith. Hebrews 11:1 in the New Testament it tells us - that FAITH is the substance of things hoped for, the evidence of things not seen. In Hebrews, 11:3 it tells us - that through FAITH we understand that the worlds were framed by the word of G-d, so that things, which are seen, were not made of things, which do appear. In addition, it was John 6:63 that said, - it's the spirit that gives life, the flesh counts for nothing.

Quantum mechanics is very perplexing! All of this might very well be impossible to grapple. But consider the "idea and thought" that the science and theory of Quantum Mechanics may be the most powerfully proven un-falsifiable tool for predicting the actions and behaviors "of our physical system" though it be through corporal or non-corporal realities or between choice and faith, including the inconceivable thought that there is no right or wrong, only "thinking" makes it so. Quantum mechanics is paramount! It is the foremost leading tool that we have ever had in our hands. It is the law!

Still the postulation of what "consciousness" is still remains. We know as humans that in order to think we need words (language)... Which brings me to an interesting question that begs the query of Genesis 2:18-20 of the Bible where Adam assigns names/words to things. It tell us that the Lord G-d had formed out of the ground all the beast of the field and all the birds of the air and then He brought them to the man to see what he would name them, and whatever the man called each living creature, that was its name. Therefore, the man gave names to all the livestock, the birds of the air and all the beasts of the field. Genesis 2:19 - Now the Lord G-d had formed out of the ground all the beasts of the field and all the birds of the air. He brought him or her to the man to see what he would name him or her; and whatever the man called each living creature, that was its name. This narrative in scripture not only must, but also in all likelihood conveys a very abounding message that gives rise to what it means to be "conscious" and also the importance of "choice among possibilities" that should be closely examined by Quantum Mechanics. It's apparent that it is presenting a message and attempting to teach us the method in which the creator built the very foundation of "human consciousness". That is to suggest that human consciousness formed a language. Which brought about verbal communication, making available the organizational advancement of civilization?

I marvel at the Bible, it is a quantum manifold of sorts, and to me a flawless masterpiece. Moses wrote the Pentateuch (first 5 books of the Hebrew Bible) centuries ago and no one to this day can dispute the fact that it is the most amazing book ever to be written - surpassing all other books. In its progression and addition to the New Testament we are presented with Jesus' thought provoking messages, which I might add are also very "quantum thinking" in nature. Such as in the parables spoken by Jesus and recorded by the disciples. Asserted statements like, the mustard seed being larger than the kingdom of heaven.

Matthew 17:20 - in which Jesus says, if you have faith as a grain of mustard seed, you will tell this mountain, move from here to there, and it will move, and nothing will be impossible for you.

Luke 11 - ask, it will be given you, seek, and you will find, knock, and it will be opened to you.

Matthew 4:4 - It is written, one does not live by bread alone but by every word that comes from the mouth of G-d.

Matthew 21:21 - Truly I tell you, if you have faith and do not doubt, not only will you do what has been done to the fig tree, but even if you say to this mountain, be lifted up and thrown into the sea, it will be done. Whatever you ask for in prayer with faith, you will receive.

Matthew 5:28 - But I say unto you, that whosoever looketh on a woman to lust after her hath committed adultery with her already in his heart.

Matthew 7 - Ask and it will be given to you, seek and you will find, knock and the door will be opened to you.

Matthew 8 - For everyone who asks receives, and the one who seeks, finds, and to the one who knocks, the door will be opened.

Matthew 8:3 - I do choose. Be made clean!
Matthew 5:13 - Go, let it be done for you according to your faith.
Matthew 8:23 - Why are you afraid, you of little faith?
Matthew 9:28 - Do you believe that I am able to do this?

Matthew 14:31 - You of little faith, why did you doubt?
Matthew 15:28 - Woman, great is your faith! Let it be done for you as you wish.

My favorite one is the one about the camel going through the eye of a needle. Matthew 19:24 - Jesus said, it is easier for a camel to go through the eye of a needle, than for a rich man to enter the kingdom of G-d. Clearly, as we all know, a camel cannot go through the eye of a needle! It is impossible and this leads to a contradictory paradox! Except, IT IS possible for a camel to go through the eye of a needle! Yes, you might want to know that in today's physics there is another branch of quantum physics called "Quantum Tunneling" and this is just not unknowable theory, but fact. Most scientists know that electronics work because of a quantum tunneling effect. In Quantum Tunneling, it is possible for an object the size of an electron to go through any obstacle - and thus if we realize that all things are constructed of atoms, this then becomes possible. Religious thinkers have also suggested that as illogical as it sounds for a camel to go through the eye of a needle - that logic is just another of G-d's creations. After all, He who created logic is permitted to disregard it.

All of these parables are implying a very important message, that our state of mind and what we are thinking is actually our true reality, and that if we ask and believe in what we ask for, we will receive. Jesus clearly exemplified the teachings of Torah to openly expose and teach the principles between science, religion and G-d. These parables were designed to be understood conceptually by non-conversant people living in a time when the terms, scientific study of matter, energy, force, and motion, and the way they relate to each other where not heard of. Mechanics, electromagnetism, optics, thermodynamics, nuclear physics, and quantum mechanics were also not known. And it is my interpretation of "Quantum Mechanics" that this is the closest science will ever come to explaining Jesus' interpretations allowing for both G-d and science to coexist together and simultaneously. In addition, when we finally awaken to the future we will have discovered that Quantum Mechanics is the only correct analogy applied for everything that exists.

So how do we truly know that G-d is in the details of Quantum Mechanics?

Herein lies the secret… Quantum mechanics is the brainchild of algorithms (mainly fractal in nature). And algorithms are mathematical and contingent to all things that exist in nature (the next few chapter will elaborate on this). They reveal themselves in self-similarity such as in a trees branches and leaves. Similarly with the repeating pattern in a shell, and in addition flowers, mountains, landscapes, shorelines, clouds, waves, currents, the DNA in all living things (anything that is structured by patterns of self-similarity). Algorithms are quantum in nature since they grow from the smallest quantity branching out to an almost infinite extreme (when small cannot get any smaller).

What algorithms show is unmistakably intelligent design - NOT evolution! When G-d spoke His utterances - which can now be understood as a series of algorithms. Psalm 119:89 - Your word, O Lord, is eternal, it stands firm in the heavens. Algorithms are possibly the one program that G-d used when He designed all that exists. But not to devalue G-d's creation, but to show its magnificence and be awed by it. Surprisingly when you get to chapter 8 (The Primordial Language) this self-similarity will additionally reveal itself in the forms of the 22 letters of the Hebrew alphabet. Oh and how surprised, pleased and liberated you will be to finally come to an understanding about this!

Moving on here… As was mentioned, Jesus' parables were expressed to be interpreted by non-conversant people, and as of today we can all see that the Bibles evolution has developed to be translated into almost all languages, and now has more copies in print then any other book in the history of mankind, no other work of history, religion or science has remained relevant for such a awe-inspiring length of time, and no one has ever proven it a fake EVER! It is another piece of un-falsifiable data. However, the most amazing thing about the Bible is its purpose and intention toward all "prophesy" - including what some have discovered as the "Bible Codes". (If you're not familiar with them, I will discuss it in the following chapters). The Bible code (words found in the Hebrew codex of the 5 books of Moses) that we have now come to recognize that foretells or have foretold of past and future events. This code is an attempt to convey a very significant and eminent message to mankind that we are participants in the change, development and shaping of the future - hence Quantum Mechanics. We have somewhat of a partial power (in the form of free will - hence consciousness) to "influence" the future and to change events. But the events we cannot change are those of prophesy found within the Bible. For all other requests, future events can only be changed if it is in rapport with G-D'S INTERESTS. Prophecy is not well received. Never has been - never will. Any feeling that we have that we can change anything or that there are "alternatives" to what is impending upon the world stage is perilous. This culmination of events and the state of world affairs is not about you or me, and this may be hard-hitting, but it is about ISREAL and G-d's chosen people. Always has been - always will be. If G-d says that the universe will melt with fervent heat, I tend to take that literally. If He says that death, hell and sin will be cast into the Lake of Fire, I take that literally too. Everything that happens(ed) in the world revolves around prophecy, Israel and the Jewish people. It is apparent that the G-d of Abraham is opposed to those who attempt to divide His covenant land.

Going off the beaten path here to elaborate about this… To know for sure that G-d's word is manifesting in the defense of Israel, all one has to do is keep their ears open to any news regarding the state of Israel and the Middle East peace meetings. Let us for a moment examine Genesis 16:11-12 which speaks about the enemies of Israel (the offspring of Ishmael). The Bible verse reads**:** Genesis 16:11-12 - And the angel of the Lord said unto her, Behold, thou art with child and shalt bear a son, and shalt call his name Ishmael; because the Lord hath heard thy affliction. In addition, he will be a wild man; his hand will be against every man, and every man's hand against him; and he shall dwell in the presence of all his brethren. This certainly coincides with what we witness in the present day with Iran and Iraq fighting against them, but nonetheless the world is trying to force a peace treaty covenant on people who aren't after peace. They're after the destruction and elimination of Israel and the Jewish people. Zechariah 12:9 inform us, (on that day I (G-d) will seek to destroy all the nations that come against Jerusalem). Zechariah 12:9 - And it shall come to pass in that day, that I will seek to destroy all the nations that come against Jerusalem

It can be argued that more has happened in Jerusalem (Israel), both spiritually, religiously, as well as governmentally, militarily, medically and even economically to change the world then anywhere else on earth, and so it ensues that the world seems to look toward Israel which is only a small fraction of the entire Middle East for a solution to peace in the region (as well as the rest of the world). All eyes, conflicts and world interest are on Israel. The news covers it almost daily. America (and other countries) based around democracy where originally influenced by Israel. Democracy and liberation (equality and freedom) stem directly from the Bible! Leviticus 25:10 - Consecrate the fiftieth year and **proclaim liberty** throughout the land to all its inhabitants. It shall be a jubilee for you; each one of you is to return to his family property and each to his own clan. Isaiah 61:1 - The Spirit of the Sovereign Lord is on me, because the Lord has anointed me to preach good news to the poor. He has sent me to bind up the brokenhearted, to **proclaim freedom** for the captives and release from darkness for the prisoners.

Now, in spite of the attempted peace process, a phenomenon can be witnessed now and again that whenever there are efforts or attempts to divide or give away the land of Israel to its enemies, certain actions take place that intervene with this effort. For instance, it has been noticed that several United States Presidents and Middle East leaders have experienced both personal and political consequences including such phenomenon as intensified weather conditions (in their states), and even loss of life for attempting to divide G-d's covenant land. Lets examine some of these instances and make a determination for whether or not they are coincidentally inclined.

For instance, United States President Franklin D. Roosevelt attempted to facilitate Middle East peace discussions between Israel and Palestine. He sent a signed letter to King Saud dated April 5th, 1945 promising that the United States would not, as long as he was president, take any measures contrary to the Arab cause or make any decision about Palestine without consulting the Arabs. One week (April 12th 1945) after he signed this letter he died of a cerebral hemorrhage and was replaced by President Harry Truman who helped Israel become a nation. To all intents and purposes, President Harry Truman was used by G-d to reestablish the state of Israel. In fact, his best friend was a Jewish man named Eddie Jacobson whom he sent to Israel (in 1949) to speak with President Chaim Weizmann and Prime Minister Ben-Gurion on his behalf. Additionally, an Eddie Jacobson Auditorium was built in Tel Aviv to commemorate the man who helped to create the state of Israel. But it was in due course, President Harry Trumans doings that helped re-gather the Jews in the Holy Land. He had fulfilled prophecy. Deuteronomy 4:27-28 - **The Lord will scatter you among the peoples**, and only a few of you will survive and the nations to which G-d will drive you. But from there you will seek the Lord your G-d, and you will find him if you search for him with all your heart and all your soul. Deuteronomy 4:30-31 - When you are in distress and all these things have happened to you, then in latter days you will return to the Lord your G-d and obey him. For the Lord your G-d is a merciful G-d; he will not abandon or destroy you or forget the covenant (the land of Israel) with your forefathers, which he confirmed to them all. Jeremiah 30:3 - For behold, days are coming declares the Lord, when I will restore the fortunes of my people Israel and Judah. The Lord says, **I will also bring them back to the land that I gave to their forefathers and they shall possess it.** Isaiah 11:11 - Then it will happen on that day that **the Lord will again recover the second time with His hand The remnant of His people,** who will remain, From Assyria, Egypt, Pathros, Cush, Elam, Shinar, Hamath, And from the islands of the sea. Joel 3:1-2 - For behold, in those days and at that time, **When I restore the fortunes of Judah and Jerusalem,** I will gather all the nations And bring them down to the valley of Jehoshaphat. Then I will enter into judgment with them there On behalf of My people and My inheritance, Israel. Ezekiel 36:24 - **For I will take you from the nations, gather you from all the lands and bring you into your own land.**

Speaking about Harry Truman's best friend... There is something about the heart of G-d relative to a Jew (G-d deals with Jews). Whether they are sinners or not, (or have broke His heart many times) they are still His chosen people whom He made a covenant with. G-d chose two Jews to fulfill His purpose of reestablishing the homeland of Israel. He first chose a man by the name of Chaim Weismann, a Russian Jewish chemist who had discovered the formula for making acetone through bacterial fermentation which then led to (TNT-Trinitrotoluene) (gunpowder) in a chemical combination with Dianisidine salts (chlorosulfonate) (sometimes called the "Weizmann Organism") which was incorporated into the TNT charges (against the Germans) during WW1 which helped us win the war. This chemical concoction (which among other things) created toxic fumes that was said to bring on violent respiratory problems. It was during WWI (World War 1) that the U.S. and Britain said to Chaim Weisman - if you will give us this formula that will help us to win this war, we will give the land of Palestine (Israel) to your people as a national homeland. This agreement was known as the "Balfour Declaration" which transpired on November 2nd 1917 when Chaim Weizmann persuaded the "British government" (who had control over Palestine at the time) to issue a statement favoring the establishment of a Jewish national home in Palestine which was thus payment to the Jews for their support of the British against the Turks during World War 1 (this was in part due to Chaim's chemical concoction of TNT that helped win the war). Chaim Weiszmann was not only a chemist. He became the Israeli President of the World Zionist Organization, and the first President of the State of Israel. He was elected on February 1st 1949 and served until he died in 1952. He also founded the Weizmann Institute of Science in Israel.

Then from 600BC to 1917 the Jews have been disbursed among the nations, (that's about 2,500 years of wondering). Twenty years later, we are now in another war. WWII (World War 2). G-d had chosen yet another Jew - Albert Einstein. It was Albert Einstein who introduced (and gave) the theory of relativity, which gave us the formula for manufacturing the atomic bomb. It was promised (guaranteed) to Einstein that if he helped produce the atomic bomb (which was the Manhattan project), Israel would become a nation after WWII. In addition, this is exactly what transpired.

Here is a list of other presidents who have attempted to intervene with efforts to divide (or give away) the land of Israel to its enemies.

United States President John F. Kennedy pressured Israel on getting rid of its nuclear plan. He was assassinated on November 22, 1963 and replaced by President Lyndon Johnson who was a great supporter of Israel. He armed Israel with weapons before the 1967 six-day war.

President Nasser of Egypt had a heart attack after the 1967 six-day Arab-Israeli war. He died on September 28th 1970 at the conclusion of a Middle East peace meeting in Cairo regarding Israel.

United States President Richard Nixon's peace plans for the Middle East were interrupted and stopped by the Watergate scandal. He was in office from January 20th, 1969 through August 9th, 1974.

United States President Gerald Ford's peace plans for the Middle East were interrupted by the Election. He was in office from August 9th, 1974 through January 20th, 1977.

United States President Jimmy Carter's peace plans for the Middle East were interrupted by the Iran hostage, interest rates and Gas prices. He was in office from January 20th, 1977 through January 20th, 1981

After President Sadat from Egypt signed the Egypt-Israeli peace treaty in March of 1979, he was condemned by the other Arab nations and excluded from the Arab League. Because of this, he was assassinated by Islamic militants in October of 1981.

United States President Ronald Reagan's peace plans for the Middle East were interrupted by the political scandal of the 1987 Iran Contra affair. He was in office from January 20th, 1981 through January 20th 1989.

On November 2nd, 1991 United States President George H.W. Bush was addressing a Middle East peace conference. He experienced within 24 hours after applying pressure on Israel to divide the land, a storm developed in the Atlantic Ocean developing the largest waves (30 feet tall) in Atlantic Ocean history flooding his vacation home in Kennebunkport Maine.

United States President George H.W. Bush was confronted with the 1992 candidacy of Ross Perot and Bill Clinton. Putting a halt on the Middle East peace talks. He was in office from January 20th, 1989 through January 20th 1993.

United States President Bill Clinton was confronted by the Monica Lewinski scandal that broke on January 17th, 1998 three days before Israeli Prime Minister Benjamin Netanyahu arrived in Washington on January 19th 1998. Clinton attempted to persuade Netanyahu to implement the Oslo agreements and withdraw from the West Bank's rural areas. The Lewinski scandal put a halt to this agreement. President Bill Clinton was in office from January 20th, 1993 through January 20th, 2001.

King Hussein of Jordan died on February 7th, 1999 of cancer shortly after the Camp David peace agreement was signed.

President Hafez Assad of Syria died on June 10th, 2000 of a heart attack while talking on the phone with the Lebanese President Emile Lahoud about the Middle East peace process.

August 27th through the 29th, 2001, Just days prior to September 11th President Bush was negotiating for Middle East peace with Crown Prince Abdullah, Prince Bandar, Colan Powel and the Saudi ambassador to the United States. The Saudi Crown Prince Abdullah warns President Bush that the administration is increasingly pro-Israel and because of this may need to reassess its relations with the United States putting an end to their alliance. Bush immediately responds by promising a new initiative for peace in the Middle East, including support of a Palestinian state. He firmly told Prince Abdullah that the Palestinian people have a right to self-determination and to live peacefully and securely in their own state and homeland, just as the Israelis have the right to live peacefully and safely in their own state. Bush corresponded that he directly supported and was in favor of a separate Palestinian state. The Saudis were appeased. On September 6th 2001 Bush holds a meeting with his top advisers and suggests a change of policy towards Palestine, including public support for a separate Palestinian state. However, days before Bush is to announce these new policies the 9-11 attacks take place and none of the planned US policy changes materializes. September 11th STOPPED THAT!

United States President George W. Bush was met head-on with the "War On Terror" which was taken to curb the spread of terrorism following the September 11th, 2001 attacks on the United States. All Middle East Peace talks are on hold.

Yasser Arafat of Palestine became comatose with an immunological decrease. It bore all the same symptoms of AIDS. Though, there were other speculations that he was poisoned. He died on November 11th, 2004. He was involved in the Middle East peace meetings to reconstitute (rename) Israel as the Palestinian state.

Prime Minister Rafik Hariri of Lebanon was assassinated by a bombing on February 14th, 2005 while trying to progress the process of the Middle East peace agreement.

Lets back up to April of 2003, the "Quartet world peace plan" consisting of a group of 4 nations that came together for the Israeli-Palestinian peace agreement under the discussion of how to divide the land. Consisting of the United States, Russia, European Union and the United Nations. 11 days after the Quartet plan broke out there, was a record breaking count of tornadoes during the first 2 weeks of May? This was the BEGINNING of the worst weather EVER in United States, Europe and Russia. 500-year heat records were broken, horrific property damage, floods, fire and heats. 4 major hurricanes in which one of them pounded President Bushes brother (Governor Jeb Bush's) state of Florida. In fact, on May 16th, 2003 President W. Bush made a public proclamation stating that May 18th through May 24th of every year forth be called "National Hurricane Awareness Week". During this year record setting United State evacuations are taking place, MILLIONS of Florida residents are ordered to evacuate an unprecedented 4 times. Whilst all this was taking place, on August 23rd, 2005 President W. Bush congratulates Israeli Prime Minister Ariel Sharon for making the "courageous decision" (in his opinion) to give the Gaza strip back to the Palestinians in the advancement of democracy. This enactment was called the "Disengagement plan" which was acted upon on August 15th 2005. The evacuation of all the Jewish residents from the area was completed by September 12th 2005. President Bush also described this mass departure of the Gaza strip as "an amazing event" (again in his opinion). Ironically Hurricane Katrina develops on August 23rd 2005 and on August 29th Katrina hits. And whilst there were thousands of businesses and homes wiped out in America, "simultaneously" hundreds of businesses and homes were lost in Gaza. Shortly after all this, Israeli Prime Minister Ariel Sharon had a massive stroke leaving him incapable of regaining office. He never adequately regained consciousness.

As if this were not enough of an eye opening indication to us, we must NOW ask ourselves, what mindful and deliberate entity is behind these workings? Are we to assert that all these instances were flukes, twists of fate or mere chance (or coincidence)? Remember - "coincidence" is not a kosher word! Would you be taken aback to find out that there is EVEN MORE observable evidence fleeting right before our eyes venerating an purposeful intelligent entity? You must keep reading to gain such knowledge. Moreover, once you've gained it, would you then be inclined to change your mind about written scripture and the world around you thus becoming more religiously attuned? I'd also like to bring another significant thought to everyone's attention regarding this mindful entity (G-d) that is behind all these workings. And this has to do with establishment of the state of Israel during the 1967 (6 day war). It may have evaded most people to consider the odds here since it was 7 Arab nations that attacked Israel with all their furry, and overnight over 800 thousand Jews found themselves at war with 50 MILLION Arabs! The Israeli army had not a single tank, while their Arabs, enemies had over 2000. Yet, amazingly Israel not only wins the war, but also establishes the only democracy in the whole Middle East. Again, on 6-5-1967, the Arab armies regroup, and the Soviet Union sells the Arabs almost 2 million dollars worth of arms. This included Tanks and fighter jets. Additionally the Soviet forces came to militarily train the Arab forces. But amazingly Israel catches wind of this and attacks, and in just 6 days Israel stuns the world - they crush the combined Arab forces of Egypt, Jordan, Syria and Lebanon, they also capture the Sinai Peninsula, the West Bank, the Gaza Strip, and the Golan heights. The factions of this are so overwhelmingly obvious that what we witnessed could have only come about by the POWER OF G-D!

Moving forth… As one can see, the fundamental problem of the U.S. (as well as the world) is our rebellion against G-d. The United States of America was founded and blessed beyond all other nations to enable us to spread the gospel all over the world and to serve as a vehicle for the reestablishment of the state of Israel (which President Harry Truman facilitated). Any betrayal against Israel is going to seal our doom as a nation. - Thus the world.

We must also watch what is impending on the world stage now. It is now the year 2009 and we have a new President in the oval office, (Barack Obama the 44th president of America). The world was watching the inauguration of President Barack Obama. In fact a tentative cease-fire between Israel and Hamas militants shooting rockets from Gaza into Israel was halted just prior and after President Barack Obama's inauguration ceremony on January 21 2009. President Barack Obama has also taken an early interest in the Middle East peace process between the Israelis and the Palestinians, with the president naming former Senator George Mitchell as his special envoy to the region who has had success with creating peace in Northern Ireland after 800 years. Barack Obama said, I do believe that the moment is "ripe" for both sides to realize that the path that they are on is one that is not going to result in prosperity and security for their people. And that instead it's time to return to the negotiating table. The wars will stop when the Palestinians love their children more than they hate us - Golda Meir (the first female Prime Minister of Israel).

We must keep a watchful eye here to see what transpires.

Finishing up here… Eventually a peace agreement will be achieved for Israel and its surrounding nations to live side by side with each other, and this peace agreement will usher in the 7-year prophetic biblical tribulation of Daniel. Daniel 9:27 - He will confirm a covenant with many for one **seven.** In the middle of the **seven,** he will put an end to sacrifice and offering. In addition, on a wing of the temple **he will set up an abomination that causes desolation**, until the end that is decreed is poured out on him. Daniel gives another description of the antichrist as a self-exalting king. Daniel 11:36-37 - And the king shall do according to his will; and he shall exalt himself, magnify himself above every G-d, shall speak marvelous things against the G-d of g-ds, and shall prosper until the indignation is accomplished: for that that is determined shall be done. Shall he neither regard the G-d of his fathers, nor the desire of women, nor regard any G-d: for he shall magnify himself above all. The book of Thessalonians gives us another description of the antichrist that tells us that the he will bring peace and safety and then sudden destruction. 1 Thessalonians 5:3 - For when they shall say, Peace and safety; then sudden destruction cometh upon them, as travail upon a woman with child; and they shall not escape. In the book of Ezekiel, we find notation that refers to the 7-year tribulation. Ezekiel 39:9 - Then those who live in the towns of Israel will go out and use the weapons for fuel and burn them up—the small and large shields, the bows and arrows, the war clubs and spears. For **seven years,** they will use them for fuel. As a cross reference so that be know that Daniels prophesy was indeed relating to the antichrist, Jesus Himself also refers to the references about the abomination and desolation from Daniel 9:27. Matthew 24:15 - When ye therefore shall see the abomination of desolation, spoken of by Daniel the prophet, stand in the holy place, (whoso readeth, let him understand). Better yet, in Mark Jesus relates this time of abomination and desolation as a time when Judea will flee to the mountains. Mark 13:14 - But when ye shall see the abomination of desolation, spoken of by Daniel the prophet, standing where it ought not, (let him that readeth understand,) then let them that be in Judaea flee to the mountains.

This peace treaty will be for 7 years and it will be facilitated by the antichrist. This is the prelude to Armageddon. Peace between the Israelis and the Palestinians WILL HAPPEN. The Bible says so. However, when it does it will bring peace, safety, and then sudden destruction. Then those that are in Judea will flee into the mountains/hills. Matthew 24:16-20 - **Then let them which be in Judaea flee into the mountains:** Let him which is on the housetop not come down to take any thing out of his house: Neither let him which is in the field return back to take his clothes. And woe unto them that are with child and to them that give suck in those days! But pray ye that your flight is not in the winter, neither on the Sabbath day.

What is G-d waiting for? He is waiting for His Bride - "the church" (the body of believers) throughout the entire world. Ephesians 5:24-27 - Therefore as the church is subject unto Christ, so let the wives be to their own husbands in every thing. Husbands, love your wives, even as Christ also loved the church, and gave himself for it; That he might sanctify and cleanse it with the washing of water by the word, That he might present it to himself a glorious church, not having spot, or wrinkle, or any such thing; but that it should be holy and without blemish. Revelation 21:1-2 - And I saw a new heaven and a new earth: for the first heaven and the first earth were passed away; and there was no more sea. In addition, I John saw the holy city, New Jerusalem, coming down from G-d out of heaven, prepared as a bride adorned for her husband.

You know what else is daunting! Even the world economy seems to be effected by prophecy. We have seen the stock market and the banks all over the world crater. Both the stock market as well as a range of cause and effect from it has put the entire world in a economic crunch... From America, Europe, to the Far East. The most interesting thing in the stock markets fall is that it occurred during what's called "the days of affliction" (from Rosh Hashanah - the New Moon of September 29th) and for the next 10 days, which are Jewish holy days called the days of Awe (basically days of affliction). It was during these days of affliction that we saw the stock market and businesses crash, and people loosing trillions of dollars. It was exactly on 9.29.08 that the stock market had it's largest one day drop in history, over 777 points.

My point in all this is that everything that happens in the world today revolves around Israel and the Jews, it always has and it always will. The Jewish holidays also revolve around prophecy, both pasts, present and future. All of this culminates to the day of atonement (the Jewish holiday of Rosh Hashanah). The earth according to the Bible has been around for

6,000 years (as aforementioned in the previous chapter). Another 1,000 would make it 7,000 years. 7 is the number of finality - it is also the year of Jubilee - the Jewish year of gathering the harvest. Could this be the end of prophecy?

"Now is the watchword of the wise" - proverbs

4 - A GLIMPSE AT CREATIONS PRIMORDIAL LANGUAGE...

Well if you thought Quantum Mechanics was something, (even though we veered off towards the end of the chapter a bit) you haven't begun to truly get the message of anything quite yet. Following this, what I write now, I don't expect you to fully understand, as its just a prelude for what's to come. The reason for this is because there is a sequence to understanding, that is why I am only giving you bits and pieces of information to read up until now so that fruition of comprehension may proceed. However, I assure you; you will begin to understand as we proceed here, and in the following chapters.

Let me start by saying that so much money, effort and dreams has gone into SETI (The Search for Extraterrestrial Intelligence) to receive a signal or an indication that someone else is out there. But suppose that we have already been contacted by G-d in a way that is so supernatural that it has become a religion or maybe INTENDED TO BE A RELIGION TO SHOW US THE WAY TO A SCIENCE that is beyond imagining. Some may still think it's a signal we should be listening for, but I'm proposing that it has already been given to us here on earth many centuries ago by our creator - G-d Himself… It is both in partnership a physical and spiritual language. IT IS THE HEBREW LANGUAGE AND ITS ALPHABET - a dialect by means of utilizing the shape, form and given/distinct position of each individual Hebrew letter. Every single one of the Hebrew letters demonstrated by their individual shape, form, and position are all a result of NATURES GUIDING FORCE (an expanding unit and growth pattern which adheres not only to our earth, but the Universe at large - which you will soon read about). The Hebrew letters also happened to be given by the finger of G-d Himself. Exodus 31:18 - When the Lord finished speaking to Moses on Mount Sinai, he gave him the two tablets of the Testimony, the tablets of stone **inscribed by the finger of G-d.**

G-d, nature and the Hebrew letters work hand-in-hand. This is a very hard concept to comprehend especially when there are no visual pictures provided (yet).

Continuing, it is these letters in geometric forms like bits and pieces of molded clay (reminiscent of atomic particles) that were used to create the whole Universe and the physical reality that we humans know and see. And I'm proposing that the Bible (Torah), which is written in these Hebrew letters, has actually created its very own letters in which it was written. Genesis 1:11-12 tells us that it's a fruit tree yielding fruit within itself.

Genesis 1:11-12 - And G-d said, Let the earth bring forth grass, the herb yielding seed, and the fruit tree yielding fruit after his kind, **whose seed is in itself**, upon the earth: and it was so. And the earth brought forth grass, and herb yielding seed after his kind, and the tree yielding fruit, **whose seed was in itself**, after his kind: and G-d saw that it was good. The word in "itself" says what it means. In addition, this gives us underlining testimony of an intelligent entity, which lays imbibed within it. The Bible itself is alive and is a living organism. It is of G-d, but written by man, since His message is to all humankind.

Delving even deeper here… as aforementioned in the past previous chapters concerning energy, matter, and light. If you can recall, light in itself is actually the ONLY thing we do see, since physics says light behaves as a particle and wave, and EVERYTHING is made up of particles that move in waves. Nothing could be closer to the truth, especially when we realize that "G-d is light" Himself. 1 John 1:5 - This then is the message, which we have heard of him, and declare unto you, **that G-d is light**, and in him is no darkness at all. Yet again, it is elevating to see the scientific evidence supporting what Scripture says. This might very well be one of the most powerful weapons that exist in the whole arsenal of scripture and science - the connection between G-d and light. Moreover, there is something even more special about the wave path that atomic light particles take (since they SPIRAL). Moreover, it is this spiraling design that forms the Hebrew letters. Briefly, IT IS THE PRIMORDIAL LANGUAGE! To sum it up, (though as I said, I don't completely expect you to understand it yet) the Hebrew letters that are reminiscent of light (since they are formed by atomic particles that move in spiraling wave patterns) are the embodiment of G-d through which He forms all things and works His ways. Please keep reading, I assure you, you will eventually come to understand this when you get to the actual chapter on the "Primordial language" (chapter 8).

Again, as I said, I don't expect you to understand this now. However, I'm sure you are in either disbelief or confusion of what you just read! Alternatively, your wondering what on G-d's green earth is this all about. You're probably thinking, how can any of this possibly be, or (more precisely) where is the evidence? Well to answer this question, you'll have to continue reading to obtain that information. But after realizing this, it will also include (part and parcel) the understanding that RELIGION ITSELF IS A SCIENCE… and you will come to understand as well as appreciate that spirituality is needed for a different kind of evolution (given in the gift of Torah). And that being, that the journey into the upper world is accompanied by and even dependant on the everyday experiences of the "material realm" but does not regard the "spiritual" as divorced from the material, but sees the material as a way to attain the spiritual.

Spirituality is to man what carbon is to steel - Me (Anita Meyer)

The answer to your query about the Hebrew letters will follow with pictographic proof in the upcoming chapter of the "primordial language". After viewing this proof, I predict that a foreseeable time will come where we will be converted from a faith-only belief to a fact-based science, and I have all the intentions of proving it as we move along in this book!

A quick introduction to the word Torah (which is the Hebrew Bible) comes from the root of the word "Horaah" which means "instruction". The Torah states this concept in the following way**:** The first word of the Ten Commandments in the Hebrew letters (Exodus 20:2) is "Anochi" which means "I am the Lord your G-d". G-d "compressed" Himself into the words of the Torah. Therefore, when a person studies Torah in this world they have a direct link and connection with G-d. This may sound esoteric and might not grab your interest or make much sense to you now, but hopefully you will be able to apply its meaning at the end culmination of this book. And for some it just might take several times of rereading this book to fully understand its entirety as well as intended message.

I would also like to cue you in, as you read this book bear in mind that what I write is counterintuitive - coming back around to beckon you. It implies retro causality. Instead of cause and effect, the effect comes before the cause, using the example of when an apple drops from a tree, is it the ground that was the cause of the apple falling in the first place?

It is reverse engineering of the infinite kind. It also entails what's called a "closed loop system" which you'll come to understand more about in the following chapters. So strike that, reverse, lets start from the beginning.

5 - SEARCHING FOR SIMILARITY...

The sequence to understanding starts here…

Ever since I can remember, the natural world has always intrigued me. From the structure of a branching tree to its similarity in a branching river network to why man made highways, expressways and roads resembled main arteries and blood vessels in the human body as well as in animals and insects and single celled creatures. This also includes all geometrical growth forms composed of algorithms (which was mentioned in an earlier chapter). I always felt (and often envisioned) that there was some simple hidden mystery and order among all the chaos. And what I soon discovered became apparent, that what seemed to be chaos was really order. I somehow had the insight that everything from my human body to the natural and material world was intertwined and connected - balanced and proportioned - symmetrical if you will. It took just finding out what that "one thing" was that everything had in common. A process of reduction in my minds eye. In addition, I knew that if I could just get to the root of all this connectedness I would uncover something of great significance and perhaps even with some scientific and mathematical credibility discover G-d in the process.

My thoughts delved deep in observing the visual world around me. I can see the association that the human body imitates in the resemblance and workings of the natural world. It takes on many similar appearances reminiscent of the rainforests and the oceans functioning as the earths lungs exchanging oxygen and carbon dioxide just like our very own human respiratory system. One can almost say that earth menstruates; mirroring bodily changes with the moons gravitational pull during the seasonal cycles that change the trees leaf colors from autumn to the perennial rebirth of spring. How candidly correlating that our rivers and streams are similar to our bodies circulatory system by way of transporting clean water and rinsing out the system. On an even broader sense of expression, even as our expressways and busy roads are likened onto main arteries and veins that carry nutrients throughout our body. The arterioles (arteries) carry oxygenated blood to the body, and the venules (veins) carry the deoxygenated blood back to the heart to be oxygenated again, and the capillaries are the thin vessels that connect them. So too does the earths "Great Oceans Conveyor Belt" circulates warm and cold water around the Pacific, Indian and Atlantic Oceans.

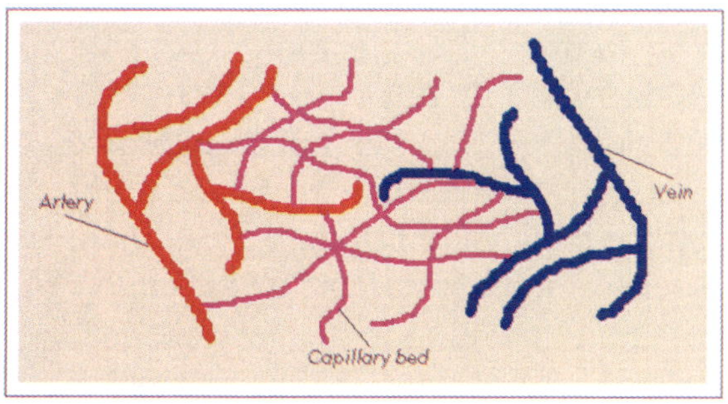

These same functions are noticed in science when the physical world is studied, since we are now learning the effects of Global Warming on the environment. The Earth whose system was fashioned by G-d has what's called perfect recycling. When we stop to consider how much un-recyclable toxic waste, we create yearly, and how we go about getting rid of those wastes, it becomes amazing when we realize that the earth recycles all its own wastes perfectly by using resourceful chemical engineering - such as the carbon, oxygen and nitrogen cycles. Some of these processes include the earths hydraulic system. It can even be witnessed occurring in the oceans where the coral (from coral reefs) and the calcium carbonate in shells help to stabilize the chemistry of the water, much in the same way as an antacid works in the human stomach thereby neutralizing acidity. Even plants, phytoplankton and algae found near the surface of lakes and oceans help to regulate carbon dioxide and oxygen levels in water and air. And in the soil, bacteria and fungi (spore-producing organisms) work as decomposers, providing inorganic nutrients for plants. What process (on its own) could duplicate such environmental harmony? Certainly not, chance processes alone!

Moving along here… Now, when we begin to expand our view to an even higher realm of understanding, it can be apprehended that the Earth from a lofty scale can be viewed as an organ within the solar system. On a majestic scale, we can view galaxies as molecules - and even in a humbling but grandiose sense, we can view planets as atoms.

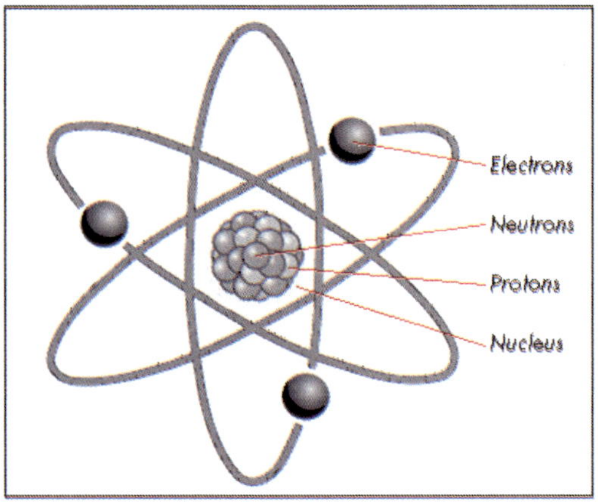

To peer even deeper, the basic building block of all matter is the atom. The center of an atom is made up of a nucleus that consists of neutrons and protons. Electrons orbit the nucleus much the same way the planets orbit the sun.

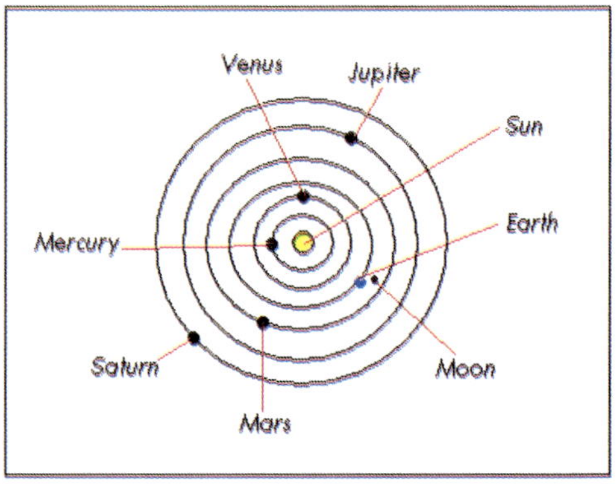

Even the human cell itself, (and notice how the nucleus is slightly off center to the side. This mathematical unit will reveal itself in the next chapter):

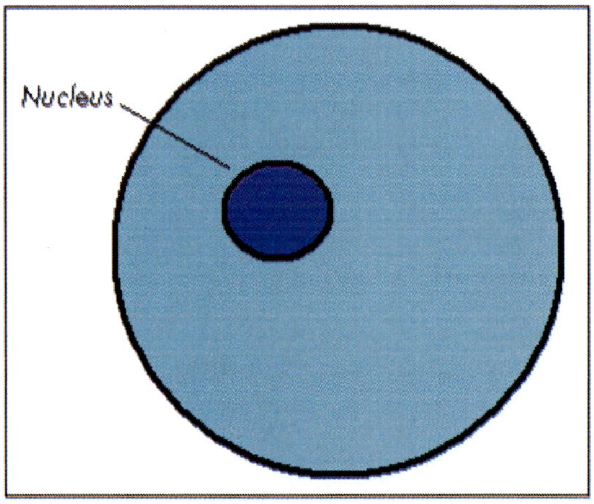

It can certainly be said that the earth can be seen as a hierarchy to our human body, just as G-d is the totality in progression of all hierarchies. If we are to understand this correctly, our bodies are built of atoms coalescing into molecules that are grouped and assembled together into cells, which then form organs and eventually unite and function together to perform the great feat in the act of operating as one working organism. Each cell is itself a conscious life responding in an individual way and having its own distinctive disposition. The organs too have their own awareness and reflect on the cells that built them. But there is a governing awareness and ingredient that informs the whole bodily structure of its totality, and it is from this grand perspective, like peas in a pod, that everything in the universe may also be a many layered being. A conscious individual entity expressing itself on various invisible levels and surrounded by innumerable various stages of development in concert with the interconnectedness of all things. ***At last, we notice G-d!***

And on behalf of the very cells in our body that serve us, so in turn do we serve something greater, perhaps unsuspectingly a G-d. A giant integrated system of systems - A subsystem of life inside of life inside of life. The full epiphany (sudden realization) of a G-d in a nutshell! An unallocated force capable of acting through, and on all things (people and objects). In view of this aspect, it can certainly be said that G-d has (in every way) full control of the whole system of systems. He is omniscient (all-knowing), omnipotent (all-powerful), omnipresent (always present everywhere). G-d is the master brain and we are the "emulations".

1 Corinthians 12:12-27 speaks about the "body of believers", however in many ways it also applies to the real "physical body" that G-d (of G-d) created for us. 1 Corinthians 12:12-27 - For as the body is one, and hath many members, and all the members of that one body, being many, are one body: so also is Christ. For by one Spirit are we all baptized into one body, whether we be Jews or Gentiles, whether we be bond or free; and have been all made to drink into one Spirit. For the body is not one member, but many. If the foot shall say, because I am not the hand, I am not of the body; is it therefore not of the body? And if the ear shall say, because I am not the eye, I am not of the body; is it therefore not of the body? If the whole body were an eye, where were the hearing? If the whole were hearing, where were the smelling? But now hath G-d set the members every one of them in the body, as it hath pleased him. And if they were all one member, where were the body? But now are they many members, yet but one body. And the eye cannot say unto the hand, I have no need of thee: nor again the head to the feet, I have no need of you. Nay, much more those members of the body, which seem to be more feeble, are necessary: And those members of the body, which we think to be less honourable, upon these we bestow more abundant honour; and our uncomely parts have more abundant comeliness. For our comely parts have no need: but G-d hath tempered the body together, having given more abundant honour to that part which lacked. That there should be no schism in the body; but that the members should have the same care one for another. And whether one member suffers, all the members suffer with it; or one member is honoured, all the members rejoice with it. Now ye are the body of Christ, and members in particular.

Now when I mentioned "emulations" above, it means to copy and imitate… When saying this, it rises to yet additional intensifying questions and relates us right back to the previous debate on the evolution theory. Do we truly invent anything new, or do we simply rediscover what is already there?

Examples of this comparative phenomenon are… Airplanes verses birds… Helicopter verses a bumblebee, hummingbird and maple seedpods… What the sun and moon is to the wheel… As a solar panel creates, energy so does a leaf… A spinning tops relationship with planetary rotation… A mirror verses a reflection on water… Or as a drop of water is to a magnifying glass… The kidneys as to a water filter… As do swimming fins and paddles, have in common with the webbed feet animals… Or the invention of the suction cup to that of octopuses… Smoke screen as a defense verses octopus ink… Fish hook and a bee stinger… Sonar associated with bats, dolphins and whales… Batteries and an Eel… Wiper fluid and tears… Windshield wiper and the eyelid… Dust filter and nostril hairs… Knife and incisor teeth... Armor and exoskeletons... Pliers to a lobster claw... A zipper and the linkages between barbs in feathers... Camouflage clothing verses a chameleon... A camera lens verses an eye... A microscope or telescope is an extension of the eye... Microphone verses an eardrum... The pump verses a heart and its valves... Chemical solution to fix punctures in tires verses coagulants found in the blood... Antibiotic medicines verses the immune system... Plumbing system verses circulatory system... Ball joint verses shoulder and knee joints... Anesthetics verses venoms and poisons... Hypodermic needle verses snake fangs... Lawn aeration verses worms and moles... Telephone cables verses spinal cord and nervous system... The shovel as an extension of the hands and feet... An automobile as an extension of the feet... The bow and arrow to the development of gunpowder and firearms... There are even such natural organisms like bacterium that can convert a toxic waste product into safe biodegradable plastic, and others that can neutralize (by eating) oil and radiation. In addition, sonoluminescence is the emission of short bursts of light from imploding bubbles in a liquid when excited by sound. Last but not least the miniature machines, turbines, sliding clamps, miniature rotary motors that are being found inside the cell.

The question is, are all of these comparisons truly the source of life's creation (evolution theory), or are they really G-d's process of creation? We must also ask the question, is anything"NEW" or is it something that has always existed from the start? In all honesty, we cannot rightfully declare that any of these creations were originally ours to begin with since the comparisons (listed above) show signs of impersonation. Furthermore, if science tries to prove the non-existence of G-d, we must then re-examine the proof. Because an "emulation" (such as one of these comparisons listed above) cannot be used to prove the non-existence of its manufacturer. Moreover, what these comparisons also substantiate (as well as corroborate) beyond any reasonable doubt is that everything that we will ever need (for life) was already there from the instant of creation.

Moving on here to yet another perspective. This perspective is termed, a "closed loop system". A closed loop system is isolated from anything coming in or going out. When I refer to a closed loop system, I am implying to the "whole" of everything that "is" and will ever be. And this includes the Earth as well as the whole gamut of space, stars, planets and universe's in existence. This also includes the four interchangeable (or interaction) forces of friction, magnetism, gravity and nuclear decay (strong, weak, electromagnetic) - the stuff that holds everything together. Outside this, nothing else exists except for G-d.

This sign is indicative of a closed loop system:

A very famous quote from King Solomon in the Bible states... What has been will be again, what has been done will be done again. There is nothing new under the sun. Is there anything of which one can say, "Look? This is something new"? It was here already, long ago; it was here before our time. Ecclesiastes 1:9-10 - What has been will be again, what has been done will be done again; there is nothing new under the sun. Is there anything of which one can say, Look! This is something new. It was here already, long ago; it was here before our time.

On a more pressing and expressive note… A closed loop system can even apply to our human immune system. Our very materialism of our auto immune system contains memories that go back to the creation of the first man and woman (Adam and Eve). G-d had created an ingenious mechanism (that rests silently) built in within us from the moment of our conception and gets passed on to the next generation. In the span of our life, these protected immune memories seem to surface intermittently (with an adaptable special ingredient) as if memories of data/information go back eons of time. For example, our body meets a viral organism for the first time, and ironically, it knows how to make a precise antibody to that virus even though our personal individual body has never met this invader before. How does it know this? It knows this because at some point in time, an ancestor (distant relation) met the virus and learned how to deal with it, and that memory is their forever in our software (this is not something that evolved over time). In fact, everything "at some point in time" replicates the entire history of everything that exists in creation. We cannot escape this closed loop system!

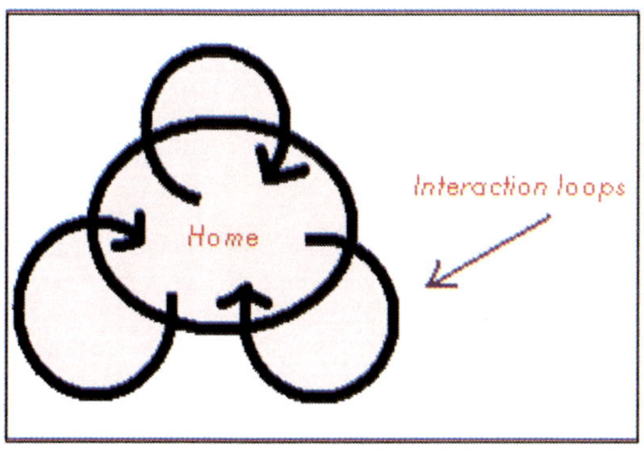

Even today, mankind is developing artificial "cells" to one-day replace our defective biological cells by means of nanotechnology. This we can also call "a closed loop system". Even though it can be seen as an "extension" of a swapping out of the biological (natural) parts with artificial (un-natural) parts within the human structure. Again we must realize that the very essence of these parts are made of silicon/quartz/gold which are also natural substances that were included in the creation of the earth, so it too is "in some branch" part of our body. For that reason, it can be argued if they are truly un-natural when we've only manipulated them. These magnificent parts/tools can extend life, thus buying us time to learn the scriptures and build a better relationship with our creator.

Here are other examples of extensions (of a closed looped system) that have progressed throughout the course of time. For example, when Samuel Morse discovered the telegraph with a tap tap tap relaying the legendary message - "what has G-d wrought?" The reply clearly didn't anticipate the social upheaval the electronic media would create from it. From the Telegraph came the telephone, radio, than the phonograph, television, photocopier, answering machine, computer, VCR, compact disc, cellular phone, FAX, internet, Mp3 player, digital technology, GPS and Satellite… ext, ext…

Some of us can remember the effect the story of H.G. Wells "War of the World" had over us by spreading wide panic. However, it also had an additional long-term effect on us. It had such a profound effect upon radio listeners and captivated people about the planet Mars, that a American pioneer named Robert Goddardt started experimenting with ways to get to Mars with ingenious designs for getting rockets from the surface of the earth up into the high atmosphere (it was regarded as silly nonsense at that time). However in Germany he had some fans who were reading his research and they then became inspired to implement his designs and improve upon them, and this led during WW2 to the development of the V2 rocket which is the predecessor for the American and Russian space programs today. This is presenting to us that there is an even more profound and peculiar connection in the course of history's events. Simply put, if we can envision it, we can create it! Cause and effect - retro causality as well as our previous discussion on Quantum Mechanics involving human consciousness. With the exception of understanding that within a closed loop system, we are still not generating or inventing anything new! We are only reinventing, implementing and manipulating something that has always been in existence from the time of creation. This can be referred to as ubiquitous in embryo! (Meaning: Existing everywhere, or present everywhere at once). In this aspect, we must realize that we cannot create something new, no not one thing, not even a new color in the rainbow! When we try to think of something that doesn't exist we can't do it! Yes, we can do amazing things with what the creator has given us, but it is only G-d alone that creates out of nothingness. In realizing this, we must now stop and consider the notion of where humankind got the conscious idea and concept of G-d itself. Where did that initial "thought" come from? We cannot solely assert that the idea of G-d was placed into our minds by early preachers or holy men, because that still leaves open the identical question of where the preachers got the idea. One cannot come up with where the existence of G-d stemmed from. We can only conclude that G-d has put the knowledge of His existence into every human being.

6 - NOTICING THE NETWORK...

The secret to understanding increase here...

In my search for similarity, I began to notice a hidden network, and it shall soon reveal itself to you not only in this chapter but also throughout this entire book. In effect, every chapter is part of the whole. In the fullness of time, it will soon be realized that there is a deep organizing principle with "ONE ORGANIZING LAW" of instruction for the world, as we know it, and the universe at large.

With that said, life just seems so full of "connections" and most of the time we don't even pay attention to the depth of life, sometimes only seeing flat surfaces. When we just look at surfaces it appears basic, but it's when we begin to look deeper that we are truly amazed by what lies embedded within. If we are to get anywhere in the hallways of understanding, we must learn to bridge the gap between things that in appearance are completely different, and yet at the same time are parallel in comparison. I could show you many corresponding pictures invoking semblance in this book but it would do no good because many eyes go through the meadow, but few see the flowers in it. That's why I am relying of the eye of the master to do more work then pictures - and by the signatures of all things, you are here to read! Besides, the soul never thinks without a mental picture, therefore ones mind is the best lens and camera.

There may be different angles of vision but they all eventually illuminate the same truth - its not what you look at, but what you see as in the minds eye. Yes, the whole of life lies in the verb seeing, and TRUE seeing is when the mind does the seeing. The eye itself sees only what the mind is prepared to comprehend. In fact, the mystery IS THE EYE ITSELF! The very "unit measurements" from the black center of the pupil to the colorful iris and then to the white of the sclera. While at the same time it is this identical unit of measurement and proportion, which forms the foundation of the material world that we are capable of seeing and finding engaging. (I will talk more about this measurement in the following chapter).

In this chapter the main goal is to contemplate (imagine, ponder, and reflect) on all pathways and channels that exist in the world around us and compare their likeness' and resemblances' to veins, capillaries, and arteries, whether it be of the human body, animal body, plants, rivers, earth, galaxy, universe, or any man made object or system. When we apply this reflection to the workings of nature and its order, we ought to come to the understanding and realization that it is always OURSELVES alone we are observing (as confusing as this may sound). Thus we cannot escape the fact that the world we know is constructed in order to see itself. It must first cut itself up into at least one state which sees, and at least on another state which is seen - hence (again) QUANTUM MECHANICS!

In the end, some may still say that the human body is exempt from being anything else other than what it is, and that (figuratively speaking) houses are perceived as houses, trees as trees, and roads as roads. However, I will tell you from my own perspective that certain forms and configurations of organization contain powerful meanings and associations for individual viewers. I will truthfully admit that it is not hard to find patterns or geometric shapes, but, it is hard to find patterns, key features, geometric shapes or configurations of organization that come together to mean something MEANINGFUL and of greater significance when it is applied to the next level of understanding that is best explained by the hallmark of an intelligent cause. It then becomes a compelling indicator or evidence of intelligence having played a role in the origin of life.

I ask you now to visualizing (in your minds eye) these veins, capillaries and arteries as "pathways" - notice that there are large and small pathways. The larger pathways are the more popular networks, which were made bigger by the smaller pathways out of necessity. The bigger pathways being the main transportation system rely heavily on the linkage and support of the smaller pathways. While the smaller less popular pathways are really the more adaptable and versatile because they are the connectors that transfer information, growth and nutrients throughout whatever system it is. Without these smaller pathways, the bigger pathways would not have a reason to exist or thrive. Keep in mind that all things come about from intent, reason and necessity. In addition, we can begin to see that the organizational system of the arteries, veins, and capillaries goes on up the scale, ad infinitum - the ever-familiar "forest from the trees" syndrome.

Now imagine that you are looking down from the hundredth floor of a large building. You will see that cars look like ants, and boats floating on water appear like platelets floating in blood. While roads and sidewalks around buildings and houses and other structures begin to look like the same similar passageways along-side organs with veins, capillaries and arteries that are within our own body.

Now we can also begin to visualize in our minds eye, the branching structure and "unit growth" of a tree and plant as well as the rooting systems, to its likeness in the splitting, diverging and forking of a river network. This same view also applies to the sprawling blood arteries, capillaries and veins in living things all the way from lungs to gills to the resemblance of branching lightning bolts. Even the skin on the palm of our hands and knuckles, in which the crease marks can be seen as main passageways formed for bending necessity. How about neighborhoods with houses and trees, shrubs and sidewalks going to and fro… from a higher elevation the shrubs begin to look like blood cells and capillaries underneath our skin dermis, and the streets and sidewalks begin to appear like blood vessels and arteries. A pond in the middle of this neighborhood can be likened to the lymphatic network or skin filled with fluid perhaps for drainage or runoff.

We often plainly see just cars in motion leaving for work, going to the gas station, grocery store and then coming home from work. But perhaps this is a system that functions along the same lines as our human circulatory system that moves blood back and forth - something akin to traffic moving down roads, streets, and highways just like blood cells moving through veins and arteries. Even the color of veins just under the surface of our skin appears blue in color, but when we bleed, it comes out red. This same phenomenon can be spectacularly witnessed at night with the color of white and red car lights (coming towards us in the opposite lanes we see white headlight, and upon stopping we see red break lights). In addition, from this perspective we can begin to see a much broader picture of semblance.

Now I'd like to bring your attention to this measuring unit of growth. Take notice of the "unit growth" in the branching limbs of a tree compared to the sprawling sparks of lightning.

Even cracked mud bears this same branching structure.

Do you suppose that either I'm looking to hard for patterns in trying to compare tree branching, Lightning, and cracked mud as to apples and oranges, or all of this is clearly showing good evidence that all living things (even nonliving) that are subject to this earth are RESPONDING TO THE SAME FORCE IN NATURE? Is it possible that a "universal formula" from a branching tree to Lightning also guides the blood vessels in pattern and path in all emergent things? And that perhaps all things though they appear different in shape, form, and category, are inextricably linked in resemblance.

Now I know that pictures are worth a thousand words as I've said before, that there are some things that I cannot show in pictures. I am also aware that all of this may be a matter of perspective, so please keep in mind that true creativity is allowing yourself to gain the loftiest perspective you can in relation to the object of your inquiry. Now I ask you to think in thought and visualize these things in your minds eye. Applying the same principles as you've just viewed above with the branching system of popular larger pathways and the smaller less popular pathways. Of course I'm just touching on a few of these examples, I know that there are many more to be conceptualized. Lets start out with our social networks. How in the world do we manage to communicate in a world consisting of billions of people? It starts out through the interaction of our immediate families, relatives, friends, work affiliates and people whom we meet around town and even through the Internet. Now lets visualize that your immediate family, relatives and close friends would be considered small blood vessels, while distant friends, work affiliates, internet connections and people who we come in contact with around town would be considered larger blood vessels. Now in order for the whole world consisting of billions of people to communicate and function as a whole organism/network, information needs to spread more effectively to a greater number of people by a process of moving through smaller vesicles in order to reach the larger arteries, and in doing so your broadcasted information has a chance to escape the boundaries of your own small immediate social group with a better chance of reaching a larger amount of people. This function is how a world consisting of billions of people can come together and soon function on an organizational level of whole purpose, necessity and reason, informing the totality of the body, which in this case would be the human organism at large. We can apply this view of how social information travels through channels and into main arteries in likeness to how a tree branches off from the main trunk into smaller branches supplying nutrients throughout the whole body of the tree, utilizing the same organizational law as show in the pictures above. Information would also flow through all living things. Amazingly, the same directorial processes would also apply to radio, television, and cell phones, the Internet, books and newspapers. Now to visualize things even more in depth we can also apply the same comparison between the working of our social networks and the functioning of our own human brains. For example, in our brains we have what is known as dendrites, these are the neuron receiving channels, and than we have the axons, which look like long wires, they are the transmission lines. The axons heading away from the neurons link up with the dendrites and other neurons, and these can be likened to communicating links up with others within the same region, and through these axons also linking regions that are on the other side of the brain. We can apply this view of how the brains neural networking travels through lobes of the brain by using similar networking lines that stem and split-off in resemblance to how a tree limb branches off from the main trunk and supplies nutrients to the smaller branches. Even the individual lobes of the brain can also be likened to the regions of transportation through a big city in its distribution of nutrients and information along the way. Unequivocally the brains chemical function is starting to look familiar in its resemblance to bolts of lightning, a river network, streets, and the underground rooting system of plants, which all mirror the same "organizational law". In addition, hark to all of this, that the circulatory system and the central nervous system of all living things also work in this same similar fashion by utilizing the same organizational branching system in its growth and distribution.

The physician's tomb - Ankhmahor in ancient Egypt. From 2330 B.C. showing a Reflexoloy treatment for both hands and feet.

In addition, other things stand in representation of this same system, but in different ways. One such body related technique is an ancient healing art called "Reflexology" that can be traced back to the Egyptians.

Reflexology is similar to acupuncture but instead of using needles, it uses the fingers to hit pressure points in the feet and hands. The theory is that when pressure is pulsed on these acupressure points it opens up the meridians so that vital nutrients and energies can pass sufficiently throughout the overall system of the body thus creating an electro chemical response. It is amazing, not only do the hands and feet reflex to all the parts of the body, but the hands and feet also mirror the body in exact reflection and representation. For instance, the toes represent the head area, and where the toes connect to the foot, we find the neck reflexes. Below that, are the shoulder and chest reflexes? And in sequence, everything above the central area of the arch of the foot reflexes to above our waistline in the body (and these areas would be the lungs, the heart, and the head). Everything below the central area of the arch of the foot will reflex to below your waistline in the body (this would include the hips, reproductive organs, and intestines). Some basic (common sense) rules apply, such as the left foot reflexes to the left side of the body, while the right foot to the right side of the body. The outer edges of each foot reflexes to the outer sides of the body including the arms and the legs. The two inner sides of the feet when put together represent the spine reflexes. And as follows, the spine reflex along the left foot for the left side have the spine, and the spine reflex for the right foot for the right side of your spine. Now along the spine reflexes we find neck, cervical, chest, thoracic, lower back, lumbar, sacral and coccyx reflexes with matching curves. It's exactly proportional! A quarter of a way down your spine will likewise have its reflex a quarter of the way down the spine reflex on the foot. If you have a long waist or a short waist, your foot will also reflect this by its structure. Incredibly, the structure of the foot matches the structure of the body almost with impeccable exactness. Similarly, equally the toes are connected to the foot, the same way your head is connected to the body by the neck. Likewise, the bony ball of the foot represents the bony chest area, whilst below that are the abdominal reflexes. Naturally, the softer part of your foot reflects the softer part of your body. Lastly, the hip and pelvic reflexes start where the structure of your foot changes at the heel line.

And it does not stop at the science of Reflexology, for it is also these same "meridians" of the body that "Acupuncture" and "Reiki" work. So pervasive are these simple practices that few think even work, or will venture to try them instead of using modern medicine.

Moreover, few people know that a field of medicinal science exists that is comparable to vaccines or allergy treatments - where the cures lie within the disease itself. This field is called "Homeopathy". Its roots can be traced back to Hippocrates, (an ancient Greek physician referred to as the father of medicine), but its true lineage stems from Qabalah (a subject of a latter discussion). In essence, medicines were used to treat, like-with-like (fighting poison with poison so to speak). For example, illnesses could be healed by taking tiny doses of the natural substance that is actually causing the suffering. For instance, foxglove is a highly poisonous plant that can stop the heart, but in a tiny dose, it can help someone whose heart is racing out of control. Theory is, by finding a plant, animal, rock or substance whose characteristics are similar to the illness, and by harnessing the energy or life force of that plant, animal, rock or substance by adding it to the person with the specific illnesses life force you get something called "resonance" which enhances ones life force. Another example of how Homeopathy works, though it may seem ridiculous to suggest as a way of getting rid of lead poisoning out of ones system, is to actually drink an almost invisible trace of it. Many of these types of medicines are noted of shamans and witch doctors, but I will tell you that such learned practices originated from the Bible and were divinely educated by G-d. The learned method in Homeopathy for how lead is removed from ones body is to actually place a minute amount of lead in water almost to the point of it being non-detectable, where only a "vibration" remains of it, or in other terms only the characteristics of the matter remains. In essence, what's actually occurring is that the "information" is copied to the water and is being used to cancel out the information of the symptoms. This is also, why holly water works when a prayer is cantered over it (as mystifying or even absurd as this may sound to some individual people). But what this is essentially teaching us (as we shall further learn) is that water has vibration as well as memory - hence Quantum Mechanics again - that substance is nothing more than vibration.

This same effect (of treating like-with-like) can also be found at work in the human body. Such as when (in some people) the body produces acid reflux (GERD - Gastroesophageal reflux disease) which comes up from the stomach. Despite what is commonly known about this chemical process, what's actually occurring is that the body or stomach is essentially "lacking" acid, and not producing enough, thus causing the body to send signaling spikes of burning acid up into the esophagus. The natural neutralizing remedy to this is drinking like-with-like. Therefore, drinking something acidic will counteract acid reflux. One of these natural chemical substances is vinegar (Apple Cider Vinegar). Vinegar is a liquid substance known as "acidic acid" - an acid made by fermenting beer, wine, cider, or in this case apples. Even an apple eaten after a meal will ferment in the stomach to produce the acid needed to neutralize any reflux (regurgitation of stomach fluid).

Now, if one thinks that all of this is just some kind of mumbo-jumbo or hogwash, I challenge your judgment against that of the Bibles. Exodus 15:22-25 - So Moses brought Israel from the Red sea, and they went out into the wilderness of Shur; and they went three days in the wilderness, and found no water. And when they came to Marah, they could not drink of **the waters of Marah, for they were bitter:** therefore the name of it was called Marah. And the people murmured against Moses, saying, What shall we drink? And he cried unto the Lord and **the Lord shewed him a tree, which when he had cast into the waters, the waters were made sweet:** there he made for them a statute and an ordinance, and there he proved them. This is a perfect example of G-d teaching Moses the principle of bitter curing bitter (like-cures-like) by casting into the waters a tree (that was bitter itself) that somehow neutralized the bitterness. There are several possibilities here of what kind of tree this could have been… One possibility is that it was an olive tree which are very bitter before ripening and contain a substance called "oleuropin", which is the acrid, bitter tasting compound found in the olives skin. A second possibility might have something in common with the process of creating tannins (usually found in Oak trees) that would interact with the water's albuminous matter to make the water drinkable. Thus, the lesson learned here is that G-d heals with the very same thing with which He smites. As aforementioned, this applies to adding a minutely invisible dose of lead to water (and consumed), which is then overturned to cancel out the information of the symptoms. Even as hazardous and unsafe as this may sound, this is still a baffling curative science that is still not completely understood yet. Science is not fully in on the verdict yet. (By the way, I don't recommend anyone try this at home).

And it does not end there. The Bible is full of these types of remedies. We can also associate this same healing effect found in the book Numbers. Numbers 21:4-9 - And they journeyed from mount Hor by the way of the Red sea, to compass the land of Edom, and the soul of the people was much discouraged because of the way. And the people spake against G-d, and against Moses, Wherefore have ye brought us up out of Egypt to die in the wilderness? For there is no bread, neither is there any water, and our soul loatheth this light bread. And the Lord sent fiery serpents among the people, and they bit the people, and much people of Israel died. Therefore the people came to Moses, and said, we have sinned, for we have spoken against the Lord, and against thee, pray unto the Lord, that he take away the serpents from us. And Moses prayed for the people. And the Lord said unto Moses, **Make thee a fiery serpent, and set it upon a pole, and it shall come to pass, that every one that is bitten, when he looketh upon it, shall live.** And Moses made a serpent of brass, and put it upon a pole, and it came to pass, that if a serpent had bitten any man, when he beheld the serpent of brass, he lived. This goes back to our original argument, that science is non-less than religion. That is why it is so important to have "religion" and to teach G-d's word in its proper context, for G-d alone teaches so much more than we could ever possibly learn in the curriculum of conventional school or science. For what is school without religion but hypocrisy. There is so much applicable scientific knowledge in the Bible that I myself cannot contain it ALL within this book. Even such sciences as psychology - the study of mind-body connection (as just pointed out above in Numbers 21:4-9).

Moving on here… Now lets continue on to the next venue, which is semi similar in its prospects. Amazing as the connections of the nerve meridians are in using "Reflexology" or the medicinal value of treating "like-with-like" are in their relationship to the human body, and to everything else we have discussed thus far. There are even things of more surmounting achievement that are man made, like the building structures of our towns and cities when viewed from a higher-level as I've pointed out earlier. However, even incredibly more superlative and profound is the persistent new growth of the Internet (the World Wide Web). It is not just merely a superficial network of computers linked together by transmission lines, but it also demonstrates this "one organizational law" by some mysterious principle of using the "same growth pattern" (revealed in the branching network of a tree) hiding in its decentralized distributed design by seeing that a few nodes such as Google, Yahoo and Amazon.com (main arteries) are more popular having a larger number of links caring more traffic information stemming from the growth of smaller hyperlinks (smaller vesicles). We can see "again" the same organizing and branching growth pattern that is so visually rendering and patently inherited in not just the tree alone, but also all living and some non living things.

It's very amazing, how something that is man made and considered nonliving is still subjected to that "one law" (organizing principle). Strangely enough, the more I fathom the idea, the Internet seems to be an ecosystem in itself. Perhaps its creation was for the purpose and extension of creating an antibody to fight off an invader. When in fact the Internet was designed for just that very intention in the first place, to implement the military with security since the telephone network was highly vulnerable. Therefore, the intention was to create a distributed communications network with the security that if a few links were wiped out it could reroute messages without the need for a central and vulnerable control point that could affect the entirety of the whole network. And the irony of the whole thing is that without even realizing it, "we" modeled this network after the very same fashion that our own human body's work, in the way that the white blood cells of our body have that same role and function in fighting off infection and foreign invaders to protect and maintain the body, and the same can also be applied to our nerves system and neurons of our brains in sending and receiving messages. Yet, we can design a system like this and not realize that even it is subject to that single "one law" source of universal energy. But in realizing that the most expansive aspect of this creation is the intention for which it was created and this "intention" is something that we share with our creator.

There is also another man made implemented structuralizing system that governs our society. The $Money System where the rich get richer. This view is looking familiar again in the same way the main arteries in the body are continually "supported" by the smaller vesicles. Once the large arteries have been established they are usually continually maintained by the smaller vesicles. Banks are such a reserve for the accumulation of money as are big corporations, casinos, restaurants, and generally any place that draws in business, and exchanges money. However, what if the richest eventually begin to struggle loosing their advantage in gathering new business and clients. Now suppose that these places of money exchange are in likeness to a main artery that becomes clogged with too much cholesterol and begins to malfunction causing the blood flow to be restricted and then eventually to shut down. Now depending on what type of main artery it is and where it transports to can be very important to the function of the main body because it could be linked to vital parts such as the brain and heart. In most cases, the body seems to be resilient and tends to repair itself and reroute rather quickly appointing the surrounding vesicles in their new role and growth of becoming main arteries. Moreover, as you can begin to see out of necessity, somebody else begins to thrive and get richer. The economy and the stock market (including everything and everyone) are also subjected to the overall health of the main body. When one fails, it usually results in a chain reaction of outcomes (possibly major illness, corruption of the system, followed by death). Drastic measures are needed (by the government) such as an economic bail out (stimulus bill) is needed (like medicine) for a recovery.

It is also well known that the US Government created a balanced, triplex form of government at the federal level (Legislative, Executive and Judicial) and it also replicates this balanced structure in three levels (Federal, State, and Local) as a symbol, marker, and indication from some of the founding fathers of our government who were religious in nature and studied the Qabalah and the diagram model of the Tree Of Life. (The subject of a later discussion).

We can even venture to point out that the electrical power grids spanning the earth are another man made system/network as well as the food web/ecosystem in the ocean as another example of a branching system. We can apply this view of how the food ecosystem works both on land and in the water as it balances from chaos to order. The animal kingdom is a good example of an ecosystem in which again everything is inextricably linked.

There's becoming something vaguely familiar about it all. Looking up at the night sky I wondered, "what mechanism is this?" Some stars brighter and bigger in the distance then others against the blackness of space. Some stars clustered in certain areas more than others did. I thought to myself looking at these clusters, could they also be main arteries, like the Milky Way and other spiral galaxies. Could we phantom the possibility that we are standing on the shoulder of a giant. Astoundingly and ironically, all of this resembles the other systems in their branching structure and networking by utilizing the "same organizational law". And that whatever the guiding force that is behind each and every one of these organizations its quite evident that they all reveal the same law of architecture and design despite their individualism, from the minute to the immense, to every growth unit found in nature and throughout the universe. I contemplated and carefully considered the idea that perhaps things that appeared to be chaotic and random with no intentional purpose were in fact very organized as can be conceptualized in the examples above, only we cannot grasp them in their entirety. Therefore, I came to the realization that things that appear to us to be random and having chaotic characteristics are in actuality either already in their ordered function or in the process of growing and adjusting to be organized with this "one law" pattern. For it is quite obvious that we are strings that are weaved within the pattern of some great design. Perhaps by some mysterious intention we will come to realize that not one thing comes to be random.

7 - EARTHS NATURAL LAW...

The secret of understanding multiplies here - where we will examine this "one law" in detail…

So I thought I had a good idea about this "one law". It is easy to see the comparisons of trees, a branching river network, lightning bolts, bronchial tree in the lungs, skin creases, a overall city view, streets, brain cells, and to see the growth and organization taking place out of necessity from smaller arteries and vesicles commingling and linking up to become main arteries, and the awesome view of our universe mirrored within our human body. But I pondered why and "how" all these things seem to adapt to grow with the same typical familiar-ness?

As I began to do more research, what I found was amazing! That what's actually going on here is in fact a fundamental organizational law of nature. It is the path of least resistance. Everything grows according to the great model of our universe. The grand shape of our universe is the whole reason and purpose for why things below and within are shaped, formed, and grow the way they do. It may be the path of least resistance, but there is no other choice, it is inescapable! As if we are all compelled.

Take notice of the spiraling shape and the distances between the spiraling arms of our universe as it sprawls out from the center. When I merged this mathematical growth pattern to things in nature I realized that it bore the same measurements. If we are built from spirals, while living in a giant spiral, then everything we put our hands to, is infused by the spiral - Maximilian Cohen

Increasingly realizing this special link, I associated it to the sequence law in nature that instructs all living things when deciding how many units to grow next, such as in a trees branches and a plants leaves. This series (of adding some quantity) can be seen everywhere in nature. This series is what has come to be called by several names; the numbers of the Fibonacci sequence also (regarded in conjunction with Phi/Pi), increase at a rate equal to oscillating around the Golden Mean Spiral.

The Fibonacci sequence goes like this: 0,1,1,2,3,5,8,13,21,34,55,89,144,… (Add the last two numbers together to get the next number 233. The Golden Section numbers are 0.61803 and 1.61803. The Golden String is 1011010110110101101... A sequence of 0's and 1's that is closely related to the Fibonacci numbers and the Golden Section.

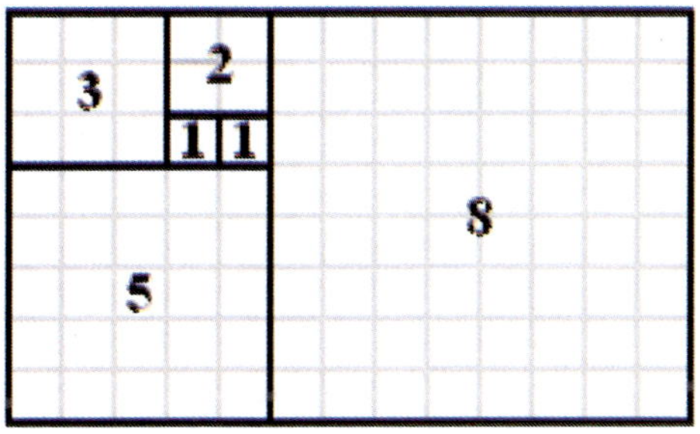

Look at the Fibonacci sequence and Golden Mean Spiral box. Notice how the numbers increase from a small squared box to a bigger sized box… (1,1,2,3,5,8...)

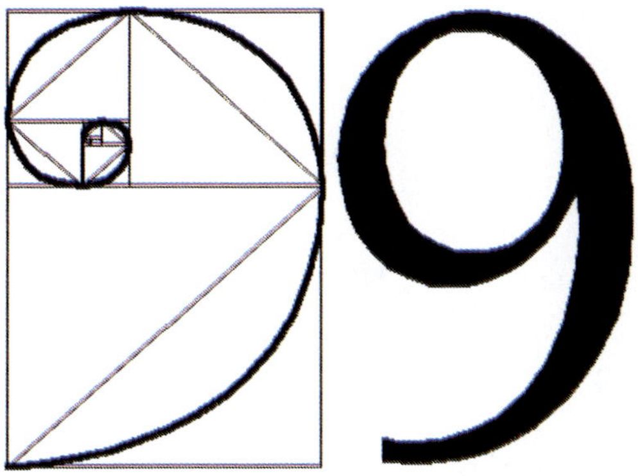

Now when we draw a flowing line intersecting these numbers, it reveals to us a spiral form in the shape of a number 9 (or even a 6 depending on how one looks at it).

There are some unmistakable forms in nature relating to this shape. Notice it here in this shells growth form.

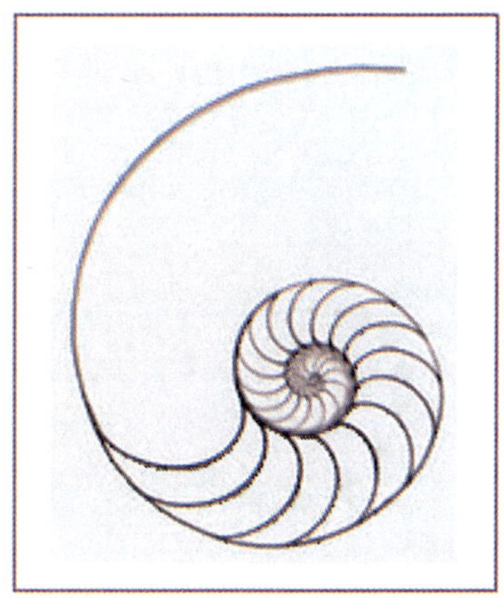

Notice it here in a leafs swirling and unfurling.

Below is a picture of a fern leaf… and a picture of what it looks like unfurling (opening up). So not only do we witness the "Fibonacci sequence" in the fractal effects in the individual growth of the leaf itself, but we also see the spiraling action of the "Golden Mean Spiral" when the leaf is unfurling. In addition, we can spot the "Golden Section" in the middle of the fern leaf. So nobody can now say that each mathematical sequence is a separate disconnected entity, as it's obvious they are all working together in something we can now call "universal law".

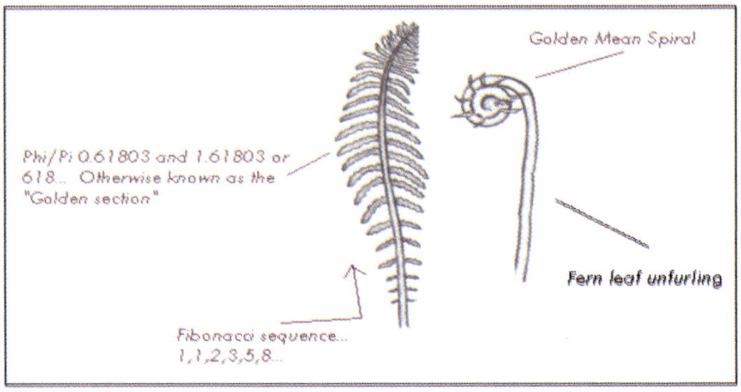

Just for reference sake, we can also use the human body in place of a fern leaf to recognize the Golden Section.

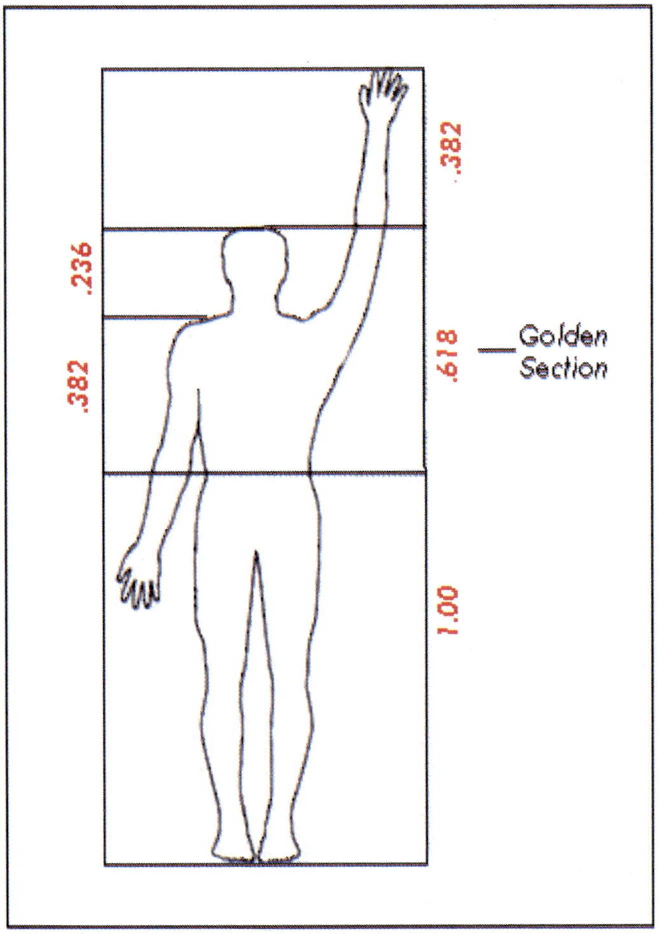

Not only is the Golden Section noticeable in the midsection of the human body, but also the Fibonacci sequence is also evident in every section of the body.

Notice it in the portions of the arm and fingers as well as leg and foot.

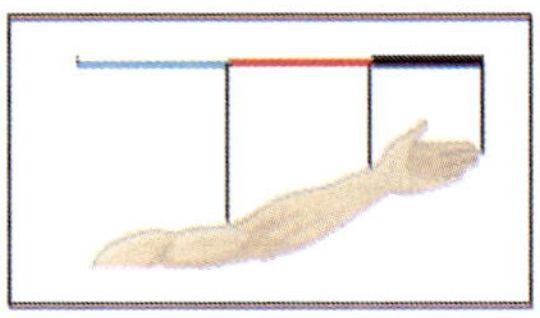

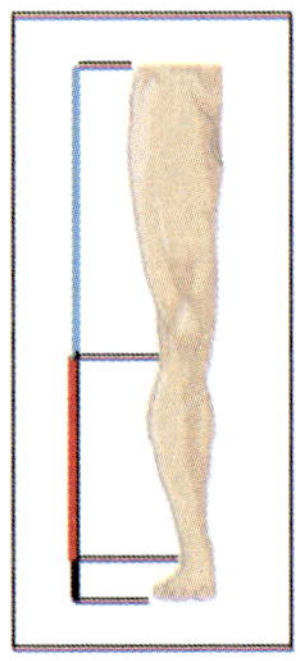

Notice it in the skeletal structure of the Finger bone.

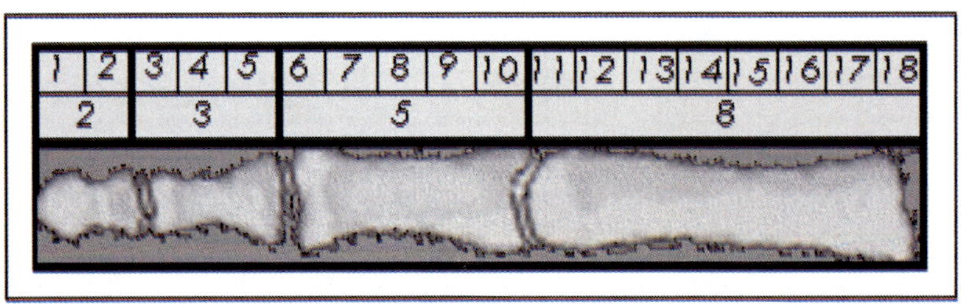

These mathematical proportions are just not prevalent in humans, but the growth pattern of EVERY LIVING THING!

It can even be noticed in the structure of nonliving things such as windblown rippled sand patterns.

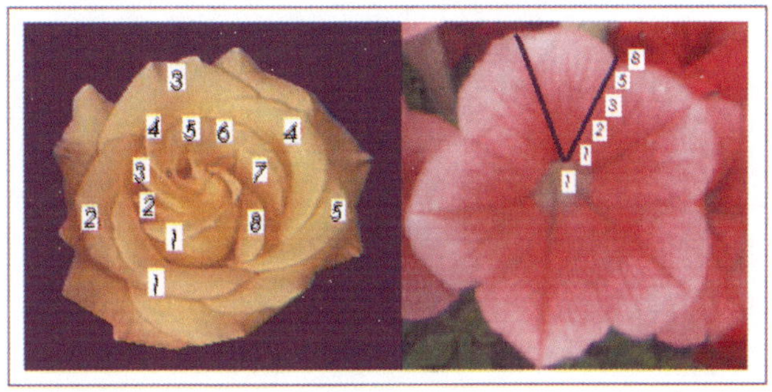

Notice it here in how the Rose blooms open, and how the Petunia flowers color opens out.

Notice it here in the sunflowers rows of seeds, and the rows in a pinecone.

Now before I precede any further, let me explain with better clarity what's happening here... For lack of a better description or word, I'd like to use a graph that best describes a unit of growth as another way of perceiving (a representation) of a shape that will latter pave the way to a higher developed understanding.

Take notice of the thorn figure outline (here:) Λ (and shown below) and how the circles within it increase by unit growth. It is this outline that I would like you to keep in mind as it illustrates that all things originate from a single point and grow (propagate) outwards from that location (DESPITE its mathematical unit of growth). And lets for a moment identify this basic structural characteristic accompanied by the motivated unmistakable feature that it grows OUTWARD away from the initial point (or tip) of the Λ as the path that all life takes in the expansion of the growth process. I'd like to refer to this structural feature as the "UNIVERAL LAW". We don't even have to attach any mathematical association to it such as Fibonacci, Phi/Pi or Golden Mean Spiral or any other named such unit of growth other than the fact that IT GROWS OUTWARD away from an initial point in a given space. You will also notice how the circles (shown below) not only increase in number, but in size.

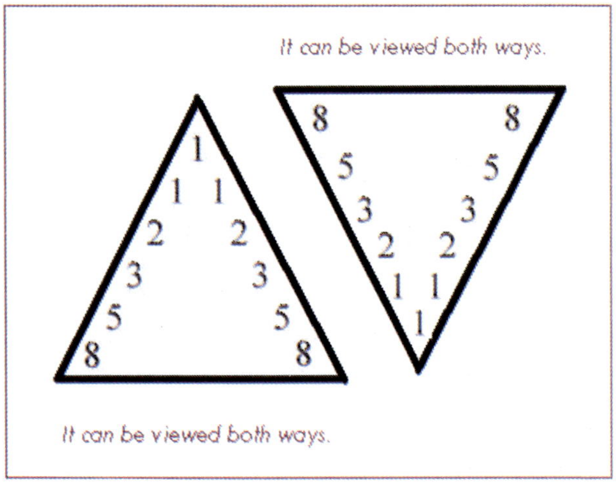

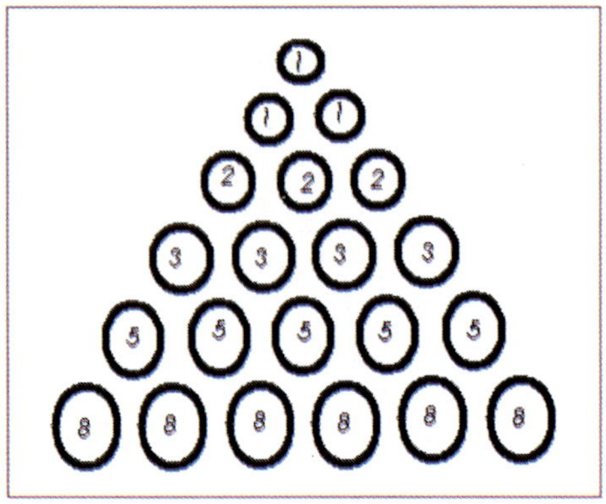

It can also be noted as the "Bell shaped curve" which is typical in many graphs (every kind of graph from heart graphs to sound waves).

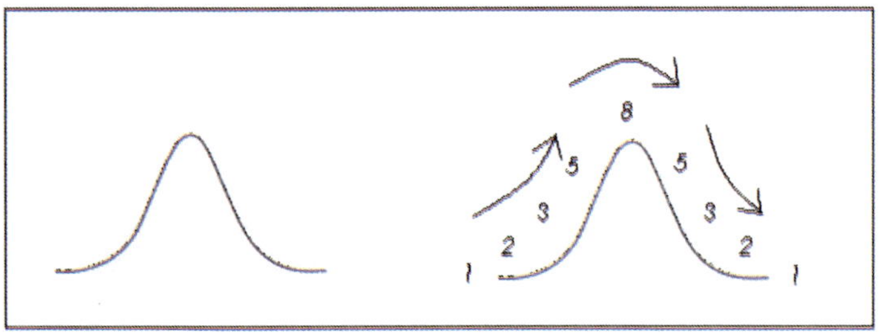

A "Bell Curve" (or simply known as curving) is a method of assigning a diagram that illustrates the distribution of quantitative or cumulative growth averages in addition to variable deviations of difficulties. Rising at the incline of interest or response and lowering at the state of residual entropy (disorder).

Lets move on here… Now I'd like you to notice it in the fractal branching structure of a classic tree.

Here it is simplified.

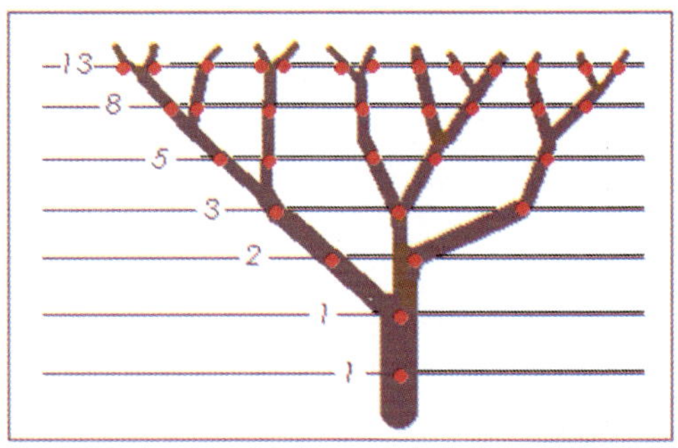

Notice it in the fractal branching of a leafs veins. One can also see Fibonacci sequence in the shape of the leaf, and the Golden Section. Psalm 1:3 - He is like a tree planted by streams of water, which yields its Fruit in season and whose leaf does not wither. Whatever he does Prospers.

Notice it in the structure of lightning branching. Psalm 97:4 - His lightning lights up the world; the earth sees and trembles. 2 Samuel 22:13 - Out of the brightness of his presence bolts of lightning blazed forth.

Notice it here in the branching growth structure of a pine trees branches.

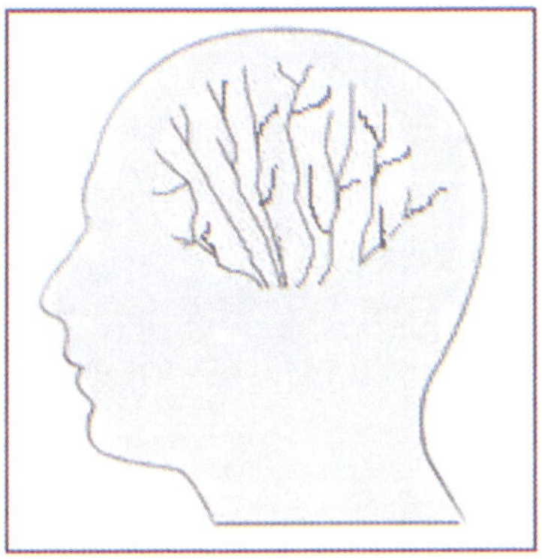

Notice this same pattern in the branching blood vessels of the brain. 1 Chronicles 28:9 - For the Lord searches every heart and understands every motive behind the thoughts. If you seek him, he will be found by you; but if you forsake him, he will reject you forever.

Notice it here in the shape and growth of the thorns. Genesis 3:18 - It will produce thorns and thistles for you, and you will eat the plants of the field.

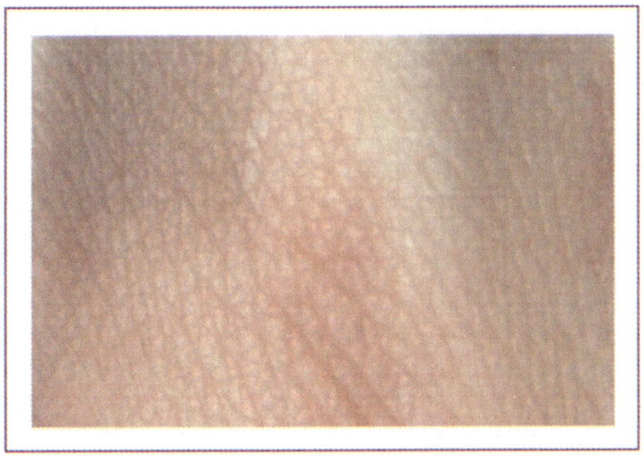

Notice it in the fractal geometry of the skin. Psalm 139:13 - For you created my inmost being; you knit me together in my mother's womb.

In fact, the skin is so organized in its woven, overlapping, and interlocking and polygon structure that it is stacked like poker chips in a casino (picture shown below). One can see that it logically has order, is prearranged, and has definite meaning with a purposeful function for the way that it is structured (which allows it to work to assist the body), thus making it easy to see that it was planned by a creator. Without the skin, we would die within seconds! One cannot live without skin. It works with all systems of the body. You need skin to protect, cool down, heat, and hold everything together.

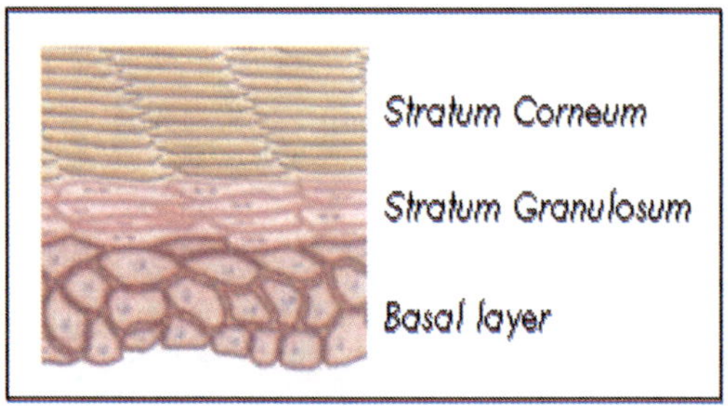

Notice it is wood grain.
(Can you see the "Bell Curve")

Exodus 25:8-10 (the Ark of the Covenant) - Then have them made a sanctuary for me, and I will dwell among them. Make this tabernacle and all its furnishings exactly like the pattern I will show you. Have them **make a chest of acacia wood**—two and a half cubits long, a cubit and a half wide, and a cubit and a half high.

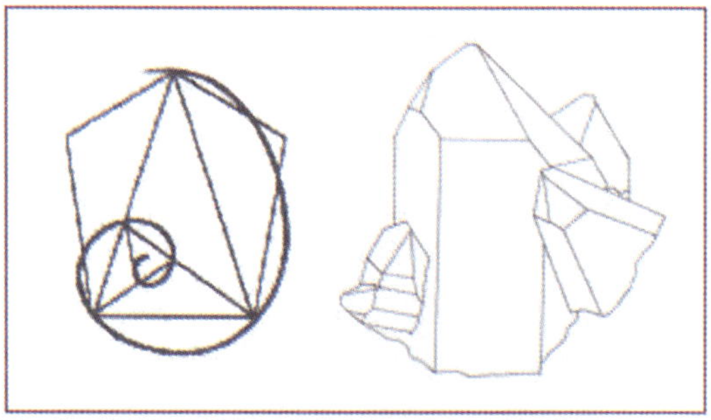

Moving right along… It can be notice it in all types of crystals. A crystal is a solid with a distinct geometric shape. The shape consists of smooth, flat surfaces that meet in sharp edges or corners and from these corner points grow outwards by a multiplying unit of measure. This unit of measure happens to be Phi/Pi as each smooth face of the crystal grows and it also happens to display the Golden Mean Spiral in its "overall" displacement of its many faces (as they grow in unit growth from large to smaller).

Notice it in the hexagonal geometry of snowflakes. Job 38:22 - Hast thou entered into the treasures of the snow?

Moreover, crystal formation should get us thinking on other lines, such as applying the mathematics of fractals geometry, not only to measure things in the natural world, but to also apply them to man made structures for maximum efficiency. Let me further explain… Fractals are everywhere in nature. In addition, it is here again that we witness a unit of expansion that is consistent with known growth patterns. Fractals not only reveal themselves in crystals, but in our lungs, kidneys, blood vessels, rhythms of the heart, weather systems - from the drainage pattern of rivers to the movements of clouds, the fertilization of an egg, the shape of a fern, a seacoast, a mountain range, or a branching tree. Fractal geometry can even be used to measure the large structure size of the Universe. Whether one looks at these shapes from far away or up close, one sees the same pattern (I don't know which has a grander view, the telescope or the microscope - someone once said). Consider a tree trunk splitting into boughs, then branches, then limbs and twigs, and stems of increasingly smaller size. As it turns out, every smaller twig and stem has the same shape as the whole. Each life form possesses the ability to reproduce itself from a portion of itself. The Bible reveals this truth to us in the record of the third day of creation when G-d created life. Genesis 1:11-12 - And G-d said, Let the earth bring forth grass, the herb yielding seed, and the fruit tree yielding fruit after his kind, whose seed is in itself, upon the earth: and it was so. And the earth brought forth grass, and herb yielding seed after his kind, and the tree yielding fruit, whose seed was in itself, after his kind: and G-d saw that it was good.

Amazingly, this very same statistical and functional pattern of fractal geometry goes into the technology of cell phone antennas (consisting of polygons and triangles) and encompassing computers, and computer graphics. These fractal patterns consists of endless repetition (self similarity and shape to the minutest size) what mathematicians call "Iteration" (repetition). Without fractals (smaller and smaller self similarity) especially for today's modern technology involving all the intricate and infinitesimal telecommunications needs we are facing would not work. Thus while observing the increasing functional importance of our technology, we begin to see within nature the identical need and commonality for fractals to exists - hence this is perhaps why nature (designed by G-d) uses them in the first place. Remember, from the earlier chapter, there is nothing new under the Sun (the entire Universe is a closed looped system). Ecclesiastes 1:9 - The thing that hath been, it is that which shall be; and that which is done is that which shall be done: and there is no new thing under the sun.

One of the new frontiers in the medical field is developing new machines that work to identify fractal formations within the body since fractals can also be used to recognize mathematical models of cancer down to the minutest blood vessels structure (vessels feeding a cancerous tumor).

Additionally by using fractal geometry, it was recently discovered that the distribution of the sizes of individual trees in a forest appears to exactly match the distribution of the sizes of individual branches within a single tree. In addition, by this, we can determine by measuring the unit growth of a single tree to determine how much oxygen a whole rainforest can produce. What does all this mean? It means that with this knowledge we can open the whole of mathematics of nature eventually allowing us a grand view into the scope of the creators works - to know that there is order, design, meaning, purpose, and a foreordained plan for all life. The colossal vista is quite apparent… G-d used both mathematics and art to design nature - Where His formulas were not just made into pictures, but physical material flesh. We must not just look at it and measure it, but to apply it to understanding it in a deeper spiritual way. And again, I would like you to notice how these things are not only growing with fractal proportions (by means of unit growth) in nature (such as crystals), but at the same time they are also spiraling (as shown in the picture below).

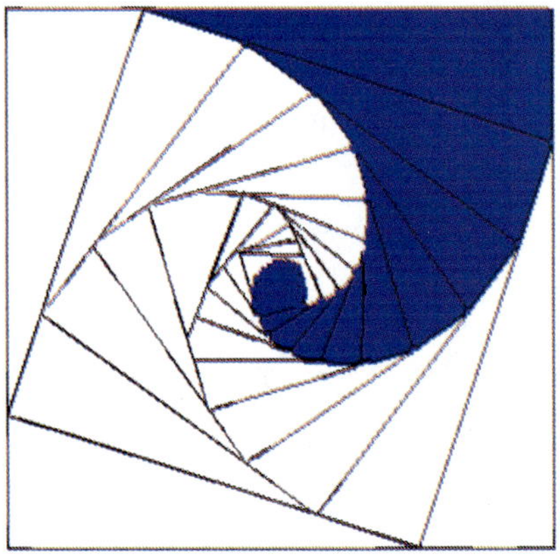

Moving right along, I'd like to give further details regarding the spiral… Now I'd like you to notice it again in the spiraling shape and pattern of a shell. Shells are exquisite little gems (in the making) that show the time progression of G-d's handiwork in action.

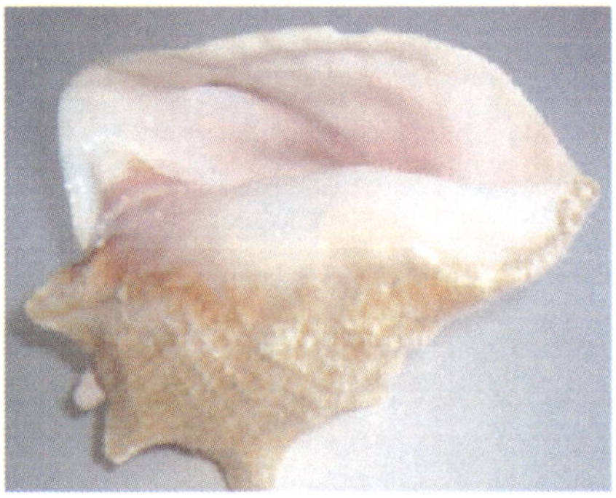

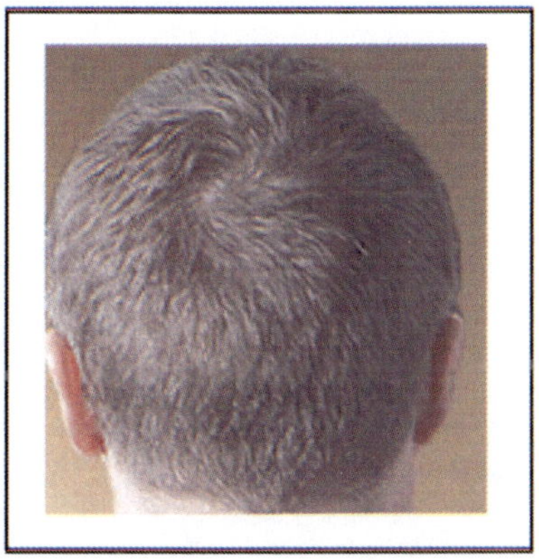

Notice it in the swirl (whorl) pattern of hair.

Notice it in fingerprints.

Notice it in waves of water.

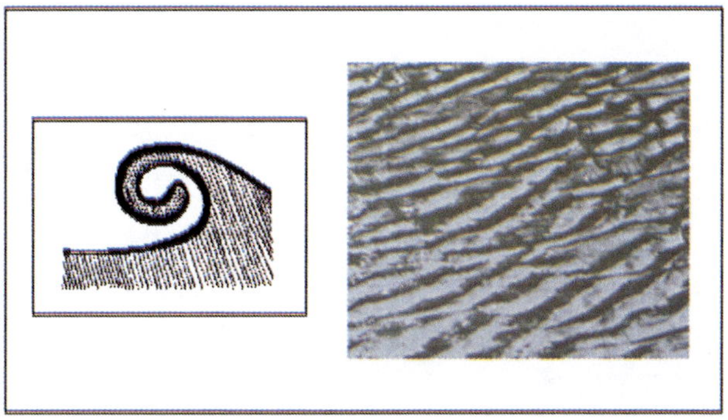

Notice it in smoke rings.

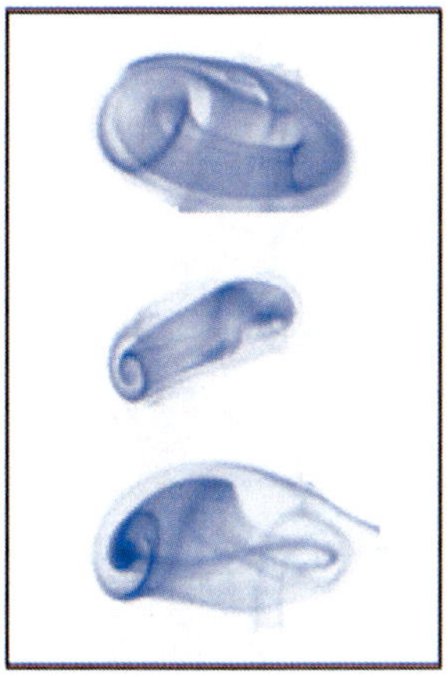

Notice it in the ear shape. Psalms 94:9 - He that planted the ear, shall he not hear?

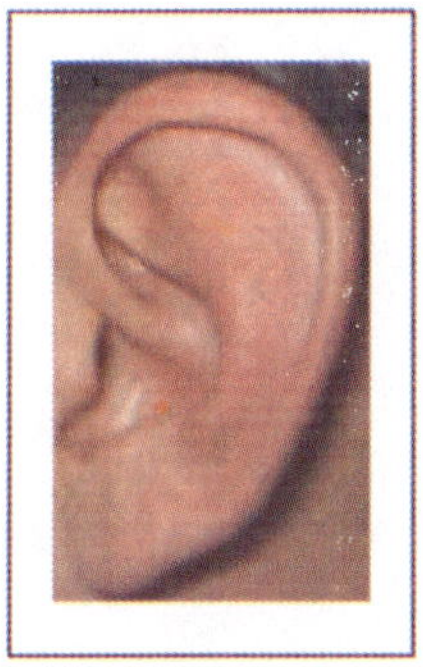

Additionally, the human ear also has an amazing function. It has a dampening muscle mechanism, which protects the eardrum and brain from loud (overloading) noises. Including hair cells that generate electrical ions to pass through the hair cells that are carried by currents to the brain and interpreted.

Also, notice it in the shape of cochlea in the inner ear.

Look at the snow capped mountains. The mountains themselves are a product of Phi/Pi, as well as the crevices of snow that appear like sprawling veins, (these too are a product of Phi/Pi.

From an even higher view the mountains look like this.

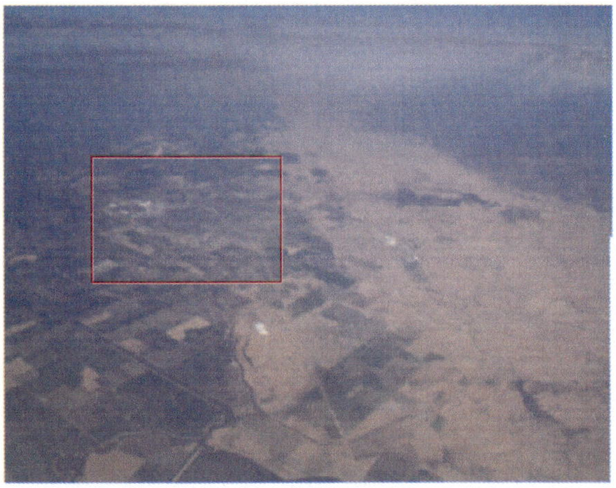

This picture was taken from the window of an airplane in which one can see a small city (epicenter) in the red square box. From the red square on out, the land segments get bigger. This is also a unit of growth pattern that goes from small to bigger.

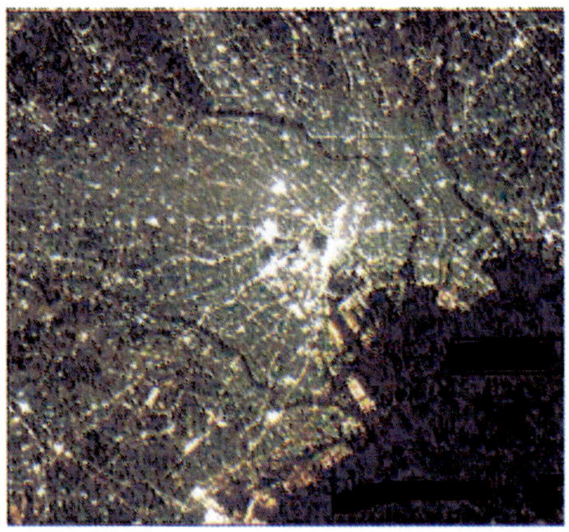

We can see the same expansion of a light-up city from a higher elevation.

Look at this beautiful beach scene. Notice how the waves are distances apart as they roll in (this is also a unit of growth pattern). Also take notice of the pattern of the Phi/Pi (the bell curve) as these waves wash up onto the shore of the sandy beach.

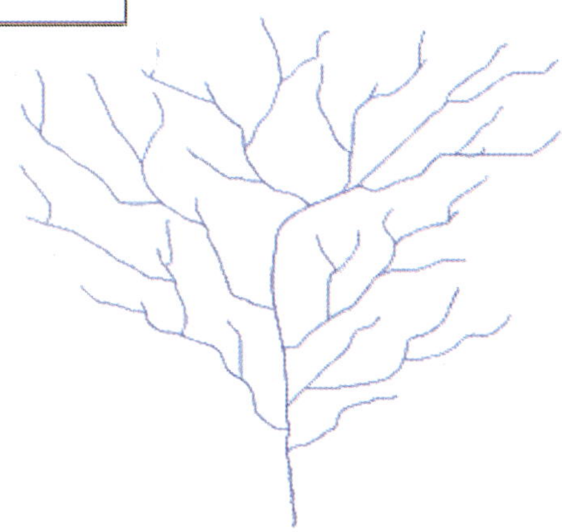

We also see this same pattern (and unit of growth) in a branching river network.

Again, these types of patterns do not only exist in all types of crystals natural and cultured. Pine cones. Flower petals. Nautilus shells (creatures exhibiting logarithmic spiral patterns such as snails and various shell fish). The way in which rivers branch. But they are also prevalent in other things, such as birds and flying insects (exhibiting Golden Mean proportions in bodies & wings). The mathematical measurements of the human heart and its beat. The human brain. The human head, eyes, nose, mouth and teeth. Even when viewed from the side, the human head illustrates the Divine Proportion. This unit of grown is also in the musical scale. Population growth. Price movement in the stock market. The geometric molecular and atomic patterns that all solid metals exhibit. The molecules of (DNA) - the foundation and guiding mechanism of all living organisms. The spiraling effect that a liquid displays when another liquid is added. The galaxy we spiral within. The way in which the wind blows, the waves roll in, and the wavy sand patterns on the beach. The eye of a hurricane. The land continents of the planet. The earths bow shaped magnetosphere. The relative distances of the 10 planets of our solar system. Saturn's rings. It has even been detected that there is an unusual energy source at the frequency of Phi/Pi found in the universe. All life forms, as we know them emerge out of enduring geometric codes. Viewing and contemplating these codes allow us to gaze directly into the subtle structure of awareness - the inner workings of the Universe and the Mind of G-d.

Such stipulations for unit growth can even be found in the Bible… In Exodus 25:10 G-d instructed Moses to build the Ark of the Covenant with these divine proportions. G-d also instructed Noah in Genesis 6:15 to build the Ark with these divine proportions. It even applies in other miraculous ways, such as when G-d gave to Moses the instructions of the Tabernacle in Exodus 26:1 - Make the tabernacle with ten curtains of finely twisted linen and blue, purple and scarlet yarn. Now if you're wondering why G-d had instructed Moses to use these colors, you can find your scientific answer in the color spectrum. If you observe the color spectrum, you will notice that the primary colors of blue and purple make scarlet. This is because colors in the spectrum are related by distances based on Phi/Pi (or the golden section). In this case, G-d used colors to mark the place where He meets with His people. And to be a place that is visually appealing, thus referring to a place that is fulfilling, faultless and perfect, (perfect with a capital P!) and free of impurities (contaminants) such as the holy Tabernacle (the place where the presents of G-d dwells).

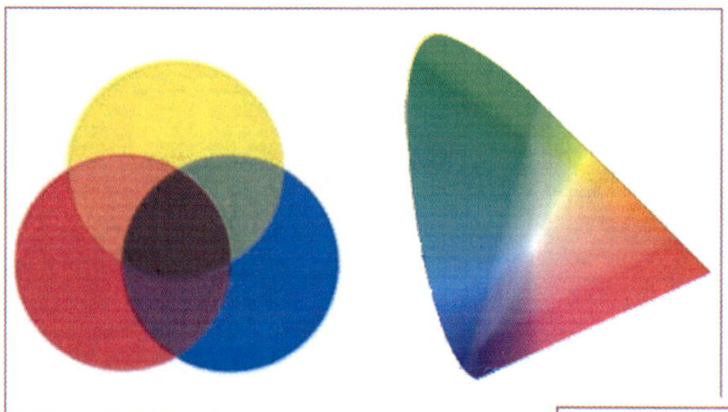

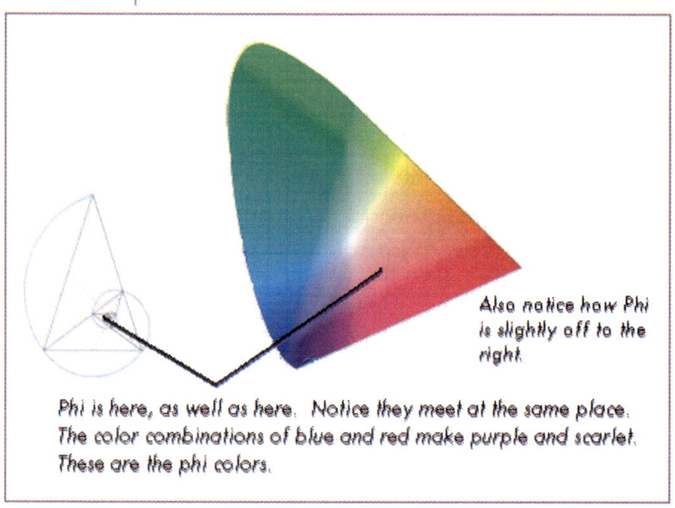

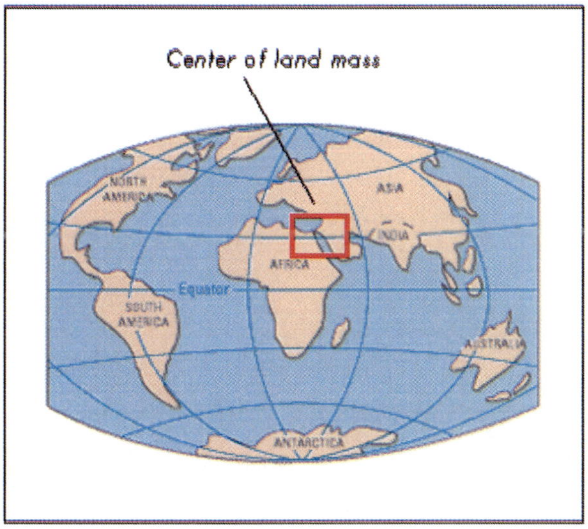

This same principle can even be applied to the holy city of Jerusalem and the place where the Jewish temple was located. Furthermore, the location of Jerusalem is 31latitude degrees by 35 longitude degrees (which is from the temple mount). However, the Great Pyramid on the Giza plateau is located at 30 latitude degrees by 30 longitude degrees, which is considered to be the "center" of the Earths land mass. (I will also talk in further detail about Egypt, Giza and the location of the Great Pyramid, and its location in a coming chapter). But concerning Jerusalem, G-d said Jerusalem is "the city which I have chosen to put my name there". 1 Kings 11:36 - 1 Kings 11:36 - And unto his son will I give one tribe, that David my servant may have a light always before me in Jerusalem, **the city which I have chosen me to put my name there.** So what was the significance of G-d selecting this location? He chose this location obviously because the city of Jerusalem is in the "Golden Section" (Phi/Pi) of the whole land mass of the Earth (Phi/Pi/Golden Section which is slightly off to the right). In addition, at the same time, Jerusalem is also located in center of the land of Israel, and within the city of Jerusalem (within the Jewish quarter), and the Wailing Wall, which is a remnant of what left of where the Jewish temple once sat. And additionally the place where the Jewish Temple once stood is a location that we also find to be the Golden Section (Phi/Pi). What's more, the Jewish Temple itself also bore the exact measurements of the Golden Section - the place of the holy of holies where the Ark of the Covenant was kept. So either way we look at it, it is Phi/Pi (the Golden Section) all around, denoting a "perfect place". (These measurements can be found in the Bible in 1 Kings, 2 Chronicles and Ezekiel).

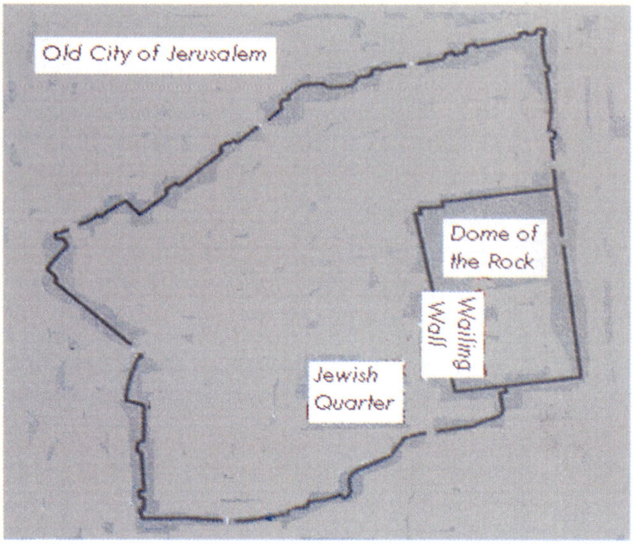

Notice how the Jewish Quarter of Jerusalem and the place where the Jewish Temple once stood (now the Dome of the Rock) is off to the right of the city.

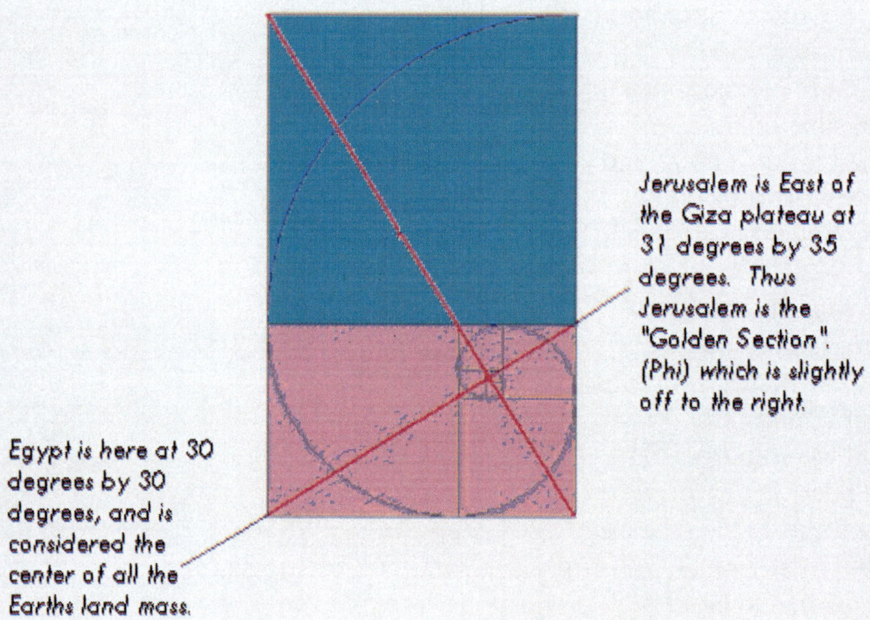

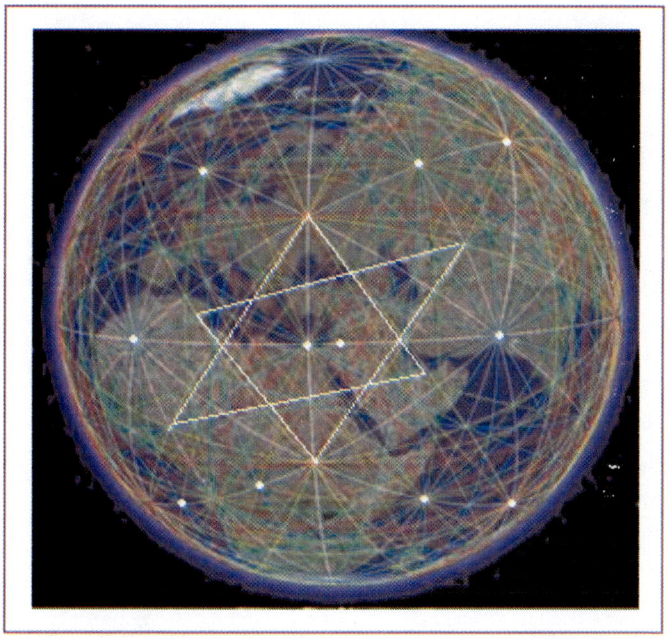

What's even more amazing is that the ley lines around the globe (the planetary grid lines) directly point to Israel as the center (slightly off to the right). It is actually Giza at 30' 30' that is the center. Look look at the picture as I've stenciled in a "Star of David" with the land of Israel found in the direct center (slightly off to the right). So, let there be no mistake for the reason G-d chose the land of Israel to be the precise center of the world. The Bible tells us that it is the location of the center of the world, and Geography can now confirm it! Ezekiel 5:5 - This is what the Sovereign Lord says: This is Jerusalem, which I have set in the center of the nations, with countries all around her.

Moving along here... Even "time" itself is also based on this proportion (Golden Ratio, or Phi), seeing that an electron has a rotational motion or spin thus stretching space and time.

And most profoundly of all, it can even be noticed in the very humblest of beginnings and endings of the atomic world. This is the life cycle of a particle. The stuff that all matter is made up of.

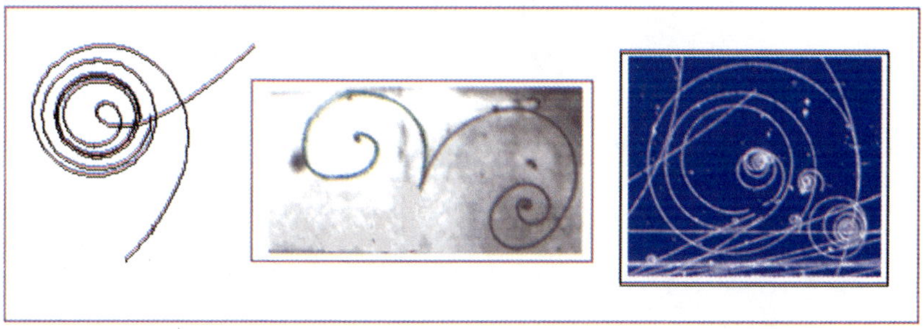

Be it as it may, that all these forms illustrated above obey the rules to real physical measurements, but there is also a side note of minimal first distinction which starts with the human eye. (picture below) In fact, the mystery is the EYE itself! It is how and why the human eye views the Phi/Pi proportion as pleasing while it is this same proportion from the black Pupil to the brown Iris to the white of the Sclera which is "Phi/Pi" in its unit of expansion that forms the foundation of the material world that we are capable of seeing. Is our brain just playing tricks on us, or is G-d a humorist? Psalms 94:9 - He that formed the eye, shall he not see?

Proverbs 20:12 - The hearing ear, and the seeing eye, the Lord hath made even both of them.

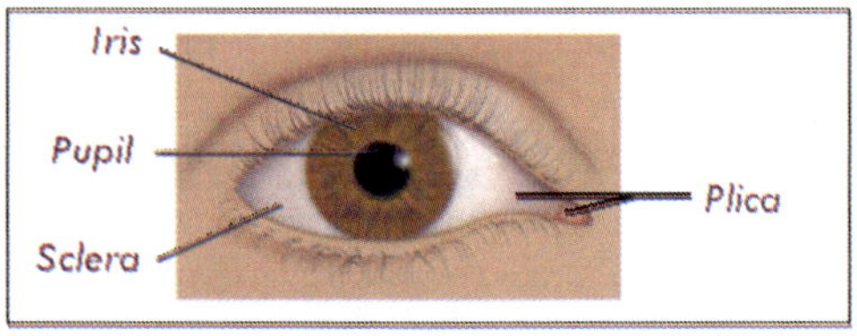

What's more, the human eye has what's called a "plica" (the piece of skin on the inside of the eye), which is "different" from any other animal on the planet (including monkeys and apes). Yes, the plica exists in several animals, but it is not anything like our human plica. Though I suppose the evolutionists see it in view of other animals having it, which suggest to them a mutation rather then seeing it as a special separate entity in humans (again, the "tool box genes" come into play). The plica is special in humans because it not only helps to keep the eye clean from debris, but also allows the human eye to turn in all directions (especially turned outwards, as well as inwards) to view things. Simply put, the human eye was specially made for us - not for, or from another animal. Additionally, no other creatures eyes match the high acuity of the human eye's central fovea region. The fovea consists of light cones that are responsible for sharp central vision, which is necessary in humans for reading, watching television, driving, and any activity where visual detail is of primary importance.

Before I move on here, I must profess that up until now I've painted a rosy picture for all to see… though not all is rosy with the Fibonacci sequence, Phi/Pi and Golden Mean Spiral. Things do not always grow "perfect" with these measurements (and they also have names for other sequences as well), but they certainly strive to take the path of least resistance. Many of us are stubborn, especially the mathematicians and scientist who are not in agreement with the confined and constricting laws of Fibonacci, Phi/Pi and Golden Mean Spiral, they make mischief of it. Yes, I will admit that they are all "individual" mathematical sequences! Some stringently say, it is Fibonacci; others say phi/Pi and other say golden ratios. They opt to segregate, isolate and dissect each one of these sequences as something different from the other. These people look at it from each "single and small" perspective and are not able to make the connection (to gap jump the bridge) and rise above the horizon to see the "fully combined perspective" of the all-engulfing mechanism. However, please recall how I have stated it above: This series is what has come to be called by several names, the numbers of the Fibonacci sequence also (regarded in conjunction with Phi/Pi), INCREASE AT A RATE EQUAL TO OSCILLATING AROUND THE GOLDEN MEAN SPIRAL. In other words, these numbers work (around) in conjunction with each other. Such as in a plants growth, as it grows by Phi/Pi it ALSO oscillates with the numbers of the Golden Mean Spiral.

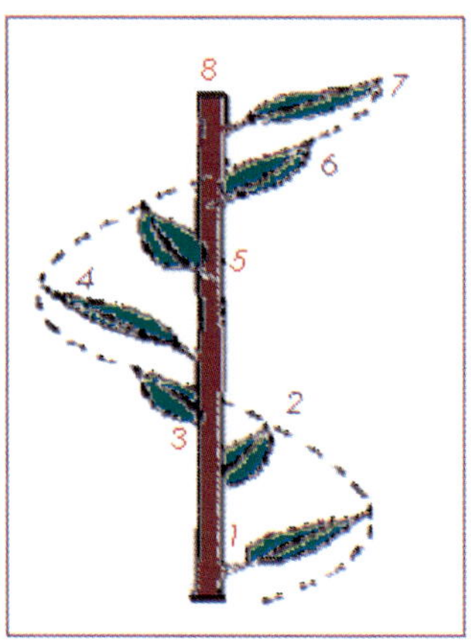

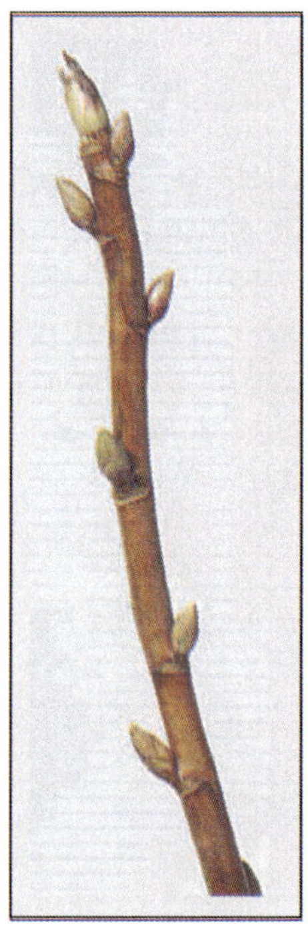

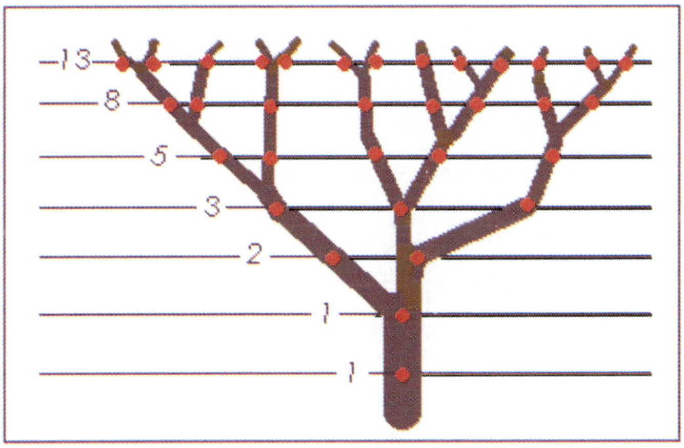

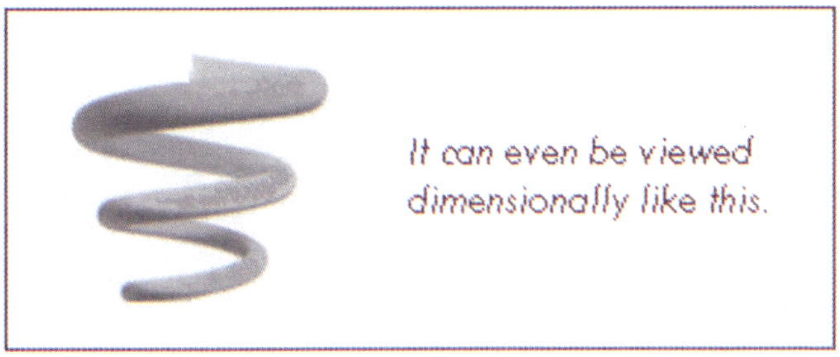

So as you can see, "at the same time", we cannot deny that these mathematical sequences form patterns that are unmistakable to the eye. In addition, there could be many other units of measure, some in which are called the Lucas sequences, Tribonacci numbers, Pentanacci, Hexanacci, and Heptanacci numbers. In addition, that all these phenomena's share a close association with extraordinary integers, each one getting closes and closer in principle to one another. The ratios work in connection with each other even though there are variations. They oscillate around one another. Commonly denoted in their approximations, the Fibonacci and Phi/Pi numbers are a kind of Golden Mean ratio in disguise - otherwise termed "generalized Fibonacci series"… They seem to be all intimately related to the Golden Mean Spiral. And we can "very nearly" see that they SUPERFICIALLY look like one another. And we cannot deny the fact that nature's patterns are built on RECIPROCALS and grow by some PROLIFERATING unit exhibiting by some chiefly means the spiral design (of our Universe) as well as in the pictures above. So its plainly obvious that these units of mathematical measurement work in conjunction with each other, therefore there is no evading that this is indeed and all-engulfing mechanism that appears to be following (while in units) "one law of order". There are no other words of explaining, or way of understanding this organizational orderliness of what's transpiring here other than characterizing it as "ONE LAW OF ORDER". There is not even a scientific word or definition given for it. So again, I am now coining (and formulating) it as the "ONE LAW OF ORDER".

It has also been noticed that these mathematical sequences start out with either the conventional Fibonacci numbers, Phi/Pi and Golden Mean Spiral, but as the sequence is extended, they seem to start converging into an erratic sequence. This should be of no surprise as we've discussed earlier the "deteriorating factor" when the law of thermodynamics and entropy (disorder) starts to set in.

We must also not ignore equally important cases that don't fit the pattern. If some things remain that ought to fit but don't, just rationalize that there is a reason for why they are "special cases". It is in my opinion that this anomaly is evidence of G-d showing the uniqueness of our planet in relationship to the whole cosmos.

A quick overview and summary at this juncture in time allows us to grasp not only the concept of fractal growth, but also the all-embracing and directing supremacy of the SPIRAL. What is so remarkable about the spiral is that all spirals in general, even of different shape and growth that progress from a smaller to larger curve is necessary for transforming higher energies into lower energies and for coalescing matter into form. Spirals are produced for the sole purpose of organizing energy into separate and distinct densities causing heavy elements of matter to clump together, and in the process slowing down time into the material form, and harmonizing it in just the right position so that we can experience physical reality, (within the elemental conditions that sustain life as we know it). As matter moves into the spiral arm, what is remarkable is that matter which appears to starts out with a completely random array of elements (but already has divine direction assigned to it) becomes a refined organized pattern where order arises from disorder. The evolutionists believe that this is how the stars and planets were formed. They confuse this viewpoint with the Universes "evolution" over millions of years. But a clear perspective of biblical doctrine informs us that G-d plainly (at a moments notice) created the universe's order. Let there be no mistake about this, there was not any chaos (disorder) ever, since everything was planned and set in motion beforehand from the mind of G-d. From this splendid order of creation everything was then "set in motion" to function perfectly. This function/force is continually directing all life. Jumping directly to a synthesis here, this is "intelligent energy" (it displays no chaos) because it has a predetermined exhibiting order to it. And even though some of us may chalk spirals off to natural occurring conditions, or the path of least resistance, or because its just the way physics works. One has to question where did the pattern of the first spiral originate from, and why does physics work the way it does? In addition, prior to that where did the mindless elements that formed the stars and planets originate? Whose order is it - is the BIG QUESTION? This spiral force "by itself" almost seems spiritual in nature with its own gusto of sophisticated workmanship. In addition, if you think it gets any easier to understand - think again. When I mention that it is the path of least resistance… Lets for a moment use the example of a plants leaves, which grow by the Fibonacci sequence regarded in conjunction with Phi/Pi and increase at a rate equal to oscillating around the Golden Mean Spiral. The plant somehow "instinctively" knows to grow this way so that each leaf gets the maximum exposure to light, casting the least shadow on the others. This also gives the best possible area exposed to falling rain so the rain is directed back along the leaf and down the stem to the roots for better physical strength to be better able to withstand wind and heavy rain. For flowers or petals, it gives the best possible exposure to insects to attract them for pollination. The whole of the plant seems to produce its leaves, flower head petals and then seeds based upon the conjunction of the Fibonacci sequence, Phi/Pi and the Golden Mean Spiral numbers. However, we must stop and think logically about how the plant "instinctively" knows this information? Just what is "instinct"? Is it something automatic, something unconsciously known, something on impulse, and something intuitively, as a reflex action, without thinking? Who incorporated this logic, and who is the thinker in a plants growth (without a brain) that we regard in something so standard as a plant? In addition, this just doesn't constitute plants; it engulfs the growth of all life contingent to this planet. It is obviously the designer (G-d) working wholeheartedly throughout all of His creation. So how do we with certainty know this?

Readers, the answer to this question is looming right before our eyes in NATURE! Please read very meticulously, what I am about to write here, as it might throw you into disarray… It turns into quite another matter when I refer to something called Gematria. Gematria plainly put is numerology of the Hebrew language and its alphabet by comparing numerical values to derive meaning or relative relationship. The word Gematria should not be a completely foreign word to us when the word itself later evolved into the Greek word "geometry" whereas the concept and system is the same as the Greek isopsephy meaning "equal" by adding up the number values of the letters in a word to form a single number. Gematria uses the Hebrew alphabet (in sequence) starting at the Hebrew letter Aleph which is letter "A" as equaling number 1, and letter "B" equaling 2 and so on. This numerical system is a way of showing how two ideas are related on a "conceptual" level by using numerology as a basis to confirm a connection. As a result of this computation it has been revealed that WHEN THE NUMERICAL VALUE FOR THE WORD OF G-D - WHICH IS SPELLED AS "ELOHIM" IN HEBREW, EQUALS THAT OF THE WORD "NATURE", WHICH IS THE HEBREW WORD "HATEVA", "both" sharing the same numerical value of the number "86". Elohim (plural) is the first Hebrew name given to G-d in the Bible. (In the beginning G-d (Elohim) created the heavens and the earth (Genesis 1:1). Low and behold, the concept that G-d is constantly within all His creations, and was He to remove Himself for a moment it would be as though nothing had ever been. So, in a roundabout way "divine guidance" (revering the spiral as the hand of G-d) is explained, as "direction from above" by a superior intelligence - for G-d comprehends all living things as His own Being. He is actually the knowledge of all "information", the knower, and the known. He is both subject and object. What G-d is in theory, the atoms, elements, and matter, are to the spiral. This is the supreme truth, which Israel proclaims in reciting the Shema, (Hear O Israel, G-d is One!). Let us also associate the Exodus from Egypt and the famous ten recorded plagues that struck Egypt using the forces of "nature" to that of G-d, which embodies Himself through nature.

The 10 plagues can be explained by nature taking its course as a chain reaction (but please do not dissect the divine from nature). Lets briefly course through them… The first plague tells of a river of blood. This leads to a number of questions; one being what specific thing could turn the water red and kill all the fish? Due to exceptionally hot weather, an outbreak of a toxic algae bloom called "Physteria" occurs. This algae kills fish and makes the water toxic. Exodus 7:14-18 - Then the Lord said to Moses, Pharaoh's heart is unyielding; he refuses to let the people go. Go to Pharaoh in the morning as he goes out to the water. Wait on the bank of the Nile to meet him, and take in your hand the staff that was changed into a snake. Then say to him, The Lord, the G-d of the Hebrews, has sent me to say to you: Let my people go, so that they may worship me in the desert. But until now, you have not listened. This is what the Lord says: By this you will know that I am the Lord: With the staff that is in my hand I will strike the water of the Nile, and it will be changed into blood. The fish in the Nile will die, and the river will stink; the Egyptians will not be able to drink its water. The second plague was that of frogs most likely caused as a chain reaction being that there were no fish alive to eat the frogs, Resulting in a much larger population explosion of frogs. Exodus 8:1-4 - Then the Lord said to Moses, Go to Pharaoh and say to him, This is what the Lord says: Let my people go, so that they may worship me. If you refuse to let them go, I will plague your whole country with frogs. The Nile will teem with frogs. They will come up into your palace and your bedroom and onto your bed, into the houses of your officials and on your people, and into your ovens and kneading troughs. The frogs will go up on you and your people and all your officials. Now because the water was also poisonous, the frogs left it, but not before gulping down and consuming a lethal dose of the toxins released by the red physteria resulting in killing the frogs, therefore because there was no frogs to eat the insets this resulting in the 3rd plague of lice or gnats. Exodus 8:16-18 - Then the Lord said to Moses, Tell Aaron, Stretch out your staff and strike the dust of the ground, and throughout the land of Egypt the dust will become gnats. They did this, and when Aaron stretched out his hand with the staff and struck the dust of the ground, gnats came upon men and animals. All the dust throughout the land of Egypt became gnats. But when the magicians tried to produce gnats by their secret arts, they could not. And the gnats were on men and animals. The 4th plague was most likely a combination of lice, flies and gnats that most likely transmitted disease to the 5th plague killing the livestock. Exodus 9:1-4 - Then the Lord said to Moses, Go to Pharaoh and say to him, This is what the Lord, the G-d of the Hebrews, says: Let my people go, so that they may worship me. If you refuse to let them go and continue to hold them back, the hand of the Lord will bring a terrible plague on your livestock in the field, on your horses and donkeys and camels and on your cattle and sheep and goats. But the Lord will make a distinction between the livestock of Israel and that of Egypt, so that no animal belonging to the Israelites will die. Plague 6 was boils carried by the flies. Exodus 9:8-9 - Then the Lord said to Moses and Aaron, Take handfuls of soot from a furnace and have Moses toss it into the air in the presence of Pharaoh. It will become fine dust over the whole land of Egypt, and festering boils will break out on men and animals throughout the land. Plague 7 was a hailstorm. Exodus 9:13-19 - Then the Lord said to Moses, Get up early in the morning, confront Pharaoh and say to him, 'This is what the Lord, the G-d of the Hebrews, says: Let my people go, so that they may worship me, or this time I will send the full force of my plagues against you and against your officials and your people, so you may know that there is no one like me in all the earth. For by now I could have stretched out my hand and struck you and your people with a plague that would have wiped you off the earth. But I have raised you up for this very purpose, that I might show you my power and that my name might be proclaimed in all the earth. You still set yourself against my people and will not let them go. Therefore, at this time tomorrow I will send the worst hailstorm that has ever fallen on Egypt, from the day it was founded till now. Give an order now to bring your livestock and everything you have in the field to a place of shelter, because the hail will fall on every man and animal that has not been brought in and is still out in the field, and they will die. Plague 8 was a locust population explosion. Exodus 10:3-6 - So Moses and Aaron went to Pharaoh and said to him, this is what the Lord, the G-d of the Hebrews, says: How long will you refuse to humble yourself before me? Let my people go, so that they may worship me. If you refuse to let them go, I will bring locusts into your country tomorrow. They will cover the face of the ground so that it cannot be seen. They will devour what little you have left after the hail, including every tree that is growing in your fields. They will fill your houses and those of all your officials and all the Egyptians, something neither your fathers nor your forefathers have ever seen from the day they settled in this land till now. Then Moses turned and left Pharaoh. The 9th Plague was a sandstorm of darkness. Exodus 10:21-23 - Then the Lord said to Moses, Stretch out your hand toward the sky so that darkness will spread over Egypt, darkness that can be felt. So Moses stretched out his hand toward the sky, and total darkness covered all Egypt for three days. No one could see anyone else or leave his place for three days. Yet, all the Israelites had light in the places where they lived. But the 10th plague is obscured with mystery. Exodus 11:4-8 - So Moses said, this is what the Lord says: About midnight I will go throughout Egypt. Every firstborn son in Egypt will die, from the firstborn son of Pharaoh, who sits on the throne, to the firstborn son of the slave girl, who is at her hand mill, and all the firstborn of the cattle as well. There will be loud wailing throughout Egypt, worse than there has ever been or ever will be again. But among the Israelites not a dog will bark at any man or animal. Then you will know that the Lord makes a distinction between Egypt and Israel. All these officials of yours will come to me, bowing down before me and saying, Go, you and all the people who follow you! After that I will leave. Then Moses, hot with anger, left Pharaoh. Nobody can logically explain what happened here. However, there is another possible explanation. The Exodus from Egypt occurred during the ancient preparation of what is known today as the Jewish Passover. Traditionally Passover lasts for 8 full days. And during that time, ancient Jews did not eat anything that had a trace of CHOMITZ (grain of leaven bread) in it. That's why they clean out their houses thoroughly in preparation - and then do a special inspection. The purpose for this has a very clever reason. The reason being that "chomitz", after it sits for a period collects moisture (especially from the plague of hail) in which a deadly poisonous black mold can grow (stachybotrys atra). This mold produces "micro-toxins" which cause hemorrhaging of the lungs. The Egyptians did not have this ritual of cleaning out their houses. As a result, during the time of drought and famine when the plagues were occurring, they fed this contaminated "chomitz" to their firstborn children as is customary of the ancient Egyptians. Children were also considered the most important asset that the family possessed; therefore, to reassure their survival they were fed first. However, this does not explain the verse about the firstborn of all the "livestock". Would they have also been fed the toxic mold? Exodus 12:29-30 - At midnight the Lord struck down all the firstborn in Egypt, from the firstborn of Pharaoh, who sat on the throne, to the firstborn of the prisoner, who was in the dungeon, and the firstborn of all the "LIVESTOCK" as well. Pharaoh, all his officials, and all the Egyptians got up during the night, and there was loud wailing in Egypt, for there was not a house without someone dead.

It's quick to assume that the Jews had the knowledge about harmful bacteria when they didn't possess a microscope in those times. What is accessible to assume is that G-d who works through nature had employed these bacteria's, which also occur naturally in nature, to work his ways. And lets not remember the most import part of the plagues was that G-d had foretold Moses of what the plagues would be IN ADVANCE and each time he went to pharaoh to profess them. In addition, each time G-d continued to harden Pharaohs heart for the mere motive of BELIEVING and acknowledging Him as the one true G-d. Exodus 10:1-2 - Then the Lord said to Moses, Go to Pharaoh, for I have hardened his heart and the hearts of his officials so that I may perform these miraculous signs of mine among them that you may tell your children and grandchildren how I dealt harshly with the Egyptians and how I performed my signs among them, and that you may know that I am the LORD G-D.

It's also quite interesting to relate each plague with that of an Egyptian G-d. This is not a coincidence that the plagues matched with one of the Egyptian G-d's. The reason for this was so that G-d could challenge the authority and ability of Pharaoh's G-d's and proved them unable to withstand the power of the one true G-d.

1 - Water to blood - this is associated with the Egyptian G-d Osiris of the Nile River.
2 - Frogs - The Egyptian god Hekt.
3- Lice - The Egyptian god Seb.
4 - Flies - The Egyptian god Kephera.
5 - Cattle - The Egyptian god Apis and Hathor.
6 - Boils - The Egyptian god Typhon.
7 - Hail - The Egyptian god Shu.
8 - Locusts - The Egyptian god Serapis.
9 - Darkness - The Egyptian god Ra.
10 - death - The Egyptian god Ptah.

Here below is the equivalence (numerical value) of the Hebrew word for G-d which is "Elohim" consisting of the Hebrew letters aleph, lamed, hey, yud and mem, its Gematria numbers are 1+30+5+10+40=86. The Hebrew word for "nature" is Hateva (aka Habeya) consisting of the Hebrew letters (hey, tet, bait and ayin, its Gematria numbers are 5+9+2+70=86. In other words, both the Hebrew words of Elohim and Hateva share the same number of "86".

Let this thought sit on you for a while.

Number	Hebrew word	Hebrew letter
1	Aleph	א
2	Bet	ב
3	Gimel	ג
4	Daled	ד
5	Hey	ה
6	Vov	ו
7	Zayin	ז
8	Heh	ח
9	Tet	ט
10	Yod	י
20	Kaph	כ, ך
30	Lamed	ל
40	Mem	מ, ם
50	Num	נ, ן
60	Samek	ס
70	Ayin	ע
80	Pe	פ, ף
90	Tsadi	צ, ץ
100	Qoph	ק
200	Resh	ר
300	Shin	ש
400	Tov	ת
500	Kaph	ך
600	Mem	ם
700	Nun	ן
800	Pe	ף
900	Tsadi	ץ

8 - THE PRIMORDIAL LANGUAGE ...

Man acts as though he were the shaper and master of language, while in fact language remains the master of man. - Martin Heidegger.

The secret to understanding intensifies here...

All right, so maybe I'm not really telling anybody anything new. However, I will be now! If at any point so far in this book you were not frightened by the realization of G-d, you will be now!

This is where I made my startling discovery. I started to think long and hard about everything that consisted of this growth pattern, which I've termed, "one universal law" and its semblance to most everything that I look at in life. As my thoughts started to converge I began to think where is G-d's connection and signature in all of this? Where is the proof that all this is "His" handiwork and architectural design? And most difficultly how to prove His existence? Is there something that I was entirely overlooking? I knew that the this basic mathematical structure of unit growth incorporated within the spiral was not only of immense importance but of a huge intelligent source, and I needed to find a way that this valuable piece of information would somehow gap the bridge between science and religion.

In my loss of understanding, I opened a holy book called the "Pentateuch and Haphtarah" which are the five books of Moses written in Hebrew of the Old Testament (Genesis, Exodus, Leviticus, Numbers, Deuteronomy). I opened it up, I focused my attention upon Genesis 1:1, and in those few seconds of seemingly eternal thought, G-d's complexity hit me! My whole purpose for writing this book, AND HIS was defined. It was immediately very easy to discern, and the answer came to me, as it eventually will to you... THAT THE "HEBREW LETTERS" APPEAR GEOMETRICALLY MATHEMATICAL IN SHAPE AND FORM SHOWING UNIT GROWTH. A growth unit "similar" to the Fibonacci sequence, Phi/Pi and Golden Mean Spiral? What I had uncovered was "intelligent design" (which will be revealed towards the end of this chapter and shown in illustrative images). In essence, what I had discovered was that G-d is "really" exchanging a few words with us, and communicating with us on a base level and simultaneously in harmonious concert with whole cosmos... a Universal Language so to speak! But before I go any further in explaining this, I first need to expand on the other aspect of it first, which reveals that there is also something else extraordinarily elusive about the VERY FIRST WORD of Genesis 1:1.

While having this tremendous revelation, I realized that it's hard to be visionary and credible at the same time. It would only stand that if the Bible was a credible source of truth, it would seem logical that the first word of the Bible would incorporate the DNA instructions that permeate all life, as we know it. And it would do so in the way that reveals and represents G-d's workings through nature "Hateva" (which was just discussed in the previous chapter). This pattern would follow and pervade throughout all life forms as a measurable unit of proliferating growth and topping it off with the signature of the spiral.

And this assumption (with confidence in G-d) turned out to be correct, that a pattern of measurable unit growth is able to be found in the first opening chapter of Genesis 1:1 in the Hebrew Bible. This was all very overpowering at the first unearthing of the FIRST 6 letter Hebrew word of Genesis 1, which is the Hebrew word "BRESHITH" (pronounced, Barrasheeth). A hunch was telling me that there was something concealed here. Something that was coded as a work of intelligent design. Therefore, like particle physics I started to pick the word apart like subatomic particles down to the molecules. The Hebrew word "Breshith" as a whole means "in the beginning", but it also has several other "conceptual" meanings attached to it starting at the first letter of the word Breshith which is B, its interpreted meaning stands for house/door/enter/in. Several others words can be derived out of it such as the word "Barra" which means to "create from nothing". It also has several other conceptual meanings listed below, but the most intriguing thing about the word is that it is composed of 6 Hebrew letters (Bet, Resh, Aleph, Shin, Yud and Tov) (בראשית) (though when you read it, it is the other way around from right to left). Not only does it contain six letters, but also the end part of the word Breshith (Shith) literally mean the number 6 (in old Aramaic) and in modern Hebrew the (ShSh) (Shin, Shin) or SHASH represents the number six. Strangely enough, the word "in itself" says exactly what it means! IT MEANS CREATED 6. As if this were not enough, the word "shith" has also been interpreted to mean a thorn… a 6 sided thorn because the whole word of Breshith (only the Hebrew, not the English version) contains 6 letters, and these SIX LETTERS are the incarnated embodiment which represents the 6 pointed Star of David (otherwise known as the Magen David) (also known as the Shield of David, and as the seal of the biblical King Solomon). Being that the word is separated into two halves (Bre-shith) we can interpret it as the two separate words of Barra (create) and Shith (six) indicating that in the process of creation G-d "created six". It can also be illustrated as two intertwined triangles. Overall, 6 letters, 6 days of creation and 6 pointed thorns that are indicative of the Star of David. In modern language, its what we call a tetrahedron, which is in the shape of a triangle. This also happens to be the "first platonic solid" in which all other life forms follow. (I will familiarize you more with platonic solids in the coming chapters). It's also important to mention that Mount Sinai (Sinai - see Exodus 19:20) is the mountain that Israel (the Jewish people) met with G-d upon leaving Egypt. The Hebrew word "Sinai" also means "thorn" because it comes from the parent root SN (shin-nun). Another word derived from this parent root is the Hebrew word "seneh" meaning "thorn bush". The bush that Moses saw burning in Exodus 3:2 is a "Seneh". It was not just a burning bush, but also a burning thorn. Like I said all of this is CONCEPTUAL (meaning vague, elusive, intangible) - hence Quantum Mechanics… for reminiscent of atomic particles (like words) they are not confined within a given space (or meaning).

Here is the 6 letter conceptual breakdown of the word BRESHITH:
The word "Breshith" (Bet, Resh, Aleph, Shin, Yud and Tov) can ALSO be broken down as such:
Bet = House/Door/enter/in
Resh = head
Aleph = One
Shin = Fire
Yud = Devine mind/first point of creation
Tov = Seal/last letter of creation

Now that I have demonstrated a CONCEPTUAL theory behind the 6 letters of the first word of Genesis 1:1, I know that many people may frown on anything that is conceptually vague because its all hearsay, they are not moved in their stance. But for the skeptics at heart who need more proof, low and behold, there is also the undeniable pattern of Phi/Pi (3.142...) which is found in Genesis 1:1 that is not just merely a conceptual theory, but now a substantial concrete one. It clearly reveals, repeats, and retells us of the eloquent pattern of Phi/Pi (3.142...)

913 = 400+10+300+1+200+2
203 = 1+200+2
86 = 40+10+5+30+1
401 = 400+1
395 = 40+10+40+300+5
407 = 400+1+6
296 = 90+200+1+5
Total 913+203+86+401+395+407+296 = 2701

Then we have 28 letters x the value of the letters which is 28 x 2 x 200 x 1 x 300 x 10 x 400 x 2 x 200 x 1 x 1 x 30 x 5 x 10 x 40 x 1 x 400 x 5 x 300 x 40 x 10 x 40 x 5 x 1 x 400 x 5 x 1 x 200 x 90 gives us 28 x 2.388

Now with 7 words x the value of the words which is 7 x 913 x 203 x 86 x 401 x 395 x 407 x 296 gives us 7 x 3.0415

Now 28 x 2.3887872 divided by 7 x 3.0415 = **3.142**... Otherwise known as Phi/Pi.

It should not surprise us that G-d would incorporate this mathematical sequence into His design, otherwise why would it publicize this in the beginning phase of the Bible, both externally that we see through nature, as well as internally though the mathematical sequence and numerical values of the Hebrew letters. The message could not be any clearer, that the Bible is clarifying (by spelling out) its genuine linkage with that of the creators.

And just to show you how conceptually correct the Hebrew language is, we can examine the two primal words that are given to us in the Bible for Adam and Eve as "man and woman". Since Adam spoke the same language as G-d, he assigned these two words in the reflected understanding of what was made. Specified in Genesis 2:23 - And Adam said, this is now bone of my bones, and flesh of my flesh: **she shall be called Woman, because she**

was taken out of Man. Surely, these two words must show some kind of intelligent design if indeed they are the language of a meaningful G-d whose purpose is all encompassing. We find this is true when we examine the Hebrew words for both man and woman… man, which is "Eeish", and woman, which is "Eeishah". Now when you look at the picture below you can see that they share two common Hebrew letters between them which is the Aleph and the Shin, and the two letters that are left are the Yud and the Hey, which is the prefix for the word G-d - hence the understanding that with G-d woman and man are brought together.

Now before I really begin to elaborate with pictures I must also confer on another aspect as well. Even more intriguing is a hierarchical array that can be found in Genesis 1:1, where each layer repeats and includes the previous layer, which clearly emphasizes a series of adding some quantity - in which once again is demonstrating this "one universal law". It does this by the use of the first word Breshith, starting with the second Hebrew letter Resh. Repeating like this: Resh, Resh, Resh ר… Tov, Tov, Tov ת… Hey, Hey, Hey ה... Yud, Yud, Yud י… (look at the diagram below, and follow the letters to view this pattern.

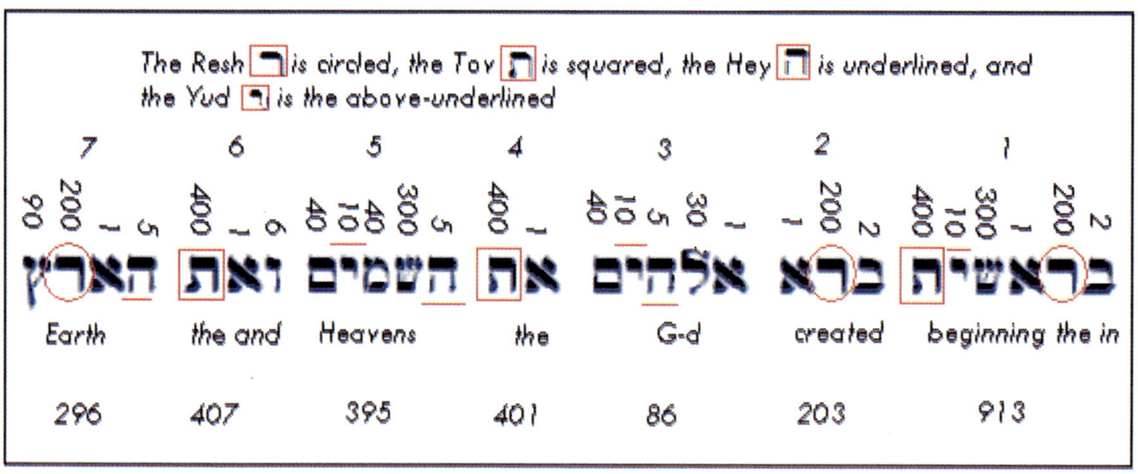

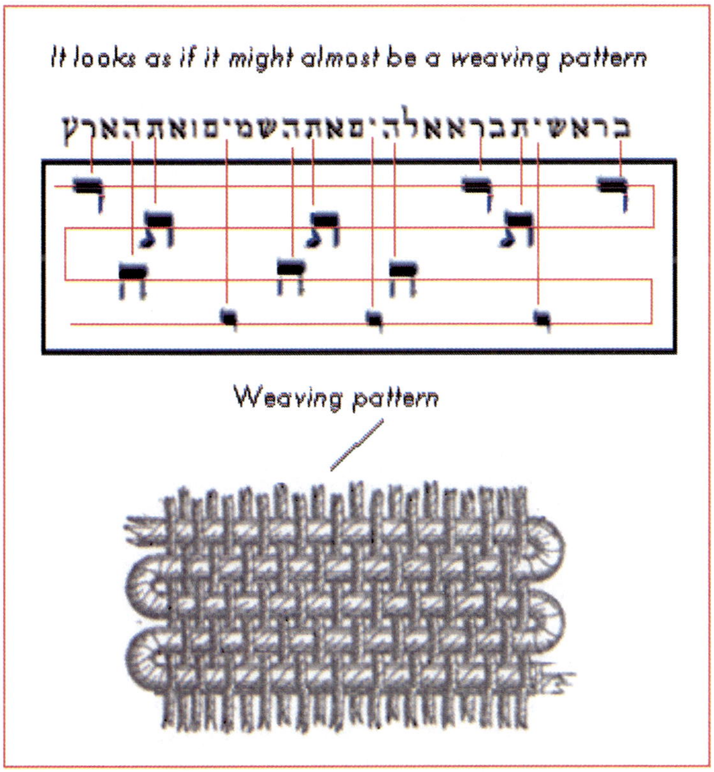

However, what is this pattern indicating? To truly understand what it is indicating we must for a moment examine something called Qabalah also known as Kabalah or Cabala. The different spellings are merely just transliterations of the word; one spelling is not necessarily right over another. The Hebrew word Qabalah means to “receive or accept” - that by studying the Torah one can gain insight and understanding about the creation of the world. It is believed that when Moses brought the Ten Commandments down from Mount Sinai he also brought with him the “ORAL LAW”. Which eventually in due course encompassed the entire spectrum of authoritative Jewish religious teachings followed by the texts of the Sefer Yesira, which include the Mishnah, Talmud, Midrash and others? They are a record of rabbinic discussions of Jewish law, ethics, customs and stories.

One of these books, the “Sefer Yesira” (which means book of formation/creations) was written to expand on the idea of how G-d used the Hebrew letters to form the material world and create all that exists. The text dates from the early centuries of this era and is actually a group of texts, which have been edited into one whole, and has sought to provide answers to questions passed down from “oral law“. Aspects of Qabalah can be traced back to the first century A.D. and later in the twelfth and thirteenth centuries to a French Jewish man from Provence named Isaac the Blind. Isaac is considered the founder of the Qabalah; or rather, he transmuted the mystical theology into the present form of the Qabalah. He alleged that this oral law was handed down from Mount Sinai from person to person until it reached him. It was Isaac who gave names to the ten Sefiroth (the ten building blocks of creation depicting the Qabalistic Tree Of Life diagram shown below).

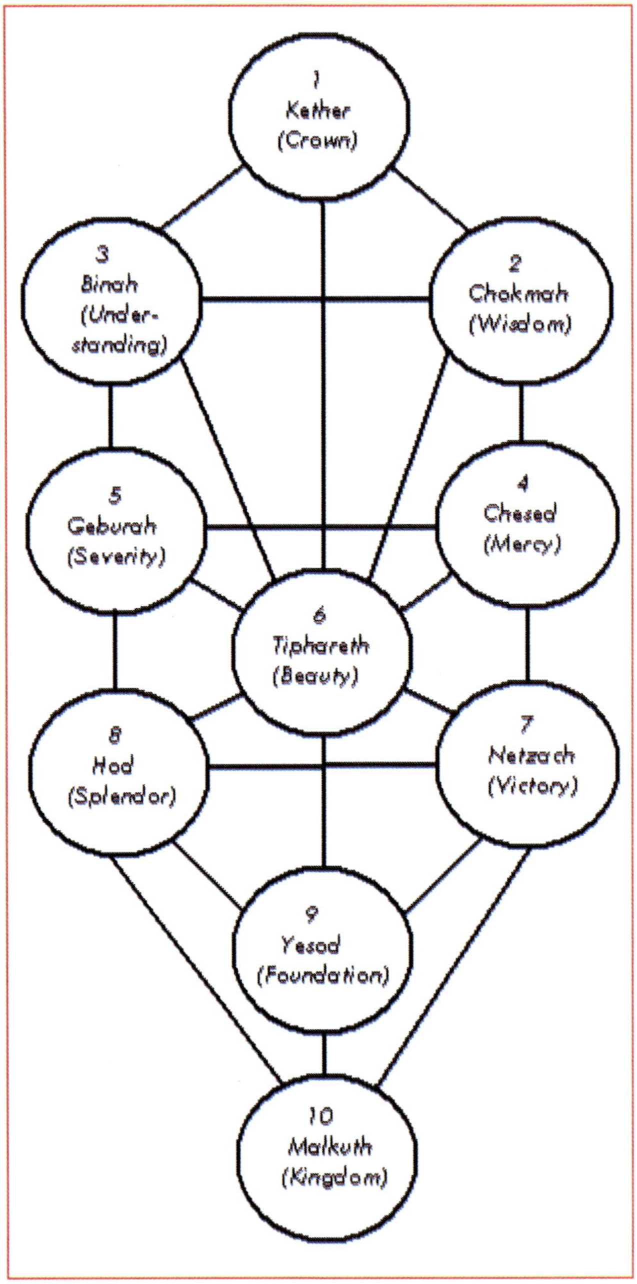

Isaac the Blind and several other centuries old religious thinkers had taken on the idea of "metempsychosis" (otherwise identified as reincarnation - but a different kind of reincarnation. A temporary reincarnation of mind and body, into spirit and soul). Passed down from "oral law" the religious thinkers were attempting to integrate an "ethereal elevating object" (boundless in space and time) that's referred to as the "MERKABAH VEHICLE", (and I will quickly get to explaining what this is). What I'm specifically aiming at here, is the hierarchical array I pointed out above which is found in Genesis 1:1 and its linkage to the "Tree of Life diagram" as well as the "Merkabah".

Overall, what this to owns up to, is that there are "layers" to reality, just as the skin is to the body. In this same sense, the conscious mind also has levels to consciousness. And the Tree of Life, to which the Merkabah vehicle comes out of, shows (teaches) us a way in which to transcend the realities of our consciousness. Look at the picture below to see how the "Star of David" comes out of the inner recesses of the "Tree of Life" diagram:

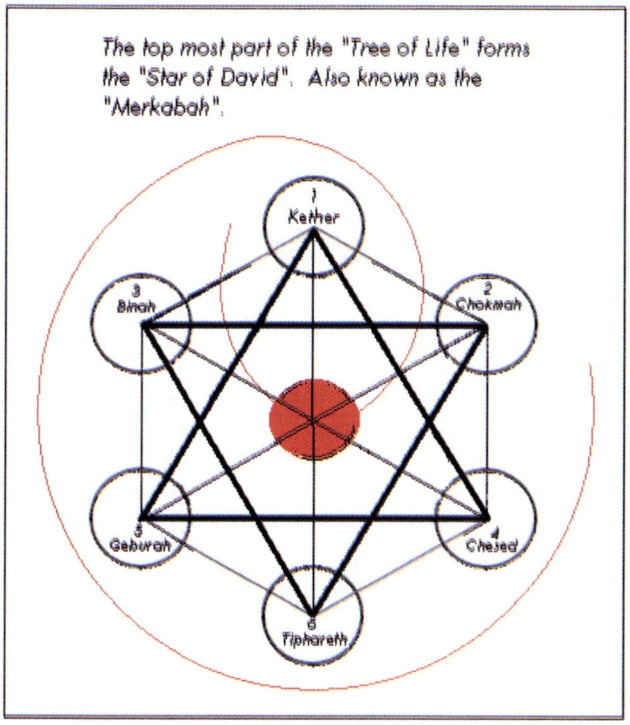

So as you can see, the Star of David is actually the personification of the Merkabah vehicle, they are the same thing. Now, to explain the Merkabah vehicle… it is simply describing an ethereal object that utilizes the "invisible spiraling forces" of nature that allows for a temporary transmigration of the souls ascent into Heaven. It was one of the first Qabalistic ideas to reach this goal by what is described as the "higher throne" of G-d through the "ten Sefiroth" which involves a process of meditation that transcends the human limitations of conscious reality and thus reaching the topmost part of the Tree Of Life, "kether" (meaning crowned head - divine throne - chariot) and sets into gear the mechanism of the Merkabah which facilitates transcendence and existence above and apart from the material world. The word Merkabah is translated to mean, "spiraling light". Such religious thinkers correlated the spiraling light (which is described as a wheel inside of a wheel) in conjunction with the 22 letters of the Hebrew alphabet forming 22 letters multiplied by 6 (representing the 6 days of creation and 6 letters in the very first word of the Torah) equaling 132 which are indicated as "partition gates to a wheel" and by transposing and permuting the Hebrew letters BACK AND FORTH, (aleph with all of them, and all of them with aleph, bet with all of them, and all of them with bet, similar to the sequence of human DNA) by way of the "wheel" (circle) GLGL (which is the Hebrew word "galgal" which means "wheel"). (gimel, lamed, gimel, lamed). However, we can also arrive at the reverse affect of 231 transmutations of the Hebrew letters if the wheel spins the other way. Now when we add 2+3+1 or 1+3+2 they both equal "6". Most wondrously AGAIN this is indicating the 6 pointed "inverse tetrahedrons" (132 opposite 231) pattern of the Star of David which is referred to in the Sefer Yesira as ONG and NGO. ONG is the Hebrew letters (ayin, nun, gimel), which means, "delight", and NGO is the Hebrew letters (nun, gimel, ayin) meaning plague (this information can be found in verse (II, 4) of the Sefer Yesira).

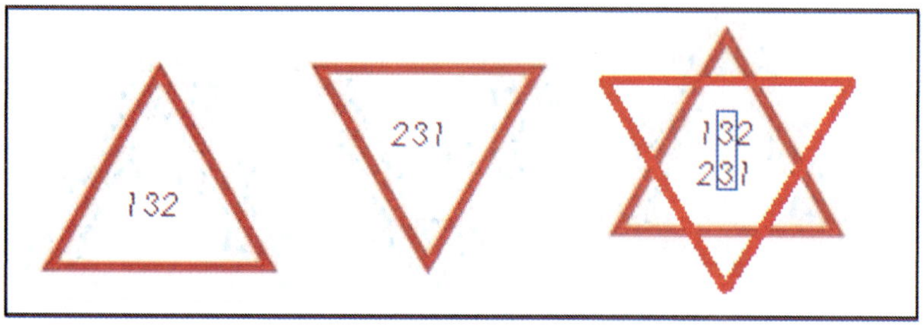

Now veering away for a moment to another aspect… oddly enough even in Genesis 1:1 sandwiched between the 28 letters is the Hebrew word “Shalosh” (Shin, Lamed, Shin, Hay) which is the Hebrew word for the number 3. We find another “partitioning separation” that is redirecting us to notice and reaffirm the Star of David pattern again.

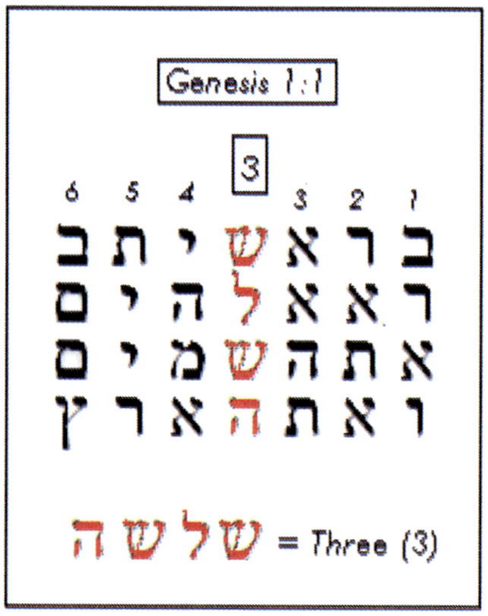

It seems that no matter how we try to break down the Genesis 1:1 pattern, it is further observed that the ubiquitous 6 is a factor of each verse, word and letter - compartmentalized by the first 6 letter word itself. There have been other commentators and religious thinkers who have tried this pattern in several different mathematical and representative ways, and who have arrived at the same ever-present 6. Six can also denote perfectionism, as it is the first “perfect“ number, in that the sum of its divisors, other than itself, is equal to itself: 6 x 1 = 6 and 3 x 2 = 6, 3+3 = 6 and 1 + 2 + 3 = 6 (the next such number is 28, and there are 28 letters in the first paragraph of Genesis 1:1. When you subtract 2 - 8 = 6).

With a spellbinding sense of purpose, all of this has been implied to draw our attention to the Star of David and its modeled design. We can certainly draw supposition like that of Isaac the Blind, but nobody can say for sure just where the Star of David pattern and design had its starting point throughout religious history? There is no doubt that something very enigmatic is hidden and sandwiched within the structured pattern of the Torah. There have been MANY ideas as to its historical origins, save for one such other hypothesis is that it resembled a fruit, the FORBIDEN FRUIT from the Tree Of Knowledge in the Garden of Eden. The Bible clearly tells us that in (Exodus 39:24-26) that they affixed “pomegranates” and shells at the High Priest’s hem. The Pomegranate is known as one of the seven holy fruits of the country enumerated in (Deuteronomy 8:8). Deuteronomy 8:8 - A land of wheat, and barley, and vines, and fig trees, and **pomegranates**; a land of oil olive, and honey. Shown below is the Pomegranate:

It is the only fruit whose image was a permanent component in the Temple. Pomegranates were also put on the pillars of Boaz and Yakhin in Solomon's Temple, (1 Kings 7:42). Solomon reared these pillars to symbolize the pillar of cloud and pillar of fire in the wilderness. Tradition also tells us that King David adorned a Magen David (Star of David) on his shield, and that King Solomon had the symbol inscribed on his ring. We may gather the reason for the Pomegranates well-regarded image for the simple reason that it was structured with 6 flower petals around the outside top of the budding Pomegranate forming a symmetric hexagram, a six-pointed geometric figure that is composed of two interwoven equilateral triangles which may symbolize creation's polarities in their crossing point and balance, including male and female, spirit and matter, heaven and earth, fire, water and spirit, which I believe is the interpretation of the Hebrew word for "Heaven" (Shamayim). The Hebrew word "Shamayim" can be broken down into two words (Sha - Mayim). The Sha is noted for its first Hebrew letter the "S" (shin). Shin = fire and Mayim = water). Now some may wonder why G-d would use fire to create Heaven when fire is notoriously know for hell. The answer to this is that not all fire is bad since fire can be used to cook, destroy, and refine things. As for Heaven G-d used fire to flavor, preserve and refine this sacred place along with everyone in it. While on the other hand hell is a place of tormenting fire and destruction.

The significance of all of this is that it reminds us of these polarities and draws our attention beyond the physical. Contemplating on the pomegranate could induce a person to reflect on all the polarities in and around them, and on G-d, the Creator. This is how it could be applied without deceiving people to deify and worship the actual symbol of the pomegranate and its successor the Star of David. Thus this "sign" is not a moralistic preachment, but a device for connecting with G-d, the Creator. Not only is this symmetrical fruit associated with the Star of David, but it is the precursor of the giving of the Torah itself and of its 613 commandments (betokened by the numbers it represents) since the Pomegranate also contains exactly 613 seeds, and if you open a pomegranate across its equator you will find that the seeds within form a Star of David pattern. It has 6 separate compartments, and if you closely examine the outside shape of a pomegranate, it is not perfectly round or oval; you can see a definite hexagon pattern on the outside exterior. Many of us are only familiar with the 10 commandments, but there are actually a total of 613 commandments in the Torah - 248 of them are positive commandments and 365 are negative commandments - totaling 613.

The Bible never overtly tells us what kind of fruit was eaten off the Tree Of Knowledge. What we do know is that this was the one tree whose fruit Adam and Eve were forbidden to eat, or even touch (Genesis 2:17, 3:3). Nevertheless, I'm guessing that it was the pomegranate that Adam and Eve partook in. The pomegranate is a small tree and even a shrub of sorts, and I don't think any one can suffer from eating them after the original sin, but its possible that G-d could have regarded this tree and its fruit as a symbol and an enigma (mystery - riddle - puzzle) for containing the information of creation - hence primordial language (as your about to find out)... thus the very first fundamental form, stage and shape containing the Tetrahedron. Its also possible, seeing that

G-d is all knowing, that he had foreseen the future hardship ahead for mankind and thus viewed the Pomegranate of containing 613 seeds as the predictable (probable choice) future commandments on account of eating from the tree and thereof sinning. You see what I'm alluding at here, is just the mere fact alone that this fruit has 613 seeds - HENCE COMMANDMENTS - is enough for us to figure out why G-d had warned us against eating from the tree of knowledge and acquainting and familiarizing ourselves with the crafty serpent. With biting witticism, G-d is teaching us the central lesson of failing to recognize that the "latter is prerequisite to the former". In addition, the question arises that can alter someone's assumptions and faith given that the pomegranate holds the "map" of scientific geometry that explains the universe (as you are about to fully find out towards the end of the chapter). Obviously, the "tree of knowledge" must have something to do with knowledge, or why should it be so named? It undoubtedly contained the key to certain divinely classified material that G-d did not want early man to possess (for reasons I will later explain in the coming chapter of Egypt's secrets). But deliberately preserved (after the Exodus from Egypt) for future generations to discover. The Pomegranate also comes in different varieties… and these different varieties may look a little different by having more or less petals, but the traditional one that grows in the Middle East has the 6 petals that resemble the Star of David.

Interestingly enough, we also seem to find another pattern in the first seven words of Genesis 1:1 (913+203+86+401+395+407+296=2701). With the value of 2701 being 37 x 73 = 2701. It is again interesting to note that both numbers consist of 3 and 7. And oddly, again if you didn't understand it from the very first 6 letter word of "Breshith" or the 132/231 (that I pointed out above), it repeats it again. And though this doesn't equal up to the value of 6, but rather 10, it is of a greater significance representing the 10 commandments. 613 (6+1+3=10) compressed from the original 10. A mystical number in its own right.

In addition, to top this off, the actual words for the Star of David known in Hebrew as the "Magen David" also shares the numerical value of the number 10. Magen is: (Mem, Gimel, Nun, 600+3+50=653, 6+5+3=14, 1+4=5) David is: (4+6+4=14, 1+4=5), therefore Magen + David is 5+5=10

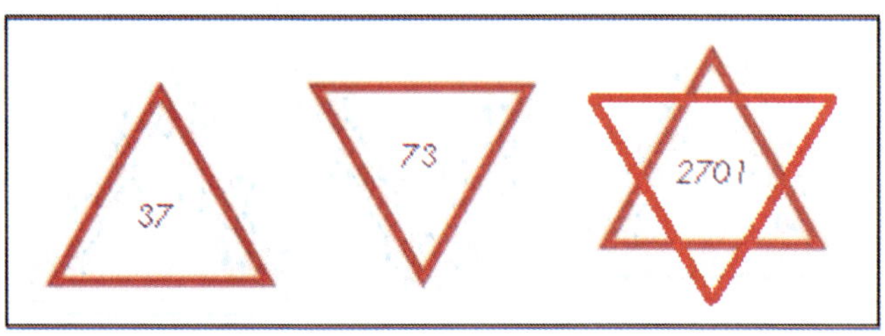

And the most strikingly apparent visual effect, over anything else I have exhibited here thus far is another pattern that can be found in Genesis 1:1. The pattern of 6 Hebrew letter "Alephs" in Genesis 1:1 forming the Star of David design. Look below at the first paragraph sentence in Genesis 1:1 and observe that there are precisely 6 Hebrew letter "Alephs" in it.

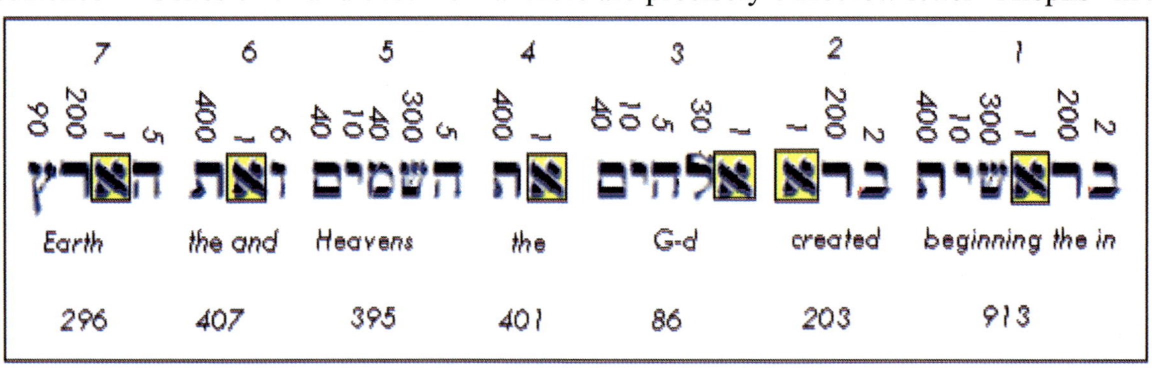

Now one might ask, what is this suggesting, or rather its significance? As if we did not have enough verification yet, G-d goes to the length of displaying this before our human eyes. As you may recall, according to "oral law" (Qabalah) passed down from Mount Sinai and Moses and migrating its way to Isaac the Blind in the texts of the Sefer Yesira, that the Hebrew letters are to be "set around a circle" working as a wheel. When we reconstruct this, starting at the very first letter of the Hebrew alphabet, which is "Aleph" (A), low and behold, a very important result ensues, and look what we arrive at!

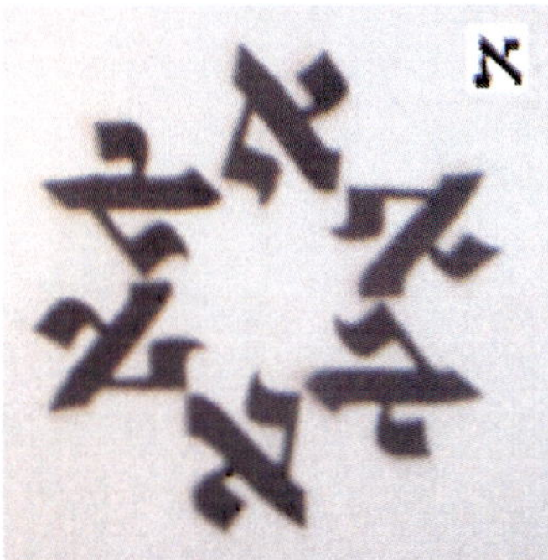

6 Hebrew letter Aleph's positioned around a circle forming the Star of David pattern.

Its significance is overwhelming - the Hebrew letter "Aleph" forms a distinguishing Star of David! As luck would have it, this also goes for "ALL" the Hebrew letters when they are set around in a circle can form the Star of David pattern. Though I did not illustrate them all, I've chosen to display two more that are both very vivid and beautifully stunning which you can see illustrated below:

6 Hebrew letter Ayin's positioned around a circle forming the Star of David pattern.

6 Hebrew letter Kaph's positioned around a circle forming the Star of David pattern.

If you cant see the Star of David pattern in them, here they are.

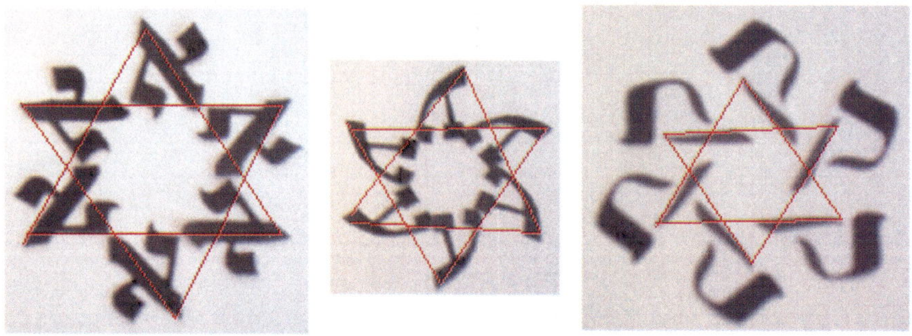

The Hebrew letters are designed in such a way as to have their tops or bottom legs geometric in nature. Now some people may think that there is nothing spectacular about this since one can also arrive at the Star of David form by using other letters (and not just Hebrew letters) and making them fit accordingly, but I can assure you that none are like that of the Hebrew letter Aleph which vividly emphasizes the Star of David form (be my guest to try for yourself). In short, the Hebrew letters (especially Aleph) are deliberately shaped for the way they appear in view of emphasizing the Star of David. One may question why the Aleph is so prominent in forming the Star of David pattern? And the answer to this will reveal itself shortly below.

However, "Aleph" is the most distinctive for forming the Star of David pattern out of all the rest of Hebrew letters. One might wonder what is so significant about the Aleph. Well, the "Aleph" as you well know is the first and foremost beginning letter of the Hebrew alphabet (the initial letter of G-d's creation). The Aleph is a representation of a "reflection of itself" consisting of two Hebrew letter "Bets" which are mirror images of each other, and thus that is why the Hebrew letter Bet (B) comes next in the Hebrew alphabet.

The "Aleph" is a partial prototype of how G-d pulled woman out of man.

ב *Hebrew letter Bet.*

Two letter Bet's can be formed out of the Aleph. The smaller one in red and the bigger one in black (which is upside-down).

Moreover, "three" letters can be made out of the "Aleph": Which are Aleph, Bet and Tsadi:

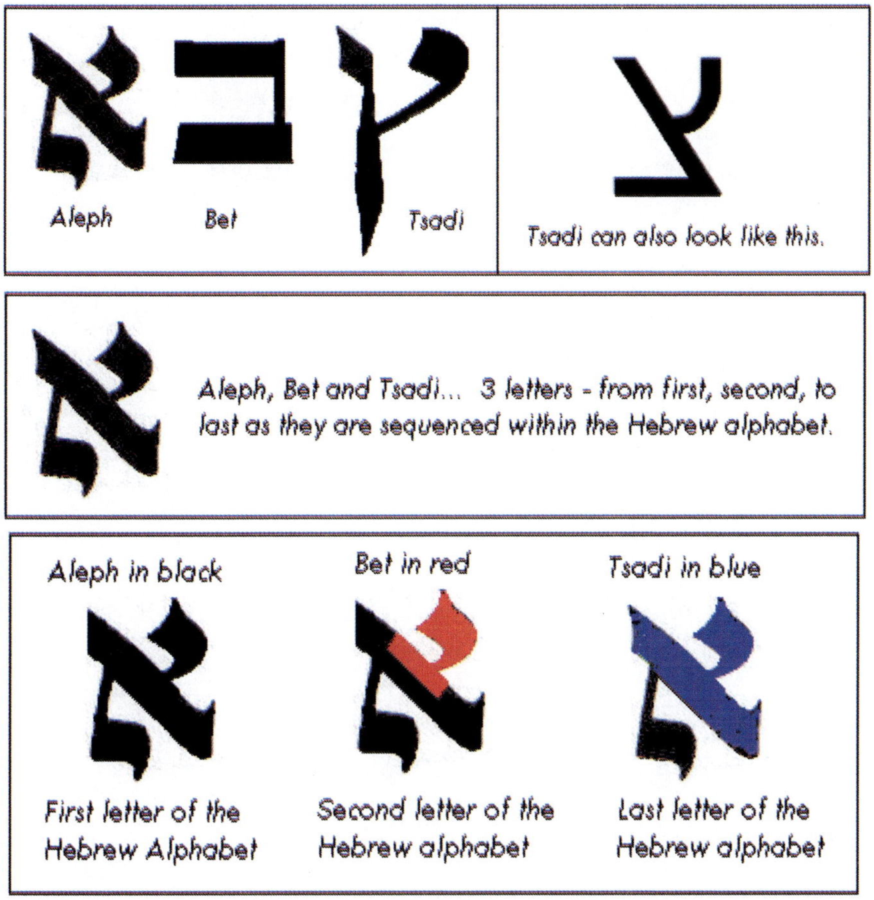

Tsadi (T)is also the very last letter (of the 5 final letters) of the Hebrew alphabet. So not only does the very first Hebrew letter contain the second letter of the alphabet, but also the last - telling us that it in itself is a complete representation of the complete Hebrew alphabet (from the first letter to that of the final last letter).

The Hebrew letter Aleph can also be viewed as a representation of the "Trinity". Since it is the very first letter, that G-d created that shows a 3-part nature to it.

There is even more mystery surrounding "Aleph". It is a mysterious letter. "Aleph" by itself it silent sounding, but may not be completely silent as it can be adhered with vowels to give it an "Ahhh" sound such as in the first word of Genesis 1:1 Breshith, (which is pronounced Barrasheeth or bAAArAAAsheeth).

The vowel A (aaahhh) in sand.

Now for the doubters at heart who are wondering to just what lengths I'm taking all this to... there is a field of study called "Cymatics" with involves the investigation and examination of wave phenomena in sound. Whereas, it has been discovered that the Hebrew letter "Aleph" and its Ahhh sound creates a identifiable pattern with sand particles (or fine powder) when subjected to vibration. As it turns out, the Ahhh sound (when sounded out by itself) creates the Star of David pattern shown here. And this includes any words (word like hallelujah or hosanna) that contain the Ahhh sounds. It seems that when Ahhh is sounded out, it creates a standing wave pattern in the form of the Star of David. (I encourage all readers to become familiar with the field of "Cymatics" and investigate this phenomenon for themselves). (The picture shown is an imitation of the original picture).

The history of the Ahhh sound can be found in some of the most powerful mantras (holy word in meditation, sacred word, chant, or sound) that are repeated during meditation to facilitate spiritual power and transformation of consciousness. Some say that this sound generates love and compassion, and have you ever noticed that the sounding of an "Ahhh" is acceptable by everyone regardless of their background, culture or religion? As it turns out, this very sound is birthed on inhalation when we are born and it is also the last sound we make on our final exhalation of air when we die. More on a biological note, this sound may be the "KEY" to the vibratory matrix of reality itself. It may be the reconstructive and repairing mechanism for living cells within all life forms. In fact, it may be the very determining mechanism in stem cells or cancer cells. It may even have an effect over the molecules in the air and water, restructuring to decontaminate the air and water. This key sound as well as its unit of growth and outward appearance (Star of David) is a supernatural (almost paranormal) power that works with the laws of nature and is vital to all life as aforementioned by the first minimal distinction of the "universal law" recognized in this book.

In the gist of all this one has to wonder, is there something to the concept of sacred language? Does a frequency sound have the power to influence and transform physical reality creating things through inherent power? Tradition tells us that through the recitation of singing Hebrew texts, such as psalms, it has the power to heal a person who has gone "out of sync". In addition, is capable of directly affecting the nervous system.

So pervasive is this message regarding the Star of David pattern as clearly witnessed... signifying (in every which way) together the number 6, 10 and 613 for the Star of David, the Pomegranate, the commandments and the combination of six Alephs around a circle, and also including its vibrating sound/form in sand. But the most pervasive thing about this structure is the main message of associating THE SPIRIT ENTERING MATTER - containing the old wisdom of - "as above so below". G-d's defining mark when one merges with Him. In other words, the double-triangle represents how one dimension pours itself out into the next dimension. It is much more than just an evasive and external symbol, as you will soon find out.

Taking a few steps back and now if you would, reexamine below the Λ diagram/picture and the "Bell Shaped Curve" that was discussed in the previous chapter.

The Λ diagram starts from a given point and grows outward away from an initial point.

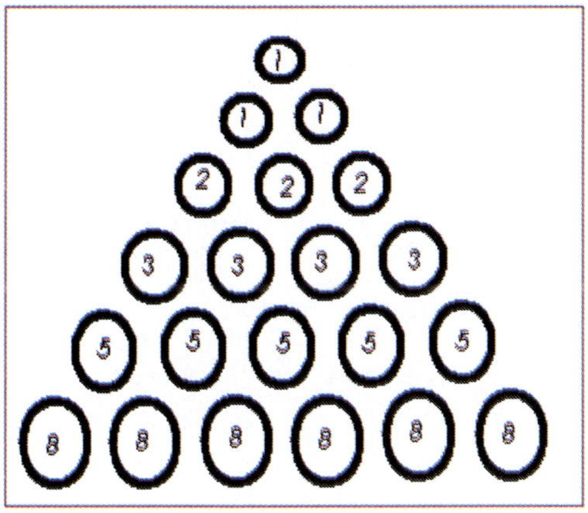

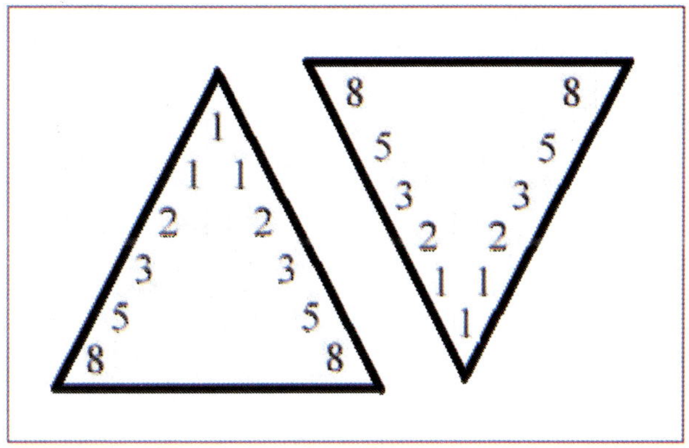

The “Bell Curve” that illustrates the distribution of quantitative or cumulative growth. Rising at the incline of interest or response and lowering at the state of residual entropy (disorder).

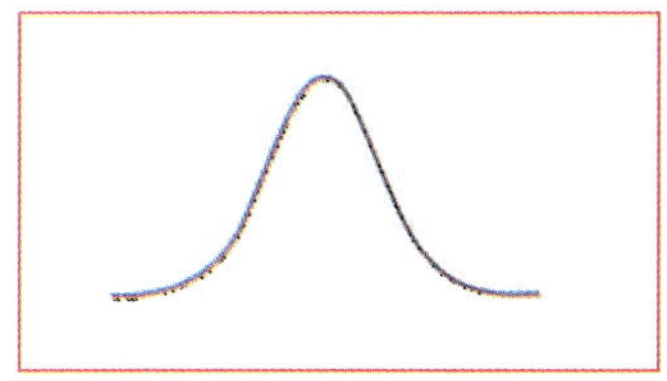

Now notice what happens when we intersect two Bell Curves, we get what called a standing wave. THERE IS NO DENYING THAT IT CLEARLY RESEMBELES THE STAR OF DAVID! Waves are a common feature in nature. A wave simply put, can be described as a disturbance that travels through a medium from one location to another location. A standing wave occurs when the interference of two waves traveling in opposite directions with the same identical equal frequency, amplitude or resonance meets its reflection. The spaces that intersect with each other are called nodes.

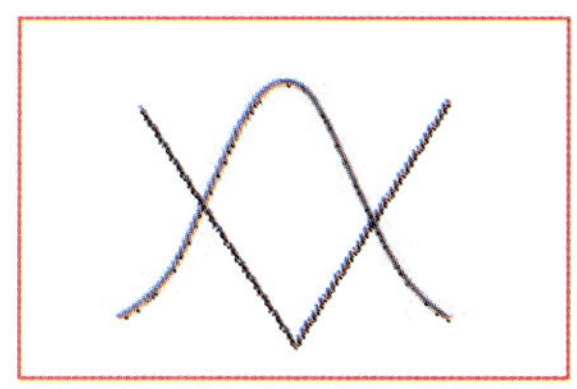

We can also unite this view with the Male and Female reproductive organs. This means that the sexual act that can lead to a new creation is actually a bringing-together of the Triangles (tetrahedrons, Bell Curves and standing wave) (Star of David, Seal of Solomon, Magen David).

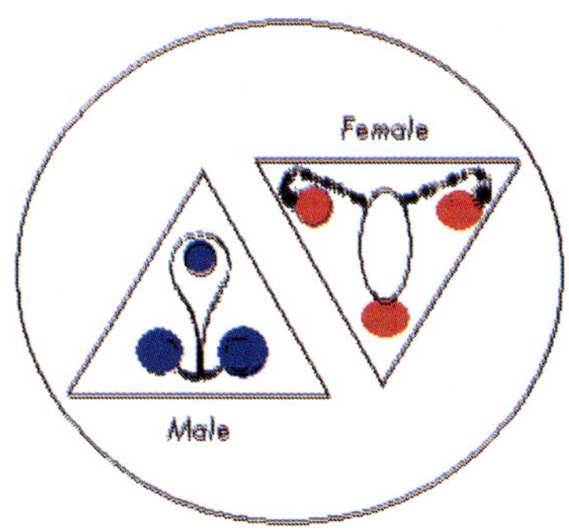

We can also see that the representation of the path of atomic particles also shares this symbolic resemblance to all of the aforementioned 6 pointed star views.

We can also relate it to the visible color spectrum.

Now here comes something of the miraculous! Not only do the Hebrew letters themselves around a circle form the Star of David pattern (as previously seem above, they also form while looking at them from a frontal view). By the way, there is no other alphabet in the ENTIRE WORLD like this that can fashion its letters from a single model pattern (prototype):

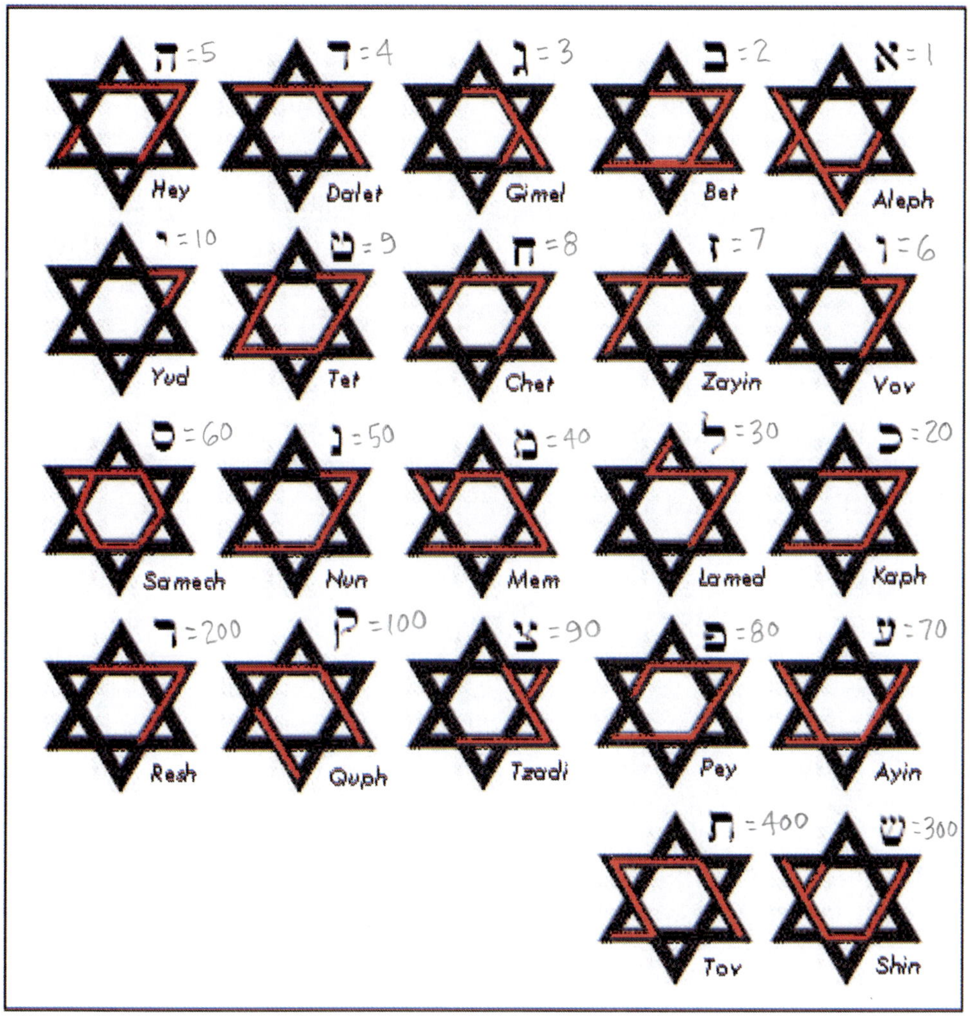

All the Hebrew letters from Aleph to Tov = "Ahtah" (G-d's accurate alphabetical name) which are all the letters of the alphabet. In addition, by this I mean, in the arranged order of alphabetic letters.

Now I know that it is not hard to find patterns of other letters within this geometric shape of the Star of David… But consequently, it is hard to find a "precise meaningful shape" which makes it of greater significance when they are attached, related or applied to any sort of organized, logical and underlying meaning - in which they do so apply as a language in exchanging information. Now you might think you are just looking at the Hebrew letters within the Star of David pattern, but the real significance (if you recall) is that the Tetrahedron displays an evenly dispensed measurable pattern of unit growth. In addition, the fact that all the Hebrew letters can be formed and found within it is almost nearly amazing!

As if this were not enough proof, IT ONLY GETS BETTER! We are getting very close here to fully understanding. This is where at the beginning of the chapter I left off telling you that the answer to the geometric shapes of the Hebrew letters will come towards the end of this chapter. We are now at that spot. So you can remember, I have cited what I said from above: THAT THE "HEBREW LETTERS" APPEAR GEOMETRICALLY MATHEMATICAL IN SHAPE AND FORM SHOWING UNIT GROWTH.

This transpires when we incorporate the "SPIRAL" into the equation. Firstly, it can just be a spiral that forms the shape of the Hebrew letters despite any unified mathematical sequence of unit growth other than the fact that it just simply resembles one of the 22 Hebrew letters, BUT IN A SPIRAL FORM. Kind of like the flowing style of cursive letters. However, THERE IS a growth sequence attached to it that expands out in a measurable pattern (where each unit is bigger than the previous one) (and you will see this).

Now I'd like you to notice below how a plant leaf opens (or unfurls). Notice how the spiral starts out from a noticeably small point and protrudes outward opening up wider as it expands (from the tip to the base). This unit of growth in it is the "universal law" of discussion. As it spirals, it expands and opens in units. Of course, this is (in part) due to the earth revolution around the sun as well as the forces of gravity, but it is also consistent with the laws of the Universe such as in spiral galaxy formations and the distances from planet to planet as well as the space distances between stars.

Now watch and see (the triple miraculous!) how the Hebrew letters are also subject to natures laws of gravity and unit growth. And for all those who believe in evolution, they will now have to come up with an explanation as to why there is absolute "order" that spells, speaks and delineates (portrays visually) a DESIGNER!

First remember, the Hebrew letter Aleph is a representation of a "reflection of itself" consisting of two Hebrew letter Bet's (which is the Hebrew letter B) which are mirror images of each other, and thus that is why the Hebrew letter "Bet" comes next in the Hebrew alphabet. You see, "Bet" (which means enter) is the "key turning (spiral) letter form" in which all the other Hebrew letters are formed from (or should I say viewed). IT IS A LITERAL KEY! How I arrived at this was by modeling with a soft metal wire an actual Hebrew letter "Bet" (into a spiral shape) and then turning it around in different positions (spinning in slow motion) until I was able to recognize the other Hebrew letters. This was truly amazing!

For starters lets examine what a letter Bet (B) looks like again**:**

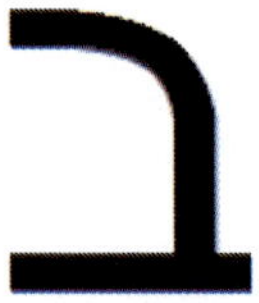

This is what the Hebrew letter Bet (B) looks like.

The Hebrew letter Bet (B) is the second letter in the Hebrew alphabet, but it just so happens to be the first initial letter that is used in the first word of the Torah in Genesis 1:1 in the word "**B**reshith", which literally means IN THE BEGINNING. But the individual letter Bet means door/enter. This is what first clued me in, seeing as it was the first initial letter I had a gut feeling that it was the KEY to unlocking all the other 22 Hebrew letters.

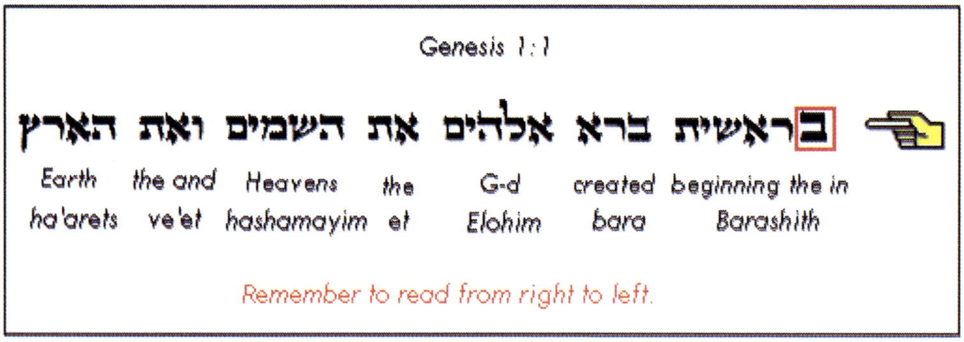

Secondly I took an actual letter "Bet" (formed out of a bendable black wire) and wound a second red bendable wire around it like so**:**

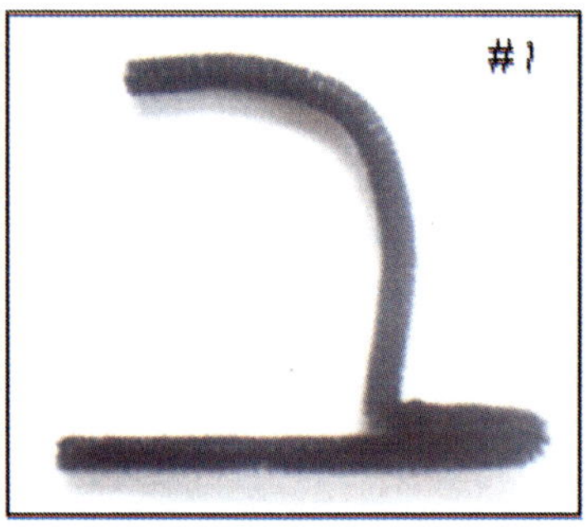

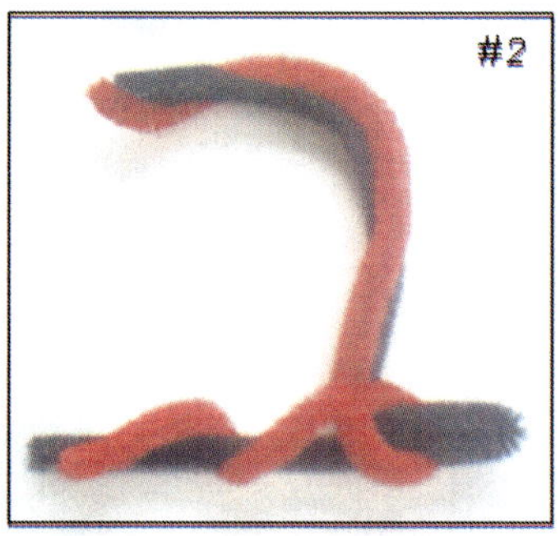

Then I took away the black wire and was left with the red wire form**:**

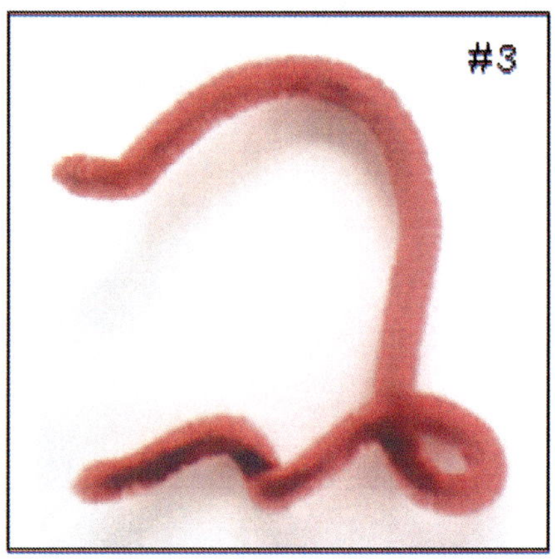

By the way, anybody can recreate these letters for themselves. The fuzzy metal wires (that I've used above) can be bought at any craft store. For the most part, any pliable/bendable wire you use will work nicely (as shown below).

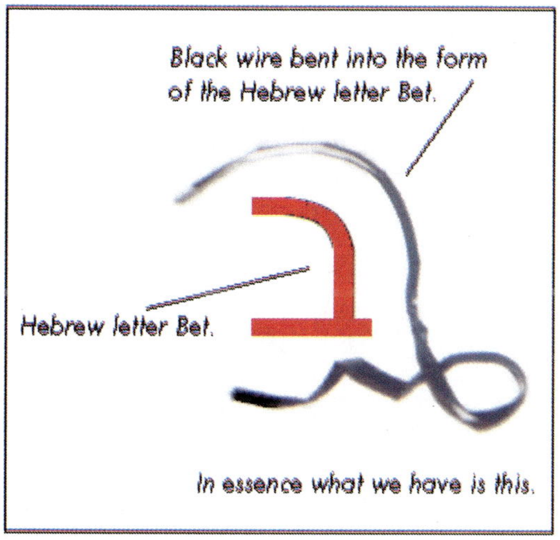

Next, I took this wire form and tilted and turned it in different positions, which enabled me to see (recognize) all of the other 21 Hebrew letters with amazing exactness. And again, like shown above where the Hebrew letters are fashioned inside the Star of David - there is no other alphabet in the ENTIRE WORLD that does this - forms its whole alphabet from one prototype letter.

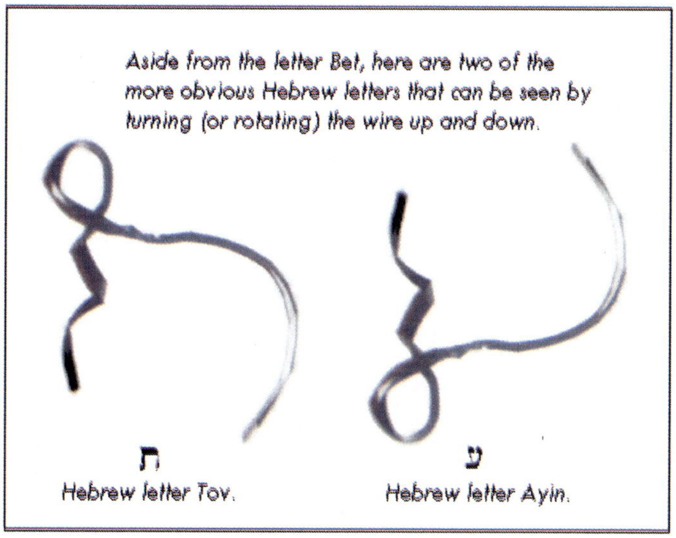

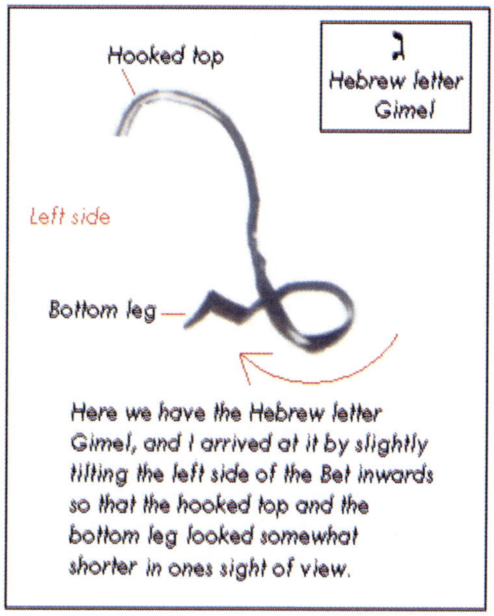

Now I'd like you to have a better look below to see how this one prototype letter, which is the first initial opening letter of the Bible (letter Bet) miraculously transforms itself into the other 21 (altogether 22) letters of the Hebrew alphabet.

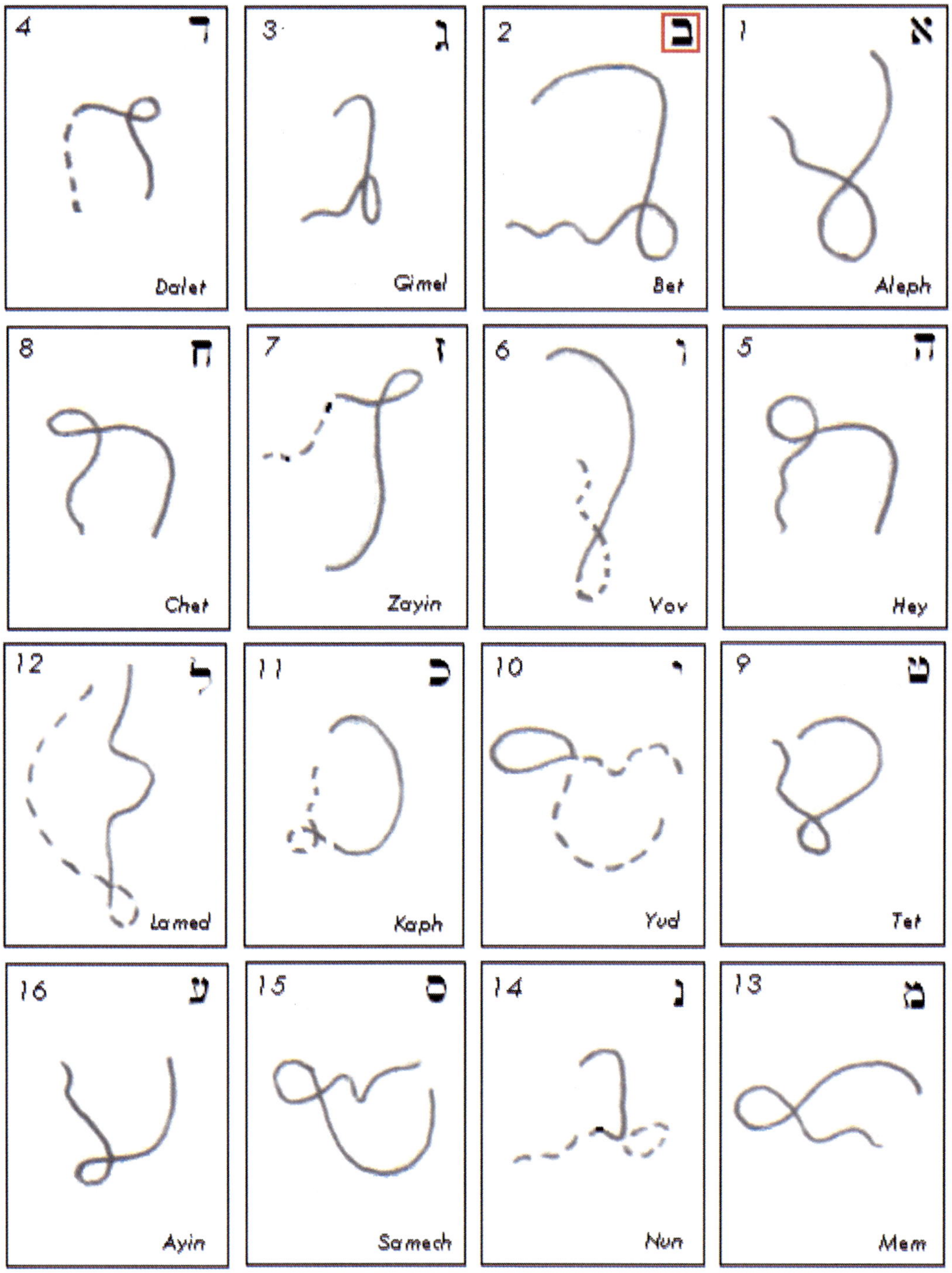
1 א Aleph
2 ב Bet
3 ג Gimel
4 ד Dalet
5 ה Hey
6 ו Vav
7 ז Zayin
8 ח Chet
9 ט Tet
10 י Yud
11 כ Kaph
12 ל Lamed
13 מ Mem
14 נ Nun
15 ס Samech
16 ע Ayin

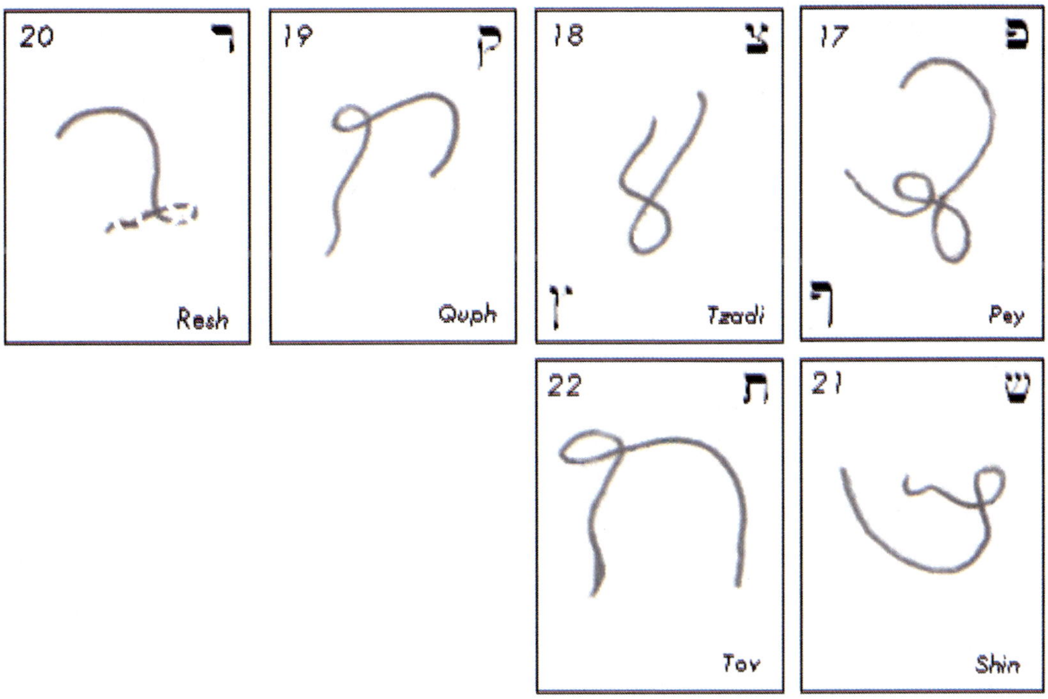

Now look below to see that these same letterforms also correspond nicely (though not exactly) to the modern "cursive" Hebrew script used today**:**

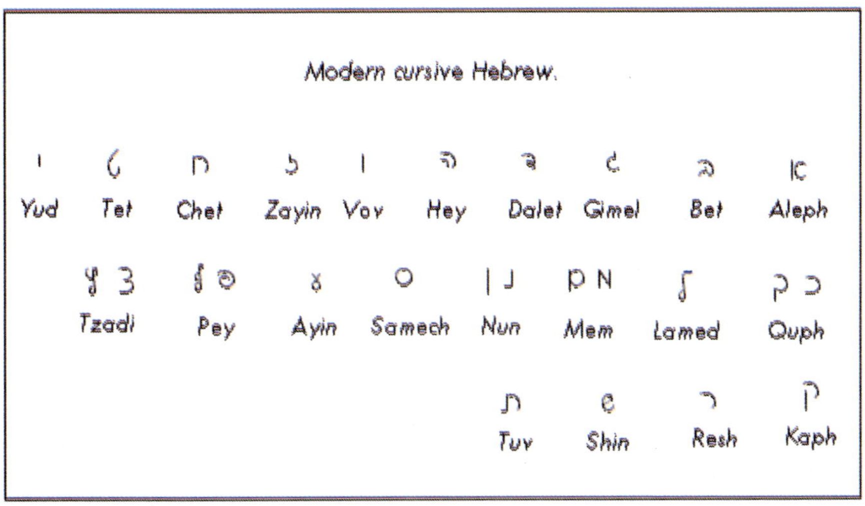

Does one understand the implications of all this? It implies a DESIGNER! - **Namely G-d** who of course is the AUTHOR of the Bible, and the designer of all life.

What is actually happening here (for those who still do not fully grasp it) is that the very model and dimensionality of the Hebrew letter Bet is formed (none other) by the "Universal Law" I've been ascribing to in this book. It does this by an increasing unit of doubly equaled growth, and spirals (or curls) this form to conform it into the Hebrew letter Bet. Here take another look**:**

These two pictures show the growth in wave graphs:

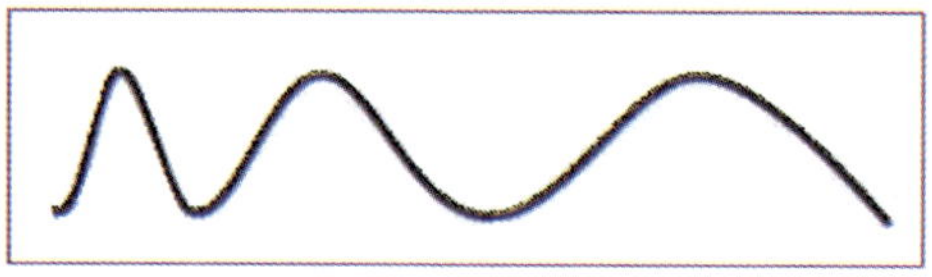

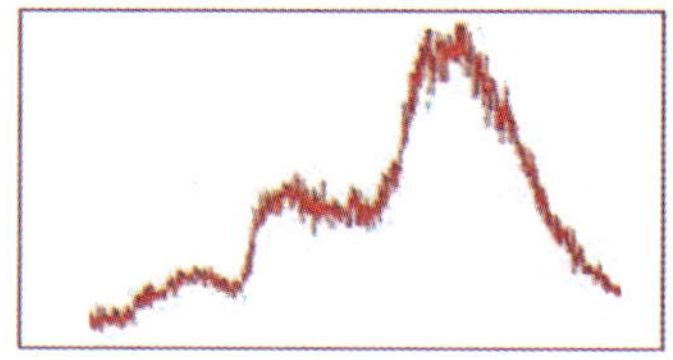

What becomes interesting is when we look at these wave graphs (in a different aspect) next to the spiral form of the Hebrew letter Bet:

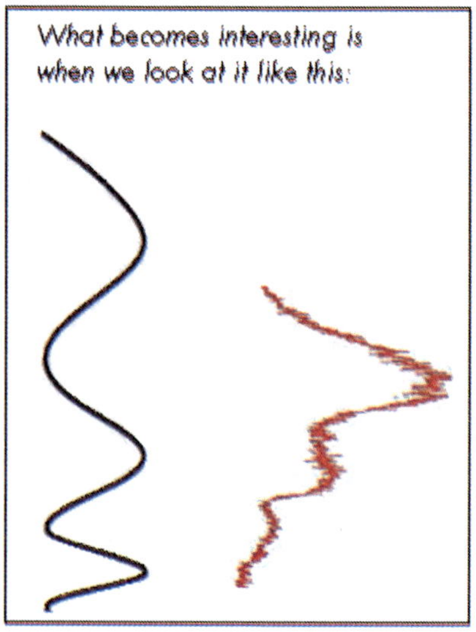

Here we can see this same growth in a living thing such as a plant when it grows from a seed into a sprout, then into a bud and lastly into a bloom:

I'd also like to bring to your attention again this same pattern in a shell. Take notice of the resemblance in the shells curvature to that of the Hebrew letters Ayin and Bet:

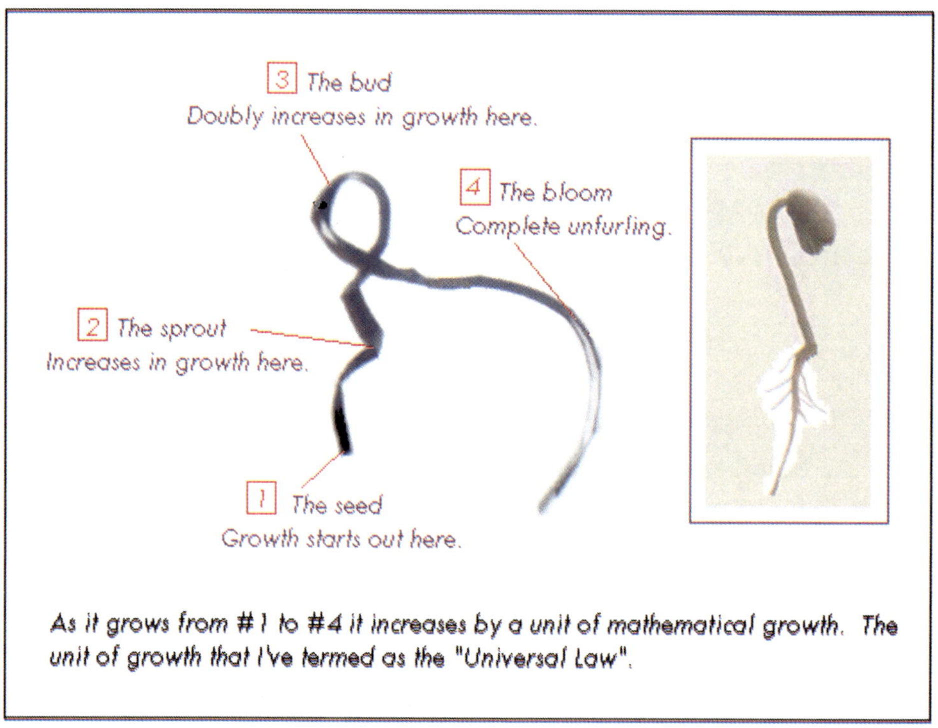

Hebrew letter Ayin.

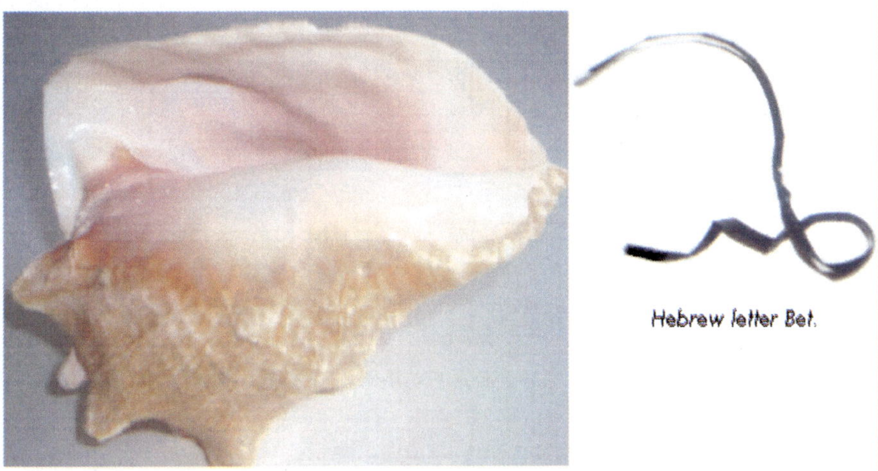

Hebrew letter Bet.

This spiral form is not meant to be frozen in perfect sturdy form, it is bendable, malleable, and in the process of continual creation. In effect, each letter in itself is a POINT OF VIEW, (a conceptual view) where Quantum Mechanics comes into play again. Remember in Quantum Mechanics (cited from a previous chapter), one cannot pinpoint the exact values for the position or momentum of a certain particle in a given space in a finite time, but, rather, it only provides a range of “probabilities” of where that particle might be. Therefore, it becomes necessary to describe these states as uncertain, unknown and unable to be measured, when it was recently discovered that the very act of just looking at these very particles in magnification, observing and measuring them caused the “atoms” to react differently and change course. A strange phenomenon occurs if you look at it in a certain way, it becomes one thing and if you turn away, it becomes something else. It seems the atomic particles have become AWARE! It has also been observed that the person performing the experiment actually influences the outcome.

Spiral thought patterns transpiring in the brain.

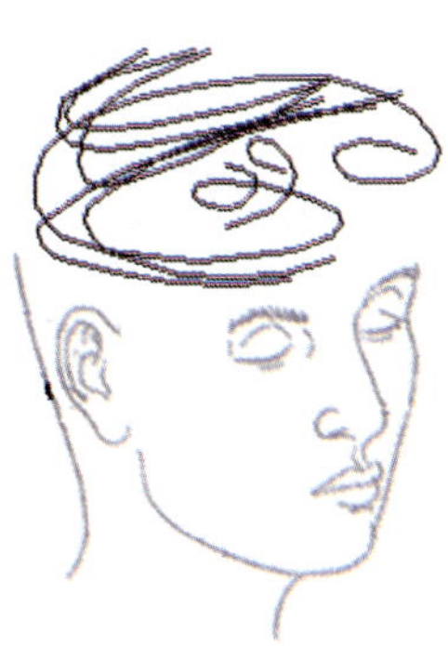

For this reason Quantum Mechanics is built on FAITH. Hebrews 11:1 in the New Testament it tells us - that FAITH is the substance of things hoped for, the evidence of things not seen. In Hebrews, 11:3 it tells us - that through FAITH we understand that the worlds were framed by the word of G-d, so that things, which are seen, were not made of things, which do appear. In addition, it was John 6:63 that said, - it’s the spirit that gives life, the flesh counts for nothing. What’s actually happening here is that these letters are "spiraling into meaning" by the conscious awareness of our thoughts. Aside from our everyday reality, they may even (to a certain extent) play a part in bringing about material form. In fact, it is even known that when the human brain is in thought the neurons are not only sparking, but also fleeting around in spirals. What starts out in the brain to appear like a somewhat rounded area of activity, (which can be seen illuminated during a brain scan) turns out to be what's called a "ring wave" that then collides with another wave and develops into spiral waves. In other words, these ring waves change into spirals. Spirals may also be the mechanism by which the brain organizes information (systemizing in growing unites). Therefore, when you have a brainstorm there may be many hurricane-like spirals occurring in your head.

Therefore, we can literally say, the Hebrew alphabet is an alphabet of FAITH! In which each letter holds its own dimensional probability (length, height, width, measurement, numerical value, meaning and unit growth pattern subjected to the laws of nature) depending on its given "spin" position. THE DIFFERENCE HERE is that these letters were intentionally "stopped" (paused by G-d within time) to communicate to mankind His divine word and His existence through language, thus confessing and baring witness that He alone is the creator. Could anything be exhibited and demonstrated more eloquently then this?! Since the Hebrew letters are based on Quantum Mechanics - hence FAITH, one can literally say that the Hebrew alphabet is not the exception, IT IS THE LAW! (As it applies to all things).

Not only that, we can now rest assure that the Bible is correct in all matters including every single aspect of it, because IT IS the literal "WORD" of G-d!

I had satisfactorily realized just how magnificent and holy the scriptures of the Torah were by the structure of the Hebrew letters (I cried out in joy), it is overwhelming! From this point on, I feared as well as respected G-d in a way I never had before. As for others whom now having the privilege of fully comprehending the significance of what they've just read and perceived with their own insightfulness - this knowledge should make a permanent impression upon you from which you are not supposed to ever recover from. From here on out, it will be a lifetime endeavor… nothing you look at will ever be seen in the same way again. You will begin to see the creators handiwork in everything around you, and ultimately the creators workings within you. This is the "Universal law and primordial language" at work! Zephaniah 3:9 - For then will I turn to the people a pure language, that they may all call upon the name of the Lord, to serve him with one consent.

And this is what is so rewarding about this hottest discovery… and this goes out to all the mathematicians, all the evolutionists, all the atheists, and to all those scientists and people who have been continually telling us all along that there is nothing significant in finding patterns. They insistently tell us that these types of things are just random and mean absolutely nothing. They go on to bulldoze the idea that just because we see a mathematical pattern in life (and in all things) does not in anyway apply to an intelligent designer. However AND HERE IS THE CLUE THAT IT DOES! As we have learned that the Hebrew letters are the result of a measurable unit of growth that is derived from Phi/Pi, Fibonacci sequence and Golden Mean Spiral (in a synergistic way). This in itself directly links the Hebrew letters to "intelligent design" that testifies of our creator. A creator that personally tells us (by way of these holy letters) that HE ALONE WAS THE ONE WHO CREATED IT ALL!

We should all be awe struck after realizing this actuality. With this information and knowledge presented before us, there is no way any of us can go on denying that a designer does not exist! What this awareness does is totally abolishes the atheist! From here on out, there will be no turning back for science. It literally means that we can tear up our notebooks and start more or less from scratch… especially for those who have a degree in mathematics or have studied calculus, number theory, real analysis, mathematical modeling, symbolic logic, physics and geometry. Additionally, having some knowledge and experience writing software coding and decoding algorithms. It should be startlingly and resplendently comprehensible that the mere notion alone that one can create all the 22 Hebrew letters from ONE PROTOTYPE letter demonstrates divine Orderliness with a capital O. Also with a capital D for Design. Then come to find out that this letter in itself (Bet - "B") has the mathematical formula that synergistically uses Phi/Pi, Fibonacci sequence and Golden mean spiral, and it does this in the shape of a spiral that furls out in unit growth. This evidence should be shell shocking to all humanity! This synergistic growth pattern can be described and labeled as several things such as "the One Universal law" (or plainly put, the Language of G-d). Hence the name of this book - The Primordial Language!

What is even more compelling to ponder on is the new research that is on the horizon to understanding just what language is. Amazingly, it has once again been found to fall into the Phi/Pi category. This is because all language type properties consist of a "tonal" language system (where changing the tone, changes the meaning), which can then be turned into visual graphics that depict a pictorial diagram of the bell curve again, (which can be a complex acoustic wave form). These acoustic wave patterns can be found in many "tonal" animals such as Dolphins, Whales, and even Prairie dogs. It can also be found in such things as the rhythm beats of the human heart, and even in the chemical compounds of insects. Subtle differences in the bell curves pictorial diagram produce letters and a language - The Holy Hebrew Letters! The Primordial Language.

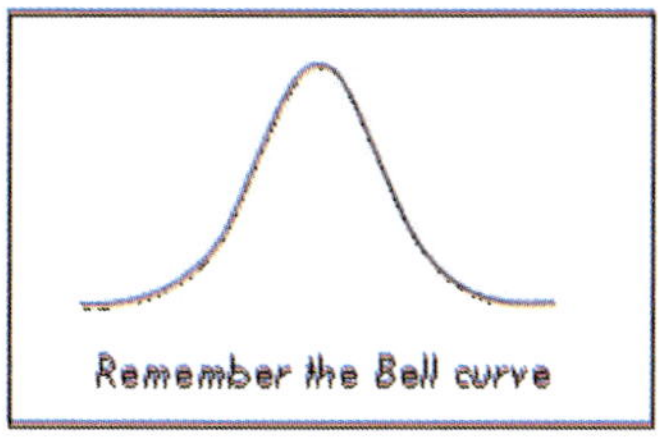

A Universal Language is all around us and we just haven't recognized it!

Before I go on to elaborate here, let me quickly add something that was left out of chapter 2 of this book (Religion verses evolution). With this chapters knowledge in mind, I can now wholly say to all those evolutionists (especially the teachers that continue to teach evolution) that their theory is mistakenly inaccurate, and flat-out blatantly wrong! Moreover, here's why. The whole premise of evolution hinges on the idea that given enough time and chance we can get life. In addition, this is brought on by insinuating that the earth is billions of years old. Oh the devil is crafty all right… he has teachers instructing students (in a little experiment) to take a hat and put all the letters of the alphabet in it from A to Z. Then the teacher, ask the students (without looking or thinking) to pull these letters out of the hat. What naturally happens on occasion is that by chance some students pull out letters in a row that spell something that they can recognize as a word, such as a letter C followed by a letter A, and followed by a letter T... The students arrived at the word CAT (a word by chance). See, (says the teacher) given enough time we could get a sequence of words by chance... and given enough time we could get sentences... and eventually given enough time we could get a whole information bank - and there it is folks, life!

We can even see the Bell curve in wood grain.

However, there is a fatal flaw in this analogy. First, who is the word CAT understood by? If it's an English speaking person, they will automatically know that it means CAT. However, if it were to a Spanish, French or Chinese person they would have no clue as to the word CAT. Therefore, this automatically tells us that it is only acknowledgeable to somebody who already knows the language. But this analogy is still incorrect even to the English speaking person that can recognize the word CAT, because there is another factor that exists in the cells of the DNA which comes into play... and these are chemicals that make up a language system that then makes up the order of molecules meaningful in the DNA. So in other words, it is "these chemicals" that read the language to make it meaningful. Therefore, you need the language to make the order of molecules and vice versa - you cannot have one without the other.

Now a chemical system that's found in the DNA consists of RNA, molecules, enzymes and proteins (down to the very atoms), which cannot even come into existence (develop, progress or flourish) without first conforming to phi/pi (3.14). This also includes all the elements in the periodic table (solids, liquids and gasses). And what is Phi/pi in terms of growth in nature? You guessed it, THE SPIRAL! And it is this magnificent "spiral" as we've come to find out, that forms the Hebrew letters and language - hence this is truly the language of G-d that makes all life possible! Therefore, without a doubt, there truly exists a G-d, and His signature (trademark) is on all things. So, as we can see life (DNA) does not really arise by chance, as the evolutionist teacher teaches. Perhaps now, the evolutionists may have second thoughts since the divine foot has stepped in the door.

Now to add the final touches... G-d in effect (as we've previously discovered) works through nature (Hatevah) to work His ways (and as you can also see the spiral in the wood grain picture above). G-d is not a physical being (though He could be, as He was through Jesus), He is a spiritual being of energy embodied within the Hebrew letters. However these letters are not the entity of G-d Himself, rather He uses these letters to facilitate His ways. The Hebrew letters existed before the Torah, even before Creation! The letters are the BUILDING BLOCKS OF CREATION. The Hebrew Bible tells us that G-d revealed Him self in several forms. In Exodus 3, G-d was the burning bush at Mount Sinai... Exodus 3:2 - And the angel of the Lord appeared unto him in a flame of fire out of the midst of a bush: and he looked, and, beholds, the bush burned with fire, and the bush was not consumed. In Exodus 14:19-20, He was the pillar of fire that stopped the Egyptians... Exodus 14:19-20 - And the angel of G-d, which went before the camp of Israel, removed and went behind them; and the pillar of the cloud went from before their face, and stood behind them: And it came between the camp of the Egyptians and the camp of Israel; and it was a cloud and darkness to them, but it gave light by night to these: so that the one came not near the other all the night. In Exodus 17:6, He stood before Moses at the rock at Horeb where Moses struck the rock and water came out… Exodus 17:6 - Behold, I will stand before thee there upon the rock in Horeb; and thou shalt smite the rock, and there shall come water out of it that the people may drink. And Moses did so in the sight of the elders of Israel. In Deuteronomy 9:10, Moses viewed G-d as a finger of Fire… Deuteronomy 9:10 - And the Lord delivered unto me two tables of stone written with the finger of G-d; and on them was written according to all the words, which the Lord spake with you in the mount out of the midst of the fire in the day of the assembly. In all our familiarity with the features of "fire" is that it too "spirals" - obviously because it has to conform to the "one universal law". We cannot literally look upon G-d and live! His actual being has no form; He is permeated within all of existence and is consciously functioning and operating through the embodiment of the SPIRAL forming the Hebrew letters that give rise to all physical material. Not only does it form, but it also speaks! The Hebrew alphabet is not merely just linguistic symbols, but the manual and blueprint for creating matter - the sole substance to all life. G-d generated substance from which He formed letters out of which He combined words, which became things - from substance out of chaos and make nonexistence into existence. This is THE PRIMORDIAL LANGUAGE! And that by definition - the one G-d - a unity underling everything. (Oh and by the way, the Hebrew letters are considered so holy and sacred that in Jewish tradition no document or book, even piece of paper with Hebrew writing on it, especially with the name of G-d on it, can be disregarded, disposed of, or thrown out into the garbage. There are certain preparations for getting rid of old written material and it's usually by burning or storing it forever in a room known as a "Genizah".

So lets quickly recap here… There are 22 Hebrew letters within the Star of David pattern shown above here in this chapter, the last 5 Hebrew letters are "finals". In addition to this, (and reconfirming) there are also 22 Hebrew letters that are formed from the prototype image of the Hebrew letter Bet. In effect, with this sort of uniformity and consistency we cannot merely just say that they are contrived by human invention or means since everything that I have revealed here implies that the Hebrew letters were created by nature, "through G-d" in utilizing this "one instructional law" which is the interweaving and unit growth mechanism that binds all of creation together. When in fact the Hebrew book of Jubilees written about 2000 years ago (found in the caves in Qumran in a settlement next to where the Dead Sea Scrolls were found mentions something about the origins of writing). Now what is profoundly special about the Jubilees manuscript is that it bears testimony that human beings did not create language or an alphabet. It distinctly tells us that Enoch was the first human being on the face of the earth to use writing, and he was given, shown, and taught this alphabet by G-d's appointed angels. Jubilees 4:17 - And he was the first among men that are born on earth who learnt writing and knowledge and wisdom and who wrote down the signs of heaven according to the order of their months in a book, that men might know the seasons of the years according to the order of their separate months. Therefore, we can confidently affirm that the Hebrew language and its alphabet is not man made, it is the language of Heaven. Additionally, the Hebrew letters are the "key sources" for any sort of unified theory, universal code, or language. Perhaps in the most eloquent and extraordinary way we end up with all there is, and will ever be. But more importantly (I want to get the strong message across), that we must be able to relate and implement the idea that an all-pervading spiritual entity is behind the whole shebang, creating an alphabet in which to communicate through us on a internally subconscious level, as well as a fully conscious external level if He so pleases... Otherwise, how can we logically explain that the universe itself speaks a language that is derived from nature, carved on stone and written upon the parchment of the Bible, openly announcing and PROCLAIMING A CREATOR!

I ran all of this past a very intelligent, scientific, and mind you ATHEISTIC person (the salt of the earth when it comes to smarts in this day and age of modern thinking and scientific knowledge). And their reaction to this discovery of the Hebrew letters was that it was from an alien source (intelligent beings from another planet). I suppose one could say that aliens could be intelligent enough to have played a role in our creation. However, the Bible tells us that G-d is of a spiritual energy - not alien being. I suppose we could probably go one step further to construe what it really means to be "alien". In other words, an alien might not be consigned to a physical body of sorts. But what doesn't make sense in this whole analogy is why an intelligent being (such as an alien) would go to the lengths of telling us through a language and its letters (compiled in a book called the Torah/Bible) PROFESSING to be the ONE-AND-ONLY BEING WHO CREATED AND MAINTAINS EVERYTHING THAT EXISTS (and along with this goes about telling us how He did that in Genesis). Lets not forget that we've just uncovered that these letters (from one prototype letter) are formed from nature (using a mathematical unit of growth) that is evident in not only things of the earth, but the distances between the planets, AND THE ENTIRE UNIVERSE. Therefore, it is of my opinion that no matter how intelligent aliens may be, they do not possess the ability to have control over the entire Universe, or to have created it all (especially only "one" particular alien who runs the whole show). I also do not believe that aliens would use a system (such as the Hebrew letters) that clearly employs "free will" which is "love" on a grandiose scale. No matter how hard we try to make an alien out to be G-d, a being of "love" is unsurpassed! Getting more to the point, the Bible speaks of NO OTHER LIFE elsewhere in the Universe except for on Earth. And we must also keep in mind that the Bible has never been found to be false, and G-d has never been found to be a liar! Thus, there are no aliens (the details of this will be discussed in a later chapter). If G-d says it, and affirms it through His word, I BELIEVE IT! To which I believe was adequately and satisfactorily revealed within this chapter! Fear only two: G-d, and the man who has no fear of G-d - Hasidic Proverb. In addition to backing this up, as to date, SETI (the search for extraterrestrial intelligence) confirms this as well - there have been NO SIGNALS received from an alien source from space, nor will there ever be (but we as humans beings will continue to search obstinately because that is our nature). In addition to this there has been NO OTHER LIFE found on any other planet (thus far) in our solar system (not even a single confirmed microbe).

Secondly, my atheist friend also threw into the mix the possibility of these Hebrew letters being a computer program that runs off a superlative imposed quantum computer. However, this analogy cannot fly either since all instructions come with intent. Someone who writes an instruction manual (or program) does so with purpose. However, my atheist friend went on to elaborate that computers in the future will have artificial intelligence and will be able to program themselves, think for themselves, and even do for themselves by repairing themselves. They will even have the ability to create by manipulating matter. In effect, they won't need humans to program them. They will also be immortal without the need of any sort of body. Nevertheless, even this does not make sense or equate to G-d adequately for the sole reason that if indeed G-d were a computer, He could eventually be manipulated. Since super computers in the future would have the ability to not only create matter, but they would also have the ability to manipulate energy, mass and time - thus "Time Travel" would be invented. And we will find out in a future chapter (the evidence for this) that there simply is no "Time Travel" (nor is it happening, or will it ever happen) because the prophesies in the Bible are being fulfilled. If "Time Travel" were possible, G-d's design for mankind would have been changed by evil forces long ago. Thus, G-d is NOT a computer or a system run by a program.

Moving forwards here... In regards to this divine alphabet, I will chillingly tell you that we have only begun to truly touch the tip of this iceberg. If you are not scared straight yet, YOU WILL BE! There is much more that remains hidden in the Bible concerning the matrix of reality, mathematics, codes, messages, images, including prophesy, we just haven't fully encroached upon it yet in this chapter, but will do so in the following chapters. Even after reading the entirety of this book, if one still has questions and if one continues to seek, I assure you, one will find the answers as I have. Matthew 7:8 - For everyone who asks receives; he who seeks finds; and to him who knocks, the door will be opened.

Adding an addendum to all this... When Moses received the stone commandment tablets written by the finger of G-d, (Deuteronomy 9:10) the letters must have been revisited and properly revised by G-d into their proper forms (gone astray from the time of Adam). As you will see the letter similarities in the following chapter 5. It is specified three times in the Torah that G-d wrote the letters and words. In Exodus 31:18: And He gave unto Moses, when He had made an end of speaking with him upon mount Sinai, the two tables of the testimony, tables of stone, written with the finger of G-d. Again in Exodus Chapter 32:16: And the tables were the work of G-d, and the writing was the writing of G-d, graven upon the tables. And again in Exodus Chapter 34:1: And the Lord said unto Moses: Hew thee two tables of stone like unto the first; and I will write upon the tables the words that were on the first tables, which thou didst break.

Now from that point on in history, according to, and in keeping with the sages, the five Hebrew books of Moses consisting of the Old Testament Bible (The Pentateuch or Havtorah) have purportedly been copied perfectly letter from letter IDENTICALLY, being preserved and kept in its original state throughout the centuries, adhering to strict observances, without any alterations or deviations, and without one letter (as rumor has it) ever being omitted, for the Jewish Torah writers were told that if they changed a letter of the text it would be the equivalent to the destruction of the universe, and so in the same respect if we took a unit of growth away from Fibonacci sequence, Phi/Pi or Golden Mean Spiral it could also calculate destruction for living things. The Torah itself is a spreadsheet outline, model and pattern for holding information together in a way that transforms it into something that is sentient. You see the "letters themselves" make the "words" which are then made into the story. When we come down to realizing it, the Bible (Torah) has created its own letters in which it was written, as Genesis 1:11-12 states, it is a fruit tree yielding fruit within itself. And we can very literally say that it is indeed imbibed with the presents of G-d.

By now, I hope that I have not fed anyone anything that is too strong to digest, but well enough to derive benefit from. I think It has been successfully demonstrated and shown, that for one thing "G-d" is certainly very real and in the mist of our reality through the use of the Hebrew letters, but more so that G-d has within his ability and power to intervene and communicate when He so desires. But when G-d chooses not to reveal himself, we as humans argue that there is no proof of His existence. We fall victim to disbelief, disconnecting ourselves from G-d. In doing so, we are therefore "separate" in a spiritual sense. Perhaps it would be best to say that from this standpoint, it is caused from man's lacking inability and refusal to acknowledge and conform to the workings of nature (Hatevah) and therefore is not in cooperation with the processes demanded by the organism to which we all belong, and therefore such a individual that does not operate in a manner that is in consonance with the intentions of its creator will be IGNORED. G-d is very real to the logical mind of those who seek. The evidence is out there and all around us, but one must learn how to integrate that evidence on their own. No one can do it for you, since the topic is the eventual source of all.

Perhaps it is the scientific exploration of Torah that G-d has wanted from us from the beginning... His will for man... What G-d is in theory, man must realize in practice.

9 - LANGUAGE SIMILARITIES...

I was faced with the dilemma that this chapter should have come before the previous one because it would have given you somewhat of a stronger disposition and respect for the Hebrew letters and language. However, if I had done so, you would have been lost wondering what I'm trying to get at. This way, I can now reinforce, and therefore build reassurance as well as accountability that Hebrew is the one and only "original" (true and G-d-given) language.

So let us begin... Many people wonder why it is important to be interested in the Hebrew language that is only spoken by a small amount of people on the planet. Although, I think all of the abovementioned has just effectively expresses why. Whilst opening this worldview up even more, I am going to expand on the Hebrew alphabets true origin and familiarize everyone on its linguistic history.

Off hand, if we were to employ the evolutionist thinking about the origin of language, it tells us that language started from primitive apelike beings making grunts and coos (which let me remind you, has never been proven in our midst). Though I don't subscribe to random events, language and speech, in my opinion is a unique attribute embedded within humans. When asked what separates humankind from all other life forms, the Sages answer is SPEECH. We find in Genesis that the world was created with a word (Speech). From Aleph to Tov, that is all the letters of the alphabet - AhTah (as aforementioned above as G-d's accurate alphabetical name. And by this, I mean, in the arranged order of alphabetic letters). In connection with this, there are 5 organs by which speech is produced which are, larynx, palate, teeth, lips and tongue. The Hebrew letter "Hey" is the "fifth" letter in the Hebrew alphabet and is numerically equal to five. Its phonic meaning stands for "behold, breath, sigh" (as one does when looking at a great sight). The sense of this passage is that AhTah (G-d) creates with His "breath" represented by the 22 divine letters of the Hebrew alphabet as they issue from the five organs of speech. Genesis 2:7 - The Lord G-d formed the man The Hebrew for man (Adam) sounds like and may be related to the Hebrew for ground (adamah) it is also the name Adam from the dust of the ground **and breathed into his nostrils the breath of life, and the man became a living being.**

It might be quite an engaging shocker to realize that all languages are rooted from one prototype language with the perspective in mind that perhaps the Bible account of Adam and Eve is not taken to far out of context - that G-d spoke a language to them in the garden (as specified in Genesis) and they understood, spoke and wrote this same conceptual language and handed it down to the following generations. Qabalistic tradition holds that Adam understood this supernatural alphabet and the formation of the universe through the powers of the 22 letters of the Hebrew alphabet, but lost much of it when he fell from Paradise. The

Bible states that when G-d spoke the universe and our earth into existence he used a language. It also states that before "Noah's flood" there was only one language. As a consequence, throughout the centuries the Hebrew letters were copied in similarity and style but through its scattered usage, it became mottled, broken and disjunctive UNTIL IT WAS PROPERLY REVISED AGAIN BY G-D (aforementioned in the previous chapter) when Moses received the stone commandment tablets written by the finger of G-d. Deuteronomy 9:10 - And the Lord delivered unto me two tables of stone written with the finger of G-d; and on them was written according to all the words, which the Lord spake with you in the mount out of the midst of the fire in the day of the assembly. Again also found In Exodus 31:18: And He gave unto Moses, when He had made an end of speaking with him upon mount Sinai, the two tables of the testimony, tables of stone, written with the finger of G-d. Again in Exodus Chapter 32:16: And the tables were the work of G-d, and the writing was the writing of G-d, graven upon the tables. And again in Exodus Chapter 34:1: And the Lord said unto Moses: Hew thee two tables of stone like unto the first; and I will write upon the tables the words that were on the first tables, which thou didst break. So the message is clear here that something supernatural took place on those stone tablets that Moses hewed out of the mountain for G-d that transformed the Hebrew letters into their square/block style that we see today reflected within the Torah. The scholars theorize that after the exile from Egypt the Jews canonized, collected and pooled together what was left of their holy scriptures. Thus, the Torah (five books of Moses) was compiled and the old Hebrew manuscripts were transcribed into a New Hebrew text. Yet, there is no evidence supporting this theory since no other known alphabetical books or multi-page manuscripts have been found from the past, only bits and pieces of inscriptions, and none resembling the square/block style that surfaced with the giving of the Torah at Mount Sinai. Bypassing the natural evolutionary trail, this style of writing seems to have appeared suddenly, and to the scholars this is an inexplicable mystery - eerily like the reporting in Genesis relating to the "Tower of Babel". Another suggestion is that the Hebrew letters are square/block in form because the Torah writers used reed/quill pens, which were distinctive of, pointed slanted and broad tips and wrote on an angle thus producing the square/block script. It is collective knowledge that the Torah writers did use reed and quill pens to write the Hebrew letters onto parchment, but this does not indicate that the letters were not originally given to Moses that way, because if one examines the reed or quill pen they would notice that its point is clipped in such a way that it also comes into play with Phi/Pi. Not that it is Phi/Pi, but that it displays the characteristic of the "universal law". (Which is classified as anything that expands from a "point", multiplies, and propagates outward like that of the Tetrahedron shape)?

Notice the unit of expansion at the tip of the Quill pen.

There is no real way for the scholars - researchers of Paleography (the study of ancient handwriting and manuscripts), or Graphologists (the study of handwriting), or even linguistic specialists to determine which alphabet (or language) "predated" the other, or the true origins of language itself - they can only surmise. That is because all alphabetic letters (and spoken words) of the modern day Hebrew alphabet seem to look and sound similar to all other alphabets (and languages) across the globe. Perhaps the scholars need to focus their attention the other way around, realizing that Hebrew was the predecessor of them all. The question should be asked, why was nobody in recorded history accredited for inventing or writing language? It is inconceivable that someone would invent an alphabet and language and his/her name would not be remembered. To me this indicates that languages and alphabetical letters were not invented by any human - but by that of G-d.

A close examination of the similarities of worldwide alphabets reveals that at a given time in history we as humans spoke and wrote one language. Starting with the oldest "suspected" alphabets all the way to the present day; one can easily see the similarities in all these letters. Though there is always room to refute their similarities, I have concluded that they are all forms of confused Hebrew that became mottled, broken, transposed and disjunctive until it was properly revised again by G-d when Moses received the stone commandment tablets written with the finger of G-d Himself.

Examples cited below from the earliest (suspected) time of writing starting approximately 3,000 to 4,000 years ago, to the recent last 100 years (though according to the Bible, writing has existed since the time of Adam and Eve which is approximately 6,000 years). Displayed below are all the written alphabetic letters (in every single dialect and written work) found documented from across the ENTIRE GLOBE! Cuneiform, Chinese and Mayan have also been displayed separately below for those who claim that they share no commonality and are completely different. And additionally Egyptian hieroglyphics, which are pictographic.

Before you look upon these letters, let me clarify again that specific dates cannot really be attached to them (for which one predates that of another) since I don't believe there was ever a time in the history of writing that it truly changed. Writing has remained analogous (meaning similar, comparable, or related) which will become evident once you view these letters. However, scholars can say, look we can associate these types of style letters with a certain group of people that lived in a specific era. However, no linguistic scholar with any credibility or certainty can say with collaborating support which writing style was the first of its kind. Scholars can say the most primitive of writings are those that are pictographic, or use dots, dashes or squiggles - but this is not so, it remains speculative. But we need not guess because the Bible is wide open about this… it clearly tells us that the Hebrew letters where given to Moses by G-d. And as we will observe are found to be comparable to other letters (of its same category and kind) they haven't changed much over a span of 3 to 4 thousand years. Now I would like you to look below and focus on each letters resemblance, similarity, and relationship to that of its original prototype letter. These original prototype letters where the ones given to Moses by G-d (they are circled in red).

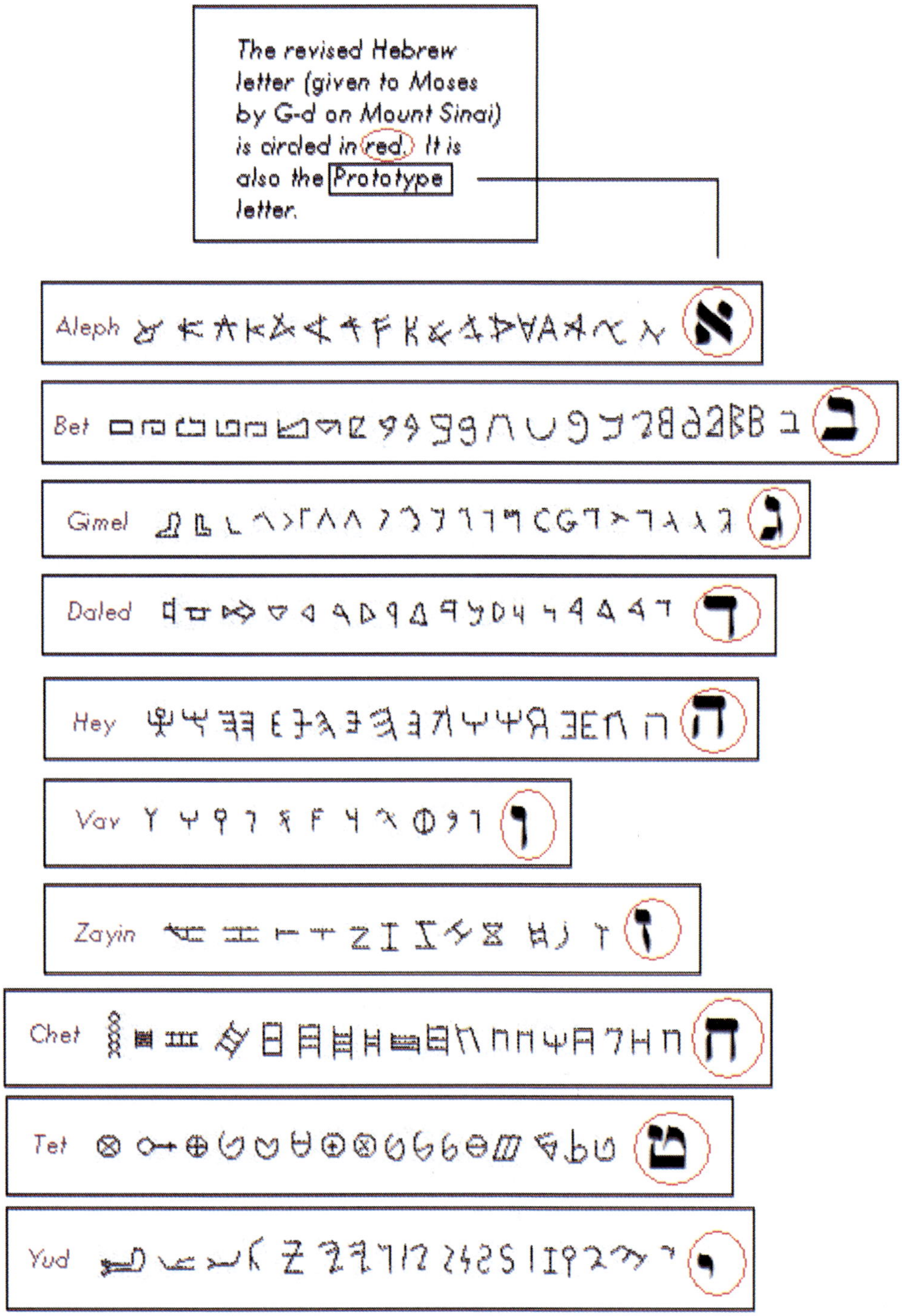
The revised Hebrew letter (given to Moses by G-d on Mount Sinai) is circled in red. It is also the Prototype letter.
Aleph
א
Bet
ב
Gimel
ג
Daled
ד
Hey
ה
Vav
ו
Zayin
ז
Chet
ח
Tet
ט
Yud
י

Now if you have carefully studied the letters above you would have noticed each letters similarity with that of the prototype Hebrew letters. Additionally you should have noticed that some of the same letter forms (shape) look like each other (or that of another letter) showing huge similarities. Which goes to prove that even though they may have transposed, swapped or switched order (one letter with that of another), or became mottled, sloppy or slap dashed through scattered usage, they are found (to this very day in history) to be comparable. Overall, they are still the same basic letter forms used over-and-over again, in excess over a large span of earths handwriting history (Graphology). Nothing really has changed, and this is significant evidence for an original language.

As you can see below, Cuneiform (also known as Ugarit and Akkadian) is usually the form of the Hebrew letter that's transpositional to that of the Hebrew letter, which is usually shown (carved or imprinted in rock) either sideways, or up or down (in other words the letters have just swapped position). Though it's not the complete alphabet, we can certainly see the unmistakable resemblance to that of several Hebrew letters:

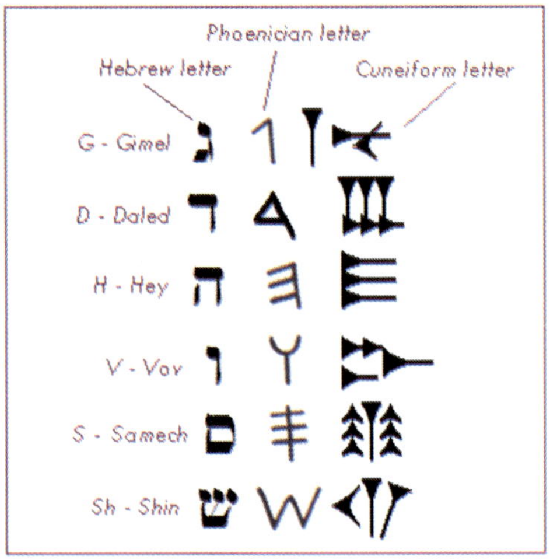

Now displayed below is several of the Chinese alphabet letters that show similarity with several of the Hebrew letters:

Similarities can even be found in the Mayan alphabet:

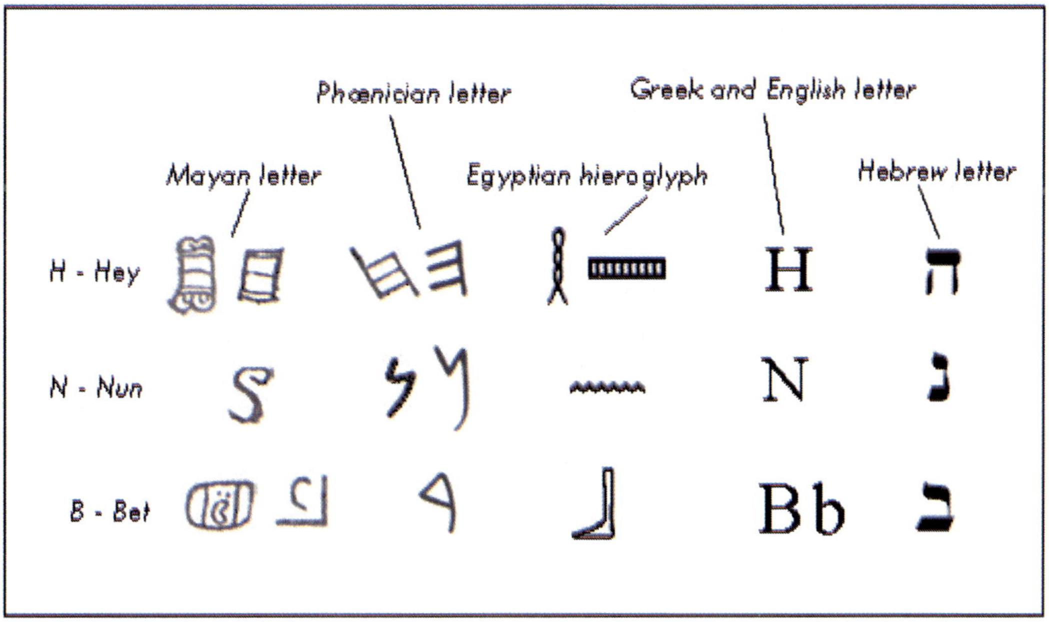

So once again as you can see, nothing has really-truly changed - NOTHING! Some letters are switched and some letters are similar, and some have swapped position, but for the most part, they remain incredibly similar. That is because they all share an unmistakable linkage to that of the original (spoken and written) language of the Earth, which was HEBREW! However, remember as was learned in the previous chapter, Hebrew is just not relegated to the Earth alone, it is relevant to the entire Universe! (Since it was these very letters that G-d used to create everything in existence).

Well with that being said, some thinkers have chosen to no longer just see the Hebrew letters as mere scribble lines or shadows, but the final clinching substance of proof.

Now lets look at the Egyptian hieroglyphics. The hieroglyphs represent the sound of the object or represent an idea associated with the object. The pictographs are figurative, symbolic, and phonetic. The scholars claim that the Egyptian hieroglyphs came about from the Phoenician alphabet, which is considered an early form of Hebrew. But there is something else very concealing (suppressed, masked and censored) about the way the Egyptians went about using this method of pictures instead of using the actual letters. The reason for this will reveal itself in the next chapter, but I will briefly tell you that when the Hebrew alphabet was introduced in Egypt the priests concealed (or obscured) what they called "hallowed characters" (or what would be considered ineffable, inexpressible, unutterable, beyond words) and thus submitted hieroglyphs in place of the Hebrew alphabetic characters that we see today for the sake of preserving what they considered holy. Just why they did this will be further explained in the following chapter.

 *An Egyptian vulture - The letter A (silent) which would be equivalent to the Hebrew letter Aleph.* א

 *An arm - Also the letter A which would be equivalent to the Hebrew letter Ayin.* ע

 *A leg - The letter B which would be equivalent to the Hebrew letter Bet.* ב

 *A quail chick - The letter W, Wu, Wuv which would be the equivalent to the Hebrew letter Vuv. (Sometimes pronounced Wuv).* ו

 *A viper - The letter F or V which would also be equivalent to the Hebrew letter Vuv.* ו

 *A reed, mat or stool - The letter P which would be equivalent to the Hebrew letter Pey.* פ

 *A reed - The letter E,Y, and sometimes which would be the equivalent to the Hebrew letter Yud.* י

 *A Owl - The letter M which would be equivalent to the Hebrew letter Mem.* מ

 Water - The letter N which would be equivalent to the Hebrew letter Nun. נ

 *A Mouth - The letter R which would be equivalent to the Hebrew letter Resh.* ר

 *A reed shelter - The letter H or Ch which would be equivalent to the Hebrew letter Hay or Chet* ה כ

 *A twisted wick - The letter H which would be equivalent to the Hebrew letter Hey.* ה

 *A placenta - The letter Kh which would be the equivalent to the Hebrew letter Kaph.* כ

 *An animal belly with tail - Also the letter Kh which would be the equivalent of the Hebrew letter Kaph.* כ

 *A folded cloth - The letter S, Sh, Ch which would be the equivalent to the Hebrew letters Shin, Samech, and Chet.* חסש

 *A door bolt - The letter S,Z,SZ which would be the equivalent to the Hebrew letter Zayin.* ז

 *A garden pool - The letter S,Sh,Ch which would be the equivalent to the Hebrew letter Shin, Samech and Chet.* חסש

 *A slope of a hill - The letter Q or K which would be the equivalent of the Hebrew letter Quph.* ק

 *A basket with a handle - Also the letter K which would be the equivalent of the Hebrew letter Quph and Chet.* כק

 *A lion - The letter L which would be the equivalent to the Hebrw letter Lamed.* ל

 *A jar stand - The letter G which would be the equivalent to the Hebrew letter Gimel.* ג

 *A bun - The letter T which would be the equivalent to the Hebrew letter Tet, Tuv and Tsadi.* צתט

 *A tethering rope - The letter T,Th,Ch,Sh which would also be the equivalent to the Hebrew letter Tet, Tuv, Chet and Shin.* שחתט

 *A hand - The letter D which would be equivalent to the Hebrew letter Daled.* ד

 *A cobra - The letter Dj which would also be the equivalent to the Hebrew letter Daled.* ד

Here is something else interesting (take a look below). Even today, the English letter A/a has a profound significance to an earlier time in history. Ever wonders where the English letters got that funny looking letter "a"? Well, it's actually a form of the Egyptian letter A/a that was originally a bird that looked like this:

Look familiar?

Now listed below is the Hebrew alphabet with 22 letters and their particular meaning, along with their number equivalent:

Hebrew letters, their meaning and their numbers values

- א A-Aleph-1- Ox, Bull, strengh, leader, first.
- ב B,V-Bet-2-Tent, House, in, into.
- ג G,Gh-Gimel-3-Camel, pride, to lift up.
- ד D-Daled-4-Door, pathway, to enter.
- ה H-Hey-5-Window, sigh, to reveal, breath.
- ו V,U,W-Vav-6-Stake, Nail, Hook, secure.
- ז Z-Zayin-7-Sword, Weapon, cut, to cut off.
- ח Ch-Chet-8-Fence, Hedge, Chamber, private, to separate.
- ט T-Tet-9-Serpent, Snake, to twist, to surround.
- י Y,I,J-Yud-10-Palm, closed hand, deed, work to make.
- כ K,Kh-Kaph-20-Arm, Wing, Palm, open hand, to cover, allow, strength.
- ל L-Lamed-30-Ox, Cattle, Goad, Staff, prod, go toward, tongue.
- מ M-Mem-40-Water, massive, overpower.
- נ N-Nun-50-Fish, activity, life.
- ס S-Samech-60-Prop, support, turn.
- ע A,Aa-Ayin-70-Eye, see, know, experience.
- פ P,Ph-Pey-80-Mouth, speak, open, word.
- צ T,Tz-Tzadi-90-Fishhook, harvest, desire.
- ק Q-Quph-100-Ear, back of the head, behind, the last, least.
- ר R-Resh-200-Head,person, highest.
- ש Sh-Shin-300-Tooth, Teeth, fire, consume, destroy.
- ת T,Th-Tov-400-Sign, Cross, covenant, to seal.

So, what happened to the Hebrew language and alphabet? How did it moderately change and become dispersed and dissimilar, (though comparable) across the entire globe? Again, the answer to this lies within the Bible. I will tell you for a second time (as was previously mentioned in an earlier chapter), it's certainly not as the evolutionists would claim that language evolved from simple grunts and coos. And we can now assess that language was not derived from genetic evolutionary means since it can be shown geographically (and now with similarity) how one language was derived from another. In other words, if linguistic evolution did take place, there WOULD NOT be any relationships at all between languages, and as you can see that is not what we find here (there are many similarities). The Bible goes out of its way to distinctly tell us about what happened to language that changed it into different languages. In addition, it all starts after the great worldwide flood with Noah and his three sons, Japheth, Shem and Ham. Genesis 9:1 - Then G-d blessed Noah and his sons, saying to them, be fruitful, increase in number, and fill the earth. Genesis 9:18-29 - The sons of Noah who came out of the ark were Shem, Ham and Japheth. These were the three sons of Noah, **and from them came the people who were scattered over the earth.** Genesis 11:1-2 - **Now the whole world had one language and a common speech.** As men moved "eastward", they found a plain in Shinar and settled there. All the descendant of Noah moved "eastward" to Shinar. This became the Tower of Babel.

Listed here are the generations and dispersions of the sons of Noah after the flood. Genesis 10:1-32 - Now these are the generations of the sons of **Noah, Shem, Ham, and Japheth**: and unto them were sons born after the flood. **The sons of Japheth**; Gomer, and Magog, and Madai, and Javan, and Tubal, and Meshech, and Tiras. And the sons of Gomer; Ashkenaz, and Riphath, and Togarmah. In addition, the sons of Javan; Elishah, and Tarshish, Kittim, and Dodanim. By these were the isles of the Gentiles divided in their lands; every one after his tongue, after their families, in their nations. **And the sons of Ham;** Cush, and Mizraim, and Phut, and Canaan. And the sons of Cush; Seba, and Havilah, and Sabtah, and Raamah, and Sabtechah: and the sons of Raamah; Sheba, and Dedan. In addition, Cush begat Nimrod: he began to be a mighty one in the earth. He was a mighty hunter before the Lord: wherefore it is said, Even as Nimrod the mighty hunter before the Lord. And the beginning of his kingdom was Babel, and Erech, and Accad, and Calneh, in the land of Shinar. Out of that land went forth Asshur, and builded Nineveh, and the city Rehoboth, and Calah, And Resen between Nineveh and Calah: the same is a great city. And Mizraim begat Ludim, and Anamim, and Lehabim, and Naphtuhim, And Pathrusim, and Casluhim, (out of whom came Philistim,) and Caphtorim. And Canaan begat Sidon his first born, and Heth, And the Jebusite, and the Amorite, and the Girgasite, And the Hivite, and the Arkite, and the Sinite, And the Arvadite, and the Zemarite, and the Hamathite: and afterward were the families of the Canaanites spread abroad. And the border of the Canaanites was from Sidon, as thou comest to Gerar, unto Gaza; as thou goest, unto Sodom, and Gomorrah, and Admah, and Zeboim, even unto Lasha. These are the sons of Ham, after their families, after their tongues, in their countries, and in their nations. **Unto Shem also**, the father of all the children of Eber, the brother of Japheth the elder, even to him were children born. The children of Shem; Elam, and Asshur, and Arphaxad, and Lud, and Aram. And the children of Aram; Uz, and Hul, and Gether, and Mash. And Arphaxad begat Salah; and Salah begat Eber. And unto Eber were born two sons: the name of one was Peleg; for in his days was the earth divided; and his brother's name was Joktan. And Joktan begat Almodad, and Sheleph, and Hazarmaveth, and Jerah, And Hadoram, and Uzal, and Diklah, And Obal, and Abimael, and Sheba, And Ophir, and Havilah, and Jobab: all these were the sons of Joktan. And their dwelling was from Mesha, as thou goest unto Sephar **a mount of the east.** These are the sons of Shem, after their families, after their tongues, in their lands, after their nations. These are the clans of Noah's sons, according to their lines of descent, within their nations. From these the nations (divided) spread out over the earth after the flood.

Because of the great iniquity (which I will soon talk about) G-d scattered all the people of Babel abroad "every one with a different language". They paired off and ventured (set out) to other places of the globe, and from these people ALL the languages and different ethnic and racial differences (as well as appearances) developed. Genesis 11:1-9 - **And the whole earth was of one language, and of one speech.** And it came to pass, as they journeyed from the east, that they found a plain in the land of Shinar; and they dwelt there. And they said one to another, Go to, let us make brick, and burn them thoroughly. And they had brick for stone, and slime had they for morter. And they said, Go to, let us build us a city and a tower, whose top may reach unto heaven; and let us make us a name, lest we be scattered abroad upon the face of the whole earth. **And the Lord came down to see the city and the tower, which the children of men builded. And the Lord said, Behold, the people is one, and they have all one language; and this they begin to do: and now nothing will be restrained from them, which they have imagined to do. Go to, let us go down, and there confound their language, that they may not understand one another's speech. So the Lord scattered**

them abroad from thence upon the face of all the earth: and they left off to build the city. Therefore is the name of it called Babel; because the Lord did there confound the language of all the earth: and from thence did the Lord scatter them abroad upon the face of all the earth.

So, what was so bad (or evil) about the Temple of Babel? The Bible is vague on this. But we do know that the actual word Babel "means" corruption and we also know that all the people on earth were in this "eastern" region and spoke only one language. What's more, if they only spoke one language, than it's logical that the whole society would function as one world order (government). This according to G-d (one world order) is not good, as we observe the opposite in nature. Nature is full of diversity as we witness this in animals, insects and many plant species in nature (as mentioned in an earlier chapter). Diversity is what makes us strong, and it insures us against extinction. Without diversity this would give Satan a perfect foot in the door to take hold, corrupt and destroy all humankind. Satan wants unity in error, and a government that consists of organized blasphemy and irreverence towards G-d. This is why G-d had to scatter, separate and disperse humankind! You see, Satan prefers an organized one world order were it's easier to control, while G-d prefers division to preserve His "truth". We know this to be true because Jesus said in Luke 12:51 - Do you think I came to bring peace (unity) on earth? No, I tell you, but division. (If you read this book again, you will appreciate a more in-depth understanding of this verse).

You see, diversity and division (though they seem harsh) is infinitely more precious than a satanic unity. Therefore, G-d's demolishment of equal unity was a blessing. It was an incredible outworking of purpose that someday we will all praise Him for, in His infinite mercy towards the human race to prevent us from plummeting into endless annihilation and depravity. G-d demolished the tower project by placing a protective barrier and stronghold within the human race against evil destruction. This was called "linguistic cultural and racial barriers". Thus, the confusion of languages put a crashing halt on any further joint effort to complete the tower. This had to be implemented by G-d so that in Satans attempt to take over the world he cannot penetrate the WHOLE human race with one sharp blow and completely destroy it instantaneously. (To appreciate the lesson of what happened at the Tower of Babel, one should marvel at how the amazing mechanisms of nature work in-sync with G-d to insure survival).

After language was confused, the people then started to drift away from each other into isolated groups that then drifted off to all other places of the world. In which each individually isolated group inbreeded with each another. When this occurred the dominant gene pool was genetically shattered allowing for the previously hidden, suppressed and recessive genes the opportunity to become visible in the physical features of the people.

This is how we have new varieties of dogs and cats. It's by isolating, inbreeding and allowing the recessive hidden genes to express themselves, and then those qualities and traits can be preserved by further inbreeding. G-d did this same exact thing to the human race. So not only do we have language barriers, but now we have double barriers because we don't even look alike anymore, and many evolutionary scientists have even gone so far as to associate black people as the closest resembling relatives to the apes. This is pure bunk! This view is adamantly a godless worldview.

At this time, I would advise, if this book has had a profound effect on you thus far, perhaps it would be wise to stop here and reflect for a while, several hours, days or months as I have. When you choose to continue, I will give you heads-up that the language similarities do not stop here, they continue in the following chapter of Egypt's secrets. It is a long story since the beginning of recorded time, man like the rival animals of this ecosystem, in his nature has always sought to achieve, advance and dominate, and it has always been a battle and power struggle among men throughout history to rule. Even man in this civilized generation, there has never been a period of history when some individual, group or race was not planning, or conspiring to take advantage of another individual or race. "We flatter ourselves by claiming to be rational and intellectual beings, but it would be a great mistake to suppose that humans are always guided by reason" - quoted by - Sir John Lubbock (1834-1913). Therefore, before you continue in this book it must be brought to your attention that this "Primordial Language" that G-d freely displays before us, may also have been skillfully constructed "as a weapon of self defense" (rather a mechanism with influential power over the earth) built within an altar (pillar) unto G-d.

10 - EGYPT'S SECRETS...

This is where things began to get very interesting! Amazed by my discovery of the Hebrew letters, in my curiosity I delved deeper... I felt that there must be some other clue and connection to something in history that might apply. I was looking for some object that also employed Phi/Pi, Fibonacci, Golden Mean Spiral and the "universal law" that is observed throughout all of nature, and that it might harbor the technology that applied the usage of the Hebrew letters.

Sure enough, I landed on another grand discovery! Among other things found in nature, a man made monument was built centuries ago in Egypt on a huge landmass consisting of limestone and granite. This structure is known as the Great Pyramid of Giza. One of the 7 wonders of the world!

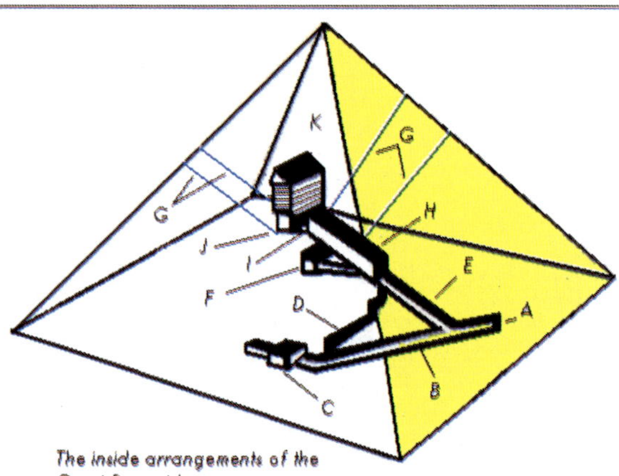

The inside arrangements of the Great Pyramid.

- A - Entrance
- B - Descending passageway
- C - Subterranean chamber
- D - Well shaft
- E - Ascending passageway
- F - Queen's chamber
- G - Ventilation shafts
- H - Grand gallery
- I - Antechamber
- J - King's chamber
- K - Relieving chambers

The Great Pyramid bares the comparable "universal law" and proportional measurements of Phi/Pi and Golden Mean Spiral. It exhibits these ratios from its exterior to its interior. I would like to draw your attention to the "interior" while focusing your attention on the passages, shafts and chambers and the "shapely outlined pattern" that all of these intersecting parts form.

Its interior shape should look familiar to you, as it bears the likeness of the Hebrew letter "Ayin". No not exactly, but if you stop to consider that "Ayin" is also derived from the "universal law" in its geometry as well as the innards of the pyramid itself, one can say it measures up quite comparably. Look at the pictures below to see this comparison.

Hebrew letter "Ayin".

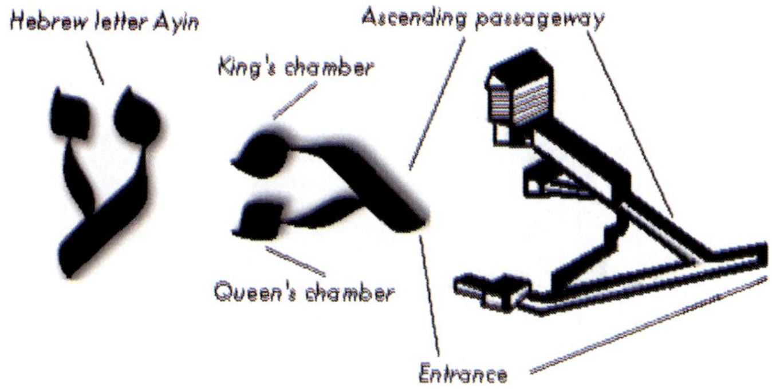

The Hebrew letter Ayin turned sideways is in resemblance to the inside arrangements of the Great Pyramid.

Great Pyramid with Hebrew letter Ayin inside.

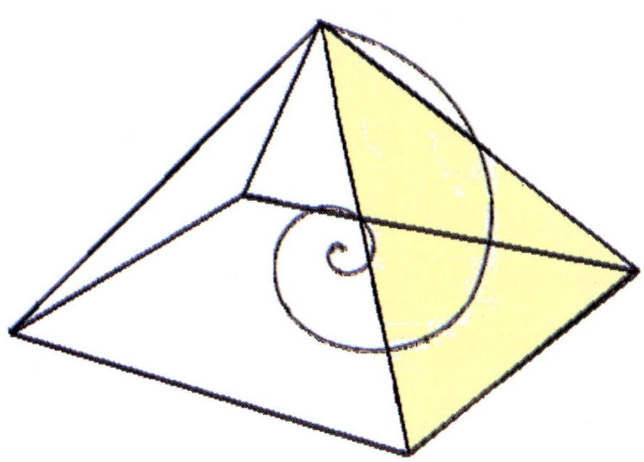

Great Pyramid with the Golden Mean Spiral inside.

Additionally notice within the Great Pyramid how the Golden spiral forms within the precise positions of the main chambers (both the King's chamber and the Queen's chamber), which are the two heads of the Hebrew, letter "Ayin". More notable the King's chamber where the Golden Mean Spiral comes to its smallest point. This precise location is Phi/Pi (the vortex) where all the culminated energy comes into play.

Now some might say this is foolhardy or impulsive to assert such glaring similarity, however, by employing the Hebrew letter "Ayin", which if you look back in the previous chapter on “language similarities” you can see for yourselves that the Hebrew letter “Ayin” represents an “actual eye“, and as we have already learned in the chapter of “the primordial language” that the Hebrew letters, (and therefore “Ayin“) is formed from “natures universal law“. The Pyramid, in fact, may really be the all Seeing Eye! (Noted as the Egyptian Eye of Ra). It’s even denoted that way on the American dollar bill within a pyramid (placed purposely on there to denote the Freemasons). Some say the founders of America were “Freemasons” who studied Qabalah. (I will mention more about the Freemasons in detail in the upcoming chapter on Qabalah). Stretching the meaning of the Hebrew letter “Ayin” a bit further… It symbolically also signifies that G-d is always “looking” from the beginning of the year to the end of the year. Thus signifying the yearly cycle from beginning to end which represents “the endless cycle“ that G-d is always in control of. (As you continue to read this will also doubly reveal why it was employed within the Great Pyramid).

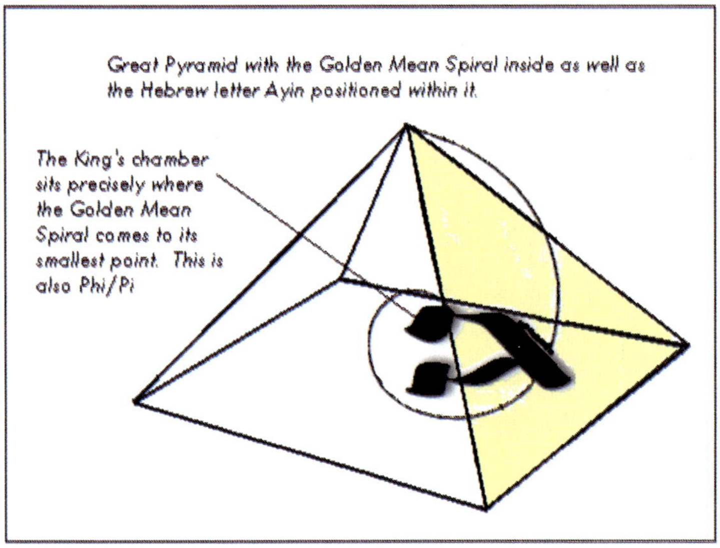

One might ask what is the significance of the Hebrew letter “Ayin” matching up to some degree, or better yet what is the significance of finding the Hebrew letter “Ayin” within the Great Pyramids innards? There are those that might laughingly chalk this off as mere a coincidence. However, in spite of this, it may soon be realized that the compelling effects of “Ayin, Phi/Pi and the Golden spiral” within the Great Pyramid might have a powerful influence over the physical world of our planet. Furthermore we will soon gather why “Ayin“ was purposely constructed and employed into the Pyramid for reasons I will eventually explain.

Case in point, the Hebrew letter “Ayin” HAS ALWAYS REPRESENTED AN “EYE” THROUGHOUT WRITTEN LANGUAGE HISTORY and has often been associated with the Great Pyramid of Giza in both an overt and covert type of way! This is suggesting several things, that the modern day pharaohs who used pictographic symbols as an alphabet (even though its roots are from Hebrew) were not the original builders of the Great Pyramid. Furthermore that the Great Pyramid is older than previously though. (I will get to that shortly).

Though few entertain the idea of just why they built all those shafts, tunnels and chambers in the LOCATIONS they did… there must have been a logical reason for their precise architecturally planned locations? I don’t think they were just built on a whim; they must have been put in those locations for functional reasons. I find it interesting that nobody has speculated on this reason yet. Perhaps they were constructed in their location for stability reasons - and those stability reasons are merely a model of “natures universal law” which durably and sturdily holds up because it is the path of least resistance (which I mentioned previously in regards to why a plants leaves grow the way they do). And it is seemingly obvious from the pictures I supplied above that Phi/Pi was a strong benefactor that took part in the building locations of the two main chambers (King and Queen chambers). Pi/Phi is actually a model of Darwin's theory at work, where the strongest survive.

But for a brief segment lets turn our attention to another arena… Researchers to this day still cannot even answer just what the Great Pyramids purpose or function was? Some say it was built as a tomb for the Pharaohs - turning them into gods in the afterworld, yet no pharaohs have been found in it. There have been MANY speculations and some I will further talk about in the paragraphs ahead, but I must emphasize that they are all speculations with no evidence to show for. Everything seems to equate to dead-ends guesswork. Aside from this, it would seem logical to me that if a nation of people were to build a immense monument like this, and have numerous generations of their populace putting their blood, sweat, heart, might and soul into it, it had better be for a good reason!

We must also realize that Egypt is a empire that has been taken over many times by wars and therefore renovated and refurbished to the new rulers likings, thus things can become very confusing to the archeologists who keep discovering new things with differing chronological dates. However, it should also be known that to this day nobody can even accurately date the age of the Great Pyramid. They try and carbon date material objects that they find like Khufu's boat that was found buried alongside the Great Pyramid. But (remember) one cannot carbon date the pyramids themselves when they're made out of stone (limestone and granite) an inorganic material (dead and non-living material). However, it can be carbon dated by its mortar (containing organic elements such as bits of charcoal, reeds and wood which were found between the mortar), but there have been a whole slew of highly anomalous different dates ranging all the way from 10,500 BC and digressing all the way down through the centuries, to the recent mortar repairs made by the Supreme Council of Antiquities in Egypt. I am literally not joking about this! There is no way to determine an accurate age for the Great Pyramid. It has even been a mockery to register a temple or pyramid with any specific pharaoh. This is because later Pharaohs often laid claim to works by earlier Pharaohs, and Pharaohs who rebuilt, refurbished, or restored structures often placed their own name on them. Not to mention, shady archeologists who were not above faking a Pharaohs cartouche to make the structure fit the theories.

Which to mention, legend tells us that the Great Pyramid did not always look the way it does to us now in this present day and age. The choppy limestone blocks that we see today were at one time covered with white marble-like polished limestone casing stones that beamed back at the sky like a reflective mirror. We find evidence of these casing stones that were once on the Great Pyramid to have been later taken off and used after the 14th century earthquake to rebuild the Cairo temples, mosques, bridges and buildings. Allegedly, on these casing stones were unexplainable "watermarks" on the stones halfway up the Great Pyramids sides. Apart from just being alleged, evidence of silt sediments containing seashells and fossils are still present at the base of the Great Pyramid. Not only are watermarks and sea fossils visible on the outside, but salt incrustations consistent with mineral salt from the sea was found within the Great Pyramid as well. All these deposits are suggestive of a sea flooding. Many esoteric thinkers have been scoffed at by the archeologist who thinks their theory alone (about the watermarks and salt) has put it to rest. But nobody has adequately explained otherwise for the substance of sea salt found on the existing limestone casing stones (even though they are found elsewhere in Egypt). Alternatively, for the existing seashells and fossils found in the region and at the base of the Great Pyramid. Again, all theories that lead to no resolute end.

The "watermarks, seashells, fossils and salt incrustations" are significant because they might hold a clue as to how old the Great Pyramid really is. It is one of many clues that put my mind into gear to a much older date in history, which led me to link the watermarks, seashells, fossils and salt incrustations to Noah's flood. Though, its just not this evidence alone that brought me to this conclusion… there are other factors involved that I am working up to. Thus with just this in mind, I am estimating that the Great Pyramid existed before the pharaohs and before and during the Great Flood of Noah's time. Therefore, if the Pyramid existed before the flood, we should be looking at pre-deluge people. One antediluvian person who was biblically recorded to live before the flood comes to mind, and his name is "Enoch" from the Bible mentioned in Genesis 5. He was the same Enoch that gave us the constellations. The Enoch in the line of Seth, and Seth who was the 3rd son of Adam and Eve. Genesis 4:25 - Adam lay with his wife again, and she gave birth to a son and named him Seth, saying, G-d has granted me another child in place of Abel, since Cain killed him. Enoch was 7th from Adam; he was the father of Methuselah and the great grandfather of Noah. Genesis 5:18-24 - Jared lived one hundred and sixty-two years, and begot Enoch. After he begot Enoch, Jared lived eight hundred years, and had sons and daughters. So all the days of Jared were nine hundred and sixty-two years; and he died. Enoch lived sixty-five years, and begot Methuselah. After he begot Methuselah, Enoch walked with G-d three hundred years, and had sons and daughters. So all the days of Enoch were three hundred and sixty-five years. And Enoch walked with G-d; and he was not, for G-d took him.

Hebrews 11:5 - By faith Enoch was taken away so that he did not see death, and was not found, because G-d had taken him, for before he was taken he had this testimony, that he pleased G-d.

However, even though Enoch is mentioned in the Bible, inharmoniously his books have been excluded from it (which I will soon deduce why). Enoch was noted as a scribe, but it was his descendants Methuselah and Noah whom this knowledge was passed down to. Some of these documented texts concerning Enoch were found among the Dead Sea Scrolls and others in Greek texts that were translated from Jewish circles. There is also an Ethiopic and Slavonic version. The book consists of five distinct sections: (The book of Watchers, the book of Parables/Similitudes, the book of the Heavenly Luminaries (also called the Astronomical book), the book of Dreams and the Epistle of Enoch). In fact so authentic are Enoch's writings that the original biblical story of Adam and Eve can be found within. Enoch 2 Chapter 30:1-16 - On the third day I commanded the earth to make grow great and fruitful trees, and hills, and seed to sow, and I planted Paradise, and enclosed it, and placed as armed (guardians) flaming angels, and thus I created renewal. Then came evening, and came morning the fourth day. On the fourth day, I commanded that there should be great lights on the heavenly circles. On the first uppermost circle I placed the stars, Kruno, and on the second Aphrodit, on the third Aris, on the fifth Zoues, on the sixth Ermis, on the seventh lesser the moon, and adorned it with the lesser stars. And on the lower I placed the sun for the illumination of day, and the moon and stars for the illumination of night. The Sun that it should go according to each constellation, twelve, and I appointed the succession of the months and their names and lives, their thundering, and their hour-markings, how they should succeed. Then evening came and morning came the fifth day. On the fifth day I commanded the sea, that it should bring forth fishes, and feathered birds of many varieties, and all animals creeping over the earth, going forth over the earth on four legs, and soaring in the air, male sex and female, and every soul breathing the spirit of life. In addition, there came evening, and there came morning the sixth day. On the sixth day I commanded my wisdom to create man from seven consistencies: one, his flesh from the earth; two, his blood from the dew; three, his eyes from the sun; four, his bones from stone; five, his intelligence from the swiftness of the angels and from cloud; six, his veins and his hair from the grass of the Earth; seven, his soul from my breath and from the wind. And I gave him seven natures: to the flesh hearing, the eyes for sight, to the soul smell, the veins for touch, the blood for taste, the bones for endurance, to the intelligence sweetness. I conceived a cunning saying to say, I created man from invisible (spiritual) and from visible (physical) nature, of both are his death and life and image, he knows speech like some created thing, small in greatness and again great in smallness, and I placed him on earth, a second angel, honourable, great and glorious, and I appointed him as ruler to rule on earth and to have my wisdom, and there was none like him of earth of all my existing creatures. And I appointed him a name, from the four component parts, from east, from west, from south, from north, and I appointed for him four special stars, and **I called his name Adam**, and showed him the two ways, the light and the darkness, and I told him: This is good, and that bad, that I should learn whether he has love towards me, or hatred, that it be clear which in his race love me. For I have seen his nature, but he has not seen his own nature, therefore (through) not seeing he will sin worse, and I said after sin (what is there) but death? In addition, I put sleep into him and he fell asleep. And I took from him a rib, and created him a wife, that death should come to him by his wife, and I took his last word and **called her name mother, that is to say, Eve.**

What's more, in these texts it describes Enoch visiting the expanses of a Heavenly realm. He describes his visits in the form of visions. In Enoch's book concerning the Heavenly Luminaries, he is taken up by G-d's appointed archangel Uriel who showed him the movements of the heavenly bodies and warned him about the future impending flood and gave him instructions for building a type of mechanism in the form of a calendar showing the machinery of the sun and moon in operation so that after the great cataclysmic flood of Noah it would help them to determine the times for sowing and reaping, and to anticipate the migration of animal and insects that also had an influence over the plants (they would need this knowledge to survive). With this understanding for the Stars, Enoch replicated this knowledge by designing something called the "HENGE" which was used as a solar calendar (like Stonehenge) (Enoch 2 Chapters 11-16) (also found in 1 Enoch 72:1-37). THIS "HENGE" IN ITS POLISHED FRUITION WAS TOPPED OFF WITH ENOCH BUILDING A 365-DAY CALENDAR - OTHERWISE KNOWN AS THE "GREAT PYRAMID OF GIZA" OR THE PILLAR OF ENOCH, which was to signify an altar unto G-d. What Enoch accomplished was of great historical significance, for as G-d directed Enoch to build the Great Pyramid; much in the same way did He direct Noah to build the Ark. However, Enoch himself may not have actually built the Great Pyramid himself or in his time, he was undoubtedly the "original" designer or early builder of the Great Pyramid. He is said to have given these instructions to his sons and their children's children (which was in the lineage of Noah). After the Great Flood Noah gave us the epact in Genesis from the antediluvian Lunisolar calendar to the modern day Lunisolar calendar. Enoch was the great-grandfather of Noah. The Book of Enoch chapter 68:1 - And after that my great-grandfather Enoch gave me all the secrets in the book and in the parables which had been given to him, and he put them together for me in the words of the book of the parables.

There is mention of this in the Bible linking Enoch's alter unto G-d as being the Great Pyramid of Giza. This notation is found in Isaiah 19:19-20 - On that day there will be an alter to the Lord in the "midst" of the land of Egypt, and a pillar to the Lord at its border. It will be a sign and a witness to the Lord of host in the land of Egypt, when they cry to the Lord because of oppressors, he will send them a savior, and will defend and deliver them. It's obvious to construe that this pillar to the Lord was the Great Pyramid of Giza, and its deliverer from the land of Egypt was later Moses. However, the Jewish slaves mentioned in the Bible did not build the Great Pyramid, "it was already existing" (having been built by pre-deluge people). The Jews of the Exodus built the storage city of Pithom and Rameses made of mud brick mentioned in Exodus 1:11. Rameses the 2nd (noted as a master builder) must have been the pharaoh of the Jews during the Exodus from Egypt because an ancient city has been found 50 miles NE of Cairo on route to "Goshen" which was at one time fertile marshlands of the Nile river. Here you can find Rameses the 2nd cartouche on pillars and statues representing him, and you can find mud brick storehouses and settlements. Many archeologists have speculated this place as being linked to the Exodus. It would make sense since the Jews lived in the region of "Goshen" that they would have naturally worked in the nearby city of Pithom and Rameses. According to the book of Genesis, 47:1-5 the land of Goshen was given to Joseph (for his family) by pharaoh. Genesis 47:1-5 - Joseph went and told Pharaoh, My father and brothers, with their flocks and herds and everything they own, have come from the land of Canaan and are now in Goshen. He chose five of his brothers and presented them before Pharaoh. Pharaoh asked the brothers, what is your occupation. Your servants are shepherds, they replied to Pharaoh, just as our fathers were. They also said to him, we have come to live here awhile, because the famine is severe in Canaan and your servants' flocks have no pasture. So now, please let your servants settle in Goshen. Pharaoh said to Joseph, Your father and your brothers have come to you.

Then a transition of power took place and a new king arose over Egypt. This knew king did not know Joseph and was afraid that the Jewish people who were multiplying in the land would grow mightier than the Egyptians and eventually overtake them in a war. This is when they became slaves. Exodus 1:8-11 - Now there arose up a new king over Egypt, which knew not Joseph. In addition, he said unto his people, Behold, the people of the children of Israel are more and mightier than we. Come on, let us deal wisely with them, lest they multiply, and it come to pass, that, when there falleth out any war, they join also unto our enemies, and fight against us, and so get them up out of the land. Therefore, they did set over them taskmasters to afflict them with their burdens. And they built for Pharaoh treasure cities, Pithom and Raamses. Some say this was (Sethi) Rameses the 1st of the 19th dynasty in Egypt. He was the father of Rameses the 2nd who was crowned king by his father Sethi (even as a child in his arms). So it's very possible that it could have been Rameses the 2nd considering that he may have also been ruling during or at the same time together with his father Rameses the 1st.

Now back to Enoch… It has been thought that the book of Enoch was excluded from the Bible because it was too esoteric in nature for the early church. Others suggest that it had been rejected by the Jews because it contained prophecies pertaining to Jesus as the Messiah. Much to our surprise by the time Jesus arrived, there is evidence that the book of Enoch was widely circulated. In fact, we even find Jesus Himself indirectly referring to and quoting Enoch (passages from the book) on numerous occasions.

Jesus quotes Enoch in several strikingly similar instances:
Jesus said: And shall inherit everlasting life (Mat. 19:29)... Enoch said: Those who will inherit eternal life. (Enoch 40:9).

Jesus said: Blessed are the meek for they shall inherit the earth (Mat. 5:5)... Enoch said: The elect shall possess light, joy and peace, and they shall inherit the earth (Enoch 5:7).

Jesus said: That ye may be called the children of light (John 12:36)... Enoch said: The good from the generation of light. (Enoch 108:11).

Jesus said: In my Father's house are many mansions (John 14:2)... Enoch said: In that day shall the Elect One sit upon a throne of glory, and shall choose their conditions and countless habitations. (Enoch 45:3).

Jesus said: The Father judgeth no man, but hath committed all judgment unto the son (John 5:22)... Enoch said: The principal part of the judgment was assigned to him, the Son of man. (Enoch 69:27).

Jesus said: Woe unto you that are rich! For ye have received your consolation. (Luke 6:24)... Enoch said: Woe to you who are rich, for in your riches have you trusted, but from your riches you shall be removed. (Enoch 94:8).

Jesus said: Woe unto that man through whom the Son of man is betrayed! It had been good for that man if he had not been born. (Mat. 26:24)... Enoch said: Where will the habitation of sinners be, who have rejected the Lord of spirits. It would have been better for them, had they never been born. (Enoch 38:2).

Jesus said: Ye also shall sit upon twelve thrones, judging the twelve tribes of Israel. (Mat. 19:28)... Enoch said: I will place each of them on a throne of glory. (Enoch 108:12).

Jesus said: Between us and you there is a great gulf fixed. (Luke 16:26)... Enoch said: By a chasm, are/their souls separated. (Enoch 22:9,11).

Jesus said: The water that I shall give him shall be in him a well of water springing up into everlasting life. (John 4:14)... Enoch said: All the thirsty drank, and were filled with wisdom, having their habitation with the righteous, the elect, and the holy. (Enoch 48:1).
There is also a direct quote in the New Testament by Jude (Jude 1:14-15, quoting Enoch 1:9) stating: And Enoch also, the seventh from Adam, prophesied of these, saying, Behold, the Lord cometh with ten thousands of his saints to execute judgment upon all, and to convince all that are ungodly among them of all their ungodly deeds which they have ungodly committed, and of all their hard speeches, which ungodly sinners have spoken against him.

In Enoch's book of Parables/Similitudes (37-71), it informs us of a figure called "the Messiah" - "the righteous one" - "the chosen one" and "the son of man". (I will discuss in detail Jesus being the Messiah in the future chapter on "expanding on the Merkabah" and "the proper religion").

A better assessment of Enoch reveals that not only did he receive knowledge from above, but also he established a dialogue where he talked with G-d and His angels, and it was most likely in the Hebrew language. All the books that have been found on Enoch were written in the Aramaic (old Hebrew) language. Some of this knowledge that Enoch possessed of the Hebrew language and its letters might have already been fairly familiar to Enoch which was most likely handed down to him from the family line of Adam, Eve, and Seth in which the knowledge of the Hebrew language was translated into written and spoken form. From Seth to Enoch, Enoch to Methuselah through Noah and his three sons, Shem, Ham and Japheth. Shem cultivated the Hebrew language and alphabet among the Jews and his son Elam to the Persians… Shem's other son Arphaxad begat Eber who begat Joktan who carried this alphabet into Arabia… Japheth carried it into Europe… Ham and his son Misraim introduced the alphabet into Egypt where in time the priests concealed (or obscured) what they called "hallowed characters" (or what would be considered ineffable, inexpressible, unutterable, beyond words) and thus submitted hieroglyphs in place of the Hebrew alphabetic characters that we see today for the sake of preserving what they considered holy.

Now I know this is hard to believe such a folklore about a man named Enoch that little is known of today. And I know its even harder to conceive that the root of all languages stem from a single source. However, together with Enoch's book there are several other biblical documents (like I already mentioned) that refer to him, such as the Dead Sea scrolls and Jesus Himself quoting Enoch. Additionally, we also have the account of Alexander the Great. Apparently when he was in India he wrote to his master Aristotle, as follows**:** When I came to such a place in India, the natives told me that they had with them the sepulcher of an ancient king that ruled over all the world, whose name was Cainan (Canaan), the son of Enos (Enoch), who foreseeing that G-d would bring a flood upon the earth, wrote his prophecy of it on tables of stone, and they are here, the writing is Hebrew writing. So we cannot merely say that it is just a single source of documentation we are hindering on.

Enoch did something even more miraculous then just being a scribe. As I said earlier, he taught us how to navigate the stars (for the seasons) with a device called a "Henge", which he later incorporated into the Great Pyramid of Giza. We also have proof of such astronomical devices like Stonehenge and the ancient Mayan Pyramid of Chichen Itzas (in Yucatan Mexico) that was used as a calendar - with four stairways, each with 91 steps and a platform at the top, making a total of 365 days in a year. In addition, the Aztec temples and the China and Japan pyramids were used as calendars as well. Even the Pyramids of the Americas were found to be used as a lunisolar calendar along with a sacred cave underneath that was used for resonating chanting sounds (this "sound" will soon be revealed). The culmination of ALL these pyramids form a worldwide pyramid system intentionally mounted on key meridians of the earth. Therefore, we must conclude that this technology was in existence and that there was such a man named Enoch that lived before the Great Flood of Noah, and he educated the world with this knowledge. To that end, it's not to far fetch to acknowledge that Enoch actually was privy to divine knowledge shown to him by G-d and His angels, which he incorporated into the Great Pyramid. In addition, in order for Enoch to have truly understood G-d and His angels, he would have had to learn how to speak and write this same language in its proper form. And with this alphabetic knowledge, Enoch recorded all his visions from G-d (prophecies) in his book. This book was then passed down to Noah. From Adam to Seth, from Seth to Enoch, Enoch to Methuselah through Noah and his three sons, Shem, Ham and Japheth. Enoch was the father of Methuselah and the great grandfather of Noah. Therefore, the book of Enoch had to have been taken on the ark with Noah for the mere fact that it exists today. However it should be clarified that Enoch (including Adam, Eve, Seth and Noah) were NOT the inventors of the alphabet, but rather only the recipients of G-d's language which was used as a communication tool between G-d and mankind.

Now lets take a better look at Enoch's "Henge" and how it relates to the Great Pyramid of Giza. In Enoch's book, it shows us how to mark solar coordinates on the horizon by creating a circular Henge observatory which would span the length of one solar year (365 days) and to keep track of time and the seasons (1 Enoch 72:1-37). This Henge was made with stones or wooden posts that are aligned to mark the astronomical movements of the planets, stars, moon and the sun. By using the sunrise and sunset of the Sun, Enoch shows us that he not only knows how to keep track of time using astronomy for the purpose of calculating the seasons, but he also incorporated the astronomical messages (prophesies) that he received from G-d (concerning the coming Messiah) which is presented through the Great Pyramid. This information was also presented to you in chapter 2 (Religion versus evolution). If you can recall I wrote: It is these constellations which mark a continuous circle around the earth today, which most of us are familiar with called the Zodiac. Today and even in the recent past these constellations have been twisted to tell a different story. Many of them being used by soothsayers, fortune-tellers and astrologers to mean many mythological things in opposition to G-d. But surprisingly we find out that the constellations bare no other meaning or purpose other than bringing the promises (prophesies) concerning the coming Messiah (Jesus the deliverer, the abolisher of sin) into fulfillment. The ancient constellations are all picture prophecies of Jesus, and are in accordance with the written word of G-d (the picture diagram can be found in chapter 2). In fact, many individual star names are recorded in the Bible as characteristic attributes and aspects of Jesus. And it was this knowledge of the constellations given to Enoch that was later passed down to Abraham, and Abraham who was latter commanded to teach the Egyptians. Moreover, the Greeks are said to have received their knowledge of the constellations from the Egyptians. In addition, which we get what's known as the modern constellations from the Egyptians.

But Enoch went above and beyond his schematics of the Great Pyramid of Giza in such a way as to verify and validate that he (Enoch) was the original designer of it. He put his mark on it for all time! He did this by leaving us clues identifying himself and the knowledge given to him by G-d (an indicating mark that no Egyptian could refurbish over). Starting with his actual name. Enoch's name is plastered all over the Great Pyramid in every single way! One can easily see this especially if they are familiar with the Hebrew language. One of the helpful hints in realizing this is recognizing that "ENOCHS" name is very similar to the Hebrew word for "INCH" (Enoch/inch). The Hebrew word for "inch" is (aleph, yud, nun, tzadi) pronounced "eench" - which is very similar, if not almost identical to Enoch's name. In Hebrew Enoch's name signifies "initiate or initiator" which is still very similar to "inch". So one might ask what is the significance of Enoch's name in relevance to the "inch". The answer to this can be found in Genesis 5:23 where it tells us: So all the days of Enoch were three hundred and sixty-five years. Now one may wonder what does 365 years have to do with the Great Pyramid. And the answer to this comes when the Great Pyramid is measured in "inches" it is equivalent to one solar year which is 365 days (to be exact one solar year is actually 36524.22) in which each year of Enoch's life was the span of one "pyramid inch" and used throughout the whole construction of the Great Pyramid. This can be derived by the length of a base (pyramid walls) from corner to corner in a straight line, which are 9,131 inches… There are four walls, so if you take 9,131 x 4 it equals 36,524 total inches (if you move the decimal over it signifies 365). The second formula for realizing that the Great Pyramid is of Enoch is by the "average" height of land above sea level on the Earth, which happens to be 5,449 inches (this can only be determined by modern day satellite technology). 5,449 also happen to be the accurate "vertical" height of the pyramid! Modern satellites have given the measurements of the highest and lowest elevations of land above sea level. The average median of the two is 5,449 inches (being that sea levels fluctuates), is also the height of the Great Pyramid in the sacred Jewish inch. The great significance here, being 5,449 inches above sea level, is that the Great Pyramid was built above sea level for the very purpose that it was meant to survive the Great Flood of Noah. Enoch's goal was to prepare his grandson Noah for the future impending flood and leave a message indicating the coming of the future messiah. Yet, something even more foundational connects Enoch with the Great Pyramid to another almost inconceivable degree… Enoch built the Great Pyramid as an altar in the sense of being a witness unto G-d and as a message to future humankind. It was explicitly built as an alter that was NOT used for sacrifice, as is declared in Joshua 22:26 - Therefore we said, let us now prepare to build us an altar, not for burnt offering, nor for sacrifice, but that it may be a witness between us, and you, and our generations after us. So we get the general idea that it was not just an ordinary altar, this was a special altar. So special that Enoch went to great lengths to communicate his divine message from G-d, and went to the extent of incorporating this message within the Great Pyramid in a most amazing and implicit way. This can be witnessed in the actual Bible phrase of Isaiah 19:19-20 which clearly states: On that day there will be an alter to the Lord in the midst of the land of Egypt, and a pillar to the Lord at its border. It will be a sign and a witness to the Lord of host in the land of Egypt, when they cry to the Lord because of oppressors, he will send them a savior, and will defend and deliver them. Now a miraculous thing happens when we apply the "gamatria" value of each Hebrew word in Isaiah 19:19-20 (broken down by each Hebrew letter). There are precisely 30 words and 126 letters. When this is added up, the total numerical value equals 5,449 which is (if you recall) the total number of height inches of the Great Pyramid "AT ITS SUMMIT PLATFORM". A note of reference: The Pyramid "sacred inch" is almost the same unit of measure that is used today - where 12 inches is equivalent to 1 foot. However, it might as well be because it's extremely close with only 1.5 centimeters difference.

The great Pyramid - a message written in stone - Enoch prophesy foretold.

Isaiah 19:19-20

ביום ההוא יהיה מזבח ליהוה בתוך ארץ מצרים ומצבה
אצל-גבולה ליהוה : והיה לאות ולעד ליהוה צבאות
בארץ מצרים כי-יצעקו אל-יהוה מפני לחצים וישלח
להם מושיע ורב והצילם :

1	2+10+6+40=58
2	5+5+6+1=17
3	10+5+10+5=30
4	40+7+2+8=57
5	30+10+5+6+5=56
6	2+400+6+20=428
7	1+200+90=291
8	40+90+200+10+40=380
9	6+40+90+2+5=143
10	1+90+30=121
11	3+2+6+30+5=46
12	30+10+5+6+5=56
13	6+5+10+5=26
14	30+1+6+400=437
15	6+30+70+4=110
16	30+10+5+6+5=56
17	90+2+1+6+400=499
18	2+1+200+90=293
19	40+90+200+10+40=380
20	20+10=30
21	10+90+70+100+6=276
22	1+30=31
23	10+5+6+5=26
24	40+80+50+10=180
25	30+8+90+10+40=178
26	6+10+300+30+8=354
27	30+5+40=75
28	40+6+300+10+70=426
29	6+200+2=208
30	6+5+90+10+30+40=181 Total 5,449

Now I know that some may raise objections on the basis alone that 5,449 is an estimative figure of the "average" level of land above sea level. But the "gamatria value" should cancel that objection out. However, it is not just the "gamatria value" alone or by itself that determines the Great Pyramids actual inches. It also works ingeniously in conjunction with another piece of evidence - THE MISSING CAPTONE!

When I capitalized in several paragraphs above, "summit platform" I am referring to a "missing capstone" that the Great Pyramid is absent of. However, if there were a capstone, then the Great Pyramids height would be 5,814 inches. This is confusing because most people tend to think that the Great Pyramid (from some pictures of it) has a pointy top to it. However, this is not the case. Pyramids in "Egypt" were usually crowned with a capstone, and in fact, several capstones have been found unattached and buried in the sandy ground nearby. However, to our query, the Great Pyramid actually has a flat truncated top to it with no capstone, and today there is a triangular antenna-looking device placed on top to demonstrate what is thought to be the approximate dimension of the capstone. Some researchers propose that there was once a capstone that rested on this top. In addition, some hold the idea that the capstone was never entirely completed. While others speculate that it was stolen or removed at some point in history. However, on the contrary some suggest otherwise, that a capstone may have never been part of the original design because there is NO REPORTS IN RECORDED HISTORY of anyone ever seeing the Great Pyramid with a capstone.

While still debating this missing capstone… to the tip of the missing apex (the antenna-looking device) we get a total of 5,814 pyramid inches. Now subtract that by the summit platform inches of 5,449 and once again, it equals up to 365 pyramid inches as well as days in a year - hence Enoch yet again. This leaves us with a valuable piece of the puzzle indicating and validating the original exact height of the Great Pyramid, which was not supposed to be 5,514 pyramid inches, but 5,449 pyramid inches (minus a capstone). Therefore, because of the missing capstone measuring 365 additional inches, it bears witness to the accurate total number of pyramid inches (5,449). Whereas if we cannot see that 365 plays a vital role in naming Enoch as its designer in all intentional purposes, it repeats it yet again with the missing capstone.

So now it leaves us with the question that if indeed the missing capstone was part of the original design, it may have also served another resourceful feature. In which the flat apex was used as a "celestial star chart" which substituted for a physical capstone - where once again we can see the wonders of the Star of David! (Shown a few picture below).

It has been suggested that the four ventilation shafts of the Great Pyramid point to the stars of Orion's belt, Sirius, Thuban, and Ursa Minor. Additionally the sun and the moon were determined by the latitude and longitude of the poles (the Sun rises in the East). These pointers may have been one of the clues that led the three wise men (Magi) to the time and location of Jesus' birth. We can also assess that a star graph (incorporated on the truncated top of the Great Pyramid) somehow foretold this same message incorporated by Enoch.

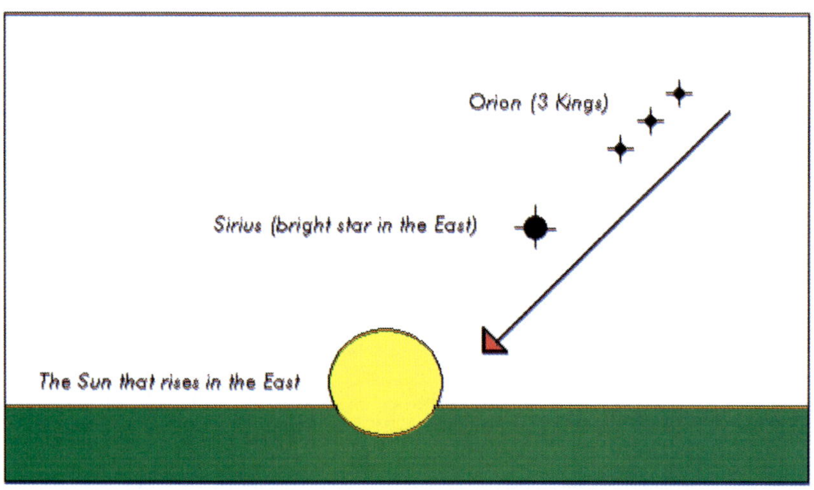

Now I'm not saying for sure, it's only a suggestion that the stars may have lined up in some way pointing eastwards. From the Sun rising in the East as well as Sirius appearing in the East, which may have also been aligned with the three stars in Orion's Belt (also noted as the 3 Kings). These stars may have all pointed (in a straight line Eastward) to the place where Jesus was born in Bethlehem. Literally, the birth of the Sun/Son.

Matthew in the New Testament tell us in 2:9-11: When they had heard the king, they departed; and, lo, the star, which they saw in the EAST, went before them, TILL IT CAME AND STOOD OVER (the house) where the young child was. When they saw the star, they rejoiced with exceeding great joy. In addition, when they came into the HOUSE, they saw the young child with Mary his mother, fell down, and worshiped him, and when they had opened their treasures, they presented unto him gifts, gold, frankincense, and myrrh. If we stop to properly analyze what Matthew was truly explaining to us, we would know that the only star that rises in the EAST is the SUN! And the only time it practically stands still (momentarily) "with the greatest duration of light" is in the northern hemisphere during the summer solstice. The word "HOUSE" is actually an astrological configuration (which sign is Cancer) and signifies the "house of the mother" - and "Cancer" as we all know IS the summer solstice. Cancer is the 4th Hebrew month of Tammuz. It also signifies "water" (the womb) in which all things come into being, and is associated with motherhood. So what the Magi were looking for, saw the signs and found, was the conjunction of all our local planetary stars forming the "Star of David" WITH THE SUN IN THE MIDDLE! What the Magi had seen was the conjunction of the NORTH EAST stars (Venus being North and Saturn being East) that would have directed them right into Bethlehem. Also, Jupiter in conjunction with Venus could have played a role having crossed paths and gleaming like one brilliant star, or even perhaps Jupiter (Jupiter - considered the royal star) in conjunction with Saturn (Saturn - considered the old ruler) moving towards the constellation Pisces. This cannot be accurately pinned down, but perhaps what the Magi saw could have been more subtle than a star or the conjunction of stars, they seemed to focus more on the movement of the "CONSTALATIONS". The Bible is inconclusive about this, but it may suggest that this was something that only the Magi saw and understood. If this is so, then perhaps this suggests something that required "INTERPRETATION" rather then something that was dazzling to the eye. If this is the case, this might also suggest that the constellation "Virgo" played a role, since the constellation "Virgo" represents the virgin birth (the Virgin Mary) (which is called Bethulah in Hebrew) and is actually a representation of a woman sitting with a child on her lap considered Jesus. And it may have been the sign of this constellation that stirred the Magi into recognizing its majestic significance - hence, and this is what Enoch would have been privy to have foreseen (as aforementioned) that G-d's ordained Angel (Uriel) revealed the constellations to Enoch along with their precise meanings, which included the coming of the Messiah! This is why the constellations (such as Virgo) were so named by Enoch as a representation (or sign) for the coming Messiah. Its also interesting to note that the word for "Virgo" is used to describe the Virgin Mary which is called "Bethulah" in Hebrew… as it turns out, the word "Bethulah" is very similar to the Hebrew town of "Bethlehem" where Jesus was born. Matthew 2:1 - After Jesus was born in **Bethlehem** in Judea, during the time of King Herod, Magi from the **east** came to Jerusalem.

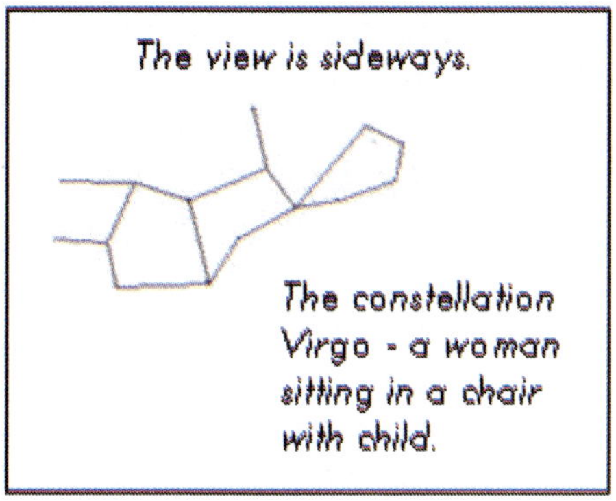

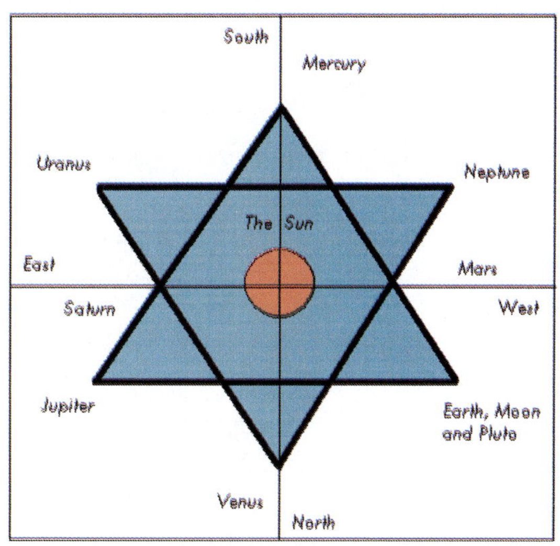

Brace yourself within this book, more is to be revealed.

To understand (more in depth) the naming of the "astrological houses" of the heavens, is firstly to understand that it was being used long before the date to which we can trace remaining evidence. Both for astrological observation and for mythical and symbolic purposes, and secondly to understand that it corresponds and shares in likeness to almost all races and religions of the world. Enoch of course was initially behind all this. To truly and thoroughly understand just how each star/planet got its name (house meaning) is by understanding the movements of the planets. For instance with the sun being born in the east, maturing at noon, and dieing at sundown - therefore East is associated with life, South with power, West with death, and North as the sun dips below the horizon, not as an end to life but as a transitory stage of birth (mother) and renewal through genealogical lineage. In addition, the "North house" is associated with "Cancer". Enoch recognized this and was privy to divine knowledge of the Messiahs coming - hence the moniker of the 4th house of Cancer, and foretelling this message by means of the Great Pyramid.

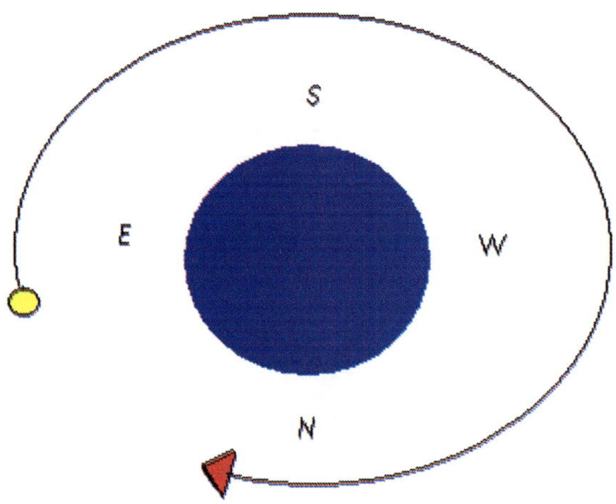

We can even see some of the apparent similarity in our modern names of the week to that of the Hebrew, and to note that each planet was named for its particular appearance, color or glint. To accentuate on this a little more... In the modern English language, like the word "Sunday" is a moniker carried over from our planetary "Sun" and the word "Sun" in Hebrew is the word "Shemesh", which literally means "light". Monday is a moniker for the "Moon" and its Hebrew word is "Lavanah" which literally means "white" the color that the Moon actually exhibits. Tuesday is in representation of "Mars" and its Hebrew word is "Maadim" which represents the color "blood red", and as we all know in this day and age with the invention of the telescope that the planet "Mars" is actually red in color. Wednesday is in representation of the planet Mercury, and Thursday represents Jupiter. Friday is a representation of the planet "Venus" which in Hebrew is called "Noga" which means to sparkle. Lastly, Saturdays moniker is the planet "Saturn" and its Hebrew name is "Shabbat" which is the holy Sabbath. Most important were the months of the equinoxes and their Hebrew names such as "Nisan" which means harvest in Hebrew with the astrological sign of "Aries". It is also the spring equinox, which means "miracle". Tammuz (a word I've mentioned briefly before) which means mother, giver of the vine and warmth with the astrological sign of "Cancer", it is also the summer solstice. "Tishrei" means, "seed time" (birth) with the astrological sign of "Libra", it is also the autumn equinox. "Tishrei" is noted as "seed time" since this is/was considered the creation/birthday of the world noted in Genesis 1:12 - And the earth brought forth grass, and herb **yielding seed** after its own kind, and trees yielding fruit with the **seed in it,** after its own kind, and G-d saw that it was good. (The month of "Tishrei" will be further explained in a future chapter). And lastly, we have "Tebheth" which means "cold" and is actually the winter solstice with the astrological sign of Capricorn.

So as we can see that most if not ALL-astronomical knowledge that we obtain today was already demonstrated in past history with its linkage to the book of Enoch 1 (72-80). Today the science of astrology and astronomy hinders on its own definitions by naming and referring new associations, and fails to realize that the "original" source of information comes from a biblical perspective. Going off on a tangent for a case in point… For instance another perspective of biblical and prophetic proportion suggests that the "casing stones" that the Great Pyramid was once said to posses was 144,000 in all. And sources tell us that these polished casing stones were so brilliant (like a reflected mirror) that they could literally be seen from hundreds of miles away. The symbolic significance of this can be found in prophecy (Revelation 7:1, 7:4, 14:1, and 14:3) where it also mentions the number of 144,000 as the future prophesied amount of people raptured and transmuted at the end time. 12,000 from each of the 12 tribes of Israel (which is 12,000x4=144,000). This prophecy was most likely carried over from Enoch and was the principal symbolic purpose for constructing the Great Pyramid with 144,000 casing stones.

The number 12 in the Bible is related to us in several different ways. Such as in separation of the hours in a day which are from 12:00 a.m. to 12:00 p.m. thus having 24 hours in a day. Another representation of the number "12" is in reference to the 12 Tribes of Israel which are spoken about in the Old Testament, and the 12 Apostles of Jesus mentioned in the New Testament, and the New Jerusalem described in Revelation 4:4, 7 and 21 - with 12 foundations, 12 gates with 12 angels, 4 walls (also 144 cubits or furlongs in height) consisting of 144,000 Jewish saints as well as the "24 Elders" whereupon they were seated about the throne of G-d with "four and twenty seats". It describes the total amount of 144,000 sectioned off by 4 (foursquare) walls: On the East three gates, on the North three gates, on the South three gates, and on the West three gates. The 12 tribes of Israel who are descended from one of the 12 sons of Jacob are: Asher, Benjamin, Dan, Gad, Issachar, Joseph, Judah, Levi, Naphtali, Reuben, Simeon and Zebulun and whose names are written on the gates. They were split up into North, East, West and South. The Northern tribes were Reuben Simeon and Gad. The Eastern tribes were Judah, Issachar and Zebulun. The Southern tribes were Reuben, Simeon and Gad. And the Western tribes were Ephraim, Manesseh and Benjamin. This may have been one of the primary reasons that Enoch had constructed the Great Pyramid to have 4 sides. In addition, its also interesting to note that Israel later in history was sectioned off to each of these tribes. Some describe the New Jerusalem to look like a square, but to me it sounds as if biblical prophecy is again referencing the description of the Great Pyramid and the "sign of the Son of Man" which is the Star of David. Matthew 24:30 - At that time the sign of the Son of Man will appear in the sky, and all the nations of the earth will mourn. They will see the Son of Man coming on the clouds of the sky, with power and great glory.

To acquire an even more enhanced understanding about astrology, we need to delve back into Genesis chapter 1:14 when G-d declares that the stars, the sun and the moon are given for signs (in the sense of navigational indicators) attributing to the Jewish holidays and incidentally human destiny. However to use astrology as a form of "divination" (foretelling) is expressly forbidden in scripture (Deuteronomy 18:10-14). James 1:5 - Our wisdom comes from G-d, not the stars. G-d also states that the other purpose of the planets and stars is for His glory (Psalms 19:1). Enoch reconfirms this for us in his book when he refers to the stars as "hosts of heaven" he correlates them to "angels". We are also told that it was G-d Himself who named the stars in Psalms 147:4 - He tells the number of the stars, He calls them all by their names. Therefore if G-d gave them their names, surely this is added evidence that they were named not just for identification but so that He may set out His plan of redemption for fallen mankind. Psalm 8:3 - When I consider thy heavens, the work of thy fingers, the moon and the stars, which thou hast ordained. Psalm 19:1 - The heavens declare the glory of G-d, and the firmament sheweth his handywork. Job 26:13 - He has garnished the heavens. Psalm 147:4 - He has all the stars numbered and named. Genesis 15:5 - And He took him outside and said, Now look toward the heavens, and count the stars, if you are able to count them. And He said to him, So shall your descendants be.

To delve even deeper into astrological mythology, the 12 constellation of the zodiac are called in Hebrew "Mazzaroth" (as aforementioned this is reference to the twelve constellations of the zodiac). Job 28:32 - Canst thou lead forth the "Mazzaroth" in their season? Or canst thou guide the bear with her sons? (the bear is Ursa Major). Job 9:9 - Who maketh the Bear, Orion, and the Pleiades, and the chambers of the south. Job 26:13 - By his spirit he hath garnished the heavens; his hand hath formed the crooked serpent. (which is Draco). Job 38:31-32 - Can you bind the cluster of the Pleiades, or loose the belt of Orion? Can you bring out Mazzaroth in its season? Or can you guide the Great Bear with its cubs? Amos 5:8 - He who made the Pleiades and Orion and changes deep darkness into morning, who also darkens day into night, who calls for the waters of the sea and pours them out on the surface of the earth, the Lord is His name.

Trying not to loose procession of thought here… I must also mention that there are also other things about the Great Pyramids innards that are "prophetic in nature" concerning the "inch" (that owns up to Enoch). This can be found throughout its chambers, passageways and shafts. There are discriminate areas where "red granite" appears in the pyramid to mark specific events in history related to the Jews, (yes the Jews!). The empty coffer in the King's chamber is also made of the red granite in which the cubic capacity of the coffer is equal to the measurements G-d gave Moses to build the Ark of the Covenant in (Exodus 25). One of the locations of red granite appears at the intersection of the ascending and descending passages in which it is suspected to correspond to the year of the Exodus where the Hebrews departed from Egypt. In addition, at the entrance to the Antechamber of the Kings room, there is a raised image on red granite with the distinguished measurements of the pyramid inch.

I must gloss over the Great Pyramid again. Cracking down on its proper height in which I specified above was a total of 5,814 pyramid inches and 5,449 inches at the summit platform. Inasmuch as all the measurements of the Great Pyramid sound credible, it flies in the face of the fact that I cannot validate these measurements. In all truthfulness, I've found that there is no firm anchor of evidence with any of the Great Pyramid measurements as they are constantly changing by some small degree (if you research this for yourself you will know what I mean). The discrepancy comes in with true measurements since the polished encasing stones which once covered the outside of the pyramid have been removed over time thus the pyramid lost approximately 30 feet off all its dimension. Whether it's labeled as cubits, the British inch, metrics, feet, Pi/Phi, golden spiral or in the "sacred Jewish inch". I am bereft of actually accruing these measurements. However, I must also conclude amid these uncertain accusations, that the Great Pyramid "today" is no longer 5,449 pyramid inches in height as weather, sedimentation and even perhaps vandalism and of course the ever pervasive law of entropy and thermodynamics having all played a role in reducing its height. HOWEVER, IT STANDS TO REASON THAT IT WAS ORIGINALLY 5,449 PYRAMID INCHES BASED ON THE CLUE OF THE MISSING CAPSTONE!

There is something else of intriguing interest… whether its relevant or not, I will leave that up to you.

I found an interesting relationship between Psalms 118:22 of the Bible - (the stone which the builders rejected is become the chief corner stone). Some say this relates to the Great Pyramids missing capstone as Jesus being the Messiah. In which it relates in two way - as the missing capstone being used as a Star of David celestial star map in which to locate the "time and place" in history of Jesus' birth by the alignments of the stars in our solar system. This is because In Psalms 118:22 there are two important Hebrew words that can be derived from this verse, which comes to the gamatria equivalent of 248. The two words that I am referring to are (rejected - corner stone). The Hebrew words are (haboneem pinnah) (hey, bet, vuv, nun, yod, mem - peh, nun, hey) (5+2+6+50+10+40+80+50+5=248). The 248 is a significant number as its a representation of the POSITIVE commandments, and if you recall the other 365 negative commandments are also representative of 365 days in a year - hence Enoch and the missing capstone, and thus equating to a grand total of 613 commandments. Perhaps just as well - the prophetic "future" commandments written in stone.

To add fuel to this supporting fire, there is so much more about the Great Pyramid and the numerous apparent mathematical coincidences corresponding in relationship with the Earth, Sun and Moon that I'm not even touching base on… in which I'm leaving open for readers to investigate for themselves.

For instance… It aligns with true north with only a slight degree of error - this is because the North Pole moves over time, but was most likely exactly aligned at one time. Its relationship with Pi/Phi is uncanny within its proportions. It calculates for us the distance to the Sun and Moon as well as the Earths radius, equatorial circumference, volume and mass, and the speed of the Earth around the Sun. Moreover, it even marks the spring equinox when it casts no shadow at noon. The spring equinox occurs on March 21st . Then on September 22nd we have the fall equinox, and then the summer solstice which occurs on June 21st, and it is on this day that the Earth sees the longest duration of daylight, and on December 22nd which is the winter solstice, is when we see the shortest amount of daylight and the longest night of the year. On these days, the Sun almost seems to pause in its orbit before resuming its course.

In addition to all this, the Great Pyramid is located "almost" at the center of the landmass of the Earth. However, at one point in time it may have been exactly at the center. There are those that will hotly deny this, saying that this can be proven wrong with a good map (some even find it at the Ukraine). However to assess this accurately by dividing the earth's land mass into approximately equal quarters. The East and West parallel that cross the MOST land, and the North and South meridian that cross the most land intersect in two places on the Earth, (30 degrees North by 30 degrees West) one in the ocean and the other at the Great Pyramid. However, what Enoch was really looking for was a solid place on land, which was "surrounded" by all the other land proximities of the earth. Suitably he chose Giza.

The importance of the Great Pyramids location will be covered in detail as you continue to read.

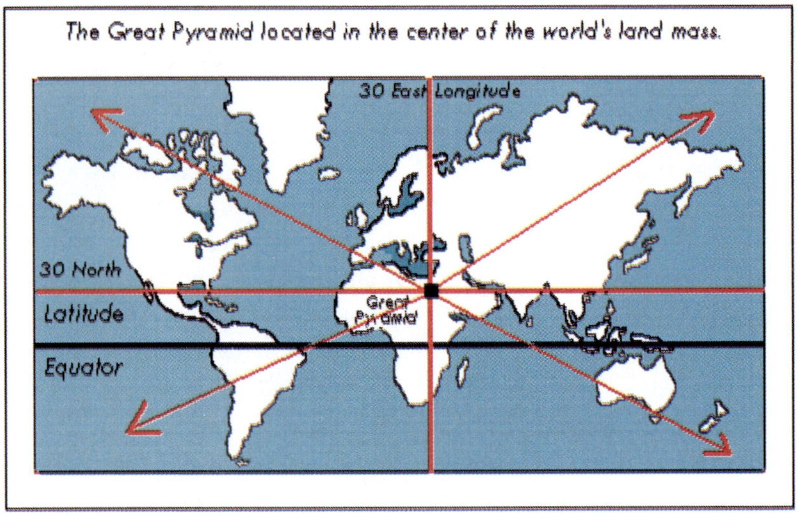

As to the location of the "altar" (Great Pyramid) in Giza, please reread Isaiah 19:19 again: On that day there will be an alter to the Lord in the midst of the land of Egypt, and a pillar to the Lord at its border.

If we closely examined the description given in Isaiah 19:19 we can evaluate the approximate location of this "altar" and peg it as the Great Pyramid of Giza. It described it as an altar at the "border" and at the same time in the "midst" of Egypt, which we can consider is the "middle" of Egypt. As it turns out the ancient word "Giza", actually means, "border"! Yet if this were not a good enough indication, we only need to take another look at the map above to realize that it is indeed at the focal point of all land mass and it is indeed at the border of "upper and lower Egypt" especially during ancient times before it was divided up into upper, middle, and lower Egypt. Incidentally, the Great Pyramid also happens to be the center of present day Egypt with Cairo being its capital. To find any other object fulfilling this description of an altar in this area is vastly unlikely.

Now we must ask ourselves the opening question to this chapter, just what was the purpose of employing the Hebrew letter "Ayin" (the all seeing eye) into the innards of the Great Pyramid? For we will soon see that the Great Pyramid speaks not only with geometry in the form of a 365 day calendar, Phi/Pi and the Golden spiral, but also with words. Just how did it do this? Well, lets explore how…

Enoch not only built a calendar, he is also acclaimed by the Bible and in his book to be the procurer of the alphabet in which he reiterated it (remember he did not create the alphabet, G-d did). In his book, he talks about being shown "weights and measures" which he recorded down. These have been suggested to be the "Hebrew letters" which were described not only as mere letters, but also as measurements in the form of vibratory modulation and control. Enoch called these frequency keys. Correspondingly, oral tradition tells us in Qabalah that these "frequency keys" are known as "ketav einayim" (eye writings).

With all this in mind, one wonders just what kind of multi-layered knowledge Enoch could have actually possessed? Could it possibly be that this "one universal law" we have been talking about all along is also a language of "frequency keys" that we witnessed in the above chapters with the Hebrew letter "Aleph" and its frequency sound forming the Star of David pattern in sand (the science of Cymatics). If this is accurate then that must also mean that each one of the 22 Hebrew letters ALSO have a specific frequency form. Similarly like the pattern of "Aleph" being formed in sand by sound vibration with the Hebrew, letter "Ayin" also having gross powers of generation. In fact the Hebrew letter "Ayin" is also silent in its sounding like "Aleph". These are the only two letters in the Hebrew alphabet that are used silently, AND "AYIN" CAN PRODUCE THE SAME SOUND AS Ahhh or Ohhhmmm. Revelation 1:8 - I am the Alpha (Ahhhlpha) and the Omega (Ohhhmega) the beginning and the end, the first and the last.

It is not beyond the realm of possibility that Enoch, with all divine intents and purposes, possessed the "keys" of the physical world, and for a skillful reason endorsed by G-d! As I also expressed earlier at some brief point in this book, and will expand upon, that this Ahhh frequency (Star of David) has biological healing powers. Therefore it is conceivable that Enoch, as suggested above, as the original designer of the Great Pyramid also placed and concealed within the Pyramid one of these frequency "keys" in the form of the Hebrew letter "Ayin". And not just physically but to function audibly. Interestingly enough, the Great Pyramid was tested for frequency sounds and it was discovered that the sarcophagus (stone coffin) in the center of the Great Pyramid is tuned to the frequency of the human heartbeat. Whereas, the human heart beat is also the Ahhh sound and is noted as the heart chakra. To understand this more in depth, the human body is built upon vibratory rates, as is everything else in the universe… The human body has seven resonance pitches on the musical scale… The lowest vibrations are found in the legs and the base of the spine, and the lower abdomen is the next higher area of vibratory resonance. Hence, the sarcophagus being in the "center" of the Great Pyramid is symbolic of its heart region and in accord shares its same frequency. The Earth itself even generates this same frequency (which besides the Ahhh sound, it also puts out 6.8 hertz). Contingent to this, it seems to be the directing mechanism in all life (as we know and experience it). It's also interesting to know that water when crystallized in snowflake form produces the hexagonal shape of the "Star of David" pattern. Thus water in its natural form is "required" to create this hexagonal form - hence when viewed in this way it is perfect in all its creation because "the frequency of love" (this same Ahhh sound) plays a part… while (on the other hand) water that is polluted (foul, nasty or vile) looses its hexagonal shape and is distorted with harmful ingredients, that if consumed by the human body or left to contaminate rivers, lakes, and the ocean could harm or destroy life, (which in a sense could be called evil, because it was caused by sin).

I'm suspecting that the Great Pyramid in its whole magnitude and immensity is a sound resonating device utilizing limestone and granite, all of which can convert sound vibrations into electromagnetism. Limestone is a rock formed essentially of quartz. Now quartz has the capability when it's subjected to mechanical pressure to produce an electrical voltage called a piezoelectric effect. Now when an alternating current is passed through the quartz it creates an oscillating back and forth resonant frequency, which in actuality is a STANDING WAVE (like the picture shown in the previous chapter)? With this divine knowledge Enoch had one thing in mind by employing the Hebrew letter "Ayin" within the vibrating quartz Pyramid… for the sole purpose of having the Great Pyramid harness this effect throughout the connecting limestone grid that spans the land continents - for reasons that most might not be able to broadly fathom at first - but exclusively for the mere purpose of reassuring the continued existence of the earths inhabitants. (Later in history after Noah, they attempted to employ this calendar technology (as well as resonating device) into the "Pyramid of the Americas" (which has a resonating cave underneath). It was recently hypothesized by researchers that the ancient city was most likely abandoned. The reason it was abandoned was not because the Pyramid calendars weren't working properly with the planting and sowing seasons, but because the city became over populated with more and more people flooding in from around the world coming to live in this lush flourishing city, and they didn't have enough harvested food to feed all the people. Therefore, the Pyramid temples were said to be set on fire and the city abandoned because the god's had failed them).

Now with Enoch's understanding that frequency is the guiding signature to all life, he recognized (with divine guidance) that the specific frequency of "Ayin" was necessary for the continual renewal of all life in the event that the earth was disturbed in any way by some sort of global disaster. Enoch could compensate for this with the Great Pyramid in such a way that the "Ayin" frequency pattern would resonate not only along the limestone grid lines, but within the atoms, restructuring the molecules and imbue G-d's original design for life... This was accomplished by simply generating the pyramid with "vibrating pressure" which sent the designated frequency pattern of the "Ayin" through the limestone quartz plateau, which then traveled all the way through the ley lines (The earth's electromagnetic grid) spanning the land and sea of the entire earth. If this resonating sound were amplified loud enough and long enough down the limestone plateau, nature would be forced to obey. Furthermore, its also interesting to know that sound travels faster through liquids and solid ground then it does through air. Enoch in a sense had a back up plan that could implement bioengineering the planet.

To our amazement, this is the internal atomic structure of quartz crystal**:**

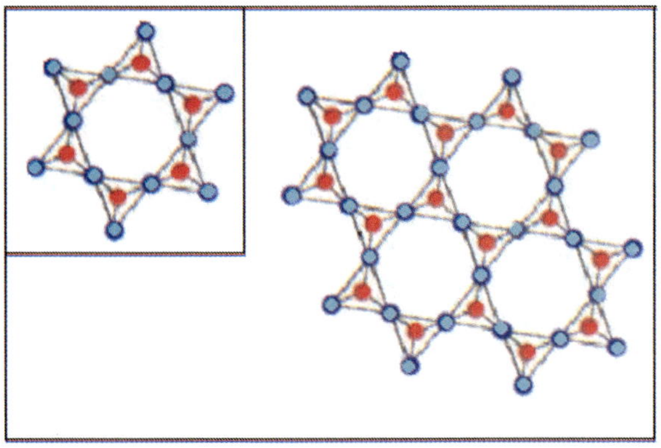

What I find ironic is the atomic structure (of each atom) of the quartz crystal itself, which displays the STAR OF DAVID PATTERN! So additionally, every ATOM within the limestone is also resonating. In addition, since we now know that the "Ahhh" sound creates the Star of David pattern, which in essence is the hexagonal shape of two intertwined triangles. We can now also associate that when it resonates (vibrates) that it ALSO produces the sound effect of the "Ahhh" sound.

Who among us does not think this is amazing!?

This was the very frequency that Enoch was harnessing from the earth itself - a kind of geo-vibratory resonance that was not audible to the human ear. He was using "low audible frequencies" that when generating and traveling through the "ground" may have an effect on nature, the environment and life. This may sound far fetched but when we stop to think of plants growing and being affected by sunlight, what we are really witnessing is the effect of a portion of the electromagnetic spectrum on plants, (the portion which includes visible light). Light in essence moves in waves similar to sound. It should not surprise us that "sound" (vibratory in nature) also influences growth because in essence it is an extension to other parts of the electromagnetic spectrum. (Readers please research this information for yourself. Plants that are subjected to certain sounds, frequencies, light and music tend to grow stronger and faster). This not only applies to plants, but all life.

Now, one may wonder just how the Great Pyramid generated the "Ayin" sound. In other words, where did the vibration come from? I have somewhat of a clue as to how this was achieved. But firstly, we must rehash over the understanding that Enoch constructively employed the Henge mechanism of the stars within the Great Pyramid which was a 365 day calendar that gave us the knowledge of knowing the proper time of planting and harvesting the soil for an abundant crop of food. And along with this (seeing that the Great Pyramid was built in Egypt) Enoch would need substantial amounts of WATER! Even though there was the abundant source of the Nile River, it was quickly realized that the river tended to move and this created a problem as to how to access water to large areas of growing crops. On top of that, they could not always rely on the Nile River to rise when they needed it to. The Nile has also been shown to dry up in urban areas that the Egyptians built cities around. Therefore, Enoch had also incorporated an additional mechanism into the Great Pyramid that would allow water to always be available, and we're not just talking locally, but also far reaching across the entire landmass of the earth. So just how did Enoch do this? Well again, we must remember that the Great Pyramids location was at the center of land mass with the ability to send a frequency signal spanning the entire land mass of the globe through the limestone grid, and it was able to do this by utilizing the Hebrew letter "Ayin's" shape and sound which was built within the internal structure of the Pyramid. When drought was in the forecast the great pyramid could give the instruction to generate water (both locally and afar)… and when an overabundance of water (rain or flood) was in the forecast the great pyramid could give the instruction to drain excess water (or cause it to stop raining). This would insure and guarantee a good harvest for the locals as well as the survival of the human race after the Great Flood of Noah. This far-reaching power may have also been preserved for an even greater future time. Possibly a time such as global warming. In addition, at a time when the Earths water source may completely dry up. In Enoch's book of Parables 1:1-4 he indicated that the book he wrote was "not for this generation, but for a remote one which is for to come".

Some may say all of this is complete bunk! However, I would carefully reconsider. Since there have been many vibratory frequency tests that have been conducted by the military and/or government as well as scientists that conclude beyond a doubt that these types of effects are very real and can have a profound effect on all life. Marine sonar from submarines and military ships at sea is one such type frequency that effects marine animals such as whales causing them to beach and even die. If there were one particular place on the Earth to broadcast such a frequency that could aid mankind in his endeavors for survival, it would logically be at a place on the Earth that is within reach to all other landmasses. Which is precisely why Enoch chose to build the Great Pyramid where he did?

Continuing with this train of thought on frequency… The microscope (as well as the human eye) shows us that each single snowflake is uniquely a symmetrical treasure. Job 38:22 - Have you entered into the treasures of the snow? What's actually making these symmetrical snowflakes is sound in the upper atmosphere as well as the water molecule itself, and as you know the "Universal Law", I have been ascribing to. So in essence, it is the sound that is forming the shape of the snowflakes, which adheres to Phi/Pi, Golden Mean Spiral and the Fibonacci sequence, which synergistically works together in cooperation to turn out "the Universal Law". This understanding follows when we stop to consider that the snowflake resembles the sound pattern of the Hebrew letter "Aleph" which forms the "Star of David" pattern in sand (shown above in the previous chapter), and as we also come to find out, the atoms within the structure of quartz (shown above). Now if we could somehow replicate (and duplicate) the pattern of water down to the exact atomic structure of the water molecule itself with its specific frequency pattern, and reverberate it back into the atmosphere of a dry environment, the vibratory sound pattern would be saying, “make water”. It is interesting because falling rain actually has a frequency sound peculiar to itself. There have even been experiments performed by certain scientists and particular people that have broadcasted these rain frequencies into the atmosphere of a air polluted city and found that these sound frequencies miraculously clear and clean the air as well as the water. Strangely, it also has a profound effect on the crime rate (lowering it) and the psychological well being of people (making them tranquil, peaceful, content and cheerful). It’s very simple really - all they did was play the waveform of the sound of rain into the air with an ordinary radio signal of music. In other words, they took the dimensional sign wave signatures of the sound of rain and converted it into a pattern that could be reproduced as a harmonic of sound and then put it into the form of music. However, these tests rendered inconclusive to the government because it could not be proved that these frequencies are what actually attributed to cleaning the air. There are also certain regulations that the EPA requires (Environmental Protection Agency).

Getting back to Enoch... So Enoch incorporated this very same resonance mechanism that worked by extrapolating these very same sound frequency patterns that would resonate within the atoms, thus restructuring (or should I say reprogramming) the molecules and creating rainwater both through the ground as well as the atmosphere. In essence, the nature of the pattern frequency determines what the resulting structure will be. This also works in a chain reaction fashion where each molecule of water influences the other molecules (like a tuning fork) that contains the original frequency or tone (picture shown below). This could give us a general idea as to the pyramids near and far reaching capabilities.

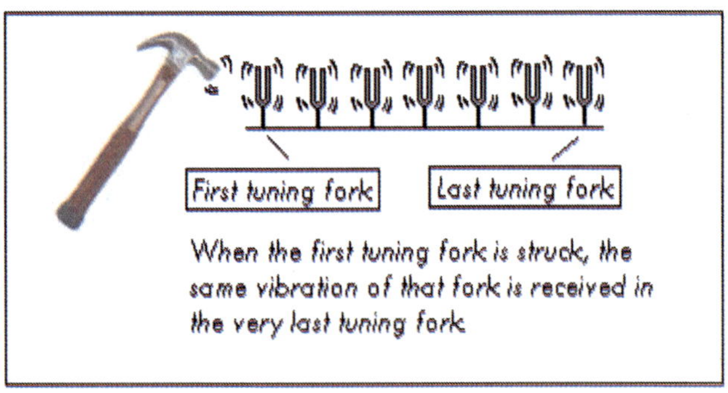

Secondly, besides its far-reaching capabilities in this respect, Enoch also devised the Great Pyramid to be a local water pump. We can begin to get a better idea of how the Hebrew letter “Ayin” was utilized if we can analyze how it was used in a local way as a water pump (as the suggested picture below demonstrates).

Egyptologist has taken notice of remnant traces of an existing rim that once surrounded and enclosed the Pyramid. It is speculative if this rim around the Great Pyramid is a “moat” which could have been used to pool water generated by the simple sound frequency (the Ayin - Ahhh sound) to pump water out the Great Pyramid in some fashion to nearby vegetation to ensure a good harvest during times of drought.

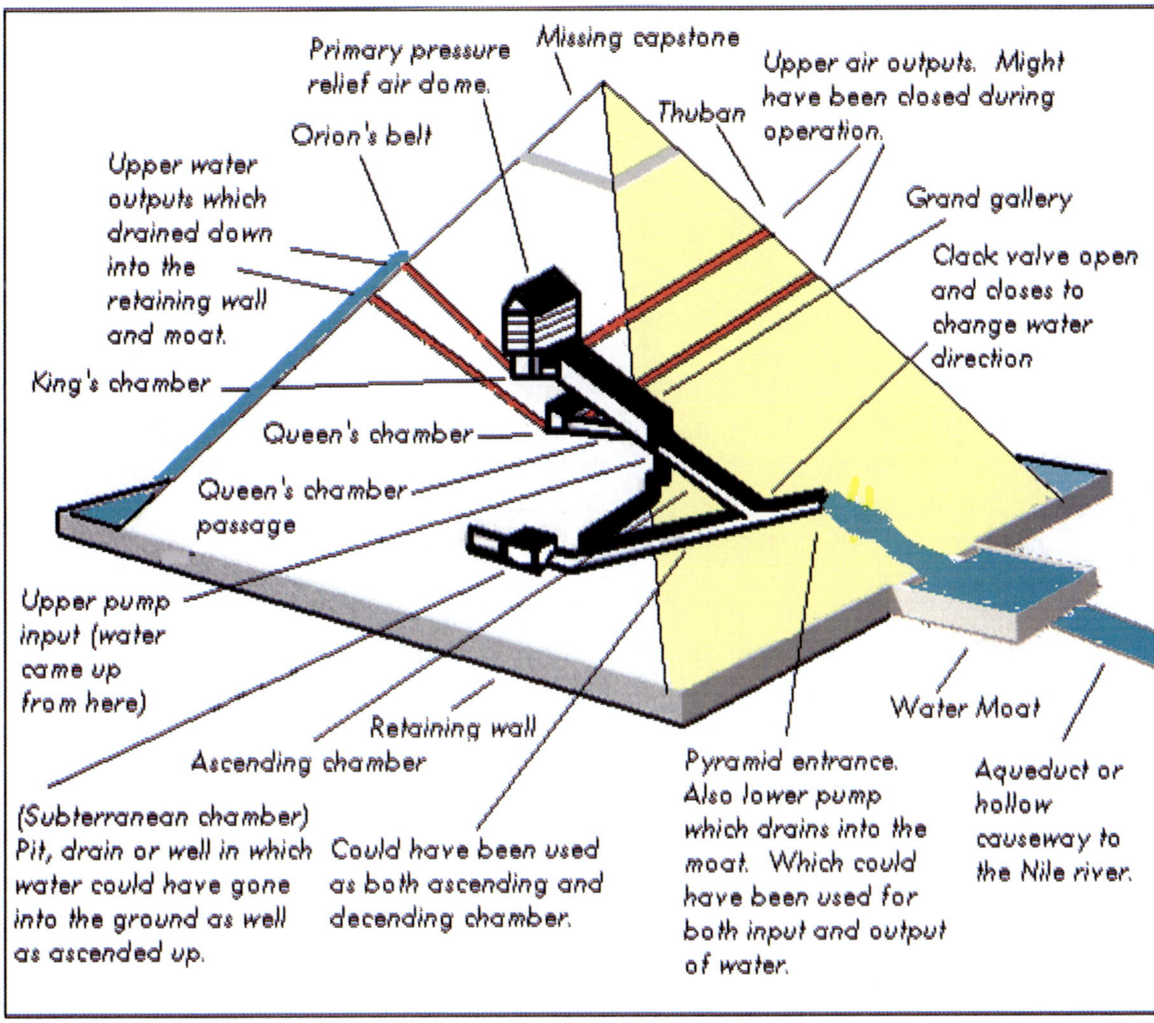

Water can be manipulated in many clever and dexterous ways. One of the ways water can be used is to create "SOUND" by fluid and acoustical dynamics such as when it is pulsated or rapidly flowing in a jet stream of water. Now when this forceful jet of water hits one of the interior limestone walls of the Great Pyramid it also creates a "sound" reverberating deflection - and thus this vibratory sound is what generated throughout the whole Pyramid operating the "Ayin". I'm assuming that this process worked in partnership between the very bottom subterranean chamber as well as the Pyramid entrance of the Great Pyramid. (Egyptologist have concluded that the very bottom subterranean chamber was left unfinished, abandoned and served no function). One way the water may have forcefully entered to create this phenomenon was either through the main entrance or the subterranean chamber pit (the well at the bottom) that was regulated by, or worked in combination with a "clack valve" that changed the directed of the water flow. As one can see from the picture above the main entrance of the Great Pyramid is above ground level for which this purpose is very telling. The purpose for this was so that the water tunnel would have been level with the moat (retaining wall) that surrounded the base of the Great Pyramid. As the moat around the Great Pyramid filled with water this allowed for water to be "gravity siphoned in" (or backwashed) with great pressure into the main entrance tunnel and up the ascending passage into the Grand Gallery that leads to both the King's chamber and the Queen's passage and chamber. However, in the picture above, I show the water running down into the water moat, but there may have also been a downward passageway that connected the water flow so that it could also be siphoned up and back into the Pyramids entrance. (?)

Modern technology concerning fluid dynamics in a hydraulic water pump reveals that water could be propelled upwards by opening valve 1 and closing valve 2. Then valve 2 opens with valve 1 closed. The resulting pumping action sends water up the top pipe to a higher level.

Curiosity then asks us how did the water drain from the Great Pyramid. The answer to this appears to be by an underground tunnel (the hollow causeway) that was built underneath the main entrance and water moat that leads to the Nile River. This aqueduct shaft is still there (found today by Egyptologists), though they claim it is no longer functional. Additionally, water was pumped up to the top chambers of both the King's and Queen's chambers and then forced out the "upper water output shafts" to trickle down the side of the pyramid into the retaining wall, which was used to store water to be distributed by some sort of irrigation system to nearby vegetation and gardens.

Its also quite possible that they even used this water system TO BUILD the Great Pyramid itself by means of elevating the huge limestone building blocks up the "well shaft" (the Subterranean chamber). This would have taken a huge load off the blocks.

The fluid dynamics of a modern hydraulic water pump. The large upper tunnel of the hydraulic pump is filled with water. First, valve 1 opens with valve 2 closed, then valve 2 opens with valve 1 closed. The resulting pumping action sends water up the top pipe to a higher level.

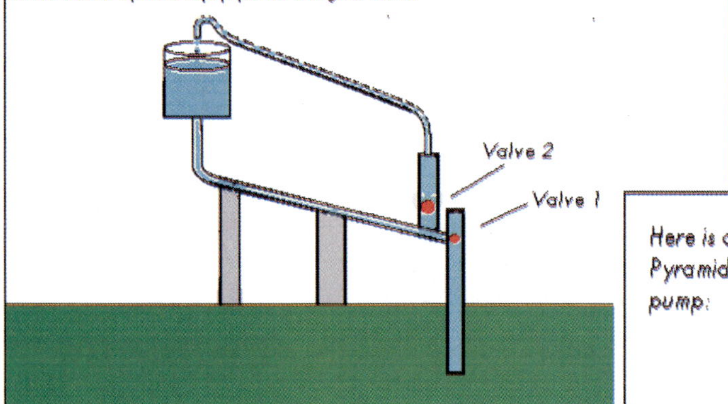

Here is another simple suggestive picture of how the Great Pyramid could have worked using frequency as a water pump:

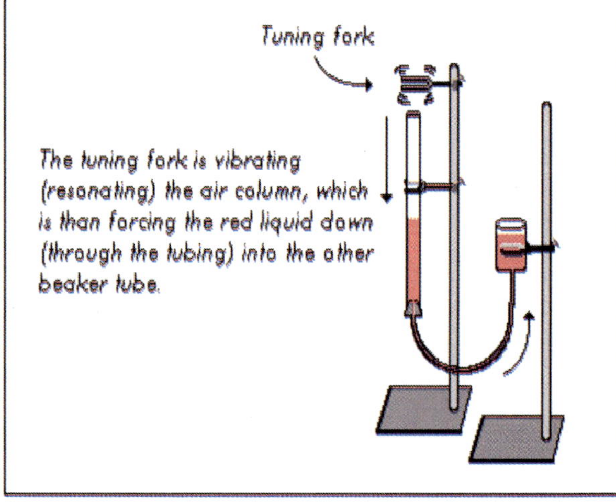

There is great physics that apply to sound (in fact sound may be the most powerful tool as well as weapon to ever be harnessed). The Indians, the Tibetan monks, and shaman healers used music that was prescribed as medicine. Likewise, the Jewish mystics taught that the Divine Word, the speech of G-d, creates and sustains both the Heaven and the Earth. They, too, believed that the various objects in our Universe are different because of a specific and unique combination of letters and sacred speech that sustains everything. Even Pythagoras believed that each planet hummed. We are now proving what these ancient groups had known all along. In fact, the field of Quantum Physics is demonstrating that all of creation is composed of vibration in the form of waves of energy or vibration. Under a microscope, we can see that everything vibrates. Which include molecules moving, and as we look further we can see the protons, neutrons, and electrons vibrating, and looking even further we can see quarks bopping around down to the nitty-gritty of photons (which was talked about in a previously chapter) are composed of the very essence of light itself.

Therefore, we know that these forces exist, and modern science has even been smart enough to employ them into modern medicine and machines. For instance, machines that use "lithotripters" (which use sound waves, or what's called "ultrasound"). They use these lithotripters (sound waves) to shatter kidney stones. In fact, ultrasound frequencies have even been discovered to kill microbes under a microscope. Additionally, ultrasound frequencies have been used to repel vermin (pests, rats, mice and cockroaches). Ultrasound frequencies have even been used to remove stains from cloths and other things. The military has even devised "frequency" into weapons of war (called sonic weapons), which use what's called "Infrasound". Infrasound is a powerful, ultra-low frequency acoustic wave in which sound waves below 16 Hertz bypass our ears but can be felt by our body in the form of vibrations. All the sound that we hear (from the lowest bass to the highest treble) is between 16 and 20,000 Hertz. Sound waves above or below these levels cannot be heard by the human ear. And it is these vibrations, (dependent upon their intensity) which some researchers say can induce a range of symptoms, from nausea, headaches, vomiting, queasiness, dizziness, disorientation, imbalance, intense irritation, sweating, and potentially a burning sensation on the skin since it heats up the body. The worst-case scenario is that it could rupture internal organs and cause death. These frequencies can also be used as a high-frequency sound, or a very high-pitched sound, which can be audible to the ear.

However, infrasound is no new invention. It reveals itself in nature. It is produced by such things as earthquakes, thunder and erupting volcanoes. The sound waves can travel vast distances and are not blocked by stone, buildings or other sounds. In fact the more we learn about balanced nature in the forest and garden we discover that certain sounds (and their frequencies) likes the birds, bees and crickets causes flowers to bloom and even ripen fruit. Research has even shown that low frequencies emitted by power lines and home appliances (even air conditioning fans, computer servers, cars and airplanes) considered "white noise" can lead to increased stress and heart palpitations.

Furthering this thought… There is nothing more beautiful than the human voice in song! When we sing, our pupils dilate and endorphins (our body's natural painkillers) surge through our bodies. It also increases oxygenation in our blood flow, tones the nervous system, heightens our immunity, and affects glandular secretions. An opera singer who sings a note that has the same resonant frequency as a crystal wine glass creates a resonance that can shatter the glass. This occurs because of the external pressure being greater than the internal pressure, collapsing it inwardly. In addition, the construction industry uses sound to break up concrete. Frequency is also used to kill and destroy bacteria and pests on fruits, vegetables and nuts (especially those that come from foreign countries). And most amazingly it has even been discovered that when radio frequencies pass through salt water it ignites into fire, and this they purpose is going to be one of the greatest inventions and investment of the 21st century allowing us to eliminate fossil fuels for organic salt water. What's actually taking place here is rather simple… the vibration of the radio waves shakes the oxygen molecules just hard enough to break the hydrogen bonds, thus freeing the hydrogen gas, which then ignites and burns. Scientist stood in wonder when they witnessed this phenomenon, but we must now ask ourselves that if a frequency can ignite a fire from salt water, we must consider what else frequency can do. The potentials are far reaching! Frequencies can also heat up and incinerate certain metals. Soon antigravity may be discovered as well as a cure for cancer. In Exodus, chapters 14:19-20 it reminds me of the pillar of fire that stopped the Egyptians at the red sea (consisting of salt water) while they crossed. Could G-d have used frequency to accomplish this? In addition, could it have been a frequency that caused the wall of Jericho to fall? Joshua 6:20 - When the trumpets sounded, the people shouted, and at the sound of the trumpet, when the people gave a loud shout, the wall collapsed; so every man charged straight in, and they took the city. How about Exodus 14:21 - When G-d instructs Moses to stretch out his hand over the sea and G-d, with a terrific east wind all night long, made the sea go back. He made the sea dry ground. The seawaters split. REMEMBER G-D IS EQUIVALENT TO NATURE (HATEVA) in which both words share the same gamatria equivalent of 86. Therefore, He used the forces of nature to work His ways.

Additionally I often wonder that if Enoch was taken and translated (converted and transformed) at 365 years of age by G-d (which is specified in the Bible). I wonder if the King's chamber sarcophagus within the Great Pyramid was his ethereal burial tomb (or where he was spectrally transmuted) somehow by the Great Pyramids resonating frequency, as his body was never found. I also have in mind to compare that the Bible states that Moses was also taken up by G-d, as well and Jesus' body just disappearing from the tomb. To this date it has never been proven that the Great Pyramid was actually a burial tomb, but there are hieroglyphics in Egypt that tell us that it was a machine for turning men into god's. Could it be that "frequency" at the right resonance can transmute human flesh?

After reading all this one, one cannot merely say that this is far fetched. It becomes clearer and clearer to ones understanding that the Great Pyramid was indeed constructed to project "resonating sound" to what effect we cannot really say for sure, but it is easy to see (as well as science rediscovering) that using vibratory frequency to reform the atomic particles in the ground as well as the atmosphere, to create whatever environment is needed, is grounded and very factual in science and physics.

Now when we accumulate all these things together we can conclude that not only did the designer of the Great Pyramid (Enoch) ingeniously incorporate it as a 365 day solar calendar, but also a water pump generated by a pumping device, which collected and siphoned water to nearby vegetation to ensure a good harvest during times of drought. Not only a solar calendar, a water pump, and a far andwide-reaching unfathomable rainmaker that utilized the Earths geo-electromagnetic limestone grid along with the frequency technology of the Hebrew letter Ayin. Additionally, we can also add to this list a SOLAR ENERGY CONVERTER - as one has to wonder what the functional purpose of the white polished limestone casing stones (mentioned earlier) were used for? It's a far cry in the dark to suggest, that aside from looking trendy, they were polished to the point of shining, kind of like today's solar panels to convert the Sun's energy into usable electricity. It has become even more interesting that the great pyramid also has four slightly curved-in sides that can only be seen from above. This could have served the purpose for water to run down efficiently. In addition, as an added feature, the Sun reflecting off the highly polished faces acted as a solar telescope or beacon!

Lets delve into frequency just a little bit more... Its also interesting to relate how the Great Pyramid assimilates with vibration and resonance when we look to other things in conjunction - like why was the "BULL" (meaning leader and head) was revered and worshiped in Egypt. The "BULL"(or Ox) has even been associated with the Hebrew letter "Aleph" (if you look back in the chapter on language similarities you will see this). We all know from Egyptian hieroglyphics that the "BULL" was worshiped, but do we truly know the real reason for WHY it was worshipped?

Funny that most don't have a clue and just except what is written in books about this stuff. Even most Egyptologist themselves don't really have a clue. I finally derived a conclusion and figured out that the sacred BULL that was revered by the Minoans from the island of Crete and Santorini (who have been "thus far" accredited with being the fabled Atlantian's we hear about). They too worshipped the BULL (as we found in their artwork) for reasons that walked, carried or rowed itself all the way to ancient Egypt.

The reason they worshiped the BULL was said to be for fertility reasons... but still one wonders why or how the BULL interacts with "fertility"? And the reason will soon astound you, that the reasons the Minoans and the Egyptians worshiped the BULL was because WHEN IT RAN IT CREATED SUCH A MOMENTUS VIBRATION ON THE GROUND THAT IT WAS FELT BY HUMANS! This "vibration" was an expression of energy. All vibration is force and energy itself - like a tuning forks that is vibrated is capable of influencing and reproducing the same exact cloned frequency in the other tuning fork (like shown in a previous picture). Likewise, this frequency (vibratory in nature) is capable of reproducing exactly the same results within us. When this frequency synchronizes in rate within us its called sympathy, attraction or love. Now when the BULL is associated with that of "fertility" this is because a person who interacts with another person can produces mental vibrations in the form of thinking, thoughts, emotion or feeling, (thus the "head of the Bull" and the Hebrew letter "Aleph"). This also ties together well with the archeological evidence of the Minoans (and their artwork of the Bull) which were found not only on the island of Crete and Santorini, but also in Egypt.

In my final analysis things started to come together quite nicely when I picked up a book about Ancient Egyptian culture and took a close look at the statues of the Pharaohs that are clasping cylinder-like object in their hands.

Found by Egyptologists, these are two metal cylinders made of copper and zinc and filled with quartz crystal, which as you now know bears the atomic structure of the Star of David. One cylinder for each hand. They are said to be tuned to the Earth's natural frequency, which dictates the energy impulse to all life on the earth (but so is the Great Pyramid also tuned to this same frequency). Egyptologists tells us that it was supposedly thought by the Egyptians that these cylinders served the purpose of stimulating the nervous and immune system and helping to fight off diseases. Though one wonders, why would the Egyptian Pharaohs hold an object that was for health purposes as a sign of power or deity unless its purpose was linked to something of far greater influential power like the Great Pyramid? Just another theory though, but is it possible that these metal and quartz objects (cylinders of resonating vibration), are receptor mechanisms that worked as an agent in conjunction with the Great Pyramid. In other words, the quartz crystal (inside the cylinders) is capable of conveying information passed from the human body to the Pyramid by relaying a signal originating from human consciousness in a similar fashion to the vibrations of radio wave frequencies striking a crystal in a radio. In other words, its quite possible that the Pharaohs used these cylinders (held in their hands) in a particular location within the Great Pyramid to resonate their conscious thoughts as a way of sending their specific vibratory frequency signal through the limestone grid lines to perhaps either control the weather or environment, as well as the emotional well-being of the human race. This particular location within the Great Pyramid could have been inside the King's Chamber sarcophagus, which is known for its peculiar acoustics as well as the precise location of Phi/Pi (the Golden Mean Spiral). Lets also not forget that the Egyptian pharaohs where looked upon as gods in the sense that they had some kind of superior influential power over the weather and the environment.

There is something else of great interest that ties in with the Great Pyramid... A limestone monument and memorial that was built near it... THE SPHINX - with the body of a cat, and the head of a pharaoh!

Though one wonders was it truly built as a CATS body with a pharaohs head, or was it wholly built as a CAT (both body and head) at one time? The smaller size of the pharaohs head attached to the cats body looks out of proportion - Like it had been sculptured down in size. Common sense tells me that it was originally sculptured as a "cat" - head and all. The Egyptians revered "cats" to the extent of making the goddess "Bastet" out of it. At one time in history the cat symbolized the biggest celebrated festival in all of Egypt which took place in the city of Bubastis which possessed Egypt's greatest temple. Some say the cat was revered because at one time in recorded history Egypt had a rat infestation that carried a dangerous plague - whereby the cats ate and chassed away the rats and kept the plague at bay, and this may also be where the superstition came in about cats chasing away ghosts or evil spirits, and it may also be that cats can see at a frequency beyond humans that can detect ghosts or evil spirits. "Bastet" by her name alone (from the research I've done) seems to be related to perfumes, alchemy and the mother of "healing".

The Sphinx with the Great Pyramid directly behind it. It almost looks like an arrow pointing the way showing that there is a direct link and connection with one another.

Bastet depicted holding a Sistrum and an Ankh.

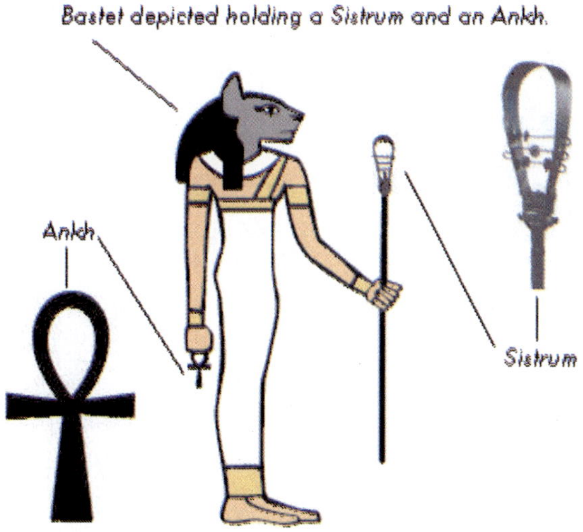

During the festival of Bastet, they had music, singing, clapping and dancing "representing lots of noise". The main objective of this festival was to produce "NOISE". Bastet is depicted holding a Sistrum (a rattling musical instrument similar to a Tambourine) and an Ankh (symbol of life, or the invisible G-d).

The word "ANKH" is a very fascinating word in itself! One can immediately hear the "Ahhh" sound associated with it and it becomes apparent that the Sistrum that Bastet is holding in the other hand (which creates a musical sounding noise) is reinforced by the "ankh" that clearly specifies that it is specifically for reciting and delivering the Ahhh sound. It becomes even more interesting with the association of words and names. For instance, the "Ankh" is an Egyptian word that is carried over from the Hebrew root word of Anochi, which is another name for G-d (it means G-d is with us). Also the Hebrew word Adonai is another name for G-d (which means master). Meanwhile the word "Ankh" is used in several of the god names of the Pharaohs such as (Tut-ankh-amun, Ankh-enaton, Sem-ankh-are). Ankh was usually incorporated into a Pharaoh's or a very important person's name likes Joseph from the Bible (Genesis 41) who was made vizier of Egypt and whose Egyptian name was "Tsopnut-pAnkh" (tSOPH nut) - "Tsophnut" is the goddess of rain or moisture. It is possible that Joseph was called Tsophnut Pankh representing "the moisture of life" because it was rain, water, and moisture, associated with the Nile river which determined the seven years of plenty or famine which he predicted in Pharaoh's dream? (Read Genesis 41). Bastet is also associated at the daughter of "Asar and Aset" (Asar and Aset are the Egyptian name for Osiris and Isis) which were the Greek names given to them. Notice how these names also contain the "Ahhh" sound.

Its also interesting that other names and words are also associated with the "Ahhh" sound such as, Amun, Atum, Atun, Amun, Aton, Amen-Ra, Anubis, Amenemhat, Ptah, Ra, Asnut, Yahovah, Yashuah ("Yashuah" which is Jesus in Hebrew) and even the all pervasive commonly used word - Amen.

This sound becomes even more reaffirmed when the Egyptians named the "cat" after its sound - MEOW! And is described as almost the same way in all languages of the world today either as the sound of "meow or cat". The Egyptian hieroglyphic word for "cat" is MIU (read: mee-u), and the Hebrew word for "cat" is "khah-tool" (notice the Ahhh sound again). Which makes it insatiable for me to reiterate that language and writing was not originally Egyptian, it was linguistically Hebrew first. In fact the word "pharaoh" is from the Hebrew interpretation of the word "per aa", which means "the great house". Does one think this could also symbolize the "house of the stars" as the Great Pyramid of Enoch?

However, one wonders what really is the true secret of the "cat"? I have a theory on this that I would like to purpose… In addition, it is really the smoking gun in all of this! Few may know this, but the "cat" has a very special quality and feature about it, and this is just not with a specific breed of cat, it is with all cats even lions and tigers. THIS SPECIAL QUALITY HAPPENS TO BE ITS PURRING NOISE! In fact this purring noise is at a frequency that has been found to heal bones, tendons, ligaments, joints, muscles, wound healing, reduction of infection, swelling, pain relief and pulmonary disease in all living things. Compared to other animals, cats don't ordinarily get chronic pulmonary disease, bone disease, muscle and tendon injuries as much as dogs - the purrr seems to be a constant strengthener. Cats rarely get cancer, but when they do, it seems to only be in their lower extremities such as the legs where the frequency is lower. (I encourage all readers to investigate this for themselves to learn the frequency of this sound range). But the residual noise admission that it puts out is none other then the Ahhh and Ohhh vowel sounds. This I believe is the secret of the "cat" and its special healing quality that Enoch meticulously elaborated on. For that reason, I'm putting across the idea that Enoch knew this healing secret and demonstrated it as the Sphinx. He not only held the "cat" in high esteem for this healing quality, but he built the Great Pyramid to include and "call attention" to this very same healing frequency sound. The resonance of the Hebrew letter "Ayin" just like "Aleph" reinforces this healing frequency and keeps the earth in tune.

The Sphinx may have also served another purpose as the constellation Leo that consists of "12 stars" and is in the shape of a Lion (or a cat) and is the symbol given to the tribe of Judah. The 12 stars are a representation of the 12 tribes of Israel as represented by the 12 sons of Jacob. (That is why I am connecting the Hebrew word for pharaoh which is "per aa", which means "the great house", but the great house of what? Obviously the stars! This is another reference to Enoch!). So as we can see the constellation Leo (consisting of 12 stars) has been used to represent the Tribe of Judah, even Jacob, as well as the tribe of King David and subsequently Jesus. In addition, remember, this was a message left by Enoch of the coming Messiah. In astrology, the constellation of Leo the Lion has always been associated with the Sun because it was common knowledge that the Sun appeared to enter during the time when the Nile River flooded its banks with water and nourished the land.

Getting close to ending this chapter… The culmination of evidence surrounding the concept of "frequency" has become very intriguing… From the cylinders to the Sphinx - working in conjunction with the Great Pyramid! One is really beginning to marvel here, that the real deal has always been about the mechanism of frequency and its near and far-reaching potentials. But one thing is unwavering from an audible standpoint! And this is the fact that the sound of "Ahhh" seems to permeate almost everything we hear with our ears in the physical world and also in the invisible world. No matter where we go on this earth, we cannot evade the "Ahhh" sound, even the ocean with the sound of the waves rushing in along with the wind blowing past our ears. It is this sound that reassures well being. It is this sound that allows for our very existence contingent to our place in the solar system and to this Earth. In fact, this same frequency sound in the range of 6.8 hertz (the Earth's balanced resonating frequency) is employed into the NASA space shuttle so that the astronauts will not become disorientated in space (thus having an effect on our human brain). It is also this frequency in part that keeps our muscles and bones from fully deteriorating in space.

In my concluding scrutiny, I will admit all of this may be conjecture, and that I may have in some degrees overstepped the boundaries of speculation. And I must also truthfully admit that the facts I found surrounding the Great Pyramid engages in all manner of vagueness and innuendo, from its accurate measurements to the evidence surrounding its true purpose is all a fistula of facts that seem to be strangely obscured. There is a lot of guessing and wondering going on here. Nothing is certain about it. In all fairness, and as esoteric as all this may sound, it leaves us open to all hypotheses and ideas concerning what purpose the Great Pyramid actually served. But why not combine all of these theories into one grand design as Enoch may have (in which he made a marriage out of all the best technology). Because in all truthfulness NONE OF THEM MAY BE WRONG!

Conceivably, I may have also overstepped myself concerning the end of days in realizing that Enoch may have created a mechanism (in the structure of the Great Pyramid) in which to bioengineer the Earth in the event of a cataclysmic event and to foretell when the Messiah will return. One wonders is the Great Pyramid still functional? The answer to this lies in the King's Chamber where structural damage has been observed. The roof beams are cracked and dislocated as well as the floor stones. Also, the walls have been moved outward. This damage might be attributed to an earthquake in past history. However, there doesn't seem to be any damage to any other chambers or passages.

As to the question of its workability? There is a still a great deal more about the Great Pyramid that none of us know, and it may be imperative that it remain that way as an element of surprise. Perhaps you are familiar with the story of the Red Sea and the Egyptians.

Exodus 14:9 - The Egyptians—all Pharaoh's horses and chariots, horsemen and troops—pursued the Israelites and overtook them as they camped by the sea near Pi Hahiroth, opposite Baal Zephon. Exodus 14:13-16 - Moses answered the people, do not be afraid. Stand firm and you will see the deliverance the Lord will bring you today. The Egyptians you see today you will never see again. The Lord will fight for you; you need only to be still. Then the Lord said to Moses, Why are you crying out to me. Tell the Israelites to move on. Raise your staff and stretch out your hand over the sea to divide the water so that the Israelites can go through the sea on dry ground. Exodus 14:23 - The Egyptians pursued them, and all Pharaoh's horses, chariots, and horsemen followed them into the sea. Exodus 14:26 - Then the Lord said to Moses, Stretch out your hand over the sea so that the waters may flow back over the Egyptians and their chariots and horsemen. Exodus 14:28 - The water flowed back and covered the chariots and horsemen—the entire army of Pharaoh that had followed the Israelites into the sea. Not one of them survived.

11 - TOWER OF BABEL...

A quick chapter before we move onto the next... Again, I was trying to make a link. This time between the Great Pyramid of Giza and the Tower of Babel. They both employed the Hebrew language and alphabet. I felt that the demise of the Tower of Babel was in part due to the usage of the Hebrew language. Therefore, I reviewed again the biblical story of the Tower of Babel of when language became confused.

Read thoroughly the following Bible story, since we all have heard or read this story many times before: (Genesis 10:32 & 11:1-9)

Genesis 10:32 - These are the families of the sons of Noah, after their generations, in their nations: and by these were the nations divided in the earth after the flood.

Genesis 11:1 - And the whole earth was of one language, and of one speech.

Genesis 11:2 - And it came to pass, as they journeyed from the east, that they found a plain in the land of Shinar; and they dwelt there.

Genesis 11:3 - And they said one to another, Go to, let us make brick, and burn them thoroughly. And they had brick for stone, and slime had they for mortar.

Genesis 11:4 - And they said, Go to, let us build us a city and a tower, whose top may reach unto heaven; and let us make us a name, lest we be scattered abroad upon the face of the whole earth.

Genesis 11:5 - And the Lord came down to see the city and the tower, which the children of men built.

Genesis 11:6 - And the Lord said, Behold, the people is one, and they have all one language; and this they begin to do: and now nothing will be restrained from them, which they have imagined to do.

Genesis 11:7 - Go to, let us go down, and there confound their language, that they may not understand one another's speech.

Genesis 11:8 - So the Lord scattered them abroad from thence upon the face of all the earth: and they left off to build the city.

Genesis 11:9 - Therefore is the name of it called Babel; because the Lord did there confuse the language of all the earth: and from thence did the Lord scatter them abroad upon the face of all the earth.

(As we‘ve already discussed in an earlier chapter), this fall had to do largely with ALL the people on Earth, (the descendants of Noah and his three sons), who speaking only one language and having only one government order were candidates for destruction simply because they lacked diversity and strength. This weakness allowed Satan to get a foothold in their government order which became corrupt by disobedient works, believing in idols and other gods other than the one true G-d Himself. If you can recall, it is diversity that makes us strong, and it insures us against extinction. Without diversity this would give Satan a perfect foot in the door to take hold, corrupt and destroy all humankind. Remember, Satan wants unity in error, and a government that consists of organized wickedness and disrespect towards G-d. This is why G-d had to scatter, separate and disperse humankind! You see, Satan prefers an organized One World Order were it’s easier to control, while G-d prefers division to preserve His truth. We know this to be true because Jesus said in Luke 12:51 - Do you think I came to bring peace (unity) on earth? No, I tell you, but division.

In addition, even though diversity and division may seem harsh, it is infinitely more precious then a satanic unity. Therefore, G-d’s demolishment of equal unity was a blessing. It was an incredible outworking of purpose that someday we will all praise Him for, in His infinite mercy towards the human race to prevent us from plummeting into endless immorality and corruption that ultimately leads to total destruction. G-d demolished the tower project by placing a protective barrier and stronghold within the human race against evil destruction. This was called “linguistic cultural and racial barriers“. Thus, the confusion of languages put a crashing halt on any further joint effort to complete the tower. This had to be implemented by G-d so that in Satans attempt to take over the world he could not (and cannot) penetrate the WHOLE human race with one sharp blow and completely destroy it instantaneously. One should marvel at how the amazing mechanisms of nature work in-sync with G-d to insure survival, not only for the human race but also within the animal, insect, and plant kingdoms.

However, is it also possible that the destruction of the Tower of Babel was also due to another cause? Perhaps the corruptness of the people involved the combination of the usage of this one language (which was Hebrew). Perhaps they may have somehow retained the knowledge from Noah and Enoch and incorporated within the tower (that ascended to great heights into the atmosphere) the technical ability to control the environment (much like the Great Pyramid). We know they certainly had the recourses to build the Tower of Babel. Genesis 11:3 - And they said one to another, Go to, let us make brick, and burn them thoroughly. And they had brick for stone, and slime had they for mortar. Science is now discovering the difference between sun baked and fired bricks. Fire baked bricks are much stronger. Firing the brick at temperatures of 1500 degrees Fahrenheit for 24 hours causes the clay particles to stick together. This consolidates into a stronger material. The strength of one of these single bricks can withstand 6,000 pounds. This would have enabled them to built the tower almost 2 miles up. This is five times taller than the worlds largest building today! So one wonders, if they had the technical ability to build a tower that far up into the atmosphere, they may have also had the knowable technology along with the inclination of employing the Hebrew letters to have some sort of colossal control over the whole environment (like Enoch employed within the Great Pyramid). The difference here is that the Tower project was not authorized by G-d. Not only was it NOT approved by G-d, but it might have also been hostile to the workings of nature by disrupting G-d’s design and cycle. In other words, they were erroneously misusing this technology for their own self-centered means. They would become spoiled to the point of not appreciating the generous kindness of the natural workings and harmonization of nature (the ecosystem) given to them by G-d. Not only that, they could seek to do whatever they wanted with such controlling power (eventually seeking to dominate and destroy, not only one another, but the environment as well). Genesis 11:6 - And the Lord said, Behold, the people is one, and they have all one language; and this they begin to do: and now nothing will be restrained from them, which they have imagined to do. It was clear to see that Satin had his foothold over the whole Earth and all its inhabitants, and this was unacceptable to G-d. This may have been the critical reason why G-d intervened.

Oh and there is something of importance that we must know so that we can link up the association between the Tower of Babel to that of the Great Pyramid. This Tower of Babel may have been very similar to the Great Pyramid(s) of Egypt as well as many other Pyramids that we find across the globe today. This Tower MAY HAVE EVEN SPIRALED IN PYRAMID FORM! We know this since archeologists have found what's been named "Ziggurats" in southern Mesopotamia, (Mesopotamia is associated with the city of Babel as well as the plain of Shinar). And if you don't know this by now, Babel is known as Babylon, which is now modern day Iraq.

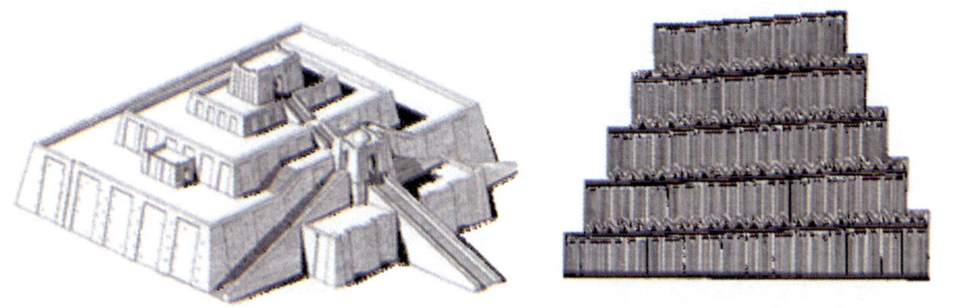

Ziggurat

It may have even tilted with a spiraling effect.

They are called Ziggurats because they zigzag in design (they follow an alternating course which gives the appearance (or impression) of a tilting effect). Though spiraling inside is not structurally the case. This spiraling effect only occurs with each stage of a staircase that reaches to the next stage (staircase). These staircases are on alternating sides so that when you have reached the top you have traveled right around the tower, thus a "spiral" staircase is descriptively true.

Ziggurat (zigzag) design.

Nearly 30 ziggurats in the area of Mesopotamia have been discovered by archaeologists. In location, they stretch from Mari and Tell-Brak in the Northwest and Dur-Sharrukin in the north, to Ur and Eridu in the South, and to Susa and Choga Zambil in the East. The structure at Eridu is the earliest observed structure that some designate a Ziggurat, is dated in its earliest level to the Ubaid period, which is estimated at around 4300-3500 years old. However, its dates are still a matter of uncertainty.

Now for the ultimate clincher in all this… So now one wonders, what could they have possibly employed within the Tower of Babel that made it as powerful and commanding like the Great Pyramid? Did the Tower also employ the Hebrew letter "Ayin" in some way? NO IT DID NOT, THEY WENT BEYOND THIS! Instead of employing the Hebrew letter Ayin; they had employed the Hebrew letter "Pey". Since the Hebrew letter "Pey" signifies a "MOUTH" by way of speech, and therefore language. Hence - we may have found the literal reason for why G-d may have chosen to confuse all languages - simply because their power ISSUED OUT OF THE MOUTH.

We should not overlook the power of the Hebrew letter "Pey" since the first words spoken by the "MOUTH" of G-d was: "let there be light". Genesis 1:3 - And G-d said, Let there be light: and there was light. This is evidence of great power, for G-d spoke and the world came into being. Can you imagine the power that one could possess to create like G-d?! Thus again, the Bible phrase of Genesis 11:6 - And the Lord said, Behold, the people is one, and they have all one language; and this they begin to do: and now nothing will be restrained from them, which they have imagined to do. This then becomes even more clearer as to the Towers ability. The people could do as they wanted and this was mocking and impersonating G-d. They in a sense were mimicking the power of G-d's mouth.

The interesting thing about the Hebrew letter "Pey" is that it comes directly after (following) the Hebrew letter "Ayin" in the Hebrew alphabet ("Ayin" being the 16th letter and "Pey" being the 17th letter). Thus, they were looking for a little more power than "Ayin" (just looking) could give them. They wanted complete and utter power!

Now the inner space of the letter "Pey" reveals the Hebrew letter "Bet". And since the first word of the Torah "Breshith" represents the "house" of creation therefore finding the letter "Bet" within the "Pey" represents the "Word of G-d" that created the Heaven and the Earth. John 1:1-3 - In the beginning was the Word, the Word was with G-d, and the Word was G-d. The same was in the beginning with G-d. All things were made by him, and without him was not any thing made that was made. In other words, the very first Hebrew letter of the Torah is a picture that the WORD of G-d is the agency behind the entire Universe. Psalm 33:6 - By the word of the Lord were the heavens made, and all the host of them by the breath of his mouth.

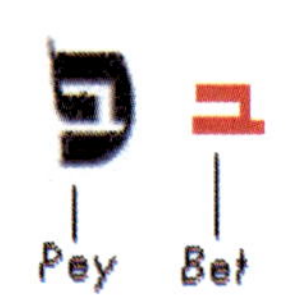

The Hebrew letter "Bet" can be found "within" the Hebrew letter Pey.

In some sense we can almost say that the Ziggurat (zigzag) design looks like a Hebrew letter "Pey".

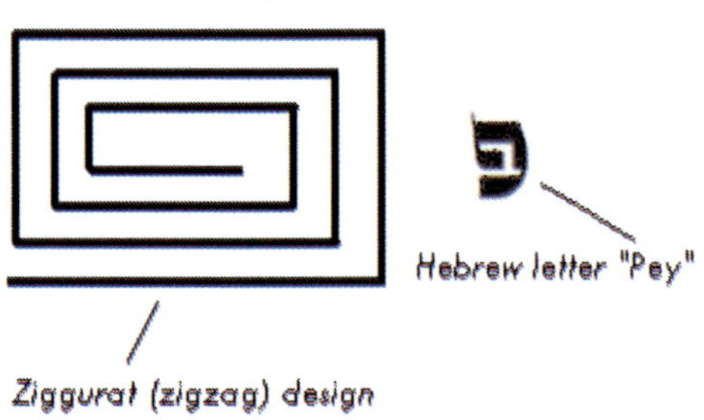

In conclusion… The sheer magnitude and scope of all of this is very overwhelming! I went from being exuberantly joyous, happy and satisfied in having uncovered the furthermost mystery of G-d by virtue of the Hebrew letters which are formed from natures law - to the extremes of feeling a sense of regret in realizing its technological dark side, giving rise to the false grandeur of Babel by corrupting what G-d had created in perfection with the Great Pyramid as well as all of creation. Sadly, I feel what started as a quest for knowledge has turned to an abusive conquest (and will continue throughout history) by being naive in not realizing that any species (such as ourselves) with a sense of curiosity and intelligence is in actuality and even perhaps unknowingly setting out to dominate and conquer! Ezekiel 28:15 - You were blameless in your ways from the day that you were created, until iniquity was found in you.

As an end in itself, this last manifestation of world domination will have its "hour", for all those that were seduced by G-ds glorious position in the awesome arrangement of His works and abused them, will break apart at the seems and collapse in a cascading procession throughout the world, as I have tremendous awe-inspiring faith that G-d's mastery over mankind is greatest and shall prevail!

12 - QABALAH...

Before we start this chapter, it is important to know that this is NOT a "complete" lesson in Qabalah. I am conferring (only touching base) on the fundamental understanding of it.

The Hebrew word Qabalah means to "receive or accept" - that by studying the Torah one can gain insight and understanding about the creation of the world. It is believed that when Moses brought the Ten Commandments down from Mount Sinai he also brought with him the "ORAL LAW". There is no earlier book or written document on it. As I've expressed Qabalah, is completely "oral tradition" which was later converted into mystical books of thought concerning it? They are a record of rabbinic discussions of Jewish law, ethics, customs and stories. Which eventually in due course encompassed the entire spectrum of authoritative Jewish religious teachings?

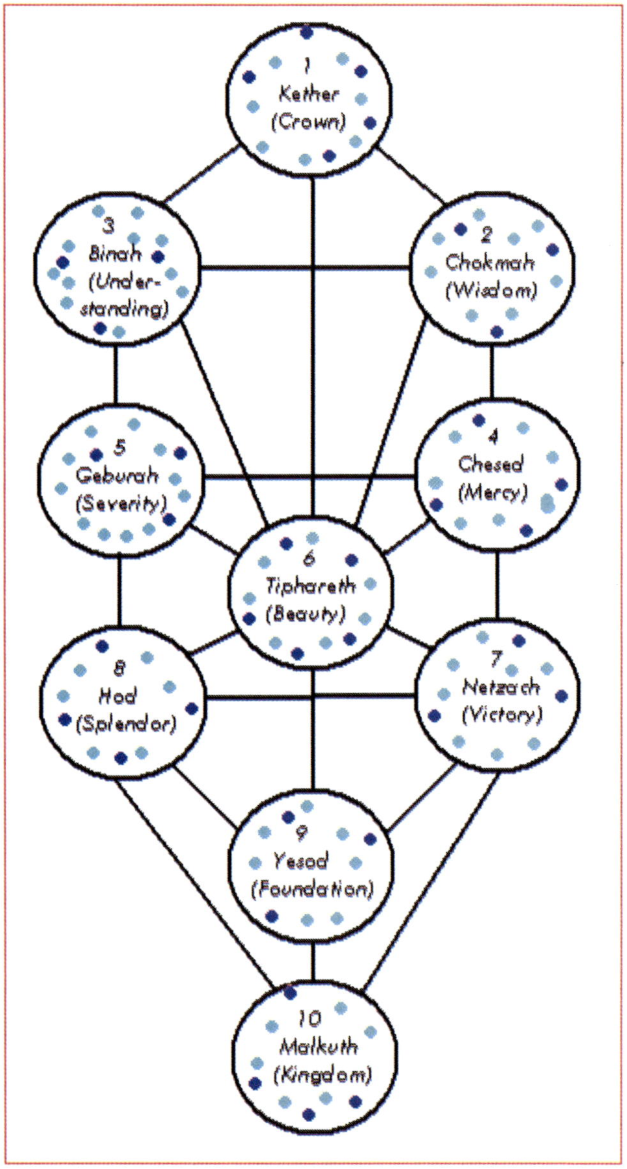

Just what Qabalah (oral law) is, one cannot merely determine by words alone but by diagrams of expressed shapes and thoughts that lead us through the physical world en-route to the spiritual so that we may come to know and understand G-d and His ways. Just what Qabalah is we are about to examine. The books about Qabalah that we have today were written to expand on the "idea" of how G-d used the Hebrew letters to form the material world and to create all that exists. Qabalah is also sought after to provide answers to questions passed down from "oral law". The books all seem to merge conscious thought with the Hebrew letters - for to understand G-d we must first understand the Hebrew letters in which He thinks, speaks and creates all things. Qabalah tries to explain this by a diagram called the "Tree Of Life" which looks like this:

The "Tree of Life" is a diagram consisting of 10 circles interconnected by 22 lines (the 22 lines are for each of the 22 letters in the Hebrew alphabet) (if you count them, there are exactly 22 lines). The circles are called sephiroth. The lines are called paths. Sephiroth literally means, "glittering sapphires". They are Spheres with the color of cosmic light and energy that literally gleams and sparkles with a neon blue (electricity) color inside of a misty fog of white (ether). It is "electromagnetic energy" (plasma like) which is blue-white in color. For those of you who are wondering what plasma actually is, it's the fourth state of matter (aside from solids, liquids and gases). Plasma is an ionized gas, a gas where adequate energy is provided to free up electrons from atoms or molecules and to permit both ions and electrons to coexist. Simply put, plasma is a gas that has been energized to the point that some of the electrons break free of their nucleus.

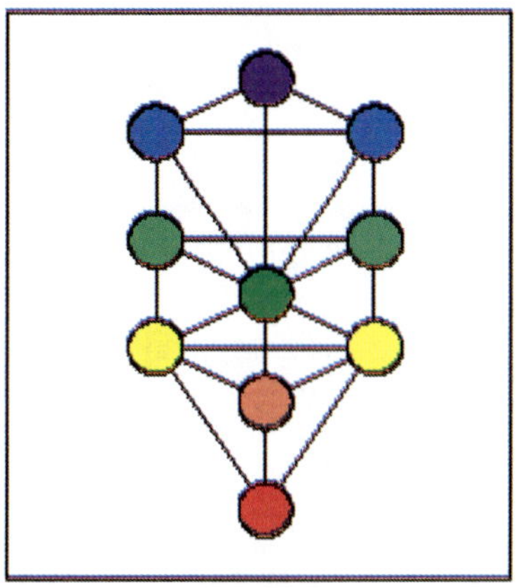

Some Qabalists even associate the frequency color spectrum to the "Tree of Life". Starting at the bottom, which is red and moving its way up to the utmost plane of violet blue. Red is considered the physical realm and violet blue is considered the spiritual plane. The color spectrum frequency goes from red (384 - 482) to violet blue at (659-769).

The names of the ten sephiroth are:

1. Kether (crown)
2. Chokmah (wisdom)
3. Binah (understanding)
4. Chesed (mercy)
5. Geburah (severity)
6. Tiphareth (beauty)
7. Netzach (victory)
8. Hod (splendor)
9. Yesod (foundation)
10. Malkuth (kingdom)

The "Tree of Life" is a map of reality, or of consciousness and the states of it. The 1st sephiroth (Kether) represents ultimate reality, ultimate consciousness, nirvana, and the source of everything. The 10th spirit (Malkuth) represents the material world (and the 1st platonic solid the Tetrahedron), which is the initial opening mechanism of the subjective individual consciousness. Sephiroth 2 through 9 represents levels of reality, manifestation or consciousness in between (Kether and Malkuth). In other words, the "Tree Of Life" can function as a chart of the relationship between ultimate reality and individual subjective reality, or, if you prefer, between divine consciousness and ordinary consciousness.

Before we go any further here, lets turn back the clock even further into past history... Good ole Enoch again is our inspired precursor in all of Qabalah long before "oral law" even came into existence at Mount Sinai. Moses was not the first to hear it as Enoch was privy to it long before. Therefore whatever "oral law" is, it seems to be the same fundamental practice that was covered with Enoch, the Angels and G-d. You might say that Qabalah is the offspring of what's known as astrology, alchemy and magic (turning copper or silver into gold by changing the atomic structure of the atoms) and also a unknown form of chemistry which seems to transform base metals into gold to produce a life-prolonging elixir and solvent called "alkahest" which can cure disease. This is also the chemical basse of medicine and poison compounds for making conventional medicine today and is the forerunner of modern pharmacology. The Bible tells us in Exodus 32:19-20 that when Moses came down from Mount Sinai with the Ten Commandments from G-d, he found that the people had taken their jewelry and fashioned themselves a god in the form of a "golden calf": (Exodus 32:19-20) tells us: And it came to pass, as soon as he came nigh unto the camp, that he saw the calf, and the dancing: and Moses' anger waxed hot, and he cast the tables out of his hands, and brake them beneath the mount. And he took the "golden calf" which they had made, and burnt it in the fire, and ground it to powder, and strawed it upon the water, and made the children of Israel drink of it.

This is the first prescription given in the history of mankind, (alkahest that is considered a universal solvent, also called liquid gold. It was prescribed for anxiety and impatience because G-d had described them as a stiff-necked people. Exodus 32:9 - And the Lord said unto Moses, I have seen this people, and, behold, it is a stiff-necked people. It still works for these and many other symptoms today to promote healing and control pain. It is even known that Tuberculosis cannot live in the presence of gold. (Tuberculosis is a bacteria that affects the lungs, the central nervous system, lymphatic system, circulatory system, bones, joints and the skin). Gold has even been injected and implanted under the skin such as in knee joints to relieve pain. It was also used to treat arthritis. Gold is also known for healing the activity of the heart and improve blood circulation, and includes treatments for skin ulcers, burns, and puncture wounds. It was also commonly used to cure "dipsomania" (uncontrollable craving for alcoholic liquors). Gold has also been used to rejuvenate the glands, stimulate the nerves and release nervous pressure. The body's temperature is also affected by gold, particularly in cases of chills, hot flashes, and night sweats. In addition, interestingly enough Gold can be used as a remedy against obesity. And even more strangely, it can be used for terminal malnutrition and starvation because it reduces pain and discomfort and the need of opiates (narcotics).

In addition to Jewish mythology, they claim to have the ability to create a Golem (a creature made of clay and brought to life by magical incantations). Qabalah also entails the capability and power to curse or hex someone. In fact people who practiced the science of alchemy (later on in history) by using magic spells and brews where considered witches and warlocks. Much of this Qabalistic knowledge has turned into pagan ways (worshiping and believing in more than one G-d) and distorted from its true origins, but real "magic" (using scientific formulas) really did and does exist, to what extent of this magic is really functional is another thing. But the ideas based behind it are from a very real standpoint between the forces of good and evil.

Behind the concept for some of these spells that were passed down from "oral law", not only included alchemical medicine but it also included the power of "CONSCIOUS THOUGHT" by means of a meditative process in combination of the Hebrew letters. Wherefore, the Hebrew letters each having a frequency and resonance sound assigned to them, and each letter and the shape of it channeling a very specific dimension of energy that we also refer to as CONSCIOUSNESS - consciousness that is the same as the life force of G-d. What these Hebrew letters are all about is what the entire revelation on Mount Sinai was for, which is to provide us with the knowledge and understanding of how to shape and form the configurations of Hebrew letters into words that manifest physical reality (Enoch described them as "weights and measures"). Also keep in mind when it was previous discussed about "conscious thought", it brings us right back to the principles of "quantum theory" and the notion that "actual thought" can and does create real physical matter starting at the atomic level. One tends to wonder if there is really something to this Qabalistic magic. Science is not to far away from spirituality and belief.

But in essence, the true meaning of Qabalah (oral law of Torah) deals with true "fulfillment". It is the theory of "thought" that emanates us to a higher level. We as humans mostly seem to be striving for something that is beyond our humanly means. As it turns out, our perception of the world is completely subjective; it fools us and says nothing about what happens outside us. Qabalah teaches us that we "see" from the inside out, and that what's actually happening in the world outside of us is all subjective to the "feelings" from within, and when we look deeply into these true desires we come to realize that we never really had a desire for the physical world. What we have really been looking for (all this time) is the true deep sense of "FULFILLMENT" and being one with G-d. However, in order to get there we must experience the physical by overcoming its illusion. For that reason is the principle behind the Qabalistic "Tree Of Life" diagram which allows for a transmigration of the souls ascent into Heaven. It was one of the first Qabalistic ideas to reach this goal by what is described as the "higher throne" of G-d through the "ten Sefiroth" which involves a process of meditation that rises from the bottom of the "Tree Of Life" (which is Malkuth) and transcending the human limitations of conscious reality and thus reaching the topmost part of the Tree Of Life (which is kether) (meaning crowned head - divine throne - chariot) and sets into gear the mechanism of what's called the "Merkabah vehicle" which facilitates transcendence of existence above and apart from the material (physical) world by way of "spiraling light". (This will be discussed in detail in the coming chapter dealing with the Merkabah).

But lets back up once again and examine in better detail the "Tree Of Life" to not only understand how this conscious process works, but to dig to the very proponents of it corresponding to the "universal law" that this book is ascribing to. To truly understand what the Qabalists were attempting to teach us through the "Tree Of Life" diagram is in essence "geometry". When we look at the "Tree Of Life" diagram, we see 3 geometric shapes:

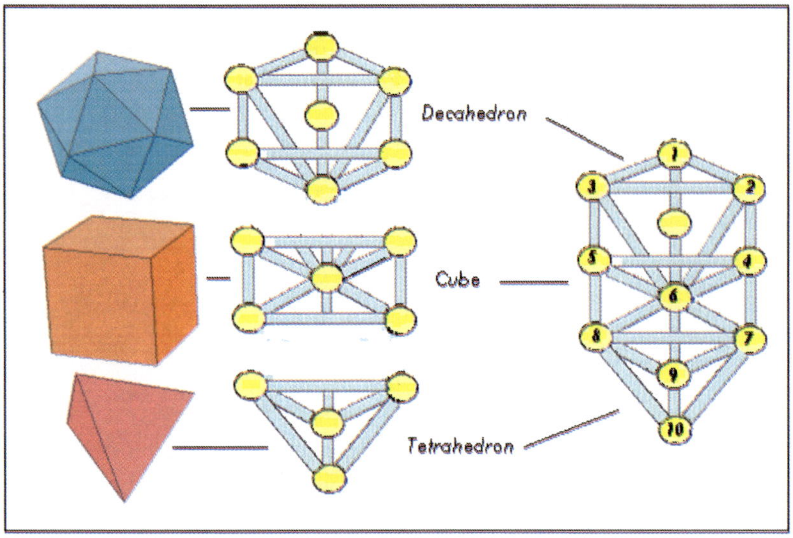

These 3 shapes are the Tetrahedron, Cube and Decahedron. And when we overlay these 3 shapes to the "Tree Of Life", we can visibly see the full body of the "Tree Of Life" diagram take form.

Sephiroth numbers 10, 9, **8, 7,** creates the Tetrahedron… numbers **8, 7,** 6, 5, 4, creates the Hexahedron/Cube and Octahedron… and numbers **6, 5, 4,** 3, 2, 1, creates the Icosahedron and the Dodecahedron. You can also see how the numbers I've underlined overlap themselves from 8,7 (the Tetrahedron) to 8,7 (Hexahedron/Cube and Octahedron) to 6,5,4 (Icosahedron and Dodecahedron). So as we can clearly see the diagram of the "Tree Of Life" shows a geometrically intelligent "unit of growth increase" of one that condenses while permeating by a unit of growth (by its distribution of one through the other) which is very similar to the Fibonacci growth sequence. (To get a better view of the shapes not shown in the picture above such as the Hexahedron (cube), Octahedron and Icosahedron look to the diagram of shapes below listed as "Euclid's Elements").

As we've discussed in earlier chapters, the Fibonacci growth sequence continues by adding the next number to the previous number in the sequence to create the one following. 2+1 = 3, 3+2 = 5, 5+3 = 8, and 8+5 = 13. Thus the sequence goes: 0,1,1,2,3,5,8,13... (Like the growth pattern of the embryo as it becomes a fetus, through adulthood, as well as the growth sequence in plants leaves). And as aforementioned, this is the growth pattern that represents all living things in nature - called the "Fibonacci sequence" (regarded in conjunction with Phi/Pi), and increases at a rate equal to oscillating around the Golden Mean Spiral. This sequence can also be found in the vibration of each musical note in the sum of the vibrations of the two notes previous resonating in sequence from 2, 3, 5, 8, and 13.

Now when we examine this growth sequences in nature and apply it to the geometric shapes of the Qabalistic "Tree Of Life" that I pointed out above with the Tetrahedron, Cube and Decahedron, we can additional back track to Pythagoras and even further to Euclid. He was the man that created (gave rebirth) to GEOMETRY some 2000 years ago, which are called the "Euclid Elements" WHOSE PRINCIPLES OF MEASURING WERE DIRECTLY TAKEN OUT OF THE HEBREW BIBLE - THE QABALAH (ORAL LAW). All along modern mathematics has failed to realize that hidden right under our noses is the obvious origin that the very basis of SCIENCE and math is none less than RELIGION. (This also goes back to our original argument of chapter 2).

Euclid was one of the most prominent mathematicians of antiquity. He is still known to every student who studies geometry. Though very little is known about Euclid's life other than he studied at Plato's Academy in Athens, he taught in Alexandria and founded the school of mathematics and wrote his prize-work, "THE ELEMENTS". The Elements are all what is known today about points, lines, angles and simple shapes. Based on Euclid's work of 10 axioms or postulates… From these 10 axioms, he built up the geometry, which is still taught today in school.

It's interesting to note that although the name "Euclid" and the word "Geometry" have often been used to mean the same thing for centuries. Centuries later the word geometry began to appear on the title page of a book. The first edition in English, published in 1970 had the words in its title: THE ELEMENTS OF GEOMETRY OF THE ANCIENT PHILOSOPHER EUCLID MENGARA.

These are Euclid's "Elements" and they are known as the 5 platonic solids:

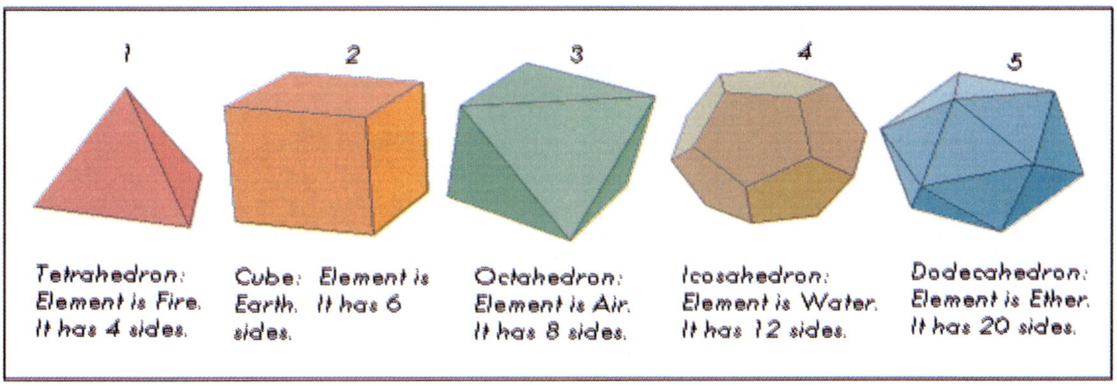

It is clear to see that Euclid carried over to science the knowledge of the Qabalistic "Tree Of Life" diagram which is the very basis of "geometry" but often without the greater context of the spiritual understanding attributed to it. There is a lot we can learn from these shapes. In fact when we look under a microscope we discover that the very first shape (or form) of "cell life" comes from a single egg which then multiplies into 2 sections and from 2 it multiplies into 4 sections, and from 4 to 8, 16, 32 and so on… Amazingly, the platonic solids multiply in a similar way starting at the "Tetrahedron" with 4 sides, and multiplies itself into a "cube" with 6 sides. In addition, from a "cube" with 6 sides, it multiplies into an "Octahedron" with 8 sides, and then it multiplies into an "Icosahedron" with 12 sides and finally and lastly it multiplies into the "Dodecahedron" with 20 sides. There are only 5 platonic solids that exist in nature; this is because each side belongs to exactly two faces and no more. (Research this for yourself). And although the sequence is not exactly the same, we can see that it multiplies itself in the same similar way by starting with a multiple number (which is a mirror reflection of itself - the number 2) and increasing by ADDING THE PREVIOUS NUMBERS. They increase by a unit of measure.

When these shapes are being formed, it is also the same principle that goes into creating matter, which turns into form with the combination of atoms forming into molecules and then into cells. The first platonic solid (the Tetrahedron) comes about by the culmination of atoms, which then turn into molecules, and then into cell form. This sequential movement, division, and expansion process is the universal rule.

Plato associated the Tetrahedron, Cube, Octahedron, Icosahedrons and Dodecahedron with the elements of Fire, Earth, Water, Air and Ether. The first of these cell divisions was the

Tetrahedron for it is the building block of the universe and crescendos into the other geometric "platonic solids". Qabalah teaches that 3 of the Hebrew letters are assigned to the principle elemental forces - Fire with the Hebrew letter "Shin", Water with the Hebrew letter "Mem", and Air with the Hebrew letter "Aleph" (and these are the exact meanings of these actual letters). For example, the Hebrew word for "Heaven" is "Shamayim" which literally means fire, water and air (Shin/fire - Aleph/air - Mem/mayim/water). This is also symbolic for the days of creation mentioned in Genesis 1... On day one light/"fire" (hydrogen) was created, day 2 Heaven and the expanses/"air" (oxygen) were created, and on day, 3 dry land and water/"water" (nitrogen) were created.

So, lets reiterate here for a moment... The "platonic solids" starting out with the Tetrahedron and growing into the Icosahedrons reveals the precise way that the energy of creation organizes itself. Every natural pattern of growth or movement conforms inevitably to the five geometric shapes, the Tetrahedron, Cube, Octahedron, Icosahedrons and Dodecahedron. Within these platonic structures we see the numerical series of the Fibonacci sequence again which is generated by adding the previous two numbers in the list together to form the next and so on: (1, 1, 2, 3, 5, 8, 13, 21, 34, 55...). Divide any number in the Fibonacci sequence by the one before it, for example 55/34, or 21/13, and the answer is always close to 1.61803. This is known as the Golden Mean Spiral (or some call it the Golden Ratio), which is reflected though nature in the form of geometry. Ironically, in the 5th (ultimate) platonic solid (the Dodecahedron) in its fullness (with 20 faces, 12 vertices and 30 edges) is equivalent to 1.618 which IS THE GOLDEN RATIO!

To expand on the 5 platonic solids, the secret here is really the atoms that compose them that give birth to their "shapes". What we are actually seeing is the manifestation of atoms. There are the electrons that travel around the nucleus, and because of the "spin angle", every time it travels around the nucleus it changes axis (orbital path). And when these electrons completely encompasses most of the space orbiting the nucleus it appears to become enclosed and gives us the illusion that the nucleus is a "solid ball like shape" of moving energy. Nevertheless, we must keep in mind that they are not "ball like structures" they are moving energy of electrons and protons. In addition, because they are in constant motion, it causes them to become adjacent with neighboring atoms, which join, or bond together, and when this happens new chemicals are made - "chemical molecular structures" that appear in one of the 5 shapes of the platonic structures. An example of this would be "carbon 6 building blocks" which are hexagonal in shape.

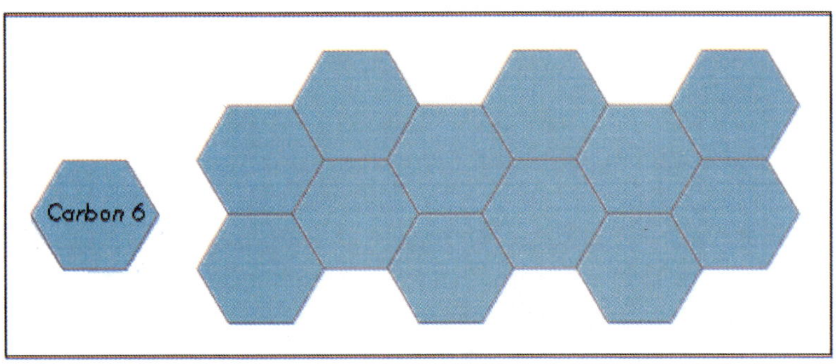

The question is how did our founding fathers of the Bible from Enoch to Moses, which later influenced Isaac the Blind when he transcribed it as the "Tree Of Life" diagram, which later influenced Euclid to convey them as the 5 platonic solids... how did they know the composure of these shapes when they didn't have microscopes in their era to see cell division? The only assumption that I can draw here for how they arrived at the 5 platonic solids long before they could be observed under a microscope is by observing things in nature and experimenting with vibration in both water and in sand (like the science of Cymatics).

However, it just may be that their true origins stemmed from a supernatural source with the purposeful intention to illustrate the fundamental truth that is woven within natures network, as well as the Qabalah (oral law of Torah) which in-turn is all part of what I've been addressing as the "universal law".

We must also examine other things regarding the Qabalah to which there are mysterious clues imbedded within it that we are yet to understand and discover, and it seems that if we look more in depth regarding religious structures we begin to make connections. For instance The "Ten Commandments"… it is not a coincidence that there are 10 commandments as well as 10 spheres in the "Tree Of Life" diagram. It can be noticed that the first 5 commandments are of a spiritual nature and deal with our relationship with G-d, while the second 5 deals with their relationship to the world.

These are the 10 commandments. Exodus 20:1-17:

These are of a spiritual nature and deal with our relationship with G-d.

1 - Thou shalt have no other g-ds before me.

2 - Thou shalt not make unto thee any graven image.

3 - Thou shalt not take the name of the Lord thy G-d in vain.

4 - Remember the Sabbath day to keep it holy.

5 - Honor thy father and thy mother.

These deal with our relationship to the world.

6 - Thou shalt not kill.

7 - Thou shalt not commit adultery.

8 - Thou shalt not steal.

9 - Thou shalt not bear false witness against thy neighbor.

10 - Thou shalt not covet thy neighbor's house nor anything that is thy neighbor's.

The "Tree Of Life" (sectional diagram which was shownearlier) also seems to be a "network type structure" of a 3x3 matrix represented by the beginning foundation of the Tetrahedron to the middle section of the Cube and the upper portion of the Dodecahedron. This is how the US government has structured its (Executive, Legislative and Judicial) with an additional three discrete levels of (Federal, State, and Local) - pending an additional future 3 levels. While textbooks do not teach it, the US government, as exemplified by our founding fathers (who were known to be "Freemasons") perfectly modeled and structured our governmental system along the lines of balance taught in Qabalistic circles.

Ironically, we also seem to find these three main branches of government in the Torah:

The Executive branch:
The Nasiy (President, Chief, Ruler or Leader)
The Nabiym (Prophets)
The Milchamah (Military or Armed Forces)

The Judicial branch:
Shaphat, Shaliysh, Shoter (The Judges and the Officers)(these are the Officials/Civil Servants)

The Legislative branch:
The Cohen (High Priests)
The Lewiym (Levites)
The elders of Tziyon (Zion) (House of Representatives)

Moreover, the first paragraph of the United States Constitution is structured in a roundabout way that is similar to the 10 commandments given to Moses by G-d at Mount Sinai:

1 - We the People of the United States, in Order to form a more perfect Union,
2 - establish Justice,
3 - insure domestic Tranquility,
4 - provide for the common defense,
5 - promote the general Welfare,
6 - and secure the Blessings of Liberty
7 - to ourselves and our Posterity,
8 - do ordain and establish
9 - this Constitution
10 - for the United States of America.

Many coincidental links can be made that the founding fathers of the United States had all the intents of applying their Masonic beliefs into the guiding development of the counrty. They felt so strongly about this that they went to the lengths of creating the US seal on the reverse side of the "One Dollar Bill", which shows a pyramid without the capstone, with a blazing eye in a triangle above it (which is symbolic for the Egyptian eye of Ra as the missing capstone of the Great Pyramid). The inscription at the top (in Latin) means, "He has looked with favor on the beginnings", and the one at the bottom mean "a new order of ages".

The Latin above the Pyramid ANNUIT COEPTIS means, "He (G-d) has looked with favor on the beginnings". The Latin below the Pyramid NOVUS ORDO SECLORUM means, "a new order of ages".

If we come to understand Qabalah and Jewish theology, it reveals to us what the founding fathers of the United States (who were "Freemasons" with Masonic beliefs) had in mind for the new nation. Once again, it entails applying the old kingdom to the new kingdom. It's a tale as old as the time. Study the past if you would define the future - Confucius. Common sense tells us to examine it with these prospects in mind, and then perhaps its true intent should reveal itself. We know (as aforementioned) that the Great Pyramid was symbolic of Enoch from the get go. Additionally we also know that the Star of David was symbolic of the Great Pyramid with its 4 sides for North, East, South and West representing the 144,000, (12,000x12, 000=144,000) as the prophesied amount of people that will be raptured and transmuted at the end time, and was specified by Enoch (by way of) the Great Pyramid having 144,000 white polished casing stones that (while glistening in the light even as an energy converter of sorts) was also symbolic for the light of souls being metamorphosed into Heaven. Whereas the Great Pyramid in conjunction with the Star of David was used as a 365 day calendar defined by the stars of the 12 astrological houses. And as also aforementioned the number 12 also represents the 12 apostles of Jesus, the 24 Elders of the New Jerusalem mentioned in Revelation 4:4, along with the mathematical descriptions given for the New Jerusalem in Revelations - wherein they all insinuate to the Star Of David pattern. This would have been one of the most important symbolic symbols to the founding fathers of the United States parallel to their Masonic beliefs. Therefore, the first place we should start is by placing a Star of David over the Pyramid and connecting its points to the surrounding letters. When we do this, an amazing word becomes visible. It spells MASON. Even though the letters are not in Hebrew (instead Latin). The first inscription at the top means, "He has looked with favor on the beginnings", and the one at the bottom mean "a new order of ages". The Masons where referring to the "ONE" G-d of the Jews with the Hebrew letter Ayin (symbolic of the all seeing eye) looking down upon the new nation of the first 13 colonies (I will get to explaining

The blue circles spell MASON.

Also notice that the Pyramid is floating, or surrounded by water.

why there are 13 stars instead of 12). This becomes apparent when we count the number of layers (or foundations) the Pyramid on the One Dollar Bill has, which is 13. And if we look closely we can also see that the pyramid is floating on water which might also be symbolic for the Great Pyramid of Giza alternately being used as a water pump supplying water both near and far. It might also be symbolic for Noah's Ark, or perhaps surviving the Great Flood of Noah.

When we look further to the right of the Pyramid on the One Dollar Bill, we see an Eagle with 13 stars over its head (forming the Star Of David pattern - again) (picture can be seen below). In addition, in the eagles left talon there are 13 arrows, and on the olive branch in the Eagles right talon there are 13 leaves. There are also 13 letters in "E Pluribus Unum" (which means "one" out of many, or from many, "one"). And 13 letters in "Annuit Coeptis" over the Pyramid (shown above). I also find it interesting that the Eagles shield (breastplate) has 13 stripes, but it could also represent the breastplate of Aaron where 12 colored Gemstones were used for protection when entering the Tabernacle in G-d's presents. Exodus 28:29 - And Aaron shall bear the names of the children of Israel in the breastplate of judgment upon his heart when he goeth in unto the holy place for a memorial before the Lord continually. The 12 gemstones where most likely used as a conduit for channeling certain high (lofty) types of energy. As we know that certain crystallized stones have a property that is naturally controlled by frequency, and can be used as a grounding plate of sorts for both electrical, electromagnetic and piezoelectricity transmissions. It may have also been used for spiritual conflict in the presents of G-d. Scientists have long known that a crystal set in an energy field will collect and focus that energy. Today gemstones are used in modern day electronics.

The Eagle in not only indigenous to America, it was also mentioned in the Torah as a central symbol of liberation: Exodus 19:4 - You have seen what I did to the Egyptians, how I bore you on eagles wings and brought you to me. Commentators have pointed out that the image of the Israelites being borne out of Egypt on eagles wings is actually "zoologically correct". Eagles are indeed known for carrying their young on the wide expanse of their wings. I'm sure this is one of the clandestine motivations that the founding fathers of the United States used as a spiritual suggestion. Perhaps true liberation can only come to our nation when we become as Eagles for one another - committed to bearing all members of our national community upon our collectively outstretched wings. The wings can also symbolize the coming together of the Old Testament and the New Testament - the 12 tribes of Israel and the 12 apostles of Jesus. An Eagle is also the sign of "Enoch". The Eagle was given the name "Phoenix" which is interpreted to mean "Pa Hanok" which means "house of Enoch" (or house of the stars) and it has always been considered the symbol for resurrection or rising again from the flame of the Sun (which is also associated with Enoch as the Sun god since the Earth revolves around the Sun 365 days, the duration of a solar year).

So why the number 13 instead of the representational 12? The message of the "Star of David" is still there, but now with a caveat to it. There seems to be a lot of mysticism associated with it both far and wide, but after all the searching I did I've deduced the reason for why?

First, to understand the number 13 we must examine its Qabalistic meaning. And lets not forget that the founding fathers of the United States were "Freemasons" who studied Qabalah (Jewish oral law).

Now since there just so happened to be 13 initial colonies at the time of the signing of the declaration of the United States, the founding fathers recognized this as a compelling sign - this was no fluke of fate! Their motto about it was that it was purposeful and intended by G-d - as all things are. With that in mind, they wanted to reflect their beliefs and values onto their descendents. Now when we delve into the understanding of Qabalah we learn that the number 13 was not associated with bad luck, but rather GOOD LUCK! Qabalistically, the number 13 indicates the ability to rise above the influence of the 12 signs of the (Zodiac) astrological houses (12+1=13), which means that one is not bound by the influences of the material physical realm. In other words, the number 13 is another way of expressing our unity with G-d. The fact that the 12 stars are all connected in the center as the 13th - 13 is the number that bonds multiplicity into oneness. For example, the twelve tribes that are adhered into their father Jacob (forming the nation of Israel). Israel is the 13th. Therefore, the meaning of the number 13 is the union of many into "ONE". In Qabalah the number 13 is among the holiest of the numbers because it is closely associated with G-d - hence Jesus being the Messiah as a symbol of Christianity and its association with the "new world order". In 1775, 13 initial colonies existed on the eve of the American Revolution. On July 4, 1776, Benjamin Franklin, John Adams, and Thomas Jefferson were given the task of creating a seal for the United States of America. In addition, on August 2nd 1776 they signed the declaration of independence. However, the seal was already in existence it was finalized and approved six years later on June 20th 1782.

Some might associate the number 13 as being bad luck because it is associated with death (Jesus' death as being included at the last supper, consisting of 13 people) (the 12 disciples plus Him). However Qabalistically thinking, this may indeed indicate death, but it also indicates transmutation, hope, faith, renewal, regeneration and rebirth - as it is at the Egyptian hour when the Sun goes down in the West on its way to North. (Aforementioned above with the Sun being born in the East, maturing at noon, and dying at sundown - therefore East is associated with life, South with power, West with death, and North as the Sun dips below the horizon, not as an end to life but as a transitory stage of birth and renewal).

In fact, to the ancient Egyptians, life was a quest for spiritual ascension, which unfolded in stages, (12 in this life and a 13th beyond - thought to be the eternal afterlife). The founders of the United States also most likely believed the American Revolution was the beginning of a "new order of the ages". In my opinion, these positive meanings surrounding the number 13 outweigh any superstition or negative association about the number. They seem to be corrupted by later cultures that came to associate 13 with bad luck, or a fear of death instead of a reverence for the spiritual aspect of it.

Though, there is more to the number 13 that must be revealed... To put a sealing cap on all of this and to be able to connect and make a certain and direct association to Qabalah, as well as the original concept that the founding fathers had for the United States of America. There is "one" piece of evidence that we cannot evade, and in my opinion, it outweighs them all! This evidence can be found on the One Dollar Bill again and found within the thirteen (13) letter of "E Pluribus Unum" (found written on the ribbon that the Eagle is holding in its beak, and under the 13 stars). And we find that something else amazing unveils itself, and that is its "GAMATRIA VALUE"! The phrase consists of 3 separate words consisting of 1, 8 and 4 letters: (E=1), (Pluribus=8), and (Unum=4), (1+8+4=13). Besides its Latin meaning, (which means "one" out of many, or from many, "one"), these numbers are also consistent to the Hebrew letters, "Aleph" which is equivalent to the number "1", "Chet" which is equivalent to "8", and Dalet which is equivalent to "4" - spelling the word "ECHAD" which is the Hebrew word for the number "ONE". To which again is symbolic for the Great Seal of the United States, that commemorates the union of 13 colonies as "ONE" nation under G-d. The additional 13th star in the "Star Of David" pattern on the "One Dollar Bill" can also be looked at as being symbolic for "Jesus" - hence a new nation under Christianity.

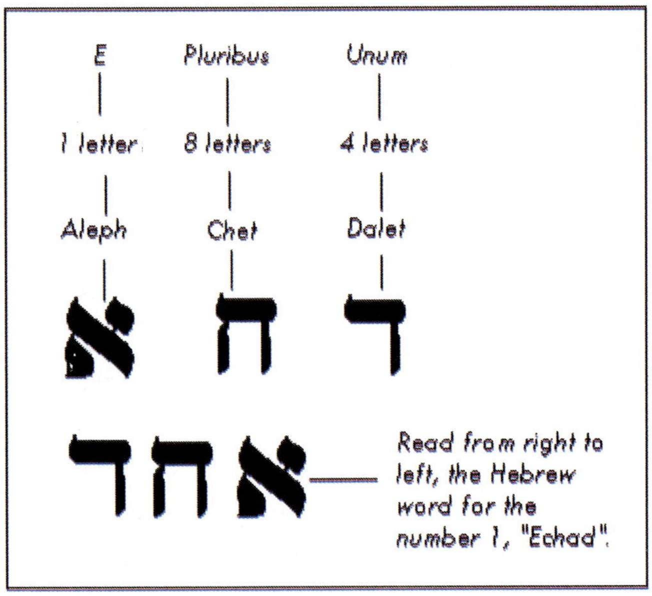

This should not surprise us, as the Romans, Greeks and Freemasons knew about the Hebrew gamatria values, BECAUSE THEY DERIVED THEIR OWN MATHEMATICS FROM IT. As aforementioned in an earlier chapter, the word "Gematria" should not be a foreign word to us when the word itself later evolved into the Greek word "geometry" whereas the concept and system is the same as the Greek isopsephy meaning "equal count" by adding up the number values of the letters in a word. As a result, this is the precise message that the founding fathers of the United States insinuated on the One Dollar Bill by coding it with the old lexicons of the original religion of humankind. In addition, as I mention earlier, a "religion" that has directed us, and taught us much about SCIENCE!

The most important prayer in all of Hebrew literature is the "Shema". It declares, "The Lord is ONE"! Deuteronomy 6:4 - Hear, O Israel: The Lord our G-d is one Lord. In Hebrew it reads: Sh'ma Yis'ra'ail Adonai Elohanu Adonai **echad.**

The question has also been posed why the individual stars within the 13 stars (above the eagles head) are not also 6 pointed stars, but rather 5 pointed stars. Its meaning becomes evidently clear when we know that the earliest five pointed stars were first seen in ancient Mesopotamia and later in ancient Egyptian hieroglyphics. But why are there 5 points instead of the traditional 6 points - what does this really symbolize? The most applicable involvement to this is again Enoch!

It may just be a happenstance that there are "5" distinct sections of Enoch book:

The book of Watchers.
The book of Parables/Similitudes.
The book of the Heavenly Luminaries (also called the Astronomical book).
The book of Dreams.
The Epistle of Enoch.

And later symbolically constructed again as the 5 books of Moses - the Pentateuch:

Genesis.
Exodus.
Leviticus.
Numbers.
Deuteronomy.

Some think they easily know the answer to why the U.S. flag uses a 5 pointed star instead of a 6 pointed star. And this has to do with the original designer who sewed the flag (Betsy Ross). According to legend, when George Washington asked Betsy Ross to sew the country's flag, the original rough copy that he showed her had a flag with 6 pointed stars. But Betsy convinced him that instead of the 6 pointed star, the 5 pointed stars would be easier for her to make because she was able to fold a piece of cloth in a way that resulted in a perfect 5 pointed star with a single snip of her scissors. This may very well be, but Betsy may have also realized that the 5 pointed star was also a Masonic sign since she designed the flag in June of 1776 before the One Dollar Bill came off the presses in 1957. But we must realize again that these Masonic symbols were well known in antiquities and floated around the entire world, even way before the first Pilgrim colony (of the United States of America) was established in 1607. Eventually it made its way to being the excepted symbolism by Congress. So it was settled, it would be a 5 pointed star that would adorn the U.S. flag.

But of course none of these insinuations are solely enough to satisfy the question. However, I can assure you that there is much more of a deliberate suggestion for the 5 pointed star than one might be able to contemplate.

If the truth were told, as aforementioned its precepts go back to Egypt and Enoch. The 5 pointed star (or the pentagram) represents the brightest glowing star in the entire sky, "Sirius" (which is known as the Dog star). We associate it with the phrase "dog days of summer" which are the hottest days of the summer in the northern hemisphere and corresponding to when the Nile river rises with water. But Sirius is also the Queen that encircles the progression of "seven stars" listed in the paragraph below. 360 degrees (a complete circle) around Sirius divided by the number of arms in a "5 pointed star" which is the number 72. (A five pointed star is the Egyptian numerical notation for the number 5). The number 72 is significant in terms of astronomy and for the "Freemasons" because 72 is the number of years it takes for the 7 stars (that revolve around Sirius) to move through one degree of their cycle (consisting of 5 cycles). Now when we multiply 360 degrees by 72 it equals 25,920. This is the cycle of precession (the conical sweep of the stars on their axis of rotation). 72 is traditionally known as the angle that the Masonic compass symbol stands at, and now we know why!

Some also see the uncanny relationship to the "Star of David":

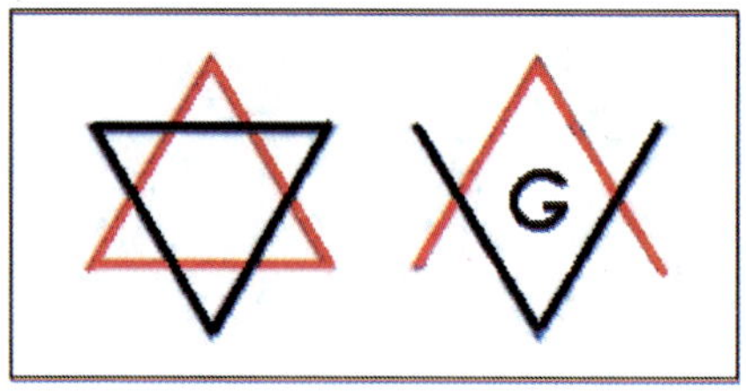

One might wonder what the letter G in the center stands for. Some suggest that it represents "geometry" or "Giza", while others suggest it stands for "G-d". However, I believe its true representation stands for the procession of the 7 stars just briefly mentioned above (which can be viewed below). From the constellation Taurus, through the Gemini twins, Castor and pollex to Procyon (in Canis minor) and from there, Sirius (which is found in Canis major) to Rigel, and from Rigel to the Sphinx passing right through the 3 stars of Orion's belt, and lastly to Betelgeuse. One can see this progression of stars making a letter "G" pattern in the northern summer sky (picture shown below) (it is also seen in the sky chart of the January constellation shown below). And it is no mistake that the Great Pyramid which has 4 descending air shafts pointing from the King and Queen's chambers to these very same constellations that reveal the "G" pattern of the summer northern sky. This is the secret path that the "Freemasons" revere, which represents RECESSION. This is also called the "sacred hoop" or the "winter hexagon".

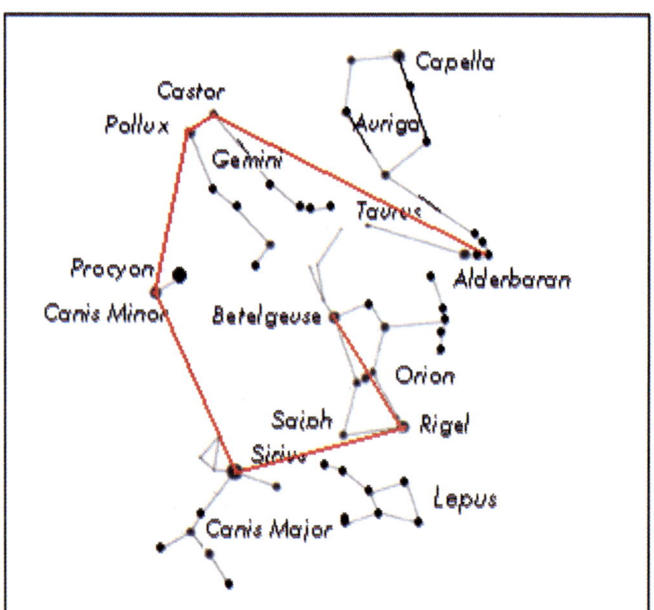

Now for the big question: Why is the phenomenon of RECESSION so important to the ancient Egyptians and the Freemasons? The answer again rests in the company of the Great Pyramid and Enoch who is a pivotal key to the whole mystery. As mentioned earlier, an Eagle is also the sign of "Enoch". The Eagle was given the name "Phoenix" which is interpreted to mean "Pa Hanok" which means "house of Enoch" (or house of the stars) and it has always been long considered in mythology the symbol for "resurrection or rising again" from the flare/flame of the precession of the Sun (which has also associated Enoch as the Sun god). It is symbolic for the start of a "great age" of complete recession mentioned earlier as 25,920 years, which is 360 divided by 5 which is 72. This is why the Phoenix has been associated with the precession of the Sun - the old Phoenix dies and is reborn, renewed, resurrecting from the ashes of its old form. This is indicative of the time of "rest" when the northern hemisphere is tilting away from the Sun followed by the "resurrection" which marks the spring equinox. The myth about the Phoenix tells us that it has a life cycle of 500 years… this is quite interesting when we examine the prophet Daniel's (Daniel 9:24) prophetic calculations of 70 weeks (70 × 7 = 490 years) its pretty close to the Phoenix's 500 year life span. These 500 year cycles appear to have been in course before the "Messiah's birth". With Abraham being born approximately 2,000 years before Jesus' birth, and nearly 500 years later around 1500 years before Jesus' birth, Israel made their Exodus from Egypt and received the Torah at Mount Sinai. Again, approximately 500 years later, (1,000 year before Jesus' birth), Solomon's temple was completed. And approximately 500 years later Zerrubabel's temple was built after Israel came out of Babylonian captivity. Finally, Jesus was born 500 years later. Among other names, Enoch was also associated and otherwise known as - the Egyptian Thoth, Hermes and Osiris.

Interestingly, the Mayans also associate this 500 year cycle with their 12-21-2012 procession ending calendar which some have interpreted as a doomsdays prophesy when the Earth will be in exact alignment with the Sun and the center of the Milky Way galaxy. This center of alignment is envisioned as a sort of entry point or entrance into rebirth, incarnation, "heavens gate", or it could simply mean just the start of a new era (such as the age of Aquarius). Some say when this alignment occurs it will cause the magnetic currents of the Earth to spin in opposite directions, thus causing a field effect reversal which will cause the Sun to melt the artic poles causing massive tsunamis (tidal waves) which will put stress on the Earth's crust (tectonic plates) causing major earthquakes all around the world. Some hypothesize that an 11 year solar storm (emanating from the sun) is in store for us in 2012 and this might have some effect on the energy grids causing major blackouts. Nevertheless, nobody knows for sure if anything will happen at all, or what effect a magnetic reversal of the Earth's poles will do? Some hypothesize that this reversal will occur instantly at the moment of the alignment, but others believe this reversal is already moving slowly into effect and we are witnessing this with "global warming". Revelation 8:7 - The first angel sounded his trumpet, and there came hail and fire mixed with blood, and it was hurled down upon the earth. A third of the earth was burned up, a third of the trees were burned up, and all the green grass was burned up.

Its also interesting to note that the constellation Sagittarius (a half man, half horse figure represented as an archer drawing a bow with an arrow). This again is a constellation shown and given to Enoch by G-d's appointed angel Uriel for a future revelation to come. What's interesting about the constellation Sagittarius is that it points (and aligns perfectly) with its bow and arrow to the exact center of the Milky Way (picture shown below). The last time it was aligned was in 1,500 BCE which is in the 500-year gap again (approximately 512 years ago). Nobody truly knows the significance of this yet. Some associate it with simply just a new era, while others may associate it with the return of Jesus. Others even say it will be a doomsday, which will bring about significant destruction of the earth. Revelation 6:1-2 - And I saw when the Lamb opened one of the seals, and I heard, as it were the noise of thunder, one of the four beasts saying, come and see. And I saw, **and behold a white horse: and he that sat on him had a bow;** and a crown was given unto him: and he went forth conquering, and to conquer. Revelation 8:1-13 - And when he had opened the seventh seal, there was silence in heaven about the space of half an hour. In addition, I saw the seven angels, which stood before G-d; and to them were given seven trumpets. And another angel came and stood at the altar, having a golden censer; and there was given unto him much incense, that he should offer it with the prayers of all saints upon the golden altar which was before the throne. And the smoke of the incense, which came with the prayers of the saints, ascended up before G-d out of the angel's hand. And the angel took the censer, and filled it with fire of the altar, and cast it into the earth: and there were voices, and thunderings, and lightnings, and an earthquake. And the seven angels who had the seven trumpets prepared themselves to sound. The first angel sounded, and there followed hail and fire mingled with blood, and they were cast upon the earth: and the third part of trees was burnt up, and all green grass was burnt up. In addition, the second angel sounded, and as it were a great mountain burning with fire was cast into the sea: and the third part of the sea became blood; And the third part of the creatures which were in the sea, and had life, died; and the third part of the ships were destroyed. And the third angel sounded, and there fell a great star from heaven, burning as it were a lamp, and it fell upon the third part of the rivers, and upon the fountains of waters; And the name of the star is called Wormwood: and the third part of the waters became wormwood; and many men died of the waters, because they were made bitter. And the fourth angel sounded, and the third part of the sun was smitten, and the third part of the moon, and the third part of the stars; so as the third part of them was darkened, and the day shone not for a third part of it, and the night likewise. And I beheld, and heard an angel flying through the midst of heaven, saying with a loud voice, Woe, woe, woe, to the inhabiters of the earth by reason of the other voices of the trumpet of the three angels, which are yet to sound!

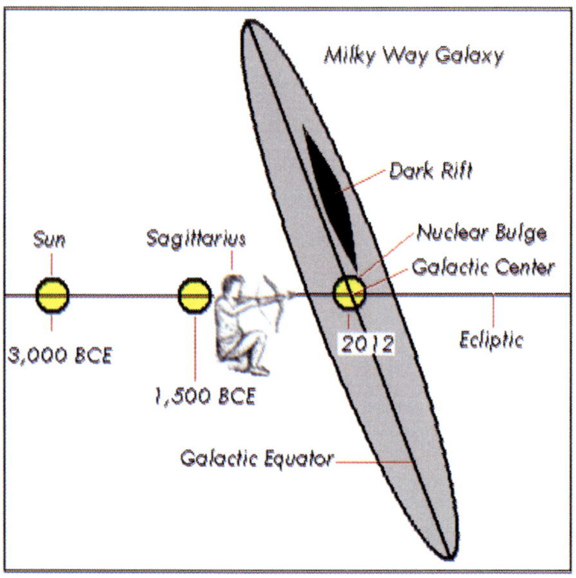

However, I will assure you that this is not the end of days scenario since all biblical prophosey has not yet been fulfilled. We are still awaiting a peace treaty between the Jews and the Arab nations. It is my opinion that 2012 ia a huge deception – the Devils day since the Mayans got this from a god they called Kukulcan which is described as a winged demon. This may indeed be a year where many bad and disturbing things happen such as power failer, satellite communications are broken down and a load of anomalous weather conditions like fires, earthquakes, tornadoes…ect… Planet alignments have also been known to happen around the times of wars…like Desert Storm. Soa war(s) could even break out.

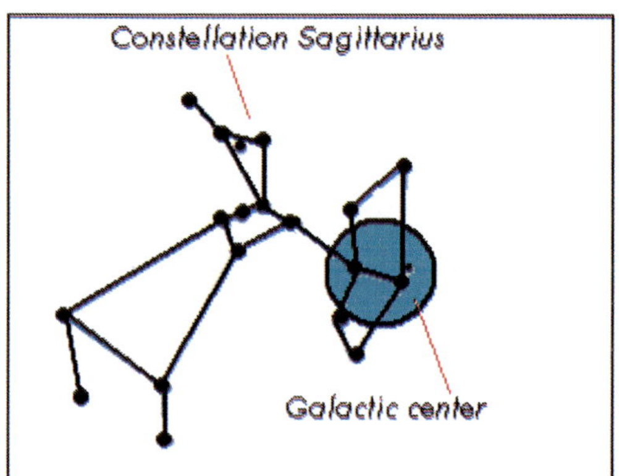

On December 21st 2012 the constellation Sagittarius will be aligned with the galactic center of the Milky Way.

Moving along here… now lets focus our attention back on the letter "G" again. There is even more to the letter "G" that appears within the Masonic symbol. As aforementioned above some suggest that it represents "geometry" or "Giza", while others suggest it stands for "G-d". However, as we have learned it also represents the procession of the 7 stars. But there may be yet another crafty as well as sly relationship to the Hebrew letter "Pey" that was discussed earlier in the previous chapter. Since the Hebrew letter, "Pey" resembles the English letter G (though in a reversed manner), and represents "mouth" and G-d's word. It may also be a representation by the "Freemasons" as a sign of the "figure of authority" over all things. Acting as a type of conical (lesser) god over a world government.

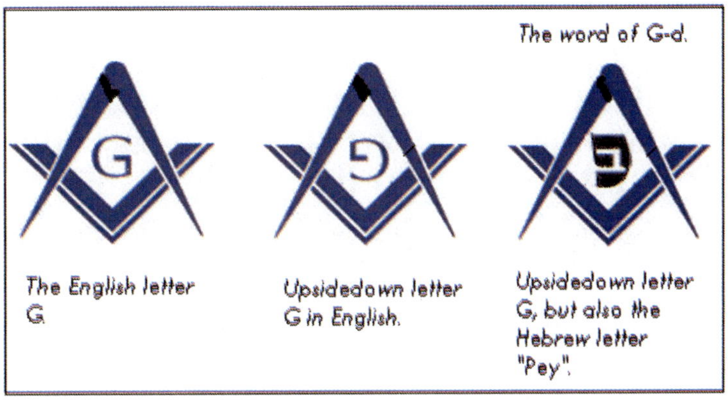

The conclusion that I derive from all this (the stars and their recession) is that we must remember what G-d tells us about it… In Genesis, 1:14 it tells us that - G-d created the stars for signs, seasons, days and years. This same message is found within Enoch 1 75:3 (which predated Moses and Genesis 1:14). Enoch 1 75:3 - And the exactness of the year is accomplished through its separate three hundred and sixty-four stations. For the signs and the times and the years and the days the angel Uriel showed to me, whom the Lord of glory hath set for ever over all the luminaries of the heaven, in the heaven and in the world, that they should rule on the face of the heaven and be seen on the earth, and be leaders for the day and the night, i.e. the sun, moon, and stars, and all the ministering creatures which make their revolution in all the chariots of the heaven. In like manner, twelve doors Uriel showed me. Come to find out, not only are these 12 doors the Zodiac that delineate (mark out) our seasons today, but this is also apparently true that the stars in the Heavens are crossed with important religious dates in Jewish past history. The precision of the stars was most likely intended by G-d for those particular events including Jesus' birth. Why else would it be necessary for an angel to show Enoch exactly how the constellations should be represented, or the number of star recessions that would occur before the Messiahs birth? I feel that Enoch was privy to know the second coming of the Messiah, which is most likely incorporated into one of the Hebrew holidays. A Jewish holiday that corresponds to this is "Rosh Hashanah" which represents the "day of judgment" (it is the Jewish New Year characterized by blowing the Shofar to awaken the listener to coming judgment) followed by 10 days of repentance just before "Yom Kippur" which is the "day of atonement". If effect, G-d gave us this holy event to teach us that we will be judged on Rosh Hashanah and sealed unto the closing of the gates on Yom Kippur. This is where the righteous are separated and rapture up to be with G-d. This holiday is mentioned in Leviticus 23:24 - Speak unto the children of Israel, saying, in the seventh month, in the first day of the month, shall ye have a Sabbath, a memorial of blowing of trumpets, an holy convocation. It also mentions something to this fact in Isaiah 26:2 - The gates of Heaven are opened on Rosh Hashanah so the righteous nation may enter. It also makes mention of this in Psalms 118:19-20 - Open to me the gates of righteousness, I will go into them, and I will praise the Lord. This gate of the Lord, into which the righteous shall enter. Because the gates of Heaven are understood to be open on Rosh Hashanah, this is further evidence that the "rapture of the believers in the Messiah" will take place between Rosh Hashanah and Yom Kippur.

Moving on here… Therefore, as not to evade the original question of what Qabalah truly is, we can see that "Egyptian culture" (borne of Enoch) played a big part in shaping the ideas for the founding fathers of the United States of America who were "Freemasons". Freemasonry shows us thriving evidence of the ancient worlds knowledge of geometry, astronomy and precession - which is the root of all RELIGION and SCIENCE today, that will yet again usher in the prophesied future. We must not lose scope of this!

And there's more, much much more… I have decided to leave no stone unturned, and to leave it up to the readers to make of it what they will. What you are now about to read will give you an in-depth open view of the stars and their linkage to understanding some of the customary ritual and religious beliefs of the ancient Egyptians. You will even be privy to such knowledge of where the Egyptians got their funky attire (such as their headdresses and crowns).

The "as above, so below" (in Earth, as it is in Heaven) principle, strikes a echoing bell… For it wasn't only Moses who coined this phrase in the Old Testament, but firstly Enoch, which followed on its course to the Egyptians. It was Enoch who first told us about his journey in which he was shown all the workings of Heaven and Earth. He was shown by the Angel in Heaven, and then replicated this knowledge on Earth. G-d even instructed Moses in Exodus 25:9-40 - According to all that I show you concerning the pattern of the tabernacle, and all of its furniture, so you shall make it. And see that you make them after the pattern for them, which is being shown you on the mountain.

Job 38:33 - Do you know the ordinances of the heavens? Can you establish their rule on the earth?

Genesis 1,11,12,21,24,25 - And G-d said, Let the earth put forth vegetation, plants yielding seed, and fruit trees bearing fruit in which is their seed, each according to its kind. The earth brought forth vegetation, plants yielding seed according to their own kinds, and trees bearing fruit in which is their seed, each according to its kind. So G-d created the great sea monsters and every living creature that moves, with which the waters swarm, according to their kinds, and every winged bird according to its kind. And G-d said, Let the earth bring forth living creatures according to their kinds: cattle and creeping things and beasts of the earth according to their kinds. And G-d made the beasts of the earth according to their kinds, and the cattle according to their kinds, and everything that creeps upon the ground according to its kind.

Proverbs 27:19 - As water reflects face, so the mind of man reflects the man.

Hebrews 9:23-24 - Thus it was necessary for the copies of the heavenly things to be purified with these rites, but the heavenly things themselves with better sacrifices than these. For Christ has entered, not into a sanctuary made with hands, a copy of the true one, but into heaven itself, now to appear in the presence of G-d on our behalf.

Moreover, Moses makes G-d's second commandment about this very implicit, that we should not worship things of this nature, but rather distinguish them by realizing that they play in harmony to reveal that they shine from "ONE" source: Exodus 20:3-5 &

Deuteronomy 5:7-9 - You shall not make for yourself a graven image, or any likeness of anything that is in heaven above, or that is in the earth beneath, or that is in the water under the earth, you shall not bow down to them or serve them, for I The Lord your G-d am a jealous G-d, visiting the iniquity of the fathers upon the children to the third and the fourth generation of those who hate Me, but showing steadfast love to thousands of those who love Me and keep My Commandments.

We cannot begin to exhaust the proof of this maxim "as above, so below" especially when it comes to the cosmos. Some even suggest that many of the Egyptian hieroglyphics we witness are actually pictures of important deities that were grafted from mirror reflections of the star constellations as we've just witnessed with the 5 pointed star. HOWEVER, THERE IS MUCH MORE TO THIS! For example, the constellations in the Northern winter skies ("Lepus" - otherwise known as Orion's chair, and "Eridanus") make up the customary attire image of an ancient Egyptian Pharaoh… We can clearly see the side view of a crown, face and beard of an Egyptian Pharaoh. (If you look at the sky chart shown below "boxed in red" with these constellations, you will be able to see the resemblance).

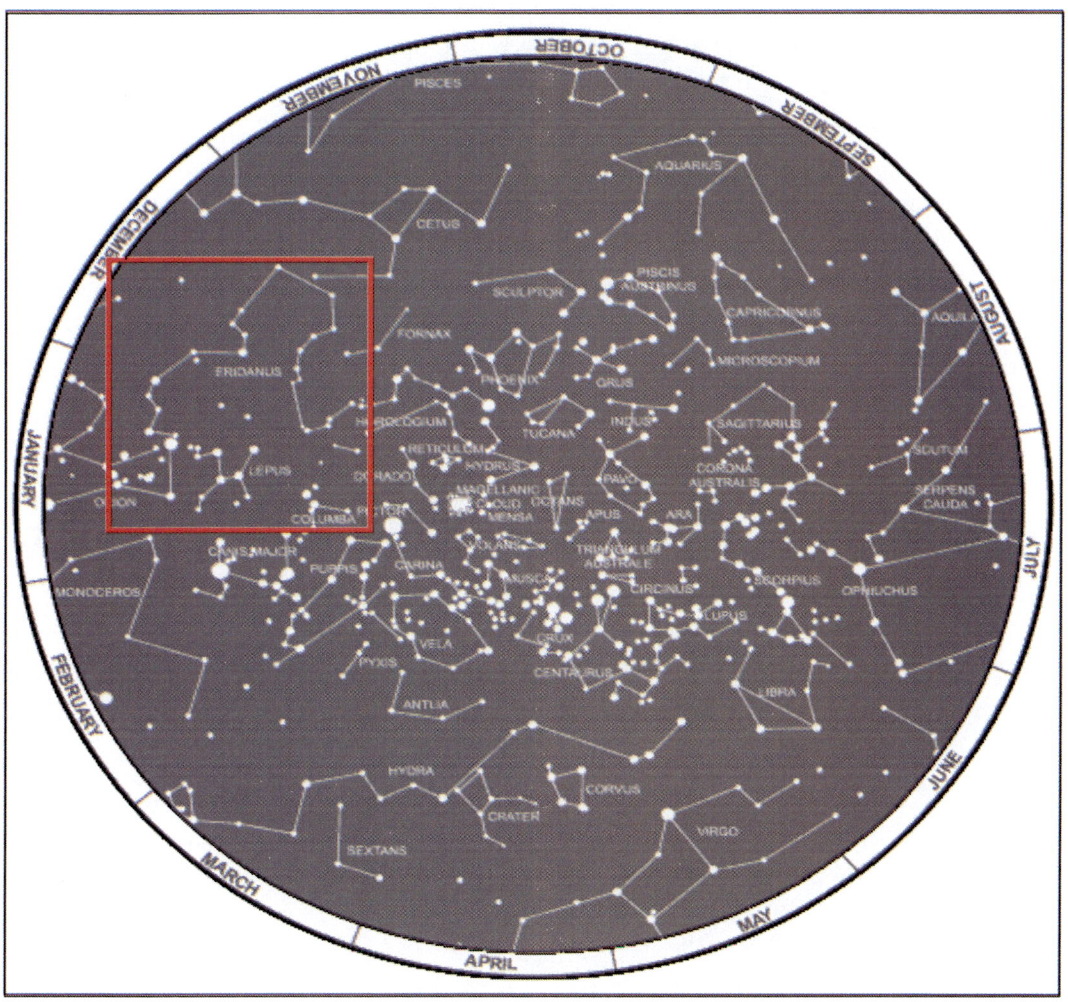

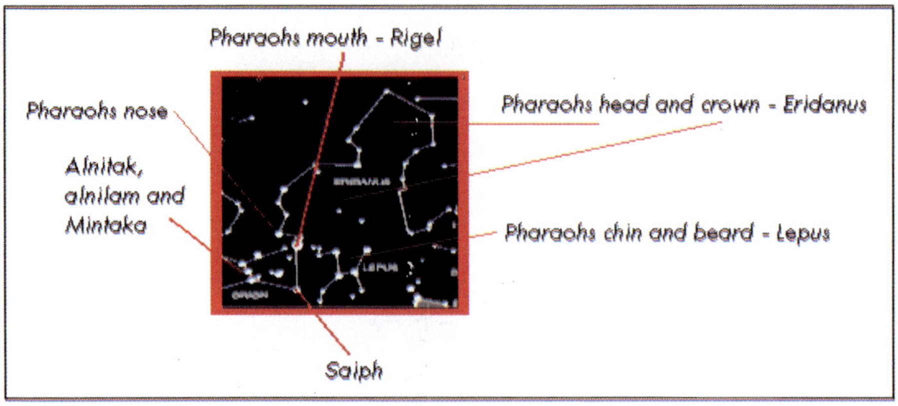

Now (shown above) directly in front of the Pharaohs mouth is the star "Rigel" (Rigel means the "left Leg"), which is found in the lower right of the constellation Orion. Ironically in HEBREW, the word "Regel" (resh, gimel, lamed) means, "leg". The star "Rigel" which is in-line directly to the 3 stars of Orion's belt which are "Alnitak, Alnilam and Mintaka", and as some suggest these 3 stars are supposedly a structural representation of "Heaven on Earth" built below as the 3 pyramids of Giza. These 3 stars emanating/uttering out of the "mouth" of the Pharaoh from the star Rigel is a figurative representation as the "word of G-d", (the Hebrew letter "Pey") suggesting that the Great Pyramid speaks the word of G-d - although not by the Hebrew "Pey", but by the Hebrew letter "Ayin". And take special notice to see that it resembles a Pyramid shape with a truncated top - from the two base legs of Rigel and Saiph to the 3 stars of Orion.

Now I'm going to show this chart again so that one can see a "full view" of all the constellations.

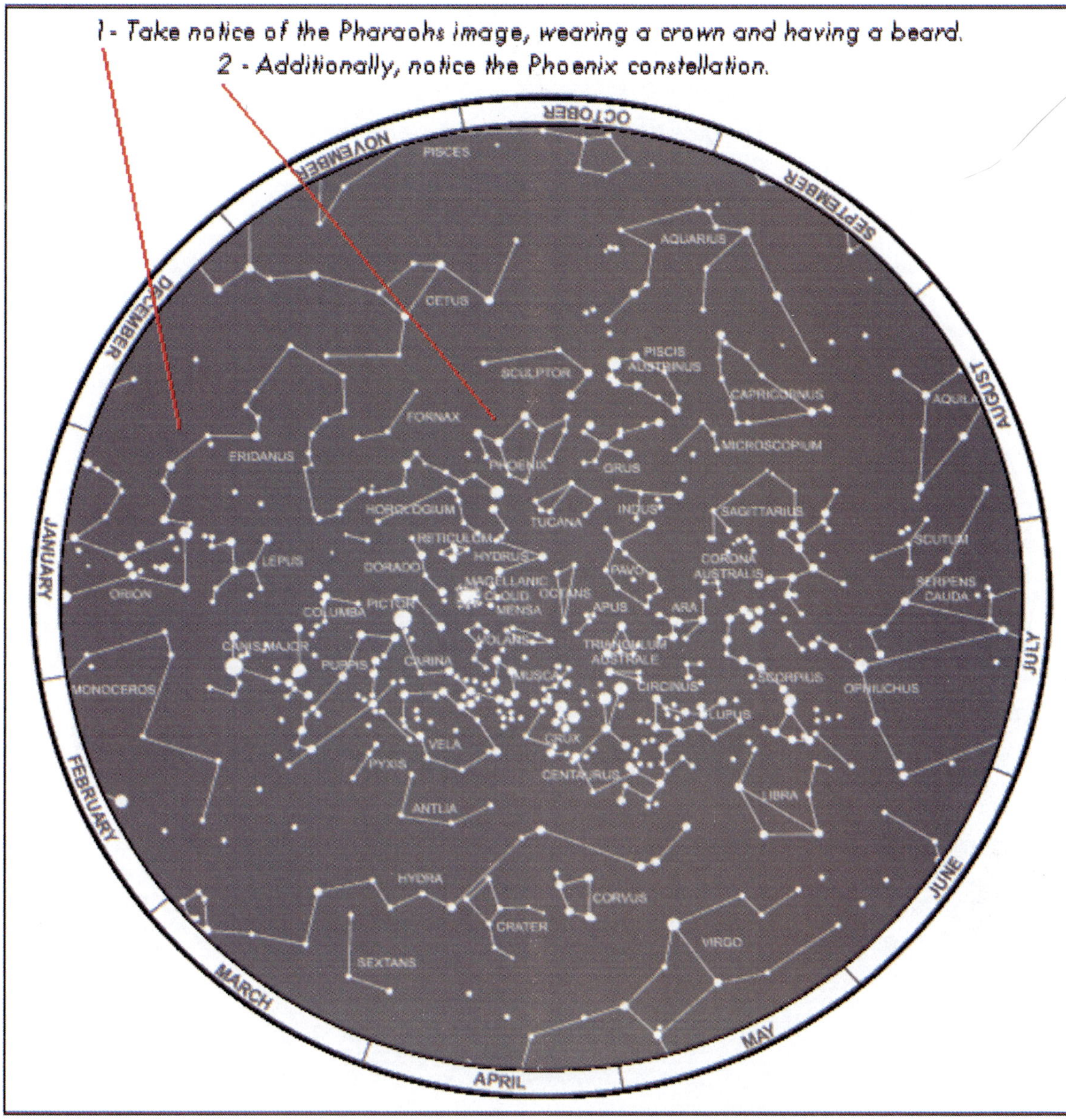
1 - Take notice of the Pharaohs image, wearing a crown and having a beard.
2 - Additionally, notice the Phoenix constellation.
OCTOBER
NOVEMBER
SEPTEMBER
DECEMBER
AUGUST
JANUARY
JULY
FEBRUARY
JUNE
MARCH
MAY
APRIL
PISCES
AQUARIUS
CETUS
PISCIS AUSTRINUS
SCULPTOR
CAPRICORNUS
AQUILA
FORNAX
MICROSCOPIUM
ERIDANUS
PHOENIX
GRUS
HOROLOGIUM
TUCANA
INDUS
SAGITTARIUS
RETICULUM
HYDRUS
SCUTUM
LEPUS
DORADO
PAVO
CORONA AUSTRALIS
ORION
MAGELLANIC CLOUD
OCTANS
SERPENS CAUDA
COLUMBA
PICTOR
MENSA
APUS
ARA
CANIS MAJOR
VOLANS
TRIANGULUM AUSTRALE
PUPPIS
CARINA
MUSCA
SCORPIUS
MONOCEROS
CIRCINUS
OPHIUCHUS
LUPUS
VELA
CRUX
PYXIS
CENTAURUS
LIBRA
ANTLIA
HYDRA
CORVUS
CRATER
VIRGO
SEXTANS

The other half of the "G" pattern showing Taurus, Gemini, Canis Minor, Orion's belt (Rigel and Betelgeuse). Canis Major is on the other side.

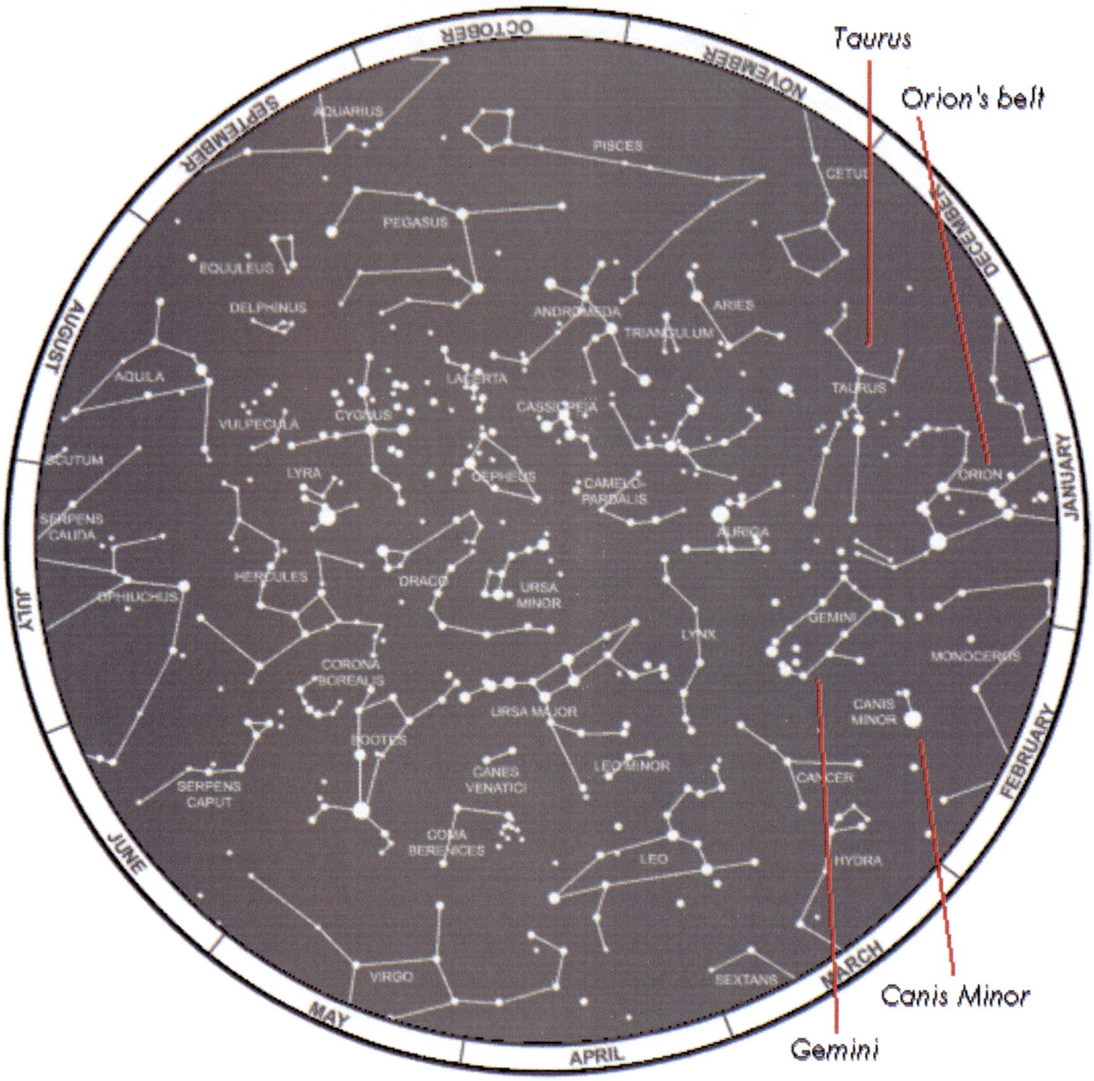

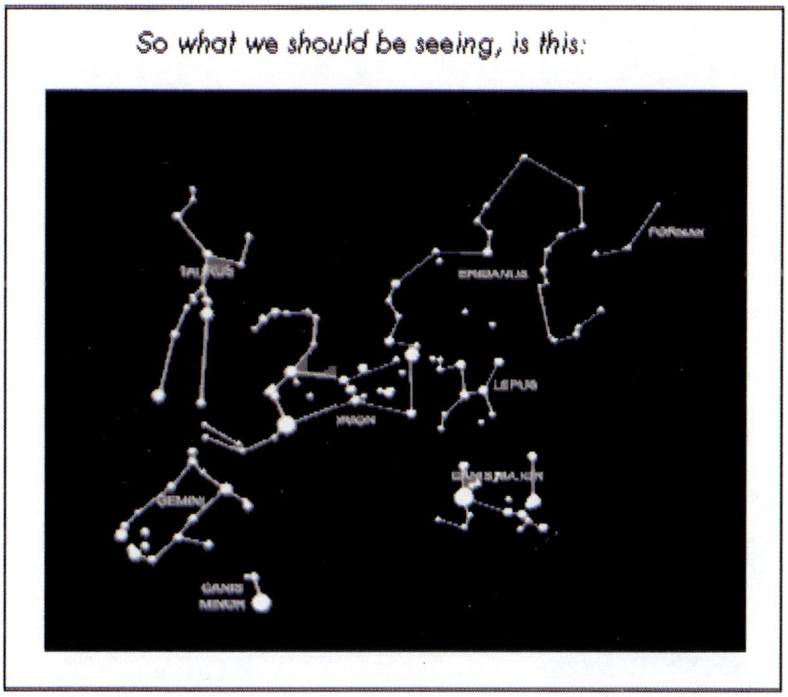

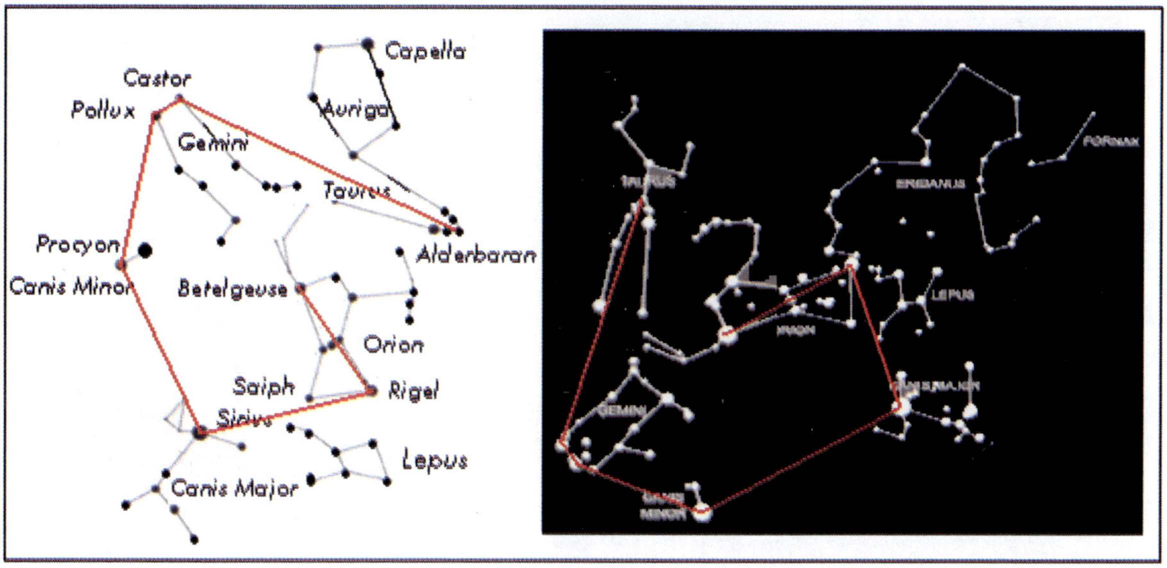

The Hebrew letter “Pey” - the MOUTH and “word of G-d”.

So not only does this look familiar to us as an Egyptian pharaoh in appropriate dress attire as well as playing a pivotal role as the Hebrew letter Pey, but we can additionally see the crown resting on the head with the back part of the crown as (Fornax) which might symbolize the combination crown of both upper and lower Egypt.

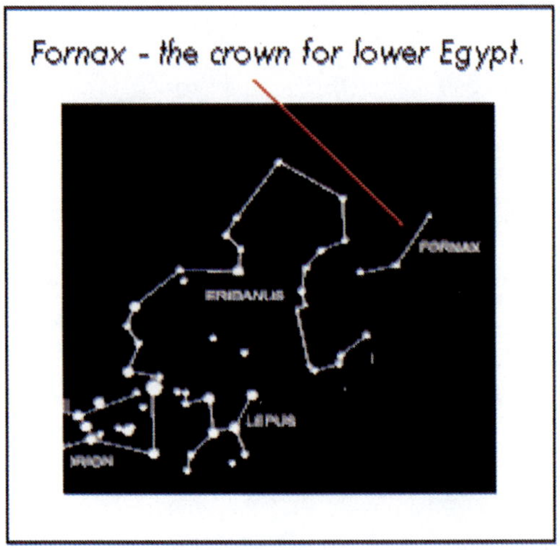

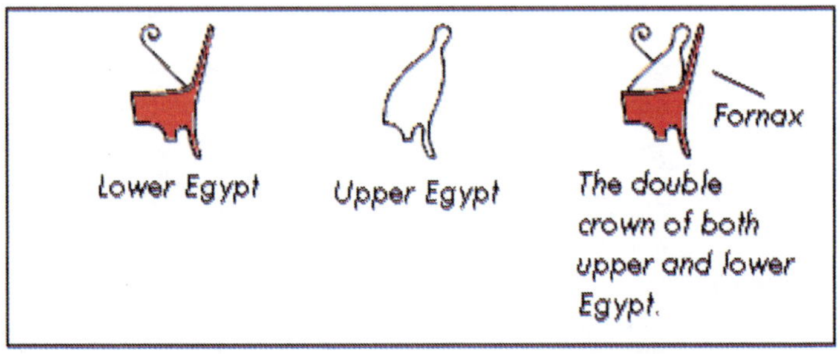

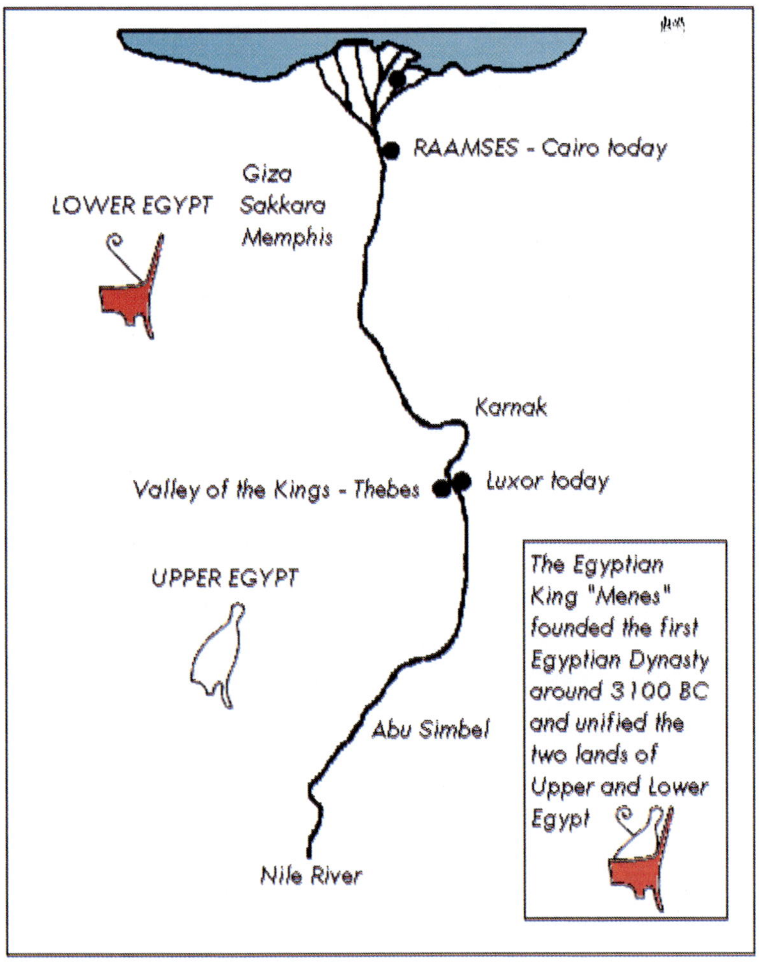

Additionally now lets look at another nearby constellation. Another representation of the constellation of Orion (the hunter) can be found behind the Pharaohs head, which is the "constellation of the Phoenix" which represents a bird with wings.

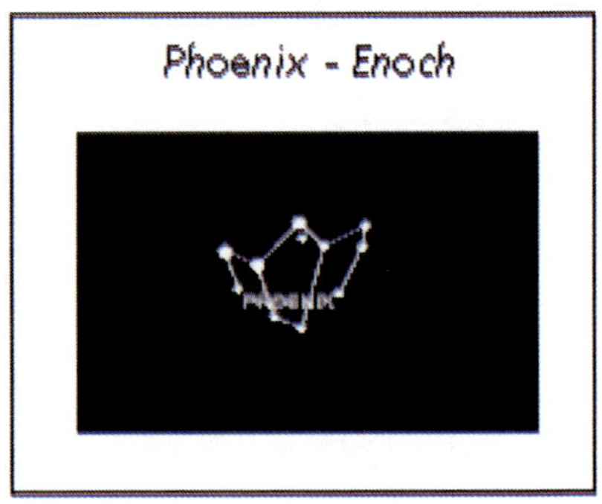

The bird with wings is also associated as the "soul of Osiris" that gives us the breath of life and death in resurrection. However, this bird is the all familiar "Phoenix" that lives for 500 years - hence birth, death and resurrection, and as we also know that the Phoenix means "house of Enoch" (or house of the stars). In addition, if you can recall, the Egyptian Hieroglyphic for a bird is ironically the phonetic sound for "Ahhh" though some associate it with the sound of Woo or Huuu. But the Egyptian hieroglyphic word and sound of Baaa (which means "soul") is traditionally represented by a bird (Osiris) and later takes its root in the Hebrew/Egyptian word Mer-Ka-Bah (which will be mentioned in its coming chapter).

So, as you can see, the Egyptians revered things of a shared nature such as with the cosmos (stars). They also observed things of a shared nature not only with the stars but also with the human body - a sort of microcosm as well as macrocosm. A "microcosm" in the sense of something being a miniature copy of something similar, especially when it represents or stands for a larger whole. And a "macrocosm" in the sense of a complex system or structure such as the world or the universe considered as a single entity that contains numerous similar smaller scale structures, and the Egyptians saw this through the role and rituals of mummification (the act of preserving a human body). Through the mummification processes they were also privy to what the inside of the human body looked like. Part of mummification requires that the human brain be extracted from the skull as well as other body parts and placed within separate jars. They were quite up on the biology of the human body and related its internal components as the symbolic structure and form of other things found in nature, as well as what we witnessed with the stars. But when it came to the human body, they highly esteemed such things like the heart and organs. Above all of these, I have found that the brain and spine were esteemed above them all, and I'll tell you just why.

After viewing these pictures (placed below) it is very compelling evidence that the crowns and objects of royalty that the Egyptians adorned themselves with and valued were replicas of not only the stars, but the human body in some outward appearance. They had seemed to revere the brain as something other than inside the skull, so they adorned it "ABOVE" their head in the form of a crown that depicted not only the pineal, pituitary and thalamus glands of the human brain but also what many hieroglyphics suggest as the "Djed column" which holds the striking resemblance of a human spinal column or backbone, but correspondingly it is also a replica of the internal structure (a microcosm) of the human brain as well. In fact the more we look at the brain, the more we begin to realize that the right and left sides (hemispheres) of the "corpus callosum" of the human brain resembles two cherubim with the Djed column between them focusing in on the center of the brain which happens to be the "pineal gland" (the pineal gland is a reddish-gray body about the size of a pea, or 8milimeters in humans). It is said to be the "third eye" which is believed to hold all the akashic knowledge of ones being. This organ is lined with fiber optic like nodules and creates the chemical melatonin that regulates rest, sleep/dreams and memories. ("Akashic" meaning, sky, space or ether. "Akashic record" is a library of knowledge that is encoded in a non-physical plane of existence, which includes all human experience. A sort of collective consciousness and sub-consciousness allowing us to reanimate our life's memories into the spiritual plane).

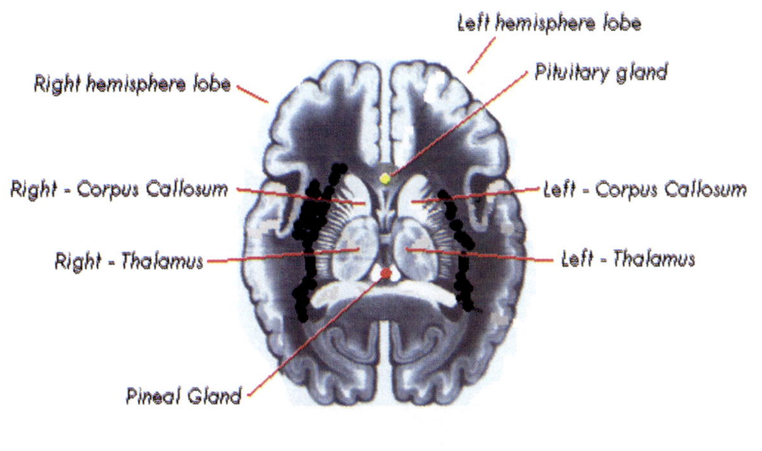

Under section of the human brain.

To see this, lets first take a look at the human brain.

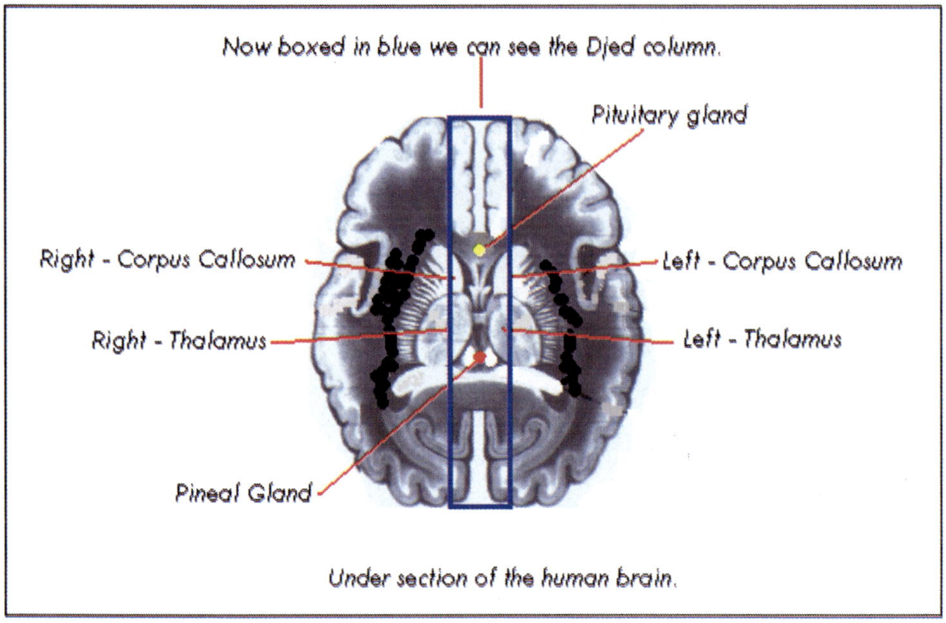

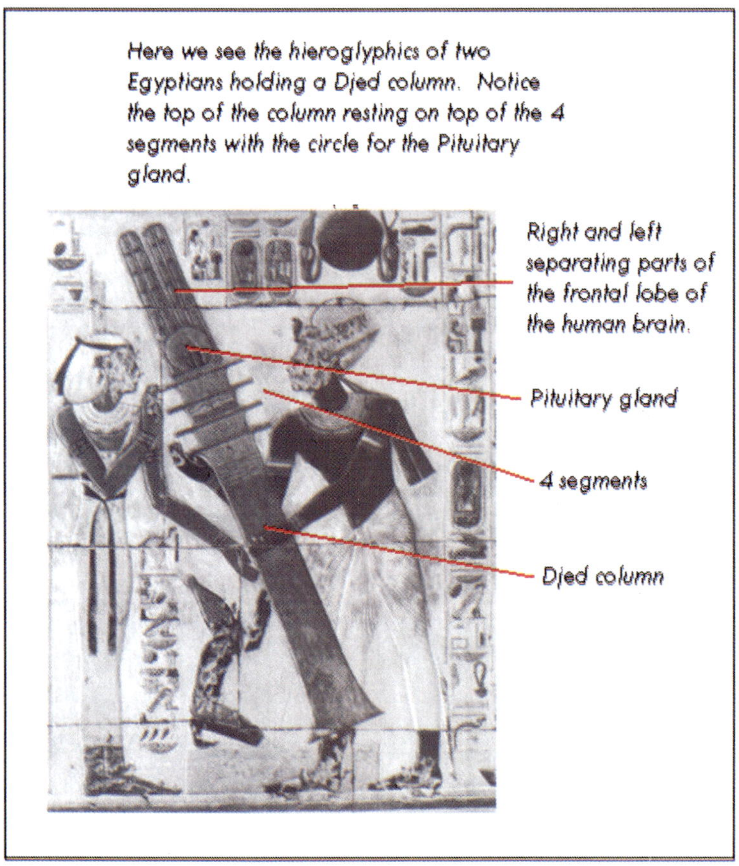

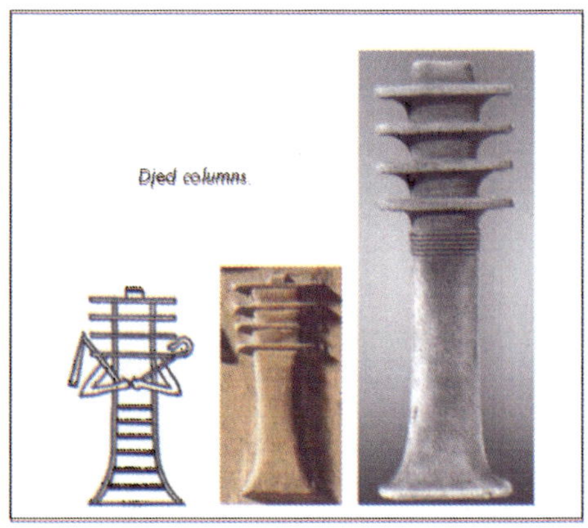

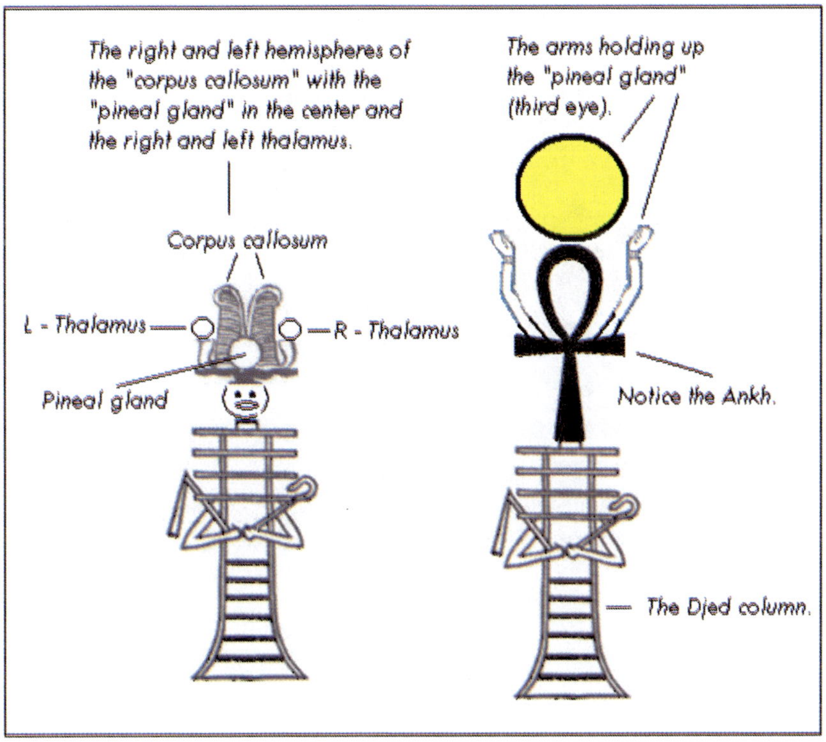

Here we can see the hieroglyphic inscriptions found within the Dendera Temple in Egypt. If we look towards the top we can see two winged female type cherubs safeguarding a Djed column.

Egyptian crown - arms holding up the Pineal gland (third eye).

This is very similar to the architecture of the Ark of the Covenant (specified in the Bible) that Moses had built to safely house the testimony of the law and the 10 commandments.

Casket (or chest) of Osiris which can be found at the Egyptian Museum in Cairo.

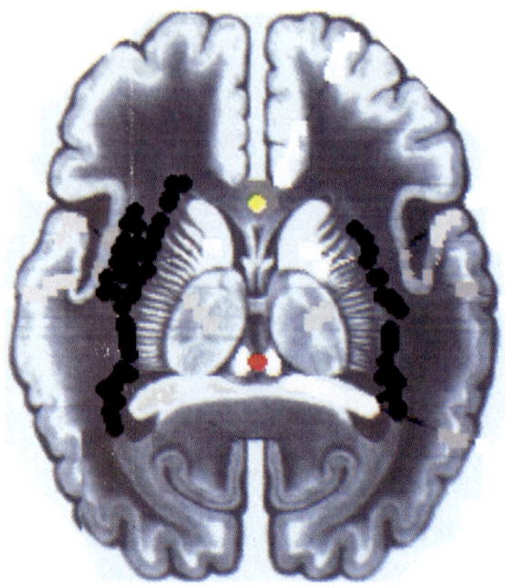

There is also a striking resemblance in the placement between the outstretched wings of the Cherubim and right and left Corpus Callosum of the human brain including the Pituitary gland and the right and left Thalamus.

Theory of what the Ark of the Covenant that the Jews built during the Exodus may have looked like:

But not that Moses had taken this architecture out of Egypt as some might think, but that it arrived there first by Enoch as (we must remember) he was the progenitor of Egyptian philosophy and culture. And the instructions were given again to Moses "BY G-D HIMSELF" to build the Ark of the Covenant. Exodus 25:8-22 - And let them make me a sanctuary; that I may dwell among them. According to all that I shew thee, after the pattern of the tabernacle, and the pattern of all the instruments thereof, even so shall ye make it. And they shall make an ark of shittim wood: two cubits and a half shall be the length thereof, and a cubit and a half the breadth thereof, and a cubit and a half the height thereof. And thou shalt overlay it with pure gold, within and without shalt thou overlay it, and shalt make upon it a crown of gold round about. And thou shalt cast four rings of gold for it, and put them in the four corners thereof; and two rings shall be in the one side of it, and two rings in the other side of it. And thou shalt make staves of shittim wood, and overlay them with gold. And thou shalt put the staves into the rings by the sides of the ark, that the ark may be borne with them. The staves shall be in the rings of the ark: they shall not be taken from it. And thou shalt put into the ark the testimony which I shall give thee. And thou shalt make a mercy seat of pure gold: two cubits and a half shall be the length thereof, and a cubit and a half the breadth thereof. And thou shalt make two cherubims of gold, of beaten work shalt thou make them, in the two ends of the mercy seat. And make one cherub on the one end, and the other cherub on the other end: even of the mercy seat shall ye make the cherubims on the two ends thereof. And the cherubims shall stretch forth their wings on high, covering the mercy seat with their wings, and their faces shall look one to another; toward the mercy seat shall the faces of the cherubims be. **And thou shalt put the mercy seat above upon the ark; and in the ark thou shalt put the testimony that I shall give thee. And there I will meet with thee, and I will commune with thee from above the mercy seat, from between the two cherubims which are upon the ark of the testimony, of all things which I will give thee in commandment unto the children of Israel.**

(Highlighted in red directly above), I find it interesting that G-d uses this architecture (a type of microcosm) to meet and communicate with us (Moses) on a subordinate level - even as abstruse as it may seem on a superficial level. G-d explicitly tells us in Exodus 33:20 - You cannot see my face, for no one may see me and live. Therefore, in order for G-d to intercede with us, He must filter Himself in a sort of subsidiary approach. When we look at the pictures above we can see that they match clearly by divine design - through and through. In addition, it is doubly amazing that this design never changes! G-d's longing has always been to be with His people in a close way. In fact, He created us in such a way as to commune with us through His spirit (possibly by Djed columns He designed and incorporated into each of our brains). The Ark of the Covenant was a part of G-d's divine design to bring us back into communion with Him. However, it was in no way the culmination of His master plan. Hebrews 8:8-10 - The day is coming, says the Lord, when I will make a new covenant with the people of Israel and Judah. This covenant will not be like the one I made with their ancestors. But this is the new covenant I will make with the people of Israel on that day, says the Lord: **I will put my laws in their minds, and I will write them on their hearts.** I will be their G-d, and they will be my people.

Now lets take an in-depth look of the crowns of Egypt:

Here are some hieroglyphic pictures of Egyptian crowns.

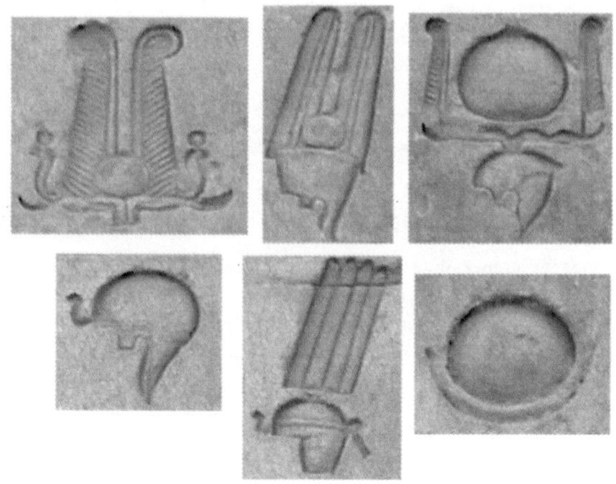

Here are the hieroglyphics of the Egyptian crowns for lower Egypt, upper Egypt and both upper and lower Egypt.

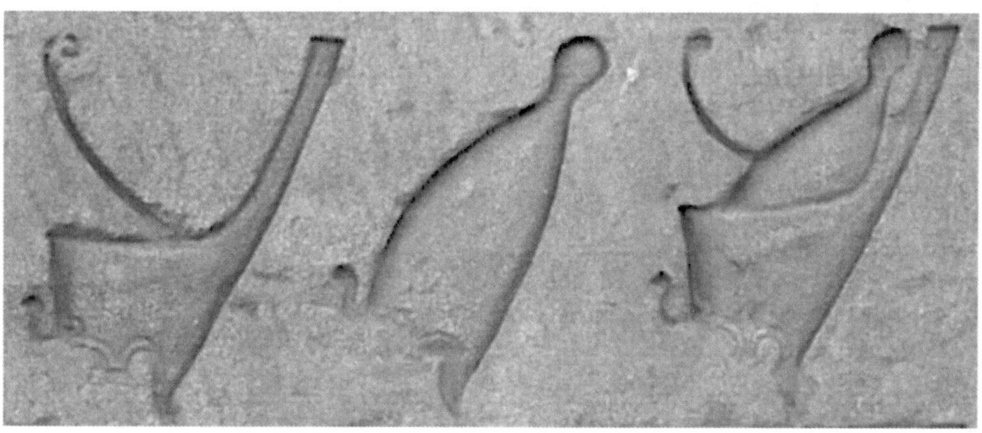

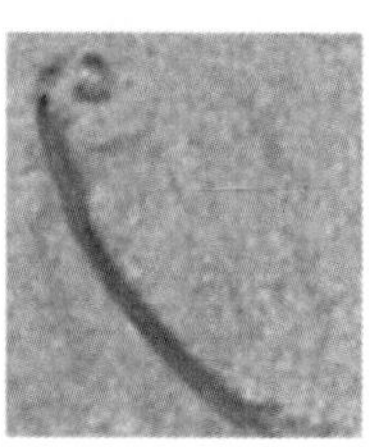

This swirl on the crown might represent the constellation Cetus.

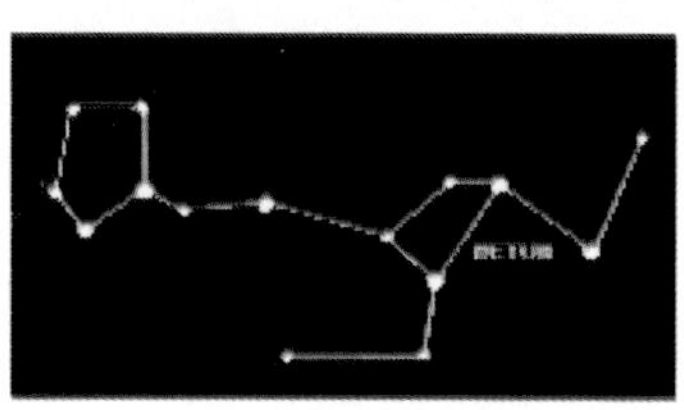

Cetus

Cetus shown here over Eridanus (the Pharaohs head).

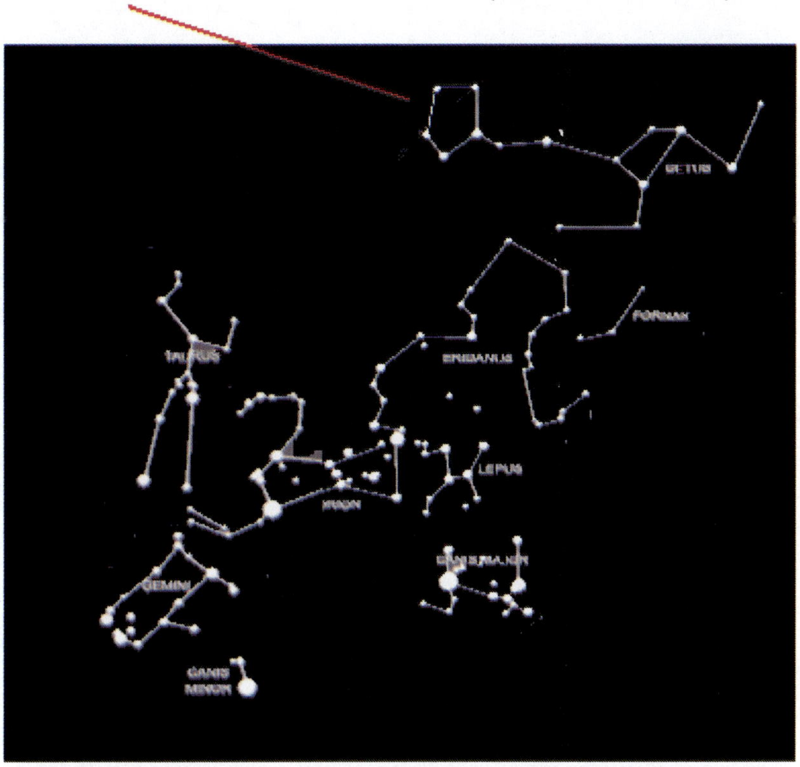

The snake in the center forehead of the crown may be a representation for the "Corpus Callosum" (also known as the "Caudate nucleus" of the brain).

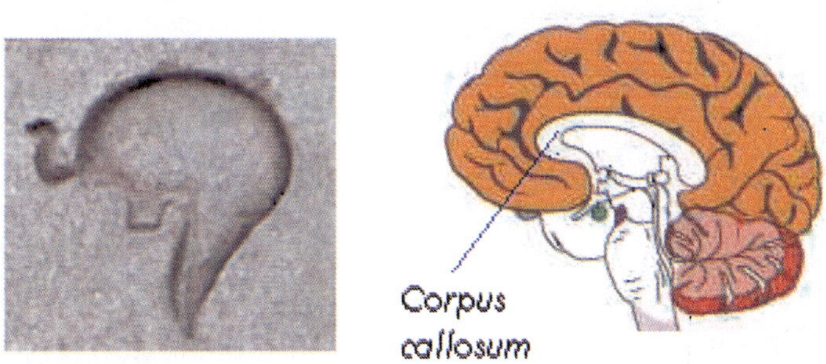

Now here is where the Qabalistic hierarchy of knowledge comes into play… It can also be said that the "cranial nerves" of the human brain are in symbiotic correspondence with the 10 sefiroth of the "Tree Of Life" diagram (another type of microcosm). These cranial nerves (which you can view in the diagram below) are nerves that originate in the brainstem and pass out of the skull to the surface of the body. There are actually 12 cranial nerve root stems in the human brain that have been termed.

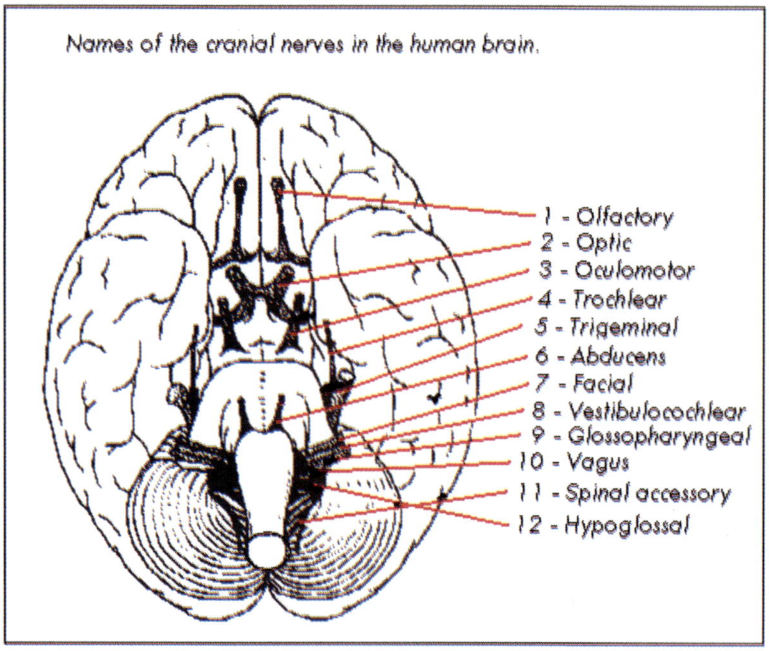

Here is more of an idea of what parts of the body these nerves control:

1 - Olfactory Nerve - smell - heavenly ether.
2 - Optic Nerve - vision - detection of light - reproduction.
3 - Oculomotor Nerve - Eye movement (upward, downward and inward) - pupil dilation - shoulders and arms.
4 - Trochlear Nerve - Eye movement (downward and inward).
5 - Trigeminal Nerve - sensory stimuli from the skin - information (touch, pain) from the face and head - muscles for chewing.
6 - Abducens Nerve - Eye movement (outward) - blinking - teeth and jaw.
7 - Facial Nerve - Taste - (tongue) - sensory stimuli from the skin - information from ear - controls muscles used in facial expression - production of saliva and tears.
8 - Vestibulocochlear Nerve - Hearing - balance.
9 - Glossopharyngeal Nerve - Taste - (tongue) - sensory stimuli from the skin - information from tongue, tonsil, and pharynx - muscles used in swallowing and for speech.
10 - Vagus Nerve - Sensory - motor and autonomic functions of internal organs (glands, digestion, heart rate).
11 - Spinal Accessory Nerve - Controls muscles used in head movement - neck turning and shoulder shrugging.
12 - Hypoglossal Nerve - Controls muscles of tongue.

Now turning our attention back to the “Tree of Life” diagram (shown below) and now we see another type of microcosm… The sefiroth of the “Tree of Life” are conjoined precisely to where the brain meets up with the cranial nerves that control that part and function of the body (just like the hands and feet overlay the body in “reflexology” as well as Reiki and Acupuncture). For example, “Kether” is the crown - divine throne - chariot (top part of the head). “Binah” (understanding) and “Chokmah” (wisdom) are the eyes on the side of the head that are essentially connected to the “Optic nerve“ that controls vision (for only in spiritual sight do we truly gain understanding and wisdom). Furthermore, even though the cranial nerves are contingent to the brain with its 5 senses of sight, smell, touch, taste and hearing, these ”nerve regions” also extend and span throughout the rest of body by way of the spinal cord. From the head to the shoulders on the “Tree Of Life” diagram is “Geburah” (severity) and “Chesed” (mercy) which is essentially controlled by the “Spinal Accessory Nerve”, for we carry severity and mercy on our shoulders - the weight of the world. Then to “Tipereth” (beauty) which is the chest and heart region (which is essentially controlled by the “Vagus nerve” (that determines heart rate), for in true beauty do we realize genuine love. “Hod” (splendor) and “Netzach” (victory) are the arm and hand regions (which is essentially controlled by the “Oculomotor nerve“), for by the arms and hands are we are able to achieve

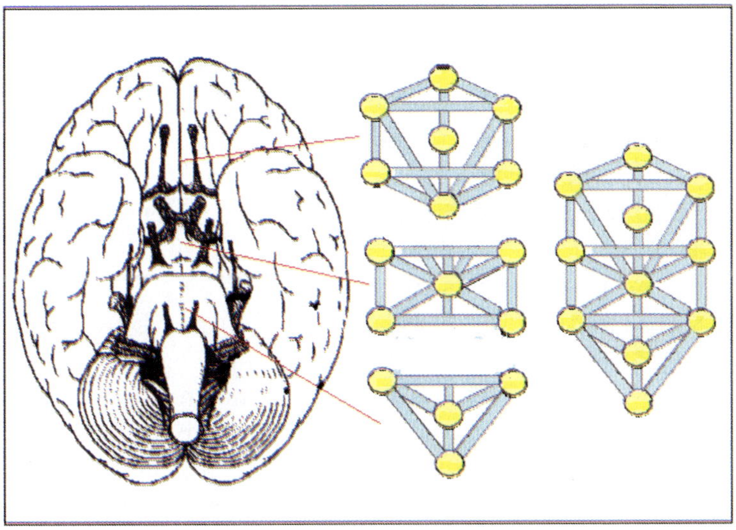

the finery of victory. Than we move to "Yesod" (foundation), which is, located in the groin area and controlled by the "Optic Nerve" for vision which also is right above the "pineal gland". This area also creates the chemical "serotonin" which is responsible for reproduction, which can be likened to the next brick (new human) added onto the foundation. Lastly "Malkuth" (kingdom), which is the feet in depiction of the "headquarters and kingdom" for which we are situated and stand on.

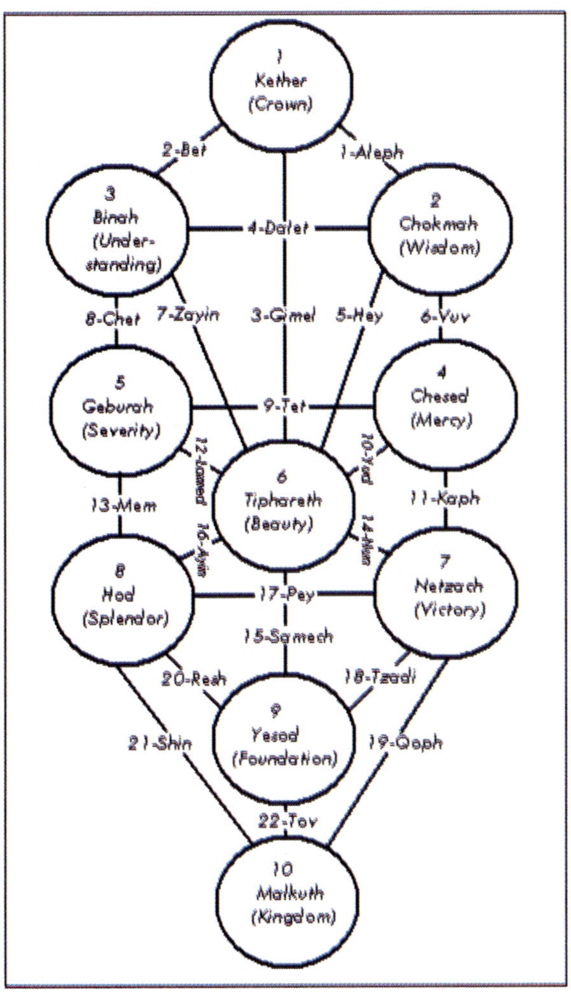

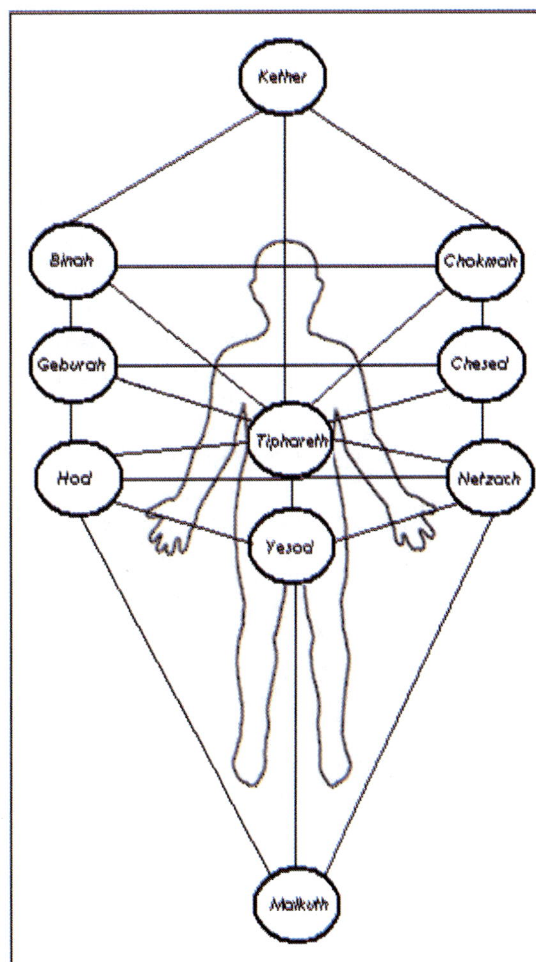

Conclusion… There are very real truths to Qabalah! Not only in Jewish rationale consisting of oral, moral and ethical laws, but its also a map of reality and states of consciousness (involving Quantum Mechanics) that incorporates the Hebrew letters not only in gamatria, but in constructively using them to play a part in shaping the probable future (which we will further read about in the next chapter). Qabalah (as we've learned earlier in the chapter) is also overwhelmingly evident in the medicinal arena, biology, geometry / mathematics/gamatria, astrology/star precession, and in "ALL" manifold aspects. Though we must remember that we are not to worship this "manifold" as something separate other than G-d - like idols or other deities. We must keep in mind that they are "ALL" from "ONE" source - G-d. Psalms 104:24 - O Lord, how manifold are thy works! In wisdom hast thou made them all: the earth is full of thy riches. As the Hebrew word Qabalah means to "receive or accept" - that by studying the Torah one can gain insight and understanding about the creation of the world, and come to realize that the very bases of all SCIENCE is non less then RELIGION! And "religion" in terms of transcending the "spiritual aspect" of all things to receive true "fulfillment". (Happiness, joy, gratification, realization, accomplishment, satisfaction and contentment).

What Qabalah holds for us is something that affixes all religions. As far as its linkage to Freemasonry which is said to be a "secret" organization, I can only pass on what Jesus said in John 18:20 - I have spoken openly to the world. I spoke nothing in secret. However, most importantly we must not loose aspect of G-d. We must start asking ourselves just who or what He actually is, and what would constitute even more objective proof of Him? Our oldest recorded documentation of who G-d may be starts before the Great Flood with Enoch. While some suggest that whoever Enoch got this knowledge from was not really G-d, but advanced alien beings who flew him away in their UFO's. However, when we really start to analyze things further, we begin to realize that this is a foolish suggestion in which the following chapter will reveal otherwise. We will also examine more unimaginable evidence in the next chapter which talks about finding meaningful codes in the Bible which will help us to determine whether we have received a genuine and sincere encounter with a divine supernatural being. Surely not an alien (creature from outer space, or another planet), but the ONE AND ONLY CREATOR (maker, inventor, originator, architect, designer, initiator), and notably "author" OF EVERYTHING - G-d!

Now before you move onto the next chapter… if this chapter was too heavily overwhelming with esoteric knowledge for you, simply put the book down for a while and ponder on what you've read. Some of you may even be feeling "anger"! This is normal as I felt it too. This is anger of secret knowledge that has been kept hidden and severed from you, but it is now knowledge that you need to connect with (not to bond or unite with, but just to connect and relate with). But please, no matter how angry you may be, please don't scorn the messenger (me)! It remains your choice to continue reading. And as aforementioned, I did say at the very beginning of this book that this information was not for the light at heart. Be forewarned, it's only going to get more intense in the next chapter!

13 - BIBLE CODES - GEMATRIA...

Like I just mentioned above, we are additionally going to discriminatingly and finely examine whether or not we have received a genuine and sincere encounter with a supernatural being. For all of you who where not touched in some way by all the above, this is the defining chapter that just may do it. It gives us even more of the substantial proof that we need to believe in G-d. Briefly, the "Bibles Codes" are a beautiful message about CHOICE! - Hence Quantum Mechanics again. Some may think these codes are all hokum, but I will let you decide that for yourself.

Simply put, the Bible Codes are words, phrases and groups of words that can be found within the Torah. The primary method by which meaningful messages have been extracted is by "equidistant letter sequence" (ELS). It is similar to finding words in a crossword puzzle (a puzzle in which numbered clues are solved corresponding to other numbers or words or phrases). And like a crossword puzzle these words and phrases, and groups of words form either horizontally, vertically, diagonally, front ways, back ways, sideways or by four corners, or even by any of the geometric shapes (square, triangle, circle) of "equal form". However not always necessarily does finding a code have to be within such strict guidelines, seeing that its even been discovered that these words, phrases and groups of words clumped together in such a way forming what appears to be "pictographic pictures" that also relate to the "context" of the code itself.

To find a code one must start with any Hebrew letter and skip over other letters by an equal amount of spacing (ignoring the spaces and punctuation), such as by the number 50.

For instance "TORAH" is the very name of the book, and a "meaningful" code can be found in it that enumerates this very message embedded within the Hebrew text itself by intervals of 50 Hebrew letter skips. What is the significance of the number 50? Because, the "Torah" was given to the Jewish people on the 50th day after they left Egypt at Mount Sinai. The giving of the Torah is celebrated by a Jewish holiday called "SHAVUOT" which marks the giving of the Torah on Mount Sinai, which was the 50th day after they left Egypt. On this holiday, the Ten Commandments are read in synagogues just as they were in the desert on Mount Sinai many thousands of years ago.

So to find this code, starting in "Genesis 1" (the first book of the Bible) the word "Torah" (circled in red by me for your convenience) is found at 50 letter intervals (which is read forwards). For example, take the last Hebrew letter Tov (T) in the first word Bereshi(th) and count 50 letters. The 50th letter is a Vuv (O). Count again 50 letters and the 50th letter is Resh (R). Repeating a count of 50 letters and the 50th letter is Hey (AH). This spells the word TORAH.

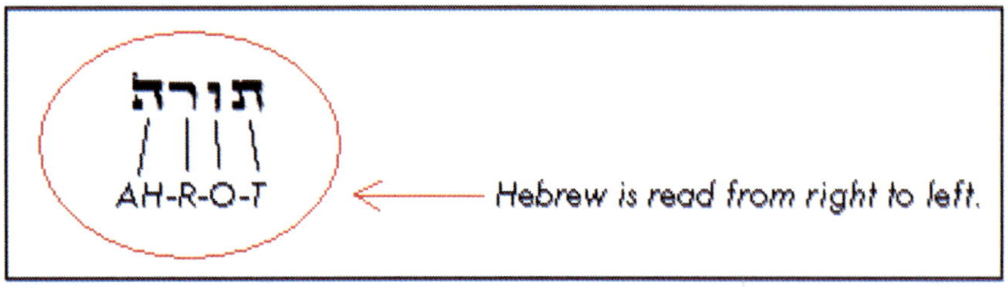

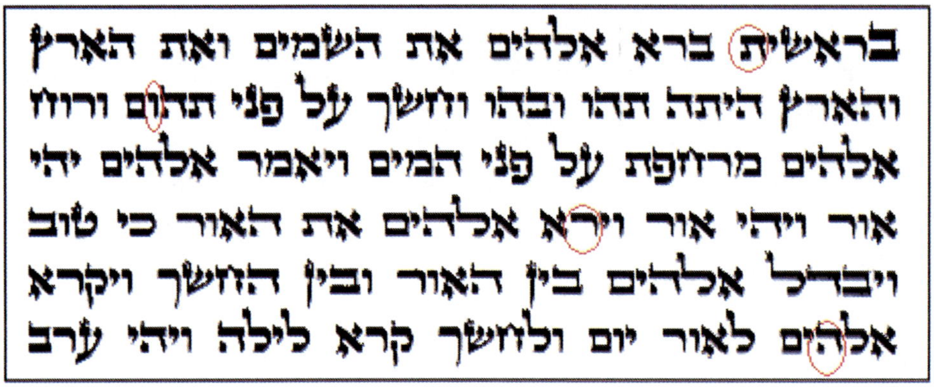
בראשית ברא אלהים את השמים ואת הארץ
והארץ היתה תהו ובהו וחשך על פני תהום ורוח
אלהים מרחפת על פני המים ויאמר אלהים יהי
אור ויהי אור וירא אלהים את האור כי טוב
ויבדל אלהים בין האור ובין החשך ויקרא
אלהים לאור יום ולחשך קרא לילה ויהי ערב

The second book of the Bible is "Exodus 1", and with astounding accuracy by using the SAME skip interval of 50, and applied to the last Hebrew letter Tov of the word Shemo(t). The search here ALSO spells TORAH (and which also reads forwards).

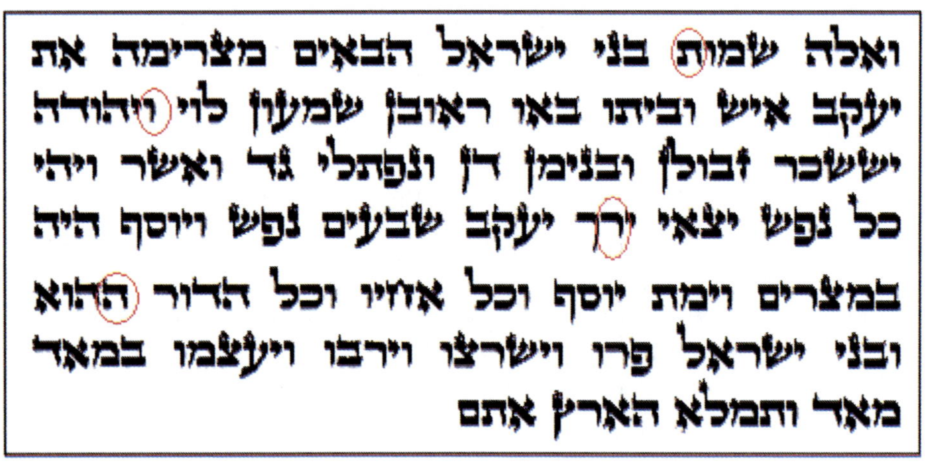
ואלה שמות בני ישראל הבאים מצרימה את
יעקב איש וביתו באו ראובן שמעון לוי ויהודה
יששכר זבולן ובנימן דן ונפתלי גד ואשר ויהי
כל נפש יצאי ירך יעקב שבעים נפש ויוסף היה
במצרים וימת יוסף וכל אחיו וכל הדור ההוא
ובני ישראל פרו וישרצו וירבו ויעצמו במאד
מאד ותמלא הארץ אתם

The 3rd book of the Bible, which is actually the "center" book of the Bible is "Leviticus 1". However, this time we do not find the word Torah, but we find something even more amazing… We find the word "Yahovah" (Y,H,V,H) (G-d's name) embedded by intervals of 8 letters. So why the number 8? It's relatively easy to assess why, and this is because the name of G-d (Yahovah) (Yud, Hey, Vuv, Hey) has the deduced gamatria equivalent of 8. Yud=10, Hey=5, Vuv=6 and Hey=5 (10+5+6+5=26)(2+6=8). So there is a loud and clear message saying that, if at first you don't believe that it is indeed G-d, check it against its numerical equivalent - which reveals itself also by the equidistant letter skipping of the number 8. There are several different Hebrew names used in the Torah when referring to G-d such as Eloheim, Yahovah and Adonai. Though some might suggest that they are referring to different G-d's, they are not… I will shortly explain why.

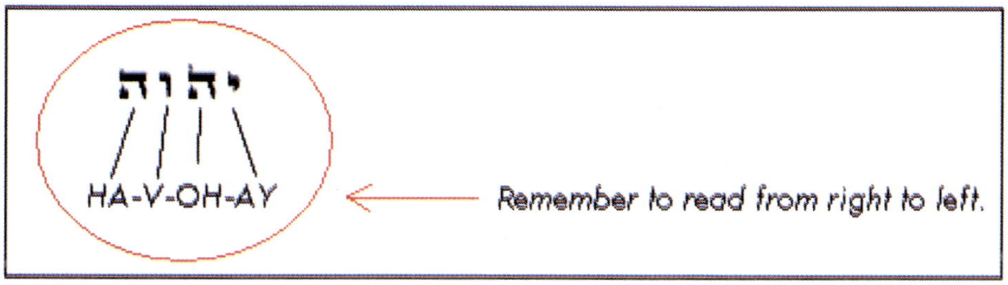

ויקרא אל משה וידבר יהוה אליו מאהל מועד
לאמר דבר אל בני ישראל ואמרת אלהם אדם
כי יקריב מכם קרבן ליהוה מן הבהמה מן הבקר

As if this were not enough, in the 4th book of the Bible, which is "Numbers 1" THIS SAME CODE by some immaculate odds of, intervals of 50 letters yet again can also be found (which this time reads backwards).

וידבר יהוה אל משה במדבר סיני באהל מועד
באחד לחדש השני בשנה השנית לצאתם מארץ
מצרים לאמר שאו את ראש כל עדת בני ישראל
למשפחתם לבית אבתם במספר שמות כל זכר
לגלגלתם מבן עשרים שנה ומעלה כל יצא צבא
בישראל תפקדו אתם לצבאתם אתה ואהרן ואתכם
יהיו איש איש למטה איש ראש לבית אבתיו

And lo and behold, in the last 5th book of the Bible "Deuteronomy 1", astoundingly again by intervals of 50 letters we find the word Torah once more which can be found (reading backwards again).

האמרי אשר יושב בחשבון ואת עוג מלך הבשן
אשר יושב בעשתרת באדרעי בעבר הירדן
בארץ מואב הואיל משה באר את התורה הזאת
לאמר יהוה אלהינו דבר אלינו בחרב לאמר רב
לכם שבת בהר הזה פנו וסעו לכם ובאו הר
האמרי ואל כל שכניו בערבה בהר ובשפלה
ובנגב ובחוף הים ארץ הכנעני והלבנון עד הנהר
הגדל נהר פרת ראה נתתי לפניכם את הארץ באו
ורשו את הארץ אשר נשבע יהוה לאבתיכם

So we have the first two books of Torah (Genesis and Exodus) pointing "FORWARDS" and in the last two books of Torah (Numbers and Deuteronomy) pointing "BACKWARDS" - so as to say its pointing to the "CENTER" (which is the book of Leviticus) - which does not have the word Torah embedded into it, but that of G-d's name in equal intervals of 8 letters spelling "Yahovah" (Y,H,V,H).

So, we have the name of the text itself (TORAH) and its AUTHOR. It appears that the Torah was specifically designed and intended to always point toward the name of G-d the proprietor. It has always been about its author revealing Himself - forwards - middle - backwards. (A closed loop system). If you recall as aforementioned in an earlier chapter, a closed loop system is isolated from anything coming in or going out. When I refer to a closed loop system, I am implying to the "whole" of everything that "is" and will ever be. In addition, this includes the whole gamut of space, stars and universes in existence. Outside this, nothing else exists.

The odds of finding these two words (Torah & G-d) in the "context" of how they are found, with a significant meaning of finding them at 50 letter intervals representing WHEN THE TORAH was given, and by letter intervals of 8 via whom it was given by, are off the charts in "probability standards". Remember, "COINCIDENCE" is not a kosher word, when it is clearly by supernatural design! There are no "coincidences" in G-d's kingdom. And this Rings True when we stop to consider what the Hebrew word "kosher" actually means? It simply means, "Prepared properly".

So incredible is the Bible Code especially with the advent of computes that so many more codes are being discovered. And it should not surprise us that the computer could be used in this way to supersede us in finding even more elusive codes - hence, remember the computer is also a man made thing, and as aforementioned in an earlier chapter in realizing that the most expansive aspect of the creation of the computer is the intention for which it was created, and this "intention" is something that we share with our creator.

Now, if you recall the second opening prolog to this book, I mentioned that I was not the official writer. I said: I didn't write this book, but that of energy working through me did. Whether you wish to call it G-d, source or spirit this energy writes all the books and delivers all the speeches. It is this source that allows the words to flow through me, through my mind and onto the paper. When we are in thought, or inspired by something, we are essentially in spirit transcending all limitations of our consciousness, as a result we are also creating! Following this thought, are we capable of realizing that there is something greater going on here?

This is how it must have also transpired with the original writers of the Bible who had no clue that encryptions (or G-d's will) were taking place between the lines of the Bible, nor did they have the technology to that extent to encrypt them by themselves, let alone in encrypting so many dimensional depths without the help of a computer. Miraculously without even knowing it "unbeknown" to them during the creating process of "conscious thought", an energy that they could not see (but feel) transpired between their physical body and the spiritual subconscious while using the mind to operate the hands to write letters upon parchment into the book of the Bible. In essence, when this was taking place the writers of Bible were in creative or inspiring thought and in connection with "source - G-d". 2 Peter 1:20-21 - Knowing this first, that no prophecy of the scripture is of any private interpretation. **For the prophecy came not in old time by the will of man: but holy men of G-d spake as they were moved by the Holy Ghost.** Acts 27:14-15 - But not long after there arose against it a tempestuous wind, called Euroclydon. And when the ship was caught, and could not bear up into the wind, **we let her drive.** However, it must also be noted that there are other sources as well that could be evil, if one connects to them in negative ways. These very real forces can manifest through the physical. They can attract and repel like atoms, and resonate like frequencies - to like minds. (This subject will also be addressed in the chapter on the "proper religion").

Now, I'd like to examine a few more codes with you that will hopefully strengthen your opinion to consider it a legitimate source of truth. I am leaning on some very "meaningful" codes that were found in the context of the Torah that completely illuminates theological thought for the purpose of WHY they were found there. Though there are many (uncountable) other codes, I chose to exemplify these.

Now we will examine why there are different names for G-d found within the Torah. Specifically the two most biblically notable names of G-d, which are Eloheim and Yahovah. Starting in the very first story of Genesis. Genesis 1:1-5 - In the beginning G-d אלהים created the heaven and the earth. And the earth was without form, and void; and darkness was upon the face of the deep. In addition, the Spirit of G-d moved upon the face of the waters. In addition, G-d said, let there be light: and there was light. In addition, G-d saw the light, that it was good: and G-d divided the light from the darkness. In addition, G-d called the light Day, and the darkness he called Night. In addition, the evening and the morning were the first day. We see that the name for G-d that is used is ELOHEIM, which is referenced 5 different times circled in "red" below (G-d refers to Himself as Eloheim). Ironically, they are also separated by 5 words each circled in "blue", and sandwiched in-between these 5 ELOHEIMS by "22 letters" we also find the name of G-d "YAHOVAH" (Y,H,V,H).

בראשית ברא אלהים את השמים ואת הארץ
והארץ היתה תהו ובהו וחשך על פני תהום ורוח
אלהים מרחפת על פני המים ויאמר אלהים יהי
אור ויהי אור וירא אלהים את האור כי טוב
ויבדל אלהים בין האור ובין החשך ויקרא
אלהים לאור יום ולחשך קרא לילה ויהי ערב
ויהי בקר יום אחד

So why do we find 5 references to "ELOHEIM", and why the name of G-d (YAHOVAH) is sandwiched in-between at intervals of 22 letters? For this answer, we must first start to analyze why there are two references to G-ds name of "Eloheim and Yahovah". We are told in Qabalah (oral tradition) that G-d created the world with His divine energy that is represented by the name "YAHOVAH", but that energy is likened onto the Sun which is to powerful for us to take the way it is. The sun has to be filtered and that's why we have the atmosphere that filters the light of the Sun. That name of G-d (Yahovah), which is responsible for creation, is to powerful on its own, and the other name of G-d "ELOHEIM" (if you recall) is equivalent to "nature" (Hatevah), and is the "natural" aspect of G-d. Therefore, we have the transcendent supernatural aspect of G-d that's filtered by the natural aspect of G-d. Whereas the message of the 22 Hebrew letters sandwiched in-between the 4 Hebrew letter of G-d's name (Y,H,V,H) is a divine representation of the energy that was used in creating the world. While the 5 uses of G-d's name (Eloheim) represents the additional 5 "final" Hebrew letters (Kaf, Mem, Nun, Peh, Tsadi) that represent the "limitations" of that divine light, (the screening and filtering of it), that's why we find 5 different occurrences of ELOHEIM in Genesis (which signifies the "filtering process" of G-d). In addition, the 22 Hebrew letters are a representation that says that they represent the divine source of "YAHOVAH".

By the way if your wondering why G-d is often referred to as "He" (a man), this has to do with the titles for all of G-d's names in the Hebrew Bible which are masculine nouns. Therefore, we naturally assert that G-d is male. However, such words as "ruach Elohim" (the spirit of G-d), or "Shekhinah" (the presence of G-d) are feminine. This again has to do with G-d's filtering process.

For your reference… The 5 additional (final) Hebrew letters are used mainly as the "last letter in a word" and therefore they take on different "forms". These are the 5 final Hebrew letters circled in red:

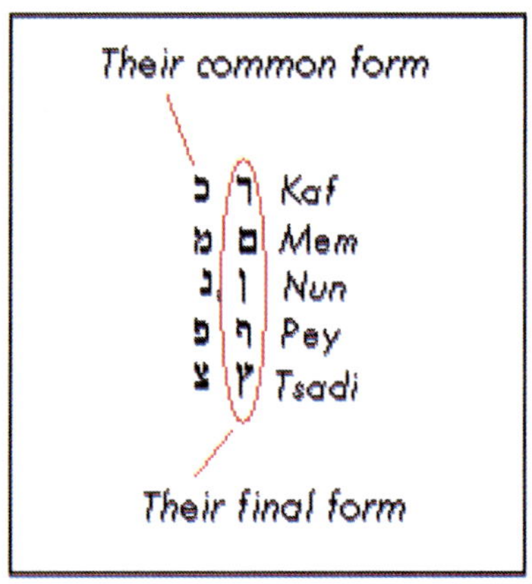

Before I go on to the next code, I would like to take a moment and elaborate on the filtering process of G-d and the operational purpose under which it works.

The Bible states insistently and repeatedly that there is only one G-d. However, this becomes confusing when it uses the names Elohim, Jehovah, and Adonai interchangeably. The conclusion may be that G-d is triune (3 in one) in nature - G-d the Father, G-d the Son, and G-d the Holy Spirit. G-d explicitly tells us in Exodus 33:20 - You cannot see my face, for no one may see me and live. Therefore, in order for G-d to intercede with us, He must filter Himself by the diminution (lessoning) of light. We had just recently witnessed this in the previous chapter with the Ark of the Covenant. But G-d also stooped to reveal Himself to us through His infallible word found within the Bible.

The proof is there that G-d has gone to elaborate lengths to explain this (3 in one) nature that was made available to us through the study of Torah. To begin to grasp and understand this triune nature all we have to do is examine the opening chapter of Genesis (the very beginning of the creation account). It is clearly emphasized for all to understand. Moreover, that we can legitimately link this (3 in one nature) up-with the claim that Jesus declared to being the Messiah. The Hebrew scholars try to avoid this at all costs! They say well that's just a plural of G-d's high-ranking majesty (referring to more than one). They assert that G-d is so great that He has to speak in plurals. But as we are about to witness He speaks in both plural and singulars.

Genesis 1:1 - In the beginning **G-d Created** the heaven and the earth.

"G-d" (In Hebrew is "Elohim" which is plural - referring to more than one).
"Created" (In Hebrew is "Bara" which is singular - referring to one).

Therefore, it can be immediately seen (by divine design) that we have both a plural noun with a singular verb. What G-d is telling us in this very first phrase is that He is a plural (Elohim), and He is also a singular (Bara)(He created)?

Then in Genesis 1:2 G-d (Elohim) tells us that He is also some kind of a moving "spirit". Genesis 1:2 - And the earth was without form, and void; and darkness was upon the face of the deep. And **the Spirit of G-d moved** upon the face of the waters. And in verse 3 G-d (Elohim) is some kind of "word" that speaks. Genesis 1:3 - **And G-d said, Let there be light**: and there was light. This G-d (Elohim) is "one", yet at the same time He is "several" - a sort of multiple persons.

Now one may ask, are there some sort of grammatical mistakes here?

Firstly, we must remember that these words were the very first words (from Genesis) that G-d dictated to Moses on Mount Sinai. Secondly, it cannot be a grammatical mistake since we've already discovered in the chapter of the "primordial language" as well as what we are discovering within this chapter (the Bible code) that every single Hebrew letter and its place within the Bible is of great significance. So much so, that if one single letter were removed, it would spell destruction for the entire Universe!

We also find this plural and singular permutational arrangement attached to other sayings of the Bible such as in Isaiah. Isaiah 6:8 - Also I hear the voice of the Lord saying whom shall I (singular) send, and who will go for us (plural)? Than said I, here am I, send me. So, as you can see again, G-d goes from the singular to the plural because He is a singular, but at the same time, He is also a plural. In addition, we witnessed this with the Ark of the Covenant as a type of microcosm in which G-d meets, communicates and intercedes with us on a subordinate level. And additionally as I said in the previous chapter, this divine design never changes! And, this is how we can authenticate it.

Here are some other New Testament passages that link-up and clarify Jesus' 3 in one triune nature: John 1:1-3 - In the beginning was the Word, and the Word was with G-d, and the Word was G-d. He was in the beginning with G-d. All things were made through Him, and without Him nothing was made that were made. John 14:6 - Jesus saith unto him, I am the way, the truth, and the life: no man cometh unto the Father, but by me. John 14:8-9 - Philip saith unto him, Lord, show us the Father, and it sufficeth us. Jesus saith unto him, Have I been so long time with you, and yet hast thou not known me, Philip? He that hath seen me hath seen the Father; and how sayest thou then, Show us the Father? Colossians 1:15 - Who is the image of the invisible G-d, the firstborn of every creature. Hebrews 1:3 - Who being the brightness of his glory, and the express image of his person, and upholding all things by the word of his power, when he had by himself purged our sins, sat down on the right hand of the Majesty on high.

Now to obtain an even better understanding of this "filtering process" that G-d uses... in an earlier chapter when it was talked about the Qabalistic 231 gates, they also come into play. The 231 gates are used for what's called a "permutation process", which uses the Hebrew letters to create all that exists. Though it is an esoteric process to truly grasp and understand, but it is through this "permutation process" that G-d created all things. This works by way of the diminution (lessoning) of light. The theory goes that things can only come into being having a physical and material body by way of light when the greater light passes down into lower degrees of light. And this can only transpire by the use of the Hebrew letters, which are in essence composed of "LIGHT" (the vibration of photons). It is by the use of the 22 Hebrew letters (along with the 5 final letters) that are what actually gives life, causing things to exist. In other words from all the 22 letters - from the first letter in the Hebrew alphabet (Aleph) to the last letter in the Hebrew alphabet (Tov). So we say "ATAH" (A-T-AH) - A (aleph)-T(tov)-AH(hey) created all that exists. The word "ATAH" represents all the Hebrew letters from Aleph to Tov and the five organs by which speech is produced (larynx, palate, teeth, lips and tongue) signified by the letter "Hey" whose gamatria value is 5. This is why during the creation process mentioned in Genesis 1 that proclaims when G-d spoke, things came into existence by the exposure (bringing to light) of the 22 Hebrew letters. The lessoning of light is a mere brightness of the living spiritual principle as a ray of sunlight is but the radiance of the sun. In essence, G-d concentrated the light, which spread from His mouth and embodied it in the combinations of the Hebrew letters, in letter changes, substitutions of the letters, gamatria values, and in placing the letters side by side. Each change and substitution indicates a descent of light from a higher to a lower degree until the desired level is reached. That is why we have all forms of life on earth such as from the lowest extremes of inert elements, rocks, to living microbes, plants, insects, animals, and humans.

Now I'm going to cast at you an intricacy of thought... From the soul flow words and letters of thought. From the letters of that thought flow the letters of speech, in which the thought is actually the life of the speech. Therefore, we can say that the speech is what brings the action into existence - hence AGAIN the "probabilities" of Quantum Mechanics.

Now for the next Bible code and its right on track with some of the subjects that have been discussed in this book so far. Now for some sort of proof beyond a reasonable doubt that it was indeed G-d who created all the stars and planets including everything in the existence of the Universe. For this proof, lets look back at the biblical phrase that tells us this. Genesis 1:14 - Genesis 1:14-19 - And G-d said, Let there be lights in the firmament of the heaven to divide the day from the night, and let them be for signs, and for seasons **(ULEMOADIM),** and for days, and years. In addition, let them be for lights in the firmament of the heaven to give light upon the earth, and it was so. Moreover, G-d made two great lights, the greater light to rule the day, and the lesser light to rule the night, he made the stars also. And G-d set them in the firmament of the heaven to give light upon the earth, and to rule over the day and over the night, and to divide the light from the darkness, and G-d saw that it was good. And the evening and the morning were the fourth day. For this proof we find embedded within the Torah text the Hebrew word "ULEMOADIM" (which means "festivals and seasons" - to which are assigned as the Hebrew festive holidays). And these are the same ones that Enoch observed with the help of the Henge to indicate the seasons.

Now, I would like to share with you and show you just how comprehensive G-d is in His reasoning. To reassure us that this is the genuine case and reason for Him to have actually created the stars for signs, seasons, days and years and that they are to be associated to the Hebrew festive holidays, He very nicely encoded it "within the actual text itself" at interval letter skips of 70. So why 70? Incidentally, because there are 70 days in the Torah that are set-aside as festive days "in which we are not to work". They are as follows listed below:

1 day Rosh Hashanah - Noted as the Jewish New Year. On this day, the Shofar is blown to awaken the listener from sleep and to alert them to coming judgment. It is also the "Day Of Judgment" for evaluating deeds before G-d.

1 day Yom Kippur - Noted as the Day of Atonement and repentance for forgiveness of sin. It is the climax of Rosh Hashanah. No eating is allowed on this day.

7 days of Sukkot - The word Sukkot in Hebrew represents the word "Sukkah" which means a "temporary shelter hut" in which the Jews built in which to eat meals during the 40 years time they wandered in the desert after the Exodus from Egypt. This eating shelter is intended to reflect G-ds generosity in providing for their needs in the desert. It is also the fruit harvest.

1 day of Shemini Atzeret and Simchat Torah - These two holidays fall on the same day. Shemini Atzeret literally means, "the assembly of the 8th day after "Sukkot" and is a more intimate celebration. Simchat Torah means "rejoicing in the Torah" in which we read the last Torah portion first, tean proceed to the first chapter of Genesis, which is to remind us that the Torah is a circle that never ends - hence a closed looped system.

7 days f Passover. "Passover" commemorates the Exodus, the liberation of the Jews from slavery in Egypt. The other meaning of the word "Passover" is the Hebrew word "Pesach" which means "skipping, or passing over", derived from the night of the 10th and last plague of the "Angel of Death" that passed over the sacrificial lambs blood on the doorpost of the Hebrew peoples houses, thus not killing their firstborn.

1 day of Shavuot - Known as the "festival of weeks" The period from Passover to Shavuot where we count each of the days from the second day of Passover to the day before Shavuot, (49 days or 7 full weeks), hence the name of the festival. In other words, the occasion is commemorated each year 50 days after the Passover holiday. The commandment for counting the "Omer" (measure of volume of grain) recorded in (Leviticus 23:15-16) - And ye shall count unto you from the morrow after the day of rest, from the day that ye brought the omer of the waving, seven weeks shall there be complete, even unto the morrow after the seventh week shall ye number fifty days, and ye shall present a new meal-offering unto the Lord. Besides the holidays significance as the day on which the Torah was given by G-d to the Jews at Mount Sinai, it is also connected to the season of the grain harvest in which lasted seven weeks (Jeremiah 5:24, Deuteronomy 16:9-11, and Isaiah 9:2). It began with the harvesting of the barley during Passover and ended with the harvesting of the wheat at Shavuot.

52 Sabbaths (Saturdays). This is derived from the average year consisting of 12 months totaling 365 days giving us 52 Sabbaths (Saturdays), plus the 17 days for the other festivals (listed above in blue) for a grand total of "70" festive days.

Shown below (circled in blue) is the Hebrew word "<u>ULE</u>MOADIM" (which means seasons/festivals). It is found in the actual text of Genesis 1:14 as a visible readable word that specifies the precise reason for why the stars were created - And G-d said, Let there be lights in the firmament of the heaven to divide the day from the night, and let them be for signs, and for seasons, and for days, and years. And in addition to that, the word "<u>HA</u>MOADIM" (which means "the holidays") is found embedded within the text at interval letter skips of 70 (circled in red) spanning throughout the text of Genesis 1:6-16. Remember the 70 letter skips is an elucidation for the exact amount of festive days.

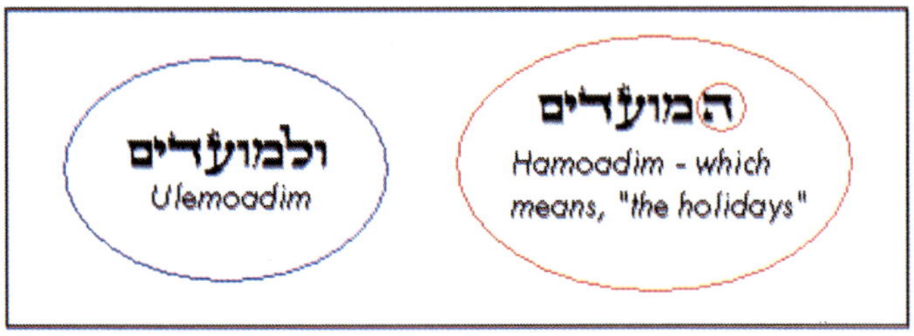

בין המים אשר מתחת לרקיע ובין המים אשר
מעל לרקיע ויהי כן ויקרא אלהים לרקיע שמים
ויהי ערב ויהי בקר יום שני
ויאמר אלהים יקוו המים מתחת השמים אל
מקום אחד ותראה היבשה ויהי כן ויקרא אלהים
ליבשה ארץ ולמקוה המים קרא ימים וירא
אלהים כי טוב ויאמר אלהים תדשא הארץ
דשא עשב מזריע זרע עץ פרי עשה פרי למינו
אשר זרעו בו על הארץ ויהי כן ותוצא הארץ
דשא עשב מזריע זרע למינהו ועץ עשה פרי
אשר זרעו בו למינהו וירא אלהים כי טוב ויהי
ערב ויהי בקר יום שלישי
ויאמר אלהים יהי מארת ברקיע השמים להבדיל
בין היום ובין הלילה והיו לאתת ולמועדים ולימים
ושנים והיו למאורת ברקיע השמים להאיר על
הארץ ויהי כן ויעש אלהים את שני המארת
הגדלים את המאור הגדל לממשלת היום ואת
המאור הקטן לממשלת הלילה ואת הכוכבים

And now for the next code of similar stature. The word ISRAEL can be found in Genesis 1:31 through Genesis 2:1-3 at letter intervals of "both" 50 and 7 within the same text. The R (Resh) in ISRAEL is the same for both words because they intersect each another. In Genesis 2:3 it clearly states - And G-d blessed the seventh day, and sanctified it, because that in it he had rested from all his work which G-d created and made. This is the 7th day (the Sabbath) in which G-d finished all His work and rested. So why can the word ISRAEL be found at intervals of both 50 and 7 within the very passage in relation to the day of rest (the Sabbath)? Because they are both numbers of "LIBERATION". The number 50 symbolizes the Exodus from Egypt as the Jews were liberated from slavery, and the number 7 symbolizes the day of rest and liberation for G-d as He rested.

Starting from the bottom of the text (circled in yellow) we have the letter "Y" (Yud) and spelling "upwards" the rest of the word Israel by 50 letter skips. Then starting from the other "Y" (Yud) (circled in blue) and spelling "downwards" the word Israel by intervals of 7 letters skips. Both words of Israel share the "R" (Resh), which are circles with both yellow and blue.

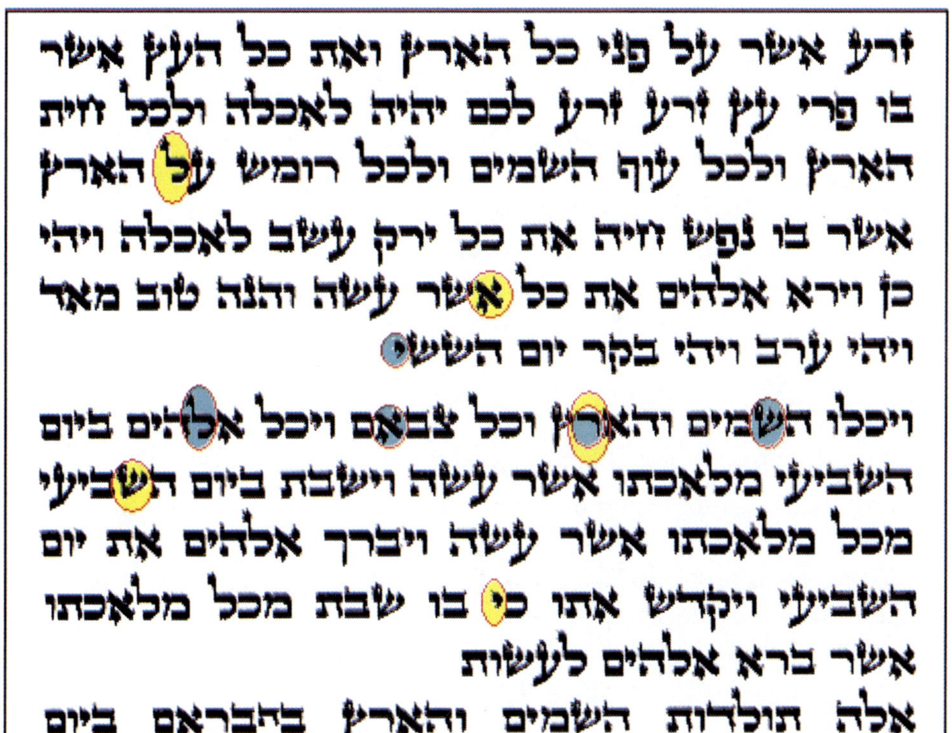
זרע אשר על פני כל הארץ ואת כל העץ אשר
בו פרי עץ זרע זרע לכם יהיה לאכלה ולכל חית
הארץ ולכל עוף השמים ולכל רומש על הארץ
אשר בו נפש חיה את כל ירק עשב לאכלה ויהי
כן וירא אלהים את כל אשר עשה והנה טוב מאד
ויהי ערב ויהי בקר יום הששי
ויכלו השמים והארץ וכל צבאם ויכל אלהים ביום
השביעי מלאכתו אשר עשה וישבת ביום השביעי
מכל מלאכתו אשר עשה ויברך אלהים את יום
השביעי ויקדש אתו כי בו שבת מכל מלאכתו
אשר ברא אלהים לעשות
אלה תולדות השמים והארץ בהבראם ביום

And this is the interesting thing about the Torah Code… its not that you find these words, but WHERE YOU FIND THEM, and that they are found in the "CONTEXT" of where YOU WOULD EXPECT to find them, and what's more, by conspiring with meaningful associated numbers. The regularity of finding these words and the "improbability" of finding them in certain places challenges the best of religious thinkers. The closer the individual letters are to one another as well as their gamatria equivalent, the smaller the odds are of it being a coincidence, (like two or more acquaintances who unexpectedly run into each other on a street corner at a certain time), as well as the chances of words being grouped so tightly together, that they have actually been found to form and illustrate GRAPHIC PICTURES that resemble something we as humans can relate to. However, just as well as finding these codes in the Bible, you can find other codes in other books and texts, but these types of "correlations" that I revealed can not be found or matched in other texts. You can find one word or another in other text, but not finding these words to the extreme of "correlating" with other words in the proper context, or to have significant meaning related to the story. Nowhere in these other books and texts does it compare with the numerous and cohesive complex codes found in the Torah. (However, I am not completely ruling it out, that it could not happen or be found).

The Bible code is a splendidly peculiar thing, as aforementioned not only do we find these codes in the Bible (mainly the Old Testament) but we can also find them in most literature (in all languages), and in most anything else if we look hard enough and assign human meaning to it. As I've mentioned earlier when we are in thought, or inspired by something, we are essentially in spirit transcending all limitations of our consciousness, and as a result WE ARE ALSO CREATING! Yet, some of the religious and scientific thinkers (as well as others of non faith) have not considered the influence of this CREATION MODE having played a part, and thus ignoring the larger picture and brushing the code aside, not giving it credit for what it truly is. We as humans have the ability to conjure up things that look familiar to us and apply meaning to them - like a cloud in the sky resembling a rabbit. However, has it evaded everyone who has studied this matter that perhaps the human mind was made in this way to perceive such a code? Or in other words, that G-d constructed the Bible Code in this manner to be "detectable" by humans - because you see, when we are looking for codes, what we are really searching for is meaning to something greater. However, even though we have the ability to "detect and perceive" these codes, we cannot fully "understand" it (I will soon get to explaining the reason why). And even though we can perceive such a code, some thinkers (that have come to except it under certain terms) exclude the divine and suggest that these codes come from a computer program by a human or alien source (perhaps a future time traveler). However, besides the Torah verifying its own authenticity and G-d as its author and proprietor (as seen in the "TORAH" code above), as well as revealing many spiritual truths hidden between the layers of the texts like the 70 festive holidays embedded in the text by intervals numbers of 70 within the very passage that talks about G-d creating the seasons for festivals, or finding the word "Israel" intersected by both numbers of liberation and in the precise passage of the Bible that tells us that G-d created the Sabbath and rested, or the two different usages of G-d's names each separated by 5 words and 22 letters - the 22 Hebrew letters as a divine representation of the energy that was used in creating the world and the additional 5 (final Hebrew letters) as a representation of the "limitations" of that divine energy. All of these things and the manner by which they are put across (communicated) suggests that G-d is a "thinking" G-d that has put much thought and consideration into His creation along with the Torah - the two play hand in hand. Some religious thinkers conclude that the Torah is beyond just being a guide for our existence, but that the Torah is actually the blueprint by which G-d used to create the world and everything that exists in it, which is somehow eluded to in the code. However, other than these type of effects, THE CODE ITSELF REFRAINS FROM MAKING ANY CONCRETE PREDICTIONS, but rather what it is revealing is PROBABILITIES AND OTHER POSSIBLE FUTURES. For example, it has even been found in the code that the world will be destroyed in 2012 by a comet. However, contrary to that, it has also been found that "it will be crumbled" with the adage: (I will tear it to pieces 2012). There are two choices (maybe more), so subsequently it does not mean that it will transpire that way or even in that year, or possibly even happen at all. Perhaps it appears in the coded text for the mere possibility that we ourselves are thinking or talking about it, or that it was mentioned somewhere in literature or at a point somewhere in time. There have been countless of other contradictory codes found. But this is only the beginning on what can be found within layers upon layers of information of names, dates, births, generations, objects, events, the list is endless. THE TORAH ALSO WARNS US ABOUT TRYING TO PREDICT THE FUTURE, OR RATHER CHANGING G-D'S PROPHECY (for prophesy is fixed). There are so many "variables" in the code that it is impossible to predict the future. FOR THAT REASON, THIS METHOD OF FINDING CODES WAS NOT INTENDED TO BE USED AS A WAY OF PREDICTING THE FUTURE, but only as a way of seeing/realizing that the Torah has "EVERYTHING IN IT". Another amazing thing about the Bible codes is that "ALL" the codes that have been found in the Torah so far have been "AFTER THE FACT", and only having meaning AFTER they occur, and this is because "MEANING" can only be obtained when/after "people" give/assign them names and labels. An example of this is the AIDS virus and Diabetes, which are found in the Bible Code, both of which were named only after they were recently discovered, named and labeled. Without knowing any new codes with new labels and names, we have no way of predicting the future. Ecclesiastes 7:14 - When times are good, be happy; but when times are bad, consider: G-d has made the one as well as the other. **Therefore, a man cannot discover anything about his future.** Ecclesiastes 8:7 - Since no man knows the future, who can tell him what is to come?

Its easy to think that perhaps a human, alien, unspecified entity, or Time Traveler went into the future, recorded past history, and came back (playing G-d) and gave the coded text to Moses. But this is not logical to assume because as aforementioned there are "contradicting codes" WITH NO REAL PREDICTIONS! The predictions are found ONLY AFTER THE FACT OF THEM OCCURING, but when we go back and look, we can find that there were also other predictions that could have transpired, but they didn't. Therefore, this leaves open the door for CHOICE. Why would a Time Traveler who has seen the outcome of the future leave us with "choices" to make? It's contradictory to suggest! Therefore, this excludes the idea or possibility of this code originating from a "Time Traveler", alien being, unspecified entity, or lifeless computer code that is devoid of determination and willpower. WHAT THIS CODE DOES REVEAL TO US is rather that it is indicative of a ruminating thinker who thinks carefully and at length about consequences and outcomes? This can be none other than a DIVINE BEING who is all knowing, all powerful and always present/everywhere, as is affirmed to in the Bible. A list of these other attributes include: Eternal, holy, unchanging, impassable, infinite, self existent, self sufficient, GOOD, LOVING, GRACIOUS, MERCIFUL, JUST, and sovereign.

This also does not means that everything is predetermined, of course, EXCEPT FOR G-D'S PROPHESY! Hebrews 3:5 - Moses was faithful as a servant in all G-d's house, testifying to what would be said in the future. Daniel 8:26 - The vision of the evenings and mornings that has been given you is true, but seal up the vision, for it concerns the distant future. Daniel 10:14 - Now I have come to explain to you what will happen to your people in the future, for the vision concerns a time yet to come.

The Bible additionally tells us that G-d listens to us in PRAYER, even though the future seems fixed. We also have free will and choice with the ability to connect to G-d and to ask for His help. Psalm 4:3 - The Lord has set apart the redeemed for Himself. Therefore, He will listen to me and answer when I call to Him.

Psalm 65:2 - And because You answer prayer, all mankind will come to You with their requests.

Psalm 66:18-19 - If I had cherished sin in my heart, the Lord would not have listened, but G-d has surely listened and heard my voice in prayer.

Proverbs 28:9 - G-d doesn't listen to the prayers of men who flout the law.

Isaiah 65:24 - I will answer them before they even call to Me. While they are still talking to Me about their needs, I will go ahead and answer their prayers.

Mark 11:24 - Therefore I tell you, whatever you ask for in prayer, believe that you have received it, and it will be yours.

Matthew 7:11 - If you, then, though you are evil, know how to give good gifts to your children, how much more will your Father in heaven give good gifts to those who ask Him.

Matthew 7:7-8 - Ask and it will be given to you, seek and you will find, knock and the door will be opened to you. For everyone who asks receives, he who seeks finds, and to him who knocks, the door will be opened.

John 14:13-14 - You can ask for anything, using My name, and I will do it, for this will bring praise to the Father because of what I, the Son, will do for you. Yes, ask anything, using My name, and I will do it.

Philippians 4:6 - Don't worry about anything, instead, pray about everything; tell G-d your needs and don't forget to thank Him for His answers.

Matthew 6:7-8 - Don't recite the same prayer over and over as the heathen do, who think prayers are answered only by repeating them again and again. Remember, your Father knows exactly what you need even before you ask Him.

Additionally, when we talk about future events being fixed we know this cannot be the case since we find G-d making exceptions. We also find that G-d does not use his power in an uncontrolled manner (He exercises self-control). For instance, Abraham pleaded with G-d not to destroy the sinful city of Sodom and Gomorrah (in which his righteous Nephew Lot and family lived). Abraham pleaded with G-d that if He could find at least 10 righteous people living there that G-d would spare the city. However, it seems that G-d had known the consequences but was willing to make exceptions, and He held back until the time was right from acting against Sodom and Gomarrah. This applies the principles of foreknowledge and foreordination in order to respect the free will that G-d gave us. This does not in any way render G-d imperfect. Rather it magnifies His greatness, and it endears him to us, for it shows that His sovereignty truly is exercised not only with omniscience and power but also with love and respect. Genesis 18:16-33 (Abraham pleads for Sodom) - When the men got up to leave, they looked down toward Sodom, and Abraham walked along with them to see them on their way. Then the Lord said, Shall I hide from Abraham what I am about to do? Abraham will surely become a great and powerful nation, and all nations on earth will be blessed through him. For I have chosen him, so that he will direct his children and his household after him to keep the way of the Lord by doing what is right and just, so that the Lord will bring about for Abraham what he has promised him. Then the Lord said, The outcry against Sodom and Gomorrah is so great and their sin so grievous that I will go down and see if what they have done is as bad as the outcry that has reached me. If not, I will know. The men turned away and went toward Sodom, but Abraham remained standing before the Lord. Then Abraham approached him and said: Will you sweep away the righteous with the wicked? What if there are fifty righteous people in the city? Will you really sweep it away and not spare the place for the sake of the fifty righteous people in it? Far be it from you to do such a thing—to kill the righteous with the wicked, treating the righteous and the wicked alike. Far be it from you! Will not the Judge of all the earth do right? The Lord said, If I find fifty righteous people in the city of Sodom, I will spare the whole place for their sake. Then Abraham spoke up again: Now that I have been so bold as to speak to the Lord, though I am nothing but dust and ashes, what if the number of the righteous is five less than fifty? Will you destroy the whole city because of five people? If I find forty-five there, he said, I will not destroy it. Once again he spoke to him, What if only forty are found there? He said, For the sake of forty, I will not do it. Then he said, May the Lord not be angry, but let me speak. What if only thirty can be found there? He answered, I will not do it if I find thirty there. Abraham said, Now that I have been so bold as to speak to the Lord, what if only twenty can be found there? He said, For the sake of twenty, I will not destroy it. Then he said, May the Lord not be angry, but let me speak just once more. What if only ten can be found there? He answered, For the sake of ten, I will not destroy it. When the Lord had finished speaking with Abraham, he left, and Abraham returned home.

Genesis 19:27-29 - Early the next morning Abraham got up and returned to the place where he had stood before the Lord. He looked down toward Sodom and Gomorrah, toward all the land of the plain, and he saw dense smoke rising from the land, like smoke from a furnace. So when G-d destroyed the cities of the plain, he remembered Abraham, and he brought Lot out of the catastrophe that overthrew the cities where Lot had lived.

So, as one can discern from Abraham's appeal, future events may not be completely fixed since we find G-d making exceptions. Jesus also knew this, that's why He attempted to pray in the Garden of Gethsemane for G-d to find another way instead of Him having to die the way He did. Matthew 26:36-45 - Then Jesus went with his disciples to a place called Gethsemane, **and he said to them, Sit here while I go over there and pray.** He took Peter and the two sons of Zebedee along with him, and he began to be sorrowful and troubled. **Then he said to them, My soul is overwhelmed with sorrow to the point of death.** Stay here and keep watch with me. Going a little farther, **he fell with his face to the ground and prayed, My Father, if it is possible, may this cup be taken from me. Yet not as I will, but as you will.** Then he returned to his disciples and found them sleeping. Could you men not keep watch with me for one hour? He asked Peter. Watch and pray so that you will not fall into temptation. The spirit is willing, but the body is weak. **He went away a second time and prayed, My Father, if it is not possible for this cup to be taken away unless I drink it, may your will be done.** When he came back, he again found them sleeping, because their eyes were heavy. **So he left them and went away once more and prayed the third time, saying the same thing.** Then he returned to the disciples and said to them, Are you still sleeping and resting? Look, the hour is near, and the Son of Man is betrayed into the hands of sinners.

Now one might also pose the question, if in fact there is nothing we can do to change our destiny, and if G-d (or fate) has already determined the outcome of a matter, what then is the point of praying? Why then should we also take measures to protect our safety? Why wear a seat belt when traveling in a car? Why wear a helmet while on a motorcycle? Why shouldn't we swim in the Ocean? Why should we avoid eating that pesticidal, sugar or lard laden food? The Bible in no way ignores or overlooks such careless behavior, in fact the Bible instructs us to be "safety conscious" about such things. Deuteronomy 22:8 - When you build a new house, make a parapet (low wall) around your roof so that you may not bring the guilt of bloodshed on your house if someone falls from the roof. The purpose of this wall was to prevent someone from accidentally falling off the roof. So, we then have to question, why would G-d give such a command if a person were already destined to fall from a roof and die. This also includes those that die from natural disaster or tragic events beyond their control. In essence, the Bible is telling us that unforeseen occurrences befall us all. So basically G-d forewarns us to be "safety conscious" because "keep in mind" we are now subjected to the laws of entropy (thermodynamics) seeing as we are under the curse of sin brought about by rebellious behavior. Therefore, this leaves us open and subjected to things of a risky and hazardous nature. This in no way suggests that we have a predetermined date with death! Now I know that many of us are familiar with the famous Bible text that reads: Ecclesiastes 3:1-8 - To every thing there is a season, and a time to every purpose under the heaven: A time to be born, and a time to die; a time to plant, and a time to pluck up that which is planted; A time to kill, and a time to heal; a time to break down, and a time to build up; A time to weep, and a time to laugh; a time to mourn, and a time to dance; A time to cast away stones, and a time to gather stones together; a time to embrace, and a time to refrain from embracing; A time to get, and a time to lose; a time to keep, and a time to cast away; A time to rend, and a time to sew; a time to keep silence, and a time to speak; A time to love, and a time to hate; a time of war, and a time of peace. Many people interpret this to mean that everything is predestined, but rather it professes only the endless cycles of things in life. Here is another texts from Ecclesiastes written by King Solomon. Ecclesiastes 9:11-12 - I returned, and saw under the sun, that the race is not to the swift, nor the battle to the strong, neither yet bread to the wise, nor yet riches to men of understanding, nor yet favour to men of skill; but **time and chance happeneth to them all. For man also knoweth not his time: as the fishes that are taken in an evil net, and as the birds that are caught in the snare; so are the sons of men snared in an evil time, when it falleth suddenly upon them.**

The solution here, is not to subject ourselves to things of a dangerous nature, or get so swept up in our daily affairs, that we overlook our creator. We must take safety measures when it comes to dangerous things that can befall us, and we must continue to pray about such things that G-d might sustain us. Our creator understands life, death, and evil completely, He does not force a destiny on us. Rather what the Bible teaches is that G-d offers all of us the option of LIVING FOREVER if we simply seek and believe in Him. However, G-d does not compel us to accept His offer. Instead, His words say: whomever wishes let them take freely of the water of life. Revelation 22:17 - The Spirit and the bride say, Come! And let him who hears say, Come! Whoever is thirsty, let him come; and **whoever wishes, let him take the free gift of the water of life.**

A small question remains… So why do small or even life taking incidences happen that can debilitate our lives in some manner? The reason for this may be that G-d is in some way protecting us by our wounds. But His purpose is only buying us the time that we need to find Him and become a workman for Him in some way. Whatever this way may be, we may never truly come to understand it. However, even as distraught, angry and defiant as we may be (especially at G-d) we must remember that all things work together for good. Romans 8:28 - And we know that all things work together for good to them that love G-d, to them who are the called according to his purpose.

Getting back to the Code… As was also mentioned earlier that the Bible Code became more accessible with the advent of the computer. This tells us that the Bible Code had a lock on it and was not assessable until the aid of the computer arose. If the Bible Code (as some thinkers suggest) is actually a computer program itself of some sort, because the Bible Code shows numerous abilities to be able to conceive of all possible outcomes of all available choices as well as arranging and programming these codes into a crossword puzzle style, they purpose that it was done with the aid of a computer as a tool to aid in human calculations, (that's why it was purposed that the Bible Code was given my a Time Traveler from the future). The thinkers ask the question: How do we as humans decipher what is really life or alive, or additionally, how do we conceive of emotions, feelings, reasoning, why cant they be part of a computer program as well as DNA? This may indeed seem like the case at first consideration, however plausibly speaking, in order to make a computer program (a future tense quantum computer) that mimics human intelligence (only appearing that way to us) with feelings, emotions and reasoning, there had to be a "LIVING ENTITY" behind it. Simply put, a computer program CANNOT create itself unless otherwise created by a like maker with mutual intentions.

Machines are the tools, not the result of G-d. It may indeed be a super intelligent machine, which exceeds human intelligence, but a machine CANNOT approach infinite intelligence! Spiritual man (with a soul) cannot be replaced by an intelligent machine, nor can G-d!

The mystery deepens, as was aforementioned earlier about "Quantum Physics", (if you recall) when it's attempted to pinpoint the exact location of atomic particles, (like electrons); it was found that it was utterly impossible because the atomic particles had no single location. This gives the particles the capability of being in more than one place at one time (a quantum phenomenon). The only explanation one could come up with is that the particles don't just exist in our physical realm; they flit into existence in other realms too (possibly by the diminution of light as discussed earlier). To this effect, the Bible Code is similar because just as we find that elections can be in numerous different places at the same time, so in turn do the Bible Codes, which also display to us that numerous opposing words that reflect meaning to us can also be found in several different places with different outcomes. However, humankinds fate is still within our control and the Bible Code is merely providing one of an infinite number of outcomes. In addition, just because G-d knows what we are going to do in advance does not in any way bind us? We still make the choices, but G-d who is not bound by any sort of time can look into the future and see what choices we have already made and therefore encode it. Either that, or G-d who has already seen it all, in every possible scenario, is so wise that He could ultimately predict an outcome. In addition, this is the beauty of choice. This very well may be the most eloquent message concerning the interpretations and implications of the Bible Codes, that there are many possibilities that could transpire, and all by choice - but that none evade G-d. When Moses dropped the first tablets (out of anger) that G-d had given him down onto the corrupted Hebrew people at Mount Sinai, did G-d foresee this? Exodus 32:19 - When Moses approached the camp and saw the calf and the dancing, his anger burned and he threw the tablets out of his hands, breaking them to pieces at the foot of the mountain. Exodus 34:1 - The Lord said to Moses, chisel out two stone tablets like the first ones, and I will write on them the words that were on the first tablets, which you broke. Or better yet; would G-d have put the effort into the tablets had He known their fate? The answer is still yes! Because there is always a purpose for everything. Perhaps this purpose was for the 3,000 that were killed at Sinai because they refused the law, and in effect latter in history, 3,000 were revived (in a manner of speaking raised from the dead) after the resurrection of Jesus when they received and accepted the Gospel. Exodus 32:28 (The Golden Calf incident) - The Levites did as Moses commanded, and **that day about three thousand of the people died.** Acts 2:41 (The holy Spirit came at Pentecost, also known as the Jewish festive of Shavuot) - Those who accepted his message were baptized, and about **three thousand were added to their number that day.** (This will be revisited in a later chapter). G-d being infinite in wisdom always has a back up measure, much as if the Bible Codes being unbounded with numerous possibilities, while (all along) hidden within lays the ultimate outcome whereupon only G-d knows.

There have also been discrepancies to finding these codes that wish to discredit them by asserting stipulations to acquiring accurate codes. Nevertheless, it doesn't matter how these codes are discovered by any sort of mathematics even if the letters are not derived equidistantly, but perhaps by some other superior equation. Not only the math in this regards, but also the data should be used with wiggle room and flexibility for the spelling of Hebrew names or words, even aliases and nicknames, thus eliminating that any such data was collected improperly (remembering, there is always a reason for everything). There should be no tightly defined rules attached to finding such codes - remember Quantum Mechanics plays a role here - the point being is that they are found however way is used, and that they are "applicable" to something of significance that we as humans can understand and relate to. G-d intended it this way. It is not irrelevant and there is no illegitimacy. Yes, G-d has bothered to play mathematical games with us so as to allow us to apply our own heart and mind to resolve the fundamental question as to the message. The message is "free choice" (quantum physics). And this point follows within all/any other literature (notes, books, journalism, writing, thoughts, prayers, building structures, artwork, speeches, daydreaming, participating ones ideas on the Internet) that was contrived by the process of "conscious thought" because from this source flows G-d. Even in our dreams (when we sleep) though as ridiculous as dreams may seem, there is always some little piece that is prophetic or visionary in nature (that has a significant meaning) perhaps in an indirect or roundabout way.

Still some find it hard to believe that G-d would go to the lengths to bury information in His Word in this way, making the Bible a seek-and-find-puzzle. Some feel this is beneath G-d to do such a thing. However there is a reason for G-d to continually play this seek-and-find game with us. The Bible Code is not there to be hidden, its there to be revealed. If G-d had made the world a complete and utter mystery, we would have no path to know Him, nor would we know that there is anything more to know. Basically, G-d wants us to believe, acknowledge and seek Him - and perhaps the prophecies and codes are a way of emphasizing this. Searching for codes may even teach us how the world works, and not only how the world works, but the thought process (that was used by G-d) in, which it was created.

Searching for codes may also be an exercise for the brain, as it may influence cognitive reasoning skills, memory, emotions, language, math, science, imagination and spiritual thought, while leaving open the door for possibilities of hope.

However, to truly understand just what the Bible Code is, one must examine the Torah in yet another degree. In addition, as aforementioned above, this leads us to answer why we can only "perceive and identify" rather than to truly "understand" the full extent and purpose of the code.

Oral law alludes to us that the Angels (who were created by G-d before us) are said to be ranked higher than us, asked to be given the Torah at Mount Sinai, but G-d chose to give the Torah to humankind instead of the Angels even though we were created lower than the Angels were. Psalm 8:5 - For thou hast made him a little lower than the angels, and hast crowned him with glory and honor. The reason G-d chose to give the Torah to humankind is because we are subjected to, or caused to undergo the physical plane of earth, thus feeling all physical attributes (5 senses including pain and suffering) that Angels cannot wholly relate to because they don't have a material body. The Angels did not readily have a need for the Torah as much as humans were in need for it. This "need" was to help us transcend the physical back into terms of spiritual aspects. Much of this loftiness was lost after the sinful fall of Adam and Eve. Genesis 3:16 - To the woman he said, I will greatly increase your pains in childbearing, with pain you will give birth to children. Your desire will be for your husband, and he will rule over you.

And G-d gave the Torah to the Jews only under one condition, that they teach it unto their children. Deuteronomy 6:7 - And thou shalt teach them diligently unto thy children, and shalt talk of them when thou sittest in thine house, and when thou walkest by the way, and when thou liest down, and when thou risest up.

So now that we understand WHY we have the Torah, we must now come to terms that it is also wrapped within a mysterious and perplexing supernatural "code" that was to be understood (or only discernable) by lofty beings who understand the full use of the HEBREW LETTERS - and this ill behooves us as humans. Therefore, we strive to understand and come to grips with the codes true purpose, for which is to elevate us back into a spiritual state, the one we lapsed from. Therefore the Bible now states that because of this "struggle" (to achieve salvation through Jesus) we are now viewed/deemed (in a certain sense) higher than the Angles, as we struggle to overcome extreme physical aspects of sin - in other words, the soul has to learn the hard way to understand the implication of what mitigates sin back into loftiness (a real challenge for a physical material body). Hebrews 1:3-5 - Who being the brightness of his glory, and the express image of his person, and upholding all things by the word of his power, when he had by himself purged our sins, sat down on the right hand of the Majesty on high. **Being made so much better than the angels, as he hath by inheritance obtained a more excellent name than they.** For unto which of the angels said he at any time, Thou art my Son, this day have I begotten thee? And again, I will be to him a Father, and he shall be to me a Son? The Torah is really about the psychology of G-d and how He relates to His creation, so when we read the Bible and find these codes what we are really doing is reading it on one level as history (past, present and future), on another level as human spiritual guidance and on the deepest level as an insight into the very nature and internal workings of G-d the creator as he cares for His creation, and ALL OF EXISTANCE.

It is the end of this chapter, and we must stop and think here for a moment to logically analyze who it is that is telling us not to kill, or steal, and not to have any other G-d's before Him? He who also tells us that He fashioned ALL of creation in Genesis 1, to the far reaches of the Heavens and ALL the stars. Who else could claim such high status unto Himself rightfully? Who else besides someone who entirely and completely cares for His creation, and for the status of souls and their moral fiber. No, He is not a Time Traveler with a computer code, or an alien being from some other distant world, as the codes were not meant to foretell the future, but rather to declare and be a witness to EVERYTHING in existence. G-d is a G-d that cares for His creation, not just on earth, but the whole universe and all of existence (however the earth is without doubt G-d's most desired, since He sent His only son here to die for us). The Bible Code is just one of the numerous wonders that confirms His unparallel authenticity.

Let us also not forget as aforementioned in an earlier chapter that the Torah actually created the very letters in which it was written in - through nature by the "universal law", Bible writing itself with letters, but these letters in both hindsight and foresight are also placed strategically and intentionally within layers of the text itself, and each letter as well having a numerical value with an attached meaning that relates to something else of significance that we as humans can relate to. Whatever this code is, it is clear that it is a living organism in itself. This alone tells us that the origin of the Hebrew letters AS WELL AS THE CODES (which work hand-in-hand) are solely from a DIVINE source that cannot be matched wits with.

And let us also not forget the prophecy of Matthew 5:18 - I tell you the truth, until heaven and earth disappear, not the smallest letter, not the least stroke of a pen, will by any means disappear from the Law until everything is accomplished.

Additionally, if there is something to be learned here about the Bible Code, it's that nothing and I do mean NOTHING happens in this world without G-d first acknowledging and allowing it. Not even one single thing in the scheme of life (even as menial or evil as it may seem). Everything has a purpose. What that purpose is - we will eventually find out. But I will tell you that abstruse as it may seem, its ultimate goal is to reunite us with G-d!

Yes, bad things happen in life! Lets not forget that entropy was brought about by sin; we are now creatures that die. Some people simply cannot understand this simple fact. Therefore when bad things happen some people cannot merely let bygones be bygones and get on with their lives. We must realize that we are physical beings, and the laws of physics (AS THEY CURRENTLY STAND) dictate that time runs in one direction only. We continue to feel responsible for what was, continue to attempt to rewrite our histories, continue to regard our past as something that somehow still "belongs" to us. Something in our nature refuses to let go, refuses to reconcile itself with the one directional flow of time and we begin to fantasize about Time Travel being possible. However, believe me when I tell you this, that G-d would never allow "Time Travel" (for us physical beings). If He did allow such a thing as Time Travel, this would create a paradox for inconsistency that would render all prophesy null and void. We know that prophesy is G-d's sovereign word in which we can trust because many prophesies of the Old Testament have already been fulfilled (which you will read about in a future chapter).

Additionally I will tell you a little secret about "Time Travel" that some have not logically considered. All processes whether they are the sending of a signal, physical, chemical, biological, geological, cosmological, time itself, you name it, they are IRREVERSIBLE (they are permanent, unalterable and irretrievable). This is because the "second law of entropy and thermodynamics" is based on this unshakable fact. It clearly tells us indisputably that we live in an irreversible universe. No human ingenuity or technology is going to outsmart the "second law" which is in absolute control of all natural processes. This is because the Universe is constantly undergoing transformations of energy/matter from one form to another IRREVERSIBLY! For instance, the simple act of just burning the gasoline in our cars is transformed to other forms of energy/matter and in the process the entropy of our world increases. Therefore, everything (in the natural process of things) is IRREVERSIBLE! Wherein can any of us see the hope of ever achieving "Time Travel"? Moreover, the universe of the past simply does not exist to go back to. In fact, the Universe would have to cease to exist in order to allow for Time Travel.

So yes, time is undoubtedly irreversible in the physical sense, but there IS something that G-d had innately put in us that CAN transcend our reality and act as a sort of Time Traveling mechanism. And this is called in Hebrew "Teshuvah", which literally means the capacity to "return" which can retroactively allows us to transform our past actions and experiences. It can so to speak allow us to change the past. In other words, when we have experiences… things don't just happen, they happen for a reason, they mean something, and they further a certain objective. We must face the facts that the tyranny of time - the "hard facts" cannot be undone, but they can "teach" us that we can make fresh starts with more fulfilling endeavors, and thus smarter in not making the same mistakes. In this respect, we have reached back in time to transform our mishaps into wake-up calls, as well as to transform distance and disharmony into a catalyst for greater things. The past is but another vista of life, open to exploration and development.

14 - BIBLICAL FOREKNOWLEDGE...

The Bible is not specified as a science book, yet it is scientifically accurate. In this chapter, I have listed several accounts on biblical foreknowledge that are consistent with known scientific facts. Many of them were recorded in the Bible thousands of years before being rediscovered. Additionally, as I've already mentioned in an earlier chapter that scientific discoveries are merely observing, understanding and using systems that G-d has already designed. All knowledge revolves around a closed loop system. Psalm 19:6 - His going forth is from the end of the heaven, and His circuit unto the ends of it, and there is nothing hid from the heat thereof. And as aforementioned in an earlier chapter, a famous quote from King Solomon in Ecclesiastes 1:9-10 - What has been will be again, what has been done will be done again. There is nothing new under the sun. Is there anything of which one can say, Look! This is something new. It was here already, long ago; it was here before our time. This also happens to be the Law of "thermodynamics and entropy" as we've discussed in previous chapters that thermodynamics is the field of physics that deals with the conversions of energy from one state to the other. In this process, there is no new energy coming in or going out. In other terms, the increase in energy of a system is equal to the amount of energy added, minus the amount lost because of any exertion. We also know that when these conversions of energy are taking place (from one state to another) there is also a deteriorating loss of energy (entropy). Romans 8:22 also endorses the law of "Thermodynamics" - For we know that the whole creation groaneth and travaileth in pain together until now. (Essentially this Bible verse is indicative that "everything is in decay").

What's more, despite everything that is going to be addressed in this chapter to provide evidence on biblical foreknowledge, we really only need to examine one verse that sums it up and puts it all into words... So colossal and complete is the Bible in its first opening verse of Genesis 1, that it contains all the physics related to our existence in a matter of 5 superseding terms: In the beginning (time) G-d (force) created (action) the Heavens (space) and the Earth (matter).

Now before we continue, lets also recap here for a moment... As aforementioned in a previous chapter we have also learned about the field called "Homeopathy" where medicines were/are taught to treat, like with like - fighting poison with poison so to speak. Exodus 15:22-25 - So Moses brought Israel from the Red sea, and they went out into the wilderness of Shur; and they went three days in the wilderness, and found no water. And when they came to Marah, they could not drink of the waters of Marah, for they were bitter: therefore the name of it was called Marah. And the people murmured against Moses, saying, What shall we drink? And he cried unto the Lord, and the Lord shewed him a tree, which when he had cast into the waters, the waters were made sweet: there he made for them a statute and an ordinance, and there he proved them. This is a perfect example of G-d teaching Moses the principle of bitter curing bitter (like cures like), in this way G-d heals with the very same thing with which He smites. We can also see this same effect in Numbers 21:4-9 - And they journeyed from mount Hor by the way of the Red sea, to compass the land of Edom, and the soul of the people was much discouraged because of the way. And the people spake against G-d, and against Moses, Wherefore have ye brought us up out of Egypt to die in the wilderness? For there is no bread, neither is there any water, and our soul loatheth this light bread. And the Lord sent fiery serpents among the people, and they bit the people, and much people of Israel died. Therefore the people came to Moses, and said, we have sinned, for we have spoken against the Lord, and against thee, pray unto the Lord, that he take away the serpents from us. And Moses prayed for the people. And the Lord said unto Moses, Make thee a fiery serpent, and set it upon a pole, and it shall come to pass, that every one that is bitten, when he looketh upon it, shall live. And Moses made a serpent of brass, and put it upon a pole, and it came to pass, that if a serpent had bitten any man, when he beheld the serpent of brass, he lived.

In addition, as was also mentioned in an earlier chapter, this is the very same method and approach that modern medicine uses today to produce vaccines that immunize the body against deadly diseases and viruses. And it was determined that vibration (Quantum Mechanics) is the underlining factor here. As aforementioned, illnesses could be healed by taking tiny doses of the natural substance that is actually causing the suffering to the point of it being non-detectable mixed in with a water, where only a "vibration" remains of it, or in other terms, only the characteristics of the matter remains. In essence, what's actually occurring is that the "information" is copied to the water (since water retains memory) and is being used to cancel out the information of the symptoms. And when we add (to the body) similar characteristics to that of the illness, we create something called "resonance" which enhances ones life force. It essentially works like the physics of vibrating tuning forks - when one is struck it causes a chain reaction to the others.

We have also learned from Exodus 32:19-20 that when Moses came down from Mount Sinai with the Ten Commandments from G-d, he found that the people had taken their jewelry and fashioned themselves a god in the form of a "golden calf", Exodus 32:19-20 - And it came to pass, as soon as he came nigh unto the camp, that he saw the calf, and the dancing: and Moses' anger waxed hot, and he cast the tables out of his hands, and brake them beneath the mount. And he took the "golden calf" which they had made, and burnt it in the fire, and ground it to powder, and strawed it upon the water, and made the children of Israel drink of it. And as we have also already learned that "gold" is considered a universal solvent, also called liquid gold. It was prescribed for anxiety and impatience because G-d had described them as a stiffnecked people (Exodus 32:9). It still works for these and many other symptoms today to promote healing and control pain. It is even known that Tuberculosis cannot live in the presence of gold. (Tuberculosis is a bacteria that affects the lungs, the central nervous system, lymphatic system, circulatory system, bones, joints and the skin). Gold has even been injected and implanted under the skin such as in knee joints to relieve pain. It was also used to treat arthritis. Gold is also known for healing the activity of the heart and improve blood circulation, and includes treatments for skin ulcers, burns, and puncture wounds. It was also commonly used to cure "dipsomania" (uncontrollable craving for alcoholic liquors). Gold has also been used to rejuvenate the glands, stimulate the nerves and release nervous pressure. The body's temperature is also affected by gold, particularly in cases of chills, hot flashes, and night sweats. And interestingly enough, Gold can be used as a remedy against obesity. And even more strangely, it can be used for terminal malnutrition and starvation because it reduces pain and discomfort and the need of opiates (narcotics).

And we have also learned in Exodus 28:29 - And Aaron shall bear the names of the children of Israel in the breastplate of judgment upon his heart when he goeth in unto the holy place for a memorial before the Lord continually. The breastplate of Aaron (with 12 colored gemstones) was used for protection when entering the tabernacle in G-d's presents. The 12 gemstones where most likely used as a conduit for channeling certain high (lofty) types of energy. As we know that, certain crystallized stones have a property that is naturally controlled by frequency, and can be used as a grounding plate of sorts for both electrical, electromagnetic and piezoelectricity transmissions as well as spiritual conflict. Scientists have long known that a crystal set in an energy field will collect and focus that energy. Today gemstones are used in modern day electronics.

These examples show significant insight into the workings of nature, physics and the human body long before they were actually rediscovered by modern science. In addition I'm sure that they also apply to the Universe at large. We are talking Quantum Physics here again where "vibration" is the underlying apparatus in which things seem to work. And we are talking about emulating things from "as above, so below" like we witnessed with the Ark of the Covenant in the previous chapter. As mentioned, divine design never changes!

Now I'm going to start bringing to your attention other things of biblical foreknowledge. One of the most significant scientific exposures that are found within the Bible (and recently brought to light) has to do with the Genesis story where G-d tells us that He created the first woman (Eve) from the "rib bone" of the first man (Adam). Genesis 2:21-23 - And the Lord G-d caused a deep sleep to fall upon Adam, and he slept: and he took one of his **ribs**, and closed up the flesh instead thereof. And the the rib, which the Lord G-d had taken from man, made he a woman, and brought her unto the man. And Adam said, this is now bone of my bones, and flesh of my flesh: she shall be called Woman, because she was taken out of Man. Now some might pose the question, why would G-d pick the rib from any other part of the human body to make another human being. And this little exploding piece of evidence alone, may be one of the biggest compellingly and powerful indicators for some people that will automatically (with 100% genuineness and dependability) confirm that the Bible is a legitimate source of all-around truth - and what's more of G-d. So why did G-d choose a rib? Was this account in Genesis simply words of mere men, as the critics claim? Come to find out, science has discovered some unique characteristics about the rib bone… it has the miraculous ability to "regenerate" itself. In fact, the rib bone is the only bone in the entire human body that has the capability of regenerating itself. We are not talking about bones having the capability of repairing themselves; we are talking about reconstruction and "REGENERATION" where the entire rib can grow back! (Much like a lizard or a mouse's tail). However, in the case of a rib, it can only re-grow itself if the rib bone is removed from what called the "periosteum" (which is a surface membrane (exterior) layer of tissue surrounding the bone). This exterior layer can be likened to the bark on a tree; or rather resembling more of a banana peel, except the peel would be peeled down just enough to remain intact. Thus, in order to build a new rib bone it is imperative that the periosteum remains intact because it contains what's called "osteoblasts" which are bone-forming cells that can rebuild new bone. Surgeons usually remove ribs for bone grafts for use in other parts of the human body because not only have they learned that ribs can re-grow themselves, but ribs in general are powerfully filled with bone marrow (the spongy tissue found inside bones). Ribs are not the only powerfully filled bone, but also the breastbone, hips, skull, and the spine, but you can't take (or use) one of these for grafting because they are obviously needed for major function. Not only are the ribs filled with marrow, but this marrow contains "stem cells" that manufacture the body's red and white blood cells, which are then recognized and excepted by the body. These white blood cells protect against disease and the red blood cells bring oxygen and nourishment to the body and remove waste products. Many of us that are familiar with "stem cells" know that they have the amazing capability to develop into many different cell types in the body. They have the choice (directed by the necessity of the body) to either remain a stem cell or become another type of cell with a more specialized function, such as a red blood cell, muscle cell, or even a brain cell if need be.

Now to examine the actual Hebrew word for "Rib" found in Genesis 2:21-22 - is **"**tsela". This word occurs nowhere else (exactly the same) in the Hebrew Bible and there is considerable uncertainly as to its true meaning. Some believe it to mean one complete side of Adam was taken (and not just a single rib bone). However, we do find variants of the word, such as tselo, tselot, ul-tsela, tsalim, batselot, la-tselot, which all clearly express the "side" of something. However, when Genesis 2:21 is read, it becomes even more defined as "one single rib" and not a whole side when it says: And he took **one of his ribs**, and closed up the flesh instead thereof.

Incidentally, this also brings rise to where the idea of anesthesia originally came from... Genesis 2:21 - And the Lord G-d **caused a deep sleep to fall upon Adam, and he slept**: and he took one of his ribs, and closed up the flesh instead thereof. Sir James Young Simpson 1847 discovered "chloroform" from the idea of G-d putting Adam into a deep sleep and taking a rib to make Eve. Therefore, James Simpson thought of the idea to put people into a state of sleep (like G-d did with Adam) where they were unaware of what was happening and couldn't feel pain. With this idea, he created "Chloroform" which is a liquid that changes to a vapor that causes unconsciousness when inhaled. It was (in earlier times) used as an anesthetic for surgical operations.

We also find such things in the Bible that clearly demonstrates the use of "biological warfare", such as with the use of "toxic honey". Proverbs 25:16 - If **you find honey, eat just enough— too much of it, and you will vomit.** Joshua 24:11-12 - Then you crossed the Jordan and came to Jericho. The citizens of Jericho fought against you, as did also the Amorites, Perizzites, Canaanites, Hittites, Girgashites, Hivites and Jebusites, but I gave them into your hands. **I sent the hornet ahead of you, which drove them out before you**—also the two Amorite kings. You did not do it with your own sword and bow. The question is what was in the honey that could cause vomiting. We find the answer to this in the "Rhododendron bush" which is a common plant in Israel and contains a toxin called "terpenoids" and are found in the nectar collected by the bees, which becomes incorporated into the honey. When the honey is eaten, this toxin affects the body with dizziness, nausea, perspiration, vomiting and effects heart rate, it may even cause coma.

Additionally we read about other medicinal sources such as "antiseptics" which reduce the growth of microorganisms, which cause infection. The First antiseptic was Hyssop oil that

G-d ordered Moses to use as a purifying agent. Hyssop oil has been shown to contain 50% antifungal and antibacterial agents. Numbers 19:18 - And a clean person shall take hyssop, and dip it in the water, and sprinkle it upon the tent, and upon all the vessels, and upon the persons that were there, and upon him that touched a bone, or one slain, or one dead, or a grave. Psalm 51:7 - Purge me with hyssop, and I shall be clean: wash me, and I shall be whiter than snow. Additionally, honey (though it has been mentioned above as a biological agent, it has also been shown to repair and even re-grow tissue. Manuka honey has the highest level of antibacterial activity because it has within it naturally occurring hydrogen peroxide. Manuka honey comes from New Zealand where beekeepers set up their hives in wild uncultivated areas in which Manuka bushes grow. The bees gather nectar from the flowers of the Manuka bush, which is indigenous only to New Zealand. Manuka honey has been shown to also have other qualities in it that heal stomach ulcers, as well as stopping the growth of the deadly MRSA bacteria (Methicillin-resistant Staphylococcus aureus) and VRE (Vancomycin-resistant Enterococci). Additionally, several studies have also shown that Manuka Honey has been effective in killing the E. coli bacteria.

Another amazing thing that I'd like to bring to your attention, that again remarkably reveals "divine design" is something called the "Warren's truss". This is a steel or wooden structural design that is used in strengthening the structure of bridges and buildings. However, this identical geometric structure can also be found in the palm bone of the bird wing (specifically the metacarpal of a vulture bone). As it turns out, birds have hollow bones designed by G-d (the Creator) to give maximum strength for minimal weight. For example, the frigate bird has a seven-foot wingspan, but the skeleton weighs only four ounces.

Here is what the internal skeleton of a metacarpal bone from the wing of a vulture looks like:

And here we see the human engineered "Warren truss" which is commonly used as a wooden or steel structural stability design in bridges and buildings.

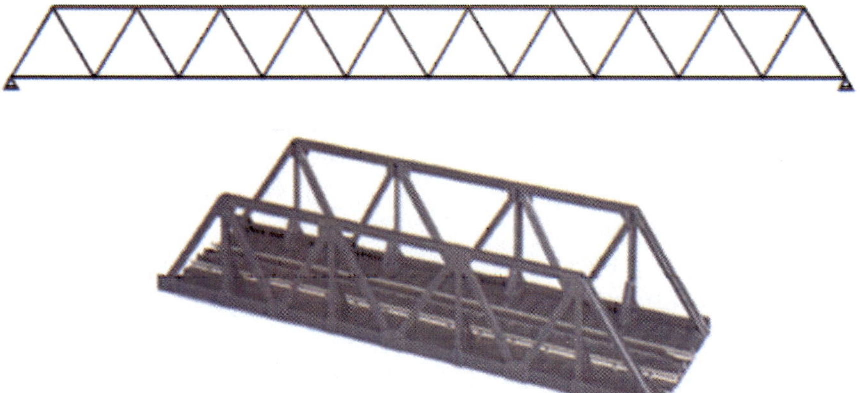

Moving on here... Now, lets examine other scientific discoveries that are found in the Bible long before modern scientists proved their existence. The patriarch "Job" in the Old Testament has some winners that defy explanation (and by the way the book of Job is thought to be one of the oldest books of the Bible). When we read some of the things Job writes about, we are left baffled by how he knew such knowledge unless otherwise privy to him from G-d! For instance, the Pleiades system of stars that is well known to astronomy today (called the seven sisters). Within this star system, we see a very bright star that is actually a huge central Sun called Alcyone. It is the brightest in the constellation Pleiades.

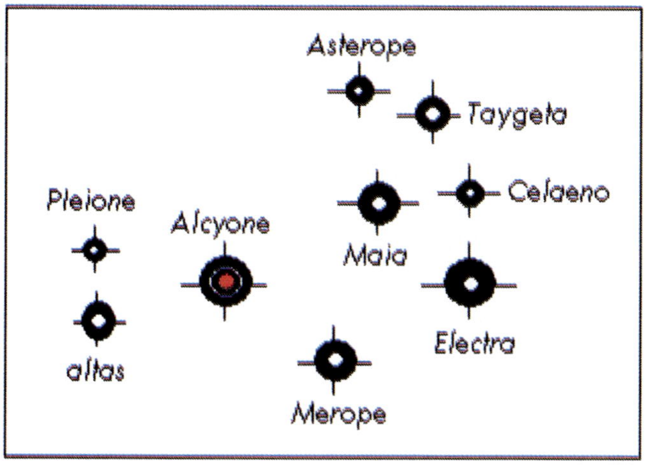

Now when we read in the book of Job we read about this same star system. Job 38:31 (G-d asks Job) - Canst thou bind the sweet influence of Pleiades, or loose the bands of Orion? It is also interesting to know that the Hebrew word for "Pleiades" is Chimah (pronounced "kee-maw") which literally means "Pivot or hinge". Incidentally, the Hebrew translation of "Polaris" (ha-tsir) also means "the pivot". What these stars are used for is to help us determine the season as well as for a navigational device as they appear to rise and set following the rotation of the Earth.

The question is, what was Job really trying to tell us (or rather G-d) when He said; canst thou bind the sweet influence of Pleiades, or loose the bands of Orion? Well we know today that the Pleiades are a small cluster of stars that make themselves more visible in the springtime. When the Pleiades becomes more visible in the night sky, it is a signal to those that specialize in studying celestial bodies that springtime is coming soon. While Orion is a constellation that announces the coming of winter. So in other words, G-d is saying, Job, there is nothing you can do to stop or prevent spring, summer, autumn or the winter from coming. So in a sense these constellations are a sort of "hinge or pivot" in determining the seasons. What's intriguing is how Job knew that these sorts of things are what keep us on our course in the solar system, and in helping us to determine our seasons here on Earth.

By some other means Job also seems to know a great many other things such as knowing that the Earth is suspended in space when he states in Job 26:7 - He stretches out the north over empty space, He hangs the earth on nothing. The fact that Job knows that the earth floats in empty space (which is a vacuum) by means of gravity is amazing!

In Job 36:27-29 it states - For He draws up drops of water, which distill as rain from the mist, which the clouds drop down and pour abundantly on man. Indeed, can anyone understand the spreading of clouds, the thunder from His canopy? How is it that Job even seems to know about "evaporation"?
Job 26:8 states - He binds up the water in His thick clouds, yet the clouds are not broken under it. Amazingly, Job even understands the process of what constitutes "condensation".

Job 37:11 states - Also with moisture He saturates the thick clouds, He scatters His bright clouds. As silly at it may sound to us in this day and age, it still begs the question of how Job knew that clouds make water, or better yet how he had a firm understanding of "Hydrology", which is the study of Earth's water system, from its distribution and circulation of the Earth and atmosphere.

These verses have remarkable scientific insight.

Even Isaiah 40:22 tells us - It is He who sits above the circle of the earth, and its inhabitants are like grasshoppers, which stretch out the heavens like a curtain, and spreads them out like a tent to dwell in.

This becomes even more interesting when we translate the Hebrew word for "circle" we get the word "chuwg" which means "circuit or something spherical".

So how did both Job and Isaiah know that the Earth was round? I find this a most intriguing question. There is another place in scripture that one can look to find the answer to this question. It tells us in an "indirect" way that the Earth is round, and if you are clever enough, you can figure it out. It tells us in Luke 17:34-36 - I tell you, in that night there shall be two men in one bed, the one shall be taken, and the other shall be left. Two women shall be grinding together, the one shall be taken, and the other left. Two men shall be in the field, the one shall be taken, and the other left. Therefore, when the Lord was describing the manner of His coming to set up His kingdom. In that night, there will be people doing the following three things: (1) sleeping (2) grinding (3) working in the fields. Now the obvious questions to ask oneself is:

When do people sleep? - At night...

When are Eastern women grinding food? - Early morning...

When are men working in the field? - Broad daylight...

This means that at the "same moment" that G-d comes, it will be night some place on Earth, early morning somewhere else, and broad daylight in yet another place. One can easily see that in order for that to be true, the world would have to be round.

Moreover, scientists have discovered something even more amazing! That Earth has an upper atmospheric layer (ionosphere), which looks like "glass" and protects us from harmful rays emanating from the Sun, which also reflects electromagnetic waves back to the Earth. However, many centuries before this was discovered, it was written in Job 37:18 - Hast thou with him spread out the sky, strong, as a molten looking glass?

Something even more amazing arises when we begin to examine the books of Enoch again (which predate the book of Job). Remarkably found within the second book of Parables chapters 37-71we are privy to some systematic factors concerning the workings between the "Earth and the Stars". Many of us think that the Universe has no real bearing (or connection) to the earth. But Enoch tells us that it most certainly does (seeing that he was privy to such workings). We find recorded in Chapter 43 and 44 **that lightning that happens down here on earth actually arises from the stars above.** Chapter 43 - **And I saw other lightning's and the stars of heaven,** and I saw how He called them all by their names and they hearkened unto Him. And I saw how they are weighed in a righteous balance according to their proportions of light: (I saw) the width of their spaces and the day of their appearing, **and**

how their revolution produces lightning: and (I saw) their revolution according to the number of the angels, and (how) they keep faith with each other. And I asked the angel who went with me who showed me what was hidden: What are these? In addition, he said to me: The Lord of Spirits hath showed thee their parabolic meaning: These are the names of the holy that dwell on the earth and believe in the name of the Lord of Spirits forever and ever. Chapter 44 - **Also another phenomenon I saw in regard to the lightning's: how some of the stars arise and become lightning's** and cannot part with their new form.

It wasn't to long ago that science learned that lightning actually stemmed from the ground upwards. However today there has been more thorough scientific investigation put into the matter. Scientists are now first discovering that the cosmic high-speed rays that come from our Milky Way are particles that are created by active stars, and that they are charged particles that are influenced by magnetic fields. Now we also know that our Sun produces a magnetic field (called the magnetosphere) that extends to the end of our solar system and is enveloped (and rallied around) the earth in solar winds. Moreover, it is a field that helps reflect many of the cosmic rays away from the Earth. However, when the Sun is going through its 11 year cycle there is a time when there are small amounts of sunspots. These sunspots are dark patches that are observed on the Sun. Then there are also cool dark patches that appear in cycles on the Sun's surface. These sunspot patches possess a powerful magnetic field. And it is during the time of minimal sunspots that the magnetosphere collapses inwardly allowing high-energy cosmic rays to penetrate (come through) more intensely into our solar system. As a result far greater cosmic rays collide with Earth and filtering down to the lower atmosphere where the ionized small particles of moisture and humidity form into water droplets that become clouds - and thus we get severe weather conditions. Therefore, it is true then… that lightning does in fact come about from the stars!

Most intriguingly to the same effect, astronauts from the space shuttle also observe a phenomenon called "Red Sprites and Blue Jets". In particular these "Red Sprites" which have been observed going fully outside of the bounds of Earths atmosphere. Ironically, these "Red Sprites" have been found to be triggered by cosmic rays, which cause lighting down on the Earth below. This was exactly what Enoch was relating too. This contributes toward the credibility of the book of Enoch being scientifically correct.

Riddle me this though… how could Enoch have possibly known what the scientists are now first discovering if his visions were incorrect or made-up?

There is more concerning Enoch's visions from G-d… Amazingly, the book of Enoch happens to be very prophetic, and it happens to agree with what the later scriptures have to say. In fact, even though Enoch's writing's predate Judaism he somehow manages to fore mention about a later day coming of the Messiah (which is basically the same future vision of the rapture - the 2nd coming of the Messiah). When comparing scripts one can witness that Enoch's visions are well matched with the book of Revelations. We can find this in chapters 45-46 where we read about the "son of man who has righteousness" (Jesus). 1 Peter 2:24 - He himself bore our sins in his body on the tree, so that we might die to sins and live for **righteousness**; by his wounds you have been healed. Philippians 1:11 - Filled with the fruit of **righteousness that comes through Jesus Christ**—to the glory and praise of G-d. 2 Corinthians 5:21 - **G-d made him who had no sin to be sin for us, so that in him we might become the righteousness of G-d.**

In this same vision, Enoch sees the Messiah, but the Messiahs name is withheld from him. Enoch then asks: where will the habitation of the sinners be? In addition, the reply that Enoch gets is: **It would be better had they not been born.** Amazingly, this can later be found quoted in the Bible when Jesus Himself said to Judas: It **would be better had he not been born.** Matthew 26:17-25 - On the first day of the Feast of Unleavened Bread, the disciples came to Jesus and asked, Where do you want us to make preparations for you to eat the Passover? He replied, Go into the city to a certain man and tell him, The Teacher says: My appointed time is near. I am going to celebrate the Passover with my disciples at your house. So the disciples did as Jesus had directed them and prepared the Passover. When evening came, Jesus was reclining at the table with the Twelve. And while they were eating, he said, I tell you the truth, one of you will betray me. They were very sad and began to say to him one after the other, Surely not I, Lord? Jesus replied, The one who has dipped his hand into the bowl with me will betray me. The Son of Man will go just as it is written about him. **But woe to that man who betrays the Son of Man! It would be better for him if he had not been born.** Then Judas, the one who would betray him, said, Surely not I, Rabbi? Jesus answered, Yes, it is you. Enoch chapter 45 - And this is the second Parable concerning those who deny the name of the dwelling of the holy ones and the Lord of Spirits. And into the heaven they shall not ascend, and on the earth they shall not come: Such shall be the lot of the sinners who have denied the name of the Lord of Spirits, who are thus preserved for the day of suffering and tribulation. On that day Mine Elect One shall sit on the throne of glory and shall try their works, and their places of rest shall be innumerable. And their souls shall grow strong within them when they see Mine Elect Ones, and those who have called upon My glorious name: Then will I cause Mine Elect One to dwell among them. And I will transform the heaven and make it an eternal blessing and light, and I will transform the earth and make it a blessing: And I will cause Mine elect ones to dwell upon it: But the sinners and evil-doers shall not set foot thereon. For I have provided and satisfied with peace My righteous ones and have caused them to dwell before Me: But for the sinners there is judgment impeding with Me, So that I shall destroy them from the face of the earth. Enoch chapter 46 - And there I saw One who had a head of days, and His head was white like wool, and with Him was another being whose countenance had the appearance of a man, and his face was full of graciousness, like one of the holy angels. And I asked the angel who went with me and showed me all the hidden things, concerning that Son of Man, who he was, and whence he was, (and) why he went with the Head of Days" And he answered and said unto me: **This is the son of man who hath righteousness,** with whom dwelleth righteousness, and who revealleth all the treasures of that which is hidden, because the Lord of Spirits hath chosen him. And whose lot hath the pre-eminence before the Lord of Spirits in uprightness for ever. And this Son of Man whom thou hast seen shall raise up the kings and the mighty from their seats (and the strong from their thrones), and shall loosen the reins of the strong, and break the teeth of the sinners. And he shall put down the kings from their thrones and kingdoms because they do not extol and praise Him, nor humbly acknowledge whence the kingdom was bestowed upon them. And he shall put down the countenance of the strong, and shall fill them with shame. And darkness shall be their dwelling, and worms shall be their bed, and they shall have no hope of rising from their beds, because they do not extol the name of the Lord of Spirits, and raise their hands against the Most High, and tread upon the earth and dwell upon it. And all their deeds manifest unrighteousness, and their power rests upon their riches, and their faith is in the gods which they have made with their hands, and they deny the name o the Lord of Spirits, and they persecute the houses of His congregations, and the faithful who hang upon the name of the Lord of Spirits.

And additionally in Enoch chapters 38 - The first Parable. When the congregation of the righteous shall appear, and sinners shall be judged for their sins, and shall be driven from the face of the earth: And when the Righteous One shall appear before the eyes of the righteous, whose elect works hand upon the Lord of Spirits, and light shall appear to the righteous and the elect who dwell on the earth, where then will be the dwelling of the sinners, and where the resting-place of those who have denied the Lord of Spirits? **It had been good for them if they had not been born.**

Even more amazingly, the book of Enoch is also in perfect accordance with the book of Daniel and John. John when he wrote the book of Revelation concerning the 2nd coming of Jesus and the casting down of Satan and his followers (Revelation 19 and 20), and Daniel when he speaks about the Messiah and the 10 kings, and how those kings will hold power and then be cast down (Daniel 7:24, Daniel 9:24-25 and Revelation 17:12).

Even in the book of Psalms, King David was actually giving a reference to the book of Enoch. Enoch 39:6-7 - And in that place mine eyes saw the Elect One of righteousness and of faith, **And I saw his dwelling-place under the wings of the Lord of Spirits.**

Psalm 91:4 - He shall cover thee with his feathers, **and under his wings shalt thou trust: his truth shall be thy shield and buckler.**

We can also find a match between the book of Enoch, Daniel and Revelations. Enoch 46:1 - And there I saw One who had a head of days, **and His head was white like wool.** Daniel 7:9 - As I looked, thrones were set in place, and the Ancient of Days took his seat. His clothing was as white as snow; **the hair of his head was white like wool. His throne was flaming with fire,** and its wheels were all ablaze. Revelation 1:14 - **His head and hair were white like wool, as white as snow, and his eyes were like blazing fire.** Revelation 1:13-16 - And among the lamp stands was someone like a son of man, dressed in a robe reaching down to his feet and with a golden sash around his chest. **His head and hair were white like wool, as white as snow, and his eyes were like blazing fire.** His feet were like bronze glowing in a furnace, and his voice was like the sound of rushing waters. In his right hand he held seven stars, and out of his mouth came a sharp double-edged sword. His face was like the sun shining in all its brilliance.

We also have an identical match found within the books of Enoch, Jude and Deuteronomy. Enoch 1:9 - And behold! **He cometh with ten thousands of His holy ones To execute judgment upon all,** And to destroy all the ungodly: And to convict all flesh Of all the works of their ungodliness which they have ungodly committed, And of all the hard things which ungodly sinners have spoken against Him. Jude 1:14 - And Enoch also, the seventh from Adam, prophesied of these, saying, Behold, **the Lord cometh with ten thousands of his saints.** Deuteronomy 33:2 - And he said, The Lord came from Sinai, and rose up from Seir unto them; he shined forth from mount Paran, **and he came with ten thousands of saints:** from his right hand went a fiery law for them.

Furthermore, with an uncanny persistence another corresponding match is found within Isaiah and 2 Enoch chapter 21 - And the cherubim and seraphim standing about the throne, the six-winged and many-eyed ones do not depart, standing before the Lord's face doing his will, and cover his whole throne, singing with gentle voice before the Lord's face: **Holy, holy,**

holy, Lord Ruler of Sabaoth, heavens and earth are full of Your glory. Isaiah 6:3 - And they were calling to one another: **Holy, holy, holy is the Lord Almighty; the whole earth is full of his glory.**

There is even more to what Enoch has to tell us concerning the Messiah. And this is something that we can all relate to concerning the familiarness of the New Testament. In the book of Enoch chapter 47:1 Enoch tells us: And in those days shall have ascended the prayer of the righteous, **And the blood of the righteous (the just one)** from the earth before the Lord of Spirits. Enoch chapter 47:4 - **And the blood of the righteous (the just one) had been required** before the Lord of Spirits. What Enoch is communicating here (in a manner of speaking) is that it was a requirement for the Messiah (Jesus) to come and shed His blood for salvation (foretold thousands of years before Jesus). Then later in history we find in the New Testament. 1 John 1:7 - ...and the blood of Jesus, His Son, purifies us from all sin.

Additionally, here in the book of Enoch chapter 48 we read that Jesus' name was not known (or revealed) to Enoch (it was hidden). The logical reason for this is because if Enoch had told about His name being Jesus, and other pertinent information about Him, such as He was to be born in Bethlehem, and His parents names were to be Mary and Joseph (from the lineage of King David) the Jews would have recognized Him as their Messiah and would not have crucified Him - thus the prophesies concerning the Messiah would not have been fulfilled. Chapter 48:6-7 - **And for this reason hath he been chosen and hidden before Him,** Before the creation of the world and for evermore. And the wisdom of the Lord of Spirits hath revealed him to the holy and righteous; For he hath preserved the lot of the righteous, Because they have hated and despised this world of unrighteousness, And have hated all its works and ways in the name of the Lord of Spirits: **For in his name they are saved,** And according to his good pleasure hath it been in regard to their life.

As you can also see highlighted above in red that "in His name we are saved". How unequivocally New Testament this all is! John 14:6 - Jesus said to him, I am the way, and the truth, and the life. No one comes to the Father except through me. Acts 4:12 - There is salvation in no one else, for there is no other name under heaven given among mortals by which we must be saved. John 1:12 - But as many as received him, to them gave he power to become the sons of G-d, even to them that believe on his name.

But even though Jesus' name was hidden from Enoch, Enoch was privy to certain details concerning Jesus' birth. For instance the virgin birth of Jesus can be found in Enoch 62:7-13 - Trouble shall come upon them, as upon a woman in travail, whose labour is severe, when her child comes to the mouth of the womb, and she finds it difficult to bring forth. One portion of them shall look upon another. They shall be astonished, and shall humble their countenance; And trouble shall seize them, **when they shall behold this Son of woman sitting upon the throne of his glory.** Then shall the kings, the princes, and all who possess the earth, glorify him who has dominion over all things, him who was concealed; for **from the beginning the Son of man existed in secret**, whom the Most High preserved in the presence of his power, and revealed to the elect. He shall sow the congregation of the saints, and of the elect; and all the elect shall stand before him in that day. **All the kings, the princes, the exalted, and those who rule over all the earth, shall fall down on their faces before him, and shall worship him. They shall fix their hopes on this Son of man, shall pray to him, and petition him for mercy.** What we have here is a canonized book that prophetically repeats itself. To this end, we have been reading the same scripture over and over again, telling the same message, and somehow in the anthology of biblical doctrine we have neglected to see it.

To prove the legitimacy of the Book of Enoch, fragments of the book have been found within the Dead Sea Scrolls, which prove that they were not penned at a latter date (as some claim). What's even more amazing about the book of Enoch (considering its authenticity) is that Enoch himself tells us in his book that he is the seventh generation from Adam. We can also asses from Enoch that his father was Jared, and that Adam was 460 years old when Jared was born, and 620 years old when Enoch was born. In addition, we know according to the Bible that Adam then lived until he was 930 years old. Therefore, Enoch was 310 years old when Adam died. Genesis 5:5 - Altogether, Adam lived 930 years, and then he died. Shockingly, this informs us that when Enoch wrote his book, both Adam and Eve were still alive! When all is said and done about Enoch, he was taken away by G-d, saw the secrets and mysteries of the Universe, which included the future of the world as well as the predetermined course of human history. Only by reading the book of Enoch do we truly grasp what is being quoted in later scriptures of the Old and New Testament. There is nothing that is contradictory! Sadly, Enoch is the greatest story never told.

Moving on here… It's not just in the books of Job, Isaiah, Luke or Enoch that we find biblical insights, but also in many of the other books of the Bible as well such as in Leviticus, Numbers and Deuteronomy. It seems that they also knew about "bacteria and parasites". Because in (Leviticus 13:45-46) it tells us to wash after handling a dead body. And in (Numbers 19:11-19) it tells us to bury excrement away from a camp. Additionally in (Deuteronomy 23:12-13) it contains instructions to dispose of human excrement away from the people, in order to keep the camp holy. This is amazing considering that people didn't understand the biological nature of germs until relatively recently (the last few hundred years). In addition, they couldn't have known this without a microscope. Yet in the Old Testament it specifies cleanliness laws and food laws, so they must have had somewhat of an understanding of germs, yet they couldn't see them, or understand how they were transmitted. So the question remains, how did they know this unless otherwise privy from G-d?

We also find other things of quirky and bizarre accounts. Such as the law of "circumcision". In Genesis G-d commanded Abraham to circumcise Isaac on the 8th day after he was born. Genesis 17:12 - And he that is eight days old shall be circumcised among you, every man child in your generations, he that is born in the house, or bought with money of any stranger, which is not of thy seed. This has since become a religious instruction ever since for covenant reasons, but has been associated throughout modern science for cleanliness reasons. However, it would have been beyond Abraham to know why it was to be done specifically on the 8th day after birth. Until recently when it was discovered that specific blood-clotting factors (vitamin K) are produced by the body at a higher concentration on the 8th day after birth. Abraham could not have known this without G-d's intervention. However, the evolutionist will tell us that when thy performed this circumcision ritual (for barbaric uncivilized reasons) they must have lost several babies and kept doing it until they discovered that a baby was stronger and better equipped for survival on or about the 8th day. However, this is a lewd assumption since the Bible specifically tells us that this "initial" ritual was commanded by G-d to circumcise Isaac precisely on the 8th day as to establish His covenant with all the descendants of Abraham. Though there are no recorded incidents of babies dying from circumcision in the Bible, there have been several recorded incidents in past history that mention babies dying from this procedure, but it was most likely because of blood loss or bacterial infection from a procedure that was not done properly. Lets not forget the Jews somehow knew about bacteria and the qualities of the blood long before it was recorded by modern science.

Additionally we also find laws for "planting and sowing". Such as in (Leviticus 25) it gives us a whole slew of instructions for leaving fields fallow (turned over and uncultivated) for a certain period of time (years). In terms of modern science this is called environmental science (Ecology), as growing the same crop repeatedly in the same place eventually robs and diminishes the soil of nutrients. Crop rotation also helps to avoid the buildup of pathogens, pests, and helps build fertility for future crops. It also helps to replenish nitrogen in the soil. How could they have known this knowledge thousands of years ago without the knowledge of what pathogens are, or even knowing that there were such things as nitrogen or nutrients in the soil? Were they that advanced those many of thousands of years ago that we are only at the verge of just discovering the importance of leaving fields fallow today to attain future bountiful yields. The evolutionist would say that we have come along way from ape men to civilized human beings over millions of years of trial and error. On the contrary, I say we were privy to such knowledge by a loving G-d who wished to share the knowledge of His creation with His children to ensure their survival.

There are many other unsolved and inexplicable Bible verses that leave us asking the question of how they knew. Another one of my favorites can be found in Genesis 22:17 - Blessing I will bless you, and multiplying I will multiply your descendants as the stars of the heaven and as the sand which is on the seashore, and your descendants shall possess the gate of their enemies. Also in (Jeremiah 33:22) - As the host of heaven cannot be numbered, nor the sand of the sea measured, so will I multiply the descendants of David my servant and the Levites who minister to me? The question remains, in those days without a telescope, how could they have possibly counted or estimated the amount of stars in the universe and possibly related it to the sands on the seashore? The fact remains besides just looking up at the night sky; they could not have estimated this amount. Amazingly, the phrase itself - I will multiply your descendants as the stars of the heaven and as the sand which is on the seashore, is coming out of the mouth of G-d Himself (the narrator) as He is telling it to Abraham.

Moving on here to other verses of biblical foreknowledge... One of the most baffling questions of religion is why a "blood sacrifice" is needed by G-d. We find the answer in the book of (Leviticus 17:11) which describes the value of blood - For the life of the flesh is in the blood, and I have given it to you upon the altar to make atonement for your soul, for it is the blood that makes atonement for the soul. We all know that blood is dependent to all living things. And when we further examine the moving parts of blood we realize that it serves several amazing functions, such as carrying water, oxygen and nutrients to every cell of the body, as well as the function to remove the "waste material" of the body. In essence, the blood is a giver of life as well as a purifier. It is indeed the atonement for the soul - as a compensator so to speak. Therefore, the question remains, besides G-d requiring it, how is it that we knew this without knowing about the moving parts of blood without a microscope. Blood also contains within it DNA which determines everything about us. DNA is the substance that carries genetic information found in all living things, which is responsible for passing along hereditary characteristics from one generation to the next. So in essence, when the Bible proclaims that it is the blood that makes atonement for the soul it is literally saying that with a sacrificial Lambs blood we are being made "perfect". And it is in the New Testament that the Bible continually proclaims to us that it is by the blood (DNA) of Jesus (a perfect sinless human being) that we are made perfect and whole again. Ephesians 1:7 - **In him we have redemption through his blood, the forgiveness of sins**, in accordance with the riches of G-d's grace. 1 Peter 1:19 - But with the precious blood of Christ, as of a lamb without blemish and without spot. John 1:29 - The next day John saw Jesus coming toward him and said, Look, the Lamb of G-d, who takes away the sin of the world! 1 Corinthians 5:7 - Get rid of the old yeast that you may be a new batch without yeast—as you really are. For Christ, our Passover lamb, has been sacrificed. Romans 8:3 - For what the law was powerless to do in that it was weakened by the sinful nature, G-d did by sending his own Son in the likeness of sinful man to be a sin offering. Therefore, he condemned sin in sinful man. What is also remarkable in this is that the Passover feast was one of the main Jewish holidays that are celebrated in remembrance of G-d's deliverance of the Israelites from their bondage in Egypt. In fact, the slaughter of the Passover lamb and applying its blood to the doorposts of the houses in Egypt is a reminiscent picture of Jesus' compensational work on the cross. It was His blood that covered and protected us from the plague of death (sin) that passed over. Exodus 12:12-13 - On that same night I will pass through Egypt and strike down every firstborn—both men and animals—and I will bring judgment on all the gods of Egypt. I am the Lord. The blood will be a sign for you on the houses where you are; and when I see the blood, I will pass over you. No destructive plague will touch you when I strike Egypt.

Let us also further examine the initial stages of blood atonement. The first account of blood atonement can be found in (Genesis 3:21) - The Lord G-d made garments of skin for Adam and his wife and clothed them. It was inferred to be a lamb that was used for its skin to cloth Adam and Eve after their initial sin in the garden. It surfaces again shortly after with "Abel" (2nd son of Adam and Eve) who is a keeper of sheep in Genesis 4:4 - But Abel brought fat portions from some of the firstborn of his flock. The Lord looked with favor on Abel and his offering. Then it resurfaces again with Abraham and Isaac as they made their way up Mount Moriah (Genesis 22:7) and Isaac asks the question to his father Abraham - Isaac spoke up and said to his father Abraham, Father? Yes, my son? Abraham replied. The fire and wood are here, Isaac said, but where is the lamb for the burnt offering? This seems to be the incarnate question of the Old Testament: "WHERE IS THE LAMB?" Peering further ahead into history, the lamb was also specified for use during the "Passover" (the Exodus from Egypt). They were instructed to place the lambs blood over the door posts of their houses in order for the "Angel of Death" (the last of 10 plagues sent by G-d on Egypt) to pass over and spare the firstborn Jews, Exodus 12:12 - On that same night I will pass through Egypt and strike down every firstborn - both men and animals - and I will bring judgment on all the G-ds of Egypt. I am the Lord. The blood will be a sign for you on the houses where you are, and when I see the blood, I will pass over you. No destructive plague will touch you when I strike Egypt. Nearly a century and a half later after Isaac begged the question, (WHERE IS THE LAMB?) and after G-d supplied the ram in the thicket, we read about John the Baptist as he was preaching and baptizing in Jordan and when he saw "Jesus", he made the greatest statement of the New Testament, John 1:29 - The next day John saw Jesus coming toward him and said, "BEHOLD THE LAMB OF G-D", who takes away the sin of the world!

Therefore, we have Genesis in the Old Testament asking the question (where is the lamb?) And in the New Testament giving us the answer (behold the Lamb of G-d who takes away the sin of the world). Interestingly enough, Isaiah in the Old Testament also foretells of a savior and liberator (the Messiah) describing Him as a lamb brought to the slaughter. Isaiah 53:7 - He was oppressed, and he was afflicted, yet he opened not his mouth. He is brought as a lamb to the slaughter, and as a sheep before her shearers is dumb, so he openeth not his mouth. Followed by Peter in the New Testament linking Jesus to Isaiah's statement, 1 Peter 2:22-23 - He committed no sin, and no deceit was found in his mouth. When they hurled their insults at him, he did not retaliate, when he suffered, he made no threats. Moreover Peter tops it off when he quotes in 1 Peter 1:19 - But with the precious blood of Christ, as of a lamb without blemish and without spot.

Lets for a moment look at the behavior and character of a lamb. A "lamb" is noted for its purity in character. Exodus 12:5 - Your lamb shall be without blemish, a male of the first year: ye shall take it out from the sheep, or from the goats. If ones has ever noticed a baby lamb in its first year of life, it is innocent, blameless, harmless, sinless and pure of heart in all its ways. Surprisingly enough, the lamb/sheep have a very special quality to them… They have a sense of direction and have been known to head back to shelter when a storm is approaching. And even more amazingly they have the ability to pursue (eat) plants that make them feel better - thereby detecting what nutrients they are deficient in. The qualities of the lamb/sheep seem to reflect certain qualities and behaviors in the life of Jesus. For instance like a lamb/sheep can heal itself, so in many ways did Jesus heal the sick, Matthew 4:23 - Jesus went throughout Galilee, teaching in their synagogues, preaching the good news of the kingdom, and healing every disease and sickness among the people. And just as lamb/sheep have a good sense of direction, Jesus too knew from what direction He came and to what direction He was going in His ministry (look where it is today). Jesus was also known to calm a storm at sea thereby escaping it. Luke 8:22 - One day Jesus said to his disciples, "Let's go over to the other side of the lake". So, they got into a boat and set out. As they sailed, He fell asleep. A squall came down on the lake, so that the boat was being swamped, and they were in great danger. (Luke 8:25) - He got up and rebuked the wind and the raging waters, the storm subsided, and all was calm. Where is your faith? He asked his disciples. In fear and amazement they asked one another, Who is this? He commands even the winds and the water, and they obey him.

There is also another special quality about the lamb/sheep that no other animal has... It pronounces the word "Baaa", which as we know is linked to the Hebrew word "Abah" (father) and it is also linked to the Egyptian word "Ba" (soul).

The lamb in character (with the pure life force of its blood being used to treat the symptoms of sin) can also be likened onto the science of "Homeopathy" - given that its "pure life force" is so strong that it overrides the attributes of sin. In fact, the life force of the blood of the lamb is none other than "VIBRATION". The blood that has retained a vibrating copy (of the animals characteristics itself) which works like a vibrating tuning fork - that when it is struck causes a chain reaction in the other, and thus overrides the erroneous vibration and replaces it with the appropriate resonating vibration thereby taking precedence over sin (a bad or sickly causing vibration).

It might seem strange for some to consider that "blood" is the established standard for atonement of sin and that a man took the surrogate position of a lamb in which to atone for all of mankind's sins. Even as ridiculous as it may seem, there is indeed a connection. Moreover, that it is an awe inspiring message that continues to reach the souls of all humankind.

Interestingly enough, this moves us right to (Leviticus 11:1-47) which tells us all about "toxic foods" that we should not eat and the purpose for keeping "kosher". Leviticus 11:1-47 - And the Lord spake unto Moses and to Aaron, saying unto them, speak unto the children of Israel, saying, these are the beasts which ye shall eat among all the beasts that are on the earth. Whatsoever parteth the hoof, and is clovenfooted, and cheweth the cud, among the beasts, that shall ye eat. Nevertheless these shall ye not eat of them that chew the cud, or of them that divide the hoof: as the camel, because he cheweth the cud, but divideth not the hoof; he is unclean unto you And the coney, because he cheweth the cud, but divideth not the hoof, he is unclean unto you. And the hare, because he cheweth the cud, but divideth not the hoof; he is unclean unto you. And the swine, though he divide the hoof, and be clovenfooted, yet he cheweth not the cud, he is unclean to you. Of their flesh shall ye not eat, and their carcase shall ye not touch, they are unclean to you. These shall ye eat of all that are in the waters: whatsoever hath fins and scales in the waters, in the seas, and in the rivers, them shall ye eat. And all that have not fins and scales in the seas, and in the rivers, of all that move in the waters, and of any living thing which is in the waters, they shall be an abomination unto you: They shall be even an abomination unto you, ye shall not eat of their flesh, but ye shall have their carcases in abomination. Whatsoever hath no fins nor scales in the waters, that shall be an abomination unto you. And these are they which ye shall have in abomination among the fowls, they shall not be eaten, they are an abomination: the eagle, and the ossifrage, and the ospray, and the vulture, and the kite after his kind, every raven after his kind, and the owl, and the night hawk, and the cuckow, and the hawk after his kind, and the little owl, and the cormorant, and the great owl, and the swan, and the pelican, and the gier eagle, and the stork, the heron after her kind, and the lapwing, and the bat. All fowls that creep, going upon all four, shall be an abomination unto you. Yet these may ye eat of every flying creeping thing that goeth upon all four, which have legs above their feet, to leap withal upon the earth. Even these of them ye may eat, the locust after his kind, and the bald locust after his kind, and the beetle after his kind, and the grasshopper after his kind. But all other flying creeping things, which have four feet, shall be an abomination unto you. And for these ye shall be unclean: whosoever toucheth the carcase of them shall be unclean until the even. And whosoever beareth ought of the carcase of them shall wash his clothes, and be unclean until the even. The carcases of every beast which divideth the hoof, and is not clovenfooted, nor cheweth the cud, are unclean unto you: every one that toucheth them shall be unclean. And whatsoever goeth upon his paws, among all manner of beasts that go on all four, those are unclean unto you: whoso toucheth their carcase shall be unclean until the even. And he that beareth the carcase of them shall wash his clothes, and be unclean until the even: they are unclean unto you. These also shall be unclean unto you among the creeping things that creep upon the earth, the weasel, and the mouse, and the tortoise after his kind, and the ferret, and the chameleon, and the lizard, and the snail, and the mole. These are unclean to you among all that creep: whosoever doth touch them, when they be dead, shall be unclean until the even. And upon whatsoever any of them, when they are dead, doth fall, it shall be unclean, whether it be any vessel of wood, or raiment, or skin, or sack, whatsoever vessel it be, wherein any work is done, it must be put into water, and it shall be unclean until the even; so it shall be cleansed. And every earthen vessel, whereinto any of them falleth, whatsoever is in it shall be unclean, and ye shall break it. Of all meat which may be eaten, that on which such water cometh shall be unclean: and all drink that may be drunk in every such vessel shall be unclean. And every thing whereupon any part of their carcase falleth shall be unclean, whether it be oven, or ranges for pots, they shall be broken down: for they are unclean and shall be unclean unto you. Nevertheless a fountain or pit, wherein there is plenty of water, shall be clean: but that which toucheth their carcase shall be unclean. And if any part of their carcase fall upon any sowing seed which is to be sown, it shall be clean. But if any water be put upon the seed, and any part of their carcase fall thereon, it shall be unclean unto you. And if any beast, of which ye may eat, die, he that toucheth the carcase thereof shall be unclean until the even. And he that eateth of the carcase of it shall wash his clothes, and be unclean until the even: he also that beareth the carcase of it shall wash his clothes, and be unclean until the even. And every creeping thing that creepeth upon the earth shall be an abomination, it shall not be eaten. Whatsoever goeth upon the belly, and whatsoever goeth upon all four, or whatsoever hath more feet among all creeping things that creep upon the earth, them ye shall not eat, for they are an abomination. Ye shall not make yourselves abominable with any creeping thing that creepeth, neither shall ye make yourselves unclean with them, that ye should be defiled thereby. For I am the Lord your G-d: ye shall therefore sanctify yourselves, and ye shall be holy, for I am holy: neither shall ye defile yourselves with any manner of creeping thing that creepeth upon the earth. For I am the Lord that bringeth you up out of the land of Egypt, to be your G-d: ye shall therefore be holy, for I am holy. This is the law of the beasts, and of the fowl, and of every living creature that moveth in the waters, and of every creature that creepeth upon the earth: To make a difference between the unclean and the clean, and between the beast that may be eaten and the beast that may not be eaten.

The whole book of Leviticus is full of laws concerning proper conduct between man and G-d (which is more than I have listed). If one is interested in knowing them all, read the whole book of Leviticus. The book of Deuteronomy contains some as well.

Deuteronomy 14 - Ye are the children of the Lord your G-d: ye shall not cut yourselves, nor make any baldness between your eyes for the dead. For thou art an holy people unto the Lord thy G-d, and the Lord hath chosen thee to be a peculiar people unto himself, above all the nations that are upon the earth. Thou shalt not eat any abominable thing. These are the beasts which ye shall eat: the ox, the sheep, and the goat, the hart, and the roebuck, and the fallow deer, and the wild goat, and the pygarg, and the wild ox, and the chamois. And every beast that parteth the hoof, and cleaveth the cleft into two claws, and cheweth the cud among the beasts, that ye shall eat. Nevertheless these ye shall not eat of them that chew the cud, or of them that divide the cloven hoof; as the camel, and the hare, and the coney: for they chew the cud, but divide not the hoof; therefore they are unclean unto you. And the swine, because it divideth the hoof, yet cheweth not the cud, it is unclean unto you: ye shall not eat of their flesh, nor touch their dead carcase. These ye shall eat of all that are in the waters: all that have fins and scales shall ye eat: And whatsoever hath not fins and scales ye may not eat, it is unclean unto you. Of all clean birds ye shall eat. But these are they of which ye shall not eat: the eagle, and the ossifrage, and the ospray, and the glede, and the kite, and the vulture after his kind, and every raven after his kind, and the owl, and the night hawk, and the cuckow, and the hawk after his kind, the little owl, and the great owl, and the swan, and the pelican, and the gier eagle, and the cormorant, and the stork, and the heron after her kind, and the lapwing, and the bat. And every creeping thing that flieth is unclean unto you: they shall not be eaten. But of all clean fowls ye may eat. Ye shall not eat of anything that dieth of itself: thou shalt give it unto the stranger that is in thy gates, that he may eat it; or thou mayest sell it unto an alien: for thou art an holy people unto the Lord thy G-d. Thou shalt not seethe a kid in his mother's milk. Thou shalt truly tithe all the increase of thy seed, that the field bringeth forth year by year. And thou shalt eat before the Lord thy G-d, in the place which he shall choose to place his name there, the tithe of thy corn, of thy wine, and of thine oil, and the firstlings of thy herds and of thy flocks; that thou mayest learn to fear the Lord thy G-d always. And if the way be too long for thee, so that thou art not able to carry it; or if the place be too far from thee, which the Lord thy G-d shall choose to set his name there, when the Lord thy G-d hath blessed thee: Then shalt thou turn it into money, and bind up the money in thine hand, and shalt go unto the place which the Lord thy G-d shall choose: And thou shalt bestow that money for whatsoever thy soul lusteth after, for oxen, or for sheep, or for wine, or for strong drink, or for whatsoever thy soul desireth: and thou shalt eat there before the Lord thy G-d, and thou shalt rejoice, thou, and thine household, and the Levite that is within thy gates, thou shalt not forsake him, for he hath no part nor inheritance with thee. At the end of three years thou shalt bring forth all the tithe of thine increase the same year, and shalt lay it up within thy gates: And the Levite, (because he hath no part nor inheritance with thee,) and the stranger, and the fatherless, and the widow, which are within thy gates, shall come, and shall eat and be satisfied, that the Lord thy G-d may bless thee in all the work of thine hand which thou doest.

As silly as all these food rules in Leviticus sound, there are reasons for their application… Some of these reasons include: Hygiene, sanitation, health reasons as well as the most misunderstood of all, for "moral" reasons. But the sole reason for keeping kosher is mainly a "reminder" of thankfulness and appreciation to G-d for taking the Jewish people out of Egypt. Therefore, without questioning the most important reason for keeping the kosher laws is based on the sole purpose of G-d entered into a covenant with the Jewish people, obligating the Jews to uphold and fulfill its commandments, and the Torah for which these Kosher laws are part of. Yes, these laws are for health reasons, but are also on a moral level of "food for thought". Some of the hygienic reasons include forbidding us to eat animals that died without a proper slaughter which requires the draining of the blood for what might seem a bizarre ritual is really to rid the blood which is a medium for the growth of bacteria. Keeping kosher also forbids eating animals that have abscesses (blemishes, sores, lumps, cancer) on their body or in their body, or other health problems. Additionally, no shellfish, mollusks, lobsters or crabs are to be eaten which have been known to spread several diseases, but are mainly known to be scavengers which eat waste material that can contain hazardous toxins, one of the primary toxins that occurs naturally in nature is "lead" which is found in soil, mud and dirt, and also the excrement (feces) of other animals (which may also contain lead and disease), thus if we eat these shellfish we are also consuming and absorbing the lead, waste materials, disease and toxins that can be harmful to our body. This is also why according to kosher law, edible fish (fit for human consumption) need to have fins and scales for the exclusive reason that scales act as an armor and protector against harmful toxins and waste material that are easily absorbed in water. Fish that have to scales or fins of armor are more susceptible to absorbing toxins more easily. This also includes insects (such as worms and snakes) and other creepy-crawlers of land and sea. Moreover, milk and meat are not to be eaten together for the mere reason that milk and meat digest at an unequal rate and are difficult for the body to break down, thus robbing it of vital nutrients and extreme ware and tear of the intestines.

Leviticus 19:23 tells us that the fruit from a tree should not be eaten before the fourth year - When you enter the land and plant any kind of fruit tree, regard its fruit as forbidden. For three years you are to consider it forbidden, it must not be eaten. In the fourth year all its fruit will be holy, an offering of praise to the Lord. However, in the fifth year you may eat its fruit. In this way, your harvest will be increased. I am the Lord your G-d. This in fact holds truth. The fruit of a tree is indeed bitter and lacks robustness and nutritional value the first few years. If eaten before its time, it has no nutritional value to us, and may actually harm us since bitter tasting fruit (if eaten in quantity) may contain toxins that cause upset stomach. Therefore, the fruits of the 5th year are effective for human consumption.

Kosher law also excludes eating certain kinds of birds, such as birds of prey even some mammals, which are predators. This is because tension, anger and hormones produced by the birds become preserved in their body and make the meat toxic for consumption, which can reflect, in the human body if eaten. Remember as in "homeopathy", even if you try to remove all remnants of the toxin or poison itself, even by cooking, a remnant or copy (vibration) still remains within the meat. Eating of the meat can change a persons disposition, (character, temperament and nature as well as health). Animal cruelty is also forbidden… when an animal is slaughtered for sacrifice or atonement, it must be done with the least amount of possible pain, that is why a special kosher knife is used that is so sharp that even the slightest nick or notch in the blade renders it unacceptable. Therefore it is the sharpness of the blade that prevents pain to the animal, and if there are visible chares found in the meat of the animal it may have been killed in pain and thus non-kosher. There are also other moral laws attributed to the humane treatment and respect of animals, such as a mother animal and her baby are forbidden to be slaughtered on the same day… additionally it is not permitted to cook a baby animal in its mothers milk. What's more, drinking or cooking milk from a non-kosher animal is also forbidden. Oral law also teaches that non-kosher food blocks the spiritual potential of the soul. However, if an animal is properly slaughtered and prepared using kosher law, it has more "sparks of holiness, purity and wholesomeness" which when eaten can be incorporated into our being. That is one of the reason we are instructed not to eat pork since a pig eats anything in its path including its own feces. In fact, whatever it eats turns to meat on its bones in a matter of just a few hours and is retained faster than it can be disposed of, therefore pork is a highly laced toxic diseased food not meant for human consumption. So not only is there a underlining health reason for us not to eat the meat of a pig, it stands that there is also a "moral" reason so that we should not absorb these qualities into ourselves. Kosher can thus be seen as "spiritual nutrition" in which there are foods that nourish the soul and foods that adversely affect it. However, it follows to reason that if we can discipline ourselves in what we eat, we can be disciplined in other areas of life as well. Self-discipline is defined as the ability to regulate one's conduct by principle and sound judgment, rather than by impulse or desire. Biblically, self-discipline can be summarized in one word: Obedience of staying within the bounds of G-d's law to evade sin and evil! Self-discipline may even indeed control consequences in the realm of "Quantum Mechanics".

Moving on here… It must also be brought to the attention of people (civilizations, nations and humanity) that there is a great deal of "borrowed knowledge" that has come out of the Bible. This has to do with the use of "sophisticated tactical warfare" that was used by many of the Old Testament characters as a means of self defense, winning territory, and wars… Upon which today the Bible has been employed as a first-rate instructional manual in establishing and securing many nations. Few people realize that the founding fathers of "DEMOCRACY" (for the free and equal representation of people) were in fact the Hebrews/Jews in their quest for liberation and establishing the state of Israel. Their weaponry and clever tactics and maneuvers have become the usage of today's modern military management, strategy and procedure. The first person in recorded history to demonstrate sophisticated tactical warfare was "Abraham" when he rescued his nephew "Lot" from the Mesopotamian army as well as freeing Israelite captives. Genesis 14:15 - And he divided himself against them, he and his servants, by night, and smote them, and pursued them unto Hobah, which is on the left hand of Damascus. Abraham's attack was a nighttime surprise raid from two different directions. Abraham strategically planned it this way so that the Mesopotamian army had no time to form into defensive position. Abraham's raid is a classical standard model nighttime maneuver, which is still taught in military academies throughout the world.

Abraham's reign did not stop there, it continued through his son Isaac, and from Isaac to Jacob and his 12 sons (the 12 tribes of Israel) who later migrated to Egypt through Jacobs younger son Joseph. They were probably a great military force known for strategy and strength, and when we stop to consider what really transpired during the time of the Jews in Egypt right before them leaving Egypt (the Exodus), and if we look at it from a slightly different angle of a "military perspective" and as a group of people with a military notability and significance, it might explain to us why they were considered "slaves". Exodus 1:13 - And the Egyptians made the children of Israel to serve with rigour. Now when we begin to examine Pharaoh's words of concern in Exodus 1:9-10 things become a whole lot clearer - And he said unto his people, Behold, the people of the children of Israel are more and mightier than we: Come on, let us deal wisely with them, lest they multiply, and it come to pass, that, when there falleth out any war, they join also unto our enemies, and fight against us, and so get them up out of the land. This verse in itself gives us several suggestions, one being that the Jews were more and mightier than the Egyptians - possible militarily, and secondly the Egyptians were afraid that they might have the military capability to overtake the land of Egypt. So to offset this, the pharaoh set them to physically required, obligatory, necessary and enforced labor, and they were reduced down to civilian workers to work on public works projects, like a military recruit that has been called up compulsorily to do military service, but they lost their status as being allied partners with Egypt and were treated like ordinary workers. The demotion from military soldier to common worker did not sit well with the Jews and was an indication to them that it was time to leave Egypt, and when they left according to Exodus 13:18, they left "militarily armed and harnessed" - But G-d led the people about, through the way of the wilderness of the Red sea: and the children of Israel went up "HARNESSED" out of the land of Egypt. However, prior to them leaving Egypt before the last plague, G-d tells Moses in Exodus 11:1-3 - Now the Lord had said to Moses, I will bring one more plague on Pharaoh and on Egypt. After that, he will let you go from here, and when he does, he will drive you out completely. Tell the people that men and women alike are to ask their neighbors for articles of silver and gold. The Lord made the Egyptians favorably disposed toward the people, and Moses himself was highly regarded in Egypt by Pharaoh's officials and by the people. Then in Exodus 12:35-36 we read - And the children of Israel did according to the word of Moses, and they borrowed of the Egyptians jewels of silver, and jewels of gold, and raiment: And the Lord gave the people favour in the sight of the Egyptians, so that they lent unto them such things as they required. And they spoiled the Egyptians. What seems to have happened here is that the Israelites are leaving Egypt and they are not equip to be in the desert, they need food, shelter, water and animals, and they take it. What hardened Pharaoh's heart again and changed his mind was that the Jews had raided and sacked Egypt for these things. Pharaoh comments about this, and said the Jews were leaving Egypt "boldly". Exodus 14:8 - The Lord hardened the heart of Pharaoh king of Egypt, so that he pursued the Israelites, who were marching out BOLDLY.

Shortly before they reach Mount Sinai we read about Joshua, who is chosen by Moses to be his military commander when the Amalakites attack the Israelites at Rephidim. Moses calls on Joshua as the military field commander to organize an army to repel them. Exodus 17:8-13 - The Amalekites came and attacked the Israelites at Rephidim. Moses said to Joshua,

Choose some of our men and go out to fight the Amalekites. Tomorrow I will stand on top of the hill with the staff of G-d in my hands. So Joshua fought the Amalekites as Moses had ordered, and Moses, Aaron and Hur went to the top of the hill. As long as Moses held up his hands, the Israelites were winning, but whenever he lowered his hands, the Amalekites were winning. When Moses' hands grew tired, they took a stone and put it under him and he sat on it. Aaron and Hur held his hands up one on one side, one on the other so that his hands remained steady till sunset. So Joshua overcame the Amalekite army with the sword.

Next we read about Joshua and his battle at Jericho (where the city's walls fall down). Joshua 6:3-4 - And ye shall compass the city, all ye men of war, and go round about the city once. Thus shalt thou do six days. And seven priests shall bear before the ark seven trumpets of rams' horns: and the seventh day ye shall compass the city seven times, and the priests shall blow with the trumpets. Today the military has developed devices that use high frequency sound waves to cause nausea and disorientate an enemy.

Then we read about Joshua when he burns the city of Ai, Joshua 8:28-29 - And Joshua burnt Ai, and made it an heap for ever, even a desolation unto this day. And the king of Ai he hanged on a tree until eventide: and as soon as the sun was down, Joshua commanded that they should take his carcase down from the tree, and cast it at the entering of the gate of the city, and raise thereon a great heap of stones, that remaineth unto this day. (He used the mountains) with the Sun rising in the East shinning right in the eyes of the Canaanites, and attacking downhill the Canaanites couldn't see fighting in the face of the Sun, nor did they eat or prepare since it was early morning. Joshua defeats the Canaanites. Joshua 10:9-10 - Joshua therefore came unto them suddenly, and went up from Gilgal all night. And the Lord discomfited them before Israel, and slew them with a great slaughter at Gibeon, and chased them along the way that goeth up to Bethhoron, and smote them to Azekah, and unto Makkedah.

The next hero of the Bible is Gideon; he rises up against the midianites. He chooses "300" elite soldiers… they carry torches, clay pots and shofers (ram horns) and at midnight Gideon strategically separates his 300 men and positions them around the midian camp on 3 sides leaving the southern part open. Judges 7:20 - And the three companies blew the trumpets, and brake the pitchers, and held the lamps in their left hands, and the trumpets in their right hands to blow withal: and they cried, The sword of the Lord, and of Gideon. The torches are seen by the midianites and in a panic they flee. The midianites are also blinded by the torches and they start fighting themselves… it's a state of confusion and complete chaos! While the remaining Israelite army in the southern part is waiting in ambush for the fleeing midianites and slaughter them.

The next military hero of the Bible is woman named "Deborah". Among her known qualities for giving advice, she is also an intelligently clever and witty woman of military command. The Philistines had taken over the Jezreel valley and she plots a strategy that uses double agents, along with her knowledge of the terrain by using Mount Tabor so that the Israelite army can see any army approaching. It also happened to be during the rainy season, which made the ground muddy so the enemy could not get there military up the muddy slopes without being bogged in the mud. Then the Israelites attack down hill with a victorious victory. Judges 4:4-7 - Deborah, a prophetess, the wife of Lappidoth, was leading Israel at that time. She held court under the Palm of Deborah between Ramah and Bethel in the hill country of Ephraim, and the Israelites came to her to have their disputes decided. She sent for Barak son of Abinoam from Kedesh in Naphtali and said to him, The Lord, the G-d of Israel, commands you: Go, take with you ten thousand men of Naphtali and Zebulun and lead the way to Mount Tabor. I will lure Sisera, the commander of Jabin's army, with his chariots and his troops to the Kishon River and give him into your hands. Then Deborah used a diversion by using a double agent to lure Sisera the Canaanite king to Mount Tabor - When they told Sisera that Barak son of Abinoam had gone up to Mount Tabor, Sisera gathered together his nine hundred iron chariots and all the men with him, from Harosheth Haggoyim to the Kishon River. However Sisera escapes and runs to the tent of the double agents wife who pounds a peg through his head, Judges 4:17-24 - Sisera, however, fled on foot to the tent of Jael, the wife of Heber the Kenite, because there were friendly relations between Jabin king of Hazor and the clan of Heber the Kenite. Jael went out to meet Sisera and said to him, Come, my Lord , come right in. Don't be afraid. So he entered her tent, and she put a covering over him. I'm thirsty, he said. Please give me some water. She opened a skin of milk, gave him a drink, and covered him up. Stand in the doorway of the tent, he told her. If someone comes by and asks you, Is anyone here? say No. But Jael, Heber's wife, picked up a tent peg and a hammer and went quietly to him while he lay fast asleep, exhausted. She drove the peg through his temple into the ground, and he died. Barak came by in pursuit of Sisera, and Jael went out to meet him. Come, she said, I will show you the man you're looking for. So he went in with her, and there lay Sisera with the tent peg through his temple-dead. On that day G-d subdued Jabin, the Canaanite king, before the Israelites. And the hand of the Israelites grew stronger and stronger against Jabin, the Canaanite king, until they destroyed him.

Saul was the next and his son Jonathan using secret ravines to work their way around the Philistines, quarantining them into a small battlefield area and triumphing over them. SAUL takes back control over the central mountain ridge of Canaan, also cutting off the trade routs in the Jezreel valley and made headway in creating a united territorial kingdom for the Israelites. 1 Samuel 13:2 - Saul chose three thousand men from Israel; two thousand were with him at Micmash and in the hill country of Bethel, and a thousand were with Jonathan at Gibeah in Benjamin. The rest of the men he sent back to their homes.

Then came King David when he took out the Philistine giant “Goliath” 1 Samuel 17:44-45 with a rock and sling as a military weapon. - And the Philistine said to David, Come to me, and I will give thy flesh unto the fowls of the air, and to the beasts of the field. Then said David to the Philistine, Thou comest to me with a sword, and with a spear, and with a shield: but I come to thee in the name of the Lord of hosts, the G-d of the armies of Israel, whom thou hast defied.

The Bible it seems was even used as a guidebook by the Greeks during the reign of Leonites. They seem to have applied the same strategy of using “300” Spartan soldiers (like Gideon did) by way of the narrow Thermopolis pass to hold back nearly 300 thousand Persian soldiers. And as mentioned earlier that many of these strategic maneuvers that were used by the Jews in ancient times are still being taught in militaries throughout the world and are still winning wars and securing territories.

As aforementioned, DEMOCRACY is the root of true liberation and equality and may even prove at some point to be a basis dependant upon our human survival. However, in this era we are locked into the greatest war in the history of the world. However, not the ordinary battle of the wars we are familiar with. The Jews always fought battles with G-d as a liaison behind them. Perhaps we could say that it has been one conglomerate war that has claimed millions of souls for the mere purpose over "ideas and philosophies" (concerning G-d) and it has become the ultimate battle for the soul of every human in the future.

Now for those who think that religion had little role in the founding of "democratic" countries like Israel, they are in for a rude awakening! America is another prime example of achieving success through democracy... for if one visit's the nations capital of Washington D.C. they would witness many biblical verses and references commemorating the name of the Jewish G-d etched within granite, marble and brass on federal institutions all over our nations capital.

The U.S. capital building has engravings that say, " What hath G-d wrought!" and "America! G-d shed his grace on thee", as well as the famous quotation that is written all over American currency that says, "In G-d we trust". When one enters the national archives building a bronze depiction of the 10 commandments gilds the entryway floor. Additionally within the capital building, there are rotunda paintings... some which include Pocahontas being baptized, and pilgrims praying for G-d's mercies before sailing to the New World. One can also view above the house chamber's main door where there are marble relief's of history's 22 greatest lawmakers, in which they are all facing Moses. It doesn't stop there, it continues on into the supreme court building above the eastern pediment (a triangular looking rock area) are histories major lawmakers (carved out in a realistic dimension setting) in which Moses is again in the center holding a copy of the 10 commandments. But to first enter the supreme court, one must pass through two oak doors which also have an engraved depiction of the 10 commandments on them, and in the court room itself, above the justices chair benches one can see another engraving of Moses holding the 10 commandments. Additionally, within the U.S. capital building's chapel there are stained glass windows, which portray George Washington praying below the phrase, "This Nation under G-d".

Additionally, the Liberty Bell (located in Philadelphia, Pennsylvania) is inscribed with Leviticus 25:10 that proclaims - Liberty throughout all the land unto all the inhabitants thereof.

Furthermore, few people may actually know that the Washington monument cornerstone contains within it a copy of the Constitution, the Declaration, and the "Kings James Bible". Additionally, the engravings and inscriptions on the monument venerate G-d, stating such things as: In G-d we trust. Moreover, on its top cap it reads "laus Deo" (which means, praise be to G-d). So let us also understand that America was founded on Christian ideals that were based off the Jewish Old Testament that "ALL MEN ARE CREATED EQUAL". Many Americans fail to realize that their "freedoms" come about because they were founded on a Judeo Christian based society, yet these rights are used against us by "atheist" when they claim its "unconstitutional" when they are subjected to religion. These people are victims of a society that is loosing its grip on the meaning of true "democracy" - for to destroy America one must simply destroy peoples faith. Hosea 4:6 - My people are destroyed for lack of knowledge: because thou hast rejected knowledge, I will also reject thee, that thou shalt be no priest to me: seeing thou hast forgotten the law of thy G-d, I will also forget thy children. The knowledge that is being referred to is the knowledge of religion that was the "first initiator" of the importance of "liberty and equality". Democracy cannot work properly without religion backing it. The reason for this is because the notion for what we as humans call "freedom" was first noticed when and only after "sin" crept into the equation (the fall of Adam and Eve), which later in history led to the giving of the 10 commandments on which "liberation" is a precondition - representing the Jews that were LIBERATED out of Egypt. Later on in history, Jesus showed us what equality means by breaking the taboos on fellowship with lepers, prostitutes, Roman soldiers and tax collectors. Even women traveled with him in his larger company of disciples. These Judeo Christian beliefs were later instilled into the 3rd president of the United States, Thomas Jefferson (in his political philosophy) that can be found in the Declaration of Independence. In one brief and fleeting paragraph, Jefferson condensed the essence of "American thinking" concerning human existence and the proper role of government. In effect, these words are the guts, backbone and spirit of America when Jefferson said: We hold these truths to be self-evident, that all men are created equal, that they are endowed by their Creator with certain unalienable Rights, that among these are Life, Liberty and the pursuit of Happiness. That to secure these rights, Governments are instituted among Men, deriving their just powers from the consent of the governed, That whenever any Form of Government becomes destructive to these ends, it is the Right of the People to alter or to abolish it, and to institute new Government.

Not just Thomas Jefferson but most if not all of the signers of the Declaration of Independence were in fact religious men. In a literal sense if it were not for the Jews, the founding fathers of the United States of America could not have appropriately made the Declaration and Constitution. But the basis for such inalienable rights and freedoms additionally stem from a source that one might not think. These rights are defined by a clause called the "laws of Nature" and of "Natures G-d". This was a law directly out of the New Testament Bible. Romans 1:19 - Because that which may be known of G-d is manifest in them; for G-d hath shewed it unto them. The bill of rights states what it considers inalienable rights based off G-d's design for nature. Such as in the right to keep and bear arms, which is a right that is based off of self defense. G-d put it in nature that we may witness every animal defending its young and know that the same is applied to us.

The first 56 words of the Declaration are noticeably G-d orientated. However, some people assume that the Constitution is separate from the Declaration because there is no mention of G-d. However, the Constitution is not godless and it is not a separate document. It may look like a secular document, but the Constitution based on article 7 states that when we ratified the Constitution that it was not done as a brand new document, but based under the Declaration. Therefore, it was not necessary for the Constitution to repeat what was already there.

However to end this argument for once and for all that the Constitution is wholeheartedly G-d orientated... When we go back to the very roots of the Constitution we specifically know from American history that the first main group of people to settle in North America were the "Puritans". The Puritans were members of the Church of England that wanted to "purify" the English Bible. At that time they were then called the "Separatists or the Pilgrims" who moved to Holland for approximately nine years, and from Holland they sailed to Virginia in 1606 in the famous ship called the "Mayflower". Now when the people traveling to the New World on this ship realized that they had no government, they assembled together and called it the "Mayflower Compact", which produced the first document written in the New World which reads: We the herby undersigned having undertaken for the glory of G-d and the advancement of the Christian faith a voyage to plant a first colony on the northern parts of Virginia do form a civil body politic (1620). This was later the blueprint of the Connecticut Constitution which was in turn the famous blueprint for the U.S. constitution which starts off - "We the people" of the United states of America. As Judeo Christian history progressed throughout the world, all of Europe was Christian, Germany - Lutheran, Switzerland - Calvinist, Holland - Dutch reform, Sweden - sweetish Lutheran, Greece - Greek orthodox, Russia - Russian orthodox, Spain, Italy, Portugal and Poland - roman catholic, England - Anglican and Scotland - Presbyterian.

And for those that continue to argue (and push) their atheist views to separate and even go to the lengths of banning religion from the schools, public places and the country… let these people remember that the founding fathers of America where Judeo Christian Masons with the original intention for a country based off of religion and G-d. This is what truly made / makes America. It will be a sad day when religion is abolished, and all those who fight for religion to be excluded from the subjection of its citizens. In effect, we will have lost the true meaning of what it means to be LIBERATED. When we loose touch of religion, we are cut off from all freedoms given to us by our loving creator, for it is only by Him that they are given. Thomas Jefferson as aforementioned: They are endowed by their Creator.

A nation, which does not remember what it was yesterday, does not know what it is today, or what it is trying to do. We are trying to do a futile thing if we do not know where we came from or what we have been about. - Woodrow Wilson the twenty-eighth President of the United States.

15 - EXPANDING ON THE MERKABAH...

Fifteen chapters later and I am off again, spinning long mesmerizing accounts from my immense research! I hope this chapter is going to be another fascinating captivator. One in which the Merkabah structure might help to clarify (and put into words) its order among the natural world (which includes divine design).

The Merkaba is by all measurable accounts - the Star of David!

To realize this, lets first start off with examining ancient scripture as well as an illustrative representation of the Merkabah, which can be traced as far back as the Phoenicians. But of course its true origin is much older as we've already discovered that it appears within the conceptual words of the very first word of Genesis 1:1 and expands throughout the opening verse which literally puts into picture (the Star of David) that it is composed of 6 (which will be reiterated about again in this chapter). Now to focus on the word "Mer-Ka-Bah" it is split up into three distinct words (body, mind and spirit). The meaning of the word "MER" is in fact a representation of a pyramid (body), in view of the fact that that the Phoenician hieroglyphics for the letter "M" resembles that of three pyramids, which might also be an intended illustration for the three pyramids of Egypt. Meanwhile the Phoenician hieroglyphics for the letter "E" are represented by a triangle/pyramid with lines going through it possibly illustrating the King and Queen's chambers within the Great Pyramid. While the Phoenician hieroglyph for the letter, "R" is illustrated by yet another form that looks like a triangle/pyramid.

The completed letters spell the word "MER":

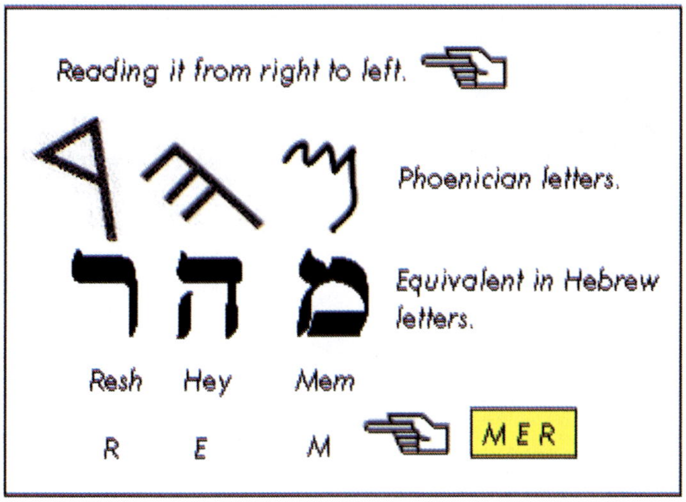

Although the word "Mer-ka-bah" is usually considered an Egyptian and Phoenician word, its usage appears first in the Hebrew language. In the English versions of the Bible, it is usually translated as "chariot" (also known as the cherubim from Ezekiel's vision and Elijah when he disappears in the whirlwind). 2 Kings 2:11-12 - And it came to pass, as they still went on, and talked, that, behold, **there appeared a chariot of fire,** and horses of fire, and parted them both asunder; **and Elijah went up by a whirlwind into heaven.** And Elisha saw it, **and he cried My father, my father, the chariot of Israel,** and the horsemen thereof. And he saw him no more: and he took hold of his own clothes, and rent them in two pieces. Keep in mind that when Elijah called it "the chariot of Israel" he was most likely referring to the "Star of David" which will soon reveal its mysterious meaning throughout this entire chapter. Now, the Hebrew word Mer-ka-bah is also related to several other conceptual root words that mean to "ride" and "wagon" since the word used in the Hebrew Bible for "chariot" is "rekeb" (RKB) and the word Merkabah (MRKBH) (as seen directly below) is made up of the same letters (thus making it a conceptual word).

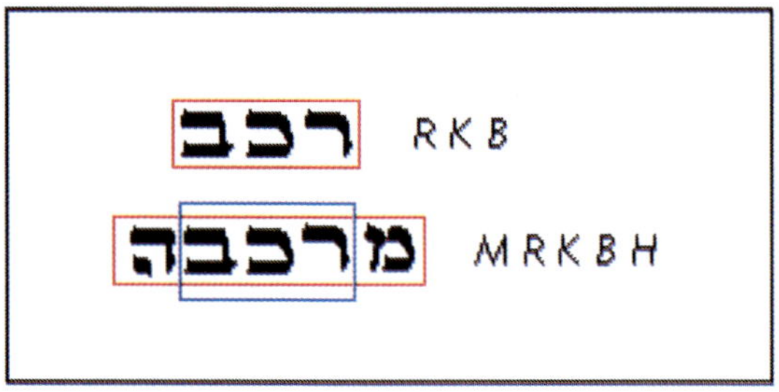

But in English terms they are all equivalent to the word "vehicle" or something that transports (or rides) your conscious being somewhere. In essence this vehicle, chariot and wagon is a counter-rotating field of circling light of plasma energy that can transport a "copy of ones being or consciousness" to other places on earth or within the universe or even to other dimensions or realms that are separated by the diminutions of light (as aforementioned that were separated by G-d in His creation process) (not to be confused though, this is NOT a Time Traveling device). This COUNTER-ROTATING field of circling light is actually two conjoined Tetrahedrons (triangles), and is "in operation" the STAR OF DAVID!

In addition, this functions by "divine design" - using the same principles as our own spiral galaxy, which is incorporated as part of the Universal Law!

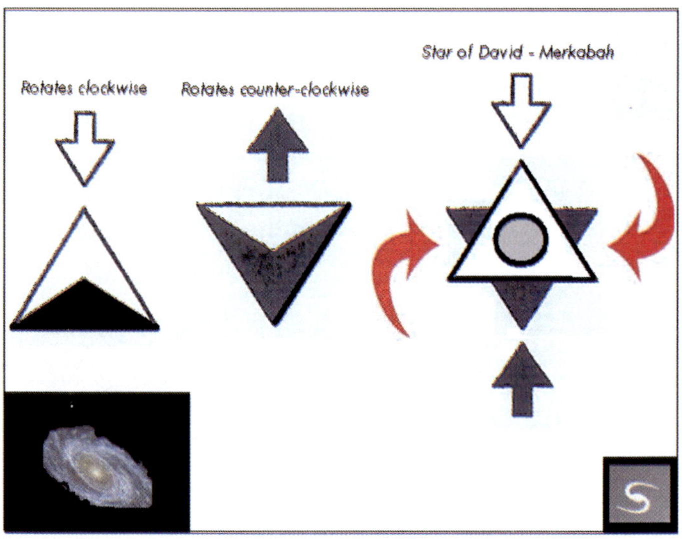

In essence we get something to this effect. Two counter-rotating tetrahedrons that continue to spin until a enormous surge of energy is released and separated from the body thus creating the Merkabah vehicle.

By the way, the "Star of David" is formed by two interlocking triangles/pyramids. The middle section in which they share is also a "Hexagon". It is noteworthy to know that the Hexagon is the strongest form in nature. However, some suggest that it is the triangle itself. I have also heard it is the sphere, but a sphere reduces itself to the atom and even atoms are not spheres but are particles held together by the electromagnetic field. Carbon is another element that forms Hexagons (and diamonds which are composed of carbon are the strongest and hardest materials known). Additionally, many living things use the form of the Hexagon (because it is structurally strong) such as a bee's honeycomb (picture shown below).

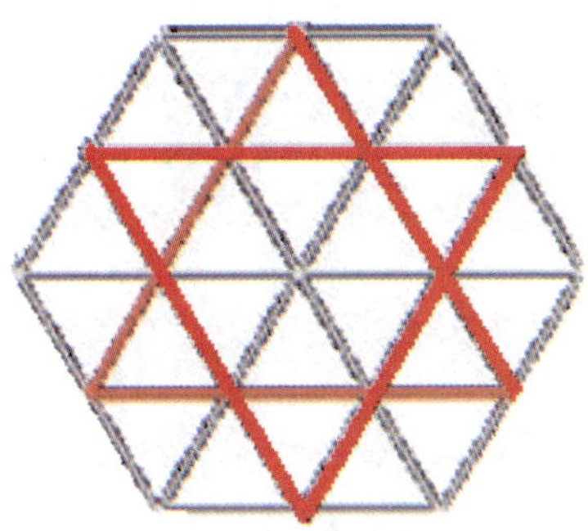

This light vehicle is actually a vibrating/frequency vehicle created by the (mind) in which the brain is stirred into coagulating hormones based on certain meditated thoughts which in-turn stimulate and engulf the body into creating electromagnetic plasma (it is a spiritual climax) - hence the three distinctive synergetic words of Mer-Ka-Bah (body, mind and spirit). This plasma usually appears invisible but sometimes depending on the intensity glows like a powdery white ball sparking with a blue twinkle (just like a sepheroth in the Tree of Life diagram). Some have been seen to float, while others dash with incredible speeds from place to place. These balls of plasma energy have been mistaken for UFOs, asteroids, and have been witnessed in creating crop circles (this is also the discussion of the next chapter). They also appear to be intelligent in the way they move and what's more seem to recognize us, when in fact they do recognize us, (as will be explained). They are also what some have categorized as OBE's (out of body experiences). For those who have experienced this phenomenon, the experience is very real, and emotions are more alive and vivid. In this state, one can see many things that the physical eyes (that are bound to the body) would not normally see. It is an enhanced vision.

It manifests to look something like this:

Now, if the Merkabah vehicle becomes visible (not all do), this is what it generally looks like:

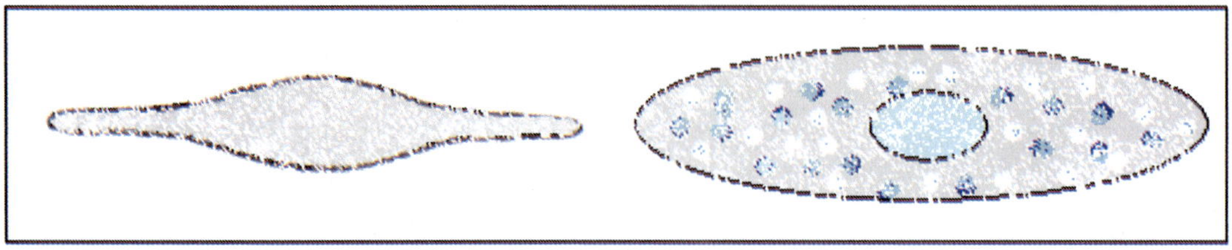

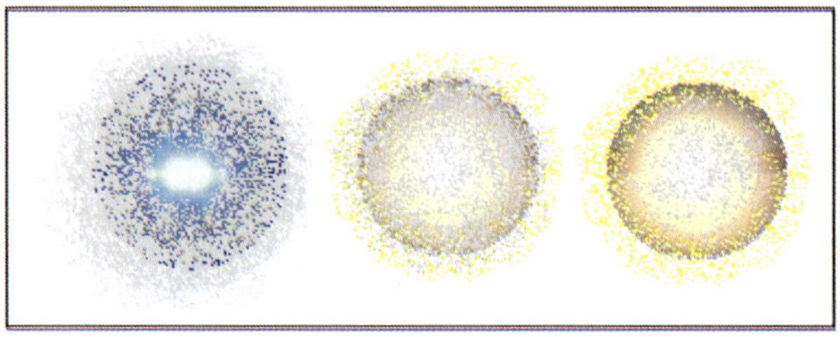

Notice how in one of the pictures it looks grey or silver (it's actually misty white) but when it reflects the light it gives off the appearance of metallic grey. Additionally, it is charged with energy that appears as spinning sparkles of neon blue combined with sparkling brilliantly white glitter (however, they can also appear in different colors which will be explained why in the next chapter). This is a common acute UFO sighting that many claim to see, but what turns out to be none less than someone's Merkabah vehicle (which is just as conscious as the individual person itself - as it is a reflected copy of their being).

By the way, "light" is the fastest thing that we know of in the universe, and the only explainable way for a vehicle to travel across the Universe with amazing speed, flitting here and there almost to the point where time is relevant, can only work by the transferring of light energy.

"Light" in itself (which consists of "protons") is an electromagnetic wave (radiation) in which Quantum Mechanics comes into play again (where protons can clone themselves). To top it all off, the photon is "mass-less" and has no electric charge. Additionally the photon (light particle) travels in the vacuum of open space at the speed of light and does not decay, however if it travels through forms of matter, it slows. The photon can be considered as a mediator because it can transfer energy from one state to another, (atoms jumping between energy states releasing radiation). Photons can also exist as a gas (photon gas) making it also a plasma - as vibrating photons are indeed expulsions of balls of plasma, (the Northern Lights also known as the Aurora Borealis are a good example of this phenomenon). This occurs when the Sun sends light (radiation and ions) it is trapped by the earths magnetosphere (a bow like shield surrounding the earth which blocks most of the radiation sent by the Sun). This radiation then become trapped and particles like electrons and protons collide with the molecules and gases of the Earths upper atmosphere and are carried by the magnetic field and solar wind (these particles are electromagnetically excited). In addition, when these collisions occur in our atmosphere they start to glow giving off a green, red, blue and violet light show across the night sky. (A cloud of such particles is called a plasma).

Therefore, it is quite possible to understand the mechanism that our human body can produce electromagnetic plasma (a gas made up of ions and electrons) as the result of over stimulation of chemical hormones produced by the brain and combined with the electricity flowing through our nervous system and spinal cord. Thus, the Merkabah vehicle is a product of this mental and spiritual state.

Lets examine this again in a different light… To clearly understand what is actually occurring here we must first realize that all material things are an illusion of empty space. Their solidity is an illusion created by the electromagnetic force field that binds atoms together. On the sub-atomic level, all interactions begin with the exchange of photons between the atoms of objects. But when we change our perception of the body (where most people think of their body as just flesh, bones and molecules), they neglect to realize that the body is also electromagnetic, and a body of light where everything that happens in our body is a result of the electromagnetic oscillation/vibration of photons. Given that, some say the laws of nature cant be broken! NOW by understanding this fact, how do we really know what the laws of nature are pertaining to the capability of transferring energy from the material body to outside its bounds? Hence Jesus' resurrection.

Reiterating this again to facilitate an even better understanding, the Merkabah is the vehicle of our soul outside the shell of our physical bodies. However, it is not the actual soul - but merely a copy or reflection of it. Releasing this plasma is a biological response involving the higher finer energies of thought and emotion, together synergistically with the biological chemical hormones produced between our body and brain, we are able to generate this power used to activate the "light vehicle / Merkabah". It is a light vehicle consisting of plasma (a gas made up of ions and electrons). As just mentioned above our Sun is plasma - and the Sun which we are all familiar with emits light and photons. The Merkabah vehicle is no different. However, in the physical world of matter a lot of power is needed to generate the fields. It is hard to move the dense particles of matter to the point needed to make the fields strong enough to detach from the body and pass thru the dimensional barriers. Although once one has broken through this barrier, they will additionally find out that they also have an impact over physical things (this includes the ability to move physical material objects and interact with the living). This sounds almost impossible or unbelievable, however, this is not something that only a few people are privileged to, or can acquire because they are gifted or psychic. I assure you that this is something that can certainly be conjured up and evoked by the chemical balance of hormones and frequency of the brain through sleep or a meditative process! It is, and I will emphasize a human potential - more commonly developed out of the need for survival. It is a state of mental confusion, depression or perhaps great exhilaration and thus a parting and separation away from the physical body occurs. Some people go through life never once experiencing this phenomenon, while others may experience this maybe only once or twice in their entire lifetime. In addition, there are others who continually experience this phenomenon. I myself have experienced it only three times in my life so far. It is much more stronger and powerful than the feeling of falling in love! It is a surge of power that is unexplainable! It moves with remarkable speed, and one experiences a sense of evanescence, ecstasy and utopia. However there is also a fear factor involved in it for those who experience it for the first time, which cuts the exploration short.

As complicated as I made it sound above, it is rather very simple to generate. Some have suggested that the beginning stages to developing the Merkabah are a natural part of our nature and start in the adolescent years. Some develop it and some don't - depending on the circumstances of ones emotional environment. It usually starts with the feeling of a "ripping sound separation" while on the verge of falling asleep or awakening during a dreaming episode (but while still asleep). It can also be achieved and witnessed during and after anesthesia or an illness. However it seems to happen more with people who go through episodes and periods in their life that consists of great turmoil or an over exhilaration of joyfulness (even spiritual-ness opens the door). The reason for this may be because it is a survival mechanism that helps us release the anguish, distress, suffering or over heightened joyfulness that is to much of an "emotional overload" for one to bare (almost like a glass of water spilling over). A release helps us cope and keep the body balanced both physically and emotionally. Aside from it happening naturally, there are numerous strategies one can use to evoke the Merkabah vehicle, but briefly, the most basic means for how to do this is as simple as "wanting it to happen" and knowing it can physically be done. Its as simple as concentrating on your body spinning around (both ways) vibrating to the point of a lightheaded wobbly dizzy spell while imagining yourself getting lighter and lighter and lifting and rising out of your body. But for some this can happen so fast that they just feel a whooshing or sucking out effect, or the ripping sound separation. Some even feel a suction sensation at the top of the head and suddenly find themselves looking down at their body. Some wander around in this state while others report seeing a tiny pinhole of light which gets bigger and bigger pulling one in. Usually the later does not happen as it means death may be looming.

Some have experienced it as a NDE (near death experience). However, science usher this experience away as a normal mechanism of the human body under such circumstances. This is of course true, IT IS A NORMAL MECHANISM OF THE HUMAN BODY! However, in order to prove the validity of its existence beyond the human body, it must be proven that what was "seen" was actually WHAT WAS WITNESSED by others. The scientist suggest that even when the entire body has shut down some can remember conversations and sounds - this may be because people might mentally reconstruct the conversation around them, or even images that are somehow internalized. Some scientists even theorize that the body may produce a flood of endorphins upon anticipation of death, (biochemical compounds produced by the pituitary gland in the brain). Endorphins are natural painkillers that produce a sense of well being which may be used by the body as a self defense mechanism to ease the body of the passage of death, (this sounds familiar in how the Merkabah vehicle is released by the synergetic stimulus of coagulated hormones). However, NDE survivors claim they can see with a birds eye view (crystal clear vision). There are many explanations that science can present, its even plausible that conversations and sounds can still be heard if the ear and brain are still somewhat functioning, but the ultimate question arises that science cannot just explain away - how one saw something (with their eyes taped shut during surgery) when they shouldn't have been able to see beyond the body? (And later verified by live witnesses to have actually happened the way the NDE saw it). Psalm 90:10 (which is talking about death) - The days of our years are threescore years and ten; and if by reason of strength they be fourscore years, yet is their strength labour and sorrow; for it is soon cut off, **and we fly away.**

Lets for a moment try an explain what "consciousness" is? Lets say (for what ever reasons) one finds themselves unconscious - the brain in no longer consciously awake. However, the body and its organs still seems to be functioning quite well and appear to have a different consciousness that directs the activity of the body. There appears to be two kinds of consciousness' occupying our body, one is a "higher consciousness" of the mind, and the other is a "lower consciousness" of the body. However there is also another form of consciousness that occurs at death when both the mind and body is beyond repair. This form of consciousness extracts itself slowly from the body and eventually all consciousnesses departs. Now when this happens, consciousness "reflects" a complete copy of itself outside the bounds of the physical body and death is final, then there is the judgment. Hebrews 9:27 - And as it is appointed unto men once to die, but after this the judgment. However, this is where things become tricky, don't forget there is also the "lower consciousness" that still remains while one is still alive and this "temporary separation" can also occur while one sleeps or during a meditative state. This reflection is also a kind of energetic echo containing a complete copy of consciousness, mind and memory outside the physical body, but will not amount to anything unless it is properly reintegrated back into the body and internalized (downloaded) into the memory banks of the brain. Now in order for the brain to properly download any memories that the reflected copy experienced while outside the bounds of the actual physical body, the conscious mind of the physical material body must stay awake in a semi-sleep state in order to be capable of recording what many of us witness as "dream memories". If these dream memories are not downloaded back into the memory banks of the brain effectively before one wakes, one will not remember their dream or OBE (out of body) experience. The reintegration, internalization and downloading result could be the reason why most people fail to remember their dreams or recognized that they had dreamed at all. Through this very concept, perhaps spirituality can teach a thing or two to science. Some of these "dream memories" are easily differentiated as nonsense while others a more real and vivid experience. However, to know the difference, it is always the ones that can be vividly remembered as flying or suddenly awakened by a falling feeling, or actually seeing oneself sleeping outside their actual body (this is the detachment stage, but doesn't always follow through to producing an actual Merkabah vehicle). Sleep specialist tell us that during sleep the familiar feeling of rising, falling and sudden jolts are just the nervous system shutting itself down for sleep. They account it all to brain chemistry and nerves. When in fact this is true to an extent! We know that "chemical hormones" play a role in releasing this state of being thus allowing us to achieve the Merkabah vehicle. But what science has not yet learned to "separate" is something that is created that goes "beyond" what the chemical hormones have induced, to a state of tangible scientific exploration of the invisible forces that allows for a state of mind/body/spirit separation. This also allows for Quantum Mechanics again to squeeze its way into the resolution (of the separations of energies, cloning of particles and connectivity with each other over vast distances) when we consider that atoms can be in multiple places at once, thanks to natures variances of "non locality" (as discussed in an earlier chapter) bio-location of ones being becomes possible outside the realm of the physical body.

However I must strongly emphasize that there is a downside to all this, even though this is a natural phenomenon, IT IS NOT SUGGESTED TO INTENTIONALLY PROVOKE OR ROUSE UP! It is not natural for the body to continually experience these episodes and fluctuations of chemical hormones. It is a transitional energy state with the central purpose of easing the passage of final death. We also know that the apostle Paul cites in the New Testament about an out of body experience. However, he tells us that there is nothing to be gained by having these experiences and that our faith should not rest on our own wisdom, but only on G-d's. In other words, it is futile to search for infinite wisdom inside (or outside) the limited mind of man. 2 Corinthians 12:1-4 - I must go on boasting. **Although there is nothing to be gained,** I will go on to visions and revelations from the Lord. I know a man in Christ who fourteen years ago was caught up to the third heaven. **Whether it was in the body**

or out of the body I do not know—G-d knows. Another reason we should not intentionally provoke or rouse up the Merkabah vehicle is that it can open and expose us to evil spirits. (This will also be addressed again in a future chapter).

Back on other matters… Additionally, many countries use psychic spies or clairvoyants called "remote viewers" to spy on other countries or individuals. In essence, the basic mechanism to how they obtain information is through the Merkabah vehicle (which can be likened unto the third eye). The process works the same way. Incidentally, many of these people have obtained their ability either through biological inheritance or an injury to the head, or life threatening injury to the body. Usually it's the head when it's knocked out of equilibrium (balance). Additionally, many of the crop circles that we see may have actually been produced by somebody's Merkabah vehicle, that's the reason why they are geometric in formation, because the Merkabah is the "Star of David" thus formed from the Tetrahedron, but to the human eye, the Merkabah vehicle appears spherical in shape. The Merkabah vehicle is also deemed as the "third eye" and is of a creative nature outside the bounds of the physically limited body. It can see and hear more vividly, in both color and shape with an extreme acuity to depth and height perception. In its own meticulous fashion, it is a mathematical/geometric genius! Moreover, since it is plasma it would be electromagnetic in nature. However, some crop circles are fakes, but they are produced by humans in a like manner out of creativity. Some fail to realize that crop circles (in past history) have been recorded all around the world (even China) since the dawn of man, which are opt opt to be the genuine ones.

Back on the direct subject of the body and Merkabah connection along with its ability to discharge a separating energy… Some say this is all bunk and a figment of ones imagination! They laugh this experience off as hallucinations and the mind playing tricks on itself. On the contrary, certainly not for those who have already experienced this phenomenon. However as laughable at it may all seem, we might begin to think otherwise when we take a closer look at what nature and science has to reveal to us concerning this geometric structure (of the "Merkabah" and the "Star of David") and its importance in the structured organization of all life.

Firstly, as I mentioned earlier above, the "Star of David" is formed by two interlocking triangles/pyramids. The middle section in which they share is also a "Hexagon", and the Hexagon as we've already learned is the strongest form in nature. However (as I also mentioned above) some suggest that it is the triangle itself. I have also heard it is the sphere, but a sphere reduces itself to the atom and even atoms are not spheres but are particles held together by the electromagnetic field. It's interesting that in all the various forms manifested in reality, the atoms themselves always line up in specific ways when they take shape. They form into a Tetrahedron first, and then that Tetrahedron 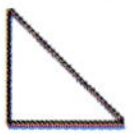 puts another Tetrahedron next to itself and it is now a Cube,  and the Cube from that stage turns into a Octahedron and then into a Icosahedron, and lastly a Dodecahedron. (As shown directly below):

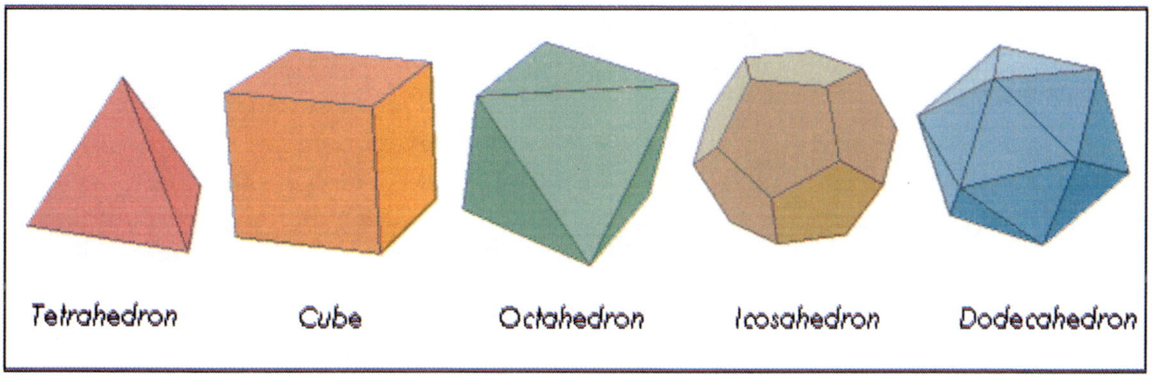

observed directly below. It becomes even more confusing when the “Octahedron” (3rd platonic solid) turns into a Hexagon when it is viewed diagonally.

Now in relations to the first stages of a Tetrahedon turning into a Cube, this is where things get tricky because the Tetrahedron can look like a Cube (even a Hexagon) depending on “ones point of view” as observed directly below. It becomes even more confusing when the “Octahedron” (3rd platonic solid) turns into a Hexagon when it is viewed diagonally.

(Traversing from Tetrahedron to Cube to Hexagon to Octahedron) depending on ones point of view:

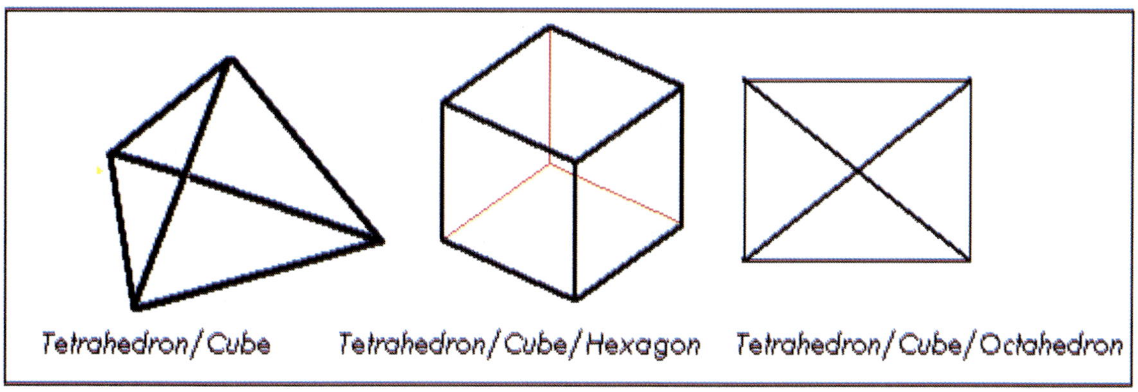

Technically from left to right it should be a Tetrahedron, a Cube and a Octahedron, and the Cube can also be seen as a Hexagon, but they are all produced from the very first platonic solid the Tetrahedron. (Point of view is very important here). However, it is especially important in understanding the very first geometric platonic solid, which is the Tetrahedron because it may actually be what is generally known about the “atom“. Nobody as yet truly knows what an atom is other than it consisting of three parts (proton (+), neutron (-) and electron (-). The atom is not even an sphere/circle (as previously thought in science), but particles held together by the electromagnetic field. The atom in its triune nature (3 in one) may actually be a standing frequency wave itself 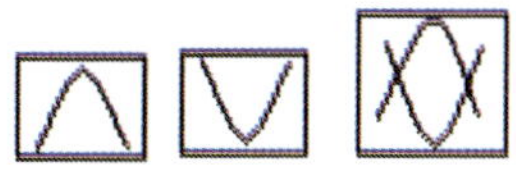 (a Tetrahedron in the beginning stages of forming the organization of material structure). And one might ask why is this? Because, a sphere has equal points of gravity (in which to keep it round). Logically, if it only had two points it would not be able to maintain being spherically round, it would also be easily destructible, but three points of gravity (where gravity balances out) makes it the strongest structure in all of creation. And the Hexagon (consisting of 6 sides) is an equal amplified growth from the main structure of a Tetrahedron.

A standing wave viewed in other ways:

Now this is where things REALLY become overwhelming… Not only can we barely grasp the concept of an atom being that of a Tetrahedron. It may even be conceivably overwhelming in realizing that the atoms true triune nature (3 in one) may actually be the synergetic forces at work of G-d the Father, the Son and the Holy Spirit working together as one.

However, it doesn't stop there! The magnificence of the surmounting organization of a Tetrahedron can be witnessed in the division of a fertilized human egg forming the first stages of life. Though this division is just not subject to humans, when in fact every single life form that we know passes through this stage. Romans 1:20 - For since the creation of the world G-d's invisible qualities—his eternal power and divine nature—have been clearly seen, being understood from what has been made, so that men are without excuse.

A 1 day old (24 hour) embryo contains 2 cells. A 3 day embryo contains 4 cells, and a 4 day old embryo contains 8 cells. On the 5th day, the embryo turns into a "blastocyst" which has so many cells that individual cells are no longer recognizable. And on the 6th day, the embryo implants itself into the uterine wall.

These are the beginning stages, starting out as an egg sphere with three gravitational balanced points thus turning into a Tetrahedron, which is the first stage of cell division.

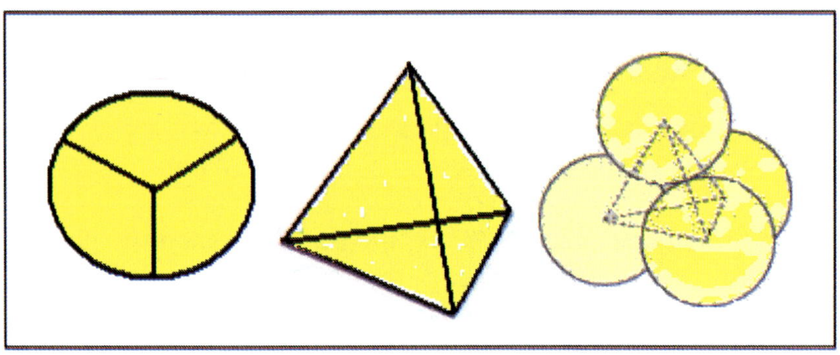

The 6th and 7th day, which form the “Star of David”, is a special stage! Because at this point in time the cell division commences (begins) soon after fertilization and proceeds rapidly into a “blastocyst“. The embryo is identified as neither a fertilized egg nor an equal multi-cellular embryo; it is now regarded as a blastocyst consisting of many cells, which indicates the beginning between the embryonic organisms to that of a human person in potency. It is also by the adhesion of the blastocyst to the uterine wall that it receives the oxygen and the nutrients that it needs to be able to grow. But before it can perform implantation, the blastocyst first needs to get rid of its strong membrane that formed around the egg that developed in the ovary. This process is called “hatching“. Following this, the 7th day is a day of rest sanctioning this interval (it is another emulated form of microcosm). Genesis 2:2 - And on the seventh day G-d ended his work which he had made, and he rested on the seventh day from all his work which he had made.

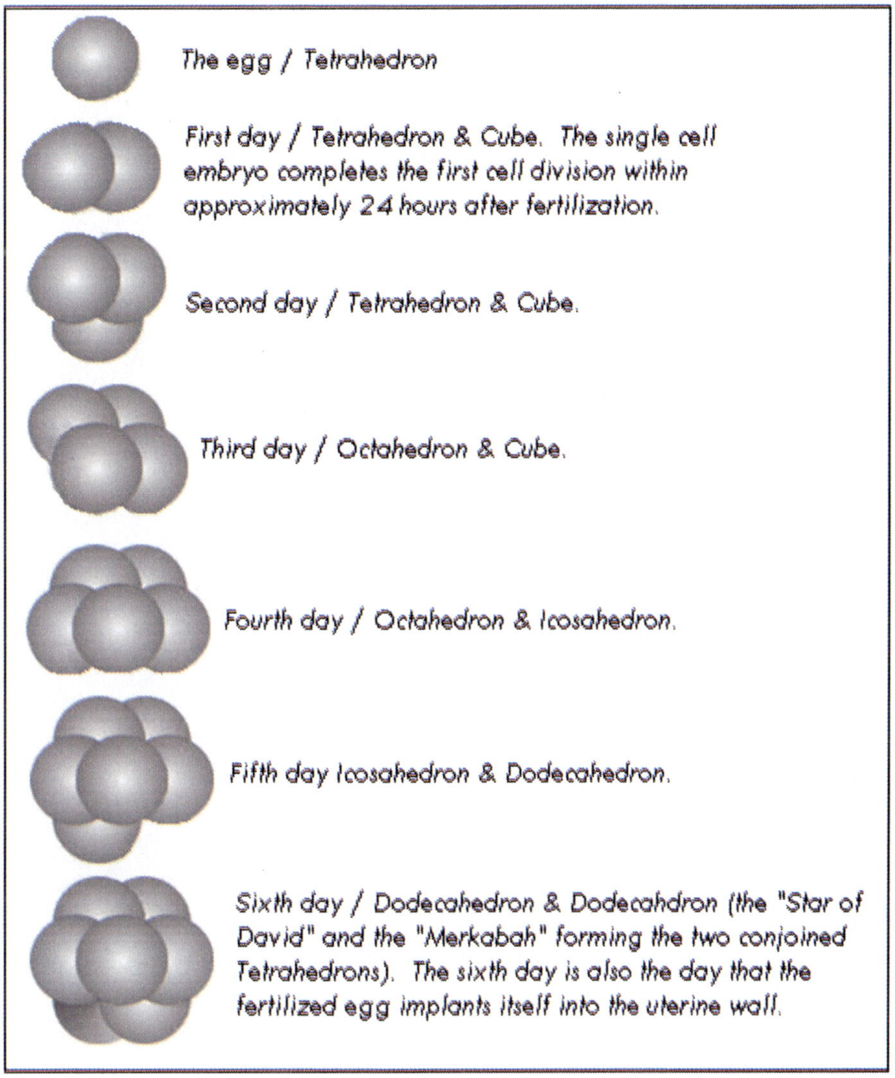

It can also be viewed like this:

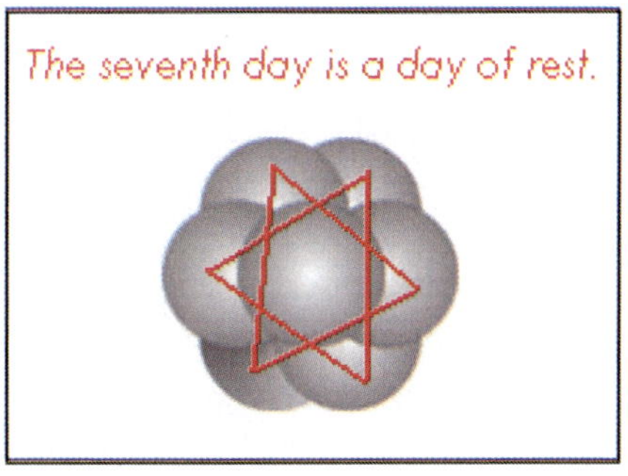

Not to distract from the subject, but underneath this whole process there is also another mysterious observable fact that science has recently discovered that was designed into the human body that clearly tells us when it comes to offspring and parenting that G-d is in favor of marriage and long lasting relationships. Yes, that right! The Bible clearly tells us this straight from the introduction of Genesis. Genesis 2:18 - And the Lord G-d said, it is not good that the man should be alone; I will make him and help meet for him. Genesis 2:21-24 - And the Lord G-d caused a deep sleep to fall upon Adam, and he slept: and he took one of his ribs, and closed up the flesh instead thereof; And the rib, which the Lord G-d had taken from man, made he a woman, and brought her unto the man. And Adam said, This is now bone of my bones, and flesh of my flesh: she shall be called Woman, because she was taken out of Man. Therefore shall a man leave his father and his mother, and shall cleave unto his wife: and they shall be one flesh. This is evidently the first wedding (jointly and in cooperation as one). We can conclude from Genesis that marriage is G-d's divine design for us. We can also make out that G-d's design for marriage is companionship and intimacy. But little do we know that G-d designed marriage to be more than just companionship and intimacy, He additionally designed marriage between a man and a woman to be an instrument to make us more "virtuous" than just happy and content. Is it possible that there is something much more profound in the heart and mind of G-d then simply to make us feel satisfied? We quickly realize that the infatuation part sooner or later starts to fade away with time and we begin to experience new emotions towards marriage. We find that we often put aside our ambitions and dreams to love and serve our partner. We also learn about unconditional love, respect, honor, obligation, commitment, responsibility, compelled and duty-bound to make ends-meet for survival, and most importantly how to forgive and be forgiven. Through all of these qualities, we can see our weaknesses and grow from that insight. We utterly develop a servants heart and draw ourselves closer to G-d. As a result, true and complete soul happiness can be achieved, and this I believe is one of G-d's ultimate desires and purposes for designing the covenant of marriage. Here are a few more Bible verses that indicate about proper and pure marriage. Deuteronomy 23:2 - No one born of a forbidden marriage nor any of his descendants may enter the assembly of the Lord, even down to the tenth generation. Hebrews 13:4 - Marriage should be honored by all, and the marriage bed kept pure, for G-d will judge the adulterer and all the sexually immoral.

Back to the blastocyst… Now an interesting thing happens, when we observe the attachment of the blastocyst to the uterine wall, it becomes a model of a compromising diplomacy. This is where everything takes root in the mother with actions so aggressive that some biologist likens it to a cancer invading the body. In order for the blastocyst to gain a supply of oxygen and nutrients it grows a placenta that adheres and attaches itself onto the blood vessels of the uterine wall and then begins shedding foreign cells into the mothers bloodstream. When this occurs, normally the mothers immune system does not accept a foreign body growing inside her, but miraculously in pregnancy it does. The reason for this actually happens "before" pregnancy occurs, from "repeated exposure" to the fathers semen, which allows the mothers immune system to learn to recognize the fathers foreign proteins, thus sending a message to the mothers immune system that this particular foreign body (protein) poses no harm. This civil proposal occurs with each act of intercourse where white blood cells from the mothers immune system cleans as well as searches her cervix for the mans foreign proteins. These proteins are then recognized by the mothers body and absorbed into the blood stream through the lymph nodes where her immune system gradually learns to recognize and tolerate them (a sort of peace negotiation). THIS IS REALLY THE KEY TO THE whole IMMUNOLOGICAL PROCESS! (The way the immune system works in the body, including allergies, resistance to disease, and acceptance or rejection of foreign tissue). This alone is scientific proof that G-d is in favor of marriage and long lasting relationships. However, some might be thinking that pregnancy can happen only after having sex "one" time. This is correct, yes, it can happen. But what's actually happening here (the first time) is that the sperm is attacked and depleted on their way to the egg (which leaves less viable sperm to fertilize the egg). Meanwhile those who have been in a long term relationship and have engaged in sex more than once thus allowing the body to build immunity to the foreign protein of the sperm have a better chance (in numbers and available choices of healthy sperm) at reaching the egg. Therefore, a greater guarantee at getting pregnant results.

Now, getting back on course to further expand on the importance of the geometric shape of the Tetrahedron. Remarkably, the Tetrahedron also emulates a water molecule. In addition, without this vital geometric structure of the Tetrahedron in the water molecule, absorption of nutrients could not take place within all living things. As the water molecule itself contains the same exact structured shape of a Tetrahedron consisting of two hydrogen atoms covalently bonded to a single oxygen atom. The hydrogen atoms bond to the oxygen atom because they share electrons. (Cited below)

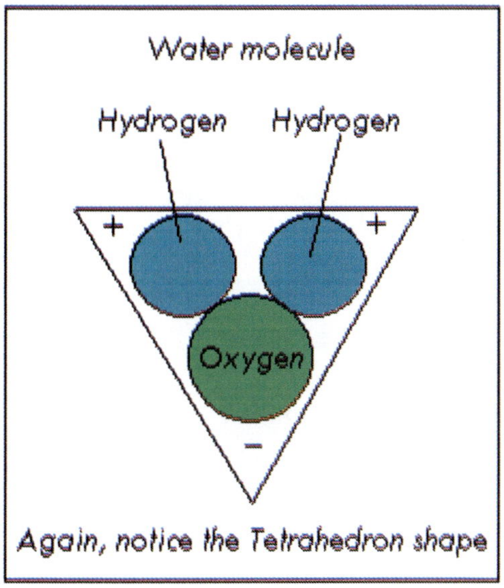

In addition, lets not forget this picture:

As we discussed in an earlier chapter this frequency and its form has the audible sound of Ahhh, which is the Hebrew letter "A" (aleph). In addition, when you make this sound, the form appears. In addition, remember, it is the "resonance" of this sound that can restructure all living cells back into its correct form. Additionally the very structure of the Tetrahedron/Star of David/Merkabah in nature is the "key" that is used to facilitate absorption of nutrients through water, and therefore in its alter ego, it is the Merkabah vehicle that our souls facilitate to cross the threshold to other regions and realms. We must remember that "divine design" never changes - it is yet but another microcosm linked in some way to the workings of G-d!

Therefore, in conceptualizing that this structure composes many things… Now, I'm going to advance all readers to an even higher level of eternal understanding of the Merkabah by unveiling its true mysterious meaning and its linkage to G-d both through nature (divine design), and the Bible.

In remembering all that was discussed in several of the first chapters regarding, "Searching for similarity" and "Noticing the network", the concept was introduced that all things despite their shape and form have the same growth pattern in common, I quoted it as the "one law" (also known as the "Universal Law"). Now we are going to go in search of that one seed, the Golden Fleece, the Chalice! The founding father to all geometric foundations. It was in my very observance of all these encompassing things and in realizing that there were so many of them - more than I could count or put in this book. I soon realized that I didn't have to count them all because everything deducted to "one". Everything was of one nature, and that one nature was G-d. I soon recognized this when I went looking for the secret of that one thing in other things, it became clear that when I could not understand the secret from the first thing, G-d, in His infinite mercy repeated the secret in greater detail in all things. I resorted to the Bible immediately once again! Wouldn't it be amazing if I could find some sort of message like this incorporated within the Hebrew words, authenticating the very nature of G-d and all of His creation? A linkage proving that He is "one in the same" with all life, as we know it. I began to think wildly - suppose the organization of all life was premeditated and carefully calculated and coded by G-d and placed within the very fundamental beginnings of Genesis 1:1 describing creation - in seeing that the first letter was like the first word and how both were like the entire first verse (as was pointed out in a previous chapter). The first letter is "Bet" symbolizing a "house" and is translated to mean "in", as if it were inviting us to enter. I focused on the first Hebrew word again of Genesis 1:1, "B'reshith" which is commonly translated to mean, "In the beginning". However this is not the only possibility as Hebrew is a "conceptual language" and if we break the word B'reshith down into smaller parts, we get the word "Bara" with means create, "Esh" which means fire, and the word "Shith" which signifies something sharp or pointed like a stem, thorn or spike, which is identifiable with a Tetrahedron, THE FIRST PLATONIC SOLID!

I had discovered that G-D HAD FLAWLESSLY PROGRAMMED ALL LIFE WITHIN THE UNIFORMITY OF A TETRAHEDRON (and science shows us the proof of this in all life forms). I was wordless to unearth this!

What does all of this prove? It not only describes G-d having a relationship with nature and creation set forth into letters which combine into words, but it really does define and describe creation in the same way that atoms form molecules, which in turn form cells, and eventually combine to form an organism, likewise the growth sequence of the platonic solids. Competently it shows us this same progression by the first word B'reshith, by forming the very first platonic solid, (the Tetrahedron) which is the foundational basis and initial instructional mechanism to all life. This powerfully indicates that the author of Genesis 1:1 is most certainly who He proclaims to be - G-d… and is literally revealing and sharing with us the mystery of all creation within the beginning verse of the Hebrew Bible. He is revealing the confirmation of His existence and handy work to all who seek Him along with the validity of the Bible. He continually strengthens the fact that we must continue to uphold His commandments and His Bible for the complete understanding that there is something more to the concept of His "sacred language of the Hebrew letters" (the Primordial Language) that allows us to connect with Him, and to know that he alone is the almighty creator.

I thought to myself, by what stretch of the imagination could I have stumbled upon something of this immense importance? The magnitude was staggering! WOW, PROOF OF G-D! This was mind-blowing and I figured it was time to write a book! This is something that science cannot just chalk off. It wholly amazed me how G-d had incorporated and integrated this working structural mechanism into His handy work, and to top it off, had left it blatantly noticeable for observation, examination and study within the penned works of the Bible and throughout all nature - and accordingly in its proper place where one would generally go looking for such information - the beginning (Genesis 1:1) - hence B'reshith!

Wait, there is more! As one reads and learns more about the "Star of David" (also known as the Merkabah) they will realize that its symbolic meaning remains universal and a representational message of G-d's plan to all humankind. As we follow its history into modern times, the "Star of David" has become a specific Jewish symbol and the flag symbol of the State of Israel. However even though the state of Israel flies this flag they neglect to realize that not only is it for the kingship of King David, but also for the coming Messiah since the prophets had written that the Messiah would come from the lineage of David, so we know that the symbol was not just specifically for David himself, but as a future symbol for the Messiah, who would come from the family line of David and completely surpass David. Jesus Himself testifies that He is this "Star of David" in Revelations 22:16 - I, Jesus, have sent my angel to give you this testimony for the churches. **I am the Root and the Offspring of David, and the bright Morning Star.**

Lets also examine what the Old Testament states about the "House and Star of David". Zechariah 12:8 - In that day shall the Lord defend the inhabitants of Jerusalem; and he that is feeble among them at that day shall be as David; **and the house of David shall be as G-d, as the angel of the Lord before them.** Zechariah 13:1 - In that day there shall be **a fountain opened to the house of David and to the inhabitants of Jerusalem for sin and for uncleanness.** Numbers 24:17-19 - I shall see him, but not now: I shall behold him, but not nigh: **there shall come a Star out of Jacob,** and a Sceptre shall rise out of Israel, and shall smite the corners of Moab, and destroy all the children of Sheth. And Edom shall be a possession, Seir also shall be a possession for his enemies; and Israel shall do valiantly. **Out of Jacob shall come he that shall have dominion,** and shall destroy him that remaineth of the city. Isaiah 22:21-22 - And I will clothe him with thy robe, and strengthen him with thy girdle, and **I will commit thy government into his hand: and he shall be a father to the inhabitants of Jerusalem, and to the house of Judah. And the key of the house of David will I lay upon his shoulder; so he shall open, and none shall shut; and he shall shut, and none shall open.** 2 Samuel 7:12-16 - And when thy days be fulfilled, and thou shalt sleep with thy fathers, I will set up thy seed after thee, which shall proceed out of thy bowels, and I will establish his kingdom. He shall build an house for my name, and I will stablish the throne of his kingdom for ever. I will be his father, and he shall be my son. If he commit iniquity, I will chasten him with the rod of men, and with the stripes of the children of men: But my mercy shall not depart away from him, as I took it from Saul, whom I put away before thee. **And thine house and thy kingdom shall be established for ever before thee: thy throne shall be established for ever.** Psalm 89:3-4 - I have made a covenant with my chosen, I have sworn unto David my servant, **Thy seed will I establish for ever, and build up thy throne to all generations.** Selah. Psalm 132:11 - **The Lord hath sworn in truth unto David; he**

will not turn from it; Of the fruit of thy body will I set upon thy throne. Luke 1:69-70 - **And hath raised up an horn of salvation for us in the house of his servant David; As he spake by the mouth of his holy prophets, which have been since the world began.** John 7:42 - **Hath not the scripture said, That Christ cometh of the seed of David, and out of the town of Bethlehem, where David was?** Hosea 3:4-5 - **For the children of Israel shall abide many days without a king, and without a prince, and without a sacrifice**, and without an image, and without an ephod, and without teraphim: **Afterward shall the children of Israel return, and seek the Lord their G-d, and David their king; and shall fear the Lord and his goodness in the latter days.**

The "Star of David" is clearly the star of Jesus the Messiah from the family line of King David. (Additional proof of Jesus' lineage to King David will follow in a future chapter). Furthermore in the New Testament Matthew tells us that the generation of the Messiah which spans 14 generation from Abraham to David, and from David to the deportation to Babylon another 14 generations and from the deportation to Babylon to the Messiah an extra 14 generations. Matthew 1:17 - So all the generations from Abraham to David are fourteen generations; and from David until the carrying away into Babylon are fourteen generations; and from the carrying away into Babylon unto Christ are fourteen generations.

Now, as illustrative a message as it may seem, an amazing thing happens when we take the "Star of David" and the Hebrew letter Vuv (which actually looks like a number 7, but also holds the Gematria number equivalent to 6) (you can look back at the Gamatra chart I supplied in a previous chapter to verify this). So we get 6 sevens (or Vuv's): (7+7) - (7+7) - (7+7) = 14 +14+14 generations… 6x7=42 generations. 42 symbolized the Messiah as well as the Star of David! (The 42 generations will reveal itself in greater detail in a coming chapter).

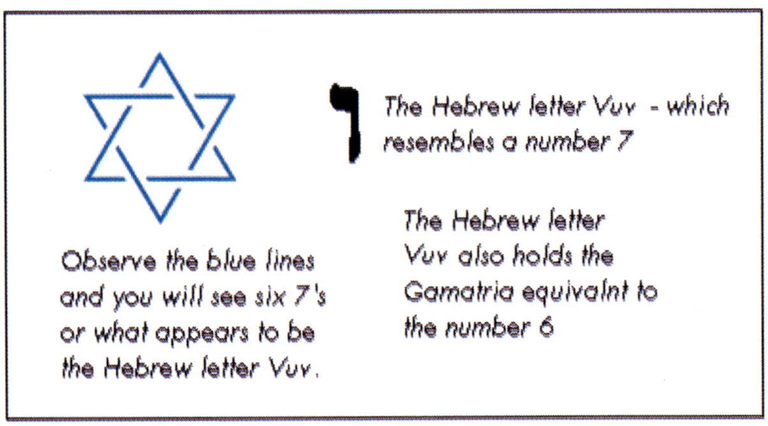

The Hebrew letter "Vuv" holds a special meaning. Its given representation is that of a "hook" because it looks like one. However, there is a more mysterious significance to it. When one studies the "order" of the Hebrew letters of the Old Testament Bible (Torah) starting at the very beginning of Genesis 1:1 they soon begin to realize that it is just not a book that unceremoniously tells a story, but a book that places each letter and each word in its precise order for an all-encompassing purpose - like pieces of a puzzle that fit together. With G-d, there is a purpose for everything, and each Hebrew letter was given a specific meaning for mysterious reasons only known to G-d. The mysterious significance to the letter "Vuv" is that it has the Gamatria value of the number 6 (which is revealed below). However, herein lies one of the greatest mysterious to "WHY" the Hebrew letters were established for representing the numbers that were assigned to them. The significant purpose for why the Hebrew letter "Vuv" has the Gamatria value of the number 6 is because it is the first and foremost initial "Vuv" of the Torah which is first sighted in Genesis 1:1 (shown below in Hebrew and English). The letter "Vuv" looks the way it does because this is how it was actually written by the finger of G-d on the first tablets of the 10 commandments (which were destroyed by Moses in his anger when he heard and saw the corrupt reveling of the Jews below) and later transcribed to Moses a second time from G-d onto the second tablets. Exodus 31:18 - When the Lord finished speaking to Moses on Mount Sinai, he gave him the two tablets of the Testimony, the tablets of stone inscribed by the finger of G-d. Deuteronomy 9:10 - The Lord gave me two stone tablets inscribed by the finger of G-d. On them were all the commandments the Lord proclaimed to you on the mountain out of the fire, on the day of the assembly. Exodus 32:16 - The tablets were the work of G-d; the writing was the writing of G-d, engraved on the tablets. Exodus 32:19 - When Moses approached the camp and saw the calf and the dancing, his anger burned and he threw the tablets out of his hands, breaking them to pieces at the foot of the mountain. Exodus 34:1 - The Lord said to Moses, Chisel out two stone tablets like the first ones, and I will write on them the words that were on the first tablets, which you broke. Exodus 34:28 - Moses was there with the Lord forty days and forty nights without eating bread or drinking water. And he wrote on the tablets the words of the covenant—the Ten Commandments.

The first "Vuv" of the Torah (as seen below) is found in Genesis 1:1 - In the beginning G-d created the heavens (and the) earth. The Hebrew word for the words (and the) is V'et. The word V'et SERVES TO JOIN EARTH AS MATTER AND HEAVEN AS SPIRIT that pervades throughout all creation. Additionally, it is not a coincidence that the letter "Vuv" first makes its appearance as the 6th word of the Torah, and it is additionally also the 22nd letter of the verse (reading from right to left) which refer to the full alphabet of the Hebrew letters - 22 in all. Even more amazing, the two letters that follow after the Hebrew letter Vuv, (which are "Aleph and Tov") is generally taken to represent ALL the 22 letters of the Hebrew alphabet, which also includes all the material and spiritual substances of creation present within Heaven and Earth (fire, water, air and earth). Moreover, if you recall from a previous chapter, all the letters of the Hebrew alphabet from Aleph to Tov = "Ahtah" (G-d's accurate alphabetical name) which are all the letters of the alphabet. By this I mean, in the arranged order of alphabetic letters.

The number 6 also applies to the six days of creation and six letters in the very first word of the Torah (B'reshit) (Bet, Resh, Aleph, Shin, Yod and Tov) as well as the six Alephs in the first verse of the Torah. And the six cubits - the dimensions of the tablets received by Moses at Mount Sinai.

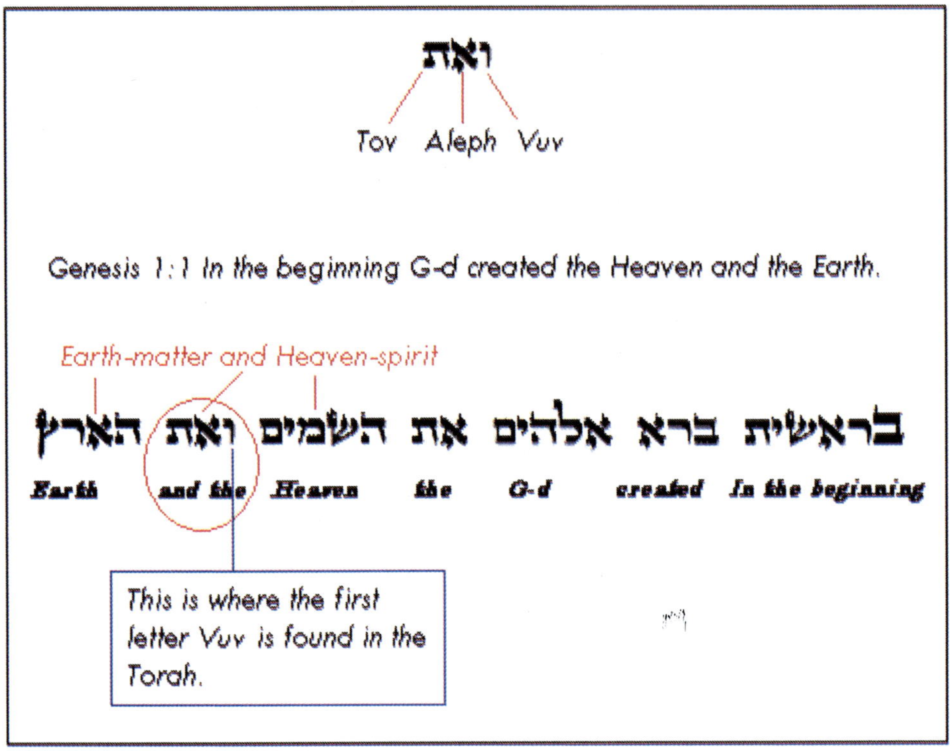

We are not stopping here! There is also another numerical observations of the "Star of David" that equates to Jesus… Something else interesting happens when we write the successive numbers next to the points and corners of the "Star of David". By adding the numbers of any one of the opposite points or corners it always gives us the sum of 14. For instance (1+13=14, 3+11=14, 6+8=14). Additionally when we add up the numbers of the "six points" it also renders 42 (1+5+12+13+9+2) = 42. And the same with the "inside corners" (3+4+8+11+10+6=42). Even if we add the numbers "diagonally" we come up with 21, but 21+21=42 again. All avenues result in 14.

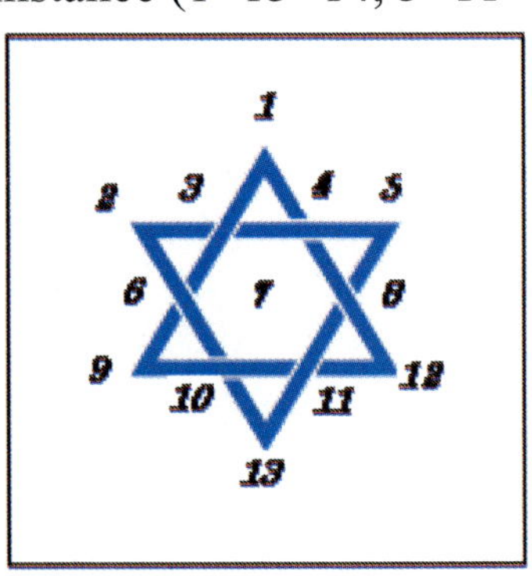

The number 14 also happens to be the numerical Gamatria value of the name David דוד (4+6+4 = 14). And remember it is also the same number 14 (in reference to King David and the coming Messiah) that turns into 42 generations (14+14+14=42). Matthew 1:17 - So all the generations from Abraham to David are **fourteen generations**; and from David until the carrying away into Babylon are **fourteen generations**; and from the carrying away into Babylon unto Christ are **fourteen generations**.

So what does all this suggest? As it turns out the "Star of David" is a numerical indicator to the inner meaning of the term "Magen David" מגן דויד "Magen" (Mem, Gimel and Nun), and "David" (Daled, Vuv and Daled) (remembering to read from right to left). The Hebrew word "Magen" means shield. Therefore the Hebrew term for the "Star of David is Magen David" which means in English "Shield of David". Psalm 115:9-11 - O Israel, trust thou in the Lord: **he is their help and their shield.** O house of Aaron, trust in the Lord: **he is their help and their shield.** Ye that fear the Lord trust in the Lord: **he is their help and their shield.**

Additionally to top it all off, the Hebrew letter "Nun" נ is the 14th letter of the Hebrew alphabet - which as we know also equals (King) David's name. King David who is the progenitor of the Kingdom of Israel, and the successor to David - the Messiah (Jesus). The Hebrew letter "Nun" typically means a "fish" (hence the symbol of the fish for Jesus), but it is also known as the kingdom of the Messiah (the successor of King David's throne). The "Nun" like the "Vuv" has revealed their mysterious reasons for their numerical values. Another meaning is attached to the Hebrew letter "Nun" because of its shape and form - as it resembles a bent over vessel (or a bent over servant). It is the receiving vessel - the Messiah.

Lastly, as was formerly and briefly mentioned (if you can recall) in an earlier chapter that the last book of the New Testament Bible (Revelation) written by the disciple John tells us about a new Heaven and a new Earth, referred to as the "New Jerusalem" coming down from Heaven. Upon reading Revelations, it becomes apparent that the description given for the new Heaven and the new Earths is actually a recognized illustration of the "Star of David". Revelation 21:1-2 - **And I saw a new heaven and a new earth: for the first heaven and the first earth were passed away;** and there was no more sea. **And I John saw the holy city, new Jerusalem, coming down from G-d out of heaven,** prepared as a bride adorned for her husband. Revelation 21:10-27 - And he carried me away in the spirit to a great and high mountain, **and shewed me that great city, the holy Jerusalem, descending out of heaven from G-d,** Having the glory of G-d: and her light was like unto a stone most precious, even like a jasper stone, clear as crystal; **And had a wall great and high, and had twelve gates, and at the gates twelve angels, and names written thereon, which are the names of the twelve tribes of the children of Israel: On the east three gates; on the north three gates; on the south three gates; and on the west three gates. And the wall of the city had twelve foundations, and in them the names of the twelve apostles of the Lamb.** And he that talked with me had a golden reed to measure the city, and the gates thereof, and the wall thereof. **And the city lieth foursquare, and the length is as large as the breadth: and he measured the city with the reed, twelve thousand furlongs. The length and the breadth and the height of it are equal. And he measured the wall thereof, an hundred and forty and four cubits, according to the measure of a man, that is, of the angel.** And the building of the wall of it was of jasper: and the city was pure gold, like unto clear glass. And the foundations of the wall of the city were garnished with all manner of precious stones. **The first foundation was jasper; the second, sapphire; the third, a chalcedony; the fourth, an emerald; The fifth, sardonyx; the sixth, sardius; the seventh, chrysolyte; the eighth, beryl; the ninth, a topaz; the tenth, a chrysoprasus; the eleventh, a jacinth; the twelfth, an amethyst. And the twelve gates were twelve pearls:** every several gate was of one pearl: and the street of the city was pure gold, as it were transparent glass. And I saw no temple therein: for the Lord G-d Almighty and the Lamb are the temple of it. And the city had no need of the sun, neither of the moon, to shine in it: for the glory of G-d did lighten it, and the

Lamb is the light thereof. And the nations of them which are saved shall walk in the light of it: and the kings of the earth do bring their glory and honour into it. And the gates of it shall not be shut at all by day: for there shall be no night there. And they shall bring the glory and honour of the nations into it. And there shall in no wise enter into it any thing that defileth, neither whatsoever worketh abomination, or maketh a lie: but they which are written in the Lamb's book of life. Revelation 22:1-5 - And he shewed me a pure river of water of life, clear as crystal, proceeding out of the throne of G-d and of the Lamb. In the midst of the street of it, and on either side of the river, was there the tree of life, **which bare twelve manner of fruits, and yielded her fruit every month: and the leaves of the tree were for the healing of the nations.** And there shall be no more curse: but the throne of G-d and of the Lamb shall be in it; and his servants shall serve him: And they shall see his face; and his name shall be in their foreheads. And there shall be no night there; and they need no candle, neither light of the sun; for the Lord G-d giveth them light: and they shall reign for ever and ever.

It can be viewed as a Tetrahedron, Cube and a Hexagon and Octahedron (remember, depending on ones point of view), and two interlocking Tetrahedrons forming a "Star of David".

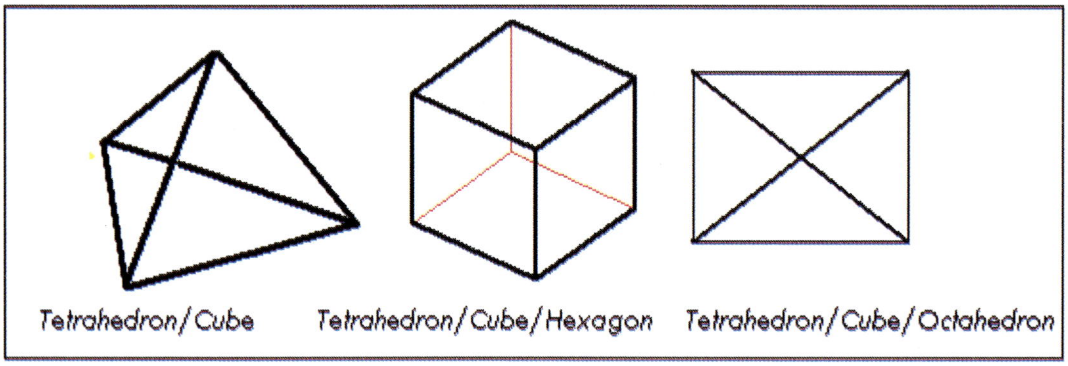

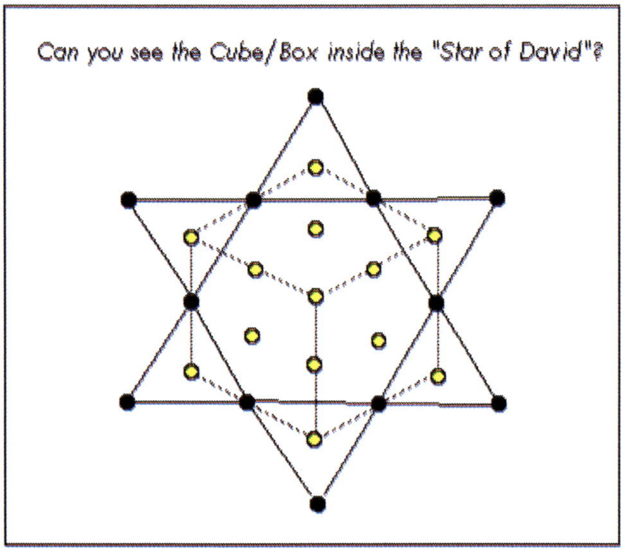

Closing notes: The Merkabah is the vehicle of our soul outside the shell of our physical body. Not only is it a real manifestation that occurs by the transference of energy from the internal to the external bounds of the body, but we also have actual proof that this "reflected copy" (separation and transference of energies) occurs, and that seeing and hearing outside the bounds of the physical body is a very real phenomenon - since during surgeries (while people have been under anesthetic) there have been actual and authentic accountable facts that have been verified by numerous witnesses of the medical field. We can additionally see this mechanism at work in the division stages of a human egg splitting in two and "making a reflection of itself" on its way to forming a living material thing - which is the full "Star of David/Merkabah" on the 6th day. The "Star of David"/Merkabah is also a mathematical representation and a symbolic illustration of the (Magen David) as the Messiah - Jesus as the biblical account of Matthew 1:17 - So all the generations from Abraham to David are **fourteen generations**; and from David until the carrying away into Babylon are **fourteen generations**; and from the carrying away into Babylon unto Christ are **fourteen generations**. Whilst the Hexagon structure (two conjoined or intertwined triangles) is the strongest form in nature (a shield). The two are one in the same. It is also spoken about, and eloquently described for us in Revelations of the New Testament, which describes it as the New Jerusalem (Revelation 21 and 22).

It was also as discussed in a previous chapter of the "Primordial Language" as a symbolic representation of the Pomegranate that contains exactly 613 seeds which is the exact total number of commandments found within the Bible.

Uncanvassed in its true light, it is indeed G-d's trademark! It is also by "divine design". It was here since time began, and unchanging until our time here on Earth ends - awaiting the enduring abode to our new home - the New Jerusalem!

16 – UFO'S AND BIBLE STORIES...

I once had a conversation with a highly intelligent person who largely believed in UFO's. Knowing that I was a religious person, this person questioned me and asked me why it is widely believed that there was a man who could walk on water, produce food and drink from thin air and rise from the grave. There are books written and stories told about Him (Jesus). And people believe 100% without contest about this man everyday. Yet, the notion that there may be life somewhere out there in the limitless reaches of space is unrealistic. (He questioned mine),Why is it that so many people will believe with unquestioning conviction the unprovable existence of a higher power, and yet call foul play due to the same lack of proof regarding extraterrestrial life? It just seemed without logic to him.

This is a very logical question to a person who is not well studied in the Bible - but who will NOW become enriched with biblical knowledge at the end of this chapter (in order to be able to satisfy the question asked above).

This chapter is not going to be easy for certain people because they cling to the belief, dream and hope that life has to exist elsewhere in the Universe other than on our own planet. They say it's only logical that such a vast Universe should have other life. Astronomers come up with something that tries to convince us that life must exists elsewhere - called the "Drake Equation", which is a mathematical equation used to estimate the number of extraterrestrial civilizations that may exist in our galaxy. Additionally organizations such as SETI (the search for extraterrestrial intelligence) continually persist to listen for alien signals or other off- worldly communications, but (to date) there have been NO genuine transmissions received. They neglect to see that we have already been contacted by the entity and spirit of G-d through His written language, (the moral fiber of the Hebrew letters, as was illustrated in the previous chapter of the Primordial Language).

People get the idea to believe that there is other life out there in the Universe because they are influenced by UFO's and strange sightings that defy explanation. Many people even believe they have actually encountered or had an intelligent experience or abduction with aliens. They swear by seeing little grey skinned aliens with dark piercing almond eyes, even giving a detailed description of the spacecraft. Then there are those people who have not even witnessed a UFO or experienced on abduction, rather they believe the stories of others. There are many things that defy explanation and I'm sure the believer does not want to here them. Even a camera (including video camcorder) does not tell the whole truth. It can take a picture (or video) of something that is distorted in both depth perception and form; this is because the camera does not have a mechanism to decipher the difference. The human eye can also deceive. Especially when exposed to weightlessness (to long) such as with many of the Astronauts who have claimed to see bright strobes of lights while in the space shuttle suspended in the upper atmosphere of our earth (in space).

Other explanations for alien abductions include sleep paralysis and the military testing hallucinogenic drugs on us as a war agent. Even the most likely of them all, the experimentations of new top-secret military crafts (some of them may actually even be old, but were stored away because they were impractical at the time). Another explanation may seem out in left field, but it may be our own government(s) experimenting with human cloning in their attempts to make an elite type of humanoid for space travel. This humanoid would of course need to be smaller and skinnier so that it could live on very little food, mostly liquids (if any at all). Its brain would have to be of a huge intelligent size along with the almond shape huge black eyes better for seeing in the dark recluses of space. And since they were cloned, their life expectancy would probably be short, (unless they have the ability to hibernate). Additionally they might also be sterile because they are a crossbreed of some kind much as if mules are the sterile offspring of mating between a horse and a donkey. Leviticus 19:19 - Keep my decrees. **Do not mate different kinds of animals.** Do not plant your field with two kinds of seed. Do not wear clothing woven of two kinds of material.

Cattle mutilations also raise a red flag. This is where many farmers find their cows cut up with precision, as if it was the work of a high-tech laser. As it turns out these cattle graze off the grass that was once a nuclear testing grounds for the military. Last I heard, it was our own government(s) testing the cattle that grazed off these radiated grounds, to analyze if there were any long-term effects absorbed into their body from radiation poisoning. The military has a special reconnaissance-flying vehicle for investigation and exploration that can maneuver silently, as well as stealthy so they won't be detected by the cattle owners. Therefore, as you can see there are explanations for these types of things other than it being associated with aliens from another planet (or even time).

Now I'd like you to look at this picture below. This is what it looks like when a jet breaks the sound barrier. Take notice of the spherical cloud that it creates. Many of these formations can be illuminated both in the day and night sky.

Oddly on certain occasions, strange light phenomenon may occur from a refraction or reflection of light. When light bounces off the ground and into the sky, or sunlight beaming through a cloud, or simply just car headlights, spotlights or a big bright light beam like the two visible light beams that were projected to symbolize the missing towers of the World Trade Centers in remembrance of the September 11th 2001 attacks. If a bird or bat happens to be flying within this projected beam of light especially at night it can appear like strobes of light (glowing, floating balls of light) darting back and forth in an intelligent pattern, when in fact it may just be an innocent bat hunting with its sonar for mosquitoes in the bright light (many insects including mosquitoes are attracted to bright light especially at night).

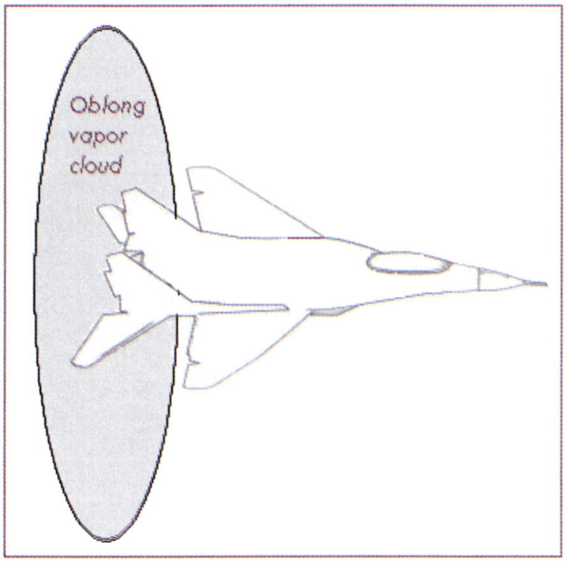

Additionally it has been discovered that swarms of bugs (such as huge masses of bees) can glow and even appear as bouncing (both) daytime and nighttime lights that repeatedly (and erratically) vanish, then reappear in a different part of the sky. It has been reported that insects (in vast numbers) could trigger electric discharges when flying beneath a storm front. This sets off a highly visible display of glowing effects resembling disco dancing in the sky, giving off various colors of blue, green, and red. What's occurring here is that the natural electric field beneath the storm front pulls away electrons from pointed, sharp or spiky exposed surfaces, (such as an insects antennae's and legs). A "sudden" large amount of these electrons IONIZES gases in the atmosphere and thus emits light. These same effects had been created with insects in several laboratories. Additionally, as the insect swarm tries to get out of the electric voltage field they dart around making a "humming sound" and then suddenly disappear from sight (and the glowing light vanishes). This phenomenon is very similar to St. Elmo's fire, which puts off a neon bluish white hue from a distance. St. Elmo's fire is actually a spark discharge (going upwards) from a high building, steeple, mast, antenna, aircraft wings, mountaintop, barbed wire fences, it can even appear on leaves and grass. It is a very rare phenomenon. When you see St. Elmo's fire near yourself, or feel your hair standing on end, this is high voltage and you must immediately leave the area as soon as possible as this is a sign that lighting in is imminent.

So, as one can see UFO's can certainly be confused with glowing insects. Incidentally, this relates to several reports of people approaching these lights, and they move away. In addition, when the people retreat, these lights move back. Because of these movements, they seem to appear to us as intelligent and curious. They may even appear to follow us (almost like a swarm of mosquitoes do in search of food/blood).

Surprisingly some may not know this, but the Earth itself produces luminous phenomena of light plasmas. These lights are produced within the Earth core. "Earthquake lights" occur before and after tremors when gasses released from within the Earth suddenly ignites and shoots sparks into the atmosphere. There is even "Ball lightning" which is an accumulation of static electricity that becomes visible and capable of moving along surfaces and floating in mid air. "Earth lights" can last long; they congregate to particular areas like a valley or mountain. This is because the tectonic stresses (movement in the Earths crust) produce tremendous force fields where plasmas are produced (luminous clusters of charged particles). At night, they will shine as lights in the sky, but in daylight, plasma can reflect light and look metallic. They can hover and shoot off and zoom out of site, they can sit on water and go under water, some come out of the water and ground where they are initially seen flashing off. They are usually seen when a volcano becomes active or before or after an earthquake. Some places on the Earth with surface tectonic faults have always been noted and associated with these light occurrences.

The common person may not be readily familiar with the science of "Earth lights" (as its rare to see them), but are more likely knowledgeable on the Northern Lights or Aurora Borealis. These lights are the plasmas put off from the Earths "magnetosphere" and being blown around by electric and magnetic forces (picture shown below). This is no different than the gasses produced by the Earths core, in which the same principles come into play here. Additionally, many UFO sightings involve the loss of electricity that can inhibit sensitive electronic components. Similarly, the earths plasma is no different (both from the magnetosphere and the Earths crust). These electro magnetically charged plasmas could disrupt radio waves and power lines, even the reception of the satellites in our upper atmosphere. They have even been known to destroy sensitive electronic equipment.

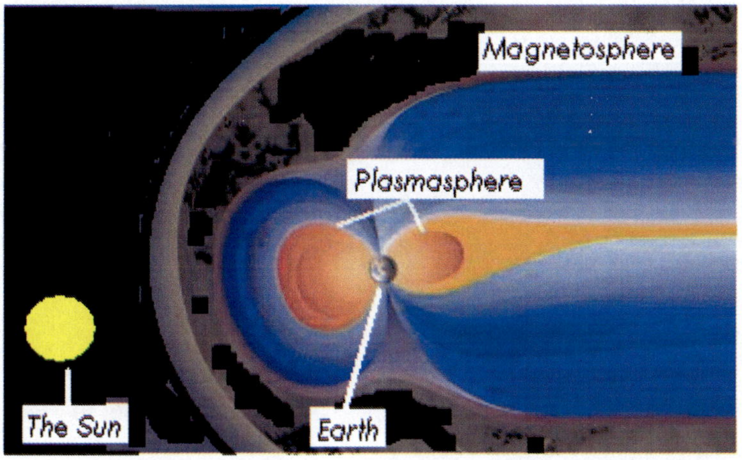

Interestingly enough we can also see a measurable unit of growth here.

Whether it be Earth lights, or earthquake lights… ball lightning… volcanoes… luminous charged particles that flash, glisten or glow as lights in the sky… the Northern Lights or Aurora Borealis… a picture taken by a camera (or video)… SETI (the search for extraterrestrial intelligence)… the Drake Equation, a mathematical equation that suggests the vast possibilities of life existing elsewhere in the Universe. This also includes those people that just want to believe, dream, wish and hope that life exists elsewhere. Then there are, those that believe they've had an actual encounter with an alien ship or have been abducted. The list goes on… sleep paralysis… the military testing hallucinogenic drugs on us as a war agent and the experimentations of new top secret military crafts… government(s) experimenting with human cloning in their attempts to make an elite type of humanoid for space travel.

Now we get to the crux of the matter, like I said, I will tell the UFO/alien believer what they want to hear, but the details will be something other than what they expected. Now that we have addressed the existence of reasonable alternate explanations, I will tell you that even these explanations are not the full truth. The full truth about the existence of alien life (or any other life existing anywhere other than on Earth) can be found in the most unexpected place - THE BIBLE! The Bible DOES NOT in any way teach that intelligent life exists anywhere else in the entire Universe except for on Earth! Although our omnipotent G-d could have created such life, had He so desired? However, it is clear from Scripture that He did not. The Bible tells us that G-d's only true celestial creations (or if you want to call them extraterrestrial) are the angels. They are the only other beings besides us that exist in the entire Universe. The angels are considered "light beings" (they do not have a body that is made up of flesh and blood). We are separated from them by the diminution (lessening) of light as we had learned earlier from the "Bible Code" about G-d's separation process. In addition, this barrier separates us from the Heavenly realm and angelic beings. The Heavenly realm or lower realm is off worldly (but angels visit us frequently here on Earth). This Heavenly realm is not visible to our living spectrum of sight - unless it wants to make itself visible by a gradation. This is because our human eyes were not made by G-d to be able to see through the separation of light barriers.

However, the argument arises from the UFO and alien believers that the Bible does indeed specifically mention such things that can be interpreted as UFO's and alien beings. However, this is where many people "misinterpret" what the Bible says. The most famous of these

Bible stories is the chariot of fire that came in a whirlwind and took the Jewish prophet Elijah into the sky never to be seen again. Many people interpret this as a UFO. 2 Kings 2:1 - And it came to pass, **when the Lord would take up Elijah into heaven by a whirlwind**, that Elijah went with Elisha from Gilgal. 2 Kings 2:11 - And it came to pass, as they still went on, and talked, that, behold, there appeared a chariot of fire, and horses of fire, and parted them both asunder; **and Elijah went up by a whirlwind into heaven.** 2 Kings 2:12 - **And Elisha saw it, and he cried, My father, my father, the chariot of Israel,** and the horsemen thereof. And he saw him no more: and he took hold of his own clothes, and rent them in two pieces. Except the Bible did not say that Elijah was taken by aliens or a UFO, rather it specifically elaborates that he was taken by G-d (the Lord, his father, the chariot of Israel), which is as we've learned from the previous chapter the "Star of David" (the Merkabah vehicle). Moses was also guided in the desert by a mysterious aerial phenomenon. Exodus 13:21 - By day the Lord went ahead of them in a pillar of cloud to guide them on their way and by night in a pillar of fire to give them light, so that they could travel by day or night. Except the cloud pillar was distinctively understood to be that of G-d, not an alien or UFO. Additionally the prophet Ezekiel was also raptured up by something in the sky. Ezekiel 1:15-18 - As I looked at the living creatures, I saw a wheel on the ground beside each creature with its four faces. This was the appearance and structure of the wheels: They sparkled like chrysolite, and all four looked alike. Each appeared to be made like a wheel intersecting a wheel. As they moved, they would go in any one of the four directions the creatures faced; the wheels did not turn about as the creatures went. Their rims were high and awesome, and all four rims were full of eyes all around. Ezekiel 1:5 - And in the fire was what looked like four living creatures. In appearance their form was that of a man. Ezekiel 1:20 - Wherever the spirit would go, they would go, and the wheels would rise along with them, because the spirit of the living creatures was in the wheels. Ezekiel 1:22 - Spread out above the heads of the living creatures was what looked like an expanse, sparkling like ice, and awesome. However, Ezekiel described these as "VISIONS". Nobody else actually saw anything to this effect.

Here we read in the Bible how angels can miraculously appear, go through walls and open locked doors like described in some alien abductions. Acts 12:7-10 - Suddenly an angel of the Lord appeared and a light shone in the cell. He struck Peter on the side and woke him up. "Quick, get up!" he said, **and the chains fell off Peter's wrists.** Then the angel said to him, "Put on your clothes and sandals." In addition, Peter did so. "Wrap your cloak around you and follow me," the angel told him. Peter followed him out of the prison, but he had no idea that what the angel was doing was really happening; he thought he was seeing a vision. They passed the first and second guards and came to the iron gate leading to the city. **It opened for them by itself, and they went through it.** When they had walked the length of one street, suddenly the angel left him.

According to the Bible, it tells us that angels can also communicate through us when we are in a dream state. The dream issue is a big part of the Judeo Christian beliefs going all the way back to Abraham in the Old Testament when G-d caused a deep sleep to come on Abraham and in this dream G-d communicated what he wanted with him. Genesis 15:12-16 - As the sun was setting, Abram fell into a deep sleep, and a thick and dreadful darkness came over him. Then the Lord said to him, Know for certain that your descendants will be strangers in a country not their own, and they will be enslaved and mistreated four hundred years. But I will punish the nation they serve as slaves, and afterward they will come out with great possessions. You, however, will go to your fathers in peace and be buried at a good old age.

In the fourth generation your descendants will come back here, for the sin of the Amorites has not yet reached its full measure.

Even Daniel when he was receiving prophesies he was in a deep sleep. Daniel 8:18 - While he was speaking to me, I was in a deep sleep, with my face to the ground. Then he touched me and raised me to my feet. In addition, when the angel appeared to Joseph in a dream. Matthew 1:20 - But after he had considered this, an angel of the Lord appeared to him in a dream and said, Joseph son of David, do not be afraid to take Mary home as your wife, because what is conceived in her is from the Holy Spirit. Matthew 2:13 - When they had gone, an angel of the Lord appeared to Joseph in a dream. Get up, he said, take the child and his mother and escape to Egypt. Stay there until I tell you, for Herod is going to search for the child to kill him. Matthew 2:19-24 - After Herod died, an angel of the Lord appeared in a dream to Joseph in Egypt and said, Get up, take the child and his mother and go to the land of Israel, for those who were trying to take the child's life are dead. So he got up, took the child and his mother and went to the land of Israel. But when he heard that Archelaus was reigning in Judea in place of his father Herod, he was afraid to go there. Having been warned in a dream, he withdrew to the district of Galilee, and he went and lived in a town called Nazareth. So was fulfilled what was said through the prophets: He will be called a Nazarene.

The question arises in the Biblical verses above, are these angels or aliens? Its quite clear that they are angels - as that is explicitly what they are expressed as. However alien encounters and abductions can be "deceiving angels" (also known as "fallen angels"). These effects are also commonly described with alien abductions, where aliens seem to put their abductees in a paralyzed state where they cannot move, and when they wake they seem to remember bits and pieces, sometimes they remember nothing at all. Again, like the mechanism of the Merkabah vehicle, the abductees body is actually a reflection of the material body. The real body is not actually "physically" taken by these aliens (fallen angels) - as they know how to induce the "reflected state" of the Merkabah in us. This is why abductions of this sort are very hard to prove physically (let alone scientifically) that they actually occur. Additionally, the reflected body has all the five senses (and then some) that the material body has - so it can see and feel pain. These "deceiving fallen angels" also know how to interfere with the memory downloading mechanism "back into body integration" of the Merkabah vehicle. They know how to do this simply because they are already in the "spiritual or etheric" state/realm. So in essence what I'm trying to put forward here is that angels (both good and bad) are not physical material aliens, they are "light beings" separated by the diminution (lessoning) of light. The diminution of light is what separates the lower worlds from higher degrees of existence. This diminution is a series of downward gradation, permutation and emanation of life forms. One of the reasons these deceiving fallen angels may interfere with the memory downloading "back into body integration process" is because they do not want the person to retain any memory of pain (abuse or exploitation), or remember any important information that they may have heard or telepathically interpreted. On the other hand, it may not be that they don't want us to remember, but rather that our own body (as a survival mechanism) does not want us to remember, and therefore the memory is not downloaded properly for memory retrieval. Their purpose with us of course is our PHYSICAL BODY! The reason for this is that they are "fallen angels" and were cast out of Heaven. Jude 1:6 - And the angels which kept not their first estate, but left their own habitation, he hath reserved in everlasting chains under darkness unto the judgment of the great day. Jesus Himself tells us that He had witnessed the fall of Satan from the estate of

Heaven mentioned in Luke 10:18 - And he said unto them, I beheld Satan as lightning fall from heaven. Additionally the Bible reiterates this fall again in Isaiah 14:12-15 - How art thou fallen from heaven, O Lucifer, son of the morning! How art thou cut down to the ground, which didst weaken the nations! For thou hast said in thine heart, I will ascend into heaven, I will exalt my throne above the stars of G-d: I will sit also upon the mount of the congregation, in the sides of the north: I will ascend above the heights of the clouds; I will be like the most High. Yet, thou shalt be brought down to hell, to the sides of the pit. The "fallen angels" yearn for the fleshly body because their holdings and resources (which they had in Heaven - the higher realm) had changed considerably when they had fallen from that estate to a lower domain. They no longer had access to the higher estates, thus causing a refinement (modification) to occur. Some religious scholars suggest that this lower estate was "the Earth". We can conclude it was the "Earth" because Revelation 12:7-9 tells us so - And there was war in heaven: Michael and his angels fought against the dragon; and the dragon fought and his angels, And prevailed not; neither was their place found any more in heaven. And the great dragon was cast out, that old serpent, called the Devil, and Satan, which deceiveth the whole world: **he was cast out into the earth, and his angels were cast out with him.** We find the "Earth" mentioned again in Job 1:7 - And the Lord said unto Satan, Whence comest thou? Then Satan answered the Lord, and said, From going to and fro **in the earth**, and from walking up and down in it. We know that 1/3 of the angels fell with Satan because the Bible tells us that he drew the third part of the stars of Heaven to the Earth. Revelation 12:4 - **And his tail drew the third part of the stars of heaven, and did cast them to the earth**: and the dragon stood before the woman which was ready to be delivered, for to devour her child as soon as it was born. We also know that when Satan and 1/3 of the angels where cast down to Earth their main purpose was to persecute women and their offspring. They did this to try to stop the fulfillment of the coming redemptive Messiah and give rise to the antichrist (this will also be the discussion of a later chapter). Revelation 12:13 - **And when the dragon saw that he was cast unto the earth, he persecuted the woman which brought forth the man-child.**

Now onto the good angels… Jesus made it clear that angels DO NOT engage in sexual activities, "at least not angels in Heaven". Matthew 22:30 - For in the resurrection they neither marry, nor are given in marriage, but are as the angels of G-d in heaven. Also angels do not die, they are perpetual (immortal, enduring and never-ending) beings. Luke 20:35-36 - But they which shall be accounted worthy to obtain that world, and the resurrection from the dead, neither marry, nor are given in marriage: **Neither can they die any more: for they are equal unto the angels;** and are the children of G-d, being the children of the resurrection. Additionally, the angels were originally made to serve G-d on a continual bases. Psalm 103:20 - Bless the Lord, ye his angels, that excel in strength, that do his commandments, hearkening unto the voice of his word. Here we read about an angel telling the prophet John not to worship him - but only G-d. Revelation 22:8-9 - And I John saw these things, and heard them. And when I had heard and seen, I fell down to worship before the feet of the angel which shewed me these things. Then saith he unto me, See thou do it not: for I am thy fellowservant, and of thy brethren the prophets, and of them which keep the sayings of this book: worship G-d. In addition to angels, they are spiritual beings that can, to a certain degree, take on physical form. Hebrews 13:2 - Do not forget to entertain strangers, for by so doing some people have entertained angels without knowing it.

Therefore, now that we know angels live forever (and do not die), they also do not marry and therefore not made to create, procreate and bare offspring. However (as mentioned) the disposition (nature) of the "fallen angels" changed when they were cast out of Heaven to a lower level on Earth (a diminution of light) to a medium state in which they became more inclined toward and yearned for the physical body (as they were drawn to it being in-between the two states). And this may be why they abduct us and try to interbreed with us. The Bible tells us of an account where these "fallen angels" looked upon the daughters of men and found them fair (beautiful) and took them as their wives and bore children to them. These children grew to be "giants".

To obtain a better understanding of these "fallen angels" who are referred to as "sons of G-d" and their offspring the "giants". We find this information recorded in the Bible. Genesis 6:1-9 - And it came to pass, when men began to multiply on the face of the earth, and daughters were born unto them, That the **sons of G-d** saw the daughters of men that they were fair; and they took them wives of all which they chose. And the Lord said, My spirit shall not always strive with man, for that he also is flesh: yet his days shall be an hundred and twenty years. **There were "giants" in the earth in those days; and also after that, when the sons of G-d came in unto the daughters of men, and they bare children to them,** the same became mighty men which were of old, men of renown. And G-d saw that the wickedness of man was great in the earth, and that every imagination of the thoughts of his heart was only evil continually. And it repented the Lord that he had made man on the earth, and it grieved him at his heart. And the Lord said, I will destroy man whom I have created from the face of the earth; both man, and beast, and the creeping thing, and the fowls of the air; for it repenteth me that I have made them. But Noah found grace in the eyes of the Lord. These are the generations of Noah: Noah was a just man and perfect in his generations, and Noah walked with G-d. (By the way, this was the reason for the flood - to rid of this evil and sin). However, even after the flood the 8 people that walked off the Ark carried (and passed on) the recessive gene for gigantism.

Furthermore, in the Old Testament we find Moses telling us that even after the flood there were still giants on the Earth. However, this did not mean that later giants were the result of fallen angel and human interbreeding. Gigantism continued down into human beings through inherited genetics. However, ALL fallen angels and human interbreeding came to a STOP at the time of the flood as these fallen angels were imprisoned by G-d in the eternal fire. In fact this interbreeding (sexual act) was so seriously disapproved by G-d that we later witness the destruction of Sodom and Gomorrah as a result of sexual interactions between human beings **who go after strange flesh.** We read about this in Jude 1:6-7 - And the angels which kept not their first estate, but left their own habitation, he hath reserved in everlasting chains under darkness unto the judgment of the great day. **Even as Sodom and Gomorrha**, and the cities about them in like manner, giving themselves over to fornication, **and going after strange flesh, are set forth for an example, suffering the vengeance of eternal fire.**

The account in the Bible about the giant "Goliath" was a prime example of one of the offspring from the "fallen angels". 1 Samuel 17:4 - A champion named Goliath, who was from Gath, came out of the Philistine camp. He was over nine feet tall. He grew to over 9 feet tall because he was a descendant of the "fallen angels". We know that he was a descendant of the "fallen angels" because when we read further we discover that Goliath had four other relatives (also giants) mentioned in the Bible who where all descendants of "Rapha" in Gath. 2 Samuel 21:16-22 - And Ishbibenob, which was of the sons of the giant (rapha), the weight of whose spear weighed three hundred shekels of brass in weight, he being girded with a new sword, thought to have slain David. But Abishai the son of Zeruiah succoured him, and smote the Philistine, and killed him. Then the men of David sware unto him, saying, Thou shalt go no more out with us to battle, that thou quench not the light of Israel. And it came to pass after this, that there was again a battle with the Philistines at Gob: then Sibbechai the Hushathite slew Saph, which was of the sons of the giant (rapha). And there was again a battle in Gob with the Philistines, where Elhanan the son of Jaareoregim, a Bethlehemite, slew the brother of Goliath the Gittite, the staff of whose spear was like a weaver's beam. And there was yet a battle in Gath, where was a man of great stature, that had on every hand six fingers, and on every foot six toes, four and twenty in number; and he also was born to the giant (rapha). And when he defied Israel, Jonathan the son of Shimeah the brother of David slew him. **These four were born to the giant (rapha) in Gath,** and fell by the hand of David, and by the hand of his servants. 1 Chronicles 20:4-8 - And it came to pass after this, that there arose war at Gezer with the Philistines; at which time Sibbechai the Hushathite slew Sippai, that was of the children of the giant (rapha): and they were subdued. And there was war again with the Philistines; and Elhanan the son of Jair slew Lahmi the brother of Goliath the Gittite, whose spear staff was like a weaver's beam. And yet again there was war at Gath, where was a man of great stature, whose fingers and toes were four and twenty, six on each hand, and six on each foot and he also was the son of the giant (rapha). But when he defied Israel, Jonathan the son of Shimea David's brother slew him. **These were born unto the giant (rapha) in Gath;** and they fell by the hand of David, and by the hand of his servants.

The Hebrew word for "giants" (or monstrous race) is called "nephilim" (plural). But the singular word found in the Bible is actually pronounced "nephiyl", נְפִיל a Hebrew word meaning "those who have fallen, or the fallen ones of iniquity" derived from the conceptual root word "naphal" which means to fall. However, the offspring (children) of these giants have also been called "Raphaym" (plural) רפאים and "Rapha" (singular) in the land of Gath, but they have also been called the sons of Anak, Anakim, Eymiym, Zamzom and Og king of Bashan. Deuteronomy 3:11 - For only Og king of Bashan remained of the remnant of giants; behold his bedstead was a bedstead of iron; is it not in Rabbath of the children of Ammon? nine cubits was the length thereof, and four cubits the breadth of it, after the cubit of a man. (The measurements of his bed equate to approximately 14 feet long by 6 feet wide). And by the way the remains of giants are being unearthed today all around the world. Some of these giant human skeleton bones measure from 8 feet all the way to 16 feet in length. (I encourage all readers to investigate this for themselves). Additionally, some suggest that these giants also possessed great knowledge given to them by their fathers, the "fallen angels" that enabled them to build great monuments of mathematical precision and align them up astronomically with the stars - possibly monuments like Stone Henge. However there is a distinct difference concerning the giants obtaining this knowledge. Remember, as we've discussed in an earlier chapter that this knowledge was "divinely" given to Enoch by G-d (who was not a giant). Therefore, this knowledge already existed with Enoch but was perverted by the giants. Enoch tells us this in book 1of the Watchers section 1 chapters 7-8 - And all the others together with them took unto themselves wives, and each chose for himself one, and they began to go in unto them and to defile themselves with them, **and they taught them charms and enchantments, and the cutting of roots, and made them acquainted with plants.** And they became pregnant, and they bare great giants, whose height was three thousand ells: Who consumed all the acquisitions of men. And when men could no longer sustain them, the giants turned against them and devoured mankind. **And they began to sin against birds, and beasts, and reptiles, and fish, and to devour one another's flesh, and drink the blood.** Then the earth laid accusation against the lawless ones. **And Azazel taught men to make swords, and knives, and shields, and breastplates, and made known to them the metals of the earth and the art of working them, and bracelets, and ornaments, and the use of antimony, and the beautifying of the eyelids, and all kinds of costly stones, and all colouring tinctures. And there arose much godlessness, and they committed fornication, and they were led astray, and became corrupt in all their ways. Semjaza taught enchantments, and root-cuttings, 'Armaros the resolving of enchantments, Baraqijal (taught) astrology, Kokabel the constellations, Ezeqeel the knowledge of the clouds, Araqiel the signs of the earth, Shamsiel the signs of the sun, and Sariel the course of the moon.** And as men perished, they cried, and their cry went up to heaven.

Enoch also testifies of these giants - the offspring of the "fallen angels" (in earlier times before the deluge) before it was recorded by Moses in Genesis. Enoch referrers to them as "Watchers". 1 Enoch 8-12 - And now, the giants, who are produced from the spirits and flesh, shall be called evil spirits upon the earth, and on the earth shall be their dwelling. Evil spirits have proceeded from their bodies, because they are born from men and from the holy **Watchers** is their beginning and primal origin, they shall be evil spirits on earth, and evil spirits shall they be called. As for the spirits of heaven, in heaven shall be their dwelling, but as for the spirits of the earth which were born upon the earth, on the earth shall be their dwelling. And the spirits of the giants afflict, oppress, destroy, attack, do battle, and work destruction on the earth, and cause trouble: they take no food, but nevertheless hunger and thirst, and cause offenses. And these spirits shall rise up against the children of men and against the women, because they have proceeded from them.

The reason these giants grew so large is that they were genetically in-between half light-beings and human flesh - thus the human flesh could not contain the embodiment or the over generous enormity of effervescence (of the light being part), therefore these children grown into big "giants". This is why G-d did not create the angels to breed and multiply because doing so would cause an overzealous pouring over of power to the human flesh, and therefore corrupt its naturally docile nature, which in effect is exactly what happened when the fallen angels interbred with earth women - their offspring where evil, cruel and corrupt. The status of their size enabled them to be strong and overpowering - thus they where looked at as deities (depicted as giants in some of the Egyptian hieroglyphics and statues).

In effect, G-d sends the Great Flood to rid the Earth of this evil. Genesis 6:4-8 - There were giants in the earth in those days; and also after that, when the sons of G-d came in unto the daughters of men, and they bare children to them, the same became mighty men which were of old, men of renown. And G-d saw that the wickedness of man was great in the earth, and that every imagination of the thoughts of his heart was only evil continually. And it repented the Lord that he had made man on the earth, and it grieved him at his heart. And the Lord said, I will destroy man whom I have created from the face of the earth; both man, and beast, and the creeping thing, and the fowls of the air; for it repenteth me that I have made them. But Noah found grace in the eyes of the Lord. Genesis 7:1 - And the Lord said unto Noah, Come thou and all thy house into the ark; for thee have I seen righteous before me in this generation. Genesis 7:4 - For yet seven days, and I will cause it to rain upon the earth forty days and forty nights; and every living substance that I have made will I destroy from off the face of the earth.

Many people have associated these giants as aliens who came in spaceships from another planet. They surmise this from the Sumerian tablets telling the story of "Gilgamesh". However, after reading the epic of Gilgamesh, one soon realizes that it is nothing more than a modified copy of the biblical flood account. Gilgamesh was a king of Uruk in Babylonia (Babylon), on the Euphrates river in modern Iraq. He lived about 4000 years ago. The story of Gilgamesh was recorded by a person named "Hannurabi" (or Hammurabi) who lived about 1000 years after Gilgamesh. This story was written in the Sumerian language on clay tablets in a script known as cuneiform (a Semitic language related to Hebrew which was illustrated in a previous chapter). There are indeed similarities between the story of Gilgamesh and the Genesis flood account in the Bible. These similarities are also found in other ancient flood accounts. My conclusion in the Gilgamesh story is that the flood was a real event in the history of humankind that was passed down through the generations of different cultures and is a "copy" of the Genesis story. The Gilgamesh story is rather a tailored saga told after the actual account of the real flood. The story seems to be fashioned to ones perspective - that lacks sincerity by the manner in which its conveyed (readers, examine this for yourself). In comparison, the Genesis account is a logical - more believable, seemingly factual account of a historical event. It is clear from the very first tablet of Gilgamesh that the story is a "modified story" which was told over. Tablet 1 tells us itself - When that man lives to tell the tale over and over… He recovered the knowledge of all the times before the Flood… And then he carved his story on stone.

Again, as pointed out above, many fabricated stories have been told about Noah's flood including its main cause "the giants" - the offspring of the "fallen angels". Nevertheless, we know from biblical scripture that it clearly implies that the offspring giants where caused by the interbreeding of "fallen angels" when it referrers to them as "sons of G-d" (Ben Elohiym) from Genesis 6:2. We can identify this from Job 38:7 - When the morning stars sang together, and all the **sons of G-d** shouted for joy. However, the "sons of G-d" can also be described in the Bible as ordinary men, leading some religious scholars to suggest that they were not angels at all, but descendants of Cain whom G-d had cursed because he killed his brother Abel. In addition, possibly because the descendants of Cain had interbred with descendants of Seth (the 3rd child of Adam and Eve). Some seem to think that this is closer to the truth. However, I tend to whole-heartedly believe the Bible word for word. Either way, it was Satan's attempt to defile the bloodline that the Messiah would descend from (which will be the subject of a later discussion).

Now the question may be asked again. Why do these "fallen angels" abduct us? They are obviously still at it, abducting us for our "physical body" to produce offspring (these are the abductions where women are impregnated by aliens that are then mysteriously without explanation taken from them, or eggs and sperm extracted, stolen and fertilized by the fallen angels). However, as I pointed out above, according to the Bible G-d does not permit these offspring to live on Earth anymore. They are now living in the half-light being realm of the fallen angels. They are producing this army for themselves. One can only surmise why? These interbred beings (half-breeds) are being created as rivalry co-conspirators in another attempt by fallen angels to corrupt mankind, and to produce an army to fight against G-d in the last battle of Armageddon (before the judgment).

The question is why would the infamous head-honcho Satan stoop to this level as disguising themselves as aliens. Probably in hopes of tricking humankind into accepting them, like at the beginning in the Garden of Eden with Eve (the first woman).

Its Amazing how this hidden agenda now comes into view! Aliens and UFO's go hand in hand with the evolutionists because they want nothing more than to have us believe that the Bible is wrong and that life evolved on other planets. Not only that, now the evolutionist has the overconfidence to claim that these intelligent aliens were the ones who created us. What's more, these aliens are now giving us the technology and the deceptive spiritual wisdom to become immorally perceived deity's like them. What these aliens really are is described for us in Ephesians 2:2 as the "Prince of the Powers of the Air". Ephesians 2:2 - Wherein in time past ye walked according to the course of this world, according to **the prince of the power of the air,** the spirit that now worketh in the children of disobedience.

Now to validate everything that I've just said, before we proceed any further in this discussion we need to back up a bit and examine something vital to the question of alien life existing "at all" (elsewhere) in the Universe. We need to first focus our attention back to our own existence here on Earth, given that we still haven't uncovered the mystery of how life on our very own planet got started. Remember the Miller experiment mentioned in a previous chapter that involved creating the building blocks of life in a lab - where Miller tried to get "amino acids" which are the building blocks of all life? Remember, he got amino acids, but he got the wrong ones that are needed for life. If you recall, amino acids come in two different kinds (right and left), and that all life on this planet is composed of 100% left amino acids. Relentlessly and insistently, the tests continue to go on and still the scientists come up with the same results. NOBODY ANYWHERE can create the right combination of amino acids to create life. In addition, this big mystery is left out of the "ALIEN and Evolution Theory". Remember the evolution theory claims that all life crawled out of the water and onto the land, but the evolution theory neglects the fact that water splits apart molecular bonds... so it couldn't have been by water that amino acids were created and thus living things. Yes, we need water to survive but until we have that "cell wall", it can't get started. In addition, remember that life cannot start with oxygen either in the atmosphere because oxygen is a corrosive that pulls apart molecular bonds - specifically amino acids! So the big question is, if life cannot get started without water or oxygen in the atmosphere, how did it get started? How did amino acids bond together to make a living cell?

In addition, herein lies the answer to life elsewhere in the Universe. Subsequently, if we CANNOT get an amino acid protein here on earth, what makes us think it can happen anywhere else in the Universe? However, the scientist, the evolutionists and UFO/alien believers consequently leave this equation out of the facts. None of these guys are telling us how hard it is for life to get started (either that, or they are not educated enough to adequately evaluate the evidence).

Aside from "amino acids", when we talk about "space travel" lets talk about (the speed of light) how fast can light travel in 1 year? It has been estimated that a light year is equivalent to "six trillion miles". From this estimation we also know that it takes approximately 800 seconds, or around 13 minutes for sunlight (coming from our Sun) to reach the planet Mars. In addition, we know that it takes light around 8 minutes to travel from the Sun to the Earth. That would mean that in order for light to travel across the Milky Way it would take approximately "one hundred thousand" years. In addition, for light to travel to the nearest galaxy it would take about two million years. Therefore, even if we could build a spaceship to travel the speed of light, it would still take us a very long time to reach other places in the Universe. Yes, we might be able to bend space by taking "wormholes" (shortcuts through gravity bent space), but keep reading to know why this is not potentially possible either.

Now, lets examine how light travels vast distances. Forget the spaceship now! If we were to send or receive a signal by radio waves, (because radio waves are the next fastest thing to light, but even they do not travel any faster than the speed of light). NOTHING KNOWN TRAVELS FASTER THAN THE SPEED OF LIGHT! (Research this for yourself). Now remember, it takes 8 minutes just for light to reach us from the Sun, that means it takes light about 4 minutes to reach the Planet Mars - that means it also takes 4 minutes for radio waves to travel back and forth to NASA from the Mars rover. Nevertheless, consider sending or receiving radio waves across the galaxy or universe, it would take years. Therefore, this whole idea for communicating across the universe is nonsense. It can only be done by the transmission of light (radio waves) or "Quantum Mechanics", non-locality or bio-location. Not even ones Merkabah vehicle can travel faster than light. However, it might be possible through "Quantum Mechanics" if particles match or correspond with each other over large distances of space - then communication can be instantaneous! However there is a glitch to this. How do the matching particles get to two opposite places (from point A to point B) without us establishing them there first? In other words, we have to arrive at the other location in the universe to place the identical particle there (this is called particle entanglement).

Now suppose we were to create a physical material spaceship that could travel in space, NASA has already done that, but that does not mean that it will literally EVER be able to travel the "speed of light" across vast distances of space and time, and I will tell you why this could never be. Yes, we could build spaceships that travel slowly through space, but when we talk about a long duration of time in space for humans - the long-term effects deteriorates our body. If we can find ways to simulate our gravity, sunlight (including frequency) and grow our food for a couple hundred years we might be able to survive, (but even these resources can only be simulated for so long, and once things like plants die out they cannot be replaced). Romans 8:22 - For we know that the whole creation groaneth and travaileth in pain together until now. Now remember, according to the laws of thermodynamics all things "even in space" will eventually deteriorate, wear-out and die. But suppose other planets, Star or Moons are ALL uninhabitable (which we will come to find out that they ALL are) we would not be able to go very far away from our home planet without eventually running out of resources that we depend on for survival.

Additionally, space is filled with deadly radiation for us humans. We fail to realize that here on Earth we are protected from our sun by the atmosphere, in which the magnetic field of the Earth gobbles up most of the radiation rays from the Sun. However, in space, we have no ozone layer or magnetic field to protect us. Even if we eventually venture out in space there is still that risk of radiation even if we get farther from the Sun, there are even such things as solar particles and galactic cosmic rays that come from the distant worlds of theoretical black holes, exploding stars and particles traveling at the speed of light. What is more, the effects of

0 gravity on our human body causes weakening (deterioration) of muscle tissue and bone mass. Additionally, lack of gravity also slows down the digestive track and eventually completely stops it. Furthermore, the effects of radiation and gravity on the human body can also combine in a threatening way. Since there is 0 gravity in space (which makes things less active) it can reek havoc on the human immune system against fighting off bacteria and viruses (which additionally feed off of radiation). This can transpire rather quickly in a matter of just a few days. Furthermore, space is also open to both extremes of radically intense heat and cold temperatures. Moreover, few actually know this, but space is also filled with toxic acids that are hazardous and detrimental to humans. In due course the more we study outer space; we will eventually find out that the Earth was made special by G-d precisely for us. And there is not one other place in the entire Universe that is fit for human habitation.

Before we continue on that thought, there is yet another problem that lurks with our physical material spaceship, if we were able to get it going up to considerable speeds in space, we then have the problem of "dust particles" in space (there are literally thousands of these particles per cubic mile of space). NASA's space shuttle is bombarded with them on a continual basis, they tear holes in the shuttle right above the Earths atmosphere when the shuttle in only floating. Now the faster ones go the greater the impact becomes with space particles. Hitting a particle the size of a kernel at great speeds would be equivalent to an atomic bomb. Followed by this, we also have the predicament of our human body against gravity. Jet fighter pilots know that 3 G's can break a blood vessel and anything over that could kill someone. Additionally, changing speeds at a fast pace would exert thousands of G's that the human body could not live through. Even such theories of wormholes that are thought to be able to bend us through space and time would tear our spaceship and physical body to pieces. Even wormholes (the warping of gravity) have space dust and everything else in-between. Wormholes twist, they twist because everything else in the Universe twists (like the spiral formations of galaxies) and when they twist with centrifugal force, speeds are faster and faster as physics takes hold. Therefore we would be pulled-apart and at the same time bombarded by space dust. Destruction is unavoidable unless we devise some kind of plasma shield that has the capability of neutralizing space particles (destroying them as our spaceship moves along). However destroying such particles at great speeds is "equivalent to light passing through matter" (which is slowed), and thus so is our spaceship. Either way, it is the impossible unless our spaceship was fully consisting of light itself (like angels, light beings, or our Merkabah vehicle). The next question is, can particles be transported from one place to the other - something like human teleportation? Yes, they can be! But these particles would not be the same particles, they would only be a "copy" of the original particles in which the originals where destroyed (by death) or go through a refinement process (for the living). This is a reflection again, like the Merkabah vehicle. And remember the Merkabah vehicle "consisting of light" cannot travel any faster than "light itself" (it is still subject to the physics of light), and if it were to travel to Mars and back it would take 8 minutes. The Merkabah vehicle itself is a teleportation device (and as I said in the previous chapter - is not meant to be roused up which will be explained why in a future chapter).

Briefly, back on the subject of human teleportation... Once you make a copy of these particles it is separate but in the same breath a replica with the same exact memories and knowledge… however, once it goes its separate way it experiences new and different things and becomes a separate individual (like twins - they are created the same, look the same, but upon separation at birth they are different personalities). However, twins can communicate with each other over vast distances because they still consist of the same cells and particles. This concept in itself is the only exception to space travel (Quantum Mechanics). This is Almost beyond comprehension or believability! However heed my word, if such a thing as human teleportation is invented it is the works of "foolery and deception" - the works of "fallen angels". (Which will be fully explained in an approaching chapter).

Now for the upcoming evidence I would ask that readers keep in mind what they've just read concerning how hard life is to get started with amino acids, as well as the forces involved in manipulating the gravity of space both with light and with a physical/material spaceship. I would like all readers to put their mind in neutral for information that is applicable and lasting. Think of yourself as a juror awaiting the evidence. After all the facts are heard, you may then reach an informed verdict that life elsewhere in the Universe does not exist at all - rather that ALL life revolves around and is centered on Earth. The Bible clearly tells us the answer to all these questions, and we find these little tidbits of information lurking within the "concordance" of the Bible if one knows how to properly search for it by comparing scripture with scripture. The answers to many mysteries and truths are available to those who patiently and diligently search and compare the concordances. But to ones dismay, just because one doesn't find what they're blatantly looking for in the Bible with the actual words staring them right in the face does not mean the answer is not there. Remember there are layers to G-d's being and to His perspective, that is why we find the Bible documented and recorded for us in a "concordance type fashion". When we search the Bible we begin to piece together some of this hidden evidence that G-d has reserved for the seeker at heart. We quickly come to realize that NASA and SETI (the search for extraterrestrial life) have deeply poisoned us in our concept of the universe compared to what the Bible/G-d has told us about its origin and purpose.

Let us start with some biblical verses that reveal to us that there are NO ALIENS lurking around on other planets ANYWHERE else in the Universe. Genesis 1:14-17 - And G-d said, Let there be lights in the firmament of the heaven to divide the day from the night; and let them be for signs, and for seasons, and for days, and years: And let them be for lights in the firmament of the heaven to give light upon the earth: and it was so. In addition, G-d made two great lights; the greater light to rule the day, and the lesser light to rule the night: he made the stars also. In addition, G-d set them in the firmament of the heaven to give light upon the earth. G-d said let there be light because our planet was meant to be inhabited - it doesn't mention anything about any other planet (or the need for light anywhere else). This verse also holds truth that G-d designed the Earth personally for us. In addition, if you recall from an earlier chapter that the "stars" where created by G-d as a calendar for the purposeful intention to serve as signs to mark our seasons (holidays), days and years. The Bible does not say anywhere that G-d created other places (in the heaven) to have inhabitable life. Some say well this only applies to our galaxy and this doesn't apply to stars that are farther out in the Universe. But be careful how you read this because G-d knew this question would arise in the future, so He specifically went out of His way to include - G-d made two great lights; the greater light to rule the day, and the lesser light to rule the night: **he made the stars also.**

So why did G-d create us? To G-d, human beings are very important because we bare His image and likeness. Genesis 1:27 - So G-d created man in his own image, in the image of G-d he created him; male and female he created them. And because the Bible says that human beings on this unique planet are "reflections to His glory" that will continue throughout all eternity to show how great a G-d He really is. Psalm 19:1 - The heavens declare the glory of G-d, the skies proclaim the work of his hands. Psalm 8:1-9 - O Lord, our Lord, how excellent is thy name in all the earth! Who has set thy glory above the heavens? Out of the mouth of babes and sucklings hast thou ordained strength because of thine enemies, that thou mightest still the enemy and the avenger. When I consider thy heavens, the work of thy fingers, the moon and the stars, which thou hast ordained; What is man, that thou art mindful of him? and the son of man, that thou visitest him? For thou hast made him a little lower than the angels, and hast crowned him with glory and honour. Thou madest him to have dominion over the works of thy hands; thou hast put all things under his feet: All sheep and oxen, yea, and the beasts of the field; The fowl of the air, and the fish of the sea, and whatsoever passeth through the paths of the seas. O Lord our Lord, how excellent is thy name in all the earth! Psalm 148:13 - Let them praise the name of the Lord, for his name alone is exalted; his splendor is above the earth and the heavens. Psalm 113:4-5 - The Lord is exalted over all the nations, his glory above the heavens. Who is like the Lord our G-d, the One who sits enthroned on high. Isaiah 55:8-9 - For my thoughts are not your thoughts, neither are your ways my ways, declares the Lord. As the heavens are higher than the earth, so are my ways higher than your ways and my thoughts than your thoughts.

Here are a few amazing Bible verses that illuminates that G-d who is transcendent and infinitely beyond the universe (the creator of all things) has humbled Himself to come down to earth and involves Himself deeply with the everyday events of humans. He is thoughtful, caring, attentive and sympathetic of man, He knows about our frustrations and concerns. He is continually concerned about us and our welfare. He is infinitely involved with everything we do. In every moment that we live, move, breath, He knows every thought, every sadness, every wrongdoing and every triumph. Psalms 113:7 - He raises the poor from the dust and lifts the needy from the ash heap. Additionally the patriarch Job tells us that G-d makes the lowly sit with princes. Job 34:19 - How much less to him that accepteth not the persons of princes, nor regardeth the rich more than the poor? for they all are the work of his hands. And in Psalms 113:9 - He settles the barren woman in her home as a happy mother of children. Praise the Lord. G-d also copes with enemies, and He has a special way of defeating and silencing them. He accomplishes this by infants and nursing babes. Jesus quotes this verse. Matthew 21:16 - And said unto him, Hearest thou what these say? And Jesus saith unto them, Yea; have ye never read, Out of the mouth of babes and sucklings thou hast perfected praise? Psalm 8:2 - Out of the mouth of babes and sucklings hast thou ordained strength because of thine enemies, that thou mightest still the enemy and the avenger. It seems to be the foolish things in this world that defeat the wise, and the week things that defeat the mighty, the proud and the arrogant. In fact, G-d's strategy in destroying the enemy is by a tiny infant in the arms of a Jewish mother, and out of that nursing babe G-d programmed the destruction of the enemy forever. His approach is to destroy the big things apparently with the little things. Isaiah 7:14 - Therefore the Lord himself shall give you a sign; Behold, a virgin shall conceive, and bear a son, and shall call his name Immanuel. Matthew 2:11 - And when they were come into the house, they saw the young child with Mary his mother, and fell down, and worshipped him: and when they had opened their treasures, they presented unto him gifts; gold, and frankincense and myrrh. Matthew 7:14 - But **small** is the gate and narrow the road that leads to life, and only a few find it. Mark 4:30-32 - Again, he said, What shall we say the kingdom of G-d is like, or what parable shall we use to describe it? **It is like a mustard seed, which is the smallest seed you plant in the ground. Yet when planted, it grows and**

becomes the largest of all garden plants, with such big branches that the birds of the air can perch in its shade.

G-d seems to be concerned only with us. This planet happens to be the center of G-d's spiritual program of the universe - His rescue (deliverance and salvation) program. This is the only problem place in the universe that's why He shed his own blood as the incarnate G-d-man in suffering (dieing on the cross) for the pardoning of the sins of people here on this planet. This act alone shows how concerned G-d is for people on this planet. According to the Bible it doesn't mention that there is any other redemptive program going on anywhere else - BECAUSE THERE IS NOBODY ANYWHERE ELSE! And lets not forget that He's coming back again for the second resurrection of the tribulation. 1 Thessalonians 4:16 - For the Lord himself shall descend from heaven with a shout, with the voice of the archangel, and with the trump of G-d: and the dead in Christ shall rise first. Revelation 7:15 - And he said, **These are they who have come out of the great tribulation;** they have washed their robes and made them white in the blood of the Lamb. Therefore, they are before the throne of G-d and serve him day and night in his temple; and he who sits on the throne will spread his tent over them. Revelation 20:4-6 - And I saw thrones, and they sat upon them, and judgment was given unto them: and I saw the souls of them that were beheaded for the witness of Jesus, and for the word of G-d, and which had not worshipped the beast, neither his image, neither had received his mark upon their foreheads, or in their hands; and they lived and reigned with Christ a thousand years. **But the rest of the dead lived not again until the thousand years were finished. This is the first resurrection.** Blessed and holy are those who have part in the first resurrection. **The second death has no power over them,** but they will be priests of G-d and of Christ and will reign with him for a thousand years. Revelation 20:12-13 - And I saw the dead, small and great, stand before G-d; and the books were opened: and another book was opened, which is the book of life: **and the dead were judged out of those things which were written in the books, according to their works.** And the sea gave up the dead which were in it; and death and hell delivered up the dead which were in them: **and they were judged every man according to their works.** John 5:29 - And shall come forth; they that have done good, unto the resurrection of life; and they that have done evil, unto the resurrection of damnation.

Now getting back on track here concerning aliens or other life forms existing elsewhere in the entire Universe, the most astounding verse that I have come across makes it deliberately known that there are no aliens or anybody else out there in the Universe except G-d and the earth that He created for us. Psalm 115:16 - The highest heavens belong to the Lord, but the earth he has given to man.

To back this up Matthew tells us in 6:9-10 - This, then, is how you should pray: Our Father in heaven, hallowed be your name, your kingdom come, **your will be done on earth as it is in heaven.**

G-d also makes it clearly known that He had created only the Earth in all of the Universe to be the one place occupied with life - all the other planets are empty. The only thing that exist in the expanses of space throughout the entire Universe is G-d (no other place is mentioned to have life). Isaiah 45:18 - For this is what the Lord says— he who created the heavens, he is G-d; he who fashioned and made the earth, he founded it; he did not create it to be empty, but formed it to be inhabited— **he says: I am the Lord, and there is no other.** Isaiah 45:21 - Declare what is to be, present it— let them take counsel together. Who foretold this long ago, who declared it from the distant past? **Was it not I, the Lord? And there is no G-d apart from me, a righteous G-d and a Savior, there is none but me.**

There are also other verses in Isaiah that proclaim that G-d alone created everything. Isaiah 44:24 - Thus saith the Lord, thy redeemer, and he that formed thee from the womb, I am the Lord that maketh all things; that stretcheth forth the heavens alone; that spreadeth abroad the earth by myself. Isaiah 45:12 - I have made the earth, and created man upon it: I, even my hands, have stretched out the heavens, and all their host have I commanded. Isaiah 48:13 - Mine hand also hath laid the foundation of the earth, and my right hand hath spanned the heavens: when I call unto them, they stand up together.

Scientists as well as many people say this view is behind the times and narrow-minded. Everyone seems to be poisoned with the idea that the Earth is just a tiny little part of a vast ocean of space, a brewing searing soup that created life by itself, and it is commonly thought that all this must have taken a time period beyond measure and power to create these other planetary bodies in space. When this view is compared to what the Bible says (that the stars were created for signs to mark seasons, as well as simply for G-d's glory) it is ridiculed, suggesting that G-d went through all this work to create billions of galaxies just for us to look at. Psalm 19:1 - The heavens declare the glory of G-d, the skies proclaim the work of his hands. Psalm 8:3 - When I consider your heavens, the work of your fingers, the moon and the stars, which you have set in place. Isaiah 48:13 - My own hand laid the foundations of the earth, and my right hand spread out the heavens, when I summon them, they all stand up together.

Lets get the record straight! Yes, G-d created billions of other galaxies just for His glory and for us to look at. But the Bible does not say that all this took G-d any work at all (especially a long drawn-out period of time). G-d spoke and it was so! Some suggest that it must have been hard, complicated, tough and tiring work because it took the seventh day for G-d to recuperate. Genesis 2:1-2 - Thus the heavens and the earth were finished, and all the host of them. In addition, on the seventh day, G-d ended his work, which he had made; and he rested on the seventh day from all his work, which he had made. However, this is not what the Bible really tells us. G-d didn't have to have six days to create anything, and He didn't have to have one day to recuperate and rest up. The reason G-d rested on the 7th day was later for Israel as a "reflection" in order to serve as a model for man's workweek. Technically, G-d could have created everything in an instant if He had wanted to. But apparently, He had us in mind even before He made us (on the sixth day) and wanted to provide an example for us to follow. According to Exodus 20:9-11 - Six days shalt thou labour, and do all thy work: But the seventh day is the sabbath of the Lord thy G-d: in it thou shalt not do any work, thou, nor thy son, nor thy daughter, thy manservant, nor thy maidservant, nor thy cattle, nor thy stranger that is within thy gates: For in six days the Lord made heaven and earth, the sea, and all that in them is, **and rested the seventh day:** wherefore the Lord blessed the sabbath day, and hallowed it. But this is the furthest thing from the truth that G-d needed to rest! Since Isaiah tells us that G-d does not become tired. Isaiah 40:28-29 - Do you not know? Have you not heard? The Lord is the everlasting G-d, the Creator of the ends of the earth. **He will not grow tired or weary,** and his understanding no one can fathom. He gives strength to the weary and increases the power of the weak.

Now onto another important focus… In so many ways, the Bible tells us that the Earth is the "spiritual center" of the Universe - the center and focus of G-d's attention and concern. We know from Genesis 1:14-15 that the Heavens exists for the benefit of us on Earth. However in the future, science (astronomy) will eventually discover that the earth IS ACTUALLY THE PHYSICAL CENTER OF THE UNIVERSE (as it is in the middle of spheres called "red-shifts" and our Milky Way galaxy is right in the middle of these spheres). Isaiah 40:22 - It is He that sitteth upon the circle of the Earth, and the inhabitants thereof are as grasshoppers; that stretcheth out the Heavens as a curtain, and spreadeth them out as a tent to dwell in.

The Earth may not readily appear to be the "physical center" of the Universe as of yet. Our Earth is also small but the Bible says that's irrelevant because G-d does not measure importance by size. Psalm 33:16 - No king is saved by the size of his army; no warrior escapes by his great strength.

There is also good evidence that indicates that the design features surrounding our Suns position in the Milky Way puts it in the ideal location for life on earth to thrive, but its position also has another advantage of keeping it at a fair distance from the galactic center. If G-d had placed the Sun closer to the Milky Way center, the dense clouds of stars, dust, and gas would have prevented us from seeing into space (the heavens). Instead, G-d put us in a most favorable and advantageous position, not at the remotest rim where the Milky Way would be dim, but far enough out to see clearly into the heights of the heavens so that we may visibly see and witness His great works (both big and wide). This may be the reason why Earth is so small and the Universe is so big. So that we can see outwards the enormity of the heavens and witness the handiwork and glory of G-d. Psalm 19:1 - The heavens declare the glory of G-d, and the firmament sheweth his handy work. For that very reason alone, we should consider ourselves fortunate and blessed for possessing prime real estate in all the vastness of space. By the way, with all our telescopes nobody has ever witnessed a star being born (aside from the Bible telling us they were made by G-d). The scientific theory of star birth tells us that when hydrogen gas comes together it compacts and then bursts, and a star is born. Strangely, though, NOBODY to this day has ever witnessed a new star being born. Rather what we do observe is stars aging and decaying and this is rather a puzzle that has bedeviled astronomers for decades. We hear the usual excuses from the evolutionary astronomers… such as the Universe is billions of years old and things take a long time to evolve, and that's why we don't have a chance to witness these things. Additionally the traveling light factor, which will be talked about shortly. In addition, another evolutionist excuse for why we don't see star birth is because they are hidden in clouds of gas and dust. This element is partially true; some stars are hidden in clusters of gas. And this is where such divergent fields as interferometry, spectrography, infrared, ultra violet, X-ray, radio, chemistry, geometry and trigonometry all come together in astronomy to help us understand what we are seeing in the world around us. However even these sciences are not not completely reliable sources, nor do they paint the full picture. What these spectral fields reveal is that (hypothetically) young disk systems show up as blue, while (hypothetically) more evolved stars, show up as red. However, this still does not equate to star birth in any shape or form. Moreover, as for the color spectrums there could be numerous other things that mimic these reactions. How scientists arrived on the color spectrum had to do with technology started here on Earth in which they test such things as lithium and beryllium from our Sun. They had observed that if you burn lithium or beryllium here on earth and spectroscope it, the readout (pattern) looks a certain way. So when the spectroscopic readout from a star has the same pattern, they surmise it is the same element - that's how they figured it out. Interestingly, science had all this spectroscopic data for about a hundred years before they knew what was causing it within the atom. Then it was explained with the model of electrons circling the nucleus of atoms. To simplify things enormously, electron leaping from orbit to orbit, and emitting or absorbing a photon in the process is what caused the spectroscopic readings and the unique pattern of each element. This same technology is what makes our cell phones that depend on Quantum Physics.

Still, this is not reliable science (remember, its only in theory). In addition, keep in mind that everything we are being told, taught and led to believe, as fact is simply not fact at all. It is only the current theory and thought of the day, which historically is/has often shown not to be true. In addition, as such, there are no such things as facts, everything is simply an opinion based on the information we have at the current time. Anyways, how can all of this positively be proven from more than several hundred light years away? This again is just theorizing. This is still a non-observable science… can that which cannot be observed be considered a science? The scientists should make that point a bit more clear to those who might draw the wrong conclusions. Red and blue colors may shed some light on some aspects, but in no way does it get a free pass to push whatever theory they are currently running with as fact. Furthermore, I would bet beyond a reasonable double that the scientists cannot really be relied upon to tell the truth - the whole truth and nothing but - if you know what I mean. It's like the Miller experiment where by using special equipment they managed to get a few amino acids to link up in the laboratory, and then the media shouts to all of us "scientists have created LIFE in a test tube!" This was not at all the full truth as we've learned in an earlier chapter… what Miller got was a mixture of both right and left amino acids, and again remember, that ALL life consists of only LEFT amino acids. Not both right and left. What Miller got was death. Therefore, we know that the experiment was grossly exaggerated, and still to this day, no ongoing test of this sort has managed to scrounge up the proper ingredients of amino acids necessary for life. Likewise, most astronomy research and retrievals from telescopic instruments that show spectral variances (or differences) fall into the same category as the Miller experiment (a lot of hype with no actual proof).

However, there is still an ultimate question mark that we did not tackle yet. And this is the idea (mentioned in an earlier chapter) that in all actuality I'm not sure if anything we see is truly "material", because what we see with our eyes is really only rays of electrical signals (energy) that is transmitted by neurons and deciphered by a tiny part in the back of the brain. Thus, the brain is also a material thing in itself, and in truthfulness the room that we are in, is really "within" us, and not the other way around. Funny how we can dream in sleep, and within our brain still have the sensations to see with our eyes and touch with our hands, yet we have neither eyes nor hands. What is left? Only an incorporeal/spiritual matter - hence faith.

Now, taking this one step further when we look to the vast stars in our Universe with telescopes (tools that we made with our own limited senses) it becomes questionable if what we are actually seeing is truly what's really out there? Keep in mind, in life, we may never truly know what is true reality when we observe nature, for what we observe is not nature itself, but nature exposed to "our" human method of questioning. Albert Einstein said: It is the theory that decides what we can observe.

Now I know that astronomy also has techniques to figure out the distance and size of stars, planets and galaxies in the Universe, but it becomes invalid when we cant see the WHOLE PICTURE (because we don't know what to compare it against). For instance, we can liken this to a camera (that is another manmade tool) that can fool the human eye with depth perception. Like a camera that views a small fly sitting on a leaf, only the camera views the fly as something huge and the tree and leaf as something small (especially if we take a close up shot of it). Though to us we can relate that it is a small fly sitting on a huge leaf attached to a tree, but the camera cannot decipher this difference. So when our telescopes look out into space we have no clue what things are really supposed to look like because this is not our known environment that we come in contact with on a daily basis. We cannot know whether or not a star is actually big or small, close or far away because we cant decipher what the background is supposed to be. Additionally we also run into the glitch of light itself from great distances away that has just reached the eyes of the telescopes here on Earth. In effect what we are actually seeing is already "past tense" since the light we are seeing in spectral readouts IS NOT RECENT and does not allow us to see what is "currently" happening! Remember (as stated from above) it takes approximately 8 minutes just for light to reach us from the Sun. Therefore light reaching us from across the galaxy or Universe would take years.

Added to all this, time is "relative" (to the speed one is traveling) - it is not something that is absolute, it only exists pertaining to ones conscious point of view or observation. This was one of Albert Einstein's greatest insights in realizing that time is relative when it speeds up or slows down depending on how fast one thing is moving relative to something else (he termed this "time dilation"). But one of the biggest monumental questions posed by some of the greatest scientific thinkers: Is time really real? Time is of your own making, its clock ticks in your head. The moment you stop thought, time too stops dead - Angelus Silesius.

Nothing keeps. There is one law in the universe: NOW - Alfred Sutro, The Open Door

So, if time is relative, and we cant believe what we are even seeing in the scope of all things, EVERYTHING literally becomes mute, and FAITH is the only thing that's left. IT ALL EQUATES TO G-D! Whether you choose to believe this or not is your own prerogative.

Add to this list the scientific labels, facts, words and technicalities… All these become mute and are drowned out in silence. Words fade and parchment crumbles. However, where bright flowers bloom, there is G-d - the pen has dropped ... And those who were seen dancing were thought to be insane by those who could not hear the music - Friedrich Nietzsche.

Interestingly I might also add… Albert Einstein had quoted one of the most notorious sayings in relation to the field of Quantum Mechanics when he said: G-d does not play dice with the universe… However, contrary to this, Stephan Hawkins another notorious scientist later quoted in rebuttal to Einstein: G-d does indeed play dice… However, it is in my opinion that when Einstein said G-d does not play dice with the Universe, he was accurate in his assertion! And the reason for this is that even though it may appear that G-d plays dice with every possible happening event (as Quantum Mechanics suggests), THERE CANNOT LOGICALLY BE ACCIDENTS IN A UNIVERSE WHICH HAS AN "ORGANIZING INTELLIGENCE" (the universal law whose growth increases in units). Some things that we are calling coincidences are much more aligned with spirit/faith then one may have ever thought them to be. Such as the mysterious Phi/Pi.

Yes, some categorize that Phi/Pi lies in the realm of the mysterious. In addition, that its not only singled out as Phi/Pi but as I stated in an earlier chapter that Phi/Pi works around Fibonacci sequence and Golden Mean Spiral. These units of measure are in cahoots and conspire with each other. They are in a state of continuity (relationship and association) with each other. As it was mentioned in a previous chapter: Increasingly realizing this special link, I associated it to the sequence law in nature that instructs all living things when deciding how many units to grow next such as in a trees branches and a plants leaves. This series (of adding some quantity) can be seen everywhere in nature. This series is what has come to be called by several names; the numbers of the Fibonacci sequence also (regarded in conjunction with Phi/Pi), increase at a rate equal to oscillating around the Golden Mean Spiral. The Fibonacci sequence goes like this: 0,1,1,2,3,5,8,13,21,34,55,89,144,… (add the last two numbers together to get the next number 233. The Golden Section numbers are 0.61803 and 1.61803. The Golden String is 1011010110110101101... A sequence of 0's and 1's that is closely related to the Fibonacci numbers and the Golden Section.

They call it mysterious because it's not an exact science that can be pinpointed (kind of like Quantum Mechanics again when trying to estimate the actions and behaviors of microscopic particles). It too is based on faith, likened also to the concept and workings of the Father, Son and Holy Spirit (which will be talked about in the follow chapters).

Continuing on star birth, the fact remains that no scientist or astronomer has ever in the history of humankind ever observed the birth or formation of a star (what they see is things that only "suggest" star birth). Maybe the reason we don't see stars being born is because they simply aren't aside from G-d originally creating them at the beginning of the Universes existence roughly only 6,000 years ago. What we are witnessing now is death and decay (stars dying). Remember, everything is in a state of decay as the laws of "thermodynamics and entrophy" apply to ALL things. We even witness that the gas in the Universe is diminishing. Observations confirm that star formation was much higher in the past due to things being closer together (as well as raw materials available) to interact more. This diminishing of gas in the Universe could certainly account for less star birth. It also mathematically calculates and spells out one inevitable fathomable fact... that the Universe will use up all its available gas and eventually all lights will go out. However, the astronomers estimate that we have billions of years left before this happens. However, billions of years or not, it is obviously the eventual as the law of thermodynamics suggest. However, in definite terms any long time duration estimation is inestimable. The answer is, we simply DO NOT KNOW THE UNIVERSES FATE. Since it was G-d's creation, He could certainly sustain His creation or create anew.

In addition to all of this, I realize that the evolutionists (as well as atheist astronomers) must maintain integrity. As a measure of control, they must mock creationists to make us look stupid. Matthew 7:6 - Give not that which is holy unto the dogs, neither cast ye your pearls before swine, lest they trample them under their feet, and turn again and rend you. What they fail to realize is that there is more evidence for creation then there is (or will ever be) for evolution of any sort. Again, everything that evolution suggests is all in THEORY. Yet, here we are on this tiny planet in a vast Universe and still nobody can PROVE otherwise how life started. With no proof, do the evolutionists realize that all their THEORIES are just based on FAITH? John 3:12 - If I have told you earthly things, and ye believe not, how shall ye believe, if I tell you of heavenly things?

Nevertheless, the most important evidence against the evolutionist in this matter is the very fact that G-d alone tells us in the opening chapter of the Bible that He single-handedly created the Heaven and the Earth as well as the stars. Genesis 1:1 - In the beginning G-d created the heaven and the earth. Genesis 1:16 - G-d made two great lights—the greater light to govern the day and the lesser light to govern the night. He also made the stars. REMEMBER, THERE IS POWER IN WORDS! If you recall, that the very sentence of "Genesis 1:1" in its Hebrew letters equates to the mathematical structure that we find in nature and the Universe (as discussed in an earlier chapter), and it is this very language (the shape and form of the Hebrew letters themselves) in the form of spirals that give rise to all physical material. It is the sole substance to all life, in which G-d generated substance from which He formed letters out of which He combined words, which became things - from substance out of chaos and make nonexistence into existence. THIS IS THE PRIMORDIAL LANGUAGE! And that by definition - the one G-d - a unity underling everything! Remember, it is this very language that the universe itself speaks that is derived from nature, carved on stone and written upon the parchment of the Bible, openly announcing and PROCLAIMING A CREATOR!

Enough said on that, now getting back on track here... Since we are fortunate to have this great view of the vastness of space, from our position in space, its only natural that we of course are even more curious about what lies beyond. Proverbs 25:2 - It is the glory of G-d to conceal a thing: but the honour of kings is to search out a matter. However, even though it is humankinds inclination to search out matters, all space programs have become a colossal failure because it bypasses, ignores and rejects G-d. Instead of reading, learning and applying what the Bible has to tell us about our origin, we are defiantly inquisitive to prove otherwise. In all the endeavors of every space program in the history of humankind that we have ever embarked on, had discovered that the Moon is uninhabited as well as the planet Mars. There is no physical life on these bodies as they've proven themselves to be uninhabitable (we cant even find bacteria). But we didn't have to go to the moon to find this information out because the Bible has already told us that the lunar origin of the moon was created by G-d in Genesis 1:16 - And G-d made two great lights; the greater light to rule the day, **and the lesser light to rule the night:** he made the stars also. Now I know that these are only two bodies that reside in our solar system, but if one wants the ultimate secret of where other planetary bodies in the Universe came from or whether or not they are inhabitable or not, you cant resolve it through restricted and limited perspectives, or through experimental science, but only through THE WORD OF G-D. There is no doubt that mankind will continue searching despite the answer already being found in the Bible. There is a reason for this that G-d has intended... and in the end I firmly believe that science will conduct us, step by step, through the whole range of astronomy, until we arrive, at length, at G-d.

It is inspiring to see how once again evidence supports what Scripture says. The Bible clearly teaches that G-d created the Heaven and the Earth and all the stars – Testifying that the heavens declare the glory of G-d as a calendar to designate the seasons (Holidays), days, and years, and additionally to know when to plant and sow (which correlates with the Jewish holidays). G-d did not say that He created the stars for other intelligent life such as aliens, or for man to travel to the vast regions of space and inhabit, nor did he mention that these other planets were even habitable. He created the earth precisely for humankind to live in. And He created all the Heavens simply to reflect His glory and for the benefit of humankind on Earth - no matter how self-centered this view may seem to the evolutionist.

Closing conclusions… As far as what UFO's or aliens truly are, as aforementioned I'd have to take several guesses… They are either somebody's Merkabah vehicle… angels (light beings)… fallen angel who are posing as aliens with the notoriousblack almond eyes - inducing the Merkabah vehicle in their victim… Earth lights or earthquake lights… ball lightning… refractive or reflecting light similes from birds or bats flying through beams of light… active volcanoes… luminous charged particles that flash, glisten or glow as lights in the sky… the Northern Lights or Aurora Borealis… the military testing hallucinogenic drugs on us as a war agent… and the experimentations of new (and old) top secret military crafts. In addition, it may also be the government involved in some kind of human crossbreeding cloning experiment gone morally and horribly wrong.

More to the point on aliens… The Bible clearly says that G-d created angels, which are Light beings, NOT ALIENS! These light beings live in Heaven, and some that rebelled where cast down to Earth into a lower spirit form (fallen angels). They do not live on other planets. Plainly put, there are no such things as aliens coming from other planets, and there is no physical/material life (of any sort) on any other planet in the entire Universe. Yes, we are all alone in this vast Universe, unique and special, a product of G-d's Magnificence and glory.

Onward…

17 - WORLDWIDE CONSCIOUSNESS...

This chapter is going to be another amazing and surprising chapter, an offshoot of the previous chapter because it is going to elaborate on the importance of humankind and our special purpose in the midst of the Universe. Upon reading this chapter, one may find that the paragraphs are somewhat out of order in terms of sentence structuring while it tries to beat around the bush to engulf the main message.

For starters, lets begin here... As was mentioned in the previous chapter, though human beings were created a little lower than the angels were, we are extremely special. Both in our place in the Universe and to the angels and especially to G-d. Hebrews 2:7 - You made him a little lower than the angels, you crowned him with glory and honor.

The Bible also tells us that we were created in the image and likeness of G-d Himself. Genesis 1:27 - So G-d created man in his own image, in the image of G-d created he him; male and female created he them.

Shortly after this we find recorded that Adam names the animals. Genesis 2:19 - And out of the ground the Lord G-d formed every beast of the field, and every fowl of the air; and brought them unto Adam to see what he would call them: and whatsoever Adam called every living creature, that was the name thereof. What nobody seems to realize by this verse alone is that we have vastly underestimated the first human being - Adam. Adam was sinless, perfect and had the full capacity of his brain. Every part of his brain was in use at the same time. It was 100% functional! Can we imagine the capability that Adam had with his brain to have the ability to name every animal only once and have the recollection to REMEMBER IT! This is astounding! Some currently may say that's impossible that Adam could have been superior than we are today (as the evolution theory claims that we evolved from monkeys and apes, and before that a singled celled organism). However, this assumption is quite WRONG. Now we know that to remember things is a human capability, but there are some people living today that have the ability to remember things beyond the norm. These people are called "savants" and they can remember most everything, from every single book (in detail) that they have ever read (this could be a whole library of books). Some savants can remember dates and events to the date and time. Others can play back music on instruments after only hearing them once. Other savants are like human calculators. They seem to remember as well as calculate all this information because their brains appear to be wired differently (where functional regions are reversed). Some regions of the brain may be reversed because of a head injury or aneurism. Some are just born with it and are labeled as hindered, underdeveloped and retarded. Can you image this! This is the furthest thing from the truth. What these people are displaying is pure genius demonstrating the capability, potential and resilience of the human brain. This alone proves that such feats of the human brain are indeed possible, and ADAM (who did not evolve according to the Bible)

POSSESSED THEM ALL. This is how he could remember all the names that he gave each animal simply because what he saw he never forgot with a photographic memory and instant recall.

So the Question is, what happened to us and why did we loose the potential that Adam had? The answer to this is that we had sinned. But even after Adam was cursed, he was still so genetically powerful with the thrust of life that G-d had given him that it took him 930 years to die! He could even produce children that were so genetically pure and devoid of mutations, that they could marry each other without any harm or defect passed down to the next generation of children. In fact, everyone mentioned in the Bible after Adam had slowly downgraded and degenerated in life expectancy and brain equivalency - from Adam down to King David. Genesis 5:1-32 - This is the book of the generations of Adam. In the day that G-d created man, in the likeness of G-d made he him; Male and female created he them; and blessed them, and called their name Adam, in the day when they were created. And Adam lived an hundred and thirty years, and begat a son in his own likeness, and after his image; and called his name Seth: And the days of Adam after he had begotten Seth were eight hundred years: and he begat sons and daughters: **And all the days that Adam lived were nine hundred and thirty years: and he died.** And Seth lived an hundred and five years, and begat Enos: And Seth lived after he begat Enos eight hundred and seven years, and begat sons and daughters: **And all the days of Seth were nine hundred and twelve years: and he died.** And Enos lived ninety years, and begat Cainan: And Enos lived after he begat Cainan eight hundred and fifteen years, and begat sons and daughters: **And all the days of Enos were nine hundred and five years: and he died.** And Cainan lived seventy years and begat Mahalaleel: And Cainan lived after he begat Mahalaleel eight hundred and forty years, and begat sons and daughters: **And all the days of Cainan were nine hundred and ten years: and he died.** And Mahalaleel lived sixty and five years, and begat Jared: And Mahalaleel lived after he begat Jared eight hundred and thirty years, and begat sons and daughters: **And all the days of Mahalaleel were eight hundred ninety and five years: and he died.** And Jared lived an hundred sixty and two years, and he begat Enoch: And Jared lived after he begat Enoch eight hundred years, and begat sons and daughters: **And all the days of Jared were nine hundred sixty and two years: and he died.** And Enoch lived sixty and five years, and begat Methuselah: And Enoch walked with G-d after he begat Methuselah three hundred years, and begat sons and daughters: **And all the days of Enoch were three hundred sixty and five years: And Enoch walked with G-d: and he was not; for G-d took him.** And Methuselah lived an hundred eighty and seven years, and begat Lamech. And Methuselah lived after he begat Lamech seven hundred eighty and two years, and begat sons and daughters: **And all the days of Methuselah were nine hundred sixty and nine years: and he died.** And Lamech lived an hundred eighty and two years, and begat a son: And he called his name Noah, saying, This same shall comfort us concerning our work and toil of our hands, because of the ground which the Lord hath cursed. And Lamech lived after he begat Noah five hundred ninety and five years, and begat sons and daughters: **And all the days of Lamech were seven hundred seventy and seven years: and he died.** And Noah was five hundred years old: and Noah begat Shem, Ham, and Japheth. Genesis 7:6 - Noah was six hundred years old when the floodwaters came on the earth. Genesis 9:28 - After the flood Noah lived 350 years. Genesis 9:29 - **Altogether, Noah lived 950 years, and then he died. From here on down the life spans after the flood dropped considerably.** Genesis 11:11 - And after he became the father of Arphaxad, Shem lived 500 years and had other sons and daughters. Genesis 11:12-32 - When Arphaxad had lived 35 years, he became the father of Shelah. And after he became the father of Shelah, **Arphaxad lived 403 years and had other sons and daughters.** When Shelah had lived 30 years, he became the father of Eber. And after he became the father of Eber, **Shelah lived 403 years and had other sons and daughters.** When Eber had lived 34 years, he became the father of Peleg. And after he became the father of Peleg, **Eber lived 430 years and had other sons and daughters.** When Peleg had lived 30 years, he became the father of Reu. And after he became the father of Reu, **Peleg lived 209 years and had other sons and daughters.** When Reu had lived 32 years, he became the father of Serug. And after he became the father of Serug, **Reu lived 207 years and had other sons and daughters.** When Serug had lived 30 years, he became the father of Nahor. And after he became the father of Nahor, **Serug lived 200 years and had other sons and daughters.** When Nahor had lived 29 years, he became the father of Terah. And after he became the father of Terah, **Nahor lived 119 years and had other sons and daughters.** After Terah had lived 70 years, he became the father of Abram, Nahor and Haran. This is the account of Terah. Terah became the father of Abram, Nahor and Haran. And Haran became the father of Lot. While his father Terah was still alive, Haran died in Ur of the Chaldeans, in the land of his birth. Abram and Nahor both married. The name of Abram's wife was Sarai, and the name of Nahor's wife was Milcah; she was the daughter of Haran, the father of both Milcah and Iscah. Now Sarai was barren; she had no children. Terah took his son Abram, his grandson Lot son of Haran, and his daughter-in-law Sarai, the wife of his son Abram, and together they set out from Ur of the Chaldeans to go to Canaan. But when they came to Haran, they settled there. **Terah lived 205 years, and he died in Haran.** Genesis 23:1 - **Sarah lived to be a hundred and twenty-seven years old. Genesis 25:7 - Altogether, Abraham lived a hundred and seventy-five years.** Genesis 35:28 - **Isaac lived a hundred and eighty years.** Genesis 47:28 - **Jacob lived in Egypt seventeen years, and the years of his life were a hundred and forty-seven.** Job 42:16-30 - After this, **Job lived a hundred and forty years**; he saw his children and their children to the fourth generation. And so he died, old and full of years. Exodus 6:16 - These were the names of the sons of Levi according to their records: Gershon, Kohath and Merari. **Levi lived 137 years.** Exodus 6:18 - The sons of Kohath were Amram, Izhar, Hebron and Uzziel. **Kohath lived 133 years.** Exodus 6:20 - Amram married his father's sister Jochebed, who bore him Aaron and Moses. **Amram lived 137 years.** Deuteronomy 34:7 - **Moses was a hundred and twenty years old when he died,** yet his eyes were not weak nor his strength gone. Joshua 24:29 - After these things, **Joshua son of Nun, the servant of the LORD, died at the age of a hundred and ten.** 1 Chronicles 29:26-28 - David son of Jesse was king over all Israel. He ruled over Israel forty years—seven in Hebron and thirty-three in Jerusalem. He died at a good old age, having enjoyed long life, wealth and honor. His son Solomon succeeded him as king. **(When we add this up, David lived to 80 years of age).**

It seems that after the flood Great Floodthe Earth was completely different from the Earth before (which will also be the discussion of a future chapter). There were large-scale global changes. These would include changes in the atmosphere including weather and temperature as well as the water cycle. This also included physical features of the Earth such as soil, rocks, even minerals such as the build up of coal. Because of the changes in the atmosphere, this would further allow for radiation from space and the Suns rays to reach the Earth thus influencing the ozone concentration and ultra violet light that changed and altered the growth of plants and other living things. Additionally genetics, diet, and a host of other chemical and physiological changes most likely took effect, wherefore the food that people ate before the flood was most likely fitter for the prolongation of life - therefore these changes caused a rapid decline of the longevity of post flood humanity (remember food is also dependant on the laws of thermodynamics). Even after Adam and Eve had sinned, and were kicked out of the garden of Eden, cursed with the providence (external circumstances) of G-d, these changes were still considered relatively good compared to what transpired after the flood where things got much worse.

Now before Adam had sinned and was cursed with the providence of G-d, he had a sinless nature about him. The question now arises, what kind of knowledge would a sinless person possess. I am going to go ahead and point this out before the actual chapter that supplies evidence that Jesus was the Messiah and G-d incarnate... When we read the Bible, it clearly paints a picture for us that the intelligence of Jesus at 12 years old was quite extraordinary and above average! We read that when Jesus was a young boy He was found in the temple in Jerusalem discussing scripture and theology with the teachers of the religious law, and apparently, they were very impressed by Him. We find this information in Luke 2:40-60 - And the child grew, and waxed strong in spirit, **filled with wisdom**: and the grace of G-d was upon him. Now his parents went to Jerusalem every year at the feast of the passover. **And when he was twelve years old**, they went up to Jerusalem after the custom of the feast. And when they had fulfilled the days, as they returned, the child Jesus tarried behind in Jerusalem; and Joseph and his mother knew not of it. But they, supposing him to have been in the company, went a day's journey; and they sought him among their kinsfolk and acquaintance. And when they found him not, they turned back again to Jerusalem, seeking him. And it came to pass, that after three days **they found him in the temple, sitting in the midst of the doctors, both hearing them, and asking them questions. And all that heard him were astonished at his understanding and answers. And when they saw him, they were amazed: and his mother said unto him, Son, why hast thou thus dealt with us? behold, thy father and I have sought thee sorrowing. And he said unto them, How is it that ye sought me? wist ye not that I must be about my Father's business?** And they understood not the saying which he spake unto them. And he went down with them, and came to Nazareth, and was subject unto them: but his mother kept all these sayings in her heart. **And Jesus increased in wisdom** and stature, and in favour with G-d and man. This confirms that even when Jesus was a kid He was remarkably intelligent, and most evidently smarter than the smartest scholars of Israel.

To accentuate on the brainpower of Jesus, we additionally find in the biblical scripture that Jesus had more than just intelligence and wisdom, He was also sinless. He additionally had total authority and dominion over the fish of the sea apparently just like G-d as well as Adam and Eve before they sinned. We find that Jesus had foresight of where fish were in the Sea of Galilee in order to catch them in nets. (The Sea of Galilee is Israel's largest freshwater lake). Genesis 1:26 - And G-d said, Let us make man in our image, after our likeness: **and let them have dominion over the fish of the sea,** and over the fowl of the air, and over the cattle, and over all the earth, and over every creeping thing that creepeth upon the earth. Luke 5:1-11 - And it came to pass, that, as the people pressed upon him to hear the word of G-d, he stood by the lake of Gennesaret, (also known as the Lake of Galilee). And saw two ships standing by the lake: but the fishermen were gone out of them, and were washing their nets. And he entered into one of the ships, which was Simon's, and prayed him that he would thrust out a little from the land. And he sat down, and taught the people out of the ship. **Now when he had left speaking, he said unto Simon, Launch out into the deep, and let down your nets for a draught. And Simon answering said unto him, Master, we have toiled all the night,**

and have taken nothing: nevertheless at thy word I will let down the net. And when they had this done, they inclosed a great multitude of fishes: and their net brake. And they beckoned unto their partners, which were in the other ship, that they should come and help them. And they came, and filled both the ships, so that they began to sink. When Simon Peter saw it, he fell down at Jesus' knees, saying, Depart from me; for I am a sinful man, O Lord. For he was astonished, and all that were with him, at the draught of the fishes which they had taken: And so was also James, and John, the sons of Zebedee, which were partners with Simon. And Jesus said unto Simon, Fear not; from henceforth thou shalt catch men. And when they had brought their ships to land, they forsook all, and followed him.

This tells us that Jesus is of a divine nature because He had dominion over the fish (He was apparently also sinless). We know this because He apparently was not under the "curse of sin" (the curse brought about by Adam and Eve sinning, "eating of the tree of knowledge" in the garden of Eden). Remember before Adam and Eve had sinned they had dominion over the fish of the sea. Jesus therefore not under the curse, (and sinless) had dominion over the fish of the sea. He had apparently commanded the fish to jump into the nets.

Here is another message that stresses that Jesus was not under the curse. Matthew 17:24 - And when they were come to Capernaum, they that received tribute money came to Peter, and said, Doth not your master pay tribute? He saith, Yes. And when he was come into the house, Jesus prevented him, saying, What thinkest thou, Simon? of whom do the kings of the earth take custom or tribute? of their own children, or of strangers? Peter saith unto him, Of strangers. Jesus saith unto him, Then are the children free. Notwithstanding, lest we should offend them, **go thou to the sea, and cast an hook, and take up the fish that first cometh up; and when thou hast opened his mouth, thou shalt find a piece of money: that take, and give unto them for me and thee.** Someone had apparently dropped a coin (shekel) into the sea of Galilee at some point in time, and in foresight (the ability to think ahead) Jesus had commanded a fish swimming by the coin to gulp it up and rush to Peters hook to be caught - that is DOMINION! Not only is it dominion, its foresight along with the knowledge and capability that none of us currently possess Today..

Therefore, the human brain had immaculate accuracy and ability before the curse took affect. I will explain this in detail as you continue to read. However, for now we need to focus on the brain capability that we currently have left. We still have quite a bit power left, enough to enable us to acknowledge G-d. We also have the ability to see our "free will" take effect over the material world that we live in. As was discussed in an earlier chapter that the knowledge of "Quantum Mechanics" has opened a loophole in the journals of science. We found that the human brain may be more than just chemicals. Remember, it can manipulate subatomic particles through electromagnetic forces. Every time we think, we are forming electrical circuits in our brain that quantumly extend out in all direction effecting the ethereal field around us. For the disbelievers at heart, a simple experiment that one can do to prove that the conscious power of the human brain can have an effect over material things is to take two canning jars and fill them less than halfway with cooked white rice (white rice enables us to see a noticeable color difference). Seal the jars good so that no air can get in or out, and label one with a sign that reads "good" and the other with a sign that reads "bad". Place them next to each other, side-by-side (with maybe a three inch difference between them). Then watch them, making a point to look at the jars filled with rice every so often throughout the day. You will begin to see a noticeable difference in the rice's color taking place between the two jars in about a couple of weeks. This usually works best if they are placed where a large majority of people can view and observe them, but it works just as well sitting on a counter top or kitchen table where they can be viewed at least 3 times a day by one or two people. To ones amazement, the one jar (filled with rice) that is labeled "good" stays preserved longer than the jar (filled with rice) labeled "bad". One should be able to see that the "bad" jar filled with white rice starts to discolor more quickly (turning yellow in color with individual pieces of rice looking black or brown).

Why does this happen? The secret lies in the water that the rice was cooked in. Remember water can retain memory through vibration (this experiment can be done with any food item that contains water). Human thoughts in essence are subtle vibrations produced by the brain during conscious thinking and with the aid of the human eyes can be directed towards and into an object. The "Anthropic Principle" comes into play here, which simply suggests that observers are necessary to bring about reality. Alternatively, in other terms, all that we are is a result of what we have thought. Believe this or not, but words have power. Using bad words can increase stress and turn the body's pH from alkaline (healthy) to acidic (unhealthy). Words can actually change the way we think and feel, and simultaneously speaking the correct form of words and thinking the correct thoughts actually has the power to change ones genetic material (DNA). Oh, I can just here the critics now! However, these types of things cannot just be turned aside from. Attention must be given to the fact that there are some unexplained physics going on here. Again, the rice experiment is proof! (I encourage all readers to do this experiment for themselves).

Furthermore, prayer works in the same way. A small but growing body of evidence suggests that prayer may indeed play a role in healing, (even from far distances). Prayer has been linked to lower blood pressure, decrease depression, build stronger immunity and even increase life longevity. But the most significant aspect of prayer reveals itself through several study's which have suggested that patients who were prayed over on a daily basis for several weeks (but did not know it) had survived, healed faster, and had fewer complications then those that were not prayed over.

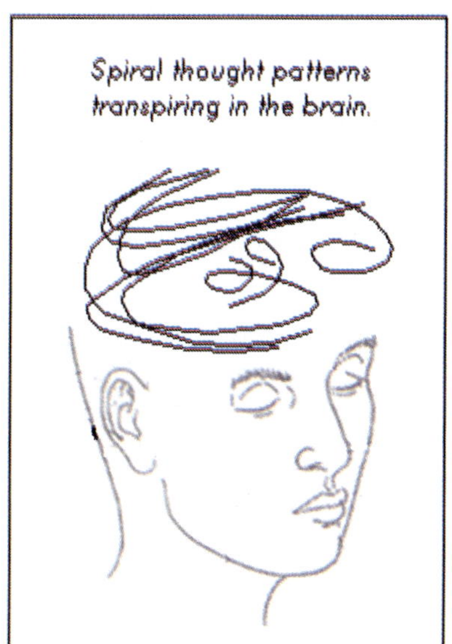

Something phenomenal is happening here! Even though we cannot fully explain this invisible force in scientific terms by the mechanism through which prayers and thoughts are propagated outwards. Somehow, these thoughts and prayers effect/direct the atomic particles much like the science of "Quantum Mechanics" which is now discovering that the person performing the experiment actually influences the outcome. Now when these thoughts are propagated outwards they are also dependent on the "one law" - in other words they span out in a "unit of increasing measure and growth" spiraling as they go (this is usually Phi/Pi). Likewise (if you can recall) any unit of measure that increases in spiraling growth can form the Hebrew letters. This is also part of the mystery of The Primordial Language. G-d somehow has programmed conscious human thought into spelling letters, and letters, which in turn correspond to words. It is a language of spiraling waves (atomic particles) that continually oscillate as they rise and fall. However G-d knows what these thoughts are at the instant we think them (even in advance) because our minds spin out the very spirals that form the Hebrew letters which makeup words, that give a detailed account and description of our innermost feelings and thoughts. We cannot evade this! None of our thoughts are personal to G-d - He literally knows all, even before we speak it!

Again, the lifespan of a particle:

This is an obvious feature of the shape and form of the "CURSIVE" Hebrew letters. Again, see for yourself:

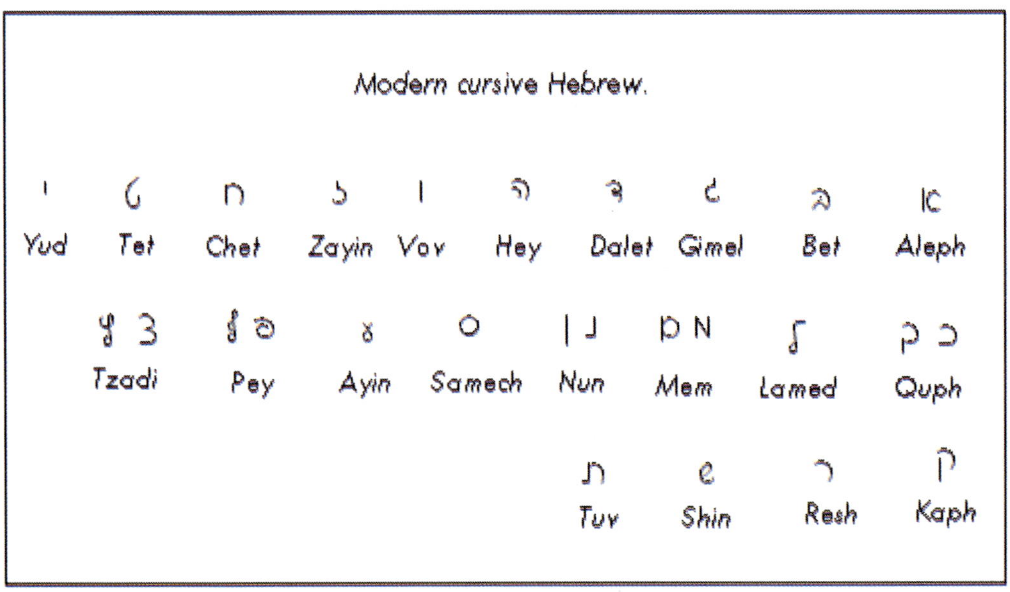

Now when we talk about human consciousness it begs the question of "what a actual though is" and where does it come from? As it turns out, a thought is a quantum fluctuation in the field of energy and information that gives rise to the material Universe (the "Antropic Principle" again). However, no matter what we say about a thought in our comprehensive human language, it is just not a thought, it is a "label" (like the good and bad signs put on the canning jars filled with rice). A label of information and energy coming out of our brain. The whole world and every living thing in it is made up of information and energy - down to the molecules and subatomic particles. Therefore, a thought is just that - information and energy. So everything is made of the same stuff, so in essence we are all part of a thinking Universe. However, come to find out that subatomic particles are not really things either! They are actually only "fluctuations" of energy and information in a huge void, and when you look further beyond this point, you see absolutely nothing! So now, what's this "nothingness" from which we all come from? Is it just an empty void or could it be the womb of creation (the infinite consciousness of G-d). Is it possible, just vaguely possible that the natural world (as it was set up by G-d) goes to the same exact place to create a thought as it does to create a material thing, or a possible future? In this bilingual sense (conjoined with our creator) we are the creators of our own future (but at the same time) endowed to us by our creator - G-d. Again, the Bible speaks boldly of itself when it states that G-d had given us this ability in the form of "free will and choice". Genesis 2:16-17 - And the Lord G-d commanded the man, saying, Of every tree of the garden thou mayest **freely** eat: But of the tree of the knowledge of good and evil, thou shalt not eat of it: for in the day that thou eatest thereof thou shalt surely die. However, because G-d is infinitely and knowledgably beyond and above all probabilities, only He alone can ultimately determine our future, which has already been prophesized for the total verdict of humanity, which according to the Bible will culminate to a horrible end for the unbeliever. However, for the individuals of faith and belief in G-d, they are "exempt" from this total. The mathematical genius that G-d is, allows Him room to calculate this in. In terms known to us, this is called "escrow" (something held independently until the condition is met).

Now moving along here in this chapter, as we have been discussing… and as was also discussed in previous chapters, that just like vibrations of energy have an influence over us, or in the same way a virus or a disease is contagious, so are IDEAS contagious. Ideas and thoughts that give rise to a "WORLDWIDE CONSCIOUSNESS". Additionally as discussed in an earlier chapter, all the "probabilities" that can be found within the Bible code or other literature, (even the advancement of technology itself) is not because somebody saw the future (other than G-d or His divinely inspired prophets). Its because it was simply thought of - this is the conditioning force (governed by free will). What the mind can think, the mind can create. The human mind (that G-d created with free will) is a powerful thing! In essence, the mind in itself is a sort of machine for the distillation of consciousness in which an ethereal energy field that has no limitations, no form, no materialness, showed up from a field of "INTENTION". All of this is coming out of the human mind. To further, extenuate on this, at this present moment in time there is a coagulation of past energy and future energy. The energy that comprises the future aspects of our reality has not yet coalesced into something real or material, it is still floating about in a super charged quantum state. Those quantum states can be affected or obstructed by thoughts. In addition, as long as humanity does not learn to work as "ONE", nothing will be accomplished. We must spread the message of love, hope and faith so that we have something to offer, something distinct, and something greater than anything else does in the universe (the hosts of Heaven). To accomplish great things we must not only act but also envisage, not only plan, but also believe - Anatole France. And it was Jesus who said: For most certainly I tell you, if you have faith as a grain of mustard seed, you will tell this mountain, "move from here to there", and it will move. Jesus also said (in so many words) that if you have thought it, you have already committed it. For that is the true purpose of the unseen in the realm of Quantum Mechanics - to bring about the thoughts that give rise to CHANGE!

We do not entirely understand the full extent and potential of how this mechanism of "intention" works. It can be perceived as a ghostly (invisible) underlining feature and linkage living within the meshwork of our environment. However, because we cannot see this force in action it has therefore eluded us. HOWEVER, IT HAS NOT ELUDED US COMPLETELY since it has come to my attention that several experiments have been done to measure the effects that human consciousness plays on worldwide random events. Without going into full detail about how this worldwide consciousness is measured, rather put in simple terms, it is measured by devices placed around the world that pick up "electronic white noise" (otherwise know as frequency). Results upon results have been accumulating to prove that when a worldwide event takes place that focuses the attention of billions of people, the random frequency sequence changes to go significantly more one way then another way, indicating a definite response outside the realm of probability! This white noise tends to peak during such events like national prayer day when large coordinated meditations focus on thoughts of peace. This also occurs with the death of an important person, significant earthquakes, and events in which we are collectively conscious of and emotionally react to. More noticeably the event that took place on September 11th when so many people worldwide were focused on the events of the World Trade Center disaster. WHAT'S EVEN MORE SIGNIFICANT IS THAT THESE RESULTS STARTED PEAKING **"BEFORE"** THE ACTUAL EVENT HAPPENED! This seems to strongly indicate that consciousness has an effect on otherwise "upcoming random" events, WHICH ARE APPARENT FUTURES! Hence, I also found it very interesting to note that "sound/frequency" is the main benefactor, which has a relating factor to most of the subjects that are talked about in this book. But its not only "sound/frequency" that gives us such results, we can also find them in other humanly made devices like the computer and the Internet by using a computer program of "Internet Web Bots " that scan the World Wide Web (WWW) using a system of cyber spiders that search for the most commonly used words. These keywords are then analyzed in a way that helps narrow down future events. Though it is not a accurate source, it is aimed at tapping into the "collective unconsciousness" of humans that interact with the Internet. We are not human beings having a spiritual experience. We are spiritual beings having a human experience - Teilhard de Chardin.

Whether these worldwide consciousness experiments are done by Government, University, or individual research… the experiments on pinning-down the actuality that such an entity truly exists brought about by the human consciousness has turned out to be quite legitimate. Although its not merely taking my word for it, but because these things can be checked out for themselves with little effort. They are public information, with the nearest results being researched and viewed on the Internet. But I'm predicting there will be many more projects like this to arise in the future, as any scientific method that demonstrates a measurable connection between the mind and the events in the physical world is worthy of further study as long as we can link it to G-d. Sadly there are many scientist involved in such studies that truly have no real understanding of what's transpiring with these experiments. They blindingly conclude the result to be wholly mans doing thus leaving out the spiritual connection aspect of G-d playing a part.

What I also found interesting about these experiments were the results of the Earth and Sun demonstrating their interactive and mutually exchanging nature. We might have some evidence to prove that both the Earth and the Sun are mutually connected (as one living being) with a shared consciousness. This estimation was based on the current solar physics when scientists received peak frequencies (from the Sun) that were observed following "nuclear testing" in which the Sun unleashed a tremendous X-Ray burst that triggered one of the largest proton storms of a peaking 11-year solar cycle. Whether the solar peak was actually caused from nuclear testing is inconclusive, however we cannot deny the fact that the Sun is a very important compromising aspects of our existence! Evidence is accumulating that shows us that how we react here on Earth (in negative ways) may also have some other ill acting effect on our Earth as well as our solar system and the rest of the Universe - if as previously mentioned in an earlier chapter that the traits of cells in our body all show signs of having a higher purpose, then it also holds true that our Sun, Earth, solar system, and the rest of the Universe is no different. In a sense, the cells of our body compared to the Universe are (again) a sort of microcosm, (a small version of something larger). And therefore, just as the stomach cells do not only think of themselves, they are also thinking of how to digest food so it can assist the heart, liver and the brain. While the liver cells somehow converses to the stomach cells that they need to operate together to help remove toxins so the rest of the body doesn't get damaged, and the heart cells asks the liver and the brain how it can pump blood so the rest of the body gets tending to, while in concert the brains cells are articulating on how to coordinate all these activities. In this awareness, every cell has a higher purpose, and every cell is conscious of what's happing in the rest of the body. Every cell also cooperates in agreement, operating with the efficiency of giving and receiving (likewise the Earth and the Sun). It is in this same respect that we as humans are interconnected with each other, and hold the same interdependence of all that exists in the Universe (including the other stars and planets). This is the infinite consciousness of G-d who is the individual entity behind the main body. In essence, the infinite consciousness of G-d can be likened as the womb of creation. And remember, from the previous chapter we just read, according to the Bible we (Earth) are the apparent center of interest and concern to G-d.

Aside from the cells of the body, some often ask the question, "where is G-d". And the answer to this must be that He exists in an unseen dimension undetectable to our human eyes. Take for example; "radio" is a method of sending out sounds for long distances by the use of electricity. The air all around us is full of radio waves broadcasted by different stations. The place where you sit as you read these lines is literally filled with radio waves, music, voices, and noises of various kinds are all around you, passing through the walls, and even your body. However, you can't feel them, hear them, or see them, but they are there, coextensive with every visible thing occupying the very SAME SPACE. Yet, each wave (or channel) exists in a different world entirely on different frequencies. Therefore they don't interfere with each other, they don't mix and become all scrambled up. This is what is also referred to by a "realm or dimension" - it is like a frequency. It can occupy the same space and yet be another place altogether. The same could apply to the place called Heaven. That's how the theory of G-d (theoretically) can be in Heaven and in the room where you are sitting at the same time. He is simply on a different frequency, in a different dimensional realm, but occupying the same space at the same time - EVERYWHERE! With the ability to permeate all of existence. However, according to the Bible G-d does not reside (live) here on Earth. Yes, He permeates here, but His home is in the outmost highest Heaven. Deuteronomy 10:14 - To the Lord your G-d belong the Heavens, even the highest heavens, the earth and everything in it. 1 Kings 8:27 - But will G-d really dwell on earth? The Heavens, even the highest heaven, cannot contain you. How much less this temple I have built! Nehemiah 9:6 - You alone are the Lord.

You made the heavens, even the highest heavens, and all their starry host, the earth and all that is on it, the seas and all that is in them. You give life to everything, and the multitudes of heaven worship you. Job 22:12 - Is not G-d in the heights of Heaven? And see how lofty are the highest stars! Psalm 102:19 - The Lord looked down from his sanctuary on high, from heaven he viewed the earth. Psalm 108:4 - For great is your love, higher than the heavens; your faithfulness reaches to the skies. Ephesians 4:10 - **He who descended is the very one who ascended higher than all the heavens, in order to fill the whole universe.**

In sharing my most profound thoughts about this ghostly (invisible) underlining feature living within the meshwork of our environment.... whether it be of nature, man, machine, cell or frequency… The Bible expresses it very clearly that G-d has assigned humankind as the acting force of this planet. Genesis 1:26 - Let them have dominion over the fish of the sea, and over the birds of the air, and over the cattle, and over all the wild animals of the earth, and over every creeping thing that creeps upon the Earth. In this regard, it has become clear that we are "INSTRUMENTS" that serve a very important purpose as the guiding mechanism of this earth, and that this ecosystem depends on not only our intelligence, but our observation and understanding of the workings of nature, but more importantly in this case, our "conscious guidance" endowed to us by our creator. Our ultimate power rests in the simple wisdom and understanding of our human brain and its co-creative effects of the clairvoyant energy of focused human consciousness, and the demonstrated physical effect of our thoughts in practical application. All of our special qualities were given to us at the moment of our creation by G-d, NOT EVOLUTION! Therefore it is necessary to recognize the potential unification and value of our connectedness with all living things, and that we have the human potential and ability "within ourselves" while collaborating as a whole to combine our collective intentions with the energy of our positive thoughts (as in the rice experiment) towards creating a better balanced environment and a more peaceful existence. All of this in obedience to His natural law. The sheer concept and idea alone, that we are equipped with the power to counter-act and balance-out these negative forces with just our "beliefs" and OUR POSSITIVE THOUGHTS is perhaps beyond human imagination! It falls into the realm of Quantum Mechanics again. Hence belief, faith, choice and prayer. So as one can see there is far more to true reality (G-d) by the invisible aspect of things.

The Bible also indicates that "each and every one of us" is important! For it was by Noah that all of humanity was spared. Genesis 6:5 - The Lord saw that the wickedness of humankind was great in the earth, and that every inclination of the thoughts of their hearts was only evil continually. And the Lord was sorry that He had made humankind on the earth, and it grieved him to His heart. So the Lord said, I will blot out from the earth the human beings I have created - people together with animals and creeping things and birds of the air, for I am sorry that I have made them. **But Noah found favor in the sight of the Lord.** Genesis 6:17 - I am going to bring a flood of waters on the earth, to destroy from under heaven all flesh. However, it may seem that G-ds habitual irony and wrath acquires a sort of tragic pathos in the face of death - Plato, it is still in view of the formidable complexity of the importance of even a single human being. Genesis 6:9 - Noah was a righteous man, blameless in his generation, Noah walked with G-d.

THIS IS WHY (INDIVIDUALLY) EVERY ONE OF US IS IMPORTANT!

In writing this book, I also wanted to come to an understanding of what evil truly is. (Which will also be the subject of a later chapter)? I wanted to understand it based on this book accompanied by nature, science and biblical scripture. What I had discovered about evil, was that it is an opposite and opposing force and that it "also" fell within the boundaries of Phi/Pi, Fibonacci sequence, and Golden Mean Spiral - in which its frequency (low thoughts) describes defiance and deformity with a degree of irregularity against the designated growth sequence of Phi/Pi, Fibonacci sequence and Golden Mean Spiral. What I found was, where Phi/Pi represents growth and life, negative Phi/Pi (anti Phi/Pi) represents decay, darkness and death.

Therefore I came to the understanding that the beast described in the Bible numbering 666 is the exact opposite process of Phi/Pi. Interesting enough, when dividing the 360 degrees of a circle by 54 degrees you get 6.66, furthermore, the other side of a 54 degree angle in a right angle is 36 degrees, and 36 divided by 54 is also .666. Revelations 13:18 - this calls for wisdom: let anyone with understanding calculate the number of the beast, for it is the number of a person. Its number is six hundred sixty six. That little revelation is quite interesting, as we know that Phi/Pi appears throughout the entire natural world and that includes every physical proportion of the human body - as was shown in the chapter on "Earths natural law". In that sense, perhaps it is the number of "mankind" in turning our thoughts against G-d's nature. This negative force can work its sneaky way through weak-minded people, using these individuals to work its way. Biblical literature tells us all about this malevolent force (which will examined in better detail in a coming chapter). In addition, as you continue to read on in the following chapters, you will gain more insight into who/what the antichrist is and his disguising and corrupting nature. However, make no mistake "evil" is very real in this sense. Some people actually think (like the evolutionists) that evil is part of the normal workings of the natural world. However on the contrary, NO IT IS NOT a natural balance of G-d, it is a "consequence" that we have chosen in sinning, for it tells us in the Bible that G-d created us "perfect" in the beginning and there was no evil or death. G-d tells us in Genesis during His creation that everything was "VERY GOOD". Genesis 1:31 - And G-d saw every thing that he had made, and, behold, it was "very good". In addition, the evening and the morning were the sixth day.

Many of us have no clue as to how a world that is structured without sin would be. It raises the logical question that if there was no sin in the beginning how could everything be "very good", when it's evident today that death, disease and extinctions are prevalent which are all part of the law of entropy and thermodynamics (as discussed earlier). This does not equate to "very good"! So how could everything have been very good? The answer is that G-d had, in effect, SUSTAINED us. And we can later witness this phenomenon in Deuteronomy 8:4 - Thy raiment waxed not old upon thee, neither did thy foot swell, these forty years. Deuteronomy 29:5 - And I have led you forty years in the wilderness: your clothes are not waxen old upon you, and thy shoe is not waxen old upon thy foot. Nehemiah 9:21 - Yea, forty years didst thou **sustain** them in the wilderness, so that they lacked nothing; their clothes waxed not old, and their feet swelled not. We can also later find in the New Testament Bible that Jesus was in effect able to "sustain" all things by the power of His word. Hebrews 1:3 - The Son is the radiance of G-d's glory and the exact representation of his being, **sustaining all things by his powerful word. After he had provided purification for sins**, he sat down at the right hand of the Majesty in heaven. And that by the power of His word we have everlasting life. John 6:47 - I tell you the truth, he who believes has everlasting life.

So, what would actually constitute as "anti Phi/Pi"? Well we now know that it physically represents decay, darkness and death. But where do we see the signs of "anti Phi/Pi" in nature and the world around us?

We can look to many examples in nature, take for example a tree that grows in a deformed, misshapen or buckled way… where we could see that some of its leaves and branches are missing or malformed, or the trunk and branches have grown in a unbalanced or different kind of way. There may be many reasons for such deformity, but to name a few, this tree may have be planted too close to an electrical cable or electrical grid, which has either stunted its growth or caused it to receive or soak up the wrong frequency signal through water which (if you recall) retains memory in the form of frequency. This frequency then confuses the tree and directs it to grow incorrectly (anti Phi/Pi). Trees can even get a type of cancer (bumps), which is supposedly very painful for the tree. Additionally, perhaps the tree may be in a place that doesn't get enough sun or water, or perhaps an area that is crowded out with other trees and plants, as well as being infested with insects, molds or fungus. These are all factors that play a part in the trees growth and survival. Not that this tree itself is evil, but that it is somehow subjected to "FAILING" (not making the mark) which in essence is considered sin, since G-d had originally created us perfect (and planted us in a perfect place - a fertile garden in a warm place between four rivers, with a river flowing through it and watering it. Genesis 2:8-10 - Now the Lord G-d had planted a garden in the east, in Eden, and there he put the man he had formed. In addition, the Lord G-d made all kinds of trees grow out of the ground—trees that were pleasing to the eye and good for food. In the middle of the garden were the tree of life and the tree of the knowledge of good and evil. A river watering the garden flowed from Eden, from there it was separated into four headwaters.

Now for those people, who think that these types of things are not evil and are just subject to cruel and unfair feats of nature, they certainly need to reconsider, thus seeing it in a different light. Lets examine what the Bible tells us about this… Matthew 13:1-9 - That same day Jesus went out of the house and sat by the lake. Such large crowds gathered around him that he got into a boat and sat in it, while all the people stood on the shore. Then he told them many things in parables, saying: A farmer went out to sow his seed. As he was scattering the seed, some fell along the path, and the birds came and ate it up. Some fell on rocky places, where it did not have much soil. It sprang up quickly, because the soil was shallow. But when the sun came up, the plants were scorched, and they withered because they had no root. Other seed fell among thorns, which grew up and choked the plants. Still other seed fell on good soil, where it produced a crop—a hundred, sixty or thirty times what was sown. He who has ears, let him hear. Matthew 13:18-23 - Listen then to what the parable of the sower means: When anyone hears the message about the kingdom and does not understand it, the evil one comes and snatches away what was sown in his heart. This is the seed sown along the path. The one who received the seed that fell on rocky places is the man who hears the word and at once receives it with joy. But since he has no root, he lasts only a short time. When trouble or persecution comes because of the word, he quickly falls away. The one who received the seed that fell among the thorns is the man who hears the word, but the worries of this life and the deceitfulness of wealth choke it, making it unfruitful. But the one who received the seed that fell on good soil is the man who hears the word and understands it. He produces a crop, yielding a hundred, sixty or thirty times what was sown. Matthew 13:24 - Jesus told them another parable: The kingdom of heaven is like a man who sowed good seed in his field. Matthew 13:27 - The owner's servants came to him and said, Sir, didn't you sow good seed in your field? Where then did the weeds come from? Matthew 13:38 - The field is the world, and the good seed stands for the sons of the kingdom. The weeds are the sons of the evil one.

Therefore, if a living thing is deformed we can interpret that sin is the culprit. Furthermore, all illnesses are sin as well, including death itself. Anything that disrupts the natural order of things is clearly and undeniably SIN!

The Bible also mentions Cain who was subjected to sin. He killed his brother Abel and was driven out from the grounds by G-d, whereupon G-d also put a unique mark on Cain that notably made him different. Genesis 4:8-15 - Now Cain said to his brother Abel, Let's go out to the field. And while they were in the field, Cain attacked his brother Abel and killed him. Then the Lord said to Cain, Where is your brother Abel? I don't know, he replied. Am I my brother's keeper? The Lord said, What have you done? Listen! Your brother's blood cries out to me from the ground. **Now you are under a curse and driven from the ground,** which opened its mouth to receive your brother's blood from your hand. When you work the ground, it will no longer yield its crops for you. You will be a restless wanderer on the earth. Cain said to the Lord, My punishment is more than I can bear. Today you are driving me from the land, and I will be hidden from your presence, I will be a restless wanderer on the earth, and whoever finds me will kill me. But the Lord said to him, Not so, if anyone kills Cain, he will suffer vengeance seven times over. **Then the Lord put a mark on Cain** so that no one who found him would kill him.

Remember there was no sin in the beginning and all things were created perfect and sustained by G-d. Therefore, because we live in a world of entropy we are no longer sustained (because of the act of sin). Romans 5:12 - Therefore, just as sin entered the world through one man, and death through sin, and in this way death came to all men, because all sinned. Therefore, when these types of things happen we can now relate them to their original source - which is ultimately the curse (cause) of sin.

Lets recap here… So its quite interesting to be able to truly understand what a world without sin would be like, and we can also understand that true evil is "anti Phi/Pi" which brings about decay, darkness and death. Now that we know this vital piece of information, we can attune our minds frequency (adjusting our thoughts in positive and optimistic ways) to quickly identify and reveal as well as REPEL (fend off) this real enemy. Ephesians 6:11-12 - Put on the whole armour of G-d, that ye may be able to stand against the wiles of the devil. For we wrestle not against flesh and blood, but against principalities, against powers, against the rulers of the darkness of this world, against spiritual wickedness in high places. Romans 7:23 - But I see another law at work in the members of my body, **waging war against the law of my mind** and making me a prisoner of the law of sin at work within my members.

If I might also add… in Genesis 1:31 where G-d tells us that everything was "very good": Genesis 1:31 - And G-d saw every thing that he had made, and, behold, it was "very good". And the evening and the morning were the sixth day. Ironically this is actually "evidence" of why the verse of Genesis 1:31 is there in the first place telling us that everything was "very good", and that evening and the morning was the "sixth" day, and why we later find it recorded in Exodus that G-d rested on the "seventh" day, AND INCLUDED THIS DAY OF REST IN THE TEN COMMANDMENTS! Because (in forethought) He knew that some day centuries later that evolutionists would try to find another explanation in favor of evolution. That is why we find it recorded in Genesis, and then in Exodus, and later on again in Deuteronomy. Obviously G-d wanted to set the record straight, that indeed He did in fact create the world in a literal 6 days, and not billions of years as the evolutionist would have us believe! Genesis 2:2-3 - And on the seventh day G-d ended his work which he had made; and he rested on the seventh day from all his work which he had made. And G-d blessed the seventh day, and sanctified it: because that in it he had rested from all his work which G-d created and made. Exodus 20:11 -Remember the sabbath day, to keep it holy. Six days shalt thou labour, and do all thy work: But the seventh day is the sabbath of the Lord thy G-d: in it thou shalt not do any work, thou, nor thy son, nor thy daughter, thy manservant, nor thy maidservant, nor thy cattle, nor thy stranger that is within thy gates: For in six days the Lord made heaven and earth, the sea, and all that in them is, and rested the seventh day: wherefore the Lord blessed the sabbath day, and hallowed it. Deuteronomy 5:12-15 - Keep the sabbath day to sanctify it, as the Lord thy G-d hath commanded thee. Six days thou shalt labour, and do all thy work: But the seventh day is the sabbath of the Lord thy G-d: in it thou shalt not do any work, thou, nor thy son, nor thy daughter, nor thy manservant, nor thy maidservant, nor thine ox, nor thine ass, nor any of thy cattle, nor thy stranger that is within thy gates; that thy manservant and thy maidservant may rest as well as thou. And remember that thou wast a servant in the land of Egypt, and that the Lord thy G-d brought thee out thence through a mighty hand and by a stretched out arm: therefore the Lord thy G-d commanded thee to keep the sabbath day.

Now I'd like to revisit and elaborate further on why the "human brain and the physical body" are downgrading because of the effects of entropy in a cursed world… Lets begin focusing on the previous chapter that determined that there was no superior or intelligent "alien" life anywhere else in the Universe signifying that the only brains that are in the Universe are the brains of humans on this planet. And as we've learned Adam, Eve and Jesus were the most superior (and down the line we find Enoch associated with the Great Pyramid and Job who knew more about the things of science then we know today). And as we also learned that we already know all there is to know about our planet and the entire Universe simply from the Bible. Additionally, today in this generation, according to everything we know about history, science, genetics and entropy all add up to one conclusion… so far from being the most advanced and brilliant of all generations that have ever lived, we are actually "the worst generation thus far". Don't misunderstand me, we have come a long way (as a whole) in science and technology, but we as "individuals" are steadily going downhill. The theory on evolution teaches otherwise, that we evolved from a single celled organism into higher complexity beings over billion of years. This is the furthest thing from the truth. The Bible tells us a different account that all things were created by G-d already fully grown, functional and perfect, and from here on down these things are slowly downgrading and deteriorating in physical quality (because of sin), even in the size of things such as living creatures that we find in the fossil record, and even amazingly today we can still see the evidence for this in the giant Redwood trees of California which are some of the oldest living things on the planet found to be 5,000 years old (a world between earth and air before the great flood of Noah). Today we take notice that the branches of trees suffer from gene depletion as apposed to the original fullness and intensity of the first genotypes of its kind.

The more inbreeding of closely related species the less and less genetic power (and possibilities) left available that are continually decreases through time to the extremity of extinction. We will eventually all entropically end. (Remember all order is relegated to disorder because of the curse).

Now before the curse and for a short time after the curse it was acceptable for brothers and sisters to marry each other. Genesis 5:4 - And the days of Adam after he had begotten Seth were eight hundred years: and he begat sons and daughters. This was not a problem "originally" because before sin and the first genetic makeup (genotypes) of humankind, there were zero mistakes (which played a role in the long life expectancy of early humans). It wasn't until Leviticus 18:6 that G-d brought in the laws against close intermarriage. Leviticus 18:6 - None of you shall approach to any that is near of kin to him, to uncover their nakedness: I am the Lord. This is the first initial mentioning in the Bible that we find concerning the rejection of marrying a close relative. The reason for this is that today after all the breeding we have been doing, mistakes (like inbreeding and over-breeding) have been adding up in our genes over time. Today we all carry these mistakes in our genes. Therefore if two people mate with the "same mistakes" in their genes, its more likely that their offspring will also inherit those mistakes. If brother and sister married today its more likely that their children would inherit their same mistakes causing deformities. This is why it's better to marry someone genetically further away.

Additionally, to some degree this is the reason why the human mind today compared to Adams is so incompetent. We cannot use full capacity. We lost full capacity because we used our G-d given brain to disobey and avert from G-d and stooped to the enemies level. Yes, we have this fantastic brain, but contingent to the laws of entropy and thermodynamics (because of the curse) it's no longer powerful enough to withstand the swap over of genes (and energy) from one generation to the next. Just because we now live in a an age of modern medicine and the high tech age of computers, including the great technological age of space exploration for sending humans to the moon, and cameras and rovers to Mars, as well as satellite cameras and telescopes out into the farthest reaches of space, even after accomplishing all this, we are still not the most brilliant or brainy. What we have obtained is the most accumulated knowledge, but not genuine brilliance. Yes, we have gathered knowledge, but in effect, we have downgraded to a lesser amount of "individual" intelligence. What we have actually discovered is that the further back we go in history the more inherent brilliance there was in humans, again such as with Enoch and Job.

Compared to the Bible, science has truly gathered ABSOLUTELY NOTHING! It has only reaffirmed for us what the Bible has already told us, and continues to do so. For instance, when science takes something as simple as a seed and tries to analyze it, cut it open and dissect it, and chemically analyzes it, this still does not really tells us anything of the seeds true nature. The answer of its true nature is incorporated into it only through the long duration of time. It is a masterpiece of arrangement - a full concerto played out in orchestra! The only way we can truly know its true nature is by planting it in the ground and watching and waiting many years to see its fulfillment… the grand finale… its grand purpose… a tree producing fruit with seed in itself to grow another tree… and in the interim supplying food for both insect, animal and human as well as supplying clean air to the environment thus making life habitable for all living things. We could have never known what the seed was for simply by scientifically analyzing it. G-d had ingeniously preprogrammed it to grow in such a way as to show us exactly what it was designed for. This is also true of you and me! G-d planed from the beginning when He brought us into existence to reflect his glory. Genesis 1:27 - So G-d created man in his own image, in the image of G-d created he him; male and female created he them. Psalm 8:4-6 - What is man that you are mindful of him, the son of man that you care for him? You made him a little lower than the heavenly beings and crowned him with glory and honor. You made him ruler over the works of your hands, you put everything under his feet. (This passage reflects G-d's original plan for mankind).

2 Corinthians 3:18 - But we all, with unveiled face, beholding as in a mirror the glory of the Lord, are being transformed into the same image from glory to glory, just as by the Spirit of the Lord. In other words, as we look at G-d continually in a focused and fixed manner, we become more like Him. The restored image of G-d in man is brought about only by G-d's work of re-creation. John 3:7 states: You must be born again. This is a spiritual rebirth and it enables us to be fashioned after G-d.

The ultimate goal of the Universe is to show the glory of G-d! I don't know any other way of saying it. Other than saying that it all starts with G-d. It's not about you or me. The purpose of our life is far greater than our own personal fulfillment, happiness, career, or even our wildest dreams. If you want to know the reason for your existence, you must begin with seeking G-d and acknowledging (from the words of the Bible) that you were born for no other reason other than G-d's purpose. Romans 11:36 - For from him, and by him, and to him [are] all things: to whom [be] glory for ever. Amen. Proverbs 16:4 - The Lord has made everything for his own purposes. The search for the purpose of life has puzzled us for thousands of years. That's because we typically begin at the wrong starting point - ourselves.

So how do we seek and get to know G-d? Moreover, how do we find His eternal purpose for us?

All we simply have to do is READ HIS BIBLE that He has programmed for us accordingly through His revealed word. We don't need to come face to face with Him to personally have repeated what He has already told the prophets of old that wrote the books of the Bible. Once G-d has done something once, He does not need to repeat Himself. In fact, a Bible phrase specifically tells us this. Luke 16:31 - He said to him, if they do not listen to **Moses and the Prophets**, they will not be convinced even if someone rises from the dead. In other words, Moses and the Prophets is the entire Bible. And everything that G-d had/has to say to us has already been recorded in the Bible. He doesn't need to reiterate Himself. Thus, if we don't read His word and come to know the truth as to our special purpose, and why we're here, we'll continue to decrease down the path of the scientist and the evolutionist. Thus like the seed, you can never find out who you are from scientific experts because its all hidden from them (and you forever) unless you open the Bible which is the pathway to G-d. The secret of our fulfillment that is programmed by G-d for you and me IS THROUGH HIS WORD! Not even highly recognized scientific specialists including those who are trained in all manners of scientific fields… (Just to name a few) such as astronomy, technology, industrial, mechanical, archeology, history, psychology, philosophy, anthropology, art, biology, chemistry, biochemistry, computer science, economics, engineering, environmental science, ethnic studies, geography, geology, marine science, mathematics, physics, sociology, music, nursing, human development and humanities.

Essentially, without the WORD OF G-D we are lost! Yes, these professions listed above are important to a certain degree, but when we are dealing with "eternal issues", these scientific studies may actually be a spiritual deterrent if we look outside the bounds of what the Bible tells us of the origin of all things in existence. One must learn by heart, mind, and spirit that the believer has divine knowledge over any unbeliever no matter how educated, accomplished, skillful, talented or brilliant they may be!

18 - EXTENDING THE QUALITY OF OUR LIFE...

1 Corinthians 6:19-20 - Do you not know that your body is a temple of the Holy Spirit, who is in you, whom you have received from G-d? You are not your own; you were bought at a price. Therefore honor G-d with your body.

Perhaps this chapter should have been last because it is the one we can more readily relate to and apply to our everyday life, but nevertheless it is here at this particular spot in the book.

We are never more alive then now at this present moment. To see beauty with the fleshly eyes and to feel with the fabric of the fingers, to smell with every breath we take, to taste with the material tongue and to hear with our ears the earthy sounds of our environment. The angels envy us because every waking moment our life is real and felt. Again let me reiterate from the previous chapter that life is short lived - Romans 8:22 - For we know that the whole creation groaneth and travaileth in pain together until now. Creation groans because we had sinned (which means death) and therefore we cannot escape the law of entropy and thermodynamics where everything is in decay. Genesis 2:17 - But you must not eat from the tree of the knowledge of good and evil, for when you eat of it you will surely die. We cannot escape "death" because it is the providence of G-d in a cursed world stemming from Adam and Eve. And as was mentioned in the previous chapter because of the fall of Adam and Eve, and secondly the flood, which caused a rapid decline in the longevity and life expectancy of humans stemming from Adam and Eve to Noah's descendants living after the flood. This may have been in part because the plants that grew food no longer had the adequate consistency of nutrients to nourish the human body since the atmospheric conditions altered plant growth. Therefore, food value was greatly diminished. Plants were also larger as well as fruits and vegetables. Therefore, a single apple or carrot would be bigger and it would contain twice as much nutrients compared to the ones that grow today (not only twice as much nutrients from the size but also in quality). Remember the soil was most likely more nutrient before Adam and Eve and the flood… this was because the solar radiation coming through the atmosphere was at different degrees (a theory for this will be found in the next chapter). Therefore, today, we would have to eat three or more carrots or apples to equal one carrot or apple before the flood. Also remember that over time the laws of entropy and thermodynamics take hold and all things from the original genotype start warring out in gene depletion as was discussed in the previous chapter. So not only do plants deplete from their original genotype from both inbreeding and over-breeding like humans and animals, but this is also subjected to all non living things as well, such as rocks and even diamonds (contrary to belief diamonds do not last forever).

Many people wonder why we live, suffer and then die. Again, the answer to is because we are living under the providence of G-d in a cursed world. IT IS THE SACRAMENT! The price (the forfeit, atonement and sacrifice) that we pay for the choice (self-punishment) we had originally made thousands of years ago to disobey G-d, which caused death (remember there was NO DEATH before sin). However, it is by the establishment of Jesus that brings grace (apology and compensation) to those that receive Him. (Things will start making a whole lot more sense for the "unbeliever", even those of another faith and religion when you reach the next chapter). Hang in there and please do not be dismayed.

This chapter is about learning how to counteract and compensate for the losses that occurred to our environment because of sin. We will once again revert to the Bible for answers to these questions starting in 1 Corinthians 15:26 - The last enemy that shall be destroyed is death. The last enemy is death because it is what originally brought about sin. It is what causes us to suffer, it is what causes us (and all things) to get old, deteriorate and die. IT IS WHAT BROUGHT ABOUT THE LAW OF ENTROPY AND THERMODYNAMICS! There is no escaping sin in a cursed world. The reason and purpose for our whole life (the remainder thereof) is now to come back to G-d. To rediscover Him, to realize that we are living in a sinful world brought about by the original sin and to ask forgiveness for the first disobedience and any succeeding sin which is in each and every one of us that has been passed down through Adam and Eve. Romans 5:14 - Nevertheless death reigned from Adam to Moses, even over them that had not sinned. After the similitude of Adams transgression. Lets face it, we are all eventually going to die no matter what we do to try and preserve ourselves to live longer, and even though we ask for forgiveness. This will not change until prophecy is fulfilled and all the faithful and true believers are raptured (raised to immortality) during the resurrection (the time of the end). Matthew 24:29-31 - Immediately after the tribulation of those days shall the sun be darkened, and the moon shall not give her light, and the stars shall fall from heaven, and the powers of the heavens shall be shaken: And then shall appear the sign of the Son of man in heaven: and then shall all the tribes of the earth mourn, and they shall see the Son of man coming in the clouds of heaven with power and great glory. And he shall send his angels with a great sound of a trumpet, and they shall gather together his elect from the four winds, from one end of heaven to the other. But in order to live long enough to learn what scripture has to teach so that we might apply it accordingly and enter into a relationship with G-d to be considered a faithful believer and to teach and pass on to the following generations the gift of immortality… we literally cannot accomplish this without obeying His written word which include the commandments of His will - all of this in obedience. Therefore we must try to sustain ourselves, care for our physical and mental being, this in turn helps us to learn how to be righteous. As we learned in a previous chapter that the Bible (Leviticus) contains many dietary kosher laws, additionally we also find laws that fall into the category of embracing emotional and mental matters.

Remember one cannot divorce the material from the spiritual because it is by the material aspects of things that we obtain the spiritual.

Now lets embark on the journey to examine what the Bible has to tell us about food, nutrition and diet… Unlike the Bible, no diet or nutritional book in the world is going to have all the answers. This is because they are all-sparse in their own ways and never truly integrate the mind, body and spirit connection. Hosea 4:6 - My people are destroyed for lack of knowledge: because thou hast rejected knowledge, I will also reject thee, that thou shalt be no priest to me: seeing thou hast forgotten the law of thy G-d, I will also forget thy children.

I hope that thus far in this book I have given you somewhat of a basis and foundation to spiritual life, now I am going to teach you how to maintain the physical - for the two play hand in hand. Firstly, one must realize pertaining to food that there is a bondage mentality and a freedom mentality. The bondage mentality says - I am a slave to food - I live to eat. And the freedom mentality says - I am a food liberator - I eat to live. In other words we should not live to eat, rather we should eat to live. We all want to be food liberators! Interestingly enough, the Bible seems to be far more concerned with how much we eat then actually, what we eat. Physical appetites are a comparison of our ability to control ourselves. If we are unable to control our eating habits we are probably also unable to control other habits such those of the physical mind and body like alcoholism, drugs, lust, money, greediness, jealousy, irritation, fury, rage, reacting with physical and mental abuse, gossip and friction. We are not to let our appetite control us; rather we must learn to control it. Because along with our appetite and hunger for food it too is brought about from that initial sin and curse that we are all subjected to. Deuteronomy 21:20 - And they shall say unto the elders of his city, this our son is stubborn and rebellious, he will not obey our voice; he is a glutton, and a drunkard. Proverbs 23:2 - And put a knife to thy throat, if thou be a man given to appetite. Secondly, G-d makes it known to us in the Bible that the environment that He supplied for us works in synchronicity, when we veer away from that balance we deteriorate at an even faster than normal rate. Our bodies (after we sinned) were designed to live to 120 years of age. Genesis 6:3 - And the Lord said, My spirit shall not always strive with man, for that he also is flesh: yet his days shall be an hundred and twenty years. Genesis was written in Moses' era, but as pointed out in the previous chapter we find recorded in the Bible the life spans of many of the old biblical patriarchs that somehow eluded the 120 year life expectancy that we have today. The reason for their long life spans was briefly discussed in the previous chapter, but will also be elaborated about in a future chapter. However, some religious thinkers suggest that many of the old patriarchs followed the kosher eating regulations and laws mentioned in Leviticus, though these food laws were not actually documented in the Bible until Moses' time.

So, the first law that should be observed is that all food should be "kosher", and abided by kosher laws of cleanliness. If not kosher then ORGANIC! All kosher food should be organic in nature and that means grown without pesticides. In addition, remember from a previous chapter that it's just not the cleanliness of keeping kosher; it's also the spiritual aspect of it.

However, before I go on to elaborate I must first revert to the Bible to the time before sin. Genesis 1:29-30 - And G-d said, Behold, I have given you every herb bearing seed, which is upon the face of all the earth, and every tree, in the which is the fruit of a tree yielding seed; to you it shall be for meat. And to every beast of the earth, and to every fowl of the air, and to every thing that creepeth upon the earth, wherein there is life, I have given every green herb for meat: and it was so. This literally means that all living things (both human and animal) were plant eaters (plants instead of meat because nothing ever died). People do not understand the immensity and importance this phrase in scripture has, and the quirk for why it is there is certainly not by mistake or coincidence… because the question arises, if we were not originally plant eaters AND NOTHING EVER DIED (for meat), why would the Bible go to the lengths of telling us this? The most logical hypothesis for the existence of this particular verse is that there was (in reality) NO DEATH BEFORE SIN! This biblical verse in itself is the masterstroke of a ringing declaration! Moreover, when we read the scripture again (before sin crept in) there is no mention of times specified to eat, such as breakfast, lunch and dinner, NOR DID THEY EVEN HAVE TO PRAY IN THANKS FOR THEIR FOOD! Food was to be eaten when hunger encroached, but its speculative that there was even hunger because hunger can fall into the pain and suffering category which would suggest sin, so maybe something more like to crave was felt. Even if they starved, G-d would sustain them because they could not die. However, we find written that after the curse took effect in Genesis that we were reduced to eating from thorns and thistles of the field… Genesis 3:18 - It will produce thorns and thistles for you, and you will eat the plants (herb) of the field. In addition, food was no longer given or supplied to us from G-d, now we had to toil for it ourselves. Genesis 3:19 - In the sweat of thy face shalt thou eat bread, until thou return unto the ground; for out of it wast thou taken: for dust thou art, and unto dust shalt thou return. (This strongly implies "death" from which there was none previously before). Now these next verses also imply that something had drastically changed! Genesis 3:23 - Therefore the Lord G-d sent him forth from the garden of Eden, to till the ground from whence he was taken. Genesis 9:3 - Everything that lives and moves will be food for you. Just as I gave you the green plants, I now give you everything. Now that there was, "death" suggests that the animals began to eat each other for food, and as they increased in number man began eating them. This is where the first food chain, ecosystem and cycle of life began. Lets slip back here for a moment and ask why G-d had chosen the Middle East region of the world for the Garden of Eden? It just so happens that the Middle East is a prime place for plants to grow and be maintained. The weather there in the Middle East (in some places) is not too hot or too cold, it is just right. There are places in the Middle East (garden valleys) that are fertile and in a semi-tropical environment. The word "Eden" literally means "fertile plain". The Bible tells us that this was a lush and fertile place with abundant water coming from 4 rivers and where the land is good, and additionally where gold, bdellium and onyx can be found. Genesis 2:8 - And the Lord G-d planted a garden eastward in Eden, and there he put the man whom he had formed. Genesis 2:10-14 - And a river went out of Eden to water the garden, and from thence it was parted, and became into four heads. The name of the first is Pison: that is it which compasseth the whole land of Havilah, where there is gold; And the gold of that land is good: there is bdellium and the onyx stone. And the name of the second river is Gihon: the same is it that compasseth the whole land of Ethiopia. And the name of the third river is Hiddekel: that is it which goeth toward the east of Assyria. And the fourth river is Euphrates. Ironically, we read that when the Jews were wondering in the desert after leaving Egypt they were once again "sustained" by G-d. Nehemiah 9:21 - Yea, forty years didst thou sustain them in the wilderness, so that they lacked nothing; their clothes waxed not old, and their feet swelled not. Psalm 78:25 - Men ate the bread of angels, he sent them all the food they could eat. But the ultimate question still remains that prior to sin and the curse and Adam and Eve being evicted from the Garden the weather still remained the same, because nothing that was mentioned suggests otherwise until the great flood of Noah which drastically changed the weather patterns around the Earth. However, for some odd reason the weather pattern in the Middle East region has not changed in climate. It has always been a semi tropical place. However, the rest of the worlds weather had drastically changed. We know that the present freezing cold artic circle (pole) was once a warm and tropical place where mammoth elephants roamed. We know this because we find food in the frozen fossilized mammoths stomachs that are from warm tropical plants. Then the flood came immediately upon them and froze them instantly into the preserved state that we find them today. It seems the world revolves around the Middle East part of the world. It always has, and it ALWAYS WILL! Psalm 132:13-14 - For the Lord hath chosen Zion, he hath desired it for his habitation. This is my rest for ever: here will I dwell; for I have desired it.

When civilization spanned out from the Garden of Eden, several great kingdoms came into existence, which were Egypt, Babylon, Assyria, Persia, Greece and Rome. What is incredible is that the small country of Israel was located right in the center of all of them, so in order to travel from east to west or north to south from those countries one had to pass through the land of Israel.

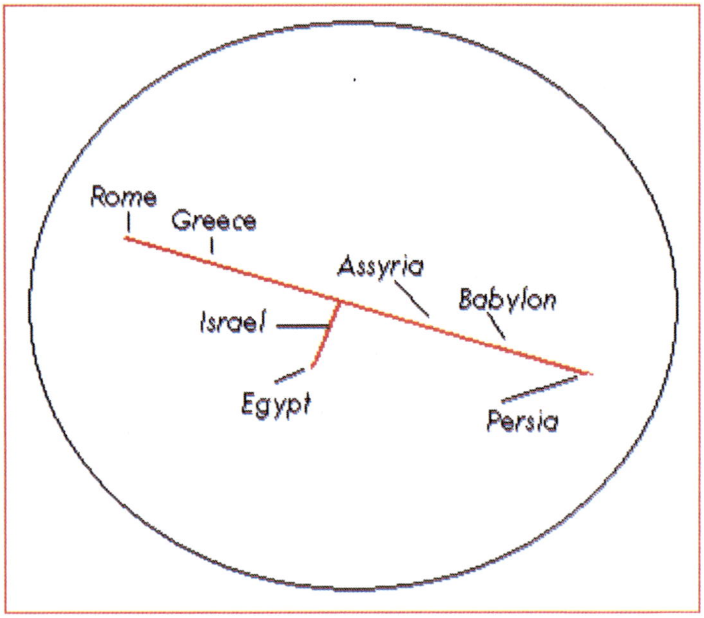

But there is an even bigger reason for why G-d chose Israel. Not only do we know that it revolves around all the land continents, and is Phi/Pi, but also G-d had chosen this precise place because from here His word could pervade in equal distributions spreading throughout the geographic center of the Earths land continents to every single human that resides on the planet. Incidentally, we find that this is even true today since many of the conflicts happening around the world involve Israel and the Middle East region of the world - and the news is center stage on this. That is the reason why the land continents are where there are today, likewise (as aforementioned) the Great Pyramid of Egypt, and the smallest but not least of all the countries, ISRAEL - the bone marrow of the world. The trading post of all the major continents is a special place where blood can be likened to nourish the world body - likewise this is a reflection of how the human body makes blood from its bones (which can be likened as another type of microcosm). From here, an antibody (a protein that fights infections) went forth around the world that purified it, sustaining it from the cancer of sin and evil. This cancer antibody was Jesus the Messiah. Leviticus 17:11 - For the life of the flesh is in the blood: and I have given it to you upon the altar to make an atonement for your souls: for it is the blood that maketh an atonement for the soul. For whoever believes in Him should not die but have everlasting life. John 6:47 - I tell you the truth, he who believes has everlasting life. Matthew 20:28 - Just as the Son of Man did not come to be served, but to serve, and to give his life as a ransom for many. John 6:54 - Whoever eats my flesh and drinks my blood has eternal life, and I will raise him up at the last day. Ezekiel 37:11-12 - Then he said unto me, Son of man, **these bones are the whole house of Israel:** behold, they say, Our bones are dried, and our hope is lost: we are cut off for our parts. Therefore prophesy and say unto them, Thus saith the Lord G-d; Behold, O my people, I will open your graves, and cause you to come up out of your graves, and bring you into the land of Israel. Then He said unto me, Son of man, these bones are the whole house of Israel, behold, they say, Our bones are dried, and our hope is lost; we are cut off on our part. Therefore, prophesy and say unto them, 'Thus saith the Lord G-d: Behold, O my people, I will open your graves, and cause you to come up out of your graves, and bring you into the land of Israel. Psalm 34:20 - He protects all his **bones,** not one of them will be broken. John 19:36 - These things happened so that the scripture would be fulfilled: Not one of his **bones** will be broken. Proverbs 3:8 - This will bring health to your body and nourishment to your **bones.**

Getting back on track here, now lets examine what else scripture has to teach us about food and sustaining us after the first sin was committed and the curse took effect... Exodus 15:26 - And said, If thou wilt diligently hearken to the voice of the Lord thy G-d, and wilt do that which is right in his sight, and wilt give ear to his commandments, and keep all his statutes, I will put none of these diseases upon thee, which I have brought upon the Egyptians: for I am the Lord that healeth thee. Exodus 23:25 - And ye shall serve the Lord your G-d, and he shall bless thy bread, and thy water; and I will take sickness away from the midst of thee. Deuteronomy 7:11-15 - Thou shalt therefore keep the commandments, and the statutes, and the judgments, which I command thee this day, to do them. Wherefore it shall come to pass, if ye hearken to these judgments, and keep, and do them, that the Lord thy G-d shall keep unto thee the covenant and the mercy which he sware unto thy fathers: And he will love thee, and bless thee, and multiply thee: he will also bless the fruit of thy womb, and the fruit of thy land, thy corn, and thy wine, and thine oil, the increase of thy kine, and the flocks of thy sheep, in the land which he sware unto thy fathers to give thee. Thou shalt be blessed above all people: there shall not be male or female barren among you, or among your cattle. And the Lord will take away from thee all sickness, and will put none of the evil diseases of Egypt, which thou knowest, upon thee; but will lay them upon all them that hate thee. Psalm 1:3 - And he shall be like a tree planted by the rivers of water, that bringeth forth his fruit in his season; his leaf also shall not wither; and whatsoever he doeth shall prosper. We can determine from these passages that if we diligently hearken to G-d, serve Him and keep His commandments and all His statutes, He will keep sickness and diseases away and heal us. He will also bless our food, animals, even the woman's womb, and the fertility of the man! THIS IS FANTASTIC, G-D WILL LITERALLY SUSTAIN US IF WE HEARKEN UNTO HIM, SERVE HIM AND KEEP THE COMMANDMENTS! It doesn't say that we will not eventually die, but it certainly says that G-d will sustain us. The word "sustained" after the curse took effect, now means to "prolong" us thereby delaying death.

Additionally G-d goes on to elaborate about the importance of the blood... Leviticus 17:11 - For the life of the flesh is in the blood: and I have given it to you upon the altar to make an atonement for your souls: for it is the blood that maketh an atonement for the soul.

Here we read about the importance of salt... Leviticus 2:13 - Season all your grain offerings with salt. Do not leave the salt of the covenant of your G-d out of your grain offerings; add salt to all your offerings. Mark 9:50 - Salt is good, but if it loses its saltiness, how can you make it salty again? Have salt in yourselves, and be at peace with each other. Job 6:6-7 - Is tasteless food eaten without salt, or is there flavor in the white of an egg? I refuse to touch it; such food makes me ill. Contrary to belief, salt is not bad for us, rather its good! Of course to much of anything is bad for us, but both salt and water is a necessary and essential ingredient for maintaining the cellular activity of all life. In fact, the body can endure periods of lack of food, but without salt and water, living cells would quickly die from dehydration. The importance of salt goes much deeper than just making food taste good. It's in our blood and all the other fluids in our body. Without it, our nerves and our muscles could not function properly, our stomachs wouldn't be able to break down food to give us energy, and our blood would have trouble carrying oxygen to our lungs. Salt also activates the salivary enzyme in the mouth - the very first enzyme in the digestive process. Additionally as I said, salt may be harmful in large quantities, though we lose a little salt every day when we perspire and urinate. Another thing regarding salt is that the body's cannot make more. We need to eat salt to replace what we lose. Perhaps that's why we crave it so much - it may be our bodies' way of making sure we eat it (animals crave salt too, such commonly known animals like dears and elephants seem to find deposits of salt in the ground to lick). Salt is also used to preserve food as it draws out water from meat and fish making them too dry for bacteria to grow, and which prevents food from rotting (remember the kosher laws in Leviticus require using salt). Preserved in salt, food can be kept safely for months or even years. Moreover, the salty amniotic fluid and environment of the mothers womb allows for the growth of the human embryo - life is dependant on this solution and from the moment of conception, we are never without a need for it.

Regrettably, the salt that most of us eat today is refined table salt, which is toxic. The high heat processing kills the salt removing essential minerals, and in addition to this, harmful additives are added to prevent moisture and caking from building up as well as corn sugar (dextrose) which is added to keep it free-flowing. Some of the additives that are used are "aluminum" (having been associated with Alzheimer's disease). To avoid this some people think that buying "organic sea salt" solves this problem. It is definitely better, but it too is essentially refined salt and generally comes from the same salt refineries. Good salt should be naturally air and sun dried and gathered by hand. This is the kind of salt spoken about in the Bible. Much to our surprise, natural salt can have healing effects over various ailments such as sinus or bronchial congestion by drawing mucous out of the body. Salt also nourishes the adrenal glands, which run on sodium. If the adrenals are weak, they cannot protect against an impending cold or flu. Yes, unbelievably salt can prevent illness. When we feel the first signs of sickness coming on we should double up on salt, which can keep at bay the impending illness, this is because salt stimulates cellular energy and enhances resistance to infections. In fact, everyone is familiar with salt causing high blood pressure… this is both true and false as refined toxic salt is an irritant to the body, but many people experience a lowering of blood pressure by using natural salt. Natural salt can also reduce fluid retention; this is because natural salt contains the proper balancing of minerals, while refined salt reverses the natural effect because it attempts to neutralize the toxins. Surprisingly, natural salt also has additional medical benefits. It can be beneficial for shock victims, hemorrhaging, physical trauma and severe internal burns caused by heat stroke or electrical shock (including lightning strikes). It can also help correct excess acidity in the body, relieve allergies, relieve skin diseases, and restore digestion through stimulating acid secretions. Interestingly, the Hebrew word for salt originated from the words for "war and peace" - this may indicate that the first wars that humankind initiated was most likely over "salt". This also follows that salt (like gold or silver) was used as money. Ironically, the English word for salt stemmed from a Roman word "salarium", which is where the word "salary" comes from. Additionally as aforementioned in Leviticus 17:11 - For the life of the flesh is in the blood: and I have given it to you upon the altar to make an atonement for your souls: for it is the blood that maketh an atonement for the soul. (Now remember, "salt" which is also found in the blood is a preservative, a sterilizer and purifier). It is in this same sense, the identical mechanism that takes over when we acknowledge Jesus as the Messiah, for it is by His blood that we are in a sense preserved, sterilized and purified of sin.

Remarkably, a salt crystal(s) looks like this:

If we look closer, it looks something like this:

Again, a salt crystal is a cube, but when we look at it from another point of view, it is also a Hexagon as well as a Octahedron, which in turn are produced from the Tetrahedron, which (as we've previous learned) is the first organized form (unit of growth) to all life.

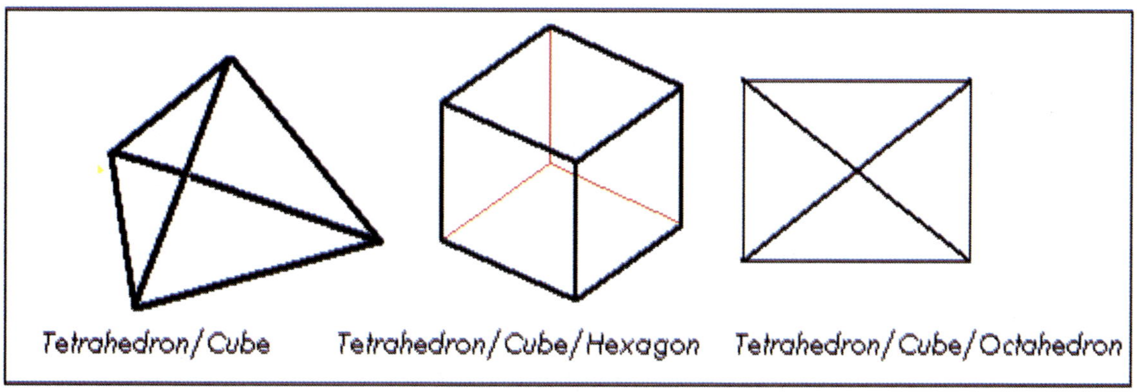

Moving on here to other biblical verses... We read in Daniel that a diet consisting of Vegetables and water keeps us healthy. Daniel 1:8 - But Daniel resolved not to defile himself with the royal food and wine, and he asked the chief official for permission not to defile himself this way. Daniel 1:11-17 - Daniel then said to the guard whom the chief official had appointed over Daniel, Hananiah, Mishael and Azariah, please test your servants for ten days: **Give us nothing but vegetables to eat and water to drink.** Then compare our appearance with that of the young men who eat the royal food, and treat your servants in accordance with what you see. So he agreed to this and tested them for ten days. At the end of the ten days they looked healthier and better nourished than any of the young men who ate the royal food. So the guard took away their choice food and the wine they were to drink and gave them vegetables instead. To these four young men G-d gave knowledge and understanding of all kinds of literature and learning. And Daniel could understand visions and dreams of all kinds. Additionally we also find mentioned in Proverbs 15:17 - Better a meal of vegetables where there is love than a fattened calf with hatred.

Now given that we know all this from just a few biblical references, I'm going to add in things that I have learned throughout my immense research and apply them to biblical scripture.

The most sought after question arises… what truly is "aging" and what's actually happening and taking place within the human body. The answer to this question in scientific terms (aside from the cause of the curse) is that ALL AGING IS A RESULT OF IMPROPER PROTEIN PRODUCTION. Our skin looses its elasticity because it has within it a substance called "elastin" which is a fibrous protein resembling collagen that happens to be the main ingredient of the elastic fibers of connective tissue. In essence, "ELASTIN IS A PROTEIN". Now when we start loosing elastin the enzymes in our body don't digest as well, our body loses fluid and becomes brittle and stiff, and our bones become thin. But the BIG question arises if it really matters what we eat as we age, and does nutrition really have an effect if the cells in our body no longer have the receptor sites after many years of abusive eating, as well as "the effects of entropy and thermodynamics" (otherwise known as the suffering of sin) to even receive or let in the nutrients that are necessary for the health of the body. According to the laws of Leviticus 11 in the Old Testament, G-d instruct us on how to keep both kosher and sanitary laws as well as affixing the importance of the spiritual aspect of it. As follows into the New Testament, these laws are again introduced, but this time with a higher spiritual aspect added to them. Matthew 6:25 - Therefore I tell you, do not worry about your life, what you will eat or drink, or about your body, what you will wear. Is not life more important than food, and the body more important than clothes? Luke 12:23 - Life is more than food, and the body more than clothes. John 4:34 - My food, said Jesus, is to do the will of him who sent me and to finish his work. 1 Corinthians 8:8 - But food does not bring us near to G- d, we are no worse if we do not eat, and no better if we do. Romans 14:14 - As one who is in the Lord Jesus, I am fully convinced that no food is unclean in itself. But if anyone regards something as unclean, then for him it is unclean. Romans 14:20 - Do not destroy the work of G- d for the sake of food. All food is clean, but it is wrong for a man to eat anything that causes someone else to stumble. Mark 7:19 - For it doesn't go into his heart but into his stomach, and then out of his body, (in saying this, Jesus declared all foods clean). When Jesus died on the cross, He put an end to the Old Testament law. Romans 10:4 - For Christ is the end of the law for righteousness to every one that believeth. Galatians 3:24-26 - Wherefore the law was our schoolmaster to bring us unto Christ, that we might be justified by faith. But after that faith is come, we are no longer under a schoolmaster. For ye are all the children of G-d by faith in Christ Jesus. Ephesians 2:15 - Having abolished in his flesh the enmity, even the law of commandments contained in ordinances, for to make in himself of twain one new man, so making peace.

However these biblical phrases (listed above) may be the spiritual case, but certainly not the physically case. However, by eating and taking care of ourselves both spiritually (firstly) and physically (secondly), one can prolong life - unfortunately, death is eventually imminent but at least we can attempt to make it last longer AND TO RESPECT G-DS ORIGINAL INTENT FOR US - relating to our earlier human body (before sin). Genesis 1:29-30 - And G-d said, Behold, I have given you every herb bearing seed, which is upon the face of all the earth, and every tree, in the which is the fruit of a tree yielding seed, to you it shall be for meat. And to every beast of the earth, and to every fowl of the air, and to every thing that creepeth upon the earth, wherein there is life, I have given every green herb for meat: and it was so. (This is a vegetarian diet).

Elaborating on this a little further… Lets first start with the most important aspect to sustaining and prolonging life… this is of course through spiritual means - having a relationship with G-d. But as for the physical aspects, we must revert to the Bible to realize what G-d's original intent was for our earlier human body (before sin). The answer to this can be found within the biblical verses specified above which include eating the proper diet that G-d had originally intended, and that was a diet rich in vegetables, fruits and plants (this would also include seeds, nuts and herbs which grow from trees and plants). Some of these herbs are used for both food and medicinal purposes. Starting in Exodus we find that "hyssop" was used as both a purifier and an important ingredient in preventing blood from coagulating. Exodus 12:22 - And ye shall take a bunch of **hyssop,** and dip it in the blood that is in the bason, and strike the lintel and the two side posts with the blood that is in the bason, and none of you shall go out at the door of his house until the morning. Leviticus 14:52 - And he shall cleanse the house with the blood of the bird, and with the running water, and with the living bird, and with the **cedar wood,** and with the hyssop, and with the scarlet. Hebrews 9:19 - For when Moses had spoken every precept to all the people according to the law, he took the blood of calves and of goats, with water, and scarlet wool, and hyssop, and sprinkled both the book, and all the people. Exodus 16:31 - And the house of Israel called the name thereof **Manna:** and it was like **coriander seed,** white, and the taste of it was like wafers made with **honey.** Numbers 11:7-9 - And the manna was as coriander seed, and the colour thereof as the colour of bdellium. And the people went about, and gathered it, and ground it in mills, or beat it in a mortar, and baked it in pans, and made cakes of it: and the taste of it was as the taste of **fresh oil.** And when the dew fell upon the camp in the night, the manna fell upon it. Hyssop was often referred to as the herb used in purification. Psalm 51:7 - Purge me with hyssop, and I shall be clean: wash me, and I shall be whiter than snow. Hyssop was also used for medicinal purposes. John 19:29-30 - Now there was set a vessel full of vinegar: and they filled a spunge with **vinegar,** and put it upon hyssop, and put it to his mouth. When Jesus therefore had received the vinegar, he said, It is finished: and he bowed his head, and gave up the ghost. Exodus 12:8 - And they shall eat the flesh in that night, roast with fire, and **unleavened bread;** and with **bitter herbs** they shall eat it. These bitter herbs most likely include, endive, chicory, lettuce, watercress, sorrel, parsley and dandelions. Matthew 23:23 - Woe unto you, scribes and Pharisees, hypocrites! for ye pay tithe of **mint** and **anise** and **cummin,** and have omitted the weightier matters of the law, judgment, mercy, and faith: these ought ye to have done, and not to leave the other undone. Jonah 4:6 - And the Lord G-d prepared a **gourd,** and made it to come up over Jonah, that it might be a shadow over his head, to deliver him from his grief. So Jonah was exceeding glad of the gourd. Job 30:4 - Who cut up **mallows** by the bushes, and **juniper roots** for their meat. Genesis 30:14 - And Reuben went in the days of wheat harvest, and found **mandrakes** in the field, and brought them unto his mother Leah. Then Rachel said to Leah, Give me, I pray thee, of thy son's mandrakes. Song of Solomon 1:14 - My beloved is unto me as a cluster of **camphire** in the vineyards of Engedi. Song of Solomon 4:13 - Thy plants are an orchard of **pomegranates,** with pleasant **fruits;** camphire, with **spikenard.** Genesis 43:11 - Then their father Israel said to them, If it must be, then do this: Put some of the best products of the land in your bags and take them down to the man as a gift—a little **balm** and a little **honey,** some **spices** and **myrrh,** some **pistachio nuts** and **almonds.** Exodus 30:34 - Then the Lord said to Moses, Take **fragrant spices—gum resin, onycha** and **galbanum**—and pure **frankincense,** all in equal amounts. Revelation 18:13 - **Cargoes of cinnamon** and spice, of incense, myrrh and frankincense, of **wine** and **olive oil,** of fine **flour** and **wheat,** cattle and sheep, horses and carriages, and bodies and souls of men. Proverbs 7:17 - I have perfumed my bed with myrrh,

aloes and cinnamon. Song of Solomon 4:14 - **Nard** and **saffron, calamus** and cinnamon, with **every kind of incense tree,** with myrrh and aloes and all the finest spices. Exodus 30:24 - 500 shekels of **cassia**—all according to the sanctuary shekel—and a hin of olive oil. 1 Samuel 25:18 - Abigail lost no time. She took two hundred loaves of **bread,** two skins of wine, five dressed sheep, five seahs of **roasted grain,** a hundred cakes of **raisins** and two hundred cakes of pressed **figs,** and loaded them on donkeys. Deuteronomy 8:8 - A land with wheat and **barley,** vines and fig trees, pomegranates, olive oil and honey. 2 Samuel 17:28 - Brought bedding and bowls and articles of pottery. They also brought wheat and barley, flour and roasted grain, **beans and lentils.** Ezekiel 4:9 - Take wheat and barley, beans and lentils, **millet and spelt,** put them in a storage jar and use them to make bread for yourself. You are to eat it during the 390 days you lie on your side. Numbers 11:5 - We remember the **fish** we ate in Egypt at no cost—also the **cucumbers, melons, leeks, onions, and garlic.** Ezekiel 17:5 - He took some of the seed of your land and put it in fertile soil. He planted it like a **willow** by abundant water. Matthew 13:31 - He told them another parable: The kingdom of heaven is like a **mustard seed,** which a man took and planted in his field. Exodus 9:31 - The **flax** and barley were destroyed, since the barley had headed and the flax was in bloom. Numbers 20:5 - Why did you bring us up out of Egypt to this terrible place? It has no grain or figs, **grapevines** or pomegranates. And there is no water to drink! Song of Solomon 6:11 - I went down to the grove of **nut trees** to look at the new growth in the valley, to see if the vines had budded or the pomegranates were in bloom. Song of Solomon 2:5 - Strengthen me with raisins, refresh me with **apples,** for I am faint with love.

However, after the fall of Adam and Eve things took a drastic change. After they had eaten from the "Tree of Knowledge" of good and bad they realized that they were naked. Genesis 3:10-11 - He answered, I heard you in the garden, and I was afraid because I was naked, so I hid. And he said, Who told you that you were naked? Have you eaten from the tree that I commanded you not to eat from? **After this G-d had to make garments of skin to cloth them, which involved the killing of the first animal.** It's not specified in the Bible, but this first animal is thought to be a lamb. The lamb of G-d otherwise recognized as Jesus in the New Testament. After this G-d had cursed them and expelled them out of the garden. Genesis 3:21 - The Lord G-d made garments of skin for Adam and his wife and clothed them. **However, remember before they were cursed nothing ever died, now, we have death and that means that animals could eat each other, and now even man can eat animals.** They had brought death into the world and now the entire creation (including animals) has to suffer the consequences. Here we find that G-d now tells Noah that along with the green plants, He has also given Noah and his family the animals to eat for meat. Genesis 9:1-3 - Then G-d blessed Noah and his sons, saying to them, Be fruitful and increase in number and fill the earth. The fear and dread of you will fall upon all the beasts of the earth and all the birds of the air, upon every creature that moves along the ground, and upon all the fish of the sea, they are given into your hands. **Everything that lives and moves will be food for you. Just as I gave you the green plants, I now give you everything.** When we read on in the Bible… it elaborates that Jesus eats fish and lamb (Jesus was not a vegetarian). Luke 24:42-43 - And they gave him a piece of a broiled fish, and of an honeycomb. And he took it, and did eat before them. Mark 14:12 - On the first day of the Feast of Unleavened Bread, when it was customary to sacrifice the Passover lamb, Jesus' disciples asked him, Where do you want us to go and make preparations for you to eat the Passover? Luke 22:7 - Then came the day of Unleavened Bread on which the Passover lamb had to be sacrificed. With all this said, there is nothing wrong with being a vegetarian. The Bible does not command us to eat meat, though it does allow for it. However, it will further be elaborated about that eating TOO MUCH meat may actually be unhealthy, as the human body was not intentionally made for it. Notice how it is only mentioned in the Bible of Jesus eating only fish and lamb. Lamb is only eaten once a year at Passover and additionally only for special or specific occasion. Therefore meat is only to be eaten moderately and most of the time refraining from it (the reason for this will follow). In regards to this, the Bible also tells us that we should not force our belief or opinion on this issue onto other people. Romans 14:2-3 - One man's faith allows him to eat everything, but another man, whose faith is weak, eats only vegetables (herbs). The man who eats everything must not look down on him who does not, and the man who does not eat everything must not condemn the man who does, for G-d has accepted him. Therefore whatever a person decides to eat is acceptable to G-d as long as we thank Him for providing. 1 Corinthians 10:31 - So whether you eat or drink or whatever you do, do it all for the glory of G-d.

Moving onwards, the Bible is overbearingly keen on cleanliness. It stresses the importance of living and dwelling in a clean environment, both physically and spiritually. Leviticus 13:58 - The clothing, or the woven or knitted material, or any leather article that has been washed and is rid of the mildew, must be washed again, and it will be clean. Leviticus 14:8 - The person to be cleansed must wash his clothes, shave off all his hair and bathe with water, then he will be ceremonially clean. After this he may come into the camp, but he must stay outside his tent for seven days. Leviticus 14:9 - On the seventh day he must shave off all his hair, he must shave his head, his beard, his eyebrows and the rest of his hair. He must wash his clothes and bathe himself with water, and he will be clean. 2 Kings 5:10 - Elisha sent a messenger to say to him, Go, wash yourself seven times in the Jordan, and your flesh will be restored and you will be cleansed. Leviticus 13:46 - As long as he has the infection he remains unclean. He must live alone, he must live outside the camp. Deuteronomy 23:12-14 - Designate a place outside the camp where you can go to relieve yourself. As part of your equipment have something to dig with, and when you relieve yourself, dig a hole and cover up your excrement. For the Lord your G-d moves about in your camp to protect you and to deliver your enemies to you. Your camp must be holy, so that he will not see among you anything indecent and turn away from you. We can apply what scripture has to teach us compared to today's standards in which things have not drastically changed. We know that fresh clean air and water is a prerequisite to sanitary conditions for staying healthy. We can also apply this to our living conditions and house dwellings. Ones house should be free of mold and bacteria that can build up in food, floors, carpets, furniture and vents (remember certain kinds of molds can actually cause death, (as the Jews learned during their time in Egypt). Its best to have floors that can be swept and washed, and leather furniture and window treatments that can be cleaned by hand or washed by machine. Change pillow cases on a weakly basis and if possible wash pillows in a washing machine with hot water and bleach, this kills bacteria. This also goes for all bedding and cloths. Additionally, we should take daily showers with clean water to wash bacteria down and away from the body. Don't take baths they harvest bacteria and get into cracks in the skin. Don't use shared bar soap, this can carry bacteria such as hepatitis, instead use liquid soap. Just before you end your shower run it cool or semi cold especially over hair as cold water stops hair from falling out and invigorates the body and spirit. Cool or cold water causes the body to generate energy from the loss of heat and sets the body clock in motion for the day. Proverbs 25:25 - Like cold water to a weary soul is good news from a distant land. And believe this or not, but people with bad breath and gum disease have a higher risk of developing pancreatic cancer.

This is because people with gum disease have elevated levels of nitrosamines (carcinogens produced by bacteria in the stomach). The solution is to prevent the buildup of bacteria and plaque on teeth so that it won't releases the chemicals that can cause inflammation and narrowing of the blood vessels, which leads to coronary artery disease. Additionally, the bacteria from the mouth may enter the blood stream and cause clumping or clotting of plaque on the inside walls of vessels, leading to coronary artery disease, heart attack (or if a clot breaks off and goes to the brain) a stroke.

Adding to the cleanliness list… the dishes, silverware and glass/cup and containers you eat and drink out of should be sterilized in extremely hot and soapy water preferable a dishwasher and not washed in a sink. This is because the sink has unwanted bacteria even with soap and hot water (because water is not hot enough for hands to tolerate) also the dishtowel you use to dry with can contains unwanted bacteria. If you use a dishwasher, make sure all phosphates are washed clean from drinking glasses; dishes and silverware… phosphates are poison! This may mean that you may have to additionally rinse your glass, dish or silverware before you drink or eat from them. (Kosher law also includes boiling all food utensils in water). For cleaning your house there is a NATUAL DISINFECTANT produced by the Earth itself. This natural disinfectant is "hydrogen peroxide" (the 3% stuff in the brown bottles at the drug store is sufficient, and best of all its cheap)! It has many uses… it can be used to sanitize most anything you can think of from toothbrushes, carpets, floor, windows, garbage cans, counter tops, stove, microwave, toilets and bathtubs, removing mold, a hand cleaner, a laundry detergent, as a mouthwash and gargle, a plant fertilizer and spray. It can be used to wash and disinfect bacteria from fruits and vegetables, but it's most notoriously known for cleaning cuts, scraps and skin infections. However, there is much more to it, you see unknown to most people hydrogen peroxide is a naturally occurring (in nature) disinfectant. Hydrogen peroxide is part of the Earth's environmental cleaning cycle. It is formed as water (H2O) passes through the ozone layer (O3) where it picks up another oxygen atom (O), becoming H202 (hydrogen peroxide). So, a small organic amount of it comes down to us in the form of rain, enters rivers and streams where it breaks down into water and oxygen, and in turn it nourishes plant and animal life. Going one step further, hydrogen peroxide is also produced naturally by the white blood cells of the human body's immune system, which help fight off infections. Other natural ingredients for cleaning your house are "olive oil and lemon juice" for polishing; deodorizing or moisturizing your floor, leather or wood furniture, countertops, stove, sink (dilute all products with water). Amazingly, olive oil shines up stainless steal to a sparkle and has been noted as a disinfectant. Tea tree oil is another natural fungicide and germicide, and is a natural antibacterial and antiseptic as well.

Again moving on here… As I've expressed earlier that "all food" should be Kosher or ORGANIC! (This means grown without pesticides or added chemicals). A connection has been found between organically grown foods and foods that have been grown with pesticides. It seems that pesticides, junk food, artificial ingredients, additives, preservatives, colors, flavors, nitrates, sugar substitutes, sweeteners such as high fructose corn syrup and MSG (monosodium glutamate) have all been found to be the cause of over stimulation of the nervous system, disrupt neurotransmitters in the brain, hyperactivity, cause migraines, behavior disorders and ADHD (attention deficit hyperactivity disorder). Some foods have even been "bioengineered" (genetically modified foods) to be resistant to insects or drought. These are not natural or healthy to eat as they cause allergies in some people. That is because they were not intentionally created by G-d in this fashion, and we learn this lesson later.

Once they introduce the plant with new genes, it now directs the production of a specific "protein" that makes the plant uniquely different. It now has the ability to produce a "new protein" that makes it resistant to certain herbicides that are used in controlling weeds, or causes it to produce a protein that is toxic to certain caterpillars, thus eliminating the need for certain conventional pesticides. Some have noted that introducing a new piece of DNA into the plant's chromosome can disrupt the function of other genes, thus crippling the plant's growth or altering the level of nutrients or toxins. For example, genetic modifications in varieties of soybeans and canola changed the fatty acid composition in the oils of those plants. Bioengineered corn has also been found and still suspected in causing nausea, introducing toxins into the blood system and causing anaphylactic shock (extreme sensitivity) in humans - as with anything else its long-term side effects are unknown. Once these new genetic organisms are introduced, they are on the planet with us ceaselessly. Pesticides also result in cancers of all kinds and children whose mothers eat/ate pesticides are found to have allergies.

So ideally, we should try to eat fresh fruits, vegetables nuts, grains, herbs and spices that are "organically grown" without pesticides, processed or bioengineered. If you buy them with pesticides, do your best to wash them off properly and additionally to kill bacteria. Fresh fruit also harvests bacteria, so make sure they are washed good, and the area you use to cut or prepare them is also washed clean. Moreover, try not to eat leftovers as they also build up and harvest bacteria even if kept in a refrigerator for 1 day. Furthermore, when we cook foods with "heat" it loses vital and important enzymes that break down food in the stomach that facilitate digestion. Foods should be eaten raw or moderately cooked. However, if food is cooked to the point where vital enzymes are lost, one can buy plant enzyme pills to counteract this (enzymes also help in breaking up fats and oils during digestion so that food can be absorbs by the body properly). However cooking vegetables does destroy their vitamin content since cooking can actually boosts your body's ability to absorb the nutrients in some vegetables. For example, the cancer-fighting phytonutrient lycopene is stronger in cooked tomato sauce then in raw tomatoes. Though, the most important rule is not to over boil vegetables! To retain the most nutrition your best bet is to either, steam or roast, with as little water as possible, and keep cooking time to a minimum. However, when it comes to fruits, "bacteria" becomes another factor in which these foods should be washed with a disinfectant (hydrogen peroxide). It should also be known that anytime you broil char or grill food (of any kind) it will produce "carcinogens" which are cancer forming chemicals. Moreover, do not microwave food because it degrades and destroys many of the essential chemical structures in the food and disrupts the hydrogen bonds, which contain the chemical structures that aid normal digestion and absorption into the body (however, a microwave is ok just for slightly warming something). Foods should also be cooked in "extra virgin olive oil". The reason for this is that olive oil comes from a greenish golden antioxidant-rich fruit that grow in the Middle East. Olive oil is a monounsaturated fatty acid (oleic acid which contains phenols) and its high in antioxidant substances which offer protection against heart disease by controlling LDL which is the bad part of cholesterol, while instead it raises the HDL which is the good cholesterol. Olive oil is beneficial for treating ulcers, gastritis, gallstones, and cardiovascular disease and even warding off colon cancer, it does this by activating the secretion of bile and pancreatic hormones. But while all olive oil is a source of monounsaturated fat, "extra virgin olive oil" should be used because this comes from the first pressing of the olives, which contains higher levels of antioxidants such as vitamin E and phenols (because it is the least processed). By least processed it means that the oil comes from the first pressing of the olives and is extracted without using heat or chemicals. The less the oil is handled, the closer it is to its natural state with containing all its beneficial nutrients. Researches also discovered that olive oil's phenolic compounds carry the same pain relieving benefits as ibuprophen. Aside from this olive oil has been used for thousands of years. Numbers 18:12 - I give you all the **finest olive oil** and all the finest new wine and grain they give the Lord as the firstfruits of their harvest. Deuteronomy 8:8 - A land with wheat and barley, vines and fig trees, pomegranates, **olive oil** and honey.

Now the quality of the water we drink is also an important matter. The finest water to drink is of course natural spring water, however even this water can contain heavy metals (even bacteria) which can be toxic to the body. Therefore, the best water to drink is boiled or steamed distilled water, even though important minerals are cooked out during the distilling process, our main source of nutrients should come from the food we eat, not the water we drink. Don't let the hyped facts get in the way… such facts that we need fluoride to protect us against cavities and brittle bones is a farce, when in fact we can get fluoride naturally from the foods we eat. Foods like tea, fresh fruits and vegetables. The reason for this is that fluoride is found naturally in rocks, volcanic ash, soil, air, and food, which then leaches its way into water sources, which eventually end up in our food. It is also found in plant and animal tissues. So we do not need the chlorinated, mercury laden and toxically over fluorinated water from the tap. Few people even realize that the tap water we drink not only contains an overabundant amount of fluoride, it also contains remnants of prescription drugs that are flushed down the drains and enter into our water supply and ground (so much for recycling). This also includes a staggering amount of burials in America that deposit gallons upon gallons of embalming fluid, formaldehyde, ethanol, and methanol into the soil each year. Additionally, the cremation process pumps all sort of poisonous chemicals into the air that we all breath in, such as dioxins, hydrochloric acid, sulfur dioxide, and carbon dioxide. Burials should be done how the Bible tells us… which is to be wrapped with a cloth and put in the ground. Ashes to ashes dust to dust. The apostle Paul appears to favor burial (which can be found in 1 Corinthians 15:35-44). It is here was Paul discusses how G-d will raise the decomposed body of a believer. The representation used is that of sowing a seed and having new life rise from the decomposing seed. Additionally, all of these people mentioned in the Bible (listed below) were buried either in the ground or in a sepulcher, which is a grave, tomb, vault, or burial chamber where the body is exposed to the air so it could decompose. Cremation was completely shunned; it was only used for the wicked. Revelation 20:15 - And whosoever was not found written in the book of life was cast into the lake of fire. Here are some referenced Bible verses of those buried either in the ground, sepulcher, grave tomb, vault or burial chamber**:** Abraham (Genesis 25:8-10), Sarah (Genesis 23:1-4), Rachael (Genesis 35:19-20), Isaac (Genesis 35:29), Jacob (Genesis 49:33 & 50:1-13), Joseph (Genesis 50:26), Moses (Deuteronomy 34:6), Joshua (24:29-30), Eleazar (Joshua 24:33), Samuel (1 Samuel 25:1), David (1 Kings 2:10), John the Baptist (Matthew 14:10-12), Ananias & Sapphira (Acts 5:5-10), and Stephen (Acts 8:2).

When it comes to drinking water, (preferably as mentioned distilled water) we can also add "tea" which is literally the juice of life! It should not be swallowed at extremely high intolerable temperatures since this can cause esophagus cancer, (this goes for any hot drinks or foods). All teas are good, however one of the best teas to drink is "green tea" which contains some of the most powerful substances found in the plant world that control inflammation, prevent cancer, help our body eliminate toxins and heavy metals, prevent cholesterol and heart disease (by thinning the blood and thereby reducing clotting), increase brain function and treat obesity by increasing our metabolism with natural caffeine. Tea also strengthens bones as it too contains fluoride found naturally in ground water and absorbed through the roots of the tree. Drinking lots of water and tea keeps the body from wrinkling and helps distribute nutrients throughout the body that facilitate organ and brain function. If you're worried about tea staining your teeth, just sip tea over the front of your teeth, or use a straw. Fruit juices are good to drink pertaining that they do not contain processed sugar such as high fructose corn syrup (or other added sugars). Juices should already be naturally sweetened with glucose. Notorious to this, we should stay away from all food that contain high fructose corn syrup, which is used as a cheap food preserver. Additionally when it comes to drinks, milk is very controversial. It is vital for a young child to receive these nutrients for fast growth up to a certain age (without question this milk should be organic)! This age would be when they stop growing, after this age milk becomes questionable. We humans seem to be the only ones that drink the milk of another animal, and we continue to do so even after we have been weaned and fully grown. However, after we are fully-grown adults be should consider obtaining calcium from other food sources.

This is for several reasons… The milk we drink today is cows milk, which is mainly made precisely for the baby cow; it is also loaded with harmful hormones that make the cows produce more milk. As I said, milk is a very controversial subject as it has been known to cause allergies, eczema, an increase in cholesterol, and one of the main causes (believe it or not) for "osteoporosis"! Contrary to belief osteoporosis IS NOT caused by a lack of calcium, it is mainly caused by a lack of activity and exercise, and to much protein (particularly animal protein) in the food you eat which causes acid (the opposite of alkaline) in your body, and in order for the body to neutralize that acid it has to take calcium OUT of your bones. Milk has also been connected to juvenile diabetes because in the milk protein (whey) which forms insulin in the pancreas to which the body produces antibodies against that "foreign" protein thus the body's reaction to this is to start destroying those cells so that the body cannot produce insulin. Milk has also been a factor in breast cancer, in heart disease, multiple sclerosis, ulcers, colitis of the intestines; female infertility and leukemia because many cows are infected with the bovine leukemia virus with can be transmitted by drinking the milk. Finally, mad cow disease, which has been otherwise known as Creutzfeldt-Jakob disease, which is virtually a form of "Alzheimer's". If one must drink milk, make sure that it is at least organic as well as hormone and antibiotic free, and drink it only in minimal amounts, instead try drinking almond milk which is high in antioxidants and calcium (vitamin D). Also virtually all green vegetables contain calcium such as all the dark green leafy vegetables like spinach, chard, leeks, artichokes, mustard greens and collard greens, as well as lettuces, like romaine, shredded cabbage, bok choy and broccoli. Even fruits contain calcium like plums, bananas and, kiwis. Additionally, virtually all nuts and seeds, especially sesame seeds, almonds and grains contain calcium. So do most beans, including navy, pinto, kidney, and black are all calcium-containing foods. In addition, try avoiding all carbonated drinks as they block calcium absorption. It should also be known that the hormones and antibiotics in milk are passed on to you through cheese and butter. Additionally cheese, butter, yogurt and even sour cream (aged foods), contain a chemical (called tyramine), which have been known to cause headaches, but what's happening here is that this chemical is actually upsetting the neurotransmitters and balance in the brain (these are the chemicals that transmit nerve impulses from one nerve to the other).

Processed sugar is another destructive agent in the food kingdom that we should veer away from. Sugar grows naturally in sugar cane, pure maple syrup from the maple tree, molasses, and raw unfiltered honey as well as beats and fruits and vegetables, even vegetables such as corn, carrots, potatoes, water melon and even rice (both brown or white). Milk even contains sugar. But these are all natural sugars, except when we start "refining" sugar it changes the molecular chemical structure so that when we consume refined sugar the body does not readily recognize it and therefore does not know how to break it down and digest it. This is because it lacks the "enzymes" that were stripped and cooked out of it. The body eventually does break it down, but damage happens in the process. This is more proof that G-d had intentionally designed natural food that grows from the Earth to work in harmony with the human body. Each chemical works hand-in-hand with the other. When we as humans disrupt or change this natural balance, it does more harm then good. That is why we should not fool around with nature! So the result we get when we refine sugar is that we strip it of its vital perfectly balanced molecular chemical structure to the point that the body no longer recognizes it in the way it should, and because it does not readily recognize it in the body, it doesn't absorb it properly and it therefore lingers around in saturation causing insulin and blood sugar to rise very quickly. Processed sugar has no nutritional value, while natural sugar is recognized by the body and therefore is absorbed more slowly keeping insulin and blood sugar stable. Processed sugar is known as "white death" and is a mild form of poison. It disturbs the natural bacteria in the stomach, which can later result in a yeast buildup and infection. Refined sugar also slows down the absorption of calcium and magnesium. It will change the metabolism so salt and water will be retained leading to high blood pressure. Sugar also affects the serotonin and endorphin hormones in the body thus effecting moods.

Refined sugar is really the nemesis in suppressing the immune system to toxins, bacteria, and viruses, by decreasing the number of white blood cells (the body's immune response to invaders). Sugar also makes your blood more acidic (bad) instead, of alkaline (good) which will be further explained in this chapter. Eating sugar throughout the day is a constant assault on the immune system. Refined sugar robs our body of vitamins, minerals, and enzymes and increases hyperactivity by preventing blood sugar levels from remaining stable. Additionally, refined sugar depletes vitamin C out of ones body, and as we know, vitamin C is used in our body to help reduce infection. It also depletes B complex (which is needed to calm the nerves). Furthermore, it also depletes zinc and chromium from the body, which are necessary for healing. It also upsets the calcium phosphorous level of your body, causes hyperglycemia, and IT PARALYZES THE IMMUNE SYSTEM FOR SEVERAL HOURS AFTER ONE HAS EATEN ANY, and the immune system therefore cannot resist disease. If you are a person that seems to get sick a lot (such as with colds) you should begin to look at the amount of your sugar intake. Substitute sugars like aspartame or saccharine also disrupts the neurotransmitters in the brain (this again is because the body was not originally programmed by G-d to recognize these processed foods for digestion). However, other sugar substitutes are natural such as "stevia" which is a herb in the sunflower family that has been used for hundreds of years. It is calorie free, no carbohydrates, no tooth decay and contains nutrients and trace minerals. It has also been noted as diabetic safe since it nourishes the pancreas and does not raise blood glucose levels. Stevia is twice as Sweet than sugar, so only a small amount is needed for sweetening. Along with stevia, there is also "xylitol". Xylitol is a naturally occurring sweetener found in the fibers of many fruits and vegetables, including various berries like raspberries and plums, corn husks, oats, mushrooms and from birch trees. Like natural sugar, xylitol can prevent the growth of bacteria. It can also heal and repair the body and build immunity. Xylitol can also be used to inhibit plaque and dental cavities and improve tooth enamel (that's why we find it as an ingredient in some gums). However, both stevia and xylitol can be refined so its important to obtain only the green or brown (whole) stevia extracts or powders, and avoid the clear extracts and white powders which are highly refined and stripped of the essential nutrient which can cause imbalance in the body as aforementioned. Additionally, sugar is not mentioned anywhere in the Bible, however the taste of something being sweet is frequently mentioned in the Bible with the use of natural things such as honey or fruits.

Honey also has other medicinal properties. It can even ease seasonal allergies. Thus transpires with eating "locally produced" honey. The theory goes: Bees in ones neighborhood feed on the same pollen that gives you itchy eyes and a runny nose. That pollen is added to the hive's honey, which when ingested it can help one build a tolerance to those allergens. In addition, as we've also learned from a previous chapter, that Manuka honey from New Zealand contains natural antibiotic properties. Of course, the same process could produce negative effects when bees visit problem plants, such as poison ivy, which can potentially cause a rash in people who ingest the ivy-tainted honey, and as previously mentioned in an earlier chapter such plants that contain terpenoids such as the Rhododendron plant can cause nausea.

Moving on here... As was expressed in previous chapters that the Sun is crucial to our existence here on Earth... additionally it is also beneficial to our health in moderation. It strengthens the immune system, kills bacteria, increases oxygen in the blood, it also calms the nervous system by relaxing you. Moreover, it lowers blood sugar by putting glucose into the cell like natural insulin. It lowers blood pressure, and it lowers cholesterol because it turns cholesterol into vitamin D when the Sun hits it in your skin. The Sun also prevents rickets (a disease that stunts growth) and osteoporosis. Despite many claims of the Sun causing skin cancer or other cancers, the Sun in "moderation" is a good agent for optimum health. In fact (despite what some experts say) the Sun can even PREVENT several types of cancers. Even viewing the bright and beautiful sunshine has good psychological effects on the brain helping it to fight depression. Ecclesiastes 11:7 - Light is sweet, and it pleases the eyes to see the sun. However, one should not sit out intentionally with the purpose of tanning the skin or overexposing it to the point of burning. Certain natural oils are good for the skin both internally and externally that can aid in protecting the skin again the Sun. Some of these natural oils are: Avocado oil, borage oil, evening primrose oil, black currant seed oil, castor oil, cocoa butter, coconut oil, grape seed oil, hemp seed oil, jojoba oil, kukui nut oil (this oil comes from the nuts of a candlenut tree which are native to Hawaii), mango butter, olive oil, palm oil, passion-fruit oil, rice bran oil, rosehip seed oil, sesame oil, shea butter and sweet almond oil. There is even organic sunscreen (Sun block) available on the market. These should be used as apposed to sunscreens that contain dangerous and harmful chemicals (which most do). Our skin is an organ too and we should avoid body lotions made with processed, unnatural or dangerous chemicals. Instead we should be rubbing into the skin natural organic moisturizing oils, and if necessary take an extra vitamin supplement that contains these essential oils, or eat foods rich in these natural oils. Some of these natural oils are: Soy bean (vitamin E), Avocado oil, borage oil, evening primrose oil, black currant seed oil, castor oil, cocoa butter, coconut oil, grape seed oil, hemp seed oil, jojoba oil, kukui nut oil, mango butter, olive oil, palm oil, passion-fruit oil, rice bran oil, rosehip seed oil, sesame oil, shea butter, sweet almond oil.

Additionally I must also include in here that a "warm" relaxing environment is essential to humans. Remember we were originally placed in a semi-tropical warm and cozy garden where clothing was not needed. Genesis 2:8 - Now the Lord G-d had planted a garden in the east, in Eden, and there he put the man he had formed. Genesis 2:25 - The man and his wife were both naked, and they felt no shame. Humans are essentially semi-tropical beings and are not equipped to deal with even mild cold. We can adapt to live in cold climates because of wearing appropriate clothing and building shelters, but ideally, we cannot endure temperatures below 70 without any clothing. Our normal body temperatures are 95 to 99°F and should be maintained as such. Scientific proof of this is the reaction of our body's responses to cold such as shivering which diverts blood away from the extremities and surface of the skin that can result in death. Additionally high extremes of heat can also result in death. These responses mean that we should maintain our core temperature wearing only light clothing.

Sleep is also extremely important to the human body… A hormone called melatonin is the thing that actually puts us to sleep. Melatonin is a hormone secreted from the pineal gland in the center of our brain. It is released when our eyes begin to register that the Sun is beginning to set and darkness begins to fall. This is what makes you go to sleep and is used in our body to regulate our sleep-wake cycles. If you wonder why older people tend to sleep less than younger people, it is because the amount of melatonin produced in their body seems to lessen as they age. Here are some biblical verses that lay emphasis on the importance of sleep and rest. Psalm 127:2 - In vain, you rise early and stay up late, toiling for food to eat— for he grants sleep to those he loves. Proverbs 6:10 - A little sleep, a little slumber, a little folding of the hands to rest. Psalm 3:5 - I lie down and sleep; I wake again, because the Lord sustains me. Psalm 4:8 - I will lie down and sleep in peace, for you alone, O Lord, make me dwell in safety. Psalm 78:65 - Then the Lord awoke as from sleep, as a man wakes from the stupor of wine. Aside from this, there is a striking connection between the amount of sleep we get and our body weight. In other words, if we loose sleep, we gain weight and the opposite effect occurs if we get plenty of sleep we loose weight. The less sleep that we get promotes the production of hormones that create cravings for food. These hormones are produced in the stomach and then send out hunger signals to the brain, which in turn commands you to be interested in food. There are also other hormones produced by fat cells that deliver similar signals to the brain. There is also a link between the right amounts of sleep in reducing the risks of ulcers, since the chemical that is responsible for healing and restoring the lining of the stomach is excreted during a restful sleep. People who do not get enough adequate sleep have symptoms of nausea, clumsiness, sluggishness, difficulty reading, unable to concentrate, suffer temporary memory lapses, trouble coping with stress, becoming irritated, instant mood swings and periods where one wont talk or communicate to others properly. A shortage of sleep can also accelerate the aging process, as opposed to an adequate restful sleep, which brings on a sense of well being and a hormone called DHEA (Dehydroepiandrosterone - an anti-aging super hormone). Your body makes most of this chemical hormone in the morning after a restful sleep. In addition, all of the hormones produced in the body are made using cholesterol as the starting material. Then out of the cholesterol the adrenal glands produce DHEA which then produces the brain chemical pregnenolone which is then converted into estrogen, progesterone, testosterone and the stress hormone cortical. People with higher DHEA levels tend to live longer, healthier lives then those with lower levels of DHEA. Now additionally, the reason we sleep is not just to produce hormones for the body, but it also has to do primarily with the brain and the connections between neurons (brain cells) called "synapses". Synapses (gap between nerve ends) allow the brain to talk to itself and the rest of the body. Therefore sleep becomes essential to the human body because it sets aside time for the fine-tuning of these synaptic connections which helps in reinforcing memory and at the same time may also erase unneeded memory. Deuteronomy 32:26 - I said I would scatter them and blot out their memory from mankind. In essence, sleep helps the brain save its superstructure. This occurs because during the wake hours of the day the brain is constantly taking in and inserting new information, when we sleep the brain miraculously shifts, rearranges and reorganizes this new and old information into their proper folders (sequential segments) so that they can be retrieved properly again and again whenever needed. This also frees up speed and efficiency. It works kind of like a "disk defragmenter" on a computer. When you do a disk defragmenter, your computer progress speeds up. It actually frees up movable space allowing your computer to work faster and you are able to visit more sites on the Internet (world wide web) at even faster speeds. Better yet, the "system restore" which can take you back to a previous time when your computer was functioning at its original optimum speed or capacity. Humorously, I find it ironic that the computers disk defragmenter can take a considerable time to run, so it's usually done at the end of the day, before going to sleep. Here are some biblical verses that lay emphasis on the importance of memory. Psalm 34:16 - The face of the Lord is against those who do evil, to cut off the **memory** of them from the earth. Psalm 45:17 - I will perpetuate your **memory** through all generations, therefore the nations will praise you for ever and ever. 2 Peter 1:13 - I think it is right to refresh your **memory** as long as I live in the tent of this body. Leviticus 26:42 - I will **remember** my covenant with Jacob and my covenant with Isaac and my covenant with Abraham, and I will remember the land.

In the midst of getting enough sleep, "exercise" is also additionally important. All of the things I've mentioned so far: Eating, sleeping, a sunny warm and clean environment, and now exercising, go hand in hand. One without the other leads to a colossal failure. So now lets discuss exercise. We must exercise at least approximately 30 minutes a day. It doesn't have to be excessive; it just has to be enough to get the heart above its normal beat. 1 Timothy 4:8 - For physical training is of some value, but godliness has value for all things, holding promise for both the present life and the life to come. 1 Corinthians 9:24-27 - Do you not know that in a race all the runners run, but only one gets the prize? Run in such a way as to get the prize. Everyone who competes in the games goes into strict training. They do it to get a crown that will not last, but we do it to get a crown that will last forever. Therefore I do not run like a man running aimlessly, I do not fight like a man beating the air. No, I beat my body and make it my slave so that after I have preached to others, I myself will not be disqualified for the prize. 2 Timothy 2:5 - Similarly, if anyone competes as an athlete, he does not receive the victor's crown unless he competes according to the rules. 2 Timothy 4:7 - I have fought the good fight, I have finished the race, I have kept the faith. In fact, the Bible is clear that we are to take good care of our bodies. 1 Corinthians 6:19-20 - Do you not know that your body is a temple of the Holy Spirit, who is in you, whom you have received from G-d? You are not your own, you were bought at a price. Therefore honor G-d with your body. Ephesians 5:29 - After all, no one ever hated his own body, but he feeds and cares for it, just as Christ does the church. Additionally remember (as we've previous discussed), the Bible also warns us about being a glutton and overeating. Deuteronomy 21:20 - And they shall say unto the elders of his city, This our son is stubborn and rebellious, he will not obey our voice, he is a glutton, and a drunkard. Proverbs 23:2 - And put a knife to thy throat, if thou be a man given to appetite.

The Bible makes it quite clear that by engaging in moderate exercise and eating reasonably, that this is G-d's instruction for our earthly body after the fall of Adam and Eve and the flood. One of the best exercises that the human body was designed to do is walk. Walking supports the whole body's health because in doing so, every part (lock, stock and barrel) of the body is in motion at the same time. When every part of the body is in motion what's actually happening is that we are stimulating the immune system by keeping our immune cells tuned up for action, ready to whip viruses and battle bacteria. A moderate walk can relieve stress that may trigger or aggravate an illness. It also aids one in overcoming illnesses such as colds (cutting their duration-time in half). It may even prevent colds and flu's from coming-on (or developing) since it gears up the immune system and keeps it on guard against invading factors. The cells activity returns to normal about 3 hours after a walk. Walking also makes one heal faster and develop fewer illnesses then people who are sedentary. It may also be noticeable to individuals that just sit around without exercising, that they frequently and continually get more colds. It also depends on WHAT one eats and HOW MUCH one eats. This is because OVEREATING can put a strained tax on the intestines, causing them work harder, and as a result shifting the body's gears (mechanism) away from fighting off bacteria and other foreign invaders. Additionally WHAT ONE EATS plays a significant role especially with foods like sugars, or high fats... seeing that this also puts a straining tax on the body forcing it to shift gears elsewhere in the body - thus setting aside the role of the immune system (postponing it and putting it on hold, or putting it on the back burner, so to speak), which gives the invaders ample time to flourish. Additionally when it comes to exercise, (on the contrary) its been found that long bouts of intense exercise like heavy running can actually "suppress" the immune system and make one more susceptible to infection, (this actually stresses the body out and runs it down).

Exercise keeps cropping up as a factor in cancer prevention and treatment. It supports weight loss and weight maintenance, it reduces the risk of heart disease and stroke, it fends off diabetes by improving the body's ability to use insulin, it reduces the pain and stiffness of arthritis, it keeps bones strong which prevents osteoporosis. In women, it relieves premenstrual and menopausal discomforts. In all people, it improves sleep, it builds strength, flexibility and stamina, it enhances mental function and it counteracts anger, depression and anxiety. Genesis 13:17 - Go, **walk** through the length and breadth of the land, for I am giving it to you. Genesis 17:1 - When Abram was ninety-nine years old, the Lord appeared to him and said, I am G-d Almighty; **walk** before me and be blameless. Exodus 17:5 - The Lord answered Moses, **Walk** on ahead of the people. Take with you some of the elders of Israel and take in your hand the staff with which you struck the Nile, and go. Leviticus 26:13 - I am the Lord your G-d, who brought you out of Egypt so that you would no longer be slaves to the Egyptians, I broke the bars of your yoke and enabled you to **walk** with heads held high. Isaiah 30:21 - Whether you turn to the right or to the left, your ears will hear a voice behind you, saying, This is the way, **walk** in it. Deuteronomy 5:33 - **Walk** in all the way that the Lord your G-d has commanded you, so that you may live and prosper and prolong your days in the land that you will possess. Isaiah 40:31 - But those who hope in the Lord will renew their strength. They will soar on wings like eagles, they will run and not grow weary, they will **walk** and not be faint. Luke 24:15 - As they talked and discussed these things with each other, Jesus himself came up and **walked** along with them. John 6:19 - When they had rowed three or three and a half miles, they saw Jesus approaching the boat, **walking** on the water, and they were terrified. Acts 14:10 - And called out, Stand up on your feet! At that, the man jumped up and began to **walk.** John 5:8 - Then Jesus said to him, Get up! Pick up your mat and **walk.**

Leviticus 26:12 - I will **walk** among you and be your G-d, and you will be my people. The Bible even says that after the believers have been resurrected we will walk in the New Jerusalem. Isaiah 35:9 - No lion will be there, nor will any ferocious beast get up on it, they will not be found there. But only the redeemed will **walk** there.

Not only should we exercise, but our brain also needs effective use. What's more, a link between exercise and intelligence has been suspected. Brain chemistry reveals an intimate relationship between mind and body. Neurons not only make contact with other neurons in the brain but they also connect with the skeletal muscles, joints and the nervous system of the body as well, by using chemical messengers. The bottom line is that physical exercise keeps the receptors that scurry in and out of synapses working and continually firing away. To stimulate and strengthen these neural connections and even create new ones there are certain cognitive exercises one can do, such as use your opposite hand to perform daily activities. To develop agility and hand-brain coordination, als try learning new tasks that involve using your five physical senses. Do things with your eyes closed, do crossword puzzles, paint, learn a foreign language, a change in environment, adventuring to new places, even the game of bingo helps in minimizing memory loss and remaining mentally sharp, it does this by bolstering hand-eye coordination. Another game that I've found that highly encourages the brain to think in marvelous ways is the Israeli game of Rummikub. It is a tile game that coordinates both color and numbers and uses them in strategic ways. The winner is the one who is able to get rid of all his/her tiles. But this game goes beyond normal thinking because it challenges (almost forces) ones brain to think of ways of substituting and manipulating numbers and colors which helps builds cognitive comprehension, plotting and planning skills, visual skills, critical and analytical thinking skills. It also develops skills in sequencing, comparing, recognizing patterns. It also allows one to develop the ability to think ahead by evaluating alternatives and generating solutions allowing one to practice, their mental mapping ability in order to outmaneuver their opponent.

For the most part ANYTHING that stimulates the brain to think is considered beneficial exercise. The more we think, the better our brains function, regardless of age. In addition, of course read, read, READ! It has even been discovered that "a muscle" (any muscle in the human body as well as the brain) can be strengthened just by "thinking" about exercising it.

Again, walking is especially good for your brain, because it increases blood circulation and oxygen that flows throughout the whole body. As explained earlier that walking uses every part of ones body that allows for all over balance distribution of blood, oxygen, nutrients, chemicals and hormones. Everything is working harmoniously with every bone, muscle, organ and cell in the body and this includes the brain. Therefore, when one walks, they are effectively oxygenating their brain. Maybe this is why it's said that walking can "clear your head" and help you to think better. Walking has also been said to improve learning ability, concentration, memory, and abstract reasoning. It is also known to help in warding off depression. Its stands to reason that movement and exercise increase breathing and heart rate (improving the hearts ability to pump blood) so that more blood flows to the brain, thus enhancing energy production in the body and also facilitating waste removal. It has even been observed that in response to exercise, cerebral blood vessels can grow. When the brain is deprived of the proper blood flow or oxygen that it needs, it begins to deteriorate and malfunction, this then sends a signal to the rest of the body. That is why all of these things I've mentioned so far: Exercising, eating, sleeping, a sunny, warm and clean environment all go hand in hand - one without the other leads to a colossal failure. This also engulfs the spiritual aspect of things - we must not leave G-d out of the equation because the Bible clearly and explicitly explains to us why we are in the DETERIORATIVE state that we are, and why we "now" need to maintain it - lest we forget, sin and the curse. Genesis 3:17 - To Adam he said, Because you listened to your wife and ate from the tree about which I commanded you, You must not eat of it, Cursed is the ground because of you; through painful toil you will eat of it all the days of your life.

Elaborating a little further here… When we exercise we breath. Just breathing alone increases the oxygen intake into the body, which in turn burns fat (thereby stimulating the metabolism that speeds up the digestive enzymes). Just by increasing oxygen to the cells, it can eliminate a multitude of diseases. Cancer, for example, cannot live in an oxygen rich environment. To explain this further, "glucose" which is primarily a form of sugar from natural plants and fruit sources? When we eat glucose, it is stored in the body for energy. Glucose combined with oxygen produces the energy that's needed for healthy cells. Without oxygen, the glucose will ferment. Cancer cells live on the fermentation of glucose caused by to the lack of oxygen. Additionally, without sufficient oxygen to the cells in the body to keep them stimulated, the cells will not be able to take in nutrients or release toxins and poisons (which cause inflammation and fluid retention) from out of the blood stream. Without oxygen and nutrients, these cells will shut down. When cells shut down (trillions of them) the body is then subjected to a whole slew of diseases such as pain, bodily weakness and fatigue, muscle aches, poor circulation, poor digestion, ulcers, dizziness, headache, memory loss, infection and bacteria, susceptibility to colds, flu's, viruses, tumors, cysts and intestinal parasites which are a result of lower immunity, anxiety, depression, stress, irritability, irrational behavior, hostility and as was mentioned earlier a result of premature aging. Therefore, as you can evaluate, oxygen is a vital cell detoxifier. Oxygen deficiency in itself is the underlying cause of cancer and other degenerative diseases, and ultimately the death of the cell, and eventually the body.

There are numerous ways to get more oxygen into our bloodstream, as mentioned from exercise and walking, but also an oxygen rich environment like the seashore, or a densely packed (with trees) woodsy area. Even exposing one self to positive ions. Positive ions can be found in the air after it rains, after cutting the grass, or by a waterfall, stream or running water, ocean waves, and large trees whose leaves bustle and blow in the wind. Getting more oxygen can be as simple as just breathing deep breaths several times a day. Job 12:10 - In his hand is the life of every creature and the **breath** of all mankind.

Additionally upon using our brains and exercising, we must also take time in life to "meditate". Meditating may sound foolish to some people but it is also biblical in nature. Joshua 1:8 - Do not let this Book of the Law depart from your mouth, **meditate** on it day and night, so that you may be careful to do everything written in it. Then you will be prosperous and successful. Psalm 1:2 - But his delight is in the law of the Lord, and on his law he **meditates** day and night. Psalm 48:9 - Within your temple, O G-d, we **meditate** on your unfailing love. Psalm 77:12 - I will **meditate** on all your works and consider all your mighty deeds. Psalm 119:15 - I **meditate** on your precepts and consider your ways. Psalm 119:27 - Let me understand the teaching of your precepts; then I will **meditate** on your wonders. Psalm 119:48 - I lift up my hands to your commands, which I love, and I **meditate** on your decrees. Psalm 119:97 - Oh, how I love your law! I **meditate** on it all day long. Psalm 119:99

- I have more insight than all my teachers, for I **meditate** on your statutes. Psalm 119:148 - My eyes stay open through the watches of the night, that I may **meditate** on your promises. Psalm 143:5 - I remember the days of long ago; I **meditate** on all your works and consider what your hands have done. Psalm 145:5 - They will speak of the glorious splendor of your majesty, and I will **meditate** on your wonderful works. But we must also be careful that we do not pray or meditate using "vain repetitions" (repeating words over and over again) since these are hypnotic in nature which allow-in (and give access to) evil inclinations. Matthew 6:7 - And when you pray, **do not use vain repetitions as the heathen do.** For they think that they will be heard for their many words.

Along with meditation, "music" is also important. What's more, the type of music we listen to is even more vitally important! Though one may not think so, I will attempt to correct this thinking. Starting with the 12th book of New Testament (Colossians), which is a letter from the apostle Paul to the church, which tells us that we are to sing (or listen) to psalms, hymns and spiritual songs. This is G-d's music! Colossians 3:16 - Let the word of Christ dwell in you richly in all wisdom; teaching and admonishing one another in **psalms and hymns and spiritual songs, singing with grace in your hearts to the Lord.** Now this may be a revolutionary new understanding for the modern person. It may even be somewhat of a turn-off for some people who are adamantly passionate to certain types of contemporary music, such as heavy metal, hard rock, or even the upbeat oldies. However, the biblical truth in this matter stands firm. The only music we should be listening to is music that endorses and glorifies G-d and is felt from the spirit. There are several factors in music to consider… One being the style of music, and the other the content of the lyrics. We must remember that the flesh is of the "carnal mind" and has fleshly desires… therefore if a certain music does not endorse G-d, and uses profane or evil lyrics and seduces you to dance in an unconventional way, it is therefore considered a fleshly desire. However, dancing is not bad under the condition that it is honoring G-d and not yourself or others (if it conjures up lust it is not acceptable). If there is any dancing to be done, it should be felt from the spirit and should not rouse any fleshly desires whatsoever. For example there are several instances mentioned in the Bible that direct us in correct dancing and singing. Such as in Exodus where Miriam was dancing to celebrate the victory of G-d's power (parting of the water) at the Red Sea. Exodus 15:19-21 - For the horse of Pharaoh went in with his chariots and with his horsemen into the sea, and the Lord brought again the waters of the sea upon them; but the children of Israel went on dry land in the midst of the sea. **And Miriam the prophetess, the sister of Aaron, took a timbrel in her hand; and all the women went out after her with timbrels and with dances.** And Miriam answered them, **Sing ye to the Lord, for he hath triumphed gloriously**; the horse and his rider hath he thrown into the sea.

Additionally in 2nd Samuel King David danced before G-d to celebrate the Ark of the Covenant that was brought back to Jerusalem. 2 Samuel 6:12-16 - And it was told king David, saying, The Lord hath blessed the house of Obededom, and all that pertaineth unto him, because of the ark of G-d. So David went and brought up the ark of G-d from the house of Obededom into the city of David with gladness. And it was so, that when they that bare the ark of the Lord had gone six paces, he sacrificed oxen and fatlings. **And David danced before the Lord with all his might;** and David was girded with a linen ephod. **So David and all the house of Israel brought up the ark of the Lord with shouting, and with the sound of the trumpet.** And as the ark of the Lord came into the city of David, Michal Saul's daughter looked through a window, and saw king David leaping and dancing before the Lord; and she despised him in her heart.

Every instance recorded in the Bible for dancing that is not considered sinful was done in worship or praise to G-d - thus glorifying Him. Here are additional guidelines to keep in mind when considering dancing. Ecclesiastes 3:4 - A time to weep, and a time to laugh; a time to mourn, and a time to dance. Psalm 149:3 - Let them praise his name in the dance: let them sing praises unto him with the timbrel and harp. Psalm 150:4 - Praise him with the timbrel and dance: praise him with stringed instruments and organs. However, as with anything else there are exceptions to the rules. Such as if lyrics in a song are true, righteous, pure and clean, pleasant, commendable, admirable, heartfelt, praiseworthy or sung to a loved one in a caring way, then there is nothing wrong with listening to a worldly song of that nature. However, I will stand firm here. The bottom line is that our bodies belong to G-d, and they are the temples of the Holy Spirit. Therefore, everything we do with our body (even think with our mind) must be in honor to Him. 1 Corinthians 6:19-20 - Do you not know that your body is a temple of the Holy Spirit, who is in you, whom you have received from G-d? You are not your own; you were bought at a price. Therefore honor G-d with your body. Philippians 4:8 - Finally, brothers, whatever is true, whatever is noble, whatever is right, whatever is pure, whatever is lovely, whatever is admirable—if anything is excellent or praiseworthy—think about such things. Colossians 3:2 - Set your minds on things above, not on earthly things.

The main thing in all this… If anything leads you to profane your body or mind or be caught up or drawn into something that does not glorify G-d, it should be avoided. “heavy metal music” is unequivocally ungodly flesh music. It is the sounds of demons that lead us away from our personal relationship with G-d. It literally makes you feel powerful, authoritative, dominant, forceful, moving, strong - to the point of possibly being violent. In fact, most people don’t know this, but this music has its roots from voodoo (from the African joojoo tribes) as a means to summon up demons (the word joojoo literally means voodoo). This thumping sound relies heavily on hypnotic drumming. If these joojoo tribes were to hear this beat (and sound) in any music they listened to, they could easily identify it with their summoning up of demons ritual.

Now let me back up here a bit and explain why (or how) this notable beating of the drum came into being for summoning up demons so that you know I‘m not just making this stuff up. Firstly, the Bible clearly informs us that Satan was/is a musical creature that while in the hierarchy of Heaven he once was a worship leader (second in command to G-d) that led billions of angels in worship. Now when he was cast out (and fell from Heaven) the Earth was corrupted and degraded with his music (it suffered the same curse he did). Ezekiel 28:13 - Thou hast been in Eden the garden of G-d; every precious stone was thy covering, the sardius, topaz, and the diamond, the beryl, the onyx, and the jasper, the sapphire, the emerald, and the carbuncle, and gold: **the workmanship of thy tabrets (timbrels which is a tambourine or small hand drum) and of thy pipes was prepared in thee in the day that thou wast created.** Secondly, the Bible also tells us that music is the “prince of the air”. Ephesians 2:2 - Wherein in time past ye walked according to the course of this world, **according to the prince of the power of the air, the spirit that now worketh in the children of disobedience.** Simply put, the “power of the air” is none other than “sound waves” which propagate through the air by compressing the molecules. Therefore, music (which in essence is sound waves) works under the same principle.

Now since these details are freely supplied to us, one can attach a sense of expansion to it. We now know that Satan was created by G-d and was made "greater" - being anointed in an expanded sense and covered by the stones insinuates a leadership role. We find this out when we compare this to other "concordances" in the Bible… we learn that as in Ezekiel 28:13 (cited above) that Satan was covered in 9 of the 12 stones (representing the 12 tribes of Israel) that the Levite priests were required to wear on their breastplates. Exodus 28:17-20 - And thou shalt set in it settings of stones, even four rows of stones: the first row shall be a **sardius,** a **topaz,** and a **carbuncle:** this shall be the first row. And the second row shall be an **emerald,** a **sapphire,** and a **diamond.** And the third row a **ligure,** an **agate,** and an **amethyst.** And the fourth row a **beryl,** and an **onyx,** and a **jasper:** they shall be set in gold in their inclosings. However, the difference here (with the breastplate) is that Satan was designed with these musical instruments physically within him. In addition to this, Ezekiel 28:14 also informs us that Satan was the cherub that "covereth", meaning that G-d had originally created him as a guarding type of filtering shield to cover and protect, (which later became defective). Ezekiel 28:14 - Thou art the anointed cherub that **covereth;** and I have set thee so: thou wast upon the holy mountain of G-d; thou hast walked up and down in the midst of the stones of fire.

Therefore we can conclusively say (without doubt now) that Satan was designed as a musical instrument for G-d and anointed to lead all of the hosts of heaven in the worship of G-d in which to create a covering for G-d to dwell in. Sadly Satan failed to do this and was cast out (he became defective because of iniquity). Ezekiel 28:15 - Thou wast perfect in thy ways from the day that thou wast created, till **iniquity** was found in thee. Luke 10:18 (Jesus tells us) - And he said unto them, I beheld Satan as lightning fall from heaven. Now because Satan was cast out of Heaven, he does not want anyone or anything to replace him. This being the case, we as G-d's creation have both the opportunity and responsibility to replace this fallen angel by calling all of G-d's people to worship and sing songs (specifically) to the Lord.

Now with this assessment… Here is something else that most people don't recognize. Some of the most beautiful music is made by the human voice itself. Additionally, there are the other musical instruments that enhance this serenity. But there will be no drums or synthesizers in Heaven since synthesizers are an electronic device that modifies sounds electronically, and drums, which are used to keep the beat, rhythm and time. As G-d would have it, there will be no drums in Heaven to keep the beat and rhythm because in Heaven there is no "time" - and therefore no drums. If drums are to be played, it will be with the uttermost softness. Now one may ask, how is it that I know there is no time in Heaven? Because the Bible clearly tells us that in Heaven there will be no need of light from the Sun or Moon which clearly implies that there is no more revolving (circling) of the planetary bodies that define day and night (in essence "time"). Revelation 22:5 - And there shall be no night there; and they need no candle, neither light of the sun; for the Lord G-d giveth them light: and they shall reign for ever and ever. Revelation 21:23 - The city does not need the sun or the moon to shine on it, for the glory of G-d gives it light, and the Lamb is its lamp. The Bible also definably tells us from the very beginning of Genesis that G-d created the Sun and the Moon to differentiate the "time and days" for us here on the Earth. Genesis 1:1-5 - In the beginning G-d created the heaven and the earth. And the earth was without form, and void; and darkness was upon the face of the deep. And the Spirit of G-d moved upon the face of the waters. And G-d said, Let there be light: and there was light. And G-d saw the light, that it was good: **and G-d divided the light from the darkness. And G-d called the light Day, and the darkness he called Night. And the evening and the morning were the "first day".**

Moving along here (and getting back on track) lets get back on the subject of eating and examine some additional eating guidelines… Now, along with the biblical eating rules found in the laws of Leviticus (as well as other places in the Bible), we see that it is "now" ok to eat the meat of an animal. However, if you recall before in the garden where no death ever occurred, that no meat was consumed because no animal ever died - therefore every living thing ate plants. This was the initial diet intended for our human bodies designed by G-d. In addition, we can still scientifically see that this is indeed the case since He created us with long winding intestines that were made to digest plant food very slowly. It digests very slowly because plants have enzymes in them that preserve them from putrefying while going through the process of the stomach. When fruits ferment, they turn into alcohol like wine, and when vegetables ferment or decay, they create gases. Meat on the other hand putrefies, which means it creates poisons that are toxic to the human body. It also allows bacteria and parasites more time to multiply. Meat also causes constipation in humans, while vegetables with fiber and roughage are the cure for constipation. Whereas, animals with small stomachs can process food more quickly, therefore food (like meat) does not have the time to putrefy and become toxic. For an example of how fast meat putrefies compared to fruits or vegetables. Take a slab of any meat (that has been defrosted) and let it sit out over night on a counter at room temperature. The time limit is 2 hours I believe, after that the bacterial count goes crazy. Leviticus 22:30 - On the same day it shall be eaten up, ye shall leave none of it until the morrow: I am the Lord. While most vegetables and fruits like apples can last almost 2 weeks after they are picked or refrigerated, and garlic can last for months. This is why animals today can eat meat or the flesh of other animals, because they have small stomachs that can process food more quickly. Whereas, the human intestines where not made to digest meat. Animals such as wolves with short elastic intestines that are smooth and designed to stretch to capacity, and push meat through quickly so that it does not sit and putrefy in the gut also lack an enzyme that is responsible for beginning carbohydrate breakdown. Instead, they have an enzyme in their saliva that destroys bacteria. However, even though all the animals were originally plant eaters at first, it follows along with the account of the curse, which speaks of GREAT CHANGES, which would follow, and each one required a genetic change and alteration. The Bible makes account for these changes in Genesis 3. Plants would grow thorns and thistles. Genesis 3:18 - Thorns also and thistles shall it bring forth to thee, and thou shalt eat the herb of the field. The serpent would slither on his belly. Genesis 3:14 - And the Lord G-d said unto the serpent, Because thou hast done this, thou art cursed above all cattle, and above every beast of the field, upon thy belly shalt thou go, and dust shalt thou eat all the days of thy life. The serpent (snake) was cursed along with the rest of the animals. Eve's body also underwent a change that made childbirth a painful experience. Genesis 3:16 - To the woman he said, I will greatly increase your pains in childbearing, with pain you will give birth to children. Your desire will be for your husband, and he will rule over you. Therefore, the Bible supports adequate evidence that some sort of genetic alteration took place, and instead of everything being "very good", the whole creation now groans and travails in pain due to the occurrence and presence of sin and its consequence - death. Romans 8:22 - For we know that the whole creation groaneth and travaileth in pain together until now. Eve's increased pain in childbirth might have something to do with the neuropathic receptors within the human body and brain. It stands to reason that after the curse indeed something fundamentally different changed within the human body. IT IS physically possible NOT TO feel pain. Since today, such marvelous techniques can be used during childbirth such as an epidural, which freezes the space around the nerves of the spine. However, other fantastic phenomenon's can take place within the human body that Geneticists have recently discovered. Such as people who have a genetic mutation that allows them to feel no pain - and it is genetic. These people have an inherited genetic mutation that stops them from feeling pain in any part of their body. The mutation is on a gene called SCN9A. It stops a sodium channel, which produces nerve impulses that convey pain signals to the brain from functioning. In fact, some drug companies are already working on a new pain relief product in preclinical development based on this genetic discovery. In hindsight, although pain hurts it is a useful sensation because it warns people of danger and injury or if something is too hot or too cold. It also has a survival benefit because when people begin to feel pain they change their behavior to avoid it. Therefore, if indeed we were in a sustained garden before the curse, we most likely didn't "NEED" to feel pain. So its not hard to imagine that in G-d's disposition and infinite wisdom He had indeed altered and adjusted creation to meet the new provision of sin, with all things feeling pain and dying, along with some animals changing temperament (to carnivorous) as well as having their intestines modified. The new sensation of feeling pain was placed there by G-d so that we might change our behavior in order to avoid it, and to ask for repentance for sin, and appreciate that G-d came later in the form of Jesus in place of our sins. In addition, today, as we constantly witness the same thing, it should be a constant reminder to our creator's condition for sin, and to readily look forward to the time when there will be no more death. 1 Corinthians 15:26 - The last enemy to be destroyed is death. And for those who believe in Jesus (the remedy and resolve of sins) shall inherit the new kingdom of G-d. Revelation 21:4 - And G-d shall wipe away all tears from their eyes, and there shall be no more death, neither sorrow, nor crying, neither shall there be any more pain: for the former things are passed away. The biggest mystery and secret of it all is that "change" will happen again! 1 Corinthians 15:51 - Listen, I tell you a mystery: We will not all sleep, but we will all be changed.

Therefore, in one breath the Bible tells us that we were originally plant eaters, then because of sin, we became meat eaters and eventually we die. I don't think that G-d changed our intestines much as we can still see the side effects of meat. And I'm sure He left it this way as a testimony and witness to the curse. In His infinite wisdom, He knew that centuries later in the age of modern science and microscopes that we would discover that meat harbored bacteria and poisonous toxins, and that we would also rediscover that our intestines were not originally designed to digest meat. This may also explain to us why we have an appendix. Not because of evolution, but because of toxins and poisons particularly from eating meat. We know this since it has been recently discovered that the appendix plays a big part in the course of fetal development. It appears that the appendix during the growth of the human embryo has a rapid development during the critical stages of bowel growth and organization. In addition, it is during this time that the development of the appendix helps to establish and maintain the bowel-blood barrier for bacteria in this area. We know this to be true when an appendix bursts it contains and releases these deadly toxins that can kill us.

When we listen to some nutritional experts, they tell us that beef contains immunity-boosting vitamins and minerals as well as valuable iron and protein that our body needs, they falsely claim that meat products are necessary for sufficient iron to ward off anemia, which is linked to weakness and fatigue. This is not so! Do not believe all that you read or hear. Yes, there are good things in meat, but the bad benefits outweigh the good. You can get all these vitamins, minerals and proteins from other natural foods such as ALL beans, leafy greens, peas, lentils, soybeans, spinach, nuts, grain, fruit and even vegetables. ONE DOES NOT NEED TO EAT MEAT! In fact, studies consistently show that meat eaters suffer more iron deficiency then vegetarians. The natural foods that I've just listed contain no cholesterol, and are high in protein, folic acid, potassium, iron and magnesium. They also have certain chemicals compounds that may help prevent diseases such as cardiovascular disease and cancer. In addition, they are a good source of fiber (unlike meat) and they contain all of the amino acids needed to make a complete protein, just like meat. In fact, all that stuff you hear about protein deprivation is nonsense! The World Health Organization found years ago that the only people on the planet who are protein-deprived are those who are actually starving.

Lets also move on to some of the other kosher laws of Leviticus, which tell us not to eat any sea creatures that don't have scales and fins. Again, we know this to be true today since modern science has discovered that fish with scales and fins do not absorb the toxins in the water (from the muddy toxic bottom) as readily as sea creatures that do not have both scales and fins (the scales and fins work like a protective armor). Additionally, we are also told in the kosher laws not to eat pork. Deuteronomy 14:8 - The pig is also unclean, although it has a split hoof, it does not chew the cud. You are not to eat their meat or touch their carcasses. As it turns out pork from a pig is a highly toxic food laced with disease and viruses because it literally eats anything in its path, including its own feces, and all of this turns to meat on its bones in a matter of just a few hours.

Additionally there is other reason why we should not eat meat. For one thing, like non-kosher fish that are subjected to mercury and lead toxins from the muddy bottom of a lake, stream, pond or ocean - animals like cattle that eat of the grass and dirt contain mercury and lead which are absorbed into the cattle's flesh. Aside from all this, meat also has unhealthy inflammatory fatty acids in it that cause our body to retain fluid, swell and bloat which encroaches upon a whole host of undesirable illnesses, the biggest one is heart disease. Additionally we've also learned that meat contains growth hormones (to make the animal grow faster), steroids (which makes the animal grow bigger), antibiotics (which helps the animal fight off bacteria that may cause it to get sick and die), herbicides and pesticides (which are chemicals used to ensure plant growth which is found in the grass and food an animal eats), viruses (which is an illness in the animal itself), as well as bacteria and parasites (which is obtained from unhealthy and unclean living conditions). Added onto this, cattle are being fed ground up animal remains. These animals are herbivores, they were meant to eat grass and therefore their stomachs do not possess the enzymes to digest animal byproducts, so in turn they retain the remains, poisons and diseases in their body, AND THEN WE EAT IT! As the old adage goes: You are what you eat... and you are what you eat EATS! Furthermore, all this goes into the milk we drink and the cheese we eat. What's more, meat is also tainted with red dyes (linked to allergies and cancer), and nitrates (a chemical radical salt preservative) that make the meat look good at the grocery store. Cancer tumors are also being found in meat products, and so we are eating "cancer itself". In addition, there is something else of biological concern when we eat meat... as soon as an animal is killed; self-destruct enzymes are released causing the formation of modified molecular protein substances, which could cause cancer in anyone who eats the meat. Not only do they cause cancer, but also on top of all this, all these hormones and steroids are affecting the development of the human female anatomy. Not only do these hormones and steroids cause females to mature at a faster rate, more women are beginning to look like men, carrying weight more in their torso.

Additionally meat-eaters have an increased risk of heart disease, colon cancer, obesity, adult-onset diabetes, gout, osteoporosis, kidney stones, gallstones, lung cancer and breast cancer. It turns out that meat and eggs are the highest sources of cholesterol and saturated fat, which are the main causes of heart disease. Not eating these foods greatly reduces the risk of heart attack. In fact, the fiber in vegetarian diets not only removes; this unneeded cholesterol and other cancer causing agents, but helps in reversing atherosclerosis (hardening of the arteries). People who consume high amounts of animal flesh have less life expectancy then people who consume little or no meat do. Not only is the quality of health and happiness better for those who eat mainly fruits, vegetables, nuts, beans, grains, and herbs, one has more energy and endurance. This is because the carbohydrates found in vegetarian foods are gradually digested and easily eliminated out of the body, thus providing a steady source of glucose to the blood. Conversely, meats contain virtually no carbohydrates and are very high in fat and protein. The fat and protein in meats are very difficult to digest, and in effect, this takes energy away from the rest of the body causing one to feel tired and sluggish after eating meat. It is also known that meats stagnate in the bowels for several days creating a toxic overload for the whole body, contributing to fatigue, headaches, excess weight, swelling and bloating, a weakened immune system, and many long-term deteriorating diseases which can result.

There are those who unrelentingly deny that meat has any ill effects over the body whatsoever. They smugly say that their great grandparent ate meat every day of their lives, and lived to be 90 years old. This might be so, but can you imagine how much longer they could have lived if they "didn't" eat meat. Yes, as I said earlier we are all going to die sooner or latter, but why not latter than sooner? If we are to be G-d fearing people we are to do whatever we can to respect and sustain our lives and the body we are given to the best of our ability. Not showing respect to the body is just another sin, since we live in retribution against sin.

There is even more to meat… Additionally, consumption of meats (flesh food) has been scientifically linked with violent and aggressive behavior (though the FDA will not openly divulge this). Correspondingly, which brings us right back to some specific laws of Leviticus, which stress that eating, certain animals like the vulture are forbidden? Leviticus 11:13 - These are the birds you are to detest and not eat because they are detestable: the eagle, the vulture, the black vulture. We find that this does not apply only to the vulture… Upon sensing death captive, frightened, helpless, and terrified cattle (animals) immediately produce and release all kinds of stress hormones and toxins. This could happen while they board the truck to the slaughterhouse and hear the screams of their fellow cattle while they stand in the slaughterhouse line. Consequently, large amounts of uric acid, adrenalin, and other hormones are found in the flesh of slaughtered animals. And it's been found that these FEAR AND STRESS HORMONES INSTIGATE VIOLENT BEHAVIOR IN HUMANS (since adrenalin and uric acid are temperament causing hormones to begin with). If you can remember from a previous chapter that such birds of prey (like the vulture) even some animals, which are predators, because of tension, anger and hormones produced, become preserved in their body and make the meat toxic for consumption which can reflect in the human body if eaten. Remember as in "homeopathy", even if you try and remove all remnants of the toxin or poison itself, even by cooking it, a remnant or copy (vibration) still remains within the meat. This happens when fright hormones (which are protein enzymes) saturate the blood and muscle cells of an animal. Some of these protein enzymes may be broken by cooking, but many remain intact recreating the anger in those who consume the flesh. Eating of the meat can change a persons disposition, (character, temperament and nature as well as health). We are in a sense ingesting anger, terror, agony, despair and depression along with causing the body to be on a constant state of unnatural alert, which may result in violent behavior. Additionally if you recall, animal cruelty is also forbidden according to kosher law… when an animal is slaughtered for sacrifice or atonement, it must be done with the least amount of possible pain, (this also might reflect that the animal senses it is going to die - which is also a sort of cruelty). That is why a special kosher knife is used that is so sharp that even the slightest nick or notch in the blade renders it unacceptable. Therefore it is the sharpness of the blade that prevents pain to the animal and if there are visible chares found in the meat of the animal it may have been killed in pain and thus non-kosher. So as one can see there are also "moral" reason for why eating the flesh of terrified or angry natured animals is forbidden, not only are they immoral but they are also harmful to the human body, and this includes unhealthy animals that were raised on hormones, steroids, antibiotics, herbicides, pesticides, and unhealthy animals that are sick with viruses, bacteria and parasites and fed ground up animal remains. WE SHOULD NOT EAT THE MEAT OF SUCH ANIMALS SO THAT WE DO NOT ABSORB THESE QUALITIES INTO OURSELVES. Leviticus 22:20-25 - But whatsoever hath a blemish, that shall ye not offer: for it shall not be acceptable for you. And whosoever offereth a sacrifice of peace offerings unto the Lord to accomplish his vow, or a freewill offering in beeves or sheep, it shall be perfect to be accepted, there shall be no blemish therein. Blind, or broken, or maimed, or having a wen, or scurvy, or scabbed, ye shall not offer these unto the Lord, nor make an offering by fire of them upon the altar unto the Lord. Either a bullock or a lamb that hath any thing superfluous or lacking in his parts, that mayest thou offer for a freewill offering, but for a vow it shall not be accepted. Ye shall not offer unto the Lord that which is bruised, or crushed, or broken, or cut, neither shall ye make any offering thereof in your land. Neither from a stranger's hand shall ye offer the bread of your G-d of any of these, because their corruption is in them, and blemishes be in them: they shall not be accepted for you.

To be frankly cynical here… Mmmm the wholesome power of cheese - milk does a body good and beef is what's for dinner! The bottom line here is to try and not eat too much meat from any living thing. If you do, it should be a very small minimal amount and in moderation. Philippians 4:5 - Let your moderation be known unto all men. The Lord is at hand. If possible, try to avoid all meats entirely. Incidentally, Jesus mainly ate a diet rich in fruits, vegetables, nuts, beans, lentils, grains, bread, water, wine and herbs, and He ate them in their natural states (not processed). He rarely ate meat, but when He did eat it, it was only mentioned to be fish and lamb, and He only ate them on special occasions. The lamb He most likely only ate during "Passover" which is celebrated only once a year. He followed the laws of Leviticus, and therefore would not have eaten pigs, rabbits or fish that did not have scales or fins, such as shellfish, crabs and shrimp (even though they have a protective armor, they are scavengers that eat off the bottom where toxins, heavy metals and poisons reside). Shellfish, crabs and shrimp are also higher in a specific acid, which is the same as the inflammatory fatty acids and saturated fats found in meat. Additionally, many shellfish, crabs and shrimp contain iodine, and a protein that causes allergies in some people.

It's also important to know the manner in which people ate in biblical times. They would eat for hours and take their time. We read that the disciples would be lounging around, relaxing and conversing while dining. This tells us that they did not eat their food in a hurry. Jesus spent his time eating and drinking. His mission strategy was a long meal, stretching into the evening. Mark 14:18 - While they were reclining at the table eating, He said, I tell you the truth, one of you will betray me—one who is eating with me. John 12:2 - Here a dinner was given in Jesus' honor. Martha served, while Lazarus was among those reclining at the table with him. There are certainly some health benefits to eating slow. The most valuable benefit of eating slowly has to do with digestion and absorption of valuable nutrients into the body. Since digestion begins in the mouth where the saliva is mixed with the food, it then starts to break it down into smaller chemicals that can be absorbed to nourish the body. Therefore chewing food thoroughly promotes smooth, complete digestion. The slower you eat, the faster and more efficiently you metabolize the food. When food is swallowed completely it makes it more difficult for the body to absorb all the important nutrients. The other significant thing about eating slowly is that it also reduces the appetite - thus reduced food intake results in weight loss. It takes the brain about 15-20 minutes to start signaling feelings of fullness, which results in eating less. Eating to fast lets you eat too much before you are fully aware of it. Additionally, when one eats slowly, they end up tasting their food more. Allowing one, the experience of more flavors, textures and smells. Eating too fast can also wreak havoc on the metabolism. Eating too fast is also associated with insulin resistance, which can lead to diabetes and heart disease, high blood pressure, obesity and high cholesterol. It may also cause acid reflux and GERD (Gastro-esophageal Reflux Disease). As a side note… there are no set times to eat meals nor should there be a specific number of meals that should be eaten within the day - only that food should be consumed before sundown. Meals should be eaten in "small quantities" and should be eaten whenever hunger calls. The reason for this is to stabilize blood sugar levels and keep then at an even keel. If this involves eating four to six small meals a day, then so be it - there should be no eating time regulations. Water should also be consumed regularly throughout the day. Additionally, it is a misconception that breakfast is needed. In fact, people who eat breakfast (a big breakfast) immediately or within several hours of waking from sleep, gain more weight. Yes, the first meal is important to get the energy and body juices flowing, but breakfast can also be at noon which is considered lunch for most people. The longer one waits to eat after waking the better the body will internalize food. In fact, it may even be better to wake up and exercise before eating anything, as this helps the body to kick into gear and burn more fuel to loose weight. Do not go to sleep on a full stomach either, its harder for the body to break the nutrients downs when its busy trying to rest and heal. Again, after an evening meal is eaten, one should exercise before going to sleep. This helps in digesting food and it also helps the body kick into repair and healing mode, and it helps in getting a better restful sleep. Exercising before bedtime also helps in releasing every bit of reserved energy that the body has left. The evening meal should be eaten several hours before going to sleep. Therefore, between ones last meal of the day, sleeping, and the next days first meal, there is a considerable time in which the body is without food. This time is ideal for the body to repair and heal itself as well as maintain body weight. I've also heard it said that eating to many meals in a day is "wear and tear" on the body and intestines, and some have even claimed that this can shorten ones lifespan. That may be so, but that is why it is important to get a restful adequate amount of sleep without food. Eating too little or too much can also have its reverse effects. Eating to little can cause the intestines to shrink up and become inoperable. It can also send mixed signals to the brain which sends a signal to the body to store whatever food rations it gets into the fat cell reserves of the body which can actually make it more difficult for some people to loose weight. While eating to much can also bloat and obstruct the bowels and intestines, but overeating at certain times can also have a good side effect that can help to kick the metabolism into higher gear causing one to actually loose weight. All of this is very confusing, it actually tells us the opposite of what some nutritional experts tell us. Nobody knows all the answers… but the bottom line in all this, is that the body should never become accustomed or reliant on its environment and provisions. Constant change and balancing is needed when it comes to keeping the body maintained. If it becomes too reliant on food at a certain time during the day, it will always expect it. Alternatively, if it becomes to accustomed to small or large amounts of food, ones stomach may shrink up or enlarge. It may also take in the same nutrients and become sedentary (and may even build a resistance to it). It's also wise not to constantly eat the same foods, but to switch off. WE DON'T WANT SEDENTARY, WE WANT ACTIVE! Additionally it also depends on what kind of message your brain is sending to your body. Both exercise along with small portions of food (or calorie restriction) keep the body functioning up to par, since its been observed in laboratory testing that "caloric restriction" has a huge effect on not only keeping weight at bay, but also suppressing aging… a process supposedly mimicked by chemicals roused up by a gene called sir2. This sir2 triggers cellular changes, and along with ones mitochondria (which is turned on by exercise) can turn nutrients such as fat, into fuel for muscle cells. And it is these sir2 (sirtuin enzymes) that have been found to reverse the aging process. Interestingly, a specific chemical called "Resveratrol" is found in the skin of both red and white grapes (and red wine), which is produced by the grapevine when it is stressed or pressured by disease such as fungal infections. When this chemical is ingested by humans or animals, it turns on the gene sir2 (the sirtuin enzymes), which have been found to reverse the aging process.

Moving on here… It is hard to veer away from tasty processed foods. But results have found that the more we veer away from eating unnaturally processed foods and repeatedly eat natural foods that grow directly from the Earth, the more our taste buds will begin to know the difference and crave its wholesome goodness (try drinking green tea without sugar, instead supplement it with lemon juice or honey - the more we drink it, the tastier it becomes). Additionally, (as was just previously mentioned) we should eat a well rounded intermittent diet, constantly changing ones selection of food and introducing them into the body at different times - the element of surprise keeps us alive! It not only keeps our intestines and internal workings active, thriving, and flourishing, but it also helps in stimulating our brain. So basically we should not continually eat the same foods, or too much of one type of food. For instance eating to many almonds (which contain cyanide) can effectively become a poison against the body, however in small minute doses of natural cyanide, almonds are good for the heart and can also be an anti-cancer compound, (it aids the health of red blood cells and also secretes cyanide compounds in the presence of cancer cells to eradicate these cells). Celery, apple seeds, apricot seeds, pear seeds, nectarine seeds, peach seeds and plum seeds also contain small traces of cyanide. Usually seeds or foods that have a bitter medicinal taste contain small traces of cyanide. Even such foods like avocadoes contain a fungicidal toxin (which is leached from the pit into the fruit) that kills breast cancer cells, but eating to many avocadoes may cause breathing problems or heart complications or palpitations. (There is no adequate scientific evidence that it causes heart problems or palpitations in humans, but it can in all animals - which suggest that it may have some ill effect on humans). Again, ones best policy is to eat a well-rounded intermittent (sporadic) diet, constantly changing ones selection of food, and eating meals in the least possible portions. Philippians 4:5 - Let your moderation be known unto all men. The Lord is at hand.

In addition to all this, it does not hurt us to abstain from food either (if the body is healthy to begin with). In fact "fasting" may actually have some health benefits. Such as cleaning ones system completely out. Cleaning out toxins (detoxification) and allowing our body's self-healing capacities to kick in. Fasting can also starve cancers in the body and may even shrink and eliminate them altogether. Cholesterol levels also drop from fasting. Stomach problems are also eliminated by fasting. Fasting can also extend ones lifespan… despite the notion that it's a starvation or wasting away. As it turns out, fasting kicks the body into survival mode, which enables it to go to the reserved fat cells for energy. During fasting, it also allows the body to create the hormone HGH (the Human Growth Hormone) produced in the pituitary gland of the brain - which is another hormone that stimulates growth and cell reproduction in both humans and animals. Fasting also allows us to become more aware (consciously), and therefore closer to G-d. The purpose of fasting should be to temporarily take our eyes off the earthly things of this world, and instead focus on G-d. It is a means through which G-d changes us. However, it is not intended to punish the flesh, or to be considered as a dieting method. Matthew 6:16-18 - Moreover when ye fast, be not, as the hypocrites, of a sad countenance: for they disfigure their faces, that they may appear unto men to fast. Verily I say unto you, They have their reward. But thou, when thou fastest, anoint thine head, and wash thy face, That thou appear not unto men to fast, but unto thy Father which is in secret: and thy Father, which seeth in secret, shall reward thee openly.

Fasting was also a sign of commitment or repentance, and for increasing ones faith, especially when accompanied by prayer. Judges 20:26 - Then all the children of Israel, and all the people, went up, and came unto the house of G-d, and wept, and sat there before the Lord, and fasted that day until even, and offered burnt offerings and peace offerings before the Lord. 1 Samuel 7:6 - And they gathered together to Mizpeh, and drew water, and poured it out before the Lord, and fasted on that day, and said there, We have sinned against the Lord. And Samuel judged the children of Israel in Mizpeh. Isaiah 58:3-10 - Wherefore have we fasted, say they, and thou seest not? wherefore have we afflicted our soul, and thou takest no knowledge? Behold, in the day of your fast ye find pleasure, and exact all your labours. Behold, ye fast for strife and debate, and to smite with the fist of wickedness: ye shall not fast as ye do this day, to make your voice to be heard on high. Is it such a fast that I have chosen? a day for a man to afflict his soul? is it to bow down his head as a bulrush, and to spread sackcloth and ashes under him? wilt thou call this a fast, and an acceptable day to the Lord? Is not this the fast that I have chosen? to loose the bands of wickedness, to undo the heavy burdens, and to let the oppressed go free, and that ye break every yoke? Is it not to deal thy bread to the hungry, and that thou bring the poor that are cast out to thy house? when thou seest the naked, that thou cover him, and that thou hide not thyself from thine own flesh? Then shall thy light break forth as the morning, and thine health shall spring forth speedily: and thy righteousness shall go before thee, the glory of the Lord shall be thy reward. Then shalt thou call, and the Lord shall answer; thou shalt cry, and he shall say, Here I am. If thou take away from the midst of thee the yoke, the putting forth of the finger, and speaking vanity, And if thou draw out thy soul to the hungry, and satisfy the afflicted soul, then shall thy light rise in obscurity, and thy darkness be as the noon day. Acts 13:3 - And when they had fasted and prayed, and laid their hands on them, they sent them away. Acts 14:23 - And when they had ordained them elders in every church, and had prayed with fasting, they commended them to the Lord, on whom they believed. 2 Corinthians 6:5 - In stripes, in imprisonments, in tumults, in labours, in watchings, in fastings. 2 Corinthians 11:27 - In weariness and painfulness, in watchings often, in hunger and thirst, in fastings often, in cold and nakedness.

Additionally fasting was used for distress, grief, mourning, sadness, remembrance and respect. Loss of appetite is a natural reaction in times of distress, grief and mourning, and it was during these times that fasting was considered appropriate. King David fasted as a sign of sorrow and heartache when Abner was killed. 2 Samuel 3:35 - And when all the people came to cause David to eat meat while it was yet day, David sware, saying, So do G-d to me, and more also, if I taste bread, or ought else, till the sun be down. There was a seven-day fast at the death of Saul. 1 Samuel 31:13 - And they took their bones, and buried them under a tree at Jabesh, and fasted seven days. Moses fasted for forty days in preparation for receiving the Ten Commandments. Exodus 34:28 - And he was there with the Lord forty days and forty nights, he did neither eat bread, nor drink water. And he wrote upon the tables the words of the covenant, the ten commandments. Daniel fasted for three weeks prior to receiving his vision. Daniel 10:2-6 - In those days I Daniel was mourning three full weeks. I ate no pleasant bread, neither came flesh nor wine in my mouth, neither did I anoint myself at all, till three whole weeks were fulfilled. And in the four and twentieth day of the first month, as I was by the side of the great river, which is Hiddekel, Then I lifted up mine eyes, and looked, and behold a certain man clothed in linen, whose loins were girded with fine gold of Uphaz: His body also was like the beryl, and his face as the appearance of lightning, and his eyes as lamps of fire, and his arms and his feet like in colour to polished brass, and the voice of his words like the voice of a multitude. Additionally, Elijah fasted forty days before speaking with G-d. 1 Kings 19:8 - And he arose, and did eat and drink, and went in the strength of that meat forty days and forty nights unto Horeb the mount of G-d. We find that Jesus also fasted for forty days in preparation for His temptation by Satan. Matthew 4:1-11 - Then was Jesus led up of the Spirit into the wilderness to be tempted of the devil. **And when he had fasted forty days and forty nights, he was afterward an hungred.** And when the tempter came to him, he said, If thou be the Son of G-d, command that these stones be made bread. But he answered and said, It is written, Man shall not live by bread alone, but by every word that proceedeth out of the mouth of G-d. Then the devil taketh him up into the holy city, and setteth him on a pinnacle of the temple, And saith unto him, If thou be the Son of G-d, cast thyself down: for it is written, He shall give his angels charge concerning thee: and in their hands they shall bear thee up, lest at any time thou dash thy foot against a stone. Jesus said unto him, It is written again, Thou shalt not tempt the Lord thy G-d. Again, the devil taketh him up into an exceeding high mountain, and sheweth him all the kingdoms of the world, and the glory of them, And saith unto him, All these things will I give thee, if thou wilt fall down and worship me. Then saith Jesus unto him, Get thee hence, Satan: for it is written, Thou shalt worship the Lord thy G-d, and him only shalt thou serve. Then the devil leaveth him, and, behold, angels came and ministered unto him.

Some may wonder how could Moses, Elijah, and Jesus have survived 40 days without food. Many of us could not go without food for even one day, let alone a month. So how did both Moses and Jesus do it? Well in the modern day of medicine, we know that we can live on intravenous fluids (the giving of liquid substances directly into a vein) for months even years thereby bypassing the natural process of eating and digestion. This liquid contains all the nutritional substances needed to sustain life such as salts, glucose, amino acids, lipids and added vitamins. So to the contrary, even though we hear of people dying from food starvation (aside from the modern day medicine of intravenous fluids), there is also another way to go without food just by drinking water and soaking up the energy of the Sun (its like running on solar energy). Sounds bizarre but its true! For starters, the Sun actually reduces the appetite. The body has a mechanism by which to receive energy directly from the Sun though the skin, eyes and brain, and stores it within the body for future nutrients. Just like plants exist on the process of photosynthesis, the human body requires energy in much the same way, and the Sun is capable of providing the body with all the energy it needs to function. What we traditionally eat is nothing but a secondary source of the Suns energy anyways. Today science knows that we get vitamin D during sun exposure as I've previously mentioned in this chapter. We also need vitamin D for the absorption of many vitamins and minerals (like calcium) before they can even be recognized or absorbed by the body. Additionally, sunlight strengthens the immune system, kills bacteria, increases oxygen in the blood, it also calms the nervous system by relaxing you, (and decreases the appetite as it is a food source itself). Moreover, it lowers blood sugar by putting glucose into the cell like natural insulin. It lowers blood pressure, and it lowers cholesterol because it turns cholesterol into vitamin D when the Sun hits it in your skin. In order to harness solar energy, one only needs to gaze at the Sun (not directly, but in the general direction) thus harnessing solar energy through the eyes and brain (which has also been known to decrease depression). This activates the endocrine glands, and the pineal gland, often considered the "third eye". The proper sunlight that one should harvest is in the morning and evening hours, one hour after sunrise and one hour before sunset. This is because these times are free from the highest UV and IR (ultraviolet and infrared) exposure and are at the minimal levels. People who fast for 40 days miraculously do not loose weight. The weight remains stable and neurons in the brain are active. Some even report that they have immensely more energy than ever before. However, no fasting should be done by children who are still in the developmental stages. Only when a person has reached full adulthood can they attempt to fast for long periods. However, I would strongly not advise fasting in such a way to anyone without doctor supervision. Most doctors would never even agree to it anyways.

Therefore, as we can see fasting has many aspects attributed to it, even though it is not recognized (or has been substantiated) by modern medicine. It is good for ones physical health as well as ones psychological health. One may also ask the question of how fasting works as a mechanism to bring us closer to G-d. The surprising and unexpected thing about it is that the Merkabah vehicle may again play a crucial role in facilitating our nearness to G-d when fasting. When the body can no longer rely on the material aspects of earthly things such as "food", the spirit begins to take over, thus allowing the Merkabah vehicle to become active. It leaves the body and roams around looking for answers (sustenance) beyond the body - thus our consciousness finds itself in an ethereal realm were angels and G-d may dwell. This may have been how Daniel (after fasting) was able to "see" the angel and received his vision, and Elijah (fasting) before being able to "speak" with G-d. Daniel 10:2-6 - In those days I Daniel was mourning three full weeks. I ate no pleasant bread, neither came flesh nor wine in my mouth, neither did I anoint myself at all, till three whole weeks were fulfilled. And in the four and twentieth day of the first month, as I was by the side of the great river, which is Hiddekel, Then I lifted up mine eyes, and looked, and behold a certain man clothed in linen, whose loins were girded with fine gold of Uphaz: His body also was like the beryl, and his face as the appearance of lightning, and his eyes as lamps of fire, and his arms and his feet like in colour to polished brass, and the voice of his words like the voice of a multitude. 1 Kings 19:8 - And he arose, and did eat and drink, and went in the strength of that meat forty days and forty nights unto Horeb the mount of G-d. However as a cautionary note, the Merkabah vehicle may also bring about evil spirits - this is another reason why it should not be intentionally roused up. If the Merkabah vehicle happens to be roused up during, "fasting" it should only be for "piety reasons" (consisting of good and faithful intentions and only with reverence for G-d).

It should also be recognized that people during times of illness, fever, injury, hunger and starvation have also been known to "hallucinate". However, these times are also injurious and negative to the body as well, thus causing the Merkabah vehicle to kick in. However as aforementioned in a previous chapter, human consciousness is far reaching, and we do not yet know the full spectrum to which it reaches. It is possible that a huge ghostly ethereal mind net exists that catches and holds all the reflected emanations, fantasies, dreams and ideas of conscious thoughts of many living and non living things (even a rock contains atoms). This ethereal mind net may be simultaneously lurking within (and about) many individuals and (nonliving things) at any given time. These include, thoughts, awareness and perceptions which may emanate from dreams, fantasies and ideas which somehow reflect, replicate, imitate, echo (and bounce off) like a mirror image, and may be viewed or received by another person or living thing when subjected to the ethereal realm (in the Merkabah form). These echoed thoughts and messages can be mixed and confusing and one must find a way of filtering them out. A person who is not of a religious mind is defenseless in determining the differences between these dimensional levels. This is because they are not developed or matured "spiritually" with the correct knowledge of G-d and scripture.

Moving on here to a another subject of importance in maintaining overall health… In an earlier chapter, we talked about reflexology (massage therapy for the feet and hands). However regular "massage" (preferably full-body massage) to any part of the body can be used to relieve stress, free anxiety, as well as stimulate the nerves, reduce muscle tension, spasms, inflammation, fluid retention, aches, stiffness, and pain. Other benefits include improved circulation, general flexibility, and increased tissue elasticity. It also builds and strengthens the immune system. Message can be used as a prelude to exercise. Massage increases circulation and warms muscles and other soft tissues like tendons and ligaments which allows the blood to circulate better in the body to flush toxins and poisons out. Massage is done by rhythmically kneading, rubbing and stroking over the muscles of the body thus stimulating circulation and blood flow which delivers oxygen and nutrients as well as eliminating waste products such as lactic acid, uric acid and other waste products that collects in the muscles (one should make sure they drink water before and after a massage to help flush these toxins out). Massage also stimulates the release of beneficial natural chemicals from the brain such as serotonin, which plays a role in mood regulation and sleep. The body also releases natural painkillers known as endorphins during massage. The release of serotonin and endorphins may be responsible for the pain reducing effect and sense of well being associated with massage. Massage can also help the body build immunity against frequent illnesses. Additionally, body scratching (commonly known as back scratching) can also help. This greatly reduces stress and anxiety and stimulates the skin as well as nervous system, thereby helping the immune system.

Just the simple act of "touching" can also have its far-reaching effects… Studies show touching and subtle massage promotes weight gain in premature babies by stimulating a part of the brain that releases food absorption hormones. The babies that are touched and massaged show a much greater behavior responses to different kinds of tests, scoring much higher than babies who are not touched or massaged.

Even as silly as this may sounds, "laughter" helps the white blood cell count and increases respiration. Job 8:21 - He will yet fill your mouth with laughter and your lips with shouts of joy. When in pain, distraction such as laughter draws attention away from pain. There is some involvement of the hypothalamus (in the brain) that releases an enzyme, which then improves the quality of life - healing the body by delighting the mind. Job 9:27 - If I say, I will forget my complaint, I will change my expression, and smile. Laughter also boosts the level of the body's natural pain killers (serotonin and endorphins) and lowers the level of stress hormones. 20 seconds of laughing gives the heart the same workout as 3 minutes of vigorous exercise. Laughter is like an aspirin; only it works twice as fast.

Additionally one should strive for happiness. There is no road, path or way to happiness, other than happiness being the way. Ecclesiastes 3:12 - I know that there is nothing better for men than to be happy and do good while they live. We should also not worry. Ecclesiastes 7:14 - When times are good, be happy, but when times are bad, consider: G- d has made the one as well as the other. Therefore, a man cannot discover anything about his future. Luke 12:22-26 - Then Jesus spoke to his disciples. He said, I tell you, do not worry. Don't worry about your life and what you will eat. In addition, don't worry about your body and what you will wear. There is more to life than eating. There are more important things for the body than clothes. Think about the ravens. They don't plant or gather crops. They don't have any storerooms at all. However, G-d feeds them. You are worth much more than birds! Can you add even one hour to your life by worrying? You can't do that very little thing. So why worry about the rest? Matthew 6:27 - Who of you by worrying can add a single hour to his life?

Additionally many people struggle throughout their life fighting against habitual habits. The secret in combating and totally eradicating any habit is by COMPLETE RESTRAINT! There is no room for excuses or exceptions, and this is NOT easier said then done. However, if one makes up their mind to do something - they can do it! For instance smoking, alcoholism, drugs or nail biting. They should be totally abstained from! There is a "bondage mentality" that tells us that we can't change, and then there is a "freedom mentality" that tells us that we do not belong to our habits. If we continue with these sorts of habits we become enslaved (or in bondage) to them - hence we are not liberated (have no freedom). The way to avoid habits is by assigning new tasks to your everyday life. Even if you are a stressful or stressed out person, keeping busy is ones best policy. When one is busy (doing other activities) the mind is elsewhere and away from the habitual habit. Whether it means spending more time in public, going to church or synagogue (study classes), taking college courses, exercising at home or at the gym, cooking, gardening, or simply reading a book. These things are very necessary for those that are susceptible to habitual habits. Constantly remember and drill into your head that total abstinence and keeping busy with other tasks takes the mind off (and away from - thus disconnecting it) from the habit. Structuring ones, time is important as to leave no room for any foolish habit to gain control over ones life and should be done in combination with all the other things I've mentioned throughout this chapter. Remember, both physical and spiritual aspects are important - one without the other is a colossal failure! When one reads this book, it should give them an incentive towards both physical and spiritual perfection. In understanding, we gain insight to fight against attacks of the flesh. Psalm 119:34 - Give me understanding, and I will keep your law and obey it with all my heart. Psalm 32:9 - Do not be like the horse or the mule, which have no understanding but must be controlled by bit and bridle or they will not come to you. 1 Chronicles 28:19 - All this, David said, I have in writing from the hand of the Lord upon me, and he gave me understanding in all the details of the plan. Proverbs 15:21 - Folly delights a man who lacks judgment, but a man of understanding keeps a straight course. Proverbs 15:32 - He who ignores discipline despises himself, but whoever heeds correction gains understanding.

Proverbs 17:27 - A man of knowledge uses words with restraint, and a man of understanding is even-tempered. Proverbs 28:2 - When a country is rebellious, it has many rulers, but a man of understanding and knowledge maintains order. Proverbs 2:11 - Discretion will protect you, and understanding will guard you. James 1:14 - But every man is tempted, when he is drawn away of his own lust, and enticed. Proverbs 4:14-15 - Enter not into the path of the wicked, and go not in the way of evil men. Avoid it, pass not by it, turn from it, and pass away. 1 Peter 5:8 - Be sober, be vigilant, because your adversary the devil, as a roaring lion, walketh about, seeking whom he may devour.

I must also not leave out that a habit that is personally attached to the physical body, “such as nail biting”, which can be a very hard habit to break as its nearest to the body. Matthew 26:41 - Watch and pray so that you will not fall into temptation. The spirit is willing, but the body is weak. Such habits are also stronger on a full or new moon, and 8 days prior to a woman’s monthly cycle (which will be addressed shortly). However, if one is determined, when the urge attacks, a simple thought should eradicate the problem. This thought is simply remembering the original temptation that strongly effected Eve that caused all humanity to fall. Every time we are tempted, think of it as picking and eating of the Tree of Knowledge. Sometimes getting to the root of the original sin helps in combating them all. James 4:7 - Submit yourselves therefore to G-d. Resist the devil, and he will flee from you. Additionally, any habit that is of the flesh is weak, and it may continually be a habit that one has to contend with for the rest of their life. One may stop for a while and then revert for whatever cause. For this reason a habit can actually help strengthen us in a battle to overcome it, thus bringing us closer to G-d since it may build ones willpower (self-control) against the malevolent desires of the flesh. This was Samson’s call. Judges 13:2-5 - A certain man of Zorah, named Manoah, from the clan of the Danites, had a wife who was sterile and remained childless. The angel of the Lord appeared to her and said, You are sterile and childless, but you are going to conceive and have a son. Now see to it that you drink no wine or other fermented drink and that you do not eat anything unclean, because you will conceive and give birth to a son. No razor may be used on his head, because the boy is to be a Nazirite, set apart to G-d from birth, and he will begin the deliverance of Israel from the hands of the Philistines. Judges 16:17-22 - So he told her everything. No razor has ever been used on my head, he said, because I have been a Nazirite set apart to G-d since birth. If my head were shaved, my strength would leave me, and I would become as weak as any other man. When Delilah saw that he had told her everything, she sent word to the rulers of the Philistines, Come back once more; he has told me everything. So the rulers of the Philistines returned with the silver in their hands. Having put him to sleep on her lap, she called a man to shave off the seven braids of his hair, and so began to subdue him. And his strength left him. Then she called, Samson, the Philistines are upon you! He awoke from his sleep and thought, I'll go out as before and shake myself free. But he did not know that the Lord had left him. Then the Philistines seized him, gouged out his eyes and took him down to Gaza. Binding him with bronze shackles, they set him to grinding in the prison. But the hair on his head began to grow again after it had been shaved. And as the story goes while Samson was in prison the Philistines unleashed him and assembled to offer him as a sacrifice to their god. Samson had prayed to G-d one last time and had regained his strength and brought down their idol temple killing more people than while he lived. Judges 16:28-30 - Then Samson prayed to the Lord, O Sovereign Lord, remember me. O G-d, please strengthen me just once more, and let me with one blow get revenge on the Philistines for my two eyes. Then Samson reached toward the two central pillars on which the temple stood. Bracing himself against them, his right hand on the one and his left hand on the other, Samson said, Let me die with the Philistines! Then he pushed with all his might, and down came the temple on the rulers and all the people in it. Thus he killed many more when he died than while he lived.

Now for the ultimate culmination, conclusion and climax of this chapter!

This is where we find scientific corroboration once again supporting what scripture says. Found in the "Ten Commandments" (the laws that G-d gave to Moses for the Jewish people to follow) is found a very specific commandment. This commandment goes as follows: Exodus 20:8-11 - Remember the sabbath day, to keep it holy. Six days shalt thou labour, and do all thy work: But the seventh day is the sabbath of the Lord thy G-d: in it thou shalt not do any work, thou, nor thy son, nor thy daughter, thy manservant, nor thy maidservant, nor thy cattle, nor thy stranger that is within thy gates: For in six days the Lord made heaven and earth, the sea, and all that in them is, and rested the seventh day: wherefore the Lord blessed the sabbath day, and hallowed it. Deuteronomy 5:12-15 - Keep the sabbath day to sanctify it, as the Lord thy G-d hath commanded thee. Six days shalt thou labour, and do all thy work: But the seventh day is the sabbath of the Lord thy G-d: in it thou shalt not do any work, thou, nor thy son, nor thy daughter, nor thy manservant, nor thy maidservant, nor thine ox, nor thine ass, nor any of thy cattle, nor thy stranger that is within thy gates, that thy manservant and thy maidservant may rest as well as thou. And remember that thou wast a servant in the land of Egypt, and that the Lord thy G-d brought thee out thence through a mighty hand and by a stretched out arm: therefore the Lord thy G-d commanded thee to keep the sabbath day. Even the Jews were delivered from Egypt on this day (the Sabbath), and given rest from slavery. Remember this was the day (in Egypt) (as well as the last plague) when the Lambs blood was placed over the doorposts of the Jewish residents so the hand of G-d would "pass over" the first-born. In remembrance of this, the Jewish holiday of "Passover" is observed.

However, this law (to remember and keep the Sabbath day) was made as a prelude before Exodus and Deuteronomy came into existence. We find the reason why it was later placed into the "Ten Commandments", because it had to do with the way in which G-d created the world, which was first, documented in Genesis. Genesis 2:2-3 - And on the seventh day G-d ended his work which he had made, and he rested on the seventh day from all his work which he had made. In addition, G-d blessed the seventh day, and sanctified it: because that in it he had rested from all his work which G-d created and made. So, this brings us to a significant question. What is the big importance about remembering and keeping the 7th day (the Sabbath), as well as not doing any work and resting on this day? And how does it apply to us in our everyday lives? The answer to this may be something of a bolt from the blue for the unbeliever, but it not only applies to us religiously, but also applies to us BIOLOGICALLY! It does so (in part) by the MOONS LUNAR CYCLE (which will be further explained). Miraculously, as it turns out, this time is the ideal time for the human body (including all other living things) to rejuvenate itself and its cells. Not only seven DAYS, but also every cell in the human body is replaced and renewed within a period of seven YEARS. The 7th day pause (for resting and rejuvenating) occurs biologically in nature for none other reason other than G-d making it so. Genesis 2:2-3 - And on the seventh day G-d ended his work which He had made, and He rested on the seventh day from all his work which he had made. And G-d blessed the seventh day, and sanctified it: because that in it He had rested from all his work which G-d created and made.

The moon has four phases (or quarters) lasting about seven days each. The first two quarters are during the waxing (or increasing light), between the new and the full moon. The third and fourth quarters are after the full moon when the light is waning, (or decreasing).

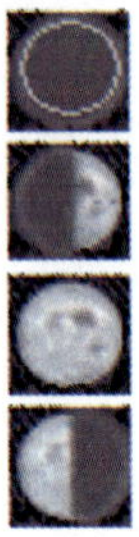

New Moon phase lasts for 1-6 days. (Rises at dawn, sets at dusk).

First quarter phase lasts for 7-13 days. (Rises at noon, sets at midnight).

Full Moon phase lasts for 14-21 days. (Rises at dusk, sets at dawn).

Last quarter Moon lasts for 22-27 days. (Rises at midnight, sets at noon).

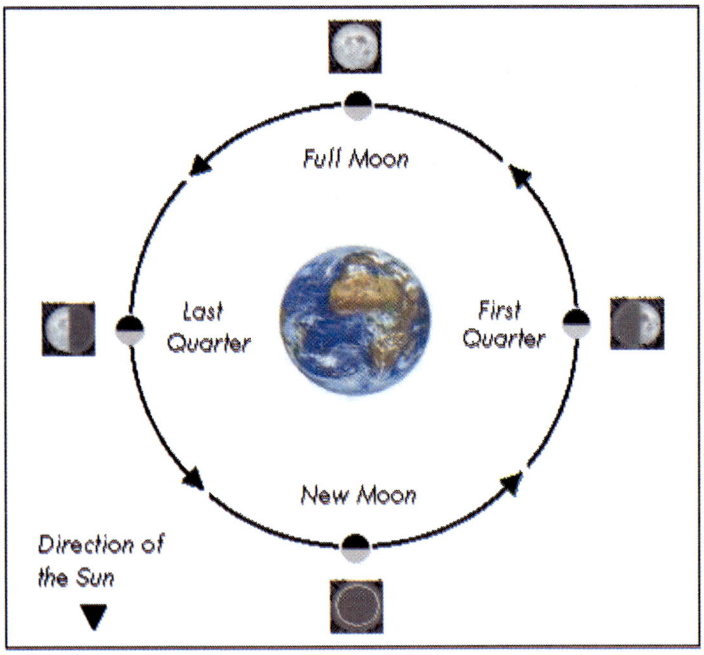

NOBODY CAN SCIENTIFICALLY EXPLAIN WHY THIS IS SO? It is not solely because of the moons cycle, or how would we then logically explain the 7 years cycle for the replacement of new cells correlating with G-d proclaiming this day as a day of rest? Because remember, it was on this 7th day that G-d had completed all that He had done and thus rested. Man could have never known this was a day of rest (BIOLOGICALLY) without G-d actually telling us so. Yes we could have observed the moon cycles, but we would never of had a way of determining the ingenious purposeful intention of rejuvenation and rest. Genesis 2:2-3 - And on the seventh day G-d ended his work which He had made, and He rested on the seventh day from all his work which he had made. And G-d blessed the seventh day, and sanctified it: because that in it He had rested from all his work which G-d created and made. Moreover was the importance of the 7th day of completion and rest, that G-d made it one of the 10 commandments found in Exodus 20:8-11 - Remember the sabbath day, to keep it holy. Six days shalt thou labour, and do all thy work: But the seventh day is the sabbath of the Lord thy G-d: in it thou shalt not do any work, thou, nor thy son, nor thy daughter, thy manservant, nor thy maidservant, nor thy cattle, nor thy stranger that is within thy gates: For in six days the Lord made heaven and earth, the sea, and all that in them is, and rested the seventh day: wherefore the Lord blessed the sabbath day, and hallowed it. And it can be found again in Deuteronomy 5:12-15 - Keep the sabbath day to sanctify it, as the Lord thy G-d hath commanded thee. Six days thou shalt labour, and do all thy work: But the seventh day is the sabbath of the Lord thy G-d: in it thou shalt not do any work, thou, nor thy son, nor thy daughter, nor thy manservant, nor thy maidservant, nor thine ox, nor thine ass, nor any of thy cattle, nor thy stranger that is within thy gates; that thy manservant and thy maidservant may rest as well as thou. And remember that thou wast a servant in the land of Egypt, and that the Lord thy G-d brought thee out thence through a mighty hand and by a stretched out arm: therefore the Lord thy G-d commanded thee to keep the sabbath day.

This 7 day cycle can be seen in primitive one-celled organisms, and are even present in bacteria. This 7 day cycle can also be witnessed in the duration days from the onset of a diseases symptoms, development and resolve. Even organ transplant patients tend to have more rejection episodes on the 7/8th , 14/15th , 21/22nd, and 28/29th days after surgery. These dates fall on days slightly before a full moon WHICH SHOULD BE A DAY OF REST! In effect, the body's immune system attacks the foreign organ. Furthermore, an even bigger rejection occurs seven days after the organ transplant operation. It is also further witnessed that medications administered at particular times may be more effective than at other times. In addition, it has been witnessed that viruses, pneumonia, and malaria infection peak at seven days. Most of us are also familiar with the common cold usually lasting a period of 7days (sometimes even two weeks).

In addition, mind you, it is just not cells, but there are numerous things that additionally arbitrate the number 7, such as: The 7 colors in a rainbow (the visible color spectrum), which are Red, Orange, Yellow, Green, Blue, Indigo, and Violet.

Additionally, the Earth has 7 layers, from the surface to the inner core that is recognized by science. These are: 1- The Crust, 2 - Asthenosphere, 3 - Mohorovicic Discontinuity, 4 - Mantle, 5 - Gutenberg Discontinuity, 6 - Outer Core and the 7 - Inner Core.

Furthermore, there are 7 layers in the Earths atmosphere. These are: 1 - Troposphere 2 - Stratosphere 3 - Ozonosphere 4 - Mesosphere 5 - Thermosphere 6 - Ionosphere 7 - Exosphere.

It has even been measured that the Earths atmosphere ends at 7 miles.

However, nowhere in the Bible is there any mention of 7 heavens. Although it has been alluded to that there are layers of Heaven such as in 2 Corinthians 12:2 - I know a man in Christ who fourteen years ago was caught up to the **third heaven.** Whether it was in the body or out of the body I do not know—G-d knows. Nonetheless, it can be found in the second book of Enoch in the first section of chapters 1-21. 2 Enoch 1-21 - And those two men lifted me up thence on to the **seventh heaven,** and I saw there a very great light, and fiery troops of great archangels, incorporeal forces, and dominions, orders and governments, Cherubim and seraphim, thrones and many-eyed ones, nine regiments, the Ioanit stations of light, and I became afraid, and began to tremble with great terror, and those men took me, and led me after them, and said to me.

Added to this, Jesus said to forgive 70 x 7. Matthew 18:21-22 - Then Peter came to Jesus and asked, Lord, how many times shall I forgive my brother when he sins against me? Up to seven times? **Jesus answered, I tell you, not seven times, but seventy-seven times.** In other words, Jesus is saying, keep on forgiving until you are complete.

Interestingly the number 7 also becomes confirmed for us when we observe the gestation periods of living things in nature. For instance, ALL bird eggs hatch in multiples of 7 day periods from laying. The hen sits three weeks (which is 21 days - 7x3), while the pigeon sits two weeks (which is 14 days - 7x2). Ducks 28 days, other Ducks 35 days, Eagles also 35 days, Owls 28 days, Penguins 49 days, (these are multiples of 7).

In addition, the list goes on. Additionally, most animals have a gestation period of multiples of seven. For instance the Mouse 21 days (3x7). The Rabbit and Rat 28 days (4x7). The Cat 56 days (8x7). And the Dog 63 days (9x7). Again as you can see all multiples of 7.

Nothing, perhaps, is more remarkable with the number 7 then the period of gestation (or pregnancy) in humans. This corresponding period is 280 days or 40x7.

But listen there is even something more ironic to all this… There are 7 colors in the light spectrum, and 7 notes in the music scale. What is the connection? Science tells us it is a matter of vibrations.

Red and red violet - F# F
Orange and red orange - G# G
Yellow and yellow orange - A# A
Yellow green - B
Green and Blue green - C C#
Blue and blue violet - D D#
Violet - E

This also brings us right back to the platonic solids since each platonic solid has a certain amount of "nodes" which forms (or outlines) its shape, but this in tern is solely determined by the frequency. A node is a geometric point (or a intersection) where parts, curves, or lines intersect or cross themselves. Therefore, when we look back at a Cymatics diagram of sound/vibration creating a form such as in the picture below, we can determine that it is an Octahedron because it has 6 nodes.

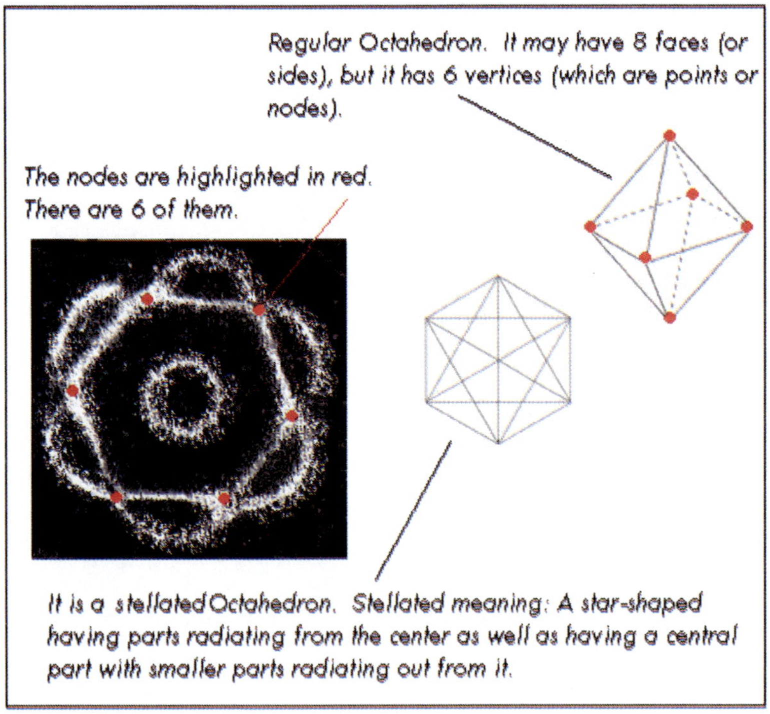

4 evenly spaced nodes form a Tetrahedron.
6 evenly spaced nodes form an Octahedron.
8 evenly spaced nodes form a Cube.
12 evenly spaced nodes form the Icosahedron
20 evenly spaced nodes form the Dodecahedron.

Now one may ask how does a regular Octahedron become a stellated Octahedron? Again, this remains a point of view in how one view an octahedron from a certain angle. Example below:

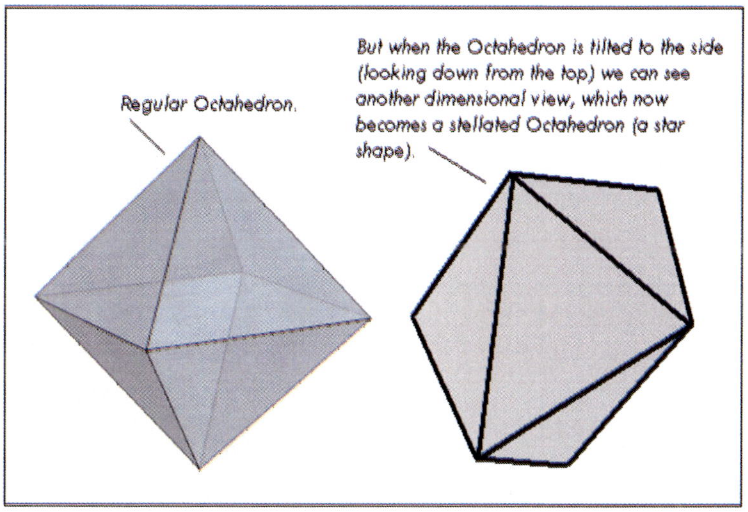

Now if one thinks a scientific explanation can still be logically found for this 7 day cycle, then one is yet left with the question of how Moses knew to write about it many centuries ago - unless he was instructed by the one true G-d that actually did create the world as He proclaims in 7 DAYS - and the cycle of living thing bear testimony to it!

There is no escaping the truth of the matter, that G-d has left His calling card bonded to living cells telling us "when" He made life in a seven day creation week (somehow these cells innately know). This cycle points directly to the life starter/giver Himself. I hope that at the end of this chapter there shouldn't be any wonder left that there is a G-d, and that He created us in the fashion in which He described directly to Moses - which was then transliterated into the book of Genesis. G-d is entitled to receive the recognition and glorification that He so deserves. Though some refuse to see it this way... since it is thought that this seven day cycle is the result of an accident, as it is commonly thought that the modern calendar is a invention which took root in the ancient world from the number of 5 visible planets in our solar system, plus the Sun and the Moon, (Sun, Moon, Mercury, Venus, Mars, Jupiter and Saturn). If there had been another visible planet, it would equate to an 8 day week. But this scenario does not explain the innate body rhythms of "seven days" found within all living things. Aside from the seven day rhythmic cycle (168 hours seven day week) there is also the more undeniable commonly known rhythm that beats in our body. This is called the "circadian rhythm" which revolves around a 24 hour period of time (one complete day). Within this day from sunrise to sunset, the heartbeat fluctuates as well as blood pressure and body temperature. Additionally, energy levels rise and fall, as well as brain waves from wake to sleep. This is derived from the exposure and duration of sunlight until dusk hitting the retina of the eye and sending a signal to the brain that produces the sleep hormone melatonin. During this 24 hour period, the body tends to become energized, hungry, tired, active and eventually lacking energy. Even a person who has been cut off from the absence of daylight is still somehow maintained by this 24 hour period. THIS RHYTHMIC CYCLE CANNOT BE DENIED! It's a funny thing that the deeper we investigate the inner workings of life from the simple to the complex we can see the marvelous display and design of the creator. To go forth claiming it to be the result of blind, random and evolutionary accidents defies the logic and support of everything we are reading (and understanding) in this book. Psalm 14:1 - The fool hath said in his heart, There is no G-d. Romans 1:20 - For the invisible things of Him from the creation of the world are clearly seen, being understood by the things that are made, even His eternal power and G-dhead, so that they are without excuse. Ironically, this 7 day cycle has drawn the interest of NASA, as it relates to further human survival in the long durations of space exploration. (I encourage all readers to further investigate this for themselves).

Before we proceed, it is also imperative that we know the biblical importance of the number "seven". The number seven has played an exceptionally large role in biblical scripture and religion. "Seven's" place is eminent among sacred numbers in scripture. For example, the creation account involving 7 days as we've just read, and thus as we've previously learned from a past chapter that the very first verse of the Hebrew Bible contains exactly "seven" words which contain a total of 28 letters. The number "seven" is typically associated with acts of completion, rest, renewal, fulfillment, spiritual importance and perfection, as you will read in some of the biblical scriptures. The number 7 (whether it be seconds, minutes, hours, days, weeks, months, or years) will almost surely culminate itself to the end of all time. The number 7 seems to be G-d's prophetic timepiece, which will be illuminated, in the next chapters. For example, one of the more prominent as well as prophetic accounts for the number 7 can be found in the biblical story of Joseph interpreting the Pharaohs dream (which consisted of sevens). This was the dream: Genesis 41:2 - when out of the river there came up **seven** cows, sleek and fat, and they grazed among the reeds. Genesis 41:3 - After them, **seven** other cows, ugly and gaunt, came up out of the Nile and stood beside those on the riverbank. Genesis 41:4 - And the cows that were ugly and gaunt ate up the **seven** sleek, fat cows. Then Pharaoh woke up. Genesis 41:5 - He fell asleep again and had a second dream: **Seven** heads of grain, healthy and good, were growing on a single stalk. Genesis 41:6 - After them, **seven** other heads of grain sprouted—thin and scorched by the east wind. Genesis 41:7 - The thin heads of grain swallowed up the **seven** healthy, full heads. Then Pharaoh woke up; it had been a dream. Genesis 41:18 - When out of the river there came up **seven** cows, fat and sleek, and they grazed among the reeds. Genesis 41:19 - After them, **seven** other cows came up—scrawny and very ugly and lean. I had never seen such ugly cows in all the land of Egypt. Genesis 41:20 - The lean, ugly cows ate up the **seven** fat cows that came up first. Genesis 41:22 - In my dreams I also saw **seven** heads of grain, full and good, growing on a single stalk. Genesis 41:23 - After them, **seven** other heads sprouted—withered and thin and scorched by the east wind. Genesis 41:24 - The thin heads of grain swallowed up the **seven** good heads. I told this to the magicians, but none could explain it to me. Genesis 41:26 - The **seven** good cows are **seven** years, and the **seven** good heads of grain are **seven** years, it is one and the same dream. Genesis 41:27 - The **seven** lean, ugly cows that came up afterward are **seven** years, and so are the **seven** worthless heads of grain scorched by the east wind: They are **seven** years of famine. Genesis 41:29 - **Seven** years of great abundance are coming throughout the land of Egypt. Genesis 41:30 - But **seven** years of famine will follow them. Then all the abundance in Egypt will be forgotten, and the famine will ravage the land. Genesis 41:34 - Let Pharaoh appoint commissioners over the land to take a fifth of the harvest of Egypt during the **seven** years of abundance. Genesis 41:36 - This food should be held in reserve for the country, to be used during the **seven** years of famine that will come upon Egypt, so that the country may not be ruined by the famine. Genesis 41:47 - During the **seven** years of abundance the land produced plentifully. Genesis 41:48 - Joseph collected all the food produced in those **seven** years of abundance in Egypt and stored it in the cities. In each city he put the food grown in the fields surrounding it. Genesis 41:53 - The **seven** years of abundance in Egypt came to an end. Genesis 41:54 - And the **seven** years of famine began, just as Joseph had said. There was famine in all the other lands, but in the whole land of Egypt there was food.

Now some may ask, why did Joseph's dream consist of seven? What was the special significance? The purpose of Josephs dream (reveals G-d working behind the scene) to insure Israel's survival. Furthermore since seven is a number of rest and renewal (also associated with the full moon). It was a symbolic representation of the foreseen future of the Exodus (mass departure) from the slavery of Egypt. Genesis 45:4-11 - And Joseph said unto his brethren, Come near to me, I pray you. And they came near. And he said, I am Joseph your brother, whom ye sold into Egypt. Now therefore be not grieved, nor angry with yourselves, that ye sold me hither: **for G-d did send me before you to preserve life.** For these two years hath the famine been in the land: and yet there are five years, in the which there shall neither be earing nor harvest. **And G-d sent me before you to preserve you a posterity in the earth, and to save your lives by a great deliverance.** So now it was not you that sent me hither, but G-d: and he hath made me a father to Pharaoh, and lord of all his house, and a ruler throughout all the land of Egypt. Haste ye, and go up to my father, and say unto him, Thus saith thy son Joseph, G-d hath made me lord of all Egypt: come down unto me, tarry not: And thou shalt dwell in the land of Goshen, and thou shalt be near unto me, thou, and thy children, and thy children's children, and thy flocks, and thy herds, and all that thou hast: And there will I nourish thee, for yet there are five years of famine, lest thou, and thy household, and all that thou hast, come to poverty. Moses was then appointed by G-d to deliver the Jews out of Egypt to the land of Israel. Exodus 5:1 - And afterward Moses and Aaron went in, and told Pharaoh, Thus saith the Lord G-d of Israel, LET MY PEOPLE GO!

Here is more correlating biblical scripture that puts emphasis on the number "seven" which are usually acts of completion, fulfillment, spiritual importance and perfection as you will read in these scriptures. Exodus 23:11 - But during the **seventh** year let the land lie unplowed and unused. Then the poor among your people may get food from it, and the wild animals may eat what they leave. Do the same with your vineyard and your olive grove. Exodus 24:16 - And the glory of the Lord settled on Mount Sinai. For six days, the cloud covered the mountain, and on the **seventh** day the Lord called to Moses from within the cloud. Leviticus 22:27 - When a calf, a lamb or a goat is born, it is to remain with its mother for **seven** days. From the eighth day on, it will be acceptable as an offering made to the Lord by fire. Leviticus 25:4 - But in the **seventh** year the land is to have a sabbath of rest, a sabbath to the Lord. Do not sow your fields or prune your vineyards. Here we find that Moses' sister Miriam was infected with a disease or virus of some kind, and G-d told Moses to remove her from the presence of the people for seven days (suggesting that the illness had a seven day incubation period). Numbers 12:14 - The Lord replied to Moses, Confine her outside the camp for **seven** days, after that she can be brought back. Again, in this next biblical phrase we find that dead flesh contains harmful bacteria lasting seven days. Numbers 19:16 - Anyone out in the open who touches someone who has been killed with a sword or someone who has died a natural death, or anyone who touches a human bone or a grave, will be unclean for **seven** days. Numbers 31:19 - All of you who have killed anyone or touched anyone who was killed must stay outside the camp **seven** days. On the third and **seventh** days you must purify yourselves and your captives. Deuteronomy 5:14 - But the **seventh** day is a Sabbath to the Lord your G-d. On it you shall not do any work, neither you, nor your son or daughter, nor your manservant or maidservant, nor your ox, your donkey or any of your animals, nor the alien within your gates, so that your manservant and maidservant may rest, as you do. Here we read that the seventh year is a year for canceling debts (ironically, government typically follows this same rule by telling us to keep 7 years of records). Deuteronomy 31:10 - Then Moses commanded them: At the end of every **seven** years, in the year for canceling debts, during the Feast of Tabernacles. Seven is also known for acts of completion and fulfillment when the walls of Jericho fell after they had marched around it seven times. Joshua 6:13 - The **seven** priests carrying the **seven** trumpets went forward, marching before the ark of the Lord and blowing the trumpets. The armed men went ahead of them and the rear guard followed the ark of the Lord, while the trumpets kept sounding.

Joshua 6:15 - On the **seventh** day, they got up at daybreak and marched around the city **seven** times in the same manner, except that on that day they circled the city **seven** times. Joshua 6:16 - The **seventh** time around, when the priests sounded the trumpet blast, Joshua commanded the people, Shout! For the Lord has given you the city! Hebrews 11:30 - By faith the walls of Jericho fell, after the people had marched around them for **seven** days. Again, seven is noted as a time of completion and fulfillment in the course of feasts. Ezekiel 45:21 - In the first month on the fourteenth day you are to observe the Passover, a feast lasting **seven** days, during which you shall eat bread made without yeast. Phenomenally, here we find Jesus replaying the importance of the number seven. Mark 8:6 - He told the crowd to sit down on the ground. When he had taken the **seven** loaves and given thanks, he broke them and gave them to his disciples to set before the people, and they did so. Mark 8:8 - The people ate and were satisfied. Afterward the disciples picked up **seven** basketfuls of broken pieces that were left over. Mark 8:20 - And when I broke the **seven** loaves for the four thousand, how many basketfuls of pieces did you pick up? They answered, **"Seven".** So important and representational is the number seven, that G-d allowed 7 other people (besides Noah) to exist in the world after the great flood. 2 Peter 2:5 - If He did not spare the ancient world when he brought the flood on its ungodly people, but protected Noah, a preacher of righteousness, and **seven** others. Furthermore, the number seven is relevant for a future time in biblical history yet to come (the opening of 7 seals). Revelation 5:5 - Then one of the elders said to me, Do not weep! See, the Lion of the tribe of Judah, the Root of David, has triumphed. He is able to open the scroll and its **seven** seals. The prophet Daniel was also told that the "Messiah" would arrive after "seventy sevens". Daniel 9:24 - **Seventy sevens** are decreed for your people and your holy city to finish transgression, to put an end to sin, to atone for wickedness, to bring in everlasting righteousness, to seal up vision and prophecy and to anoint the most holy. And in the first chapter of Matthew, "the genealogy of Jesus" is organized into three groups of fourteen generations (7+7=14). Matthew 1:17 - Thus there were **fourteen** generations in all from Abraham to David, **fourteen** from David to the exile to Babylon, and **fourteen** from the exile to the Christ. (Remember the "Star of David" example from the previous chapter discussing the Merkabah).

There are many more sacred numbers mentioned in the Bible (each with a special meaning), but the number seven seems to hold precedence above them all as the CYCLIC ACTION of G-d in the affairs of humankind. The seventh day of creation remains a day of pause and worship. In addition, not only pause and worship, but completion, fulfillment and **FINALIZATION!** Therefore, we can conclude that the number "seven" is deliberately commemorating G-d's original creation of us.

In addition, there may be something even more spectacular to the number 7 yet… and this has to do with recorded history (in other words, "measured time"). It is my estimation that there will be exactly only 7 thousand years of measured time - as this is G-d's great culminating plan.

We are now nearing the 7 thousand years of measured time - which is the Grand Jubilee?

Adam - 930
Lamech - 874-1651
Noah - 1056-2006
Terah - 1879-2084

Abraham - 2009-2184
Flood - 1650

So if we estimate the timeline today from Jesus it was approximately 2,000 years, and from Jesus to Abraham was 2,000 years, and from Abraham back to Adam is about 2,000 years, which gives us an approximate age of 6,000 years of recorded (and measured) time.

(Now from 6,000 years), if we estimate that Jesus died in 33AD, and this is now the year 2009 AD... that means that 33 AD was exactly 1,976 (one thousand nine hundred and seventy six) years ago. Therefore, if we subtract 1,976 from the 2,000 year mark, we are roughly 24 years short of the 6,000th year, which would bring us to the year 2033. After these 24 years, we have another 1,000 years left to make the Grand Jubilee of 7,000 years! We may very well be in (or nearing) the 1,000 years specified in Revelation 20:1-11 - And I saw an angel come down from heaven, having the key of the bottomless pit and a great chain in his hand. And he laid hold on the dragon, that old serpent, which is the Devil, and Satan, and bound him **a thousand years,** And cast him into the bottomless pit, and shut him up, and set a seal upon him, that he should deceive the nations no more, **till the thousand years should be fulfilled:** and after that he must be loosed a little season. And I saw thrones, and they sat upon them, and judgment was given unto them: and I saw the souls of them that were beheaded for the witness of Jesus, and for the word of G-d, and which had not worshipped the beast, neither his image, neither had received his mark upon their foreheads, or in their hands; and they lived and reigned with Christ **a thousand years.** But the rest of the dead lived not again until **the thousand years** were finished. This is the first resurrection. Blessed and holy is he that hath part in the first resurrection: on such the second death hath no power, but they shall be priests of G-d and of Christ, and shall reign with him **a thousand years.** And when **the thousand years** are expired, Satan shall be loosed out of his prison, And shall go out to deceive the nations which are in the four quarters of the earth, Gog, and Magog, to gather them together to battle: the number of whom is as the sand of the sea. And they went up on the breadth of the earth, and compassed the camp of the saints about, and the beloved city: and fire came down from G-d out of heaven, and devoured them. And the devil that deceived them was cast into the lake of fire and brimstone, where the beast and the false prophet are, and shall be tormented day and night for ever and ever. And I saw a great white throne, and him that sat on it, from whose face the earth and the heaven fled away; and there was found no place for them.

Revelation 20:7-10 talks about satan's doom - And when **the thousand years are expired, Satan shall be loosed out of his prison, And shall go out to deceive the nations which are in the four quarters of the earth, Gog, and Magog,** to gather them together to battle: the number of whom is as the sand of the sea. And they went up on the breadth of the earth, and compassed the camp of the saints about, and the beloved city: and fire came down from G-d out of heaven, and devoured them. And the devil that deceived them was cast into the lake of fire and brimstone, where the beast and the false prophet are, and shall be tormented day and night for ever and ever. However what is not known is wheather or not this extra 1,000 years takes place before the rapture of the church (which is the body of believers) or takes place afterwards as the remaining 1,000 years after the rapture when Jesus dwells with those left on the Earth (which I believe will be the Jews when they finally realize their Messiah). I also am personally in favor of believing that the extra 1,000 years will take place before the rapture.

Now when the Bible talks about the four quarters of the Earth as "Gog and Magog" (highlighted directly above), it is referring to Russia. Ezekiel 38:2 - Son of man, set thy face against Gog, the land of Magog, the chief prince of Meshech and Tubal, and prophesy against him. We know from secular history that the Magog's were named Magagiles, but who are known (and called) by the Greeks as Scythians, and this was told to us by the historian Josephus (antiquities of the Jews). We can also find in the New Columbia Encyclopedia that during the 6th century BC the Scythians actually dominated the land that was later to become Russia. Furthermore, John Lawrence on his history of Russia tells us that modern Russians sometimes call themselves Scythians. The Scythians who were the Magagiles can actually be traced all the way back from a grandson of Noah since Meshech is the 6th son of Japheth, son of Noah, to which out of him come a family, then a tribe, then a nation.

Now we must also remember (and calculate in) what we had discussed in the previous chapter about the importance of RECESSION. Remember, it was of great importance to both the ancient Egyptians and the Freemasons. In addition, if you recall the Great Pyramid and Enoch was the pivotal key to the whole mystery of the ages. And as mentioned earlier, the Eagle was used as a sign for "Enoch". The Eagle was given the name "Phoenix" which is interpreted to mean "Pa Hanok" which means "house of Enoch" (or house of the stars) and it has always been long considered in mythology the symbol for "resurrection or rising again" from the flare/flame of the precession of the Sun (which has also associated Enoch as the Sun god). It is symbolic for the start of a "great age" of complete recession mentioned earlier as 25,920 years, which is 360 divided by 5 which is 72. This is why the Phoenix has been associated with the precession of the Sun - the old Phoenix dies and is reborn, renewed, resurrecting from the ashes of its old form. This is indicative of the time of "rest" when the northern hemisphere is tilting away from the Sun followed by the "resurrection" which marks the spring equinox. The myth about the Phoenix tells us that it has a life cycle of 500 years… this is quite interesting when we examine the prophet Daniel's (Daniel 9:24) prophetic calculations of 70 weeks ($70 \times 7 = 490$ years) its pretty close to the Phoenix's 500 year life span. These 500 year cycles appear to have been in course before the "Messiah's birth". With Abraham being born approximately 2,000 years before Jesus' birth, and nearly 500 years later around 1,500 years before Jesus' birth, Israel made their Exodus from Egypt and received the Torah at Mount Sinai. Again, approximately 500 years later, (1,000 year before Jesus' birth), Solomon's temple was completed. And approximately 500 years later Zerrubabel's temple was built after Israel came out of Babylonian captivity. Finally, Jesus was born 500 years later.

And as was also previously discussed in a earlier chapter that Enoch was privy to know the second coming of the Messiah which is most likely incorporated into one of the Hebrew holidays. A Jewish holiday that corresponds to this is "Rosh Hashanah" which represents the "day of judgment" (it is the Jewish New Year characterized by blowing the Shofar (a rams horn - to awaken the listener to coming judgment) followed by 10 days of repentance just before "Yom Kippur" which is the "day of atonement". In effect, G-d gave us this holy event to teach us that we will be judged on Rosh Hashanah and sealed unto the closing of the gates on Yom Kippur. This is where the righteous are separated and raptured up to be with G-d. This holiday is mentioned in Leviticus 23:24 - Speak unto the children of Israel, saying, in the seventh month, in the first day of the month, shall ye have a Sabbath, a memorial of blowing of trumpets, an holy convocation. It also mentions something to this fact in Isaiah 26:2 - The gates of Heaven are opened on Rosh Hashanah so the righteous nation may enter. It also makes mention of this in Psalms 118:19-20 - Open to me the gates of righteousness, I will go into them, and I will praise the Lord. This gate of the Lord, into which the righteous shall enter. Because the gates of Heaven are understood to be open on Rosh Hashanah, this is further evidence that the "rapture of the believers in the Messiah" will take place between Rosh Hashanah and Yom Kippur.

Additionally, this will be on the Jubilee. Leviticus 25:9-19 - Then shalt thou cause the trumpet of the jubile to sound on the tenth day of the seventh month, in the day of atonement shall ye make the trumpet sound throughout all your land. And ye shall hallow the fiftieth year, and proclaim liberty throughout all the land unto all the inhabitants thereof: it shall be a **jubile** unto you; and ye shall return every man unto his possession, and ye shall return every man unto his family. A jubile shall that fiftieth year be unto you: ye shall not sow, neither reap that which groweth of itself in it, nor gather the grapes in it of thy vine undressed. For it is the **jubile;** it shall be holy unto you: ye shall eat the increase thereof out of the field.

In the year of this **jubile** ye shall return every man unto his possession. And if thou sell ought unto thy neighbour, or buyest ought of thy neighbour's hand, ye shall not oppress one another: According to the number of years after the **jubile** thou shalt buy of thy neighbour, and according unto the number of years of the fruits he shall sell unto thee: According to the multitude of years thou shalt increase the price thereof, and according to the fewness of years thou shalt diminish the price of it: for according to the number of the years of the fruits doth he sell unto thee. Ye shall not therefore oppress one another; but thou shalt fear thy G-d: for I am the Lord your G-d. Wherefore ye shall do my statutes, and keep my judgments, and do them; and ye shall dwell in the land in safety. And the land shall yield her fruit, and ye shall eat your fill, and dwell therein in safety.

However, let me clarify something. Though I had given a calculation of 24 years being short of the full 6,000th year, we also have the Jewish calendar to contend with, which does not match the modern calculation that I purposed. According to the Rabbinical computation for the current date since the creation of the world, it has only been 5,769 years following the Hebrew year 5,770 which will begin at sundown on the evening of the 18th of September 2009. This gives us a longer time period left of 230 years (or 1,230). As you can see, there is a discrepancy. And even though I have calculated this in my own way does not mean I am anywhere near being correct since the Bible tells us that the hour and time is unknown, however we do know the season (the holiday of Rosh Hashanah). The Bible also tells us that despite not knowing this exact hour and time, TO BE READY! Matthew 24:36 - No one knows about that day or hour, not even the angels in heaven, nor the Son, but only the Father. Mark 13:32-37 - No one knows about that day or hour, not even the angels in heaven, nor the Son, but only the Father. Be on guard! Be alert! You do not know when that time will come. It's like a man going away: He leaves his house and puts his servants in charge, each with his assigned task, and tells the one at the door to keep watch. Therefore keep watch because you do not know when the owner of the house will come back—**whether in the evening, or at midnight, or when the rooster crows, or at dawn.** If he comes suddenly, do not let him find you sleeping. What I say to you, I say to everyone: Watch! The reason that we do not know the day or the hour is because this solely depends on the "New Moon", since it is the New Moon that determines the starting day of "Rosh Hashanah". (This will be explained again).

(Trying not to make things to complicated)... Let me also quickly add that the "Jewish calendar system" for the end of days according to some Rabbinical Rabbis is for the year

5777, which is eerily very close to our present day. The number 777 is G-d's number (which you will read and learn about in the coming chapter). And with this in mind, there is strong speculation that the number 777 also suggests that it is the final hour, day and year. Now, since it is currently the Hebrew year 5770, (which equates to the modern calendar year of 2009-2010) + 7 years is 2017 - thus the Hebrew year 5777. Therefore, according to some religious thinkers who study scripture (put in terms that we can understand according to the modern year calendar), the year 2017 is the end of days for those that are left unbelieving! This they correlate with the "rapture" (the second coming of Jesus).

Now before we proceed, we need to back-up and revisit the details concerning the MOONS LUNAR CYCLE. Not only will we find that the Moons cycle is supported by biblical scripture, but we will additionally find that it corroborates with known scientific data, which will ultimately lead to confirmation of G-d-linked truth. Now, as I mentioned earlier, these 7 day cycles are determined by the MOONS LUNAR CYCLE. The moon was created on the 4th day. And as you may recall, G-d created the Moon, the Sun and the stars to serve as signs to mark season, days and years. Genesis 1:14-19 - And G-d said, Let there be lights in the expanse of the sky to separate the day from the night, and let them serve as signs to mark seasons and days and years, and let them be lights in the expanse of the sky to give light on the earth. And it was so. **G-d made two great lights—the greater light to govern the day and the lesser light to govern the night (the sun and the moon).** He also made the stars.
G-d set them in the expanse of the sky to give light on the earth, to govern the day and the night, and to separate light from darkness. And G-d saw that it was good. And there was evening, and there was morning—**the fourth day.** Psalm 104:19 - The moon marks off the seasons, and the sun knows when to go down. Isaiah 66:23 - From one New Moon to another and from one Sabbath to another, all mankind will come and bow down before me, says the Lord.

Other than G-d creating the MOON to serve for both governing the night and marking off the seasons, science has come to learn that the moon has a vast effect and array over all living and nonliving things on Earth. The moon affects tides (both ocean, lake, river and stream). One can visually see the effects of high and low tides throughout the day. We can observe this if we sit and watch the ocean shore throughout the day as the tide moves farther away or closer to where you are sitting. One will find that when the seashore line (tide) moves up to their chair (or sitting spot) they have to relocate farther up on shore. This is evidence happening before our very eyes that such a phenomenon is occurring. Additionally the gravitational effects of the Moon cause volcanoes to erupt, (earthquakes, shifting and buckling of the Earths plates, tsunamis, tornados and the flow of liquid magma from below the Earths surface). Revelation 6:12 - I watched as he opened the sixth seal. There was a great earthquake. The sun turned black like sackcloth made of goat hair, the whole moon turned blood red. Gravity is the main cause of this, because as the moon orbits the Earth the gravitational pull tugs at everything on Earth (to the Moons direction). The Earth itself has a rotational pull as well as the Sun. In effect, all these forces culminated together are strongest on a full Moon. This is because when we can see the full Moon it is aligned with that part of the Earth. Luke 21:25 - There will be signs in the sun, moon and stars. On the earth, nations will be in anguish and perplexity at the roaring and tossing of the sea.

The effects of the Moon also include all biological life. This stands to reason that the human body as well as all life is susceptible to the gravitational pull of the moon (in combination of the Earth and Sun) since most living things are made up of mostly water, fluid and blood that flow like the oceans. The brains of insects, animals and humans may also be susceptible to this gravitational pull since the brain of any living things is composed of mostly liquid and semi-solid material to the likes of which a full Moons gravitational effect may disrupt the normal flow of fluids, which in return could have an effect over behavior and mood. The Moon can also affect hormones, organ function, and the skin of living things. Fluids in the body play a major role in the onset of disease and psychologically disturbed behavior, as glands and hormones produce chemicals that promote well being, lust, passion, anxiety, fear, worry and depression of various mental conditions and illnesses. If glands become imbalanced, their hormonal secretions become abnormal which can throw the body out of whack. During extreme tides, (from the pull of the Moon) there is also an increase in the surrounding electromagnetic activity, which may stir the nervous system. It may also cause an increase in depression, restlessness, insomnia, have an influence over organ function and may promote the cellular growth of cells. Additionally as evidence of this phenomenon, a rapid heartbeat occurs just prior to and during periods of a full Moon. However, to advert these bad tendencies and promote the good tendencies we must "not work". Instead we must relax, sleep and rest up - both physically as well as mentally (which greatly supports and corroborates what scripture says again - to not work on the 7th day of the week, which is the Sabbath). Even fasting can help in averting these tendencies along with cleaning the blood and promoting cellular growth. Ironically, the best days for fasting are on the new and full Moon days (several days before, and several days after). It's not actually the full or new Moon itself, it is the gravitational pulling actions that transpire before the actual full or new Moon… the full or new Moon is actually the desired achieved result and "a pause of rest" occurs upon reaching the ultimate goal. Then after this relieving pause (of the Moon reaching its ultimate goal), the downward depreciation or pulling away (of the Moon) stirs the Earth with stress again. This is like the workweek, where we rest the weekend and start again on Monday working through to Friday. There are various strengths throughout the month. Science tells us of high tide on new Moon and full Moon days, and low tide on the seventh day of the lunar cycle as the Sun comes nearer to the Earth. The Earth, Moon and Suns attraction to each other is greater on certain days because of the bodies' gravitational effect on each other. This is due in part to the Earth having an elliptical rotation with the moon lagging minutely behind (as shown in the picture below). When the Earth and the Moon are aligned in this elliptical sequence it tugs at the Earths crust, which is what causes all chaotic weather conditions, as well as influencing all living and nonliving things on the planet. Additionally, other strange phenomenon occur which collaborates along with the Sun, Earth and Moons alignment along with the other planets in our solar system. In August of 1990, there was a 6-planet alignment, which was the onslaught of the Desert Storm war.

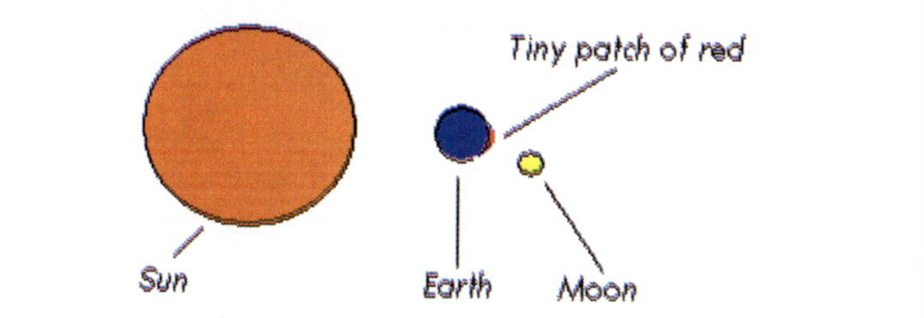

Notice the blue earth with a tiny patch of red facing the moon. This is the power of the gravitational pull and tug of the moon over the earth. It is usually stronger when the Sun, Earth and Moon are in alignment.

The difference between a new Moon and a full Moon is that during a full Moon it rises at the same time the sunsets, while the new moon is barely visible and is very close to the position of the Sun. The full Moon rises roughly 14 days after the new Moon.

Interestingly, during this time, just as the Moon affects the tides of the Earth's oceans, it also has an influence on plants to make them grow faster and stronger. Certain experiments have revealed that after the new Moon and before the full Moon, there is a gravitational tug-of-war. This happens when the gravitational pull of the Moon is at its strongest for the month, which makes it easier for seeds to germinate faster. Once they germinate, the roots grow upward and at the same time grow deeper into the soil. What's even more interesting to witness, is mushroom harvesting where flushes appear at approximately 7 day intervals, (one can buy a mushroom kit to witness this). Additionally during this time (of the gravitational pull of the Moon) there is also a slight increase in the Earths "electricity". This electrical energy has its different effects on all living things. In humans this electrical charge stimulates our body such as muscles and organ function. However, in plants, it promotes growth by delivering a slight tingling of electricity, which awakens seeds to growth. Genesis 8:22 - As long as the earth endures, seedtime and harvest, cold and heat, summer and winter, day and night will never cease. In other words what this biblical verse is explaining to us, is that the Earth will continue to endure from season to season, summer and winter, and day and night simply because of the revolutions of the Earth, Sun and Moon.

In fact so important is the Earth, Moon and Sun connection (as aforementioned in a previous chapter), which serves to shed light on the amount of "daylight" that we have in a single 24 hour day, which is also extremely necessary for producing the hormone "melatonin" in the human brain that brings on sleep. And so important is the hormone melatonin that it is linked between the brain, the pituitary gland, and the gonads (testes and ovary), which in turn produce a hormone called gonadotropin, which is the key hormone that primes the body for sex. Simply put, if we don't sleep we simply could not reproduce offspring or have survived. This also brings us back to another issue... the issue of existing in long term zero gravity space, and this includes possible pregnancies both in humans and other living things, since there have been tests on rat embryos that had produced decreased skeletal and brain development. Bottom line, we simply were not made to survive in a space environment, which begs the question if any other similar life out that could exist either? The answer to this is that - there simply isn't. At least not on this realm of existence).

Next, before we continue reading, we are to know that the Sabbath was indeed determined by the Moon. Biblical scripture freely and liberally tells us this. All one has to do to find this proof (and know the truth of this) is to look within the Bible, (The Bible continually links the new Moon with the Sabbath). G-d, in His infinite wisdom and mercy, has planted clues in His word the Bible - clues that those with an understanding mind can uncover and apply. 2 Chronicles 2:4 - Behold, I build an house to the name of the Lord my G-d, to dedicate it to him, and to burn before him sweet incense, and for the continual shewbread, and for the burnt offerings morning and evening, **on the sabbaths, and on the new moons,** and on the solemn feasts of the Lord our G-d. This is an ordinance for ever to Israel. 2 Chronicles 8:13 - Even after a certain rate every day, offering according to the commandment of Moses, **on the sabbaths, and on the new moons,** and on the solemn feasts, three times in the year, even in the feast of unleavened bread, and in the feast of weeks, and in the feast of tabernacles. Nehemiah 10:33 - For the shewbread, and for the continual meat offering, and for the continual burnt offering, **of the sabbaths, of the new moons,** for the set feasts, and for the holy things, and for the sin offerings to make an atonement for Israel, and for all the work of the house of our G-d. Isaiah 1:13 - Bring no more vain oblations, incense is an abomination unto me, **the new moons and sabbaths,** the calling of assemblies, I cannot away with; it is iniquity, even the solemn meeting. Isaiah 66:23 - And it shall come to pass, that **from one new moon to another, and from one sabbath to another,** shall all flesh come to worship before me, saith the Lord. Ezekiel 46:3 - Likewise the people of the land shall worship at the door of this gate before the Lord in the **sabbaths and in the new moons.** Amos 8:5 - Saying, **When will the new moon be gone, that we may sell corn? and the sabbath, that we may set forth wheat,** making the ephah small, and the shekel great, and falsifying the balances by deceit?

So how do we know which day is the exact day of rest? It is the 15th day of the second month. We know this again from scripture. It is very interesting that Moses should mention the EXACT DAY that the Israelites murmured against him and Aaron in the wilderness - nowhere else in the account of the Israelites wanderings from Egypt to the Promised Land is an exact date mentioned! This day - the 15TH DAY OF THE SECOND MONTH was so important to G-d that He appeared in the cloud before the Israelites and then caused scores of quail to be blown into the Israelites camp. This event occurred between the two evenings of the 15th. Exodus 16:1 - The whole Israelite community set out from Elim and came to the Desert of Sin, which is between Elim and Sinai, on the fifteenth day of the second month after they had come out of Egypt. Leviticus 23:6 - On the fifteenth day of that month the Lord's Feast of Unleavened Bread begins, for seven days you must eat bread made without yeast. Leviticus 23:34 - Say to the Israelites: On the fifteenth day of the seventh month the Lord's Feast of Tabernacles begins, and it lasts for seven days. Leviticus 23:39 - So beginning with the fifteenth day of the seventh month, after you have gathered the crops of the land, celebrate the festival to the Lord for seven days; the first day is a day of rest, and the eighth day also is a day of rest.

Why was this day so important to G-d? Because in foresight of the future G-d Knew what was to come. It was a day that G-d Himself was speaking to the Jews ON A SABBATH in the wilderness on their way to the Promised Land of Israel! It is evident that G-d (who in human flesh later became Jesus) first preached to Adam on the original Sabbath. Jesus declared Himself "Lord even of the Sabbath". Mark 2:27-28 - And He said unto them, the Sabbath was made for man, and not man for the Sabbath: Therefore, the Son of man is Lord also of the Sabbath. Adam was created on the sixth day of creation sometime in the late afternoon. We can deduce this because the making of man was the last act of creation on that day. In addition, it was on this day that G-d had instructed Adam that the wages of sin is death. Genesis 2:15-17 - And the Lord G-d took the man, and put him into the Garden of Eden to dress it and to keep it. And the Lord G-d commanded the man, saying, Of every tree of the garden thou mayest freely eat: But of the tree of the knowledge of good and evil, thou shalt not eat of it: for in the day that thou eatest thereof thou shalt surely die. And here in the wilderness after leaving Egypt G-d is again preaching to Israel, through Moses, ON THE SABBATH! In addition, in the New Testament we find Jesus preaching on the Sabbath. Luke 4:16 - And he came to Nazareth, where he had been brought up: and, as his custom was, he went into the synagogue on the sabbath day, and stood up for to read. Matthew 12:8 - **For the Son of man is Lord even of the Sabbath day.**

Again, here we read that in the "afternoon" before the Sabbath (on the 15th) G-d sent the Jews in the wilderness quail for food, and the next morning manna was found on the ground. But something special happened to the manna that was found on the ground on the 7th day. This special incident was that the manna was preserved from rotting. Miraculously the manna was kept fresh and preserved, but directly after the Sabbath day, it rotted. If we link the Sabbath to that of a full moon (which scripture clearly indicates to, as we've just read above). This would mean that several days "before" the actual full moon, when the earth is being tugged by the moon which is lagging minutely behind, the gravitational forces are prime for the rejuvenation of all living things - which would include the bacteria forming in the manna that did not rot, go moldy, decay or decompose. It is also (as mentioned before) the ideal time for the human body to rejuvenate itself and its cells. Additionally (as aforementioned) in light of recent medical science, it has been discovered that patients may reject an organ transplant if it is done before the 7th day, or after the 7th day. In addition, as aforementioned, this happens to be the case with immune response to infections such as the common cold, which usually lasts for about a week.

Now after this time things are more susceptible to decay when the Moon passes past the elliptical, (the egg shaped oval of the Earth as shown in the picture above). Again, we find the Bible being scientifically correct about this. Read what the Bible teaches about it: Exodus 16:11-30 - The Lord said to Moses, I have heard the people of Israel talking about how unhappy they are. Tell them, When the sun goes down, you will eat meat. In the morning you will be filled with bread. Then you will know that I am the Lord your G-d. That evening quail came and covered the camp. In the morning the ground around the camp was covered with dew. When the dew was gone, thin flakes appeared on the desert floor. They looked like frost on the ground. The people of Israel saw the flakes. They asked each other, What's that? They didn't know what it was. Moses said to them, It's the bread the Lord has given you to eat. Here is what the Lord has commanded. He has said, Each one of you should gather as much as you need. Take two quarts for each person who lives in your tent. The people of Israel did as they were told. Some gathered a lot, and some gathered a little. When they measured it out, those who gathered a lot didn't have too much. And those who gathered a little had enough. All of them gathered only what they needed. **Then Moses said to them, Don't keep any of it until morning. Some of them didn't pay any attention to Moses. They kept part of it until morning. But it was full of maggots and began to stink.** So Moses became angry with them. Each morning all of them gathered as much as they needed. But by the hottest time of the day, the thin flakes had melted away. On the sixth day, the people gathered twice as much. It amounted to four quarts for each person. The leaders of the community came and reported that to Moses. He said to them, Here is what the Lord commanded. He said, **Tomorrow will be a day of rest. It will be a holy Sabbath day. It will be set apart for the Lord. So bake what you want to bake. Boil what you want to boil. Save what is left. Keep it until morning. So they saved it until morning, just as Moses commanded. It didn't stink or get maggots in it. Eat it today, Moses said. Today is a Sabbath day in the Lord's honor. You won't find any flakes on the ground today. Gather them for six days. But on the seventh day there won't be any. It's the Sabbath. In spite of what Moses said, some of the people went out on the seventh day to gather the flakes. But they didn't find any. Then the Lord spoke to Moses. He said, How long will all of you refuse to obey my commands and my teachings? Keep in mind that I have given you the Sabbath day. That is why on the sixth day I give you bread for two days.**

All of you must stay where you are on the seventh day. No one can go out. So the people rested on the seventh day.

This was a miracle from G-d showing us, which is the right 7th day to observe the Sabbath! Starting on the 15th day of the month, G-d made available the quails and manna for food for the Jews in the wilderness, then on the seventh day after the 15th, He did not provide any more manna - thereby confirming that this day (now the 22nd day of the month) was also a Sabbath. Evidently, if the 15th was seven days before the 22nd, it as well was a Sabbath. And if the 15th and the 22nd were Sabbath days, then it stands to reason that the 8th and the 29th of the month were also Sabbaths. So here we see a pattern 8th, 15th, 22nd, and the 29th (each 7 days apart). These dates are all central to the established calendar because they correspond to the phases of the Moon. In effect, we can now see that G-d was instituting His weekly Sabbath cycle for the Jews by the Moon phases. This is a rhythm tuned to G-d's own activity of work and rest. The days should be like such:

1 - Sunday - creation - birth - resurrection.
2 - Monday
3 - Tuesday
4 - Wednesday
5 - Thursday
6 - Friday - Sundown (the start of the Sabbath).
7 - Saturday - Sabbath - rest - break - renewal - cell regeneration.

Corresponding to all this, a woman's menstrual cycle takes place every 28 days and lasts for approximately seven days. However, seven days prior to the onset of a cycle, certain hormonal chemicals come into play and a woman can start to feel her breasts hurting. And after the seven-day cycle, the hormones begin to decrease within the next seven days. The menstrual cycle is broken down and deduced within a month (4 weeks in a month x 7 days in a week, equals 28 days). The reason a woman gets her menstrual cycle on the 28th day of each month, and why the menstrual cycle last for 7 days, is because Eve had to have been created on the 6th day late in the evening possibly around the early hours of the approaching 7th day. Which in the scheme of things may have some relevance to why a woman's menstrual cycle last for 7 days (4 weeks in a month x 7 days in a week, equals 28 days). It may also have something to do with the fact that the Moon makes a complete turn (or cycle) around the Earth in approximately 29 days. From new Moon to new Moon is a synodic time lasting 29.53059 days (precisely 29 days, 12 hours, 44 minutes, 2.8 seconds). In addition, the sidereal cycle, the measure taken from where the Moon appears at the same place in the sky, which is 27.33 days? Since neither of these cycles accurately measures the solar cycle the 28th day is an average of the two main lunar cycles. Genesis 1:27 - So G-d created man in his own image, in the image of G-d created he him, male and female created he them. Genesis 1:31 - And G-d saw every thing that he had made, and, behold, it was very good. And the evening and the morning were the sixth day. Genesis 2:18 - **The Lord G-d said, It is not good for the man to be alone. I will make a helper suitable for him.**

Moving on here... Now that the seven-day cycle has been made clear, and a basic dietary plan brought into proper perspective (from the first part of this chapter), we can now move on to the next increment in spiritual growth. This next stage helps us in advancing our survival skills. A skill in which one learns the art of self-sufficiency, (which means being dependant upon ourselves). 1 Thessalonians 4:11-12 - Make it your ambition to lead a quiet life, to mind your own business and to work with your hands, just as we told you, so that your daily life may win the respect of outsiders and so that you will not be dependent on anybody. Thessalonians 3:8 - Nor did we eat anyone's food without paying for it. On the contrary, we worked night and day, labouring and toiling so that we would not be a burden to any of you. This includes learning to live in harmony with nature and make use of the environment and our resourceful ingenuity (inventiveness, cleverness and creativity) that G-d has given to us. This includes that the home in which one lives in should be self sufficient on its own and not needing to pay for utilities such as gas, electricity and water from another source. This house should also be in a place that does not harm the environment or destroy the natural habitat that is around it, such as cutting down trees and depleting wildlife. Electricity should come from an energy-generated source on ones own property such as a wind mill (powerful modern one) or solar power (panels) of some kind. This technology is advancing in leaps and bounds powerful enough to run all electrical appliances in ones house. Waterpower can also be included as well as any other ingenious method that can be thought of that generates energy. Ones water source should also come from a well on the property, or a nearby lake, pond or stream. Sterilization of water should be done anyways at all times, which includes boiling the water and distilling it (as I mentioned earlier that distilled water is probably the healthiest water to drink). Remarkably and miraculously, "water" can actually be harvested without a water source from just the air itself. A method, which turns air into water. This method is nothing new and has been used by survivalists for centuries. It's done by using a method of harvesting the moisture right out of the air itself by collecting tiny water droplets from humid air and early morning dew. These devices are called "fog fences" (or solar still or distillery) which can be used to collect water from the air in any climate, given that enough sunlight and moisture are present.

Here is what some of them look like:

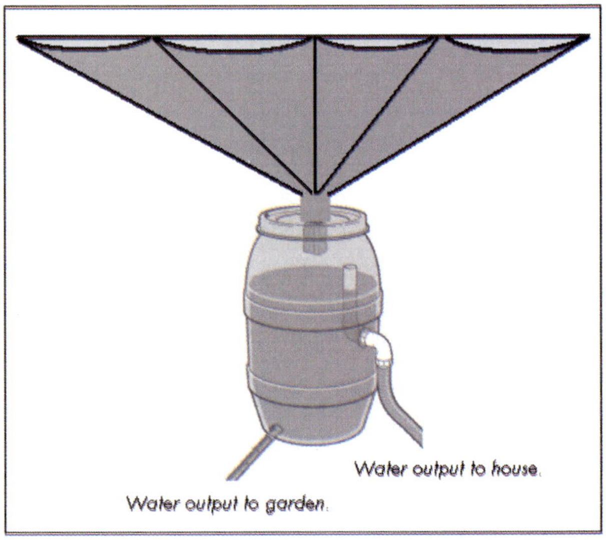

Water moisture can even be drawn out of the ground. Judges 6:38 - And that is what happened. Gideon rose early the next day, he squeezed the fleece and wrung out the dew—a bowlful of water.

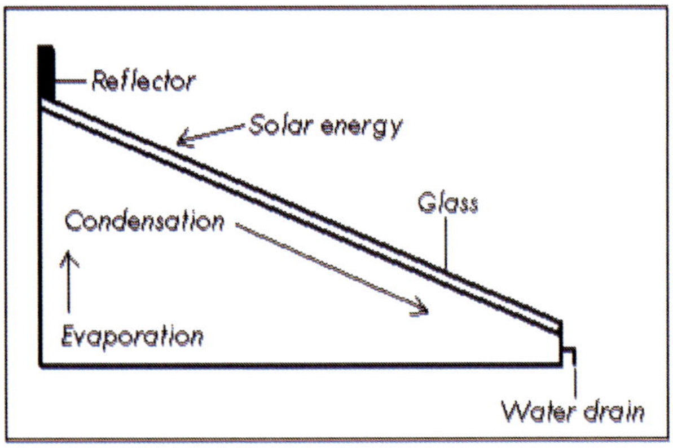

For beginners, start out with a small model to prove to yourself that IT CAN BE DONE! Then once you have acquired water with this small test, it's then just a matter of scale to generate the amount of water that you want. The more surface area, the more you will collect (we're talking gallons of water). Sooner or later one begins to realize that it would be more parctical to use ones roof of their house in which to both dew and rain which can then be directed and channeled into storage tanks or into the house or garden for direct use in water pipes for the sinks, showers, washing machine and toilets. Such a device placed elsewhere on ones property may actually provide shade in a garden (if needed for certain plants that require shade). Deuteronomy 32:2 - Let my teaching fall like rain and my words descend like dew, like showers on new grass, like abundant rain on tender plants.

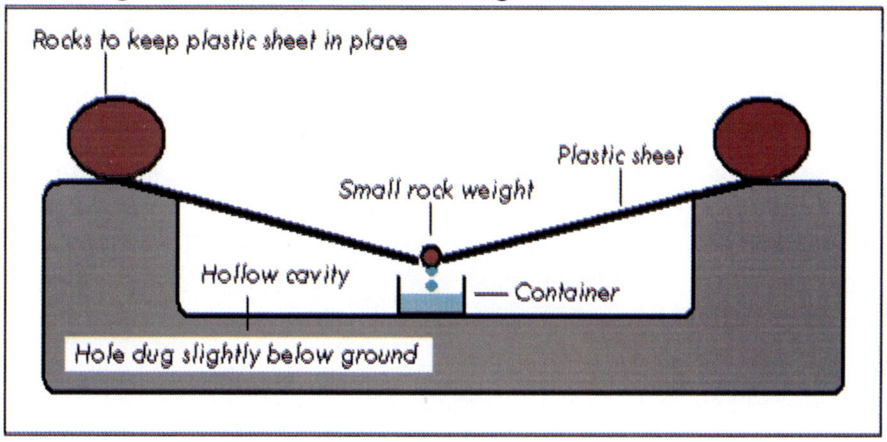

Additionally, to conserve water one can save rainwater in a rain barrel(s). This can easily be done by routing the gutters on the roof of a house directly into a rain barrel(s). Then with more gutters or hosing (and a shut off valve attached to the barrel for when water is needed) one can direct the water from the rain barrels into a garden, or can even be routed into the plumbing of the house. Heavy rains come every year around the world and its only common sense to reserve this water. Water can be saved in rain barrels for a full year (even longer) and used when needed.

Now some people think that the use of electricity is a modern invention. But as was brought up in an earlier chapter that electricity was also used by ancient people who learned that electricity could be channeled when certain metals (iron and copper) come in contact with a chemical acid (lemon juice, grape juice or vinegar) like the Baghdad Battery found in the Middle East. However, the Bible speaks about "electricity" in Job and Jeremiah's day long before Alexander Graham Bell sent the first "electrically transmitted" message in 1844. Ironically, Bells first message was: What hath G-d wrought! However, in the biblical book of Job we find that G-d is amazingly describing messages transmitted by electricity much to the similarity of Bells invention of the telephone. Job 38:35 - Do you send the lightning bolts on their way? Do they report to you, "Here we are"? In addition, in Jeremiah it tells us that G-d speaks through thunder, lightning and wind. Jeremiah 10:13 - When He thunders, the waters in the heavens roar, He makes clouds rise from the ends of the earth. He sends lightning with the rain and brings out the wind from His storehouses. Alexander Bell also invented a device greater than the telephone that enabled sound to be transmitted on a beam of light using a selenium crystal and a mirror that vibrated in response to sound. This invention is built upon the modern principle today that use lasers and fiber optics for communications.

Coming to a close here, I'd like to briefly revisit the subject of food again... It escapes my understanding why we willfully eat poisoned and processed food. Food that is laden with cancer causing pesticides, and taste enhancing and preserving chemicals, as well as bioengineered. Organic growers seem to somehow manage growing healthy foods in large quantities due largely to the fact that they are grown without pesticides; they are actually naturally stronger at resisting pests. Nevertheless, the truth of the matter is that we really don't need to grow fruits and vegetables in large quantities beyond ones immediate means. In fact we need very little to survive on (as I mentioned earlier). It is my reasoning and rational thinking that submits me to suggest that all people (young and old) should be educated on agriculture (it should be implemented into the school curriculum). Every household should have a greenhouse or a small lot of land in which to grow crops for their immediate family. Food can even be grown indoors with just a single window that allows natural sunlight in. Ultra violet sun lamps can also be used. Any excess food should be canned, dehydrated or frozen for future use, or shared or traded amongst the immediate community for other food sources in which one lacks. Money should be OBSOLETE and should not have any value to it what so ever! Living in such a community would be hugely beneficial. Life would go on as it normally does, only these people would be more self-reliant, and in a sense living a life more wealthier than some billionaires. Deuteronomy 8:17 - You may say to yourself, my power and the strength of my hands have produced this wealth for me. Matthew 6:24 - No one can serve two masters. Either he will hate the one and love the other, or he will be devoted to the one and despise the other. You cannot serve both G-d and Money. Revelation 3:17 - You say, I am rich, I have acquired wealth and do not need a thing. But you do not realize that you are wretched, pitiful, poor, blind and naked. 1 Timothy 6:17 - Command those who are rich in this present world not to be arrogant nor to put their hope in wealth, which is so uncertain, but to put their hope in G-d, who richly provides us with everything for our enjoyment.

Included in all this is knowing how to live on as little as possible. Less is more - Robert Browning (1812-1889). This includes both food as well as material things. Things like manufactured products that no one truly needs (useless items that take up space, harvest dust, and add to clutter). These useless things are usually made of plastic and end up in garbage dumps anyways. In regards to food, if we really want to stretch its limits all we need to do is go to any forest or field area and there are plenty of wild weeds, even flowers and edible foliage (leaves) that we could eat. Many people associate food with either being a fruit, vegetable or meat that we buy at the supermarket, but if one is educated enough to know what's edible and what's not, they could easily survive and thrive on any nearby residential forest. For instance, the leaves of most trees are edible including their seeds. Grass, clover, dandelion leafs, marigold flowers, roses, geraniums, clover, mint, watercress, mustard, chickweed, cedar berries, wild onion, oregano, violets, daisies, chives, thyme, daffodil flower and leaves, lilacs, wild mushrooms, plant roots, and tree sap (maple syrup) are also edible. With all these edible things, one could die of infection before starvation. The Indians found out what was edible by simply rubbing the plant on their skin. If no rash or side effect appeared, they then proceed to eat tiny bits of it, and if there were still no bad reaction, they would proceed to eat increasingly more until they discovered that it was an edible plant. The medicinal factors of many plants were also discovered in this same manner. Today many people are more interested in destroying these plants with pesticides (like dandelions) because they look bad in a lawn. However, what people don't realize is that by killing these valuable plants they are destroying the ecological environment in which they naturally grow to the point of disappearance and eventually extinction. Jeremiah 12:4 - How long will the land lie parched and the grass in every field be withered? Because those who live in it are wicked, the animals and birds have perished. Moreover, the people are saying, He will not see what happens to us. So, the next time we stroll through the woods, take a good look at all the plant life around us and appreciate every single one of them from the ugliest to the prettiest. We should not take these plants for granted, remembering that everything was put here for a reason.

Moving onward here… Transportation is also another hindrance that we must learn to get around. Aside from walking, the bicycle is the best, cheapest and healthiest means to use. However, bio-fuels, electric, water and solar cars are the things of the future. The more we veer away from gas and oil, and invest in electric, water and solar energy cars, the more technology will advance in making these cars more durable and powerful. Future technology will eventually lead in this direction especially with the threat of global warming, pollution, threatened wildlife and using up all the Earths natural resources. Additionally, it would be wise just for security purposes to stock up on bicycle tires. If in the event, a revolution ever breaks out, or an EMP (electric magnetic pulse) bomb hits, we will be helpless, rendered back to the pioneer days.

The places where we reside or live (inside our houses) should be plain and simple absent from clutter and knickknacks (idols) that are all dust collectors. 1 Samuel 12:21 - Do not turn away after useless idols. They can do you no good, nor can they rescue you, because they are useless. And as aforementioned, ones house should be self-sufficient. It should also be made of natural earthy products and with abundant windows to let in natural light. Air vents should have cheesecloth (a cotton material) within or underneath the actual vent to catch stagnant bacteria and dust particles. Inside the house, one should also have plants and a running source of water like a small fountain; this helps in cleaning and filtering the air as well as adding oxygen to it. Furniture in the house (like couches or chairs) should not be made of fabric unless it can be taken off and washed regularly. Rather furniture should be made of leather or natural woods that can be cleaned easily and efficiently with just water, hydrogen peroxide, lemon juice or olive oil diluted with water in a spray. Chemicals should not be used as they are absorbed right through the skin (which is also an organ) and into the body. All floors and bedding should be washed regularly including pillows. Carpets should be avoided as they harbor bacteria, dust mites and even mercury and lead which we track in with our shoes from the dirt, grass, or ground we walk on before we enter the house. Few people actually realize that dirt (and dust) contains enormous amounts of mercury and lead. These are natural occurring metals that were thrown into the environment by volcanic eruptions. Lead harms almost everything in our bodies from the central nervous system, the brain, the kidneys, the blood and even the reproductive system. Much in the same way, mercury causes mental retardation, developmental disabilities, brain damage, and organ damage. It is even suspected as an agent in causing cancer. That is why we must try our best to keep these metals out of our home.

In addition to our self-sufficiency, it might come in handy just to learn how to knit, crochet, weave and sew clothing (not just associated among women, but men as well). As well as the familiarity and method of knowing where the material comes from (such as cotton and wool) and learning how to spindle it into usable thread. This is an important survival skill since biblical times and is still in force today. Exodus 35:35 - He has filled them with skill to do all kinds of work as engravers, designers, embroiderers in blue, purple and scarlet yarn and fine linen, and weavers—all of them skilled workers and designers. Exodus 28:8 - Its skillfully woven waistband is to be like it—of one piece with the ephod and made with gold, and with blue, purple and scarlet yarn, and with finely twisted linen. Though most clothing manufactures do the job, it is not a lost art as some might think. One should pick up a magazine on knitting and look at all the beautifully modern styles that can be made by hand. If one has the means available to plant cotton or have several sheep as pets, they can harvest and sheer their own material knitting it into the desired fashionable clothing and never paying a penny for manufactured store bought cloths. Along with clothing, many people might be surprised at the natural elements that were once used for survival to manufacture things. In fact, there are still people of the world today (like the Amish people) who still use some of these valuable skills.

Another thing that most people frown upon is weapons (such as guns). Weapons are needed for self defense and as expressed in an earlier chapter, it is our G-d given right to protect ourselves, such as it is innately built into nature, (like a mother bear protecting her cubs). This is a right that is based off G-d's design for nature - and the same applies to us as humans. At no time should we sit by ignorantly watching our freedom being torn away from us (the right to bear arms). WEAPONS are our G-d given right to protect ourselves! Romans 1:19 - Because that which may be known of G-d is manifest in them, for G-d hath shewed it unto them. Hosea 13:8 - Like a bear robbed of her cubs, I will attack them and rip them open. Like a lion I will devour them, a wild animal will tear them apart. Job 12:7 - But ask the animals, and they will teach you, or the birds of the air, and they will tell you.

If I might also add to all of this... In a previous chapter I brought up the idea that evil was anti-Phi/Pi in nature. But evil is much more proliferated than just through nature. Evil is so cunning that it has us trapped in modern society and leaves us with almost no other option to survive without money. Money is the culprit that leads to all evil. 1 Timothy 6:10 - For the love of money is the root of all evil: which while some coveted after, they have erred from the faith, and pierced themselves through with many sorrows. Truly, I tell you the mark of the beast deals with ALL MONETARY MONEY! This includes, coins, bills, currency, savings, credit, wages, salary, assets, property (basically anything that can be converted into cash). In some cases this may even include a commodity like gold or silver that can be used to purchase goods and services. However even exchanges can be acceptable such as for food and essential commodities. Revelation 13:16-18 - He also forced everyone, small and great, rich and poor, free and slave, to receive a mark on his right hand or on his forehead, so that no one could buy or sell unless he had the mark, which is the name of the beast or the number of his name. This calls for wisdom. If anyone has insight, let him calculate the number of the beast, for it is man's number. His number is 666.

Now some people think this will be a chip (implanted in the skin somewhere), or a number (such as social security number), or credit card that is linked to a computerized on-line world banking system. But I now tell you that it is ANYTHING that deals with money! THE MARK OF THE BEAST IS ALREADY HERE! It is that number, that credit card, that check or monetary money that gives us a "FREE TICKET" to go buy things. And these are things we truly do not need. Things like that dishwasher, washing machine or drier, microwave or conventional stove, air conditioning or heating units, television, car, or processed foods that are made in a factory and sold at the grocery store. We truly do not need these luxuries. These are luxuries that Satan tells us we absolutely DO NEED to live in comfort and make things easier for ourselves. Satan tells us that these are things we need in order to exist in today's society, not only to exist and survive, but additionally to blend in, and look normal (it also involves coveting). The plan of Satan is to enslave people in any which way he can. Therefore, anything that deals with money is the mark of the beast. Satan is considered the god of Moloch (materialism). The only way we are ever going to completely escape from the mark of the beast is to be very self-sufficient where money is not needed, and has no stronghold in our lives. Money is foolishness! All we truly need is food, clothing and shelter to survive. If we look around in nature, these things are freely supplied by G-d. Job 12:7 - But ask the animals, and they will teach you, or the birds of the air, and they will tell you. Matthew 6:26 - Look at the birds of the air, they do not sow or reap or store away in barns, and yet your heavenly Father feeds them. Are you not much more valuable than they? This does not mean that we literally have to live in a cave-like dwelling, similar to animals in a undignified or demeaning type way. It simply means that we should live as plainly as possible and not above our needs (less is actually better, it is also MORE). The house we live in should be simple, not to big, and not to small (depending on occupants). There should not be an excess of furniture or useless things like knickknacks lying around (or displayed), or things that add to clutter and dust. We should not have an excess amount of cloths either. A small wardrobe is all we need. Lets think sensibly here! All we truly need is our daily essentials.

Now for those who show off, display, and selfishly and greedily live in luxury, these luxuries will not last long anyways, since G-d tells us (in the Bible) that in the end days, we will again be driven back to the ground and the caves, as happened to men after the fall of Babel. Job 30:6 - They were forced to live in the dry stream beds, among the rocks and in holes in the ground. G-d will once again bring humankind back to the ground. He will humble the proud and lofty, the arrogant and the over-confidents of luxury and laziness. Isaiah 2:9-12 - So man will be brought low and mankind humbled—do not forgive them. Go into the rocks, hide in the ground from dread of the Lord and the splendor of his majesty! The eyes of the arrogant man will be humbled and the pride of men brought low, the Lord alone will be exalted in that day. The Lord Almighty has a day in store for all the proud and lofty, for all that is exalted (and they will be humbled). Isaiah 2:17 - The arrogance of man will be brought low and the pride of men humbled, the Lord alone will be exalted in that day. Isaiah 2:19 - And they shall go into the holes of the rocks, and into the caves of the earth, for fear of the Lord, and for the glory of his majesty, when he ariseth to shake terribly the earth. Isaiah 2:21 - To go into the clefts of the rocks, and into the tops of the ragged rocks, for fear of the Lord, and for the glory of his majesty, when he ariseth to shake terribly the earth.

However, that still leaves open the question of how we are to live today without the use of money, if indeed we still need to reside in a specific place (land) or live in a dwelling (house). Can one truly escape ALL MONEY?! Even if the house is paid off, we still have to pay TAXES on the land that is owned by the county, country or the state. What does biblical scripture have to say about this? This is what Jesus said: Matthew 17:24-27 - When they came to Capernaum, those who collected the two-drachma tax came to Peter and said, Does your teacher not pay the two-drachma tax? He said, Yes. And when he came into the house, Jesus spoke to him first, saying, What do you think, Simon? From whom do the kings of the earth collect customs or poll-tax, from their sons or from strangers? When Peter said, From strangers, Jesus said to him, Then the sons are exempt. However, so that we do not offend them, go to the sea and throw in a hook, and take the first fish that comes up, and when you open its mouth, you will find a shekel. Take that and give it to them for you and Me.

The texts of Matthew, Mark and Luke each tell of another account, when the Pharisees tried to trap Jesus in his words, and find a reason to accuse him. In Matthew 22:15-22 - Then the Pharisees went and plotted together how they might trap Him in what He said. And they sent their disciples to Him, along with the Herodians, saying, Teacher, we know that You are truthful and teach the way of G-d in truth, and defer to no one, for You are not partial to any. Tell us then, what do You think? Is it lawful to give a poll-tax to Caesar, or not? But Jesus perceived their malice, and said, Why are you testing Me, you hypocrites? Show Me the coin used for the poll-tax. And they brought Him a denarius. And He said to them, Whose likeness and inscription is this? They said to Him, Caesar's. Then He said to them, Then render to Caesar the things that are Caesar's, and to G-d the things that are G-d's. And hearing this, they were amazed, and leaving Him, they went away.

In Romans 13:1 the apostle Paul brings additional clarification to the concept of paying taxes. Romans 13:1 - Every person is to be in subjection to the governing authorities For there is no authority except from G-d, and those which exist are established by G-d. Therefore, we can conclude from this biblical verse, that if we don't pay taxes we are rebelling against the authorities established by G-d. The Bible gives us this warning if we don't pay taxes and rebel against the authority, this person is thereby rebelling against what G-d has instituted, and those who do so will bring judgment onto themselves. Romans 13:2 - Therefore whoever resists authority has opposed the ordinance of G-d, and they who have opposed will receive condemnation upon themselves.

And lastly, regarding the paying of taxes, the apostle Paul couldn't make it anymore understandable in Romans 13:5-7 - So you must submit to them, not only to avoid punishment, but also to keep a clear conscience. Pay your taxes, too, for these same reasons. For government workers need to be paid. They are serving G-d in what they do. Give to everyone what you owe them: Pay your taxes and government fees to those who collect them, and give respect and honor to those who are in authority.

Therefore, the Bible leaves no doubt that Jesus taught his followers to give to the government any taxes that are owed. However, at the same time even though we must pay taxes, we should ultimately strive to be self-sufficient without the need for money. And ultimately to trust and have confidence in the resources (of nature) that the Lord G-d gives us - just as Jesus did with tax money that was found in the mouth of the fish that first came up. It is simply ironic how many of us can relate to finding the divine provisions needed at the moment. However, even though Jesus got the tax money by divine sources there is a lesson for all of us to learn here. Matthew 17:25-27 - (Jesus was declaring Himself as King as well as the son). What do you think, Simon Jesus asked? From whom do the **kings of the earth** collect duty and taxes—from their own sons or from others? From others, Peter answered. **Then the sons are exempt, Jesus said to him.** But so that we may not offend them, go to the lake and throw out your line. Take the first fish you catch; open its mouth and you will find a four-drachma coin. Take it and give it to them for my tax and yours. Additionally before the curse, humankind had dominion over the fish. Apparently, Jesus still had this power because He was sinless. Genesis 1:28 - G-d blessed them and said to them, Be fruitful and increase in number, fill the earth and subdue it. Rule over the **fish** of the sea and the birds of the air and over every living creature that moves on the ground. Time and again, Jesus defied conventional wisdom by turning completely and exclusively to His heavenly father for all of

His needs. Through this example, Jesus taught his followers not to rely on human ways and means for their needs, but instead to rely wholeheartedly on G-d (even if this entails money).

There is also something else that must be brought forth concerning "the mark of the beast" and the great lengths that Satan will go to pull the wool over our eyes. There is a very fine deceiving line here, which must be regarded with suspicion… Satan is so good at twisting and spinning the truth, so much so that he even has us believing in "ecology" which has us worshiping the Earth in this day and age. Ecology, labeled as environmental science, environmentalism, or in other terms known as "green politics". He has the scientists up in arms, coming and going to meetings concerning global warming telling us that the earths ecosystem is being destroyed. He has us literally living in fear! Amidst all this whether its true or not, we should focus on what scripture says… Hebrews 1:3 - Who being the brightness of his glory, and the express image of his person, **and upholding all things by the word of his power,** when he had by himself purged our sins, sat down on the right hand of the Majesty on high.

In other words, G-d/Jesus is holding this planet together until sin/Satan runs its course. There is an appointed time for this mentioned in the book of Revelations wherein there will come a time when G-d will restore all things. Isaiah 65:17 - Behold, I will create new heavens and a new earth. The former things will not be remembered, nor will they come to mind. Revelation 21:1 - Then I saw a new heaven and a new earth, for the first heaven and the first earth had passed away, and there was no longer any sea. Mark 9:12 - Jesus replied, "To be sure, Elijah does come first, and restores all things. Why then is it written that the Son of Man must suffer much and be rejected? In the mean time, G-d is constantly shifting and equalizing the ecosystem as necessary to cause the environment to be satisfied in balance. These are compensated by natural cycles that are often misunderstood by scientists. Yes, there are droughts, pollutants, oil spills, even nuclear testing sites, but these are nothing the earth cannot clean up by itself. The birds are still chirping and the clouds and sunshine are still hovering and gleaming in radiance. However, we must be careful that we don't worship secular ecology; this in itself can also be a religion in which Satan is programming us into worshipping mother Earth. Search the real facts on this matter… global warming temperatures are not increasing, they are actually decreasing, or in other terms - evening out (being compensated for, or upheld by G-d). Even though scientific research leads us to believe that warming temperatures of the Earth is in fact occurring, we also have available data from past millennia that temperature changes were frequently severe and rapid long before man allegedly had any impact at all (such as in producing industrial emissions). Additionally, there are other plausible explanations for global warming temperatures that can be caused by other factors, such as volcanic activity on the earth itself, solar system effect such as sunspot activity, the Earths orbital movement, the Moons influence, etc… Therefore, we can eliminate the overrated hype that the alarmists put forth of current climate change. However, one thing we know for sure, that the Earth is indeed "wearing out" and has been continually decaying since the fall of Adam and Eve. But we must also keep in mind that while not worshiping the earth, we must also respect it. Yes, environmental issues are important, but the most important pressing issue facing mankind is finding faith in G-d.

We must also keep in mind that this whole world system (the Earth) is a system of wickedness. It is evil, satanic, corrupt, lustful and sinful in all its concepts and ways. Sadly, the whole world system is in the power of Satan. His evil ways are contrary to the word of

G-d. He hinders, subdues, and continually curtails the ways of our daily lives with G-d. We have one foot in the Church and the other in the world. 1 John 5:19 - And we know that we are of G-d, **and the whole world lieth in wickedness.** 1 John 2:15-17 - Love not the world, neither the things that are in the world. If any man love the world, the love of the Father is not in him. **For all that is in the world, the lust of the flesh, and the lust of the eyes, and the pride of life, is not of the Father, but is of the world.** And the world passeth away, and the lust thereof: but he that doeth the will of G-d abideth for ever. John 18:33-37 - Then Pilate entered into the judgment hall again, and called Jesus, and said unto him, Art thou the King of the Jews? Jesus answered him, Sayest thou this thing of thyself, or did others tell it thee of me? Pilate answered, Am I a Jew? Thine own nation and the chief priests have delivered thee unto me: what hast thou done? **Jesus answered, My kingdom is not of this world: if my kingdom were of this world, then would my servants fight, that I should not be delivered to the Jews: but now is my kingdom not from hence.** Pilate therefore said unto him, Art thou a king then? Jesus answered, **Thou sayest that I am a king. To this end was I born, and for this cause came I into the world, that I should bear witness unto the truth.** Every one that is of the truth heareth my voice.

We must become conscious of the fact that we cannot serve both G-d and Satan. Luke 16:13 - No servant can serve two masters: for either he will hate the one, and love the other; or else he will hold to the one, and despise the other. Ye cannot serve both G-d and Money.

The world is a rat race where the struggle of individuals to survive (and make progress) in the competitive environment of modern life has become dehumanizing and an ultimately futile activity. It is pure and simple foolishness. How is it that we can separate ourselves from money when it seems to be the way of the world? We get up in the morning and go to a money making job to support our family that perhaps deals with the stock market, communications, manufacturing, banking, trading, shipping, exporting, purchasing, real estate, lawyer, accountant, economist, computer technician, engineer, fashion designer, clerk, cashier, librarian, school principle, zoologist, even doctors, dentist, nurses and psychologists since these too deal with money, etc… if your going to call it a job - it deals with money, and you rely on it for survival, (there is no way around it). The simple truth is that G-d never intended us to live like this in the first place. He put us in a "garden to be farmers" and to dwell with Him!

Coming to a close here… All who have now read this chapter and have obtained this information, (and who have attempted to apply it even in some small way to their life), are now potentially and partially accountable to the next person. For that reason, before I end this chapter I would like to point out the two most important commandments that a person can do… The very first commandment that a person can do is to love G-d with all their heart, soul, mind and strength. And the second most important commandment "according to Jesus" is to love thy neighbor as thyself. Mark 12:28-34 - One of the teachers of the law came and heard the Sadducees arguing. He noticed that Jesus had given the Sadducees a good answer. So he asked him, Which is the most important of all the commandments? Jesus answered, Here is the most important one. Moses said, Israel, listen to me. The Lord is our G-d. The Lord is one. **Love the Lord your G-d with all your heart and with all your soul. Love him with all your mind and with all your strength (which is also mentioned in Deuteronomy 6:4-5). And here is the second one. Love your neighbor as you love yourself (which is also mentioned in Leviticus 19:18).** There is no commandment more important than these.

You have spoken well, teacher, the man replied. You are right in saying that G-d is one. There is no other G-d but him. To love G-d with all your heart and mind and strength is very important. So is loving your neighbor as you love yourself. These things are more important than all burnt offerings and sacrifices. Jesus saw that the man had answered wisely. He said to him, You are not far from G-d's kingdom. From then on, no one dared to ask Jesus any more questions.

Therefore, “firstly” we must always trust in the Lord. Secondly, we must learn how to become self-sufficient, thus sustaining ourselves as well as encouraging and assisting others in obtaining the skill of self-sufficiency. We must also **apply and share** all of the knowledge that we have gained and learned from this chapter (as well as this entire book). This includes caring for ourselves and our body in three specific ways - physically, mentally and spiritually (one without the other is a colossal failure). Lastly, whatever knowledge and ability a person has, they should share in helping others. This should be the Universal Law that we ALL subscribe to.

This should be our fulltime job - moneyless and self-rewarding!

19 - THE PROPER RELIGION...

This is the culminating that chapter melds all the other chapters together, and ends everything in perfect harmony!

Perhaps this should have been one of my first chapters, but I have chosen it as second to the last as to leave no reasonable doubt left. Though truly it doesn't matter in what order you read this book, whether you read it from the first chapter to the last chapter, or last to first. Matthew 19:30 - But many who are first will be last, and many who are last will be first.

Some may find this chapter difficult, challenging and intense. However, I would like to make it very clear that there is no ill intentions towards any other religion or race - nor is it intended to be of a preachy nature especially in an irritatingly or overbearingly way. I am merely summing up the knowledge and information that has accumulated over the length of my investigation into writing this book.

Lets start by examining the many religions of the world such as: (Animism, Judaism, Christianity, Hinduism, Islam, Atheism, Bahai, Buddhism, Candomble, Jainism, Jehovah's Witness, Mormon, Paganism, Rastafari, Santeria, Shinto, Sikhism, Taoism, Unitarianism, Scientology, Zoroastrianism). While each somewhat different from the other, they are ALL offshoots of the very first religion in the history of humankind - "JUDAISM". Some furiously argue this point refusing to investigate that all of Earths combined religions seem to convey the "same story" in similar ways, especially concerning the story of a worldwide flood and a Garden of Eden. Though each religion basically serves the same purpose, we find that each and every religion answers questions which all humans seem to be programmed to ask: Who is my maker, or the source of my energy from which my life flows? For this reason alone religion seems to exists and continues to do so. However, inasmuch as all these religions of the world beg the insatiable question of a maker or creator, it has been found that in all the religions of the world that ONLY "Judaism and its off-shoot Christianity" have been found to be of a confirmed "PROPHETIC NATURE". Things of a prophetic nature include passages or segments from historical biblical documents that are farsighted and foretelling of future events. More so that the script itself has a life of its own resonating in a field of dimensional quality, which I feel has been adequately demonstrated thus far in the previous chapters, but it only gets better.

Prophesy means history written in advance. Isaiah 46:9-10 - Remember the former things of old: for I am G-d, and there is none else; I am G-d, and there is none like me, **Declaring the end from the beginning**, and from ancient times the **things that are not yet done**, saying, My counsel shall stand, and I will do all my pleasure. The reason why the end is mentioned before the begging is because there were no human witnesses in the beginning, but all are witnesses in the end.

In order to provide evidence of "prophetic nature" of something that was farsighted or predicted one must stop and think here for a moment and consider the Jewish Old Testament Bible and what it claims to be? It seems that the texts themselves happen to be a "message" written by the creator to His creation. The question now arises as to how the creator authenticates His message, and how He goes about proving to a creation that He is indeed the one conveying the message. If indeed G-d is all knowing, then the primary way in which the Creator would chose to convey His message is to tell us the future before it happens. When in fact the Creator has done just that as we've witnessed in the previous chapter on the Bible Code. Though if you remember the Bible Code itself does not actually predict the future but serves to give us a number of probable outcomes, and only "after the fact" is the predicted future then seen and acknowledged. But the Creator went one step further and appointed "prophets" who He had divinely inspired to write down their message from Him of future events that HE WANTED KNOWN. (He chose man to receive His message since He Himself came in the human form of Jesus, which we will soon discuss). So not only are there hidden messages in the Bible Code, but that these codes are somehow alluded to, connected, backed, and reinforced by the predictions declared to the prophets (by G-d) in the Bible. We clearly saw some of these predictions brought forth and defined for us in the chapter of the Bible Code such as the 70 Hebrew letter skips which is an elucidation for the exact amount of Jewish festive days, and to which the Hebrew word for "the holidays" can be found within that precise location that talks about the purpose for G-d creating the stars. (Look back in the chapter if you must). In addition, lest you forget, it is by the Sun, Moon and Stars that these Jewish holidays are determined. Genesis 1:14 - And G-d said, Let there be lights in the expanse of the sky to separate the day from the night, and let them serve as signs to mark seasons and days and years. Added to this, in the New Testament, Jesus supports and attaches to the meaning of this by alluding to the number 70 as a measure of completeness. Jesus said to forgive 70 x 7 in Matthew 18:21-22 - Then Peter came to Jesus and asked, Lord, how many times shall I forgive my brother when he sins against me? Up to seven times? **Jesus answered, I tell you, not seven times, but seventy-seven times.** In other words, Jesus is saying, keep on forgiving until you are complete. Again, the number "seven" is typically associated with acts of completion, rest, renewal, fulfillment, spiritual importance and perfection whether it be seconds, minutes, hours, days, weeks, months, or years, it will almost surely culminate itself to the end of all time. The number 7 is G-d's prophetic timepiece. Jesus also goes one step further in expressing this... He proclaims Himself to be the year of the Lord's favor - which is the Grand Jubilee! Luke 4:16-21 - He (Jesus) went to Nazareth, where he had been brought up, and on the Sabbath day he went into the synagogue, as was his custom. And he stood up to read. The scroll of the prophet Isaiah was handed to him. Unrolling it, he found the place where it is written: The Spirit of the Lord is on me, because he has anointed me to **preach good news to the poor.** He has sent me to **proclaim freedom for the prisoners and recovery of sight for the blind,** to **release the oppressed,** to **PROCLAIM THE YEAR OF THE LORD'S FAVOR.** Then he rolled up the scroll, gave it back to the attendant and sat down. The eyes of everyone in the synagogue were fastened on him, and he began by saying to them, **Today this scripture is fulfilled in your hearing.** In Luke 4:16-21 Jesus was quoting the scriptures of Isaiah 61:1-2 (from the Old Testament) to fully summarize His ministry on earth, which addresses the Jubilee (which is the year of the Lord's favor) in reference to the future Messiah - Himself! Isaiah 61:1-2 - The Spirit of the Sovereign Lord is on me, because the Lord has anointed me to **preach good news to the poor.** He has sent me to **bind up the brokenhearted,** to **proclaim freedom for the captives**

and release from darkness for the prisoners, to **proclaim the year of the Lord's favor** and the day of vengeance of our G-d, to comfort all who mourn.

Additionally we must also focus our attention on who **"the Spirit of the Lord"** is. (Old Testament) Isaiah 11:2 - **And the spirit of the Lord shall rest upon him**, the spirit of wisdom and understanding, the spirit of counsel and might, the spirit of knowledge and of the fear of the Lord. (Old Testament) Isaiah 61:1-2 - **The Spirit of the Lord G-D is upon me**; because the Lord hath anointed me to preach good tidings unto the meek; he hath sent me to bind up the brokenhearted, to proclaim liberty to the captives, and the opening of the prison to them that are bound; To proclaim the acceptable year of the Lord, and the day of vengeance of our G-d; to comfort all that mourn. (New Testament) Luke 4:14-21 - **And Jesus returned in the power of the Spirit into Galilee**: and there went out a fame of him through all the region round about. And he taught in their synagogues, being glorified of all. And he came to Nazareth, where he had been brought up: and, as his custom was, he went into the synagogue on the sabbath day, and stood up for to read. And there was delivered unto him the book of the prophet Isaiah. **And when he had opened the book, he found the place where it was written, The Spirit of the Lord is upon me,** because he hath anointed me to preach the gospel to the poor; he hath sent me to heal the brokenhearted, to preach deliverance to the captives, and recovering of sight to the blind, to set at liberty them that are bruised, To preach the acceptable year of the Lord. And he closed the book, and he gave it again to the minister, and sat down. And the eyes of all them that were in the synagogue were fastened on him. **And he began to say unto them, This day is this scripture fulfilled in your ears.**

The "Spirit of the Lord" is clearly Jesus!

Now to fully understand what exactly the Jubilee is, lets revisit Leviticus 25:8-17 - Count off seven Sabbaths of years—seven times seven years—so that the seven Sabbaths of years amount to a period of forty-nine years. Then have the trumpet sounded everywhere on the tenth day of the seventh month; on the Day of Atonement sound the trumpet throughout your land. **Consecrate the fiftieth year and proclaim liberty throughout the land to all its inhabitants.** It shall be **a jubilee** for you; **each one of you is to return to his family property and each to his own clan.** The fiftieth year shall be a jubilee for you; do not sow and do not reap what grows of itself or harvest the untended vines. For it is a jubilee and is to be holy for you; eat only what is taken directly from the fields. In this Year of Jubilee **everyone is to return to his own property.** If you sell land to one of your countrymen or buy any from him, do not take advantage of each other. You are to buy from your countryman on the basis of the number of years since the Jubilee. And he is to sell to you on the basis of the number of years left for harvesting crops. When the years are many, you are to increase the price, and when the years are few, you are to decrease the price, because what he is really selling you is the number of crops. Do not take advantage of each other, but fear your G-d. I am the Lord your G-d.

So, as one can clearly see all three separate texts match up with each other (Leviticus 25:8-17, Isaiah 61:1-2 and Luke 4:16-21). In other words, the Jubilee is the Sabbath of Sabbaths, and Jesus is the Sabbath personified, (the ultimate Sabbath in person)! And with great confidence He went on proclaiming in Matthew 11:28 - Come unto me, all ye that labour and are heavy laden, and I will give you rest.

Now, as you continue to read you will discover that in many of these divinely inspired messages recorded in the Old Testament Bible, shockingly they have indeed foretold of impending events that had happened. These are not only encoded within the Bible Codes, but they are outwardly mentioned/written and seen in plain sight within the Bible. One of the most triumphant and successful prophetic prophecies is concerning the Jewish Messiah - the birth, death and resurrection of Jesus, (though the Jews refute Jesus being the Messiah, we will shortly examine the reason for why towards the end of this chapter).

The Jewish Messiah seems to be the centering and pivoting hinge in which all prophecies concerning Jesus in the Old Testament have happened in perfect fulfillment. Jesus had somehow managed to flawlessly synchronize with many of the old and new messianic prophecies. Not only did He synchronize, He is the only one who has seen to it that the prophecies were fulfilled (who but G-d can proclaim the future, let alone fulfill it - Jesus did just that! Some might think that many of these prophecies are from the New Testament that tells His story, but contrary to this we find this largely in the forecasted prophesies of the Old Testament regarding His birth, life, death, and resurrection. In addition, we find that the New Testament backs-up what the Old Testament says. While additionally within the Old Testament we are given a wealth of information of detailed descriptions that the coming Messiah will possess. However, some today still refute these prophecies and relegate them to fiction. However if you keep reading you will acknowledge that the laws of probability in this case cannot give us a reasonable explanation for how these prophecies were fulfilled, other than them being divinely inspired by G-d. Many people know nothing of what I speak about yet, while others may already have the advantage. I ask those who do not, to hold fast, because I will soon shed light on these divinely foretold predictions, and at the end of this chapter, there will be no excuse for denying that there is evidence of a supernatural being transpiring throughout the history of humankind.

Now before I begin to elate here on such "foretold prophecies" I must first bring attention to some contradicting disinformation that has surrounded the speculation of Jesus' existence. Some have even gone so far as to believe that Jesus was a fictitious story and even debate whether such a man ever existed in history. However it should be known that almost all historical figures are known by "document evidence", NOT PHYSICAL PROOF, this can also be likened to other people spoken of in literature such as Shakespeare, Julius Caesar, Alexander the Great, and even Santa Clause can be applicable here. We also have such people like Albert Einstein the great theoretical physicist of our age, and mind you a JEW, had (with great confidence) acknowledged the historical existence of Jesus. As a child I received instruction both in the Bible and in the Talmud. I am a Jew, but I am enthralled by the luminous figure of the Nazarene.... No one can read the Gospels without feeling the actual presence of Jesus. His personality pulsates in every word. No myth is filled with such life - Albert Einstein. Even the book of Isaiah in the Old Testament was known to be composed by Isaiah himself, with additional material added by later disciples of the prophet who where committed to continuing Isaiah's work. We also have to realize that throughout history, anytime a prophet showed up and testified to bad news, the people didn't want to hear it, but afterwards followers began to remember and record their experiences, in effect this is when the fulfillment of earlier prophesy can be witnessed. Now, the disciples actually wrote and completed the New Testament, which was said to be in 70AD, which is approximately 37 years later after the crucifixion of Jesus. However, the gospel was "preached" IMEDIATLLY after Jesus' death on the cross. Evidence of this can be found in 1 Corinthians 15:3 which is the earliest report of Jesus written by the apostle Paul - For I delivered unto you first of all that which I also received, how that Christ died for our sins according to the scriptures. Then three years later Paul tells us in Galatians that he went to Jerusalem to see Peter and James (Jesus' brother) for fifteen days to talk about the gospel. Galatians 1:18-19 - Then after three years I went up to Jerusalem to see Peter, and abode with him fifteen days. But other of the apostles saw I none, save James the Lord's brother. Aside from the New Testament disciples that recorded the story of Jesus, we also have recorded documents by other known historians, writers and philosophers who lived during or not long after the time of Jesus (33AD) that supply us with a document of evidence. One of these historians is a person by the name of Cornelius Tacitus who lived after Jesus from 55AD to 120AD. He was a senator and Roman historian. He writes about a man called Chrestus (a common connotation of Christ) who was executed by Pilate the Roman official of Judea during the reign of the Roman emperor Tiberias. Ironically, Tacitus was hostile toward Christianity, therefore it would not have been in his interest to record that Jesus existed, but he did anyways because it would have been disloyal to the historical record to not do so. Another Greek writer named Lucian of Samosata who is recorded to have lived within two hundred years of Jesus additionally referred to Jesus. Another Roman historian and court official was a man by the name of Suetonius. He refers to the expelling of Jews from Rome because of their activities on behalf of Jesus (called Chrestus again). Another governor by the name of Pliny in 112AD was responsible for executing Christians for not worshipping or bowing down to a statue of the Roman emperor. Another philosopher by the name of Mara Bar-Serapion who lived some time after 70 AD wrote a letter to his son describing how the Jews executed their King. However, perhaps the most famous of all, happens to be a Jewish historian named Josephus who was born 4 years after the crucifixion of Jesus in 37AD. He wrote the Jewish Antiquities describing Jesus as a wise man who did wonderful works and calling Him the Messiah. Josephus also affirmed that Jesus was executed by Pilate and had shortly after, rose from the dead. There are also two other ancient historians of indisputable origins... One was named Thallus and the other Phlegon. Both confirmed the biblical account of an earthquake and an eclipse of the Sun when Jesus took His last breath and gave up His spirit on the cross and died. We find a record of this in the New Testament of Luke 23:45 - And the sun was darkened, and the veil of the temple was rent in the midst. Also, in Matthew 27:51 - At that moment the curtain of the temple was torn in two from top to bottom. The earth shook and the rocks split. Moreover, additionally in Mark 15:38 - The curtain of the temple was torn in two from top to bottom. However, the Roman historian Thallus wrote in his journals in 52AD (before the disciples wrote the New Testament, which was said to be in 70AD) - On the whole world, there pressed a most fearful darkness, and the rocks were rent by an earthquake, and many places in Judea and other districts were thrown down. Thallus also writes - As appears to me without reason, an eclipse of the sun. Additionally, the Roman historian Phlegon born in 80AD records in 140AD - That, in the time of Tiberius Caesar, at full moon, there was a full eclipse of the sun from the sixth hour to the ninth - manifestly that one of which we speak. Phlegon also writes -The darkness that occurred in the 4th year of the 202nd Olympiad (corresponding to 32/33 A.D), There was the greatest eclipse of the sun. It became as night in the sixth hour of the day (noon) so that the stars even appeared in the heavens. There was a great earthquake in Bithynia and many things were overturned in Nicaea. The evidence of a recorded earthquake in history clearly shows that this event occurred on a date and time that agrees with the scriptures, which makes a strong case for Jesus' authenticity. It is also well documented that within the lifetime of people who knew Jesus and His disciples, that they were persecuted for believing in Him. This tells us several truths, the biggest one of all is that they believed He was real and chose to be tortured to death rather than deny Him as the Messiah. And the history of people who chose to defend this belief from the beginning of His resurrection is continuous from the first century to the present. I think you'll soon begin to see what I mean that there is a wide source of reporting as well as numerous accounts of Old Testament fulfillment with such a wide degree of uniformity that no one person could have made it up.

Now lets examine the dating of the Crucifixion... The dating has to do with understanding the Jewish calendar. Our first clue to the date of the crucifixion has to do with Jewish law and custom. Since the earth was created in 6 days and on the 7th day G-d rested, the 7th day was considered the Sabbath. The Sabbath was considered a Saturday, considering that G-d started the creation on a Sunday. The Sabbath (Saturday) was a day of complete rest where no work could be done; therefore, Friday was known as a day of preparation. It was a day when food and other things needed for Saturday were prepared in advance.

All four Gospels state that Jesus was crucified on Preparation Day, which was a Friday. Mark 15:42 - And now when the even was come, because it was the preparation, that is, the day before the sabbath. The New Testament also records that the crucifixion occurred the day before the Passover festival. This is a also an important clue because the Jewish Passover (which occurs at sundown) always begins on the 14th day of the Jewish lunar month of Nisan, which is the season of spring, which is why Easter is celebrated at that time. Therefore, the crucifixion must have occurred in a year when the 14th of Nisan fell on a Friday (Preparation Day).

Now to narrow down the year in which Jesus lived. Ancient non-Biblical historians record that Jesus was put to death by the Roman governor Pilate whom we know to have ruled from 26AD through 36AD. We also find documented in the New Testament book of John that there were 3 annual Passovers during Jesus' ministry (which incidentally tells us that Jesus was 33 years old when He died on the cross because Jewish men traditionally become Rabbis at age 30). We can then narrow down this date to two dates, which fell on a "Friday" the 14th of Nisan (according to the Jewish calendar). One, which falls on April 7th of 30AD and the other April 3rd of 33AD. It's becoming clearer that April 3rd, 33 AD is our date with the documentation of an ancient Jewish prophecy concerning the Messiah.

This prophecy took place about a half a century BEFORE the birth of Jesus, when a Jewish Old Testament prophet by the name of Daniel existed. At this time, Daniel is in captivity in Babylon when the angel Gabriel brings him a message from G-d. He tells him that the order is soon going to come forth ordering the rebuilding of Jerusalem and He informs Daniel of the coming Messiah.

Here is Daniels prophecy concerning the coming Messiah... Daniel 9:22-24 - And he informed me, and talked with me, and said, O Daniel, I am now come forth to give thee skill and understanding. At the beginning of thy supplications the commandment came forth, and I am come to shew thee; for thou art greatly beloved: therefore understand the matter, and consider the vision. **SEVENTY WEEKS** are determined upon thy people and upon thy holy city, to finish the transgression, and to make an end of sins, and to make reconciliation for iniquity, and to bring in everlasting righteousness, and to seal up the vision and prophecy, and to anoint the most Holy. The Hebrew word for "weeks" is actually plural for the word "seven", thus the term "seventy sevens" we can relate from Jewish history are "weeks of years". Then it goes on to tell us in Daniel 9:25-26 - Know therefore and understand, that from the going forth of the commandment to restore and to build Jerusalem unto the Messiah the Prince shall be **SEVEN WEEKS, AND THREESCORE AND TWO WEEKS:** the street shall be built again, and the wall, even in troublous times. And after **THREESCORE AND TWO WEEKS** shall Messiah be cut off, but not for himself: and the people of the prince that shall come shall destroy the city and the sanctuary; and the end thereof shall be with a flood, and unto the end of the war desolations are determined.

The angel Gabriel told Daniel that the rebuilding and restoring of Jerusalem would begin with a period of seven sevens (weeks), threescore sevens, plus two sevens (two weeks). The Hebrew word for "weeks" is actually plural for the word "seven", thus, the term "seventy sevens" that we can relate from Jewish history are "weeks of years". This is 69 sevens of "weeks of years" or, 69x7 = 483 years. Now 483x360 (360 degrees in a complete circle - the revolution of the stars for one year, as calculated by some) = 173,880 days.

So in essence the angel Gabriel tells Daniel that exactly 483 years from that day, (the day that the commandment is given to restore and build Jerusalem), the Messiah will present Himself to Israel (this is the first time that Jesus presented Himself to the world as the Messiah, the second time will occur at the rapture). Well, that comes out to exactly 173,880 days. This dated timeline is well known in secular history. The day was March 30th 33AD (which was exactly 5 days before Passover on April 3rd, 33 AD), the day Jesus rode into Jerusalem on a donkey presenting Himself as the Messiah. For that date in history to be contrived by Jesus beforehand is impossible! This is again stressed in the New Testament of Matthew 21:4-11 - All this was done, that it might be fulfilled which was spoken by the prophet, saying, Tell ye the daughter of Sion, Behold, thy King cometh unto thee, meek, and sitting upon an ass, and a colt the foal of an ass. And the disciples went, and did as Jesus commanded them, And brought the ass, and the colt, and put on them their clothes, and they set him thereon. And a very great multitude spread their garments in the way; others cut down branches from the trees, and strawed them in the way. And the multitudes that went before, and that followed, cried, saying, Hosanna to the son of David: Blessed is he that cometh in the name of the Lord; Hosanna in the highest. And when he was come into Jerusalem, all the city was moved, saying, Who is this? And the multitude said, This is Jesus the prophet of Nazareth of Galilee.

We also know from scripture (according to John's Gospel in the New Testament) that 6 days before, Jesus rode into Jerusalem on a Donkey presenting Himself as the Messiah on March 30th 33 AD, He went to Bethany to visit Lazarus whom He had raised from the dead. John 12:1 - Then Jesus **six days before** the Passover came to Bethany, where Lazarus was, which had been dead, whom he raised from the dead. John 12:9-13 - Much people of the Jews therefore knew that he was there: and they came not for Jesus' sake only, but that they might see Lazarus also, whom he had raised from the dead. But the chief priests consulted that they might put Lazarus also to death; Because that by reason of him many of the Jews went away, and believed on Jesus. On the next day much people that were come to the feast, when they heard that Jesus was coming to Jerusalem, **Took branches of palm trees,** and went forth to meet him, and cried, Hosanna: Blessed is the King of Israel that cometh in the name of the Lord. (This is noted today as Palm Sunday). Therefore is was on the 6th day before the Passover when Jesus came to Bethany where Lazarus was, and it was on the 5th day before the Passover that He rode into Jerusalem exactly on March 30th (which is Palm Sunday). Therefore, it is estimated that He was crucified four days later, the evening before the Passover. Based on this data, it is also suggested that Jesus' ministry lasted about 3 ½ years from roughly around 29AD until the spring of 33AD. We can estimate this based on Luke 3:1. John the Baptist began his ministry in the 15th year of Tiberious' reign in 29AD. Luke 3:1 - **In the fifteenth year of the reign of Tiberius Caesar**—when Pontius Pilate was governor of Judea, Herod tetrarch of Galilee, his brother Philip tetrarch of Iturea and Traconitis, and Lysanias tetrarch of Abilene. Jesus began His ministry shortly thereafter at the age of thirty. Luke 3:23 - Now **Jesus himself was about thirty years old when he began his ministry.** He was the son, so it was thought, of Joseph.

Nevertheless, I must stress that there is a distinction here. The Jewish calendar is based on the lunar cycle, and our modern calendar is based on the solar cycle. Therefore, in the eyes of the modern calculated calendar (which is NOT the ancient Hebrew equivalent) - the Hebrew equivalent to this occurred on Friday the 14th (Preparation day) 33AD, which is at the sundown (and the first full moon) before the morning of the 15th (which is the Jewish Passover). Mark 15:42 - And now when the even was come, **because it was the preparation, that is, the day before the Sabbath.** But the date of Friday the 14th 33AD in the eyes of the modern calculated calendar falls on April the 3rd 33AD. This can become very confusing for some.

However to simplify things, we may conclude - without absolute certainly though, but with an extremely high degree of probability that the crucifixion of Jesus fell on Friday April the 14th in the evening (at sundown) which is the Hebrew holiday of Passover (Nissan 1). Which also coincided with the first full Moon. Either way, both dates (April the 14th and April the 3rd) are accurate since they both fall on APRIL 33AD, which is the matching date for both month and year! (We can also see this if we look on a modern day calendar that the Jewish Passover occurs in the month of April). Exodus 12:2-6 - This month shall be unto you the beginning of months: it shall be the first month of the year to you. Speak ye unto all the congregation of Israel, saying, In the tenth day of this month they shall take to them every man a lamb, according to the house of their fathers, a lamb for an house: And if the household be too little for the lamb, let him and his neighbour next unto his house take it according to the number of the souls; every man according to his eating shall make your count for the lamb. Your lamb shall be without blemish, a male of the first year: ye shall take it out from the sheep, or from the goats: And ye shall keep it up until the fourteenth day of the same month: and the whole assembly of the congregation of Israel shall kill it in the evening. 2 Chronicles 8:13 - According to the daily requirement for offerings commanded by Moses for Sabbaths, New Moons and the three annual feasts—the Feast of Unleavened Bread, the Feast of Weeks and the Feast of Tabernacles.

There is also another way of figuring out the date to Daniels prophecy. Since we also know from the Bible that the Persian Emperor Artaxerxes who ruled Persia from 464-424BC issued the edict to rebuild Jerusalem on the 1st of Nissan in the 20th year of his reign (that comes out to March 5th 444BC). Daniel 9:25-26 - **Know therefore and understand, that from the going forth of the commandment to restore and to build Jerusalem unto the Messiah** the Prince shall be seven weeks, and threescore and two weeks: the street shall be built again, and the wall, even in troublous times. And after threescore and two weeks shall Messiah be cut off, but not for himself: and the people of the prince that shall come shall destroy the city and the sanctuary; and the end thereof shall be with a flood, and unto the end of the war desolations are determined. And we can also calculate this from Nehemiah 2:1-8 - And it came to pass in **the month Nisan, in the twentieth year of Artaxerxes the king,** that wine was before him: and I took up the wine, and gave it unto the king. Now I had not been beforetime sad in his presence. Wherefore the king said unto me, Why is thy countenance sad, seeing thou art not sick? this is nothing else but sorrow of heart. Then I was very sore afraid, And said unto the king, Let the king live for ever: why should not my countenance be sad, when **the city, the place of my fathers' sepulchres, lieth waste, and the gates thereof are consumed with fire?** Then the king said unto me, For what dost thou make request? So I prayed to the G-d of heaven. And I said unto the king, If it please the king, and if thy servant have found favour in thy sight, that thou wouldest send me unto Judah, unto the city of my fathers' sepulchres, **that I may build it.** And the king said unto me, (the queen also sitting by him,) For how long shall thy journey be? and when wilt thou return? So it pleased the king to send me; and I set him a time. Moreover I said unto the king, If it please the king, let letters be given me to the governors beyond the river, that they may convey me over till I come into Judah; And a letter unto Asaph the keeper of the king's forest, that he may give me timber to make beams for the gates of the palace which appertained to the house, and for the wall of the city, and for the house that I shall enter into. And the king granted me, according to the good hand of my G-d upon me. Nehemiah 2:17-18 - Then said I unto them, **Ye see the distress that we are in, how Jerusalem lieth waste, and the gates thereof are burned with fire: come, and let us build up the wall of Jerusalem,** that we be no more a reproach. Then I told them of the hand of my G-d which was good upon me; as also the king's words that he had spoken unto me. And they said, Let us rise up and build. So they strengthened their hands for this good work.

Therefore 173,880 days from March 5th 444BC ends at March 30th 33AD (which is Palm Sunday, the day Jesus rode into Jerusalem presenting Himself as the Messiah). We can also look at it this way… March 5th 444BC to March 5th 33AD = 476 years (this is because 1BC to 1AD is one year - THERE IS NO YEAR ZERO). Now, 476 x 365 days (which is the length of one calendar year) = 173,855 days. And from March 5th to March 30th is another 25 days. 173,855 days + 25 days = 173,880 days. To refine this even further for ones better understanding, lets start at March 444BC. Here is where the calendar differs… again, the Jewish calendar is based on the lunar cycle, and our modern calendar is based on the solar cycle. Therefore, in our calendar system there is 365 days a year. Therefore, if we divide our days (365) into 173,880 days we get 476 years. Now when we subtract 476 years from 444BC we arrive at 33AD (remember there is no year zero therefore 32AD is 33AD).

So, we have several valid estimation dates. One from the prophecy of Daniel, one from the disciple Matthew, and the other from the Persian Emperor Artaxerxes. This is all very convincing and compelling evidence! Though some may stringently argue over such small discrepancies, however, the time period in history is close enough. Job 6:25 - How forcible are right words! But what doth your arguing reprove?

But perhaps the most powerful evidence yet, is astronomical (an eclipse that took place). The gospels recount that the Sun was darkened on the day of the crucifixion from noon until 3:00pm in the afternoon. Ancient non-Biblical sources also confirm this. As aforementioned, Phlegon Trallianus records in his history of 32-33AD - A failure of the Sun took place greater than any previously known, and night came on at the sixth hour of the day (noon), so that stars actually appeared in the sky; and a great earthquake took place in Bithynia and overthrew the greater part of Niceaea. However, the evidence does not stop here. Amazingly to correlate with this, a final modern day discovery answers the questionable date of the crucifixion with precision. In today's scientific advancements in astronomy, we can determine exactly when historical eclipses occurred. This was found with special computer software that could pier back into the past to look below the horizon and see Earths shadow begin the eclipse. When they did this, scientists had discovered that only one Passover lunar eclipse was visible from Jerusalem while Pilate was in office. And it occurred on April 3rd, 33 AD at precisely 3:00pm in the evening (calculated using the modern day calendar). The New Testament gospels also tell the order of events. Nails were spiked through Jesus' hands and feet into the cross at 9:00am. As He was raised up. At noon and for 3 hours the sky was darkened. In the holy Temple in Jerusalem, a double lined curtain excluded men from the holy of holies. At the end of the crucifixion (Jesus' last breath when He gave up the spirit), the veil in the holy temple was torn apart from top to bottom, as an earthquake split rocks and broke open tombs. Jesus was recorded (in the New Testament) to have died at 3:00pm. After that, His body was removed from the cross before nightfall to observe the approaching Passover. Mark 15 - **And straightway in the morning** the chief priests held a consultation with the elders and scribes and the whole council, and bound Jesus, and carried him away, and delivered him to Pilate. Mark 15:25 - **And it was the third hour, and they crucified him.** Mark 15:33-39 - And when the **sixth hour** was come, there was darkness over the whole land **until the ninth hour.** And **at the ninth hour Jesus cried with a loud voice,** saying, Eloi, Eloi, lama sabachthani? which is, being interpreted, My G-d, my G-d, why hast thou forsaken me? And some of them that stood by, when they heard it, said, Behold, he calleth Elijah. And one ran and filled a sponge full of vinegar, and put it on a reed, and gave him to drink, saying, Let alone; let us see whether Elias will come to take him down. And Jesus cried with a loud voice, and gave up the ghost. And the veil of the temple was rent in twain from the top to the bottom. And when the centurion, which stood over against him, saw that he so cried out, and gave up the ghost, he said, Truly this man was the Son of G-d.

Lets get back on track here… You've briefly had the advantaged of reading (and confirming) one of the most prophetic prophecies of Daniel concerning Jesus, but before we get to exploring other "prophetic prophecies" I'd like to additionally examine another source of speculation concerning Jesus' existence "as a real person" to put to rest the skeptics at heart. This includes the vague speculation that during the alleged time that Jesus was said to live, no one bothered to write down "at that exact juncture in time" anything about the most extraordinary person in history. Perhaps the reason nobody wielded a pen in His support is that they feared ridicule and prosecution from both the Jews and the Roman government. This was a very controversial and commotional time, a period of tumult. The New Testament also elaborates on how eventually Jesus' ministry caused so much upheaval that the Romans were threatened by Him and the Jews demanded his execution. However, everything seemed to change after His crucifixion. The New Testament tells us that the disciples ran away in fear at the garden of Gethsemane documented in Matthew 26:55-56 - At that time Jesus said to the crowd, Am I leading a rebellion, that you have come out with swords and clubs to capture me? Every day I sat in the temple courts teaching, and you did not arrest me. However, this has all taken place that the writings of the prophets might be fulfilled. **Then all the disciples deserted him and fled.** However, amazingly a week after the crucifixion and resurrection of Jesus, the disciples who went into hiding are now out confidently preaching.

Afterwards, we also read about many of the disciples who were persecuted for preaching about Jesus. One such person was Jesus' brother James who was stoned to death.

In fact, this information can be found documented OUTSIDE of the Bible - for those that doubt that a man named Jesus, or His brother James and disciples truly existed. It can be found in the Josephus's Jewish Antiquities (Antiquities 20:9), which clearly states that the brother of Jesus whose name was James met his death in Jerusalem 62AD as having occurred after the death of the procurator Porcius Festus, yet before Clodius Alinus took office. It was the high priest Ananus ben Artanus who condemned James on the charge of breaking the law, and then had him executed by stoning. We also have Eusebius who quotes Josephus' account as well as Hegesippus and Clement of Alexandria saying: The Pharisees were upset at James' teachings and threw him from the summit of the Temple in Jerusalem and then stoned him as well as breaking his skull with a fuller's club. Additionally, James the brother of John was also killed. Acts 12:1-2 - Now about that time Herod the king stretched forth his hands to vex certain of the church. And he killed James the brother of John with the sword. So as we can see it's quite obviously that the disciples (and believers) were convinced that He (Jesus) was resurrected after seeing His ethereal body appear before them, they no longer doubted - they had obsolete proof! And it was now that they believed more than ever. So much so, that they figured they might also be killed, but if Jesus could beat death, then so could they.

What is also a firm-rock-solid testimonial of Jesus' death was when His resurrected body was seen and witnessed by more than 500 people, and all at the same time. 1 Corinthians 15:3-7 - For what I received I passed on to you as of first importance that Christ died for our sins according to the Scriptures, that he was buried, that he was raised on the third day according to the Scriptures, and that he appeared to Peter, and then to the Twelve. After that, **he appeared to more than five hundred of the brothers at the same time**, most of whom are still living, though some have fallen asleep. Then he appeared to James, then to all the apostles. The testimony of one or two eyewitness may have been easy to dismiss, but the testimony of more than five hundred people provides more than enough compelling evidence that Jesus was indeed raised from the dead.

Surprisingly there are even more evidence of Jesus' existence and crucifixion, and most Christians; even Jews don't know this - that Jesus was also mentioned about (and recorded) in the Jewish Talmud. The Talmud is a book of Jewish law (as aforementioned in a previous chapter) that consists of a collection of ancient Jewish writings that make up the basis of Jewish religious law. However, the Talmud is not a part of the "word of G-d", it is only a collection of rabbinical statements and arguments. Many of these legal discussions go back to the first century BC and are recorded in the Talmud from Rabbis that lived in Israel during the time of Jesus. In addition, it is within these recorded discussions that we find the Rabbis discussing Jesus. Not only in the Talmud is the documentation of Jesus found, but also in documents called the Baraitha and Tosefta, which are supplements to the Talmud.

The Talmud actually mentions Jesus by name, and even mentions the crucifixion, (Sanhedrin page 43a). Within this passage, the Rabbis are discussing how a convicted criminal (Jesus) was to be executed. It says, for 40 days before the execution of Jesus took place a Harold went out delineating that if they find Him innocent, they discharge him, but if not, He goes forth to be stoned seeing that He committed the offense. Jesus' charge was that He practiced sorcery, was a beguiler, and enticed Israel to apostasy. Apostasy as in driving Israel to the renunciation or abandonment of the Jewish faith. The Talmud authorities do not deny that Jesus worked signs and wonders, but they looked upon them as acts of sorcery. We find the same thing mentioned in the New Testament where the Pharisees speak in similar terms about Him. Mark 3:22 - And the scribes which came down from Jerusalem said, He hath Beelzebub, and by the prince of the devils casteth he out devils. Matthew 9:34 - But the Pharisees said, He casteth out devils through the prince of the devils. Matthew 12:24 - But when the Pharisees heard it, they said, This fellow doth not cast out devils, but by Beelzebub the prince of the devils. Therefore, Jesus did work signs and wonders, but they had to explain it away by saying that He did it by the powers of sorcery.

And this is the interesting thing... here we have even more proof that there indeed did exist a man named Jesus, and He was crucified - clearly documented within the rabbinical sources of the Jewish Talmud that plainly gives an account and admits with clear admission that He was indeed hung (by crucifixion) by a decision of a Jewish court and on the eve of Passover (which was a Friday - Preparation day). Not only does the Talmud admit to this, but it also clearly refers to Jesus being connected to royalty. The Hebrew phrase "mekurav le malcut" means connected to royalty. It admits that He was influential; He was connected / related with/to the government / kingdom specifically from the household of David, (the Davidic lineage). What's even more interesting concerning the Jewish Talmud... found within Sanhedrin 97a we find a chronological timeline for the Earths age (of 6,000 years old) as well as the an affirmation that the Messiah has already come some 1900 years ago. The Talmud reads something like this: The world is to stand for 6,000 years... 2,000 in confusion and void, 2,000 with the law, and for 2,000 the time of the Messiah (that was already some 1,900 years ago).

In conclusion... the Talmud contains an important cross-reference that shows the biblical account of Jesus and furthermore that it is consistent to what the New Testament tells us, which cannot be denied by the Jews (or anyone else for that matter) even in the darkest hours of Christianity.

So as is thought, the earliest reference (as aforementioned) that we have on Jesus came from the Jewish historian Josephus who was born 4 years after the crucifixion of Jesus in 37AD (though some have even claimed Josephus was a forgery). In this case, the "Talmud" may even be EARLIER credible evidence, since it was documented before, during and after Jesus' crucifixion took place. Some even claim that the Jesus the Talmud is talking about is not the Jesus we know, rather it was another Jesus since Jesus was a common name back then. The other argument is that a mixture of stories of crucified saviors or redeemers seemed to have previously existed in several religions prior to Jesus. However, as we already discovered that all religions of the world originated from one basic source, so it should not surprise us that we find similarities of the story. The other and most ludicrous profound suggestion is that the person called Jesus was never really a living breathing physical person but was rather a "metaphoric story" for the description of the Sun passing through the 12 constellations during the 4 seasons of 1 year - from the Suns death in autumn to its resurrection in spring. We can account this knowledge of symbolic rotation of the planets and stars to our first patriarchal father "Enoch" again. Some say the word for "Heaven" is described "as above" in the Stars. And it is in Heaven where some have concluded that the story of Jesus actually took place - not on earth, but in the Stars above (which is a ridiculous and laughable postulation). Yes, it is true the story is foretold by Enoch in the Stars of the coming Messiah, but it did not physically happen in the Heaven, but rather on Earth.

This "metaphoric story" goes something like this: It all begins in the constellation of Aquarius when the Sun is just below the horizon which is the January rainy season and the first harvest cycle in the Middle East region of the world. It is also correlated in reference to John the Baptist who baptized Jesus with water. That is the beginning stages of the story. John the Baptist is a representation of Aquarius, and Jesus "the Sun" whom makes His way to see His cousin John the Baptist. Shortly after we read in the New Testament scriptures that John the Baptist is put in prison, some suggest this is a parable for the constellation Aquarius moving below the horizon until it cant be seen. Come February, the Sun moves into the constellation Pisces that is a symbolic representation of two fish. We then read that Jesus visits Simon and his brother Andrew, who are two fishermen and the representation of constellation Pisces. In the month of March, the Sun moves into the constellation Aries (the Lamb or Ram). Ironically, Jesus is referred to as the Lamb of G-d. In March the sunrises directly due east which is the spring Equinox. For Christians the spring Equinox when the sun rises directly due "east" is noted as Easter - hence East-er. For Jews when the Sun passes over directly due east, this is when "Passover" is celebrated. In April, the sun passes through the constellation Taurus (the Bull). Then in May, the sun passes across the Milky Way, and afterwards follows on its path into the constellation Gemini (the twins) (the two brightest stars, Pollux and Castor). This might have been when Jesus healed two demon-possessed men (Pollux and Castor). In June, the Sun passes through Cancer (the Crab). This is the summer solstice - the longest day of the year when the days of sunlight start getter shorter. Then in July, the Sun passes through the constellation of Leo the Lion. These are the hottest days of summer. Its during this time that we find recorded that John the Baptist reappears and is beheaded in Mark 6:14 - And king Herod heard of him; (for his name was spread abroad:) and he said, That John the Baptist was risen from the dead, and therefore mighty works do shew forth themselves in him. When Aquarius went below the horizon as the Sun moved into the month of January, its then after 6 months that the constellation Aquarius begins to rise again on the other side of the horizon. As this transpires, it appears that the head of Aquarius is cut off by the horizon. And as aforementioned in an earlier chapter with the sun being born in the east, maturing at noon, and dieing at sundown - therefore east is associated with life, south with power, west with death, and north as the Sun dips below the horizon, not as an end to life but as a transitory stage of birth. In the Middle East region Aquarius moves below the western horizon in January and reappears on the eastern horizon, this was symbolic for John the Baptist rising from the dead, but not an end to life, but as a transitory stage of birth (or in this case rebirth).

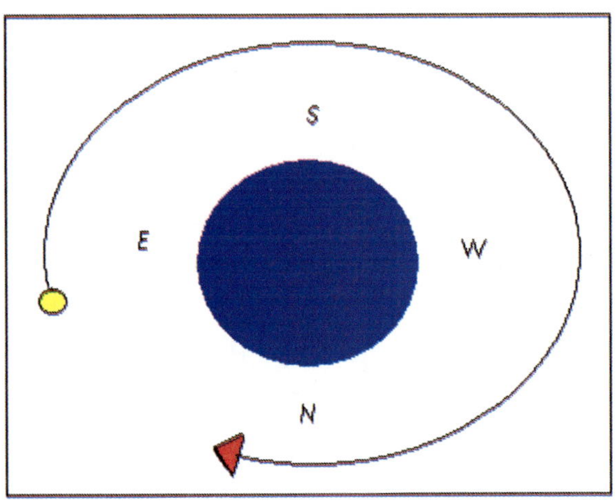

It is now the month of August when the Sun is in the constellation of Virgo (the Virgin). This is the harvest time. This is the food, which grew out of the virgin soil of the earth. Harvest time is associated with "The Kingdom of Heaven". Jesus refers to the Kingdom of Heaven coming soon in several New Testament passages. Matthew 10:7 - As you go, preach this message: The kingdom of heaven is near. Matthew 13:24 - Jesus told them another parable: The kingdom of heaven is like a man who sowed good seed in his field. Come September, October and November, after the harvest and everything has been gathered, the

Sun moves into the constellation of Libra (Libra symbolizes the scales of righteousness). It is now time to sell the harvest, and the scales of Libra are then used like money in buying and selling the harvest. When there is nothing left to sell, the money changers pack up and leave. This is also in accordance to when Jesus entered the temple in Jerusalem, overturned the tables, and drove out those who were purchasing and selling with the scales of rectitude there. This is thought to have been when the season of the harvest had passed. Mark 11:12-19 - And on the morrow, when they were come from Bethany, he was hungry: And seeing a fig tree afar off having leaves, he came, if haply he might find any thing thereon: and when he came to it, he found nothing but leaves; for the time of figs was not yet. In addition, Jesus answered and said unto it, No man eat fruit of thee hereafter for ever. And his disciples heard it. And they come to Jerusalem: and Jesus went into the temple, and began to cast out them that sold and bought in the temple, and overthrew the tables of the moneychangers, and the seats of them that sold doves; And would not suffer that any man should carry any vessel through the temple. And he taught, saying unto them, Is it not written, My house shall be called of all nations the house of prayer? but ye have made it a den of thieves. And the scribes and chief priests heard it, and sought how they might destroy him: for they feared him, because all the people was astonished at his doctrine. And when even was come, he went out of the city. Marks' recording of Jesus' parable concerning the fig tree is a very import depiction of the season of Autumn when the trees begin to dry up and the leaves fall from them. Jesus therefore said "no man eat fruit of thee hereafter for ever". Many conclude that it is simply Autumn when Jesus, (who is referred to as the Sun), causes the foliage to fall from the plants. Matthew in the New Testament also makes note of this. Matthew 21:18-22 - Early in the morning, as he was on his way back to the city, he was hungry. Seeing a fig tree by the road, he went up to, it but found nothing on it except leaves. Then he said to it, May you never bear fruit again! Immediately the tree withered. When the disciples saw this, they were amazed. How did the fig tree wither so quickly? they asked. Jesus replied, I tell you the truth, if you have faith and do not doubt, not only can you do what was done to the fig tree, but also you can say to this mountain, Go, throw yourself into the sea, and it will be done. If you believe, you will receive whatever you ask for in prayer.

So we have 3 months of Spring, 3 months of Summer, 3 months of Autumn, and 3 months of Winter, totaling 12 months in a year, and all in which is determined by the Earth revolution around the Sun. Some conclude that the Last Supper with Jesus and His twelve disciples is a reference to the 12 months and eating the harvest, which was produced by the Sun (Jesus). Matthew 26:17-30 - On the first day of the Feast of Unleavened Bread, the disciples came to Jesus and asked, Where do you want us to make preparations for you to eat the Passover? He replied, Go into the city to a certain man and tell him, The Teacher says: My appointed time is near. I am going to celebrate the Passover with my disciples at your house. So the disciples did as Jesus had directed them and prepared the Passover. When evening came, Jesus was reclining at the table with the Twelve. And while they were eating, he said, I tell you the truth, one of you will betray me. They were very sad and began to say to him one after the other, Surely not I, Lord? Jesus replied, The one who has dipped his hand into the bowl with me will betray me. The Son of Man will go just as it is written about him. But woe to that man who betrays the Son of Man! It would be better for him if he had not been born. Then Judas, the one who would betray him, said, Surely not I, Rabbi? Jesus answered, Yes, it is you. While they were eating, Jesus took bread, gave thanks and broke it, and gave it to his disciples, saying, Take and eat; this is my body. Then he took the cup, gave thanks and offered it to them, saying, drink from it, all of you. This is my blood of the covenant, which is poured out for many for the forgiveness of sins. I tell you, I will not drink of this fruit of the vine from now on until that day when I drink it anew with you in my Father's kingdom. When they had sung a hymn, they went out to the Mount of Olives. It was Judas who sold Jesus for 30 pieces of silver; he was the scale of Libra. Matthew 27:3 - When Judas, who had betrayed him, saw that Jesus was condemned, he was seized with remorse and returned the thirty silver coins to the chief priests and the elders. Matthew 27:9 - Then what was spoken by Jeremiah the prophet was fulfilled: They took the thirty silver coins, the price set on him by the people of Israel. It's actually in Zechariah that we find reference to Jeremiah. Zechariah 11:12-13 - I told them, If you think it best, give me my pay; but if not, keep it. So they paid me thirty pieces of silver. And the Lord said to me, Throw it to the potter - the handsome price at which they priced me! So I took the thirty pieces of silver and threw them into the house of the Lord to the potter.

In effect after the harvest the Sun rises lower each day, the days become shorter, colder and darker. Jesus is therefore handed over to Pilate and crucified. But it is prophecy that Jesus (the Sun) will rise again three days later. Mark 8:31 - He then began to teach them that the Son of Man must suffer many things and be rejected by the elders, chief priests and teachers of the law, and that he must be killed and after three days rise again. December 22nd is the Winter Solstice (the shortest day of the year). Three days later on December 25th (we know this to be Christmas) the Sun rises faintly (barely by a whisker) in the north representing the rising of the Sun again as the beginning cycle of the New Year.

So one might ask what does transfiguration, resurrection or rising from the dead really mean? The explanation for this can be found in Mark and Matthew when Jesus goes to a high mountain and gleams like the incredibly bright Sun. The transfiguration would have occurred during the warm summer months, but the resurrection occurred during the Winter Solstice and the rising from the dead occurred three days later when the Sun rises faintly in the north. Mark 9:2-13 - After six days Jesus took Peter, James and John with him and led them up a high mountain, where they were all alone. There **he was transfigured** before them. His clothes became dazzling white, whiter than anyone in the world could bleach them. In addition, there appeared before them Elijah and Moses, who were talking with Jesus. Peter said to Jesus, Rabbi, it is good for us to be here. Let us put up three shelters—one for you, one for Moses and one for Elijah. (He did not know what to say, they were so frightened.) Then a cloud appeared and enveloped them, and a voice came from the cloud: "This is my Son, whom I love. Listen to him!" Suddenly, when they looked around, they no longer saw anyone with them except Jesus. As they were coming down the mountain, Jesus gave them orders not to tell anyone what they had seen until the Son of Man had risen from the dead. They kept the matter to themselves, **discussing what "rising from the dead" meant.** In addition, they asked him, "Why do the teachers of the law say that Elijah must come first?" Jesus replied, "To be sure, Elijah does come first, and restores all things. Why then is it written that the Son of Man must suffer much and be rejected? But I tell you, Elijah has come, and they have done to him everything they wished, just as it is written about him. Matthew 17:1-3 - After six days Jesus took with him Peter, James and John the brother of James, and led them up a high mountain by themselves. There he was transfigured before them. **His face shone like the sun**, and his clothes became as white as the light. Just then there appeared before them Moses and Elijah, talking with Jesus.

Conclusion… For some, after reading that the story of Jesus is a "metaphoric story" of the Sun passing through the 12 constellations, and understanding what the transfiguration, resurrection and rising from the dead really symbolize, I will admit it can surely make one wonder that the real person of Jesus was merely just a metaphoric story. It seems to make considerable sense. Nevertheless, I hope that it does not in any way sway ones beliefs to the point that they should abandoned what they believe. If anything, it should be a positive note that can lead one to a more spiritual and respectful understanding of life concerning the seasons. However, I will stress how important it is that we do not forget everything that we have read or gathered thus far. We should not forget that G-d and nature are "ONE IN THE SAME" as demonstrated several times in past chapters. In rational opinion, this only leads to more credence of Jesus in His approach AS THE PERSONIFICATION OF G-D HIMSELF! Lets also not forget that the Heavens (which were created by G-d to represent (Hamoadim) the seasonal holidays) are a manifold of "AS ABOVE, SO BELOW" and serve as a type of microcosm (a small version of something larger)… Additionally, a type of reincarnation in a closed loop system. Strangely the Bible also makes note of this in Hebrews 1:3 - **The Son is the radiance of G-d's glory "and the exact representation of his being",** sustaining all things by his powerful word. After he had provided purification for sins, he sat down at the right hand of the Majesty in heaven. And let us not forget that the "initial" and foremost oldest documentation of the stars and their revealed representations to humankind were all incorporated into Enoch's divine message (from G-d) to us, and have held accuracy throughout all of biblical and cosmological history onto the present day. I'd also like to make clear here that when I just recently said above that G-d and nature are "ONE IN THE SAME"… G-d is NOT entirely nature itself; He is a separate individual entity that WORKS His ways through the workings of nature (His designed program, which is a filtering system as was discussed in a previous chapter). And this can become confusing to some people who are inclined to think in terms of the "new age/new spirituality movement" that has been sweeping the era that serves to reverse our thinking and put into our minds that "G-d is in everything one sees", and in everyone and everything, which leads directly to wrongly thinking that the recognition of G-d is the recognition of oneself (and this is Satan's reverse psychology of deception). Do not get confused here… G-D IS NOT MAN! For in this new age, spirituality teaches man that he is a god and therefore does not sin or do wrong and thus has no need of salvation. The Bible is clear that we are not god(s). Ezekiel 28:2 - Son of man, say to the ruler of Tyre, This is what the Sovereign Lord says: In the pride of your heart you say, I am a god, I sit on the throne of a G-d in the heart of the seas. **But you are a man and not a god,** though you think you are as wise as a god.

Now lets finally get to explore those other "prophetic prophecies"… Like I mentioned there are many religions of the world, but only Judaism and its off-shoot Christianity musters up enough evidence to stand up to the test of actually being "prophetic in nature". Lets start by first examining the best piece of "prophetic" evidence that we have available - the Old Testament - the text that I have continually been elaborating about. It will shortly become apparent that Jesus is the central theme, as well as the sum and substance of all biblical prophecy of the entire Bible (both Old and New Testament), (and as aforementioned also the texts of Enoch). Some say Jesus was a pretender that knew that He was not G-d. The Jews describe Him as a blaspheming rebel and a heretic who practiced sorcery. While others, say He was a "self deceived" man who sincerely believed He was the Messiah, savior and the son of G-d. However, prophecy will prove this to be otherwise, that He WAS INDEED who He claimed to be! I also find it ironical that Jewish rabbi's know the New Testament better than most Christians do. They acknowledge Jesus (especially from the Talmud), but they say they don't believe that He was the Messiah. They say Jesus was a great teacher, a brilliant and honorable Jew that should have never been crucified. But I will soon open the blockage of scripture leaving those religious men sitting there looking at the Torah scrolls and saying, oh maybe we got this wrong! The Jews for years have misinterpreted the last words spoken by Jesus on the cross when He said: My G-d, my G-d why hast thou forsaken me? Which in the Jewish mind, was said because He could not save Himself from dying on the cross and therefore was not the Messiah nor did He have the power of G-d behind Him? However, they neglect to realize that Jesus was reciting the LAST prophesy about Himself. He was quoting the prophecy about Himself in Psalms 22 surrounding the crucifixion of His death, which declares that they pierced His hands and His feet and that they cast lots upon His garments. Psalms 22:1-31 - **My G-d, my G-d, why hast thou forsaken me?** why art thou so far from helping me, and from the words of my roaring? My G-d, I cry in the day time, but thou hearest not; and in the night season, and am not silent. But thou art holy, O thou that inhabitest the praises of Israel. Our fathers trusted in thee: they trusted, and thou didst deliver them. They cried unto thee, and were delivered: they trusted in thee, and were not confounded. But I am a worm, and no man; a reproach of men, and despised of the people. All they that see me laugh me to scorn: they shoot out the lip, they shake the head, saying, He trusted on the Lord that he would deliver him: let him deliver him, seeing he delighted in him. But thou art he that took me out of the womb: thou didst make me hope when I was upon my mother's breasts. I was cast upon thee from the womb: thou art my G-d from my mother's belly. Be not far from me; for trouble is near; for there is none to help. Many bulls have compassed me: strong bulls of Bashan have beset me round. They gaped upon me with their mouths, as a ravening and a roaring lion. I am poured out like water, and all my bones are out of joint: my heart is like wax; it is melted in the midst of my bowels. My strength is dried up like a potsherd; and my tongue cleaveth to my jaws; and thou hast brought me into the dust of death. For dogs have compassed me: the assembly of the wicked have inclosed me: **THEY PIERCED MY HANDS AND MY FEET.** I may tell all my bones: they look and stare upon me. **THEY PART MY GARMENTS AMONG THEM, AND CAST LOTS UPON MY VESTURE.** But be not thou far from me, O Lord: O my strength, haste thee to help me. Deliver my soul from the sword; my darling from the power of the dog. Save me from the lion's mouth: for thou hast heard me from the horns of the unicorns. I will declare thy name unto my brethren: in the midst of the congregation will I praise thee. Ye that fear the Lord, praise him; all ye the seed of Jacob, glorify him; and fear him, all ye the seed of Israel. For he hath not despised nor abhorred the affliction of the afflicted; neither hath he hid his face from him; but when he cried unto him, he heard. My praise shall be of thee in the great congregation: I will pay my vows before them that fear him. The meek shall eat and be satisfied: they shall praise the Lord that seek him: your heart shall live for ever. All the ends of the world shall remember and turn unto the Lord: and all the kindreds of the nations shall worship before thee. For the kingdom is the Lord's: and he is the governor among the nations. All they that be fat upon earth shall eat and worship: all they that go down to the dust shall bow before him: and none can keep alive his own soul. A seed shall serve him; it shall be accounted to the Lord for a generation. They shall come, and shall declare his righteousness unto a people that shall be born, that he hath done this.

The gambling of His garments is indeed a documented event that took place with Jesus during the crucifixion which was recorded in John 19:23-24 - Then the soldiers, when they had crucified Jesus, took his garments, and made four parts, to every soldier a part; and also his coat: now the coat was without seam, woven from the top throughout. They said therefore among themselves, Let us not rend it, but cast lots for it, whose it shall be: that the scripture might be fulfilled, which saith, They parted my raiment among them, and for my vesture they did cast lots. These things therefore the soldiers did.

But if one read the whole segment of Psalms 22:1-31 and compares it to Isaiah 53, there are several similar phrases that ring out at us about what happened to Jesus in His last hours and the purpose He served. Such as: A reproach of men, and despised of the people... All they that see me laugh me to scorn… They shoot out the lip… They shake the head, saying, He trusted on the Lord that he would deliver him… Let him deliver him, seeing he delighted in him… But thou art he that took me out of the womb… They gaped upon me with their mouths, as a ravening and a roaring lion… I am poured out like water, and all my bones are out of joint… My heart is like wax, it is melted in the midst of my bowels… My strength is dried up like a potsherd, and my tongue cleaveth to my jaws; and thou hast brought me into the dust of death… For dogs have compassed me: the assembly of the wicked have inclosed me… Deliver my soul from the sword… I will declare thy name unto my brethren… All ye the seed of Jacob, glorify him… For he hath not despised nor abhorred the affliction of the afflicted; neither hath he hid his face from him… The meek shall eat and be satisfied… All the ends of the world shall remember and turn unto the Lord… All the kindreds of the nations shall worship before thee… He is the governor among the nations... All they that go down to the dust shall bow before him… None can keep alive his own soul... They shall come, and shall declare his righteousness unto a people that shall be born, that he hath done this. Isaiah 53 - Who hath believed our report? and to whom is the arm of the Lord revealed? For he shall grow up before him as a tender plant, and as a root out of a dry ground: he hath no form nor comeliness; and when we shall see him, there is no beauty that we should desire him. He is despised and rejected of men; a man of sorrows, and acquainted with grief: and we hid as it were our faces from him; he was despised, and we esteemed him not. Surely he hath borne our griefs, and carried our sorrows: yet we did esteem him stricken, smitten of G-d, and afflicted.

But he was wounded for our transgressions, he was bruised for our iniquities: the chastisement of our peace was upon him; and with his stripes we are healed. All we like sheep have gone astray; we have turned every one to his own way; and the Lord hath laid on him the iniquity of us all. He was oppressed, and he was afflicted, yet he opened not his mouth: he is brought as a lamb to the slaughter, and as a sheep before her shearers is dumb, so he openeth not his mouth. He was taken from prison and from judgment: and who shall declare his generation? for he was cut off out of the land of the living: for the transgression of my people was he stricken. And he made his grave with the wicked, and with the rich in his death; because he had done no violence, neither was any deceit in his mouth. Yet it pleased the Lord to bruise him; he hath put him to grief: when thou shalt make his soul an offering for sin, he shall see his seed, he shall prolong his days, and the pleasure of the Lord shall prosper in his hand. He shall see of the travail of his soul, and shall be satisfied: by his knowledge shall my righteous servant justify many; for he shall bear their iniquities. Therefore will I divide him a portion with the great, and he shall divide the spoil with the strong; because he hath poured out his soul unto death: and he was numbered with the transgressors; and he bare the sin of many, and made intercession for the transgressors.

However, this is only the beginning… Here are more biblical correspondences that tie in with Jesus as the foretold Messiah… In the New Testament in John 1:10-11 - He was in the world, and the world was made by him, and the world knew him not. He came unto his own, and his own received him not. John 7:5 -For neither did his brethren believe in him. John

7:48 - Have any of the rulers or of the Pharisees believed on him? John 13:18 - I speak not of you all: I know whom I have chosen: **but that the scripture may be fulfilled, He that eateth bread with me hath lifted up his heel against me.** Then we find foretold in the Old Testament in Psalm 41:9 - Yea, mine own familiar friend, in whom I trusted, which did eat of my bread, **hath lifted up his heel against me.**

There is even more prophecy to contend with... In the Old Testament of Zechariah it is foretold that the Messiah will be sold for 30 pieces of silver. Zechariah 11:12 - And I said unto them, If ye think good, give me my price; and if not, forbear. So they weighed for my price **thirty pieces of silver.** Then in the New Testament Matthew recounts the 30 pieces of silver. Matthew 26:14-15 - Then one of the twelve, called Judas Iscariot, went unto the chief priests, And said unto them, What will ye give me, and I will deliver him unto you? And they covenanted with him for **thirty pieces of silver.**

Again, we find in the Old Testament of Isaiah that the Messiah will be silenced before His accusers. Isaiah 53:7 - He was oppressed, and he was afflicted, yet he opened not his mouth: he is brought as a lamb to the slaughter, and as a sheep before her shearers is dumb, so he openeth not his mouth. Again in the New Testament Matthew recounts this foretold prophecy. Matthew 27:12-14 - And when he was accused of the chief priests and elders, he answered nothing. Then said Pilate unto him, Hearest thou not how many things they witness against thee? And he answered him to never a word; insomuch that the governor marvelled greatly.

And yet again it is foretold first in the Old Testament and then again in the New Testament that the Messiah will be mocked... Psalm 22:7-8 - All they that see me laugh me to scorn: they shoot out the lip, they shake the head, saying, He trusted on the Lord that he would deliver him: let him deliver him, seeing he delighted in him. Matthew 27:31 - And after that they had mocked him, they took the robe off from him, and put his own raiment on him, and led him away to crucify him.

It was additionally foretold in the Old Testament, and then again in the New Testament that the Messiah will be beaten... Isaiah 52:14 - As many were astonied at thee; his visage was so marred more than any man, and his form more than the sons of men. Matthew 27:26 - Then released he Barabbas unto them: and when he had scourged Jesus, he delivered him to be crucified.

And additionally it is mentioned that He will also be spit upon in the Old Testament and then again in the New Testament... Isaiah 50:6 - I gave my back to the smiters, and my cheeks to them that plucked off the hair: I hid not my face from shame and spitting. Matthew 27:30 - And they spit upon him, and took the reed, and smote him on the head.

It is also mentioned that the Messiah will have His hands and feet pieced which was foretold in the Old Testament and then again in the New Testament... Psalm 22:16 - For dogs have compassed me: the assembly of the wicked have inclosed me: they pierced my hands and my feet. Matthew 27:31 - And after that they had mocked him, they took the robe off from him, and put his own raiment on him, and led him away to crucify him.

The Bible also makes mention of the Messiah being crucified with "thieves" in both the Old Testament as well as the New Testament... Isaiah 53:12 - Therefore will I divide him a portion with the great, and he shall divide the spoil with the strong; because he hath poured out his soul unto death: and he was numbered with the transgressors; and he bare the sin of many, and made intercession for the transgressors. Matthew 27:38 - Then were there two thieves crucified with him, one on the right hand, and another on the left.

It also tells us both in the Old Testament as well as the New Testament that the Messiah will pray for His persecutors… Isaiah 53:12 - Therefore will I divide him a portion with the great, and he shall divide the spoil with the strong; because he hath poured out his soul unto death: and he was numbered with the transgressors; and he bare the sin of many, and made intercession for the transgressors. Luke 23:34 - Then said Jesus, Father, forgive them; for they know not what they do. And they parted his raiment, and cast lots.

During Jesus crucifixion it makes mention that He was pierced in His side. This was both foretold in the Old Testament as well as the New Testament… Zechariah 12:10 - And I will pour upon the house of David, and upon the inhabitants of Jerusalem, the spirit of grace and of supplications: and they shall look upon me whom they have pierced, and they shall mourn for him, as one mourneth for his only son, and shall be in bitterness for him, as one that is in bitterness for his firstborn. John 19:34 - But one of the soldiers with a spear pierced his side, and forthwith came there out blood and water. There is also significant scientific evidence for the account of Jesus' pierced side spewing both blood and water. This happened when the Roman soldier verified that Jesus was dead and pierced His side John told us that the blood was separated from the water when it poured out. The only way that blood in the body could divide from water is if circulation has stopped and the blood begins to clot and separate. It is questionable whether or not John knew this to be evidence of death. He most likely didn't and only reported on what he saw. Therefore, I don't think he had fabricated what had actually happened.

The Bible also tells us that Jesus was given gall and vinegar to drink during His crucifixion, but this was also foretold in the Old Testament, first in Psalms and then we find it noted again in the New Testament of Matthew and Luke. Psalm 69:21 - They gave me also gall for my meat; and in my thirst they gave me vinegar to drink. Matthew 27:34 - They gave him vinegar to drink mingled with gall: and when he had tasted thereof, he would not drink. Luke 23:36 - And the soldiers also mocked him, coming to him, and offering him vinegar.

The Bible also makes mention that the Messiah will have no broken bones. This was found already foretold in the Old Testament of Psalms and reiterated again in the New Testament of John. Psalm 34:20 - He keepeth all his bones: not one of them is broken. John 19:32-36 - Then came the soldiers, and brake the legs of the first, and of the other which was crucified with him. But when they came to Jesus, and saw that he was dead already, they brake not his legs: But one of the soldiers with a spear pierced his side, and forthwith came there out blood and water. And he that saw it bare record, and his record is true: and he knoweth that he saith true, that ye might believe. For these things were done, that the scripture should be fulfilled, A bone of him shall not be broken.

The Messiah was foretold to have been buried in a rich man's tomb. Documented in the Old Testament of Isaiah and then later in the New Testament by Matthew. Isaiah 53:9 - And he made his grave with the wicked, and with the rich in his death; because he had done no violence, neither was any deceit in his mouth. Matthew 27:57-60 - When the even was come, there came a rich man of Arimathaea, named Joseph, who also himself was Jesus' disciple: He went to Pilate, and begged the body of Jesus. Then Pilate commanded the body to be delivered. And when Joseph had taken the body, he wrapped it in a clean linen cloth, And laid it in his own new tomb, which he had hewn out in the rock: and he rolled a great stone to the door of the sepulchre, and departed.

It was also foretold in the New Testament of Psalms that the Messiah would rise from the dead, and this can be found recited again in the New Testament of Mark and Acts. Psalm 16:10 - For thou wilt not leave my soul in hell; neither wilt thou suffer thine Holy One to see corruption. Mark 16:6 - And he saith unto them, Be not affrighted: Ye seek Jesus of Nazareth, which was crucified: he is risen; he is not here: behold the place where they laid him. Acts 2:31 - He seeing this before spake of the resurrection of Christ, that his soul was not left in hell, neither his flesh did see corruption.

It was also foretold that the Messiah would ascend into Heaven both in the Old Testament of Psalms and found recited again in the New Testament in Acts. Psalm 68:18 - Thou hast ascended on high, thou hast led captivity captive: thou hast received gifts for men; yea, for the rebellious also, that the Lord G-d might dwell among them. Acts 1:9 - And when he had spoken these things, while they beheld, he was taken up; and a cloud received him out of their sight.

It was also foretold that the Messiah sits at the right hand of G-d both in the Old Testament of Psalms, and recited again in the New Testament in Hebrews. Psalm 110:1 - The Lord said unto my Lord, Sit thou at my right hand, until I make thine enemies thy footstool. Hebrews 1:3 - Who being the brightness of his glory, and the express image of his person, and upholding all things by the word of his power, when he had by himself purged our sins, sat down on the right hand of the Majesty on high.

As we can see, a great deal of Old Testament prophecies were made by Isaiah, but who really was Isaiah and how and why was he able to be privy to such prophecies concerning the coming Messiah? Isaiah lived before Ezekiel and Jeremiah. He was a prophet of the southern kingdom of Judah and Jerusalem. He became a prophet in the year that King Uzziah died in 740 BC, during a critical period in the history of the Jewish nation. G-d appointed Isaiah to warn the people in Judah and to give reference to the coming Messiah (the anointed one). The vision that he interpreted as his call to service is recorded in chapter 6 of the Book of Isaiah in which he responds with the words "Here am I. Send me!" Isaiah 6 - In the year that king Uzziah died I saw also the Lord sitting upon a throne, high and lifted up, and his train filled the temple. Above it stood the seraphims: each one had six wings; with twain he covered his face, and with twain he covered his feet, and with twain he did fly. And one cried unto another, and said, Holy, holy, holy, is the Lord of hosts: the whole earth is full of his glory. And the posts of the door moved at the voice of him that cried, and the house was filled with smoke. Then said I, Woe is me! for I am undone; because I am a man of unclean lips, and I dwell in the midst of a people of unclean lips: for mine eyes have seen the King, the Lord of hosts. Then flew one of the seraphims unto me, having a live coal in his hand, which he had taken with the tongs from off the altar: And he laid it upon my mouth, and said, Lo, this hath touched thy lips; and thine iniquity is taken away, and thy sin purged. Also I heard the voice of the Lord, saying, Whom shall I send, and who will go for us? **Then said I, Here am I; send me.** And he said, Go, and tell this people, Hear ye indeed, but understand not; and see ye indeed, but perceive not. Make the heart of this people fat, and make their ears heavy, and shut their eyes; lest they see with their eyes, and hear with their ears, and understand with their heart, and convert, and be healed. Then said I, Lord, how long? And he answered, Until the cities be wasted without inhabitant, and the houses without man, and the land be utterly desolate, And the Lord have removed men far away, and there be a great forsaking in the midst of the land. But yet in it shall be a tenth, and it shall return, and shall be eaten: as a teil tree, and as an oak, whose substance is in them, when they cast their leaves: so the holy seed shall be the substance thereof. Incidentally, Bible writers like Isaiah did not actually know what they were writing - they were guided by the Holy Spirit. In effect, G-d wrote through Isaiah dictating these words to him - Isaiah was just the transcriber. In addition, when Isaiah wrote something, he most likely looked at what he wrote in amazement. What is remarkable is that Isaiah wrote about the Jewish Messiah (Jesus) hundreds of year before He was even born. And we have scripture to back up that the Bible writers were effectively guided by the Holy Spirit. 2 Peter 1:16 - For we have not followed cunningly devised fables, when we made known unto you the power and coming of our Lord Jesus Christ, but were eyewitnesses of his majesty. 2 Peter 1:20-21 - Knowing this first, that **no prophecy of the scripture is of any private interpretation. For the prophecy came not in old time by the will of man: but holy men of G-d spake as they were moved by the Holy Ghost.** Acts 27:14-15 - But not long after there arose against it a tempestuous wind, called Euroclydon. And when the ship was caught, and could not bear up into the wind, **we let her drive.**

Now lets turn our attention to one of the most "prophetic prophecies" that Jesus quoted in hindsight about "Himself", which can be found in both the Old Testament as well as the New Testament. Deuteronomy 18:15 (Moses speaking) - The Lord your G-d will raise up for you a prophet like me from among your own brothers. You must listen to him. And now found within the New Testament starting in Luke 24:44 - And He said unto them, These are the words which I spake unto you, while I was yet with you, that all things must be fulfilled, which were written in the law of Moses, and in the prophets, and in the psalms, concerning me. Matthew follows up with this and tells us in Matthew 5:18 - I tell you the truth, until heaven and earth disappear, not the smallest letter, not the least stroke of a pen, will by any means disappear from the Law until everything is accomplished. Again, we find recorded in John that Moses wrote of Him. John 5:46 - For had ye believed Moses, ye would have believed me; for he wrote of me.

Still, added onto all this documentation, this is the one that wont let sleeping actors lie. Jesus' divinity becomes immediately apparent that it was Him (in trinity) who talked to Moses. John 8:58 - Jesus said unto them, Verily, verily, I say unto you, Before Abraham was, **I AM.** Exodus 3:14 - And G-d said unto Moses, **I AM THAT I AM:** and He said, Thus shalt thou say unto the children of Israel, I AM hath sent me unto you.

It can be substantiated that in fact Jesus supported the truth of the Old Testament, as well as quoting the book of Enoch as aforementioned in an earlier chapter. As I mentioned earlier the book of Enoch was later rejected and abandoned by the Jews because it contained prophecies pertaining to Jesus as the Messiah, which they considered blasphemous and improper that a man could possess the embodiment of G-d. We know this because it was found documented by many of the early church fathers (Arthenagoras, Clement of Alexandra, Irenaeus and Tertullian) who wrote that the book of Enoch had been rejected by the Jews because it contained prophecies pertaining to Jesus. Here we find in the New Testament Jude reviewing us about Enoch and substantiating Jesus as the Messiah. Jude 1:14-15, (which is also the exact words (quoting) from the book of Enoch 1:9) - And Enoch also, the seventh from Adam, prophesied of these, saying, Behold, the Lord cometh with ten thousands of his saints to execute judgment upon all, and to convince all that are ungodly among them of all their ungodly deeds which they have ungodly committed, and of all their hard speeches, which ungodly sinners have spoken against him. (Enoch 1:9 has also been found quoted within the Qumran Dead Sea Scrolls). It has been estimated that the book of Enoch is quoted at least 128 times in the New Testament. Additionally, in Enoch's book of Parables / Similitude's (37-71) it informs us of a figure called "the Messiah" - "the Holy One" - "the righteous one" - "the chosen one" and "the son of man". Parables/Similitude's 37.1 - The second vision which he saw, the vision of wisdom which Enoch the son of Jared, the son of Mahalalel, the son of Cainan, the son of Enos, the son of Seth, the son of Adam, saw. And this is the beginning of the words of wisdom which I lifted up my voice to speak and say to those which dwell on earth: Hear, you men of old time, and see, you that come after, the words of **the HOLY ONE** which I will speak before the Lord of Spirits.

Let us not forget or loose scope of the knowledge that Enoch was given a divine message from G-d in which he later encompassed into the Great Pyramid, which was a calendar that foretold the coming of the Messiah. Therefore, it is fitting for Jesus to have quoted what Enoch said about Him.

In the book of Daniel (from the Old Testament) it also denotes the Messiah as "the son of man". Daniel 7:13-14 - I saw in the night visions, and, behold, one like the **SON OF MAN** came with the clouds of heaven, and came to the Ancient of days, and they brought him near before him. And there was given him dominion, and glory, and a kingdom, that all people, nations, and languages, should serve him: his dominion is an everlasting dominion, which shall not pass away, and his kingdom that which shall not be destroyed. Not only is He mentioned as the "son of man" (the Messiah) by Daniel but He is also noted as coming in the clouds and having dominion and glory over the kingdom. This is also reiterated in the New Testament. Matthew 24:30 - At that time the sign of the **Son of Man** will appear in the sky, and all the nations of the earth will mourn. They will see the **Son of Man** coming on the clouds of the sky, with power and great glory. Matthew 26:64 - Yes, it is as you say, Jesus replied. But I say to all of you: In the future you will see the **Son of Man** sitting at the right hand of the Mighty One and coming on the clouds of heaven. Mark 13:26 - At that time men will see the **Son of Man** coming in clouds with great power and glory. Mark 14:62 - I am, said Jesus. And you will see the **Son of Man** sitting at the right hand of the Mighty One and coming on the clouds of heaven. 1 Thessalonians 4:17 - After that, we who are still alive and are left will be caught up together with them in the clouds to meet the Lord in the air. And so we will be with the Lord forever. Jude 1:12 - These men are blemishes at your love feasts, eating with you without the slightest qualm—shepherds who feed only themselves. They are clouds without rain, blown along by the wind; autumn trees, without fruit and uprooted—twice dead. Revelation 1:7 - Look, he is coming with the clouds, and every eye will see him, even those who pierced him; and all the peoples of the earth will mourn because of him. So shall it be! Amen.

Here is another passage that I found extremely moving and harmonized precisely in place. It shows how appropriate it was that Jesus fulfills prophecy. It must have also been very frustrating, and in the same breath, overwhelmingly compelling that it broke Him to tears.

John 12:37-50 - But though he had done so many miracles before them, yet they believed not on him: That the saying of Esaias (Isaiah) the prophet might be fulfilled, which he spake, Lord, who hath believed our report? and to whom hath the arm of the Lord been revealed? Therefore they could not believe, because that Esaias (Isaiah) said again, He hath blinded their eyes, and hardened their heart; that they should not see with their eyes, nor understand with their heart, and be converted, and I should heal them. These things said Esaias, when he saw his glory, and spake of him. Nevertheless among the chief rulers also many believed on him; but because of the Pharisees they did not confess him, lest they should be put out of the synagogue: For they loved the praise of men more than the praise of G-d. **JESUS CRIED AND SAID,** He that believeth on me, believeth not on me, but on him that sent me. And he that seeth me seeth him that sent me. I am come a light into the world, that whosoever believeth on me should not abide in darkness. And if any man hear my words, and believe not, I judge him not: for I came not to judge the world, but to save the world. He that rejecteth me, and receiveth not my words, hath one that judgeth him: the word that I have spoken, the same shall judge him in the last day. For I have not spoken of myself; but the Father which sent me, he gave me a commandment, what I should say, and what I should speak. And I know that his commandment is life everlasting: whatsoever I speak therefore, even as the Father said unto me, so I speak.

The list goes on… The New Testament leaps off the pages with quotes that Jesus is of G-d!.. John 6:35 - And Jesus said unto them, I am the bread of life: he that cometh to me shall never hunger; and he that believeth on me shall never thirst. John 8:12 - Then spake Jesus again unto them, saying, I am the light of the world: he that followeth me shall not walk in darkness, but shall have the light of life. This is actually a repeated insinuation of Genesis 1:3 - And G-d said, let there be light, and there was light. John 10:7 - Therefore Jesus said again, I tell you the truth, I am the gate for the sheep. He is identifying Himself as the door for all people into Heaven. John 10:11 - I am the good shepherd. The good shepherd lays down his life for the sheep. He is telling us that He is laying down His life for our sins in place of the sacrificial lamb. This is another repeated insinuation from the Exodus during Egypt when the lambs blood was put on the doorposts. He is the Passover lamb whose blood is shed for us, the blood of atonement so that deaths angel shall pass over us. John 11:25 - Jesus said to her, I am the resurrection and the life. He who believes in me will live, even though he dies. John 14:6 - Jesus answered, I am the way and the truth and the life. No one comes to the Father except through me. The Old Testament also rises up off the pages with quotes designating Jesus as the Messiah. Isaiah 41:4 - Who hath wrought and done it, calling the generations from the beginning? I the Lord, the **first,** and with the **last**; I am he. Isaiah 44:5-7 - One shall say, I am the Lord's; and another shall call himself by the name of Jacob; and another shall subscribe with his hand unto the Lord, and surname himself by the name of Israel. Thus saith the Lord the King of Israel, and his redeemer the Lord of hosts; I am the first, and I am the last; and beside me there is no G-d. And who, as I, shall call, and shall declare it, and set it in order for me, since I appointed the ancient people? and the things that are coming, and shall come, let them shew unto them. Isaiah 48:12 - Hearken unto me, O Jacob and Israel, my called; I am he; I am the first, I also am the last. Then in the New Testament in Revelation 22:13 Jesus declares again - I am the Alpha and the Omega, the **First** and the **Last**, the Beginning and the End. (This is also an indicator to us that no other prophets will come after Him, not even the Muslim prophet Muhammad that will be covered in detail towards the end of this chapter).

In the Old Testament opening in Genesis 22, He (Jesus) was the promised one. Abrahams son (Isaac) that had been promised to Abraham years ago, and who had finally arrived, despite all the odds. This is the son who would go on to have his own children (Jacob), who would in time become a great nation, which would bless the whole world.

He is the seed of the woman (which in superior logic is basically the whole reason for the gospel of Jesus in a nutshell, which will be explained more in depth shortly)… Genesis 3:15 - And I will put enmity between you and the woman, and between your offspring and hers; he will crush your head, and you will strike his heel. Genesis 21:12 - But G-d said to him, Do not be so distressed about the boy and your maidservant. Listen to whatever Sarah tells you, because it is through Isaac that your offspring will be reckoned. Ruth 4:17 - The women living there said, Naomi has a son. And they named him Obed. He was the father of Jesse, the father of David (and from David came Jesus). Isaiah 11 - A shoot will come up from the stump of Jesse; from his roots a Branch will bear fruit. Isaiah 11:10 - In that day the Root of Jesse will stand as a banner for the peoples; the nations will rally to him, and his place of rest will be glorious. Luke 3:32 - The son of David, the son of Jesse, the son of Obed, the son of Boaz, the son of Salmon, the son of Nahshon. Romans 15:12 - And again, Isaiah says, the Root of Jesse will spring up, one who will arise to rule over the nations; the Gentiles will hope in him. Recorded in Matthew 1-25 is the genealogy of Jesus - The book of the generation of Jesus Christ, the son of David, the son of Abraham. Abraham begat Isaac; and Isaac begat Jacob; and Jacob begat Judas and his brethren; And Judas begat Phares and Zara of Thamar; and Phares begat Esrom; and Esrom begat Aram; And Aram begat Aminadab; and Aminadab begat Naasson; and Naasson begat Salmon; And Salmon begat Booz of Rachab; and Booz begat Obed of Ruth; and Obed begat Jesse; And Jesse begat David the king; and David the king begat Solomon of her that had been the wife of Urias; And Solomon begat Roboam; and Roboam begat Abia; and Abia begat Asa; And Asa begat Josaphat; and Josaphat begat Joram; and Joram begat Ozias; And Ozias begat Joatham; and Joatham begat Achaz; and Achaz begat Ezekias; And Ezekias begat Manasses; and Manasses begat Amon; and Amon begat Josias; And Josias begat Jechonias and his brethren, about the time they were carried away to Babylon: And after they were brought to Babylon, Jechonias begat Salathiel; and Salathiel begat Zorobabel; And Zorobabel begat Abiud; and Abiud begat Eliakim; and Eliakim begat Azor; And Azor begat Sadoc; and Sadoc begat Achim; and Achim begat Eliud; And Eliud begat Eleazar; and Eleazar begat Matthan; and Matthan begat Jacob; And Jacob begat Joseph the husband of Mary, of whom was born Jesus, who is called Christ. **So all the generations from Abraham to David are fourteen generations; and from David until the carrying away into Babylon are fourteen generations; and from the carrying away into Babylon unto Christ are fourteen generations.** Now the birth of Jesus Christ was on this wise: When as his mother Mary was espoused to Joseph, before they came together, she was found with child of the Holy Ghost. Then Joseph her husband, being a just man, and not willing to make her a public example, was minded to put her away privily. But while he thought on these things, behold, the angel of the Lord appeared unto him in a dream, saying, Joseph, thou son of David, fear not to take unto thee Mary thy wife: for that which is conceived in her is of the Holy Ghost. And she shall bring forth a son, and thou shalt call his name JESUS: for he shall save his people from their sins. **Now all this was done, that it might be fulfilled which was spoken of the Lord by the prophet, saying, Behold, a virgin shall be with child, and shall bring forth a son, and they shall call his name Emmanuel, which being interpreted is, G-d with us**. Then Joseph being raised from sleep did as the angel of the

Lord had bidden him, and took unto him his wife: And knew her not till she had brought forth her firstborn son: and he called his name **JESUS.**

Here is the prophetic prophesy found in the Old Testament foretold by Isaiah 7:14 - Therefore the Lord himself shall give you a sign; Behold, a virgin shall conceive, and bear a son, and shall call his name Immanuel. And found backed-up within the New Testament of Matthew 1:22-23 - All this took place to fulfill what the Lord had said through the prophet: The virgin will be with child and will give birth to a son, and they will call him Immanuel—which means, G-d with us.

He is the lion of the tribe of Judah… In the New Testament book of Revelation 5:5 - Then one of the elders said to me, Do not weep! See, the Lion of the tribe of Judah, the Root of David, has triumphed. He is able to open the scroll and its seven seals. In the Old Testament in the book of Genesis the patriarch Jacob (the great grandfather of Jesus) refers to his son Judah as a young lion. Genesis 49:9 - Judah is a lion's whelp: from the prey, my son, thou art gone up: he stooped down, he couched as a lion, and as an old lion; who shall rouse him up?

In the Old Testament book of Leviticus He is the high priest, the sacrifice and the offering for our sins… Leviticus 16:31-33 - It is a sabbath of rest, and you must deny yourselves; it is a lasting ordinance. The priest who is anointed and ordained to succeed his father as high priest is to make atonement. He is to put on the sacred linen garments and make atonement for the Most Holy Place, for the Tent of Meeting and the altar, and for the priests and all the people of the community.

In the Old Testament book of Numbers, He is the star that shall rise out of Jacob - He's the scepter of Israel… Numbers 24:17 - I see him, but not now; I behold him, but not near. A star will come out of Jacob; a scepter will rise out of Israel. He will crush the foreheads of Moab, the skulls of all the sons of Sheth. Genesis 49:10 - The sceptre shall not depart from Judah, nor a lawgiver from between his feet, until Shiloh come; and unto him shall the gathering of the people be.

In Deuteronomy, Moses tells us that He's that prophet that shall come in the future that is greater than he is, and that we should listen to Him for He will supersede him. Deuteronomy 18:15-22 (Moses speaking to the whole congregation of Israel) - The Lord your G-d will raise up for you a prophet like me from among your own brothers. You must listen to him. For this is what you asked of the Lord your G-d at Horeb on the day of the assembly when you said, Let us not hear the voice of the Lord our G-d nor see this great fire anymore, or we will die. The Lord said to me: What they say is good. I will raise up for them a prophet like you from among their brothers; I will put my words in his mouth, and he will tell them everything I command him. If anyone does not listen to my words that the prophet speaks in my name, I myself will call him to account. But a prophet who presumes to speak in my name anything I have not commanded him to say, or a prophet who speaks in the name of other gods, must be put to death. You may say to yourselves, How can we know when a message has not been spoken by the Lord? If what a prophet proclaims in the name of the Lord does not take place or come true, that is a message the Lord has not spoken. That prophet has spoken presumptuously. Do not be afraid of him. So we can certainly asses from what Moses tells us that the way to tell if a message is spoken by G-d is if what the message proclaims comes to pass or takes place. If the message does not come to pass or take place then obviously, it is not true or from G-d. If we put Jesus to this test, it clearly reveals to us that what He said (and what the prophesies foretold about Him) in effect has happened; therefore, what He spoke was TRUE! (We will continue to visit these prophesies).

Its also interesting to note such similarities between both Moses and Jesus... for instance, like Moses, the Messiah would be the one appointed by G-d, a leader (a person in charge), a teacher (rabbi), a prophet, a lawgiver, a deliverer, and a human being that was to be one of G-d's chosen people (specifically a Jew) that would perform the role of mediator between G-d and man. This included speaking the words of G-d. Like Moses, the Messiah would offer Himself to die for the sins of the people. Both Moses and Jesus performed many miracles confirming their message.

As newborn baby's, both their lives were threatened by malevolent kings (Moses by Pharaoh and Jesus by Caesar), and both were supernaturally protected from harm.

Both spent their early years in Egypt.

Both cured lepers (Num 12:10-15; Matt. 8:2-3) and confronted demonic powers.

Moses lifted up the brazen serpent to heal all his people who had faith; Jesus was lifted up on the cross to heal all who would have faith in Him.

Moses appointed 70 elders to rule Israel (Num. 11:16-17); Jesus appointed 70 disciples to teach the nations (Luke 10:1-17).

Both Moses and Jesus died on a mountain. Please continue to read on.

It was also predicted in the Old Testament of Micah that the Messiah (from the lineage of David) is to be born in Bethlehem. Micah 5:2 - But thou, Bethlehem Ephratah, though thou be little among the thousands of Judah, yet out of thee shall he come forth unto me that is to be ruler in Israel; whose goings forth have been from of old, from everlasting. Then in the New Testament of Matthew and Luke it chimes in with this same message. Matthew 2:1 - Now when Jesus was born in Bethlehem of Judaea in the days of Herod the king, behold, there came wise men from the east to Jerusalem. Luke 2:4-7 - And Joseph also went up from Galilee, out of the city of Nazareth, into Judaea, unto the city of David, which is called Bethlehem; (because he was of the house and lineage of David:) To be taxed with Mary his espoused wife, being great with child. And so it was, that, while they were there, the days were accomplished that she should be delivered. And she brought forth her firstborn son, and wrapped him in swaddling clothes, and laid him in a manger; because there was no room for them in the inn.

It was also predicted in the Old Testament in the book of Hosea that Jesus would be called out of Egypt. Hosea 11:1 - When Israel was a child, then I loved him, and called my son out of Egypt. And it is insinuated again in Matthew of the New Testament. Matthew 2:14-15 - When he arose, he took the young child and his mother by night, and departed into Egypt: And was there until the death of Herod: that it might be fulfilled which was spoken of the Lord by the prophet, saying, Out of Egypt have I called my son.

When Herod (King Herod of Judea) heard about the prophecy of a new ruler over Israel, he ordered the killing of the infants… Jeremiah 31:15 - Thus saith the Lord; A voice was heard in Ramah, lamentation, and bitter weeping; Rachel weeping for her children refused to be comforted for her children, because they were not. Then we read in the New Testament in Matthew 2:16-18 - Then Herod, when he saw that he was mocked of the wise men, was exceeding wroth, and sent forth, and slew all the children that were in Bethlehem, and in all the coasts thereof, from two years old and under, according to the time which he had diligently inquired of the wise men. Then was fulfilled that which was spoken by Jeremiah the prophet, saying, In Rama was there a voice heard, lamentation, and weeping, and great mourning, Rachel weeping for her children, and would not be comforted, because they are not.

In Isaiah 40:3-5 of the Old Testament John the Baptist was foreseen leading the way of Jesus - The voice of him that crieth in the wilderness, Prepare ye the way of the Lord, make straight in the desert a highway for our G-d. Moreover, in the New Testament Matthew recounts this message again. Matthew 3:1-3 - In those days came John the Baptist, preaching in the wilderness of Judaea, And saying, Repent ye: for the kingdom of heaven is at hand. For this is he that was spoken of by the prophet Esaias (Isaiah), saying, The voice of one crying in the wilderness, Prepare ye the way of the Lord, make his paths straight.

Zechariah of the Old Testament foresees that the Messiah would enter Jerusalem as a king on a donkey. Zechariah 9:9 - Rejoice greatly, O daughter of Zion; shout, O daughter of Jerusalem: behold, thy King cometh unto thee: he is just, and having salvation; lowly, and riding upon an ass, and upon a colt the foal of an ass. In addition, in the New Testament, Matthew gives an account of this message again. Matthew 21:4-9 - All this was done, that it might be fulfilled which was spoken by the prophet, saying, Tell ye the daughter of Sion, Behold, thy King cometh unto thee, meek, and sitting upon an ass, and a colt the foal of an ass. And the disciples went, and did as Jesus commanded them, And brought the ass, and the colt, and put on them their clothes, and they set him thereon. And a very great multitude spread their garments in the way; others cut down branches from the trees, and strawed them in the way. And the multitudes that went before, and that followed, cried, saying, Hosanna to the son of David: Blessed is he that cometh in the name of the Lord; Hosanna in the highest.

Now pay careful attention to what you are about to read because it is the key element of conclusion and resolution. Moses told us that the way to tell if a message has been spoken by G-d is if what the message proclaims is true and does in fact come to pass. Deuteronomy 18:15-22 - The Lord your G-d will raise up for you a prophet like me from among your own brothers. You must listen to him. For this is what you asked of the Lord your G-d at Horeb on the day of the assembly when you said, Let us not hear the voice of the Lord our G-d nor see this great fire anymore, or we will die. The Lord said to me: What they say is good. I will raise up for them a prophet like you from among their brothers; I will put my words in his mouth, and he will tell them everything I command him. If anyone does not listen to my words that the prophet speaks in my name, I myself will call him to account. But a prophet who presumes to speak in my name anything I have not commanded him to say, or a prophet who speaks in the name of other gods, must be put to death. You may say to yourselves, How can we know when a message has not been spoken by the Lord? **If what a prophet proclaims in the name of the Lord does not take place or come true, that is a message the Lord has not spoken.** That prophet has spoken presumptuously. Do not be afraid of him.

What we find is that Jesus DID INDEED put the prophecies to the test which clearly exhibits to us that what He said (and what the prophesies foretold about Him) in effect have truly come to pass. However, some continue to refute the authenticity of these prophecies. Some say Jesus knew these prophesy of the Old Testament extremely well and that He conspired with the disciples to fulfill them. However, how does one explain how Jesus knew who would be on duty the day He was crucified so He could be nailed between two thieves? In addition, fulfilling the scriptures so that He could have them gambling for His cloths, or made them feed Him vinegar laced with gall when He said I thirst, and finally the resurrection. These are things He had no control over. To thwart this view the disbelievers argue that the scriptures were written after the fact. However, even that argument was defeated in the discovery of the Dead Sea Scrolls, which recorded the words of Isaiah at least 250 years before the coming of Jesus. But as aforementioned earlier in this chapter one of the most stunning and indisputable prophesies concerning the arrival of Jesus in Israel as the foretold Messiah had to do with the exact year, the exact month, and in fact the exact day when Jesus presented Himself to Israel as the Messiah. This was Daniels prophecy as foretold - mentioned earlier, which came out to be 173,880 days. And for those who also want to refute the book of Daniel - the entire book was found within the Dead Sea Scrolls and it matches-up with the Bibles versions that we have today.

There is no way to refute this, as the dates are well known in secular history. And as was determined, this date was April the 6th 33AD the day Jesus rode into Jerusalem on a donkey presenting Himself as the Messiah. For that date in history to be contrived is impossible! This is again stressed in the New Testament in Matthew 21:4-11 - All this was done, that it might be fulfilled which was spoken by the prophet, saying, Tell ye the daughter of Sion, Behold, thy King cometh unto thee, meek, and sitting upon an ass (Donkey), and a colt the foal of an ass. And the disciples went, and did as Jesus commanded them, And brought the ass, and the colt, and put on them their clothes, and they set him thereon. And a very great multitude spread their garments in the way; others cut down branches from the trees, and strawed them in the way. And the multitudes that went before, and that followed, cried, saying, Hosanna to the son of David: Blessed is he that cometh in the name of the Lord; Hosanna in the highest. And when he was come into Jerusalem, all the city was moved, saying, Who is this? And the multitude said, This is Jesus the prophet of Nazareth of Galilee. Zechariah 9:9 - Rejoice greatly, O daughter of Zion; shout, O daughter of Jerusalem: behold, thy King cometh unto thee: he is just, and having salvation; lowly, and riding upon an ass, and upon a colt the foal of an ass.

There is also another prophecy that comes out of Daniel's prophecy concerning the Messiah, and is linked to what Peter cites Jesus saying. In the New Testament Peter asks Jesus how many times should he forgive, and Jesus "prophetically" answers. Matthew 18:21-23 - Then came Peter to him, and said, Lord, how oft shall my brother sin against me, and I forgive him? till seven times? Jesus saith unto him, I say not unto thee, Until seven times: but, **Until seventy times seven.** Therefore is the kingdom of heaven likened unto a certain king, which would take account of his servants. This is also a prophetic prophecy that not only specifies Jesus as the Grand Jubilee as we had discussed earlier, but also that Jesus was referring to Himself as "Daniels Message" concerning the prophesized Messiah - 70x7=490. G-d had given the Jews only 490 years (483 from the decree to rebuild Jerusalem plus 7 more years to acknowledge, accept and receive their Messiah).

Now the unsurpassable question arises... How is it that Jesus conspired the time and place where He was to be born in Bethlehem so that He could ride into Jerusalem at exactly the right time of Daniels prophecy? Yes, He knew prophecy had to be fulfilled, and yes, He knew He was the Messiah, but the events that fell into place are certainly something in which He had no control over. Therefore, we must come to terms that the Old Testament along with the New Testament is a solitary framework of the foretold Messiah. It was G-d's whole agenda to make the Bible a masterpiece of "PROPHETIC KNOWLEDGE" that allows it to authenticates itself. No other religion or text (religious or non-religious) in the history of humankind is of this underlining nature. And this is what makes it a compelling piece of standing and lasting scientific evidence! By the way, while we're on the subject... these types of prophecies can also attest to people about "fictitious claims" that Jesus was married and had children. We can know from the scriptures that Jesus was not married, nor did He have children because the Bible talks exclusively about the life of Jesus in the Old Testament, and when Jesus came onto the scene in the New Testament He fulfilled the prewritten prophesies from the Old Testament. Such prophesies like... how many pieces of silver was He betrayed with. Well we know that He was betrayed with 30 pieces of silver. But why not 35 or even 50 pieces of silver? This is because Zachariah in the Old Testament said that 30 pieces of silver would be His price. Additionally when Jesus was on the cross why didn't they break His legs? They didn't break His legs because the Old Testament book of Psalms says that none of His bones will be broken. Everything that happened in the life of Jesus was foreordained in the scriptures. Nowhere in the Old Testament or the New Testament scriptures does it, (or did it) make mention that Jesus was to marry Mary Magdalene and have children with her. In addition, this is how we know it's a lie of deception - basically, false doctrine.

Sadly, what I often find is that even Christians are very misinformed about the Bible (G-d's word). They tend to twist (or spin) the Hebrew words around to fit their own premises. For instance, some Christians misinterpret that when G-d created the Earth it was not "perfect" but "very good". They attempt to change "very good" into "not perfect". They also persist to change around the true meaning of other things as well, such as in the Genesis story of creation they change an actual 24 hour day into long drawn out years, or change the meaning of the Noah's worldwide flood into just a local flood. If one allows this type of thinking, it then opens the door to evolutionary precepts. This then turns the Bible into a doctrine of devils. When we do this, we are no longer serving G-d, but man.

Folks, we must not discredit or harm the reputation of the word of G-d in this way. Some of us have one foot in the Church and the other in the world. The Bible plainly tells us that we serve either G-d or man. Matthew 6:24 - No one can serve two masters. Either he will hate the one and love the other, or he will be devoted to the one and despise the other. You cannot serve both G-d and Money. And do not think for one moment that there are not evil dis-informants that infiltrate into peoples lives with the main purpose of deceiving those away from G-d by trying to discredit His word in some way. By changing G-d's words one is then serving man. Matthew 7:13 - Enter through the narrow gate. For wide is the gate and broad is the road that leads to destruction, and many enter through it.

Lets recap here for a moment... Who is the one who is said to be our redeemer and savior, the son of man, the anointed one, the King of Kings, the Lord of Lords, the servant of the Lord and the great I AM? To Enoch (the architect of the Great Pyramid), He's the chief cornerstone. Wise men showed up saying where is He King of the Jews. Matthew immediately declares Him as the king of the Jews. The Jews rejected Him. The Roman governor Pontius Pilate condemned Him. The Romans crucified Him. Our sins killed Him. Pontius Pilate writes the declaration and places it on the cross declaring, Jesus of Nazareth as "the King of the Jews". He was deity on foot! G-d wroth in human flesh! The embodiment of the expression of G-d Himself in human form. He rose triumphant on the 3rd day and lives forever at the right hand of G-d the father. 1 Peter 3:22 - Who has gone into heaven and is at G-d's right hand—with angels, authorities and powers in submission to him. To know Jesus is to know G-d.

Aside from Jesus being our advocate and intercessor, He is also the loftiest idea in literature and philosophy - known worldwide… So much so, that the most phenomenal piece of evidence of Jesus' existence and purpose is actually the mere realization and observation that He continues to persists as the focal point of civilization. What greater proof is there then that? Jesus Christ is to me the outstanding personality of all time, all history, both as Son of G-d and as Son of Man. Everything he ever said or did has value for us today and that is something you can say of no other man, dead or alive. There is no easy middle ground to stroll upon. You either accept Jesus or reject him - Sholem Asch, Jewish author (1880-1957). We can even go one step further in realizing the crucial proof that the Bible itself explains its divine origin, for prophecy never had its origin in the will of man, but men spoke from G-d as they were carried along by the Holy Spirit which is described for us in 2 Peter 1:21 - For the prophecy came not in old time by the will of man: but holy men of G-d spake as they were moved by the Holy Ghost. Numbers 23:19 - G-d is not a man, that he should lie; neither the son of man, that he should repent: hath he said, and shall he not do it? or hath he spoken, and shall he not make it good?

And as an end to this segment of the chapter, the most startling proof that Jesus was indeed the Messiah can be found incorporated within the "Bible Code". As expected, it can be found in the most conspicuous place one would expect to find it. In the chapter itself, which declares that, the Messiah was stricken with our transgressions. Isaiah 53:6-10 - He was taken from prison and from judgment: and who shall declare his generation? for he was cut off out of the land of the living: for the transgression of my people was he stricken. And he made his grave with the wicked, and with the rich in his death; because he had done no violence, neither was any deceit in his mouth. Yet, it pleased the Lord to bruise him; he hath put him to grief: when thou shalt make his soul an offering for sin, he shall see his seed, he shall prolong his days, and the pleasure of the Lord shall prosper in his hand.

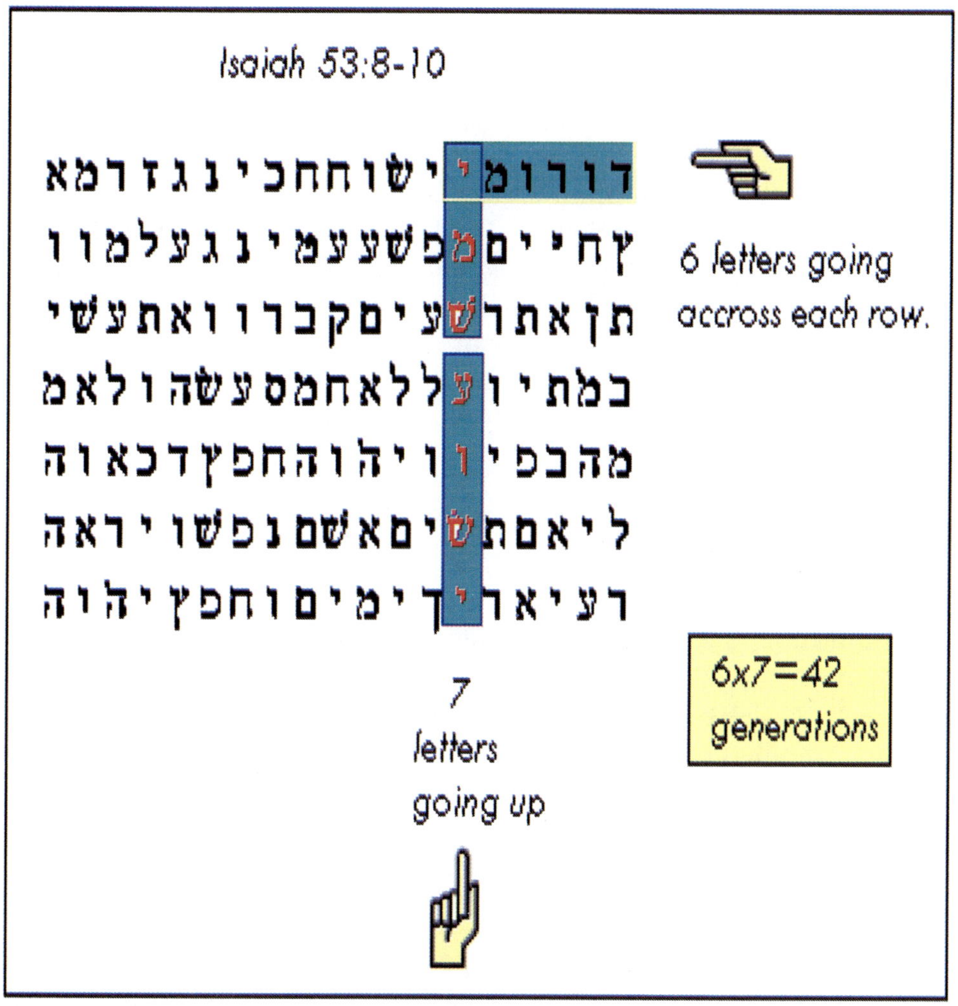

Therefore, who was the person that was stricken with our transgressions (sins)? We can find the name of this person in the "Bible Code" which can be found at every 20th Hebrew Letter starting in Isaiah 53:10 and going in reverse to Isaiah 53:8. It spells: **JESUS IS MY NAME!**

So why the number 20? The number 20 is the Gamatria equivalent to the Hebrew letter "Kaph" which means "holding or reaching hand" expressing the purpose of "SALVATION", deliverance, liberation, redemption, or rescue. Jesus' name in Hebrew literally means "salvation", so a holding or reaching hand is indicative of Jesus. The Hebrew letter "Kaph" is also positioned on the "RIGHT HAND" side between Chesed/Mercy and Netzach/victory on the "Tree Of Life" diagram. The "right hand" is also indicative of both G-d and Jesus mentioned in prophesy throughout both the Old and New Testament. Matthew 26:64 - Yes, it is as you say, Jesus replied. But I say to all of you: In the future you will see the Son of Man sitting at the **right hand** of the Mighty One and coming on the clouds of heaven. Hebrews 12:2 - Let us fix our eyes on Jesus, the author and perfecter of our faith, who for the joy set before him endured the cross, scorning its shame, and sat down at the **right hand** of the throne of G-d. 1 Peter 3:22 - who has gone into heaven and is at

G-d's **right hand**—with angels, authorities and powers in submission to him. Exodus 15:6 - Your **right hand**, O Lord, was majestic in power. Your right hand, O Lord, shattered the enemy. Isaiah 41:10 - So do not fear, for I am with you; do not be dismayed, for I am your G-d. I will strengthen you and help you; I will uphold you with my righteous **right hand.** Luke 22:69 - But from now on, the Son of Man will be seated at the **right hand** of the mighty G-d.

Off hand, while looking at the script above I also took notice of another phenomenon. Which I do not believe to be a coincidence, as I've expressed earlier that coincidence is not a kosher word, that everything is prepared properly by G-d especially His divine word. Nothing has been changed with the script or the structure or placement of the letters; they are taken directly out of the Pentateuch in the same order. What's interesting is that we can see from this same script above and shown below that there are 6 Hebrew letters before each letter of the words of "Yeshua Shme" (which means Jesus is my name), and there are 7 Hebrew letters (reading upwards) that are composed of the Hebrew words "Yeshua Shme". This appears to be another reaffirming echo of Matthews prophecy of the 6x7=42 generations of Jesus' line of descent as the coming Messiah, (which if you recall is also represented as the Star of David). Matthew 1:17 - So all the generations from Abraham to David are **fourteen generations**; and from David until the carrying away into Babylon are **fourteen generations**; and from the carrying away into Babylon unto Jesus are **fourteen generations**. (7+7) - (7+7) - (7+7) = 14 +14+14 generations... 6x7=**42** generations.

Now let us examine why the Jews do not except their Messiah...

Firstly, lets bring to light where the notion of "Messiah" actually originated. Where does the Jewish concept of Messiah come from? One of the crucial promises of biblical prophecy is the promise of a future time in history demonstrated by universal peace. Listed below in red are some of the Old Testament prophesies that the Jews sternly cling to concerning the Messiah. Isaiah 2:1-4 - The word that Isaiah the son of Amoz saw concerning Judah and Jerusalem. And it shall come to pass in the last days, that the mountain of the Lord's house shall be established in the top of the mountains, and shall be exalted above the hills; and all nations shall flow unto it. And many people shall go and say, Come ye, and let us go up to the mountain of the Lord, to the house of the G-d of Jacob; and he will teach us of his ways, and we will walk in his paths: for out of Zion shall go forth the law, and the word of the Lord from Jerusalem. And he shall judge among the nations, and shall rebuke many people: and they shall beat their swords into plowshares, and their spears into pruninghooks: nation shall not lift up sword against nation, neither shall they learn war any more. Isaiah 11:1-9 - And there shall come forth a rod out of the stem of Jesse, and a Branch shall grow out of his roots: And the spirit of the Lord shall rest upon him, the spirit of wisdom and understanding, the spirit of counsel and might, the spirit of knowledge and of the fear of the Lord; And shall make him of quick understanding in the fear of the Lord: and he shall not judge after the sight of his eyes, neither reprove after the hearing of his ears: But with righteousness shall he judge the poor, and reprove with equity for the meek of the earth: and he shall smite the earth: with the rod of his mouth, and with the breath of his lips shall he slay the wicked. And righteousness shall be the girdle of his loins, and faithfulness the girdle of his reins. The wolf also shall dwell with the lamb, and the leopard shall lie down with the kid; and the calf and the young lion and the fatling together; and a little child shall lead them. And the cow and the bear shall feed; their young ones shall lie down together: and the lion shall eat straw like the ox. And the sucking child shall play on the hole of the asp, and the weaned child shall put his hand on the cockatrice' den. They shall not hurt nor destroy in all my holy mountain: for the earth shall be full of the knowledge of the Lord, as the waters cover the sea. Isaiah 32:15-18 - Until the spirit be poured upon us from on high, and the wilderness be a fruitful field, and the fruitful field be counted for a forest. Then judgment shall dwell in the wilderness, and righteousness remain in the fruitful field. And the work of righteousness shall be peace; and the effect of righteousness quietness and assurance for ever. And my people shall dwell in a peaceable habitation, and in sure dwellings, and in quiet resting places. Isaiah 60:15-18 - Whereas thou has been forsaken and hated, so that no man went through thee, I will make thee an eternal excellency, a joy of many generations. Thou shalt also suck the milk of the Gentiles, and shalt suck the breast of kings: and thou shalt know that I the Lord am thy Saviour and thy Redeemer, the mighty One of Jacob. For brass I will bring gold, and for iron I will bring silver, and for wood brass, and for stones iron: I will also make thy officers peace, and thine exactors righteousness. Violence shall no more be heard in thy land, wasting nor destruction within thy borders; but thou shalt call thy walls Salvation, and thy gates Praise. Jeremiah 31:33-34 - But this shall be the covenant that I will make with the house of Israel; After those days, saith the Lord, I will put my law in their inward parts, and write it in their hearts; and will be their G-d, and they shall be my people. And they shall teach no more every man his neighbour, and every man his brother, saying, Know the Lord: for they shall all know me, from the least of them unto the greatest of them, saith the Lord: for I will forgive their iniquity, and I will remember their sin no more. Jeremiah 23:5-6 - Behold, the days come, saith the Lord, that I will raise unto David a righteous Branch, and a King shall reign and prosper, and shall execute judgment and justice in the earth. In his days Judah shall be saved, and Israel shall dwell safely: and this is his name whereby he shall be called, the Lord our righteousness. Jeremiah 30:7-10 - Alas! for that day is great, so that none is like it: it is even the time of Jacob's trouble, but he shall be saved out of it. For it shall come to pass in that day, saith the Lord of hosts, that I will break his yoke from off thy neck, and will burst thy bonds, and strangers shall no more serve themselves of him: But they shall serve the Lord their G-d, and David their king, whom I will raise up unto them. Therefore fear thou not, O my servant Jacob, saith the Lord; neither be dismayed, O Israel: for, lo, I will save thee from afar, and thy seed from the land of their captivity; and Jacob shall return, and shall be in rest, and be quiet, and none shall make him afraid. Jeremiah 33:14-16 - Behold, the days come, saith the Lord, that I will perform that good thing which I have promised unto the house of Israel and to the house of Judah. In those days, and at that time, will I cause the Branch of righteousness to grow up unto David; and he shall execute judgment and righteousness in the land. In those days shall Judah be saved, and Jerusalem shall dwell safely: and this is the name wherewith she shall be called, The Lord our righteousness. Zephaniah 3:9 - For then will I turn to the people a pure language, that they may all call upon the name of the Lord, to serve him with one consent. Hosea 2:20-22 - I will even betroth thee unto me in faithfulness: and thou shalt know the Lord. And it shall come to pass in that day, I will hear, saith the Lord, I will hear the heavens, and they shall hear the earth; And the earth shall hear the corn, and the wine, and the oil; and they shall hear Jezreel. Hosea 3:4-5 - For the children of Israel shall abide many days without a king, and without a prince, and without a sacrifice, and without an image, and without an ephod, and without teraphim: Afterward shall the children of Israel return, and seek the Lord their G-d, and David their king; and shall fear the Lord and his goodness in the latter days. Amos 9:13-15 - Behold, the days come, saith the Lord, that the plowman shall overtake the reaper, and the treader of grapes him that soweth seed; and the mountains shall drop sweet wine, and all the hills shall melt. And I will bring again the captivity of my people of Israel, and they shall build the waste cities, and inhabit them; and they shall plant vineyards, and drink the wine thereof; they shall also make gardens, and eat the fruit of them. And I will plant them upon their land, and they shall no more be pulled up out of their land which I have given them, saith the Lord thy G-d. Micah 4:1-4 - But in the last days it shall come to pass, that the mountain of the house of the Lord shall be established in the top of the mountains, and it shall be exalted above the hills; and people shall flow unto it. And many nations shall come, and say, Come, and let us go up to the mountain of the Lord, and to the house of the G-d of Jacob; and he will teach us of his ways, and we will walk in his paths: for the law shall go forth of Zion, and the word of the Lord from Jerusalem. And he shall judge among many people, and rebuke strong nations afar off; and they shall beat their swords into plowshares, and their spears into pruninghooks: nation shall not lift up a sword against nation, neither shall they learn war any more. But they shall sit every man under his vine and under his fig tree; and none shall make them afraid: for the mouth of the Lord of hosts hath spoken it. Zechariah 8:23 - Thus saith the Lord of hosts; In those days it shall come to pass, that ten men shall take hold out of all languages of the nations, even shall take hold of the skirt of him that is a Jew, saying, We will go with you: for we have heard that G-d is with you. Zechariah 14:9 - And the Lord shall be king over all the earth: in that day shall there be one Lord, and his name one. Ezekiel 34:11-31 - For thus saith the Lord G-d; Behold, I, even I, will both search my sheep, and seek them out. As a shepherd seeketh out his flock in the day that he is among his sheep that are scattered; so will I seek out my sheep, and will deliver them out of all places where they have been scattered in the cloudy and dark day. And I will bring them out from the people, and gather them from the countries, and will bring them to their own land, and feed them upon the mountains of Israel by the rivers, and in all the inhabited places of the country. I will feed them in a good pasture, and upon the high mountains of Israel shall their fold be: there shall they lie in a good fold, and in a fat pasture shall they feed upon the mountains of Israel. I will feed my flock, and I will cause them to lie down, saith the Lord G-d. I will seek that which was lost, and bring again that which was driven away, and will bind up that which was broken, and will strengthen that which was sick: but I will destroy the fat and the strong; I will feed them with judgment. And as for you, O my flock, thus saith the Lord G-d; Behold, I judge between cattle and cattle, between the rams and the he goats. Seemeth it a small thing unto you to have eaten up the good pasture, but ye must tread down with your feet the residue of your pastures? and to have drunk of the deep waters, but ye must foul the residue with your feet? And as for my flock, they eat that which ye have trodden with your feet; and they drink that which ye have fouled with your feet. Therefore thus saith the Lord G-d unto them; Behold, I, even I, will judge between the fat cattle and between the lean cattle. Because ye have thrust with side and with shoulder, and pushed all the diseased with your horns, till ye have scattered them abroad; Therefore will I save my flock, and they shall no more be a prey; and I will judge between cattle and cattle. And I will set up one shepherd over them, and he shall feed them, even my servant David; he shall feed them, and he shall be their shepherd. And I the Lord will be their G-d, and my servant David a prince among them; I the Lord have spoken it. And I will make with them a covenant of peace, and will cause the evil beasts to cease out of the land: and they shall dwell safely in the wilderness, and sleep in the woods. And I will make them and the places round about my hill a blessing; and I will cause the shower to come down in his season; there shall be showers of blessing. And the tree of the field shall yield her fruit, and the earth shall yield her increase, and they shall be safe in their land, and shall know that I am the Lord, when I have broken the bands of their yoke, and delivered them out of the hand of those that served themselves of them. And they shall no more be a prey to the heathen, neither shall the beast of the land devour them; but they shall dwell safely, and none shall make them afraid. And I will raise up for them a plant of renown, and they shall be no more consumed with hunger in the land, neither bear the shame of the heathen any more. Thus shall they know that I the Lord their G-d am with them, and that they, even the house of Israel, are my people, saith the Lord G-d. And ye my flock, the flock of my pasture, are men, and I am your G-d, saith the Lord G-d. Ezekiel 37:21-28 - And say unto them, Thus saith the Lord G-d; Behold, I will take the children of Israel from among the heathen, whither they be gone, and will gather them on every side, and bring them into their own land: And I will make them one nation in the land upon the mountains of Israel; and one king shall be king to them all: and they shall be no more two nations, neither shall they be divided into two kingdoms any more at all. Neither shall they defile themselves any more with their idols, nor with their detestable things, nor with any of their transgressions: but I will save them out of all their dwelling places, wherein they have sinned, and will cleanse them: so shall they be my people, and I will be their G-d. And David my servant shall be king over them; and they all shall have one shepherd: they shall also walk in my judgments, and observe my statutes, and do them. And they shall dwell in the land that I have given unto Jacob my servant, wherein your fathers have dwelt; and they shall dwell therein, even they, and their children, and their children's children for ever: and my servant David shall be their prince for ever. Moreover I will make a covenant of peace with them; it shall be an everlasting covenant with them: and I will place them, and multiply them, and will set my sanctuary in the midst of them for evermore. My tabernacle also shall be with them: yea, I will be their G-d, and they shall be my people. And the heathen shall know that I the Lord do sanctify Israel, when my sanctuary shall be in the midst of them for evermore.

Jews do not accept Jesus as the Messiah for several principle reasons… These principles reasons include that Jesus did not fulfill the messianic prophecies of what is expected of the Messiah in scripture. These messianic fulfillments include that He did not build the third temple mentioned in Ezekiel 37:26-28, and He did not gather all the Jews back to the land of Israel mentioned in Isaiah 43:5-6. Additionally, He did not usher in an age of world peace, and end all war mentioned in Isaiah 2:4. They also claim that He did not spread worldwide knowledge of the G-d of Israel, which will unite all humanity under one G-d mentioned in Zechariah 14:9. Jews quote scripture proclaiming that when the Messiah comes He will fulfill these prophecies "outright". This however can be argued that Jesus in fact DID, HAS, and IS "at present" doing all these things - as they have been fulfilled throughout the centuries and are happening in this present day and age.

Lets again examine some of these… In Matthew Jesus, prophesized the destruction of the second temple in Jerusalem that was renovated by Herod in 20AD and was later destroyed by the Romans in 70AD. Matthew 24 - And Jesus went out, and departed from the temple: and his disciples came to him for to shew him the buildings of the temple. **And Jesus said unto them, See ye not all these things? verily I say unto you, There shall not be left here one stone upon another, that shall not be thrown down.** And as he sat upon the mount of Olives, the disciples came unto him privately, saying, Tell us, when shall these things be? and what shall be the sign of thy coming, and of the end of the world?

This indeed DID HAPPEN! The holy temple as prophesized by Jesus fell 40 years later in 70AD. And in fact there was even a prophesy made around 572BC about the shutting of Jerusalem's east gate. Ezekiel 44:1-2 - Ezekiel 44:1-2 - Then he brought me back the way of **the outer gate of the sanctuary, the one facing east,** and it was shut. The Lord said to me, **This gate is to remain shut.** It must not be opened; no one may enter through it. **It is to**

remain shut because the Lord, the G-d of Israel, has entered through it. This prophecy was fulfilled when Jesus made His triumphal entry into Jerusalem on the back of a donkey announcing Himself as the Messiah (this day is known in history as Palm Sunday). His path entailed coming from the Mount of Olives and entering into the east gate of Jerusalem. Matthew 21:1-2 - As they approached Jerusalem and came to Bethphage **on the Mount of Olives**, Jesus sent two disciples, saying to them, Go to the village ahead of you, and at once you will find a donkey tied there, with her colt by her. Untie them and bring them to me. This gate led into the temple courts as it reads in the texts "the outer gate of the sanctuary which faces toward the east". Matthew 21:12 - **Jesus entered the temple area** and drove out all who were buying and selling there. He overturned the tables of the money changers and the benches of those selling doves. Following this, the city of Jerusalem was destroyed in 70AD and it remained in ruins for centuries later. After this in 1530AD the Muslim's rebuilt a wall around the city and completely walled shut the "East Gate", and thus without knowing it fulfilled the ancient prophecy of Ezekiel, "This gate is to remain shut because the Lord, the G-d of Israel, has entered through it". To this very day, the East Gate REMAINS sealed shut! This is a huge contradictory prophecy for the Jews since they believe that their Messiah has not yet come. I don't understand how it is that they can evade these two fulfilled historical facts - the destruction of Jerusalem in 70AD, along with the enclosure of the Eastern Gate. What these two prophecies prove (beyond a shadow of a doubt) is that the Messiah has already come!

Jesus also prophesized that Jews will be dispersed among the nations, and then will eventually come back to inhabit and reclaim the land again. Luke 21:24 - And they shall fall by the edge of the sword, and shall be led away captive into all nations: and Jerusalem shall be trodden down of the Gentiles, until the times of the Gentiles be fulfilled. Amos 9:14-15 - And I will bring again the captivity of my people of Israel, and they shall build the waste cities, and inhabit them; and they shall plant vineyards, and drink the wine thereof; they shall also make gardens, and eat the fruit of them. And I will plant them upon their land, and they shall no more be pulled up out of their land which I have given them, saith the Lord thy G-d.

This indeed has ALSO HAPPENED (and is currently still happening)! Israel was reborn as a nation on May 14th 1948. In addition, many Jews flooded back into the homeland, especially after the Holocaust.

It would not make sense that a Messiah would magically come (at first), and as a result in an instant change the nature of every human being on the planet. However, what does make sense is that this should transpire over a period of time as to be a WITNESS TO PROPHECY in order to "gain the belief" of the unmoving, skeptical, and disbelieving - in case we forget, the Bible has always been about prophesy! It seems to be G-d's method of teaching us. And from what we know of G-d's nature, there seems to always be a purposeful driven intent towards a major outcome that serves to restore, perfect, and complete us - just like the fulfillment of 7's.

Jesus also fulfilled the Old Testament in Jeremiah 31:31-33 (which the Jews should be adherent to), which tells us about a coming prophesy of a "New Covenant" - Jeremiah 31:31-33 - Behold, the days are coming, declares the Lord, when I will make a NEW COVENANT with the house of Israel and the house of Judah, not like the covenant that I made with their fathers on the day when I took them by the hand to bring them out of the land of Egypt, my covenant that they broke, though I was their husband, declares the Lord. But this is the covenant that I will make with the house of Israel after those days, declares the Lord: I will put my law within them, and I will write it on their hearts. And I will be their G-d, and they shall be my people.

The book of Hebrews of the New Testament reiterates this same message with a qualification to it that tells us that the "New Covenant" will be for the remission of sins. Hebrews 8:8-12 - For he finds fault with them when he says: Behold, the days are coming, declares the Lord, when I will establish a new covenant with the house of Israel and with the house of Judah, not like the covenant that I made with their fathers on the day when I took them by the hand to bring them out of the land of Egypt. For they did not continue in my covenant, and so I showed no concern for them, declares the Lord. For this is the covenant that I will make with the house of Israel after those days, declares the Lord: I will put my laws into their minds, and write them on their hearts, and I will be their G-d, and they shall be my people. And they shall not teach, each one his neighbor and each one his brother, saying, Know the Lord, for they shall all know me, from the least of them to the greatest. **For I will be merciful toward their iniquities, and I will remember their sins no more.**

However, the Jews refute this by saying that the Torah states that all religious duties and commandments remain binding forever, and anyone coming to change the Torah is immediately identified as a false prophet. Deuteronomy 13:1-4 - If there arise among you a prophet, or a dreamer of dreams, and giveth thee a sign or a wonder, And the sign or the wonder come to pass, whereof he spake unto thee, saying, let us go after other gods, which thou hast not known, and let us serve them; Thou shalt not hearken unto the words of that prophet, or that dreamer of dreams: for the Lord your G-d proveth you, to know whether ye love the Lord your G-d with all your heart and with all your soul. Ye shall walk after the Lord your G-d, and fear him, and keep his commandments, and obey his voice, and ye shall serve him, and cleave unto him. Jews who do not believe that Jesus was the Messiah cleave to this commandment, and they cite that throughout the New Testament, Jesus contradicts the Torah and states that its commandments are no longer applicable. For example in the New Testament of John which records that Jesus made a alteration in violation of Shabbat, which caused the Pharisees to say that "He does not observe Shabbat!" John 9:1-41 - And as Jesus passed by, he saw a man which was blind from his birth. And his disciples asked him, saying, Master, who did sin, this man, or his parents, that he was born blind? Jesus answered, Neither hath this man sinned, nor his parents: but that the works of G-d should be made manifest in him. **I must work the works of him that sent me, while it is day: the night cometh, when no man can work. As long as I am in the world, I am the light of the world.** When he had thus spoken, he spat on the ground, and made clay of the spittle, and he anointed the eyes of the blind man with the clay, And said unto him, Go, wash in the pool of Siloam, (which is by interpretation, Sent.) He went his way therefore, and washed, and came seeing. The neighbours therefore, and they which before had seen him that he was blind, said, Is not this he that sat and begged? Some said, This is he: others said, He is like him: but he said, I am he. Therefore said they unto him, How were thine eyes opened? He answered and said, A man that is called Jesus made clay, and anointed mine eyes, and said unto me, Go to the pool of Siloam, and wash: and I went and washed, and I received sight. Then said they unto him, Where is he? He said, I know not. They brought to the Pharisees him that aforetime was blind. **And it was the sabbath day when Jesus made the clay, and opened his eyes.** Then again the Pharisees also asked him how he had received his sight. He said unto them, He put clay upon mine eyes, and I washed, and do see. **Therefore said some of the Pharisees, This man**

is not of G-d, because he keepeth not the Sabbath day. Others said, How can a man that is a sinner do such miracles? And there was a division among them. They say unto the blind man again, What sayest thou of him, that he hath opened thine eyes? He said, He is a prophet. But the Jews did not believe concerning him, that he had been blind, and received his sight, until they called the parents of him that had received his sight. And they asked them, saying, Is this your son, who ye say was born blind? how then doth he now see? His parents answered them and said, We know that this is our son, and that he was born blind: But by what means he now seeth, we know not; or who hath opened his eyes, we know not: he is of age; ask him: he shall speak for himself. These words spake his parents, because they feared the Jews: for the Jews had agreed already, that if any man did confess that he was Christ, he should be put out of the synagogue. Therefore said his parents, He is of age; ask him. Then again called they the man that was blind, and said unto him, Give G-d the praise: we know that this man is a sinner. He answered and said, Whether he be a sinner or no, I know not: one thing I know, that, whereas I was blind, now I see. Then said they to him again, What did he to thee? how opened he thine eyes? He answered them, I have told you already, and ye did not hear: wherefore would ye hear it again? will ye also be his disciples? Then they reviled him, and said, Thou art his disciple; but we are Moses' disciples. We know that G-d spake unto Moses: as for this fellow, we know not from whence he is. The man answered and said unto them, Why herein is a marvellous thing, that ye know not from whence he is, and yet he hath opened mine eyes. Now we know that G-d heareth not sinners: but if any man be a worshipper of G-d, and doeth his will, him he heareth. Since the world began was it not heard that any man opened the eyes of one that was born blind. If this man were not of G-d, he could do nothing. They answered and said unto him, Thou wast altogether born in sins, and dost thou teach us? And they cast him out. Jesus heard that they had cast him out; and when he had found him, he said unto him, Dost thou believe on the Son of G-d? He answered and said, Who is he, Lord, that I might believe on him? And Jesus said unto him, Thou hast both seen him, and it is he that talketh with thee. And he said, Lord, I believe. And he worshipped him. And Jesus said, For judgment I am come into this world, that they which see not might see; and that they which see might be made blind. And some of the Pharisees which were with him heard these words, and said unto him, Are we blind also? Jesus said unto them, If ye were blind, ye should have no sin: but now ye say, We see; therefore your sin remaineth.

The Torah given by Moses should therefore not be exclusively linked to the conditional undeviating covenant, since the deeper covenantal purpose of G-d has always been to remove the curse (sin) from humanity. Yet, the Jews are also overlooking another irrefutable factor when they dispute that Jesus healed the blind man on the Sabbath day. Since G-d is all-powerful and unbounded, logic tells us that He who created this law is permitted to disregard it. In this instance, Jesus tells us that He IS THE LAW and had come into the world for judgment (He is obviously above the law). G-d is also above absolutes! Mankind cannot equal G-d's compassion, mercy and ability to love. In this respect, G-d is not feigning His own high principles, though the Pharisees in scriptures would have it look that way.

The list goes on … The Jews say that the Messiah has to be a descendent from King David specified in Genesis 49:10 and Isaiah 11:1, and this must be from His fathers side. Therefore, if Jesus was the product of a virgin birth as biblically claimed, he had no human father and therefore could not have fulfilled the messianic prophesy of being a descendent of his fathers side from King David. This however is ridiculous to claim such determinacy when we start to consider who is true blood or an orphan we must remember who was orphaned.

The Jews should remember that they were once slaves in Egypt. Deuteronomy 24:17-22 - Do not deprive the alien or the fatherless of justice, or take the cloak of the widow as a pledge. Remember that you were slaves in Egypt and the Lord your G-d redeemed you from there. That is why I command you to do this.

We must also remember who was really of Jewish royal blood or gentile. In addition, who was raised by whom, such as Moses, Josephs children, Esther and Ruth. Exodus 2:5-10 - Then Pharaoh's daughter went down to the Nile to bathe, and her attendants were walking along the river bank. She saw the basket among the reeds and sent her slave girl to get it. She opened it and saw the baby. He was crying, and she felt sorry for him. This is one of the Hebrew babies, she said. Then his sister asked Pharaoh's daughter, Shall I go and get one of the Hebrew women to nurse the baby for you? Yes, go, she answered. And the girl went and got the baby's mother. Pharaoh's daughter said to her, Take this baby and nurse him for me, and I will pay you. So the woman took the baby and nursed him. When the child grew older, she took him to Pharaoh's daughter and he became her son. She named him Moses, saying, I drew him out of the water. Hebrews 11:24 - By faith Moses, when he had grown up, refused to be known as the son of Pharaoh's daughter. Genesis 46:20 and then later elaborated about in the apocryphal of Joseph - In Egypt, Manasseh and Ephraim were born to Joseph by Asenath daughter of Potiphera, priest of On (Asenath was a pagan girl who converted to Judaism to marry Joseph. She bore him two sons, "Manasseh and Ephraim" who became two of the patriarchs of the Israelite tribes). Esther 2:7 - Mordecai had a cousin named Hadassah, whom he had brought up because she had neither father nor mother. This girl, who was also known as Esther, was lovely in form and features, and Mordecai had taken her as his own daughter when her father and mother died. Ruth 1:4 - They married Moabite women, one named Orpah and the other Ruth. After they had lived there about ten years. Ruth 1:16 - But Ruth replied, Don't urge me to leave you or to turn back from you. Where you go I will go, and where you stay I will stay. Your people will be my people and your G-d my G-d. Ruth 4:13 - So Boaz took Ruth and she became his wife. Then he went to her, and the Lord enabled her to conceive, and she gave birth to a son.

In the Old Testament of Ruth, the author states that Ruth and Boaz' child is named Obed and that Obed became the father of Jesse, and Jesse who was the father of King David. Therefore, we can see that the message of the Bible clearly shows acceptance of the Israelites marrying converts to Judaism, which demonstrates that a marriage to a foreigner is acceptable to G-d when the foreigner follows G-d. Ruth 4:17-22 - And the women her neighbours gave it a name, saying, There is a son born to Naomi; and they called his name Obed: he is the father of Jesse, the father of David. Now these are the generations of Pharez: Pharez begat Hezron, And Hezron begat Ram, and Ram begat Amminadab, And Amminadab begat Nahshon, and Nahshon begat Salmon, And Salmon begat Boaz, and Boaz begat Obed, And Obed begat Jesse, and Jesse begat David.

The connection between Ruth and King David is very important because Jesus was born of Mary and adopted by Joseph, but the Jews disregard this because Joseph was not Jesus' blood father. But if we examine scripture a little closer we soon realize moreover that "BOTH" (Mary and Joseph) according to the New Testament of Luke and Matthew are of the lineage of David. Luke 3:23-38 - And Jesus himself began to be about thirty years of age, being (as was supposed) the son of Joseph, which was the son of Heli, Which was the son of Matthat, which was the son of Levi, which was the son of Melchi, which was the son of Janna, which was the son of Joseph, Which was the son of Mattathias, which was the son of Amos, which was the son of Naum, which was the son of Esli, which was the son of Nagge, Which was the son of Maath, which was the son of Mattathias, which was the son of Semei, which was the son of Joseph, which was the son of Juda, Which was the son of Joanna, which was the son of Rhesa, which was the son of Zorobabel, which was the son of Salathiel, which was the son of Neri, Which was the son of Melchi, which was the son of Addi, which was the son of Cosam, which was the son of Elmodam, which was the son of Er, Which was the son of Jose, which was the son of Eliezer, which was the son of Jorim, which was the son of Matthat, which was the son of Levi, Which was the son of Simeon, which was the son of Juda, which was the son of Joseph, which was the son of Jonan, which was the son of Eliakim, Which was the son of Melea, which was the son of Menan, **which was the son of Mattatha, which was the son of Nathan, which was the son of David, Which was the son of Jesse, which was the son of Obed,** which was the son of Booz, which was the son of Salmon, which was the son of Naasson, Which was the son of Aminadab, which was the son of Aram, which was the son of Esrom, which was the son of Phares, which was the son of Juda, Which was the son of Jacob, which was the son of Isaac, which was the son of Abraham, which was the son of Thara, which was the son of Nachor, Which was the son of Saruch, which was the son of Ragau, which was the son of Phalec, which was the son of Heber, which was the son of Sala, Which was the son of Cainan, which was the son of Arphaxad, which was the son of Sem, which was the son of Noe, which was the son of Lamech, Which was the son of Mathusala, which was the son of Enoch, which was the son of Jared, which was the son of Maleleel, which was the son of Cainan, Which was the son of Enos, which was the son of Seth, which was the son of Adam, which was the son of G-d. This can also be evaluated (with an associated provision) to it from Matthew 1:1-16 - The book of the generation of Jesus Christ, the son of David, the son of Abraham. Abraham begat Isaac; and Isaac begat Jacob; and Jacob begat Judas and his brethren; And Judas begat Phares and Zara of Thamar; and Phares begat Esrom; and Esrom begat Aram; And Aram begat Aminadab; and Aminadab begat Naasson; and Naasson begat Salmon; And Salmon begat Booz of Rachab; and **Booz begat Obed of Ruth; and Obed begat Jesse; And Jesse begat David the king; and David the king begat Solomon** of her that had been the wife of Urias; And Solomon begat Roboam; and Roboam begat Abia; and Abia begat Asa; And Asa begat Josaphat; and Josaphat begat Joram; and Joram begat Ozias; And Ozias begat Joatham; and Joatham begat Achaz; and Achaz begat Ezekias; And Ezekias begat Manasses; and Manasses begat Amon; and Amon begat Josias; And Josias begat Jechonias and his brethren, about the time they were carried away to Babylon: And after they were brought to Babylon, Jechonias begat Salathiel; and Salathiel begat Zorobabel; And Zorobabel begat Abiud; and Abiud begat Eliakim; and Eliakim begat Azor; And Azor begat Sadoc; and Sadoc begat Achim; and Achim begat Eliud; And Eliud begat Eleazar; and Eleazar begat Matthan; and Matthan begat Jacob; And Jacob begat Joseph the husband of Mary, of whom was born Jesus, who is called Christ.

Thus, Ruth is the foremother of Jesus found in (Matthew 1:5). The line can be traced as: Boaz, father of Obed, Obed, father of Jesse, Jesse, father of David, David, the eventual father of Joseph, and Joseph the adoptive father of Jesus. However, the New Testament lists two "DIFFERENT" genealogies for Jesus, one in Matthew and one in Luke (as shown above highlighted). Matthew shows a lineage from David, father of Solomon, and Luke shows a lineage through Nathan, a son of David. Many people don't know that King David had several other full-blooded sons from his wife Bathsheba, which were Shammua, Shobab, Solomon and Nathan. 2 Samuel 5:13-14 - After he left Hebron, David took more concubines and wives in Jerusalem, and more sons and daughters were born to him. These are the names of the children born to him there: Shammua, Shobab, **Nathan,** Solomon. 1 Chronicles 3:5 - And these were the children born to him there: Shammua, Shobab, **Nathan** and Solomon. These four were by Bathsheba daughter of Ammiel. The explanation for this is that Matthew is stating Joseph's line, and Luke is stating Mary's line. Under this interpretation, Jesus would ALSO be a biological descendant of David through his mother Mary who was in the line of lineage from Nathan (King David's other son).

So as you can see Jesus' lineage from "BOTH" His parents where of the Davidic line. However, all of this really becomes silly when one begins to realize that whether one is of royal blood or raised by one who is a descendent becomes null and void. All that truly matters is one that was raised under the "Judaic law". This is evidently, what truly makes a Jew a royal ancestor. Bickering about this will not suffice. The Lord G-d takes mercy on all peoples of the faith. Therefore if Jesus was born of the faith and found raised among those in the royal line of King David despite enforced guidelines proposed by the Jews, He was indeed a descendent as well as the successor of the royal Davidic line. Additionally, when the prophecy was made that the Messiah would be of the "Davidic line" it didn't specifically mention "flesh and blood" parents (this is only assumed).

Quickly before we move on... there is also something else pertinent to Ruth and Boaz whom were the parents of Obed, who became the father of Jesse, and from Jesse came King David, Solomon, and Jesus. Ruth was a Moabite that came with her mother-in-law back to Bethlehem. Now Ruth was without a husband and Naomi sent her into the threshing fields (called the threshing floor) to gather wheat during harvesting time. The threshing floor a is a flattened surfaced or a hard flat piece of ground (even rock) where the farmers tread and separate the grain from the straw. This particular parcel of land that Ruth threshed was owned by Boaz, and this is where Ruth met Boaz. Ruth 2:5-7 - Boaz asked the foreman of his harvesters, Whose young woman is that? The foreman replied, She is the Moabitess who came back from Moab with Naomi. She said, Please let me glean and gather among the sheaves behind the harvesters. She went into the field and has worked steadily from morning till now, except for a short rest in the shelter. But it was later on in history that the 1st Jewish temple in Jerusalem was built on a threshing floor (the temple mount). 2 Samuel 24:18 - On that day Gad (David's prophet) went to David and said to him, Go up and build an altar to the Lord on the threshing floor of Araunah the Jebusite. This is also found recorded in 1 Chronicles 21:18 - Then the angel of the Lord ordered Gad (David's prophet) to tell David to go up and build an altar to the Lord on the threshing floor of Araunah the Jebusite. Then we find in 2 Chronicles 3:1 - Then Solomon began to build the temple of the Lord in Jerusalem on Mount Moriah, where the Lord had appeared to his father David. It was on the threshing floor of Araunah the Jebusite, the place provided by David. This is the same exact parcel of land that G-d revealed to Abraham to sacrifice Isaac (which was Mount Moriah) that was later bought by King David and thereafter the site where his son Solomon built the 1st Jewish temple in Jerusalem (which is now where the Arab Dome of the Rock sits). To correspond with this, a scripture can be found in the New Testament that tells us that Jesus is the one who comes to clean "His threshing floor". Matthew 3:12 - Whose fan is in his hand, **and he will thoroughly purge his floor, and gather his wheat into the garner;** but he will burn up the chaff with unquenchable fire.

Moving on here… Now the Jews also firmly stick to Hebrew scripture and claim that biblical verses can only be understood by studying the "original Hebrew texts". They claim that there are many discrepancies and mistranslation in Hebrew words. One of the most commonly known words of mistranslation is between the words "young girl and virgin" found in Isaiah 7:14 - Therefore the Lord himself shall give you a sign; Behold, a virgin shall conceive, and bear a son, and shall call his name Immanuel. The conflict comes from the idea that a virgin is otherwise known as a young girl since the Hebrew word "alma" has always meant a young woman. However, we are forgetting that Hebrew is a "conceptual language" so when we encounter two semi different words that are close in meaning we should not exclude them as a mistranslations. For instance "alma" (young girl) and virgin. The word "virgin" is simply the feminine form of the corresponding masculine noun that pertains to age, not sexual experience. They are relating to the same thing. For one to argue such a close proximity of words is really inefficient and ridiculous. And should not be used as an excuse by Jews to inflexibly deceive oneself of true prophecy. Additionally, another verse that is picked over is found in Psalms when it makes mention that, "they pierced my hands and feet". The Jewish commentators claim this is not what it "really" says. In Hebrew, the world Ki-ari (like a lion) is grammatically similar to the word "gouged or gashed", and it does not mean pierced hands or feet, but rather like a lion, they gouge or gash me. Again relating to the same thing. Psalm 22:16 - For dogs have compassed me: the assembly of the wicked have inclosed me: they pierced my hands and my feet.

Another rationalism that the Jews adamantly stick to is based on several biblical passages that Moses conveyed to the Israelites in Deuteronomy concerning proof of G-d. This proof of G-d has been termed as a "national revelation". A national revelation is an event that is witnessed by a nation of people. This national revelation occurred when G-d spoke "outright" to a large population of Jews at Mount Sinai. Deuteronomy 4:9-13 - Only take heed to thyself, and keep thy soul diligently, lest thou forget the things which thine eyes have seen, and lest they depart from thy heart all the days of thy life: but teach them thy sons, and thy sons' sons; Specially the day that thou stoodest before the Lord thy G-d in Horeb, when the Lord said unto me, Gather me the people together, and I will make them hear my words, that they may learn to fear me all the days that they shall live upon the earth, and that they may teach their children. And ye came near and stood under the mountain; and the mountain burned with fire unto the midst of heaven, with darkness, clouds, and thick darkness. And the Lord spake unto you out of the midst of the fire: ye heard the voice of the words, but saw no similitude; only ye heard a voice. And he declared unto you his covenant, which he commanded you to perform, even ten commandments; and he wrote them upon two tables of stone.

What's more, Judaism is the only religion in the history of the world that claims that everyone heard G-d speak with their own eyes and ears, and makes the claim that the Torah also predicts that there will never be another official claim of national revelation again in history. They go on to contest against Jesus being the Messiah because they cant get past why G-d would establish His entire relationship with a nation through one man based only on blind faith - a personal revelation by a man named Jesus, which has turned out to be dependent on the testimony of others. Therefore, they exclude Him as being the Messiah on these grounds. However, lets not so quickly forget that the Torah "predicted" that there would never be another claim of "national revelation" in history. The reason for this is obvious, that G-d knew He would speak publicly only "on one occasion" at Mount Sinai some 3,000 years ago.

Therefore, Jesus does not have to conform to these regulations, but as you read on you will see that in many ways He actually did. Deuteronomy 4:32-33 - For ask now of the days that are past, which were before thee, since the day that G-d created man upon the earth, and ask from the one side of heaven unto the other, whether there hath been any such thing as this great thing is, or hath been heard like it? Did ever people hear the voice of G-d speaking out of the midst of the fire, as thou hast heard, and live? Or hath G-d assayed to go and take him a nation from the midst of another nation, by temptations, by signs, and by wonders, and by war, and by a mighty hand, and by a stretched out arm, and by great terrors, according to all that the Lord your G-d did for you in Egypt before your eyes? Unto thee it was shewed, that thou mightest know that the Lord He is G-d; there is none else beside him. Out of heaven he made thee to hear his voice, that he might instruct thee: and upon earth he shewed thee his great fire; and thou heardest his words out of the midst of the fire. Deuteronomy 5:1-4 - And Moses called all Israel, and said unto them, Hear, O Israel, the statutes and judgments which I speak in your ears this day, that ye may learn them, and keep, and do them. The Lord our G-d made a covenant with us in Horeb. The Lord made not this covenant with our fathers, but with us, even us, who are all of us here alive this day. The Lord talked with you face to face in the mount out of the midst of the fire. Deuteronomy 18:14-2 (tells us about the Prophet yet to come) - The nations you will dispossess listen to those who practice sorcery or divination. But as for you, the Lord your G-d has not permitted you to do so. The Lord your G-d will raise up for you a prophet like me from among your own brothers. You must listen to him. **For this is what you asked of the Lord your G-d at Horeb on the day of the assembly when you said, Let us not hear the voice of the Lord our G-d nor see this great fire anymore, or we will die.** The Lord said to me: What they say is good. I will raise up for them a prophet like you from among their brothers; I will put my words in his mouth, and he will tell them everything I command him. If anyone does not listen to my words that the prophet speaks in my name, I myself will call him to account. But a prophet who presumes to speak in my name anything I have not commanded him to say, or a prophet who speaks in the name of other gods, must be put to death. You may say to yourselves, How can we know when a message has not been spoken by the Lord?

Based on this, the Jews believe (rather assume) that when G-d or the Messiah comes He will immediately be acknowledged, the world will recognize Him, and this will proceed with immediate peace. **Again, how can this be if indeed it was predicted that G-d would never speak outright in this incorporeal way again?** Deuteronomy 18:16-18 - For this is what you asked of the Lord your G-d at Horeb on the day of the assembly when you said, Let us not hear the voice of the Lord our G-d nor see this great fire anymore, or we will die. The Lord said to me: What they say is good. I will raise up for them a prophet like you from among their brothers; I will put my words in his mouth, and he will tell them everything I command him. You will begin to see that this in itself is very contradicting. When Jesus came, He did indeed proclaim to be the Messiah, and many did hear and see with their own personal eyes and ears - both in His message and miracles (which were also recorded in the Jewish Talmud). And additionally, (like at Mount Sinai), there were also witnesses that gave testimony to hearing Jesus' word, seeing miracles, and there are even many people recorded that claimed seeing His resurrected body after the crucifixion. Incidentally, at Sinai when Moses threw down the first two tablets of stone onto the people below who were sinning (as well in unbelief), the Bible states that 3,000 Jewish people were killed by this incident, but miraculously after Jesus' resurrection the new law was being preached, and the word of G-d went out (spoken in different languages) and 3,000 people were saved and came to faith (as almost to say that the 3,000 that were slain for unbelief at Sinai during the giving of the law are now 3,000 raised back up again in salvation from sin). Exodus 32:19 - And it came to pass, as soon as he came nigh unto the camp, that he saw the calf, and the dancing: and Moses' anger waxed hot, and he cast the tables out of his hands, and brake them beneath the mount. Exodus 32:28 - The Levites did as Moses commanded, and that day about **three thousand of the people died.** Acts 2:1-8 -And when the day of Pentecost was fully come, they were all with one accord in one place. And suddenly there came a sound from heaven as of a rushing mighty wind, and it filled all the house where they were sitting. And there appeared unto them **cloven tongues like as of fire, and it sat upon each of them.** And they were all filled with the Holy Ghost, **and began to speak with other tongues,** as the Spirit gave them utterance. And there were dwelling at Jerusalem Jews, devout men, out of every nation under heaven. Now when this was noised abroad, the multitude came together, and were confounded, because that **every man heard them speak in his own language.** And they were all amazed and marvelled, saying one to another, Behold, are not all these which speak Galilaeans? And how hear we every man in our own tongue, wherein we were born? Acts 2:41 - Those who accepted his message were baptized, and about **three thousand were added to their number that day.** On top of this, the Jews at Sinai said let us not hear the voice of the Lord G-d (or His fire) again because they feared they would die, so it was agreed upon that instead G-d would raise up a prophet like Moses, whom would profess the word of G-d - this was Jesus, but the Jews neglect to see this. But He did indeed appear before us (as the man Jesus), but in such a miraculous filtered way that we were not subjected to the supersonic spirit voice of G-d along with His great fire - like what was witnessed at Sinai.

The Jews also cite from the Ten Commandments which states: "You shall have no other G-d's before me", meaning that it is forbidden to set up a mediator between G-d and man. Therefore, the Jews reject Jesus because He serves as an intermediary. However, Jesus rectified this when He said in the New Testament: "No man cometh unto the Father but by me". John 14:6 - Jesus saith unto him, I am the way, the truth, and the life: no man cometh unto the Father, but by me. Again, the Jews refute this and hold fast to the "Shema" which is the basis of Jewish belief. The Shema is the most important prayer in Hebrew scripture, which states: **"Hear O Israel, the Lord our G-d is ONE".** To the Jews (and understandably) this is contradictory to the Christian theology that breaks G-d up into 3 separate beings: The father, the Son, and the Holy Ghost, which is considered idolatry to the Jews. Matthew 28:19 - Go ye therefore, and teach all nations, baptizing them in the name of the **Father, and of the Son, and of the Holy Ghost.** But the true meaning of this is rectified by Jesus in John 10:30 - I and my Father are one. This concept can be very confusing for one to grasp, so to clarify this is terms that one can understand, the triad (three parts) position of the G-dhead can be explained in these terms. Back in the beginning of Genesis, we come upon a certain scripture that is a mystery. Genesis 1:26-27 - And G-d said, Let us make man in our image, after our likeness: and let them have dominion over the fish of the sea, and over the fowl of the air, and over the cattle, and over all the earth, and over every creeping thing that creepeth upon the earth. So G-d created man in his own image, in the image of G-d created he him; male and female created he them. The Hebrew scholars are baffled at this; they want to know how a singular G-d can speak with a plural pronoun? Some suggest that G-d was talking to the angels whom He created somewhere between the 3rd and the 6th day of creation. However the mystery of G-d talking in plurals deepens when we additionally come across other scriptural passages such as in Isaiah 6:8 - Also I heard the voice of the Lord, saying, Whom shall I send, and who will go for us? Then said I, Here am I; send me.

However, the answer to why G-d has spoken in plural form in both these passages is to our very amazement (especially to the Jews) can be found within the New Testament. Howbeit fittingly, the New Testament is the linking piece of evidence that sheds light on the Old Testament and answers many mysterious and unknown question for us. You see, there happens to be within the "one single G-dhead" three distinct persons who have existed together forever - the Father, the Son and the Holy Spirit (Ghost). Some say, how can it possibly be that three individual people can be G-d? The Jews cannot even begin to conceive of such a thing as it is blasphemous to suggest G-d as something other than "ONE". However, the concept becomes understood only upon realizing that each of the three G-dhead's have a distinct function within the administrative structure of what G-d does in everything that take place. Nothing here suggests anything is "contradictory". All three are set into place to work together to perform a certain function of organization. For instance, Hebrew scholars often say well that's just a plural of majesty that G-d is so great He has to speak in plurals (plural means more than one, and singular means referring to one). Nevertheless, as we are also about to witness in the paragraph below that G-d can also speak in singulars in Genesis 1:1.

Genesis 1:1 - "in the beginning G-d (Elohim) Created" the Heaven and the earth. As we can see the Hebrew word for "G-d" used in Genesis 1:1 is "Elohim" which is plural (G-d refers to Himself using this word - meaning more than one) (also remember, the filtering process of G-d's name "Elohim" was also unveiled for us in the chapter of the Bible Code). However, in this same phrase the Hebrew word "created" (Bara-shith) is a singular word with literally means to create. So we have a plural noun with a singular verb used together in the very first sentence of the Hebrew Bible. Then in Genesis 1:2 G-d (Elohim) tells us that He is also some kind of a moving "spirit". Genesis 1:2 - And the earth was without form, and void; and darkness was upon the face of the deep. And **the Spirit of G-d moved** upon the face of the waters. And in verse 3 (G-d Elohim) is some kind of "word" that speaks. Genesis 1:3 - **And G-d said, Let there be light**: and there was light. This G-d (Elohim) is "one", yet at the same time He is "several" - a sort of multiple persons. In essence, what G-d is telling us in His very first phrase to us, is that He is both a plural (Elohim), and a singular (He created).

There is clearly definition between the wordings. Some suggest especially the Jewish scholars that this is a grammatical mistake and a huge misinterpretation. However, it cannot be a grammatical mistake since we've already discovered in the chapter of the "primordial language" as well as the "Bible Code" that every single Hebrew letter and its place within the Bible is of great significance. So much so, that if one single letter was removed it would spell destruction of the Universe! There are no grammatical mistakes, or coincidences, the Torah is very kosher! (Meaning prepared properly).

This is something that the Jews, especially the old scholars of the Hebrew Bible never fathomed concerning the Old Testament, especially from the most holiest of Torah sections - at the very beginning of Genesis.

But it doesn't stop there… as I said; it paves its way to the New Testament. In fact, this acknowledgement is so overwhelming to the Jew, especially to the old scholars of the Hebrew Bible, that the "trinity" is a time bomb waiting to explode in the New Testament! John in the New Testament tells us, John 1:1-3 - In the beginning was the Word, and the Word was with G-d, and the Word was G-d. The same was in the beginning with G-d. All things were made by him; and without him was not any thing made that was made. However, this is the proper way we should be reading and understanding it: In the beginning was the Word, (the second person - the Son) and the Word was with G-d, (the first person - the Father) and the Word was G-d (both the Father and the Son together, despite the fact that He's dissimilar than the Father). The same was in the beginning with G-d. All things were made by him (the second person - the Son) and without him (the Son) was not any thing made that was made. John 1:10 - He was in the world, and the world was made by him, and the world knew him not. John 1:16-17 - And of his fullness have all we received, and grace for grace. For the law was given by Moses, but grace and truth came by Jesus Christ. Hebrews 1:2-3 - Hath in these last days spoken unto us by his Son, whom he hath appointed heir of all things, by whom also he made the worlds; Who being the brightness of his glory, and the express image of his person, and upholding all things by the word of his power, when he had by himself purged our sins, sat down on the right hand of the Majesty on high.

So the G-d of the Bible is three (3), yet He is one (3 in one). And as we can see this also shows up in the very creation story itself (Genesis 1:1), and in His creation of the Father, the Son, and the Holy Spirit in the New Testament, which again equals one G-d.

Isaiah 6:3 - And one cried unto another, and said, **Holy, holy, holy,** is the Lord of hosts: the whole earth is full of his glory. Revelation 4:8-11 - And the four beasts had each of them six wings about him; and they were full of eyes within: and they rest not day and night, saying, **Holy, holy, holy,** Lord G-d Almighty, which was, and is, and is to come. And when those beasts give glory and honour and thanks to him that sat on the throne, who liveth for ever and ever, The four and twenty elders fall down before him that sat on the throne, and worship him that liveth for ever and ever, and cast their crowns before the throne, saying, Thou art worthy, O Lord, to receive glory and honour and power: for thou hast created all things, and for thy pleasure they are and were created. In Hebrew when a word is repeated three (3) times, it means it is absolute. Additionally, Jesus rose on the 3rd day as a sign of completeness.

What the trinity does is "stress unity while recognizing diversity within that oneness". What is so disparaging in all this is that Jesus walked the streets of Israel and nobody could see with their limited eyes the one who created the Universe. Without even being known or seen, (unbeknownst) to the people living at that time that His word would eventually became the **invisible** "Holy Spirit". The book of Colossian bares witness to this fact. Colossians 1:15-17 - Who is the image of the invisible G-d, (the Son) the firstborn of every creature: For by him were all things created, that are in heaven, and that are in earth, **visible and invisible,** whether they be thrones, or dominions, or principalities, or powers: all things were created by him, and for him: And he is before all things, and by him all things consist. And he is the head of the body, the church: who is the beginning, the firstborn from the dead; that in all things he might have the preeminence. For it pleased the Father that in him should all fullness dwell. Hebrews 1:1 - Hath in these last days spoken unto us by his Son, whom He hath appointed heir of all things, by whom also He made the worlds.

The divine orderliness of things becomes even more clear to this fact when we keep reading. John 17 - These words spake Jesus, and lifted up his eyes to heaven, and said, Father, the hour is come; glorify thy Son, that thy Son also may glorify thee: As thou hast given him power over all flesh, that he should give eternal life to as many as thou hast given him. And this is life eternal, that they might know thee the only true G-d, and Jesus Christ, whom thou hast sent. I have glorified thee on the earth: I have finished the work which thou gavest me to do. And now, O Father, glorify thou me with thine own self with the glory which

I had with thee before the world was. I have manifested thy name unto the men which thou gavest me out of the world: thine they were, and thou gavest them me; and they have kept thy word. Now they have known that all things whatsoever thou hast given me are of thee. For I have given unto them the words which thou gavest me; and they have received them, and have known surely that I came out from thee, and they have believed that thou didst send me. I pray for them: I pray not for the world, but for them which thou hast given me; for they are thine. And all mine are thine, and thine are mine; and I am glorified in them. And now I am no more in the world, but these are in the world, and I come to thee. Holy Father, keep through thine own name those whom thou hast given me, that they may be one, as we are. While I was with them in the world, I kept them in thy name: those that thou gavest me I have kept, and none of them is lost, but the son of perdition; that the scripture might be fulfilled. And now come I to thee; and these things I speak in the world, that they might have my joy fulfilled in themselves. I have given them thy word; and the world hath hated them, because they are not of the world, even as I am not of the world. I pray not that thou shouldest take them out of the world, but that thou shouldest keep them from the evil. They are not of the world, even as I am not of the world. Sanctify them through thy truth: thy word is truth. As thou hast sent me into the world, even so have I also sent them into the world. And for their sakes I sanctify myself, that they also might be sanctified through the truth. Neither pray I for these alone, but for them also which shall believe on me through their word; That they all may be one; as thou, Father, art in me, and I in thee, that they also may be one in us: that the world may believe that thou hast sent me. And the glory which thou gavest me I have given them; that they may be one, even as we are one: I in them, and thou in me, that they may be made perfect in one; and that the world may know that thou hast sent me, and hast loved them, as thou hast loved me. Father, I will that they also, whom thou hast given me, be with me where I am; that they may behold my glory, which thou hast given me: for thou lovedst me before the foundation of the world. O righteous Father, the world hath not known thee: but I have known thee, and these have known that thou hast sent me. And I have declared unto them thy name, and will declare it: that the love wherewith thou hast loved me may be in them, and I in them. John 10:29 - My Father, who has given them to me, is greater than all, no one can snatch them out of my Father's hand. John 14:28 - You heard me say, I am going away and I am coming back to you. If you loved me, you would be glad that I am going to the Father, for the Father is greater than I.

In all this, Jesus is clearly emphasizing that He and the Father are the same. He is explaining the concept of the three-part G-dhead. In addition, in this, He explains that the Father is greater than He is, ONLY in administration, organization and management of the world. This is thoroughly a process of "regeneration", as this is how the world works. It was by the death of the second person (the Son, as an incarnation of the Father) that our sins were compensated for. However, the Father or the Son could not have accomplished anything until the 3rd person was administered as the "Holy Spirit". Therefore, Jesus (the Son) was not on different levels, He just happens to interchange as "different persons" within the administrative G-dhead. This is how it HAD to be… since the only way anything could conclude would be as a result of total self-sacrifice-just as Jesus went willingly when they came for Him in the garden of Gethsemane. Mark 14:48-50 - Am I leading a rebellion, said Jesus, that you have come out with swords and clubs to capture me? Every day I was with you, teaching in the temple courts, and you did not arrest me. But the Scriptures must be fulfilled. Then everyone deserted him and fled.

In effect, there is only one G-d, not three. However, G-d played three different roles during three different ages for our salvation. Understanding that G-d is One - expressed in 3 persons, Father, Son and Holy Spirit. Jesus testifies as being G-d. He tells us "Before Abraham was, I am". In the New Testament of John 8:58 - I tell you the truth, Jesus answered, **before Abraham was born, I am!** And what happens here is that divinity is immediately apparent that it was Jesus (in trinity) who talked to Moses in the Old Testament. Exodus 3:14 - And G-d said unto Moses, **I AM THAT I AM:** and he said, Thus shalt thou say unto the children of Israel, **I AM** hath sent me unto you. He also tells us that He is the same as the Holy Spirit when He says, "I am with you always". Matthew 28:20 - And teaching them to obey everything I have commanded you. And surely I am with you always, to the very end of the age. In the prophecy of Isaiah, we read, that He is a "Wonderful Counselor, Mighty G-d, the Everlasting Father". Isaiah 9:6 - For to us a child is born, to us a son is given, and the government will be on his shoulders. And he will be called Wonderful Counselor, Mighty G-d, Everlasting Father, Prince of Peace. Also, Jesus was called Emmanuel, which means, G-d with us. Now take notice of how it was first prophesized in Isaiah, and then later restated again, word for word, and fulfilled in the New Testament by Matthew. Isaiah 7:14 - Therefore the Lord himself will give you a sign: The virgin will be with child and will give birth to a son, and will call him Immanuel. Matthew 1:23 - The virgin will be with child and will give birth to a son, and they will call him Immanuel—which means, G-d with us. We also read in the opening book of the Gospel of John, "the Word (Jesus) was with G-d and the Word was G-d". In that same passage, we see that He (Jesus) was there in the beginning and that nothing was created without Him. This in itself is a very hard concept to grasp. John 1:1-3 - In the beginning was the Word, the Word was with G-d, and the Word was G-d. The same was in the beginning with G-d. All things were made by him; and without him was not any thing made that was made.

Surprisingly, the Hebrew Old Testament Bible itself DOES supports the view that G-d can become a man, without ceasing to be G-d, since there are places where G-d appeared in human form in the Bible. We can find this exclusive information in Genesis 18:1-33 when "three" men came to visit Abraham, and they spoke, ate, and drank with him, but one remained behind with Abraham while the other two went to Sodom. The one that stayed behind was G-d Himself/Jesus. Genesis 18:1-33 - And the Lord appeared unto him in the plains of Mamre: and he sat in the tent door in the heat of the day; And he lift up his eyes and looked, and, lo, **three men stood by him**: and when he saw them, he ran to meet them from the tent door, and bowed himself toward the ground, And said, My Lord, if now I have found favour in thy sight, pass not away, I pray thee, from thy servant: Let a little water, I pray you, be fetched, and wash your feet, and rest yourselves under the tree: And I will fetch a morsel of bread, and comfort ye your hearts; after that ye shall pass on: for therefore are ye come to your servant. And they said, So do, as thou hast said. And Abraham hastened into the tent unto Sarah, and said, Make ready quickly three measures of fine meal, knead it, and make cakes upon the hearth. And Abraham ran unto the herd, and fetcht a calf tender and good, and gave it unto a young man; and he hasted to dress it. And he took butter, and milk, and the calf which he had dressed, and set it before them; and he stood by them under the tree, and they did eat. And they said unto him, Where is Sarah thy wife? And he said, Behold, in the tent. And he said, I will certainly return unto thee according to the time of life; and, lo, Sarah thy wife shall have a son. And Sarah heard it in the tent door, which was behind him. Now Abraham and Sarah were old and well stricken in age; and it ceased to be with Sarah after the manner of women. Therefore Sarah laughed within herself, saying, After I am waxed old shall I have pleasure, my lord being old also? And the Lord said unto Abraham, Wherefore did

Sarah laugh, saying, Shall I of a surety bear a child, which am old? Is any thing too hard for the Lord? At the time appointed I will return unto thee, according to the time of life, and Sarah shall have a son. Then Sarah denied, saying, I laughed not; for she was afraid. And he said, Nay; but thou didst laugh. And the men rose up from thence, and looked toward Sodom: and Abraham went with them to bring them on the way. And the Lord said, Shall I hide from Abraham that thing which I do; Seeing that Abraham shall surely become a great and mighty nation, and all the nations of the earth shall be blessed in him? For I know him, that he will command his children and his household after him, and they shall keep the way of the Lord, to do justice and judgment; that the Lord may bring upon Abraham that which he hath spoken of him. And the Lord said, Because the cry of Sodom and Gomorrah is great, and because their sin is very grievous; I will go down now, and see whether they have done altogether according to the cry of it, which is come unto me; and if not, I will know. **(22) And the men turned their faces from thence, and went toward Sodom: but Abraham stood yet before the Lord.** And Abraham drew near, and said, Wilt thou also destroy the righteous with the wicked? Peradventure there be fifty righteous within the city: wilt thou also destroy and not spare the place for the fifty righteous that are therein? That be far from thee to do after this manner, to slay the righteous with the wicked: and that the righteous should be as the wicked, that be far from thee: Shall not the Judge of all the earth do right? And the Lord said, If I find in Sodom fifty righteous within the city, then I will spare all the place for their sakes. And Abraham answered and said, Behold now, I have taken upon me to speak unto the Lord, which am but dust and ashes: Peradventure there shall lack five of the fifty righteous: wilt thou destroy all the city for lack of five? And he said, If I find there forty and five, I will not destroy it. And he spake unto him yet again, and said, Peradventure there shall be forty found there. And he said, I will not do it for forty's sake. And he said unto him, Oh let not the Lord be angry, and I will speak: Peradventure there shall thirty be found there. And he said, I will not do it, if I find thirty there. And he said, Behold now, I have taken upon me to speak unto the Lord: Peradventure there shall be twenty found there. And he said, I will not destroy it for twenty's sake. And he said, Oh let not the Lord be angry, and I will speak yet but this once: Peradventure ten shall be found there. And he said, I will not destroy it for ten's sake. And the Lord went his way, as soon as he had left communing with Abraham: and Abraham returned unto his place. Genesis 19:1 - **And there came two angels to Sodom at even**; and Lot sat in the gate of Sodom: and Lot seeing them rose up to meet them; and he bowed himself with his face toward the ground.

To confirm the basis that G-d Himself (in human form) actually appeared as one of the three men, note that Genesis 18:22 (highlighted above) says the men got up and headed towards Sodom and Gomorrah, while the Lord remained behind with Abraham. Then in Genesis 19:1 (also highlighted above), we read that the two angels arrived at Sodom. Now if G-d wasn't one of the three men then we would expect to find that all three men went ahead to Sodom. However, this is not what the text says. Therefore, the third man was obviously G-d Himself/Jesus who had remained behind to talk to Abraham. The funny thing about this is that the Rabbis willfully admit that Abraham did in fact see G-d. However, what they don't believe is that He appeared as a man (or was one of the three men) a view that clearly "contradicts" the plain reading of the text.

There is also something else of great symbolic significance here when we read the beginning verse of Genesis 18:1-2 - **And the Lord appeared unto him** in the plains of Mamre: and he sat in the tent door in the heat of the day; And **he lift up his eyes and looked,**

and, lo, three men stood by him. It is not my intention to upset the non-believers in Jesus (such as the Jews) who think that believing in Jesus is idolatry, especially the idea alone of 3 persons in the form of G-d. But if I may put forth that to truly know, G-d is to know G-d as "triune" (3 - in one). The eternal G-d - the spirit who leads us through the son to the father (for the overall purpose of working the Fathers redemptive plan). True Idolatry is only when we associate something with G-d that is not G-d. Worshiping Jesus does not apply to idolatry since He is of G-d (leading us the way to G-d). For what ever reasons the Jews do not seem to want to focus on this symbolic precept, let alone acknowledge the clear concept that the trinity (G-d the Father, the Son, and the holy spirit) existed in the Old Testament long before it was reintroduced into the New Testament. The 3 divine visitors who visited Abraham can be seen as G-d's revelation of His 3-fold nature.

There are also other places in the Bible that support the fact that G-d can also be in the form of a man (as a mortal). Just as well, the Jews condone it, and quote the text word-for-word that states that G-D IS NOT A MAN. However, none of these biblical texts really-truly say that G-d CANNOT be a man, but rather what they do say is that G-d IS not a man (that…). There is a distinguishing difference here… along with saying that G-D IS NOT A MAN, these verses are attached with contingencies (or clauses)… and what these contingencies say is that G-d is not fallible like man is (flawed or imperfect because of sin), rather if He was a man He would be sinless (such as Jesus was). Nowhere in any of these verses (as you are about to read) does it rightly say that G-d could not come in the form of a man. Numbers 23:19 - G-d is not a man, that he should lie, nor a son of man, that he should change his mind. Does he speak and then not act? Does he promise and not fulfill? Samuel 15:29 - He who is the Glory of Israel does not lie or change his mind; for he is not a man, that he should change his mind. 1 Samuel 15:29 - He who is the Glory of Israel does not lie or change his mind; for he is not a man, that he should change his mind. Hosea 11:9 - I will not execute the fierceness of mine anger, I will not return to destroy Ephraim: for I am G-d, and not man; the Holy One in the midst of thee: and I will not enter into the city.

What remains remarkable are the documented details recorded by the disciples about Jesus' nature (character, temperament and personality)… Jesus was very distinctive of G-d. Indeed Jesus did not lie, He did not repent, and He hath spoken, clarified, and made good of His word. Not one word that Jesus spoke was in error, He said them with infinite accuracy and authority. In addition, in no way did He lie, contaminate, minimize, or remise scripture. Jesus was like no other man, that is to say He was absolutely pure and sinless. Psalms 119:89 - Your word, O Lord, is eternal; it stands firm in the heavens. Luke 21:33 - Heaven and earth will pass away, but my words will never pass away. In fact so important was written scripture - especially to Jesus "when He quotes" the biblical passages and replies to Satan's temptation to deny and pervert the Fathers word, He said "IT IS WRITTEN" three times, and accurately quoted appropriate passages from the Old Testament. Matthew 4:1-11- Then was Jesus led up of the Spirit into the wilderness to be tempted of the devil. And when he had fasted forty days and forty nights, he was afterward an hungred. And when the tempter came to him, he said, If thou be the Son of G-d, command that these stones be made bread. But **He answered and said, It is written, Man shall not live by bread alone, but by every word that proceedeth out of the mouth of G-d.** Then the devil taketh him up into the holy city, and setteth him on a pinnacle of the temple, And saith unto him, If thou be the Son of G-d, cast thyself down: for it is written, He shall give his angels charge concerning thee: and in their hands they shall bear thee up, lest at any time thou dash thy foot against a stone. **Jesus said unto him, It is written**

again, Thou shalt not tempt the Lord thy G-d. Again, the devil taketh him up into an exceeding high mountain, and sheweth him all the kingdoms of the world, and the glory of them; And saith unto him, All these things will I give thee, if thou wilt fall down and worship me. **Then saith Jesus unto him, Get thee hence, Satan: for it is written, Thou shalt worship the Lord thy G-d, and him only shalt thou serve.** Then the devil leaveth him, and, behold, angels came and ministered unto him.

There are also other key indicators that strongly supports that Jesus was extraordinarily archetypal of G-d and sinless. 2 Corinthians 5:21 - For he hath made him to be sin for us, who knew no sin; that we might be made the righteousness of G-d in him. Hebrews 4:15 - For we have not an high priest which cannot be touched with the feeling of our infirmities; but was in all points tempted like as we are, yet without sin. Hebrews 7:26 - For such an high priest became us, who is holy, harmless (innocent), undefiled, separate from sinners, and made higher than the heavens. Hebrews 9:14 - How much more shall the blood of Christ, who through the eternal Spirit offered himself without spot to G-d, purge your conscience from dead works to serve the living G-d? 1 Peter 1:18-19 - Forasmuch as ye know that ye were not redeemed with corruptible things, as silver and gold, from your vain conversation received by tradition from your fathers; But with the precious blood of Christ, as of a lamb without blemish and without spot. 1 Peter 2:22 - He committed no sin, and no deceit was found in his mouth. 1 Peter 3:18 - For Christ also hath once suffered for sins, the just for the unjust, that he might bring us to G-d, being put to death in the flesh, but quickened by the Spirit. 1 John 2:1 - My little children, these things write I unto you, that ye sin not. And if any man sin, we have an advocate with the Father, Jesus Christ the righteous. 1 John 3:3 - He is pure. All who hope to be like him make themselves pure. 1 John 3:5 - But you know that Christ came to take our sins away. And there is no sin in him.

However, the best one of all (that is totally contradicting) is when Judaism acknowledges and understands that the Messiah will be born of human parents, and will possess normal physical attributes like any ordinary person. They also believe that the Messiah could be alive in every generation with the capacity to step into the role of the Messiah at any given time. They say He will not even need to possess any divinely supernatural qualities.

However, (in their rational understanding) they completely "separate" the entity of the Messiah from being joined, united, or coupled to/with G-d. The reason for this is based on the Jewish conviction that G-d is so essentially different from and beyond humanity that he could never become a human. 1 Kings 11:3 - Howbeit I will not rend away all the kingdom; but will give one tribe to thy son for David my servant's sake, and for Jerusalem's sake which I have chosen.

So as you read from the paragraph above from the Old Testament that the Jews clearly acknowledge the fact that the Messiah will come from a human descendant of King David, but conversely this would have to mean that it also unequivocally predicts that the Messiah of King David is actually G-D IN THE FLESH! Isaiah 9:6-7 - **For unto us a child is born,** unto us a son is given: and the government shall be upon his shoulder: and his name shall be called Wonderful, Counsellor, **The mighty G-d, The everlasting Father,** The Prince of Peace. Of the increase of his government and peace there shall be no end, **upon the throne of David,** and upon his kingdom, to order it, and to establish it with judgment and with justice from henceforth even for ever. The zeal of the Lord of hosts will perform this. Jeremiah 23:5-

6 - Behold, the days come, saith the Lord, that I will raise unto David a righteous Branch, and a King shall reign and prosper, and shall execute judgment and justice in the earth. In his days Judah shall be saved, and Israel shall dwell safely: and this is his name whereby he shall be called, **THE LORD OUR RIGHTEOUSNESS.** These texts unmistakably indicate that G-d is to become man in the Person of the Messiah. And as we clearly read here, the Messiah is a child who is born (from the lineage of King David, showing that He is truly human, while also being the Mighty G-d and everlasting Father at the same time - whom the New Testament says is Jesus.

By the way, there is also another prophetic prophecy that beautifully ties-in with Jesus as being the "RIGHTEOUS ONE" (as highlighted right above in all CAPS). In addition, this first starts out with the patriarch Abraham. Genesis 15:6 (Old Testament) - Abram believed the Lord, **and he credited it to him as righteousness.**

The Old Testament also says that G-d would send the Messiah as the source through which righteousness would come to mankind. Jeremiah 23:5-6 - The days are coming, declares the Lord, when **I will raise up to David a righteous Branch,** a King who will reign wisely and do what is just and right in the land. In his days Judah will be saved and Israel will live in safety. This is the name by which he will be called: **The Lord Our Righteousness.** And it can be found again in Jeremiah 33:15-16 - In those days and at that time I will make a righteous Branch sprout from David's line; he will do what is just and right in the land. In those days Judah will be saved and Jerusalem will live in safety. This is the name by which it will be called: The Lord Our Righteousness. Now the New Testament goes on to says that Jesus is that Messiah, and that righteousness comes only through faith in Him. 2 Corinthians 5:21 (New Testament) - **G-d made him who had no sin to be sin for us, so that in him we might become the righteousness of G-d.**

Conclusion... in other words, a person is declared righteous, based upon faith by the sacrifice of Jesus. Romans 10:4 (New Testament) - **Christ is the end of the law so that there may be righteousness for everyone who believes.** What is prophetic here is that both the Old Testament and the New Testament indicate that Abraham was considered righteous because He believed G-d. Romans 4:3 (New Testament) - What does the Scripture say? **Abraham believed G-d, and it was credited to him as righteousness.** The New Testament puts forward that one does not become righteous by simply obeying G-d's law, but by believing - and by the mercy of G-d through the forfeiting sacrifice of His Son (just like Abraham, when G-d asked him to sacrifice his son Isaac).

Now getting back to the triune (three in one) nature of G-d... For those who are still distant at heart regarding this matter, there is still another way to properly determine the three-part nature of the G-dhead, and authenticate Jesus as that person. Currently at this stage in the book I am going to give away yet another one of my great secrets that sheds even more evidence of Jesus' G-dly incarnate nature - that He in fact WAS G-D! Like I have mention so many times before that all the answers lie within the Bible itself - where scripture backs-up scripture. This is the way G-d made it. He seems to wants us to delve into the Bible and diligently search and compare "concordances" for the answers. The answer to another great secret regarding Jesus' authenticity and inclined nature with G-d (as G-d) lies at the very beginning of the Bible in Genesis. It involves the creation story itself where life was created in a literal six days. It tells us that G-d put trees on dry land before He put life in the ocean.

Genesis 1:11-12 - And G-d said, Let the earth bring forth grass, the herb yielding seed, and the fruit tree yielding fruit after his kind, **whose seed is in itself**, upon the earth: and it was so. And the earth brought forth grass, and herb yielding seed after his kind, and the tree yielding fruit, whose seed was in itself, after his kind: and G-d saw that it was good. And we read it again in Genesis 2 (so that it is clearly understandable) that plants were made before rain. Genesis 2:4-5 - These are the generations of the heavens and of the earth when they were created, in the day that the Lord G-d made the earth and the heavens, **And every plant of the field before it was in the earth, and every herb of the field before it grew: for the Lord G-d had not caused it to rain upon the earth,** and there was not a man to till the ground.

Then we read in Genesis that G-d created the great whales and birds. Genesis 1:20-21 - And G-d said, Let the waters bring forth abundantly the moving creature that hath life, and fowl that may fly above the earth in the open firmament of heaven. And G-d created **great whales,** and every living creature that moveth, which the waters brought forth abundantly, after their kind, and every winged fowl after his kind: and G-d saw that it was good.

G-d created life in this sequence to implicitly show us that He didn't need any water to do His work of creating life. He also created whales and birds (fowl) first, and then waited a day to create the land animals to prove that He didn't need their help to initially give rise to the whales. Genesis 1:24-25 - And G-d said, Let the earth bring forth the living creature after his kind, **cattle, and creeping thing, and beast of the earth after his kind**: and it was so. And G-d made the beast of the earth after his kind, and cattle after their kind, and every thing that creepeth upon the earth after his kind: and G-d saw that it was good.

Its very ironical here because the evolution theory says that whales developed from existing land mammal's as they have what looks like remains of skeletal feet and bones, (but it has been recently discovered that these appendages (vestigial legs) are actually used for reproduction. Furthermore, these vestigial legs are only found in the "male" and are used to anchor the muscles attached to the penis. Furthermore, the evolution theory says that birds must have come from previously existing reptiles, so G-d in foresight of this, deliberately, intentionally and purposely created the birds on the fifth day, and waited one day to create reptiles to prove He didn't need their help in giving rise to the birds either. Genesis 1:20 - And G-d said, Let the waters bring forth abundantly the moving creature that hath life, **and fowl that may fly above the earth in the open firmament of heaven.** Genesis 1:24-25 - And G-d said, Let the earth bring forth the living creature after his kind, **cattle, and creeping thing, and beast of the earth after his kind**: and it was so. And G-d made the beast of the earth after his kind, and cattle after their kind, and every thing that creepeth upon the earth after his kind: and G-d saw that it was good.

The message behind these order of events will eventually reveal to us who is really and truly the creator. Allowing us to see that Jesus is of G-d.

So how do we really know that Jesus was of G-d? Well if Genesis is correct that G-d created everything FULLY GROWN AND MATURE, these same attributes would also be prevalent in Jesus' nature (which indeed they are). Lets first examine what the disciples wrote about Him when they were describing Jesus' three-part nature, (G-d the Father, G-d the Son, and G-d the Holy Spirit). John 1:3 - All things were made by Him; and without Him was not any thing made that was made. John 1:10 - He was in the world, and the world was made by Him, and the world knew Him not. Colossians 1:17 - And He is before all things, and by Him all things consist. Hebrews 1:2 - Hath in these last days spoken unto us by his Son, whom He hath appointed heir of all things, by whom also He made the worlds. Now lets look at some of the miracles that Jesus did… During Jesus' time here on Earth He performed certain unexplainable miracles. He most likely performed these miracles because they were in some way suggestive, indicative (or representative) of prophecy. Such as when during the wedding feast, He made wine FULLY MADE (pressed and aged). John 2:1-11 - On the third day a wedding took place at Cana in Galilee. Jesus' mother was there, and Jesus and his disciples had also been invited to the wedding. When the wine was gone, Jesus' mother said to him, They have no more wine. Dear woman, why do you involve me? Jesus replied, My time has not yet come. His mother said to the servants, Do whatever he tells you. Nearby stood six stone water jars, the kind used by the Jews for ceremonial washing, each holding from twenty to thirty gallons. **Jesus said to the servants, Fill the jars with water; so they filled them to the brim. Then he told them, Now draw some out and take it to the master of the banquet. They did so, and the master of the banquet tasted the water that had been turned into wine.** He did not realize where it had come from, though the servants who had drawn the water knew. Then he called the bridegroom aside and said, Everyone brings out the choice wine first and then the cheaper wine after the guests have had too much to drink; but you have saved the best till now. This, the first of his miraculous signs, Jesus performed in Cana of Galilee. He thus revealed his glory, and his disciples put their faith in him.

Another miraculous miracle that Jesus did involved feeding a mass of hungry people. Before the Passover hour and with no money, and a great crowd of people following, Jesus took a boys lunch (consisting of five small barley loaves and two small fishes) and miraculously created a lot more loaves of bread ALREADY BAKED (fit to be eaten) from barley and wheat that was not grown or harvested, BUT FULLY GROWN, DRIED AND BAKED. And fishes that DIDN'T HATCH IN THE SEA, OR COME FROM THE WATER, OR CAUGHT BY MEN. John 6:1-15 - Some time after this, Jesus crossed to the far shore of the Sea of Galilee (that is, the Sea of Tiberias), and a great crowd of people followed him because they saw the miraculous signs he had performed on the sick. Then Jesus went up on a mountainside and sat down with his disciples. The Jewish Passover Feast was near. When Jesus looked up and saw a great crowd coming toward him, he said to Philip, Where shall we buy bread for these people to eat? He asked this only to test him, for he already had in mind what he was going to do. Philip answered him, Eight months wages would not buy enough bread for each one to have a bite! Another of his disciples, Andrew, Simon Peter's brother, spoke up, Here is a boy with five small barley loaves and two small fish, but how far will they go among so many? Jesus said, Have the people sit down. There was plenty of grass in that place, and the men sat down, about five thousand of them. Jesus then took the loaves, gave thanks, and distributed to those who were seated as much as they wanted. He did the same with the fish. When they had all had enough to eat, he said to his disciples, Gather the pieces that are left over. Let nothing be wasted. So they gathered them and filled twelve baskets with the pieces of the five barley loaves left over by those who had eaten. After the people saw the miraculous sign that Jesus did, they began to say, Surely this is the Prophet who is to come into the world. Jesus, knowing that they intended to come and make him king by force, withdrew again to a mountain by himself. These entire things were produced by HIS MERE WORD! Can we see the linkage and the relationship? What this tells us is that "both" G-d and Jesus can create things FULLY GROWN simply by mere word. Just as G-d created things FULLY GROWN in the Garden of Eden, so could Jesus when He made wine, bread and fish FULLY MADE. This indicates that He shares the same divine nature with G-d.

Additionally if you recall that Jesus had dominion over the fish because He was sinless. And He was sinless because He was not under the curse (the curse brought about from Adam and Eve sinning, "eating of the tree of knowledge" in the garden of Eden). Remember before Adam and Eve had sinned they had dominion over the fish of the sea. Jesus therefore not under the curse (and sinless) had dominion over the fish of the sea. He also had foresight of where fish were in the Sea of Galilee in order to catch them in nets. Genesis 1:26 - And G-d said, Let us make man in our image, after our likeness: **and let them have dominion over the fish of the sea,** and over the fowl of the air, and over the cattle, and over all the earth, and over every creeping thing that creepeth upon the earth. Luke 5:1-11 - And it came to pass, that, as the people pressed upon him to hear the word of G-d, he stood by the lake of Gennesaret, (also known as the Lake of Galilee). And saw two ships standing by the lake: but the fishermen were gone out of them, and were washing their nets. And he entered into one of the ships, which was Simon's, and prayed him that he would thrust out a little from the land. And he sat down, and taught the people out of the ship. **Now when he had left speaking, he said unto Simon, Launch out into the deep, and let down your nets for a draught. And Simon answering said unto him, Master, we have toiled all the night, and have taken nothing: nevertheless at thy word I will let down the net. And when they had this done, they enclosed a great multitude of fishes: and their net brake. And they beckoned unto their partners, which were in the other ship, that they should come and help them.** And they came, and filled both the ships, so that they began to sink. When Simon Peter saw it, he fell down at Jesus' knees, saying, Depart from me; for I am a sinful man, O Lord. For he was astonished, and all that were with him, at the draught of the fishes which they had taken: And so was also James, and John, the sons of Zebedee, which were partners with Simon. And Jesus said unto Simon, Fear not; from henceforth thou shalt catch men. And when they had brought their ships to land, they forsook all, and followed him.

Here is another scriptural message that stresses that Jesus was not under the curse. Matthew 17:24 - And when they were come to Capernaum, they that received tribute money came to Peter, and said, Doth not your master pay tribute? He saith, Yes. And when he was come into the house, Jesus prevented him, saying, What thinkest thou, Simon? of whom do the kings of the earth take custom or tribute? of their own children, or of strangers? Peter saith unto him, Of strangers. Jesus saith unto him, Then are the children free. Notwithstanding, lest we should offend them, **go thou to the sea, and cast an hook, and take up the fish that first cometh up; and when thou hast opened his mouth, thou shalt find a piece of money: that take, and give unto them for me and thee.** As I had explained in the prior chapter, that someone had apparently dropped a coin (shekel) into the Sea of Galilee, and in foresight, Jesus had commanded a fish swimming by the coin to gulp it up and rush to Peters hook to be caught - that is DOMINION! Not only is it dominion, its foresight along with the foreknowledge and capability that none of us currently possess.

Now for those who are still doubtful about Jesus' authenticity as the Messiah, I'm going to reach over the heads of the Jewish scholars (who refute Him), to reveal a vital component that's been left out of the questionnaires, and left for the scholars to decide by a secret ballot. This vital factor has to do with who is rightfully KING! It has been recorded throughout the Old Testament that all Israel's Kings were crowned AT THE SAME TIME. And this took place on one specific (and significant) time of year, which was during Passover - the 15th of Nissan! (Which takes place in the spring and is the first full Moon). This is also called the

New Year for Kings! This is when Israel had coronate(d) all her Kings. This holiday correlates to Passover because it was in this month that Moses brought the Jews out of Egypt. It was a day of LIBERATION! Even Moses became the leader of Israel in that month, so it makes perfect sense that Kings were crowned at this specific time. In other words, the greatest date in the Jewish calendar is Nissan the 15th (which is the coronation of Kings). The book of Kings tells us that King Solomon was crowned in the first month (which was Nissan the 15th) and it tells us his reign was from Nissan the 15th (the day his reign started), to Nissan the 15th (the day his reign ended). 1 Kings 6:1- And it came to pass in the four hundred and eightieth year after the children of Israel were come out of the land of Egypt, in the fourth year of Solomon's reign over Israel, **in the month of Zif (which is late April early May) (after Passover), which is the second month that he began to build the house of the Lord.** This tells us that King Solomon was crowned the previous month. This is repeated in 2 Chronicles 3:2 - And he began to build in the second day of the second month, in the fourth year of his reign. This is extremely important because if all Israel's Kings were crowned at that time, that takes us up to the crowning of Jesus (the crown of thorns). Jesus was crowned on Passover when they put the crown of thorns on His head during His crucifixion. This was actually His coronation! He was coronate(d) as King of the Jews while He was hanging and being put to death upon the cross. John 19:2-3 - And the soldiers platted a crown of thorns, and put it on his head, and they put on him a purple robe, And said, Hail, King of the Jews! and they smote him with their hands. Not only was Jesus crowned here, but they also clothed Him in a royal garment which is also customary for Jewish Kings. John 19:4-5 - Pilate therefore went forth again, and saith unto them, Behold, I bring him forth to you, that ye may know that I find no fault in him. **Then came Jesus forth, wearing the crown of thorns, and the purple robe. And Pilate saith unto them, Behold the man! (which basically means, "the one you've been waiting for").** John 19:6-8 - When the chief priests therefore and officers saw him, they cried out, saying, Crucify him, crucify him. Pilate saith unto them, Take ye him, and crucify him: for I find no fault in him. The Jews answered him, We have a law, and by our law he ought to die, because he made himself the Son of G-d. When Pilate therefore heard that saying, he was the more afraid. John 19:14 - And it was the preparation of the Passover, and about the sixth hour: **and he saith unto the Jews, Behold your King!** Along with this Pilate wrote a sign that read "King of the Jews". When the Jews read this they soon realized the implications of this (after the fact) and wanted that sign taken down, but Pilate refused and said, "I've written what I've written". John 19:19-22 - And **Pilate wrote a title, and put it on the cross. And the writing was JESUS OF NAZARETH THE KING OF THE JEWS.** This title then read many of the Jews: for the place where Jesus was crucified was nigh to the city: and it was written in Hebrew, and Greek, and Latin. **Then said the chief priests of the Jews to Pilate, Write not, The King of the Jews; but that he said, I am King of the Jews. Pilate answered, What I have written I have written.**

Many Jews do not except the crowning, death, burial and resurrection of Jesus' crucifixion as being a coronation (ceremony of crowning) - BUT NONETHELESS IT IS! It took place at the appropriate time - the New Year for Kings during the most important holiday of the Jewish calendar - the starting day of Passover - the day that Moses liberated the Jews from Egypt. (Incidentally April 16th is Emancipation Day in the United States which celebrates the "liberation of slavery" that President Abraham Lincoln abolished in 1862). So, as one can see this date in history is commemorated (though some people are unaware of it) as a day of "liberation". Make no mistake about this; it was April Nissan 1 on the 15th day! Exodus 16:1 - The whole Israelite community set out from Elim and came to the Desert of Sin, which is between Elim and Sinai, **on the fifteenth day of the second month after they had come out of Egypt.** Numbers 33:3 - The Israelites set out from Rameses **on the fifteenth day of the first month, the day after the Passover. They marched out boldly in full view of all the Egyptians.** Remember, there are no coincidences, since coincidence is not a kosher word. There is a reason and a time for everything that takes place in the scheme of things. G-d has immaculate foresight and is foretelling in this arena. It is not too hard to see the quirk of fate that the Messiah (in the line of King David) should follow in the footsteps of all other Kings. This is actually the masterstroke of congruity (great consistency), which commemorates an important secular event in Jewish history. Is He not a G-d that backs His word? Numbers 23:19 - G-d is not a man, that he should lie, nor a son of man, that he should change his mind. Does he speak and then not act? Does he promise and not fulfill? Luke 24:44 - And he said unto them, These are the words which I spake unto you, while I was yet with you, that all things must be **fulfilled,** which were written in the law of Moses, and in the prophets, and in the psalms, concerning me.

Even G-d (the Father) recognized Jesus as King. Psalms 2:6 - **I have set my king upon my holy hill of Zion.**

Furthermore we also have an astounding future prophesy concerning Jesus as being the reigning King of Kings in the last and final book of the New Testament which describes His second coming. Revelations 19:12 - His eyes are like blazing fire, and **on his head are many crowns.** He has a name written on him that no one knows but he himself. Revelations 19:16 - On his robe and on his thigh **he has this name written: KING OF KINGS AND LORD OF LORDS.**

So some may wonder why a "crown of thorns" was placed upon Jesus' head instead of a typical crown of gold. Well there happens to be something very significant about thorns, and this has to do with the time when Adam and Eve sinned back in the Garden. When G-d came down and confronted them with what they had done (eating of the Tree of good and evil, which G-d told them not to), G-d said cursed is the ground for your sake - from now on, it will bring forth thorns. "Thorns" were the first thing mentioned because of the curse. Genesis 3:17-18 - To Adam he said, Because you listened to your wife and ate from the tree about which I commanded you, You must not eat of it, Cursed is the ground because of you; through painful toil you will eat of it all the days of your life. **It will produce thorns and thistles for you,** and you will eat the plants of the field. Therefore, when Jesus had the crown of thorns placed upon His head (and forehead), this symbolized the first of the curses upon the human race. In addition to this, when the blood of Jesus dripped out from His forehead and covered those thorns, it can be said that Jesus was literally paying the price for the curse of sin.

Again, another unsurpassable question arises… How is it that Jesus' crowning "fittingly happened" in addition to Daniels prophesies? How is it that Jesus conspired the time and place where He was to be born in Bethlehem so that He could ride into Jerusalem at exactly the right time of Daniels prophecy? Yes, He knew prophecy had to be fulfilled, and yes, He knew He was the Messiah, but the events that fell into place are certainly something in which He had no control over. So now, we also have to contend with the fact that He was also crowned at the proper time. There is no way on G-d's green earth that Jesus could have had perfect control over all these things in addition to being made King with a crown of thorns and on the anniversary of all other Jewish Kings. As well as the symbolic representation of thorns being purified with blameless blood. The secret ballot is in - this is what totally authenticates (endorses and confirms) Jesus as the true Messiah.

Oh the maladies of the whole saga… One would think it would end there, but it doesn't. The Jews that don't acknowledge Jesus as their Messiah have even gone so far as to use the excuse that He was not a proper or valid sacrificial sacrifice according to the requirements of the Old Testament laws. They put forth that His death by crucifixion was an unacceptable process for the atonement of sins. The reason for this they conclude was that He was not slaughtered in the proper way.

The rules for a sacrificial animal (lamb or male-goat) have to be without any physical defects or blemishes. It has to be slaughtered and its blood placed on the doorposts of houses. It has to be roasted, burned and eaten. There can be no leftovers. Human sacrifice is strictly prohibited! Furthermore, the Jews proclaim that Jesus was beaten before being crucified which left His body blemished and scared, and therefore unacceptable for a sacrificial offering.

But all of this preposterous! Jesus was not a meal offering, He was a spiritual giving. A "blood offering" for the "atonement" of sin. However, the Jews claim that the Paschal Lamb was a festive and commemorating offering and not a sin offering. But they conveniently forget that before it became a festive and commemorating holiday, the original purpose of putting blood on the doorposts in Egypt (before the Exodus) was for the mere purpose of G-d passing over. Some can conclude that the blood was used to cover up sin as He passed over. And we must also remember that the killing of the first animal by G-d (suggested to be a lamb) was recorded in Genesis, and was used to cover the shameful nude bodies of Adam and Eve after they had committed sin (in other words, the lambs skin (along with its blood in death) was used as a covering for their sin). Genesis 3:21 - The Lord G-d made garments of skin for Adam and his wife and clothed them. And it was also used again in Exodus 12:12 - On that same night I will pass through Egypt and strike down every firstborn—both men and animals—and I will bring judgment on all the g-ds of Egypt. I am the Lord. The blood will be a sign for you on the houses where you are; and when I see the blood, I will pass over you. No destructive plague will touch you when I strike Egypt. In contrast to this, Peter in the New Testament (a Jew in those times who clearly understanding the religious rules concerning a sacrificial lamb) remedies this for us that Jesus WAS INDEED a lamb without blemish or spot when he tells us in 1 Peter 1:19 - But with the precious blood of Christ, as of a lamb without blemish and without spot.

So how do we rightly know that Jesus was an acceptable sacrifice onto G-d? We know this because He rose on the 3rd day Just as prophecy said He would. Exodus 19:11 - And be ready by the third day, because on that day the Lord will come down on Mount Sinai in the sight of all the people. (Old Testament) Hosea 6:2 - **After two days He will revive us; on the third day He will restore us, that we may live in His presence.** Matthew 16:21 - From that time on Jesus began to explain to his disciples that he must go to Jerusalem and suffer many things at the hands of the elders, chief priests and teachers of the law, and that he must be killed and on the third day be raised to life. Matthew 17:23 - They will kill him, and on the third day he will be raised to life. And the disciples were filled with grief. Matthew 20:19 - And will turn him over to the Gentiles to be mocked and flogged and crucified. On the third day he will be raised to life! Matthew 27:64 - So give the order for the tomb to be made secure until the third day. Otherwise, his disciples may come and steal the body and tell the people that he has been raised from the dead. This last deception will be worse than the first.

Luke 9:22 - And he said, The Son of Man must suffer many things and be rejected by the elders, chief priests and teachers of the law, and he must be killed and on the third day be raised to life. Luke 13:32 - He replied, Go tell that fox, I will drive out demons and heal people today and tomorrow, and on the third day I will reach my goal. Luke 18:33 - On the third day he will rise again. Luke 24:7 - The Son of Man must be delivered into the hands of sinful men, be crucified and on the third day be raised again. (New Testament) 1 Corinthians 15:4 - For I delivered unto you first of all that which I also received, how that **Christ died for our sins according to the scriptures; And that he was buried, and that he rose again the third day according to the scriptures.** Jonah 1:17 - Now the Lord had prepared a great fish to swallow up Jonah. And Jonah was in the belly of the fish three days and three nights. Matthew 12:40 - **For as Jonas was three days and three nights in the whale's belly; so shall the Son of man be three days and three nights in the heart of the earth.** John 2:19 - Jesus answered and said unto them, Destroy this temple, and in three days I will raise it up.

Again the Jews assert that killing a human being in place of a sacrificial sacrifice unto G-d is exclusively forbidden since it's one of the 10 commandments listed in Exodus 20:13 and again in Deuteronomy 5:17 - Thou shalt not kill. However, this again can be refuted when G-d told Abraham that He was to offer his son Isaac for a burnt offering. Genesis 22:2 - And he said, Take now thy son, thine only son Isaac, whom thou lovest, and get thee into the land of Moriah; and offer him there for a burnt offering upon one of the mountains which I will tell thee of.

The contradicting question that arises is why would G-d even ask such a thing if it were not permitted in His logic and reasoning? Perhaps the difference here is that even though it was just a test of faith, it did not surpass G-d to request such a demoralizing act as a test of faith. The answer might come as a surprise since G-d Himself (incarnate) came down to Earth, was sacrificed and died in the form of Jesus. Here is the full scripture of Genesis 22:1-18 - Some time after these things G-d tested Abraham. He said to him, "Abraham!" "Here I am!" Abraham replied. G-d said, "Take your son – your only son, whom you love, Isaac and go to the land of Moriah! Offer him up there as a burnt offering on one of the mountains which I will indicate to you." Early in the morning Abraham got up and saddled his donkey. He took two of his young servants with him, along with his son Isaac. When he had cut the wood for the burnt offering, he started out for the place G-d had spoken to him about. On the third day Abraham caught sight of the place in the distance. So he said to his servants, "You two stay here with the donkey while the boy and I go up there. We will worship and then return to you." Abraham took the wood for the burnt offering and put it on his son Isaac. Then he took the fire and the knife in his hand, and the two of them walked on together. Isaac said to his father Abraham, "My father?" "What is it, my son?" he replied. "Here is the fire and the wood," Isaac said, "but where is the lamb for the burnt offering?" "G-d will provide for himself the lamb for the burnt offering, my son," Abraham replied. The two of them continued on together. When they came to the place G-d had told him about, Abraham built the altar there and arranged the wood on it. Next he tied up his son Isaac and placed him on the altar on top of the wood. Then Abraham reached out his hand, took the knife, and prepared to slaughter his son. **But the Lord's angel called to him from heaven,** "Abraham! Abraham!" "Here I am!" he answered. "Do not harm the boy!" the angel said. "Do not do anything to him, **for now I know that you fear G-d because you did not withhold your son, your only son, from me."** Abraham looked up and saw behind him a ram caught in the bushes by its horns. So he went over and got the ram and offered it up as a burnt offering instead of his son. And Abraham called the name of that place "The Lord provides." It is said to this day, "In the mountain of the Lord provision will be made." **The Lord's angel called to Abraham a second time from heaven** and said, **"'I solemnly swear by my own name,' decrees the Lord, 'that because you have done this and have not withheld your son, your only son, I will indeed bless you,** and I will greatly multiply your descendants so that they will be as countless as the stars in the sky or the grains of sand on the seashore. Your descendants will take possession of the strongholds of their enemies. **Because you have obeyed me,** all the nations of the earth will pronounce blessings on one another using the name of your descendants.

The question remains, is G-d contradicting Himself when He declared the 6th commandment that thou shalt not kill? If the atheists had their way, they would claim that G-d violated His own commandment in ordering Abraham to kill his only son Isaac. There are also other places in scripture that G-d had ordered the Jews to kill thousands of people in order to obtain a homeland in the Middle East. This almost sounds like and open and shut case for the hypocrisy of the G-d of the Bible. So to answer the question adequately of whether or not G-d breaks His own commandments, we need to determine if G-d actually committed the crime of "murder" (such as, killing people without cause). When reading certain parts of the Bible we find the answer to this. The most notorious example of G-d killing is the Great Flood. Another example is when G-d ordered the Jewish people to kill ALL the people in Canaan. If indeed G-d ordered or participated in the killing of innocent people, then He would be guilty of murder. Let's look at two of the most well known examples.

According to the Bible, G-d killed every human except Noah, his wife and his three sons, and their wives in the great flood. The question is, were the other people alive at that time killed unjustly besides Noah and his family? The Bible specifically tells us that ALL people (except Noah and his family) had become corrupted. Not only had all people become corrupted, but also they were continually evil. Genesis 6:5 - And G-d saw that the wickedness of man was great in the earth, and that every imagination of the thoughts of his heart was only evil continually. Genesis 6:11-12 - Now the earth was corrupt in G-d's sight and was full of violence. G-d saw how corrupt the earth had become, for all the people on earth had corrupted their ways. Therefore, the Bible indicates that no "innocent" people were killed in the flood, as they were all corrupt and evil.

There was another time in Biblical texts that G-d orders the killing of every man, woman and child in Canaan. Joshua 10:40 - So Joshua smote all the country of the hills, and of the south, and of the vale, and of the springs, and all their kings: he left none remaining, but utterly destroyed all that breathed, as the Lord G-d of Israel commanded. Joshua was given this command from G-d because these people were burning their own sons and daughters in sacrifices to their god's. Deuteronomy 12:31 - Thou shalt not do so unto the Lord thy G-d: for every abomination to the Lord, which he hateth, have they done unto their gods; for even their sons and their daughters they have burnt in the fire to their gods. So, we can see that these people were not innocent either. Some ask why the children of these corrupt people have to suffer the same fate of death as well. This is because some people tend to assume that children are innocent, even if their parents are doing bad things. However, these children are already indoctrinated with the same evil tendency as their parents. Corruption literally does breed corruption, which is why G-d could not chance a tainting of this sort.

Conclusion… The Hebrew translation of "thou shalt not kill" translates as "do not put anyone to death without cause" which nearly always refers to intentional killing without cause. Therefore, the commandment "thou shalt not kill" is rather referring to premeditated, unjustified killing, which is considered "murder". Additionally, there is a difference between killing and murder, and this is differentiated by the two separate Hebrew words. One that can be found in the Ten Commandments, which reads "Lo Tirtzach" (which means thou shalt not murder). While the other (not to kill) reads "Lo Teherog". The two different (but similar) words determine that there is a difference between killing in war and murdering someone. Even though G-d ordered the extermination of entire cities, He did so in righteous judgment on a people whose corruption had led to extreme wickedness. Therefore, G-d is not a murderer. Addendum to this, we must not forget that "logic" (for killing or murder) is also another part of G-d's creation. In addition, He who created logic is permitted to disregard it, perhaps for reasons that we as humans do not yet understand. Nevertheless, as far as we know, G-d has not broken any of His commandments.

Now, concluding on the Abraham sacrificing Isaac issue… There arose another problem in conversing with several Rabbis on this matter. They strongly asserted that I was not interpreting my Hebrew correctly. They all seemed to hinge on the notion that G-d never truly asked Abraham to sacrifice Isaac. According to the Hebrew interpretation, G-d only asked him to raise him up as a sacrifice. They say G-d wanted Abraham to say no! And they conclude that Abraham flunked this test of G-d because he was tempted. And they go on to make another assumption that after G-d had tempted Abraham He never spoke directly to him again on the basis that it was always an angel of G-d, and not G-d directly who spoke to Abraham, as if to say he flunked the test. And an even more twisted assumption on their part is that G-d used this as an example of "what not to do". Thus, the Jews theorize that it was a trial example that humans were never to be used as a sacrifice for sin - thereby eliminating the acceptance of Jesus dying as an acceptable sacrifice for sin. But these are all lame excuses since this may have actually been an "introduction" for what was to come. And this introduction was the leading concept behind the sacrifice of G-d's own son (Jesus) who was the sacrificial lamb that was substituted in the thicket. Genesis 22:13 - And Abraham lifted up his eyes, and looked, **and behold behind him a <u>ram</u> caught in a thicket by his horns:** and Abraham went and took the ram, and offered him up for a burnt offering in the stead of his son. Now there is a difference between a lamb and a ram. A lamb is a young sheep less than 1 year old, and a ram is specifically an older male sheep. This is telling us that it is an adult male sheep that was used for the sacrifice, instead of a young sheep in place of young Isaac. Though Isaac wasn't really that young, he was approximately about 30 years of age and could have easily overpowered his father Abraham if he truly wanted to, but instead went along willingly with the sacrifice upon his life. He was a son who willingly submitted himself to his father much as Jesus also later willingly submitted Himself to G-d on the cross. The "ram" was in fact a prelude to an adult male that was to be sacrificed. In addition, here is another startling realization. Jesus was sacrificed as an older lamb, which may have symbolically been in relationship to Isaac who was also around 30 years of age (just like Jesus). Moreover, the significance of the actual day that the lamb is eaten is of a HUGE importance! It will help us to further understand the significance of sacrifice. The lamb was eaten on the 15th of Nissan, since the 14th (preparation day) is a day that is specifically mentioned for "observance". Because it is this night that the lamb is killed in the evening. This observance was specifically set aside in "remembrance" for the time when the spirit of G-d (as the destroying plague) passed over the houses of the Jews in Egypt that had the lambs blood on the doorposts. While this last plague of death was passing over, the Jews sat together safely in their houses enumerating over the special hallowed significance of this day as they ate the Passover lamb. This was the sacrificial lamb whose blood enabled Israel to escape the judgment of G-d. The Talmud tells us that the Passover lamb had to be eaten by "midnight", and recorded in Antiquities, Josephus, (who was a Pharisee living in Jesus' day) tells us that the law for the Passover lamb called for it to be eaten during the night with nothing left for morning.

So, the 15th is the actual day that the Lamb is eaten, but it is the evening/night of the 14th that the lamb is actually killed. And as we can see there are HUGE conflicting ideas floating around here between the two days even though they go hand-in-hand. The 14th is for spiritual reasons, and the 15th is for the flesh, which recalls both slavery and redemption (here again we witness the twofold dimensional nature of G-d). Without question, the sacrificial lamb of the Exodus Passover was eaten on the 15th day of Nissan, but it just wasn't the eating of the lamb that saved the Jews in Egypt from the destroyer that "passed over" them that night. I am bereaved (at a loss) that the Jews (with a mis-focused perspective) cannot grasp the concept that this same sacrificial lamb (as Jesus) was killed and its blood shed and painted on the doorposts the night of the 14th day of Nissan, and that this lamb's blood was what protected them from the death of their first born, (they were saved from their sins of death). In effect, sin literally passed over! All G-d had to see was the lamb's blood on the doorposts (the night of the 14th) that caused the "passing over" of the plague (sin that brought death). In essence, **on this particular night of the 14th, the Exodus lamb was not yet meant to be a fleshly eating, but rather a spiritual blessing for the protection of the flesh.** This was the demonstrating factor of Jesus in place of the sacrificial lamb. He indeed was our "true spiritual sacrifice" who gave His body, blood and life as the sin-bearing lamb for all humankind. Exodus 12:13 - And the blood shall be to you for a token upon the houses where ye are: and when I see the blood, I will pass over you, and the plague shall not be upon you to destroy you, when I smite the land of Egypt. Exodus 12:21-27 - Then Moses called for all the elders of Israel, and said unto them, Draw out and take you a lamb according to your families, and kill the Passover. And ye shall take a bunch of hyssop, and dip it in the blood that is in the bason, and strike the lintel and the two side posts with the blood that is in the bason; and none of you shall go out at the door of his house until the morning. For the Lord will pass through to smite the Egyptians; and when he seeth the blood upon the lintel, and on the two side posts, the Lord will pass over the door, and will not suffer the destroyer to come in unto your houses to smite you. And ye shall observe this thing for an ordinance to thee and to thy sons for ever. And it shall come to pass, when ye be come to the land which the Lord will give you, according as he hath promised, that ye shall keep this service. And it shall come to pass, when your children shall say unto you, What mean ye by this service? That ye shall say, It is the sacrifice of the Lord's Passover, who passed over the houses of the children of Israel in Egypt, when he smote the Egyptians, and delivered our houses. And the people bowed the head and worshipped.

In essence Jesus clarified for us in a very specific way that He was the "true spiritual sacrifice" that took place on the 14th of Nissan and was presented before G-d as the protection from death (sin). This is the actual day that He died on the cross, in the evening on this day and was put in His burial place minutes before the 15th Sabbath (none of the crucifixion took place on the actual day of the 15th).

The lesson to be learned here… The eating of the lamb that night along with the bitter herbs and unleavened bread was significant for the Jews to remember the great miracle that G-d had done for them THAT NIGHT in causing the plague of death (sin) to pass over them.

This day is also significant for the Jews as a day of great joy and liberation because they were finally able to leave Egypt after experiencing great bondage. This was a lesson for all generations to learn and remember. Numbers 9:11- They are to celebrate it on the fourteenth day of the second month at twilight. They are to eat the lamb, together with unleavened bread and bitter herbs. In conclusion, our true required spiritual Passover is on the 14th (the time Jesus' sacrifice actually began). In addition, the bona fide day of the 15th is what G-d calls "a day that is to be observed, remembered and kept". Exodus 12:42 - Because the Lord kept vigil that night to bring them out of Egypt, on this night all the Israelites are to keep vigil to honor the Lord for the generations to come. It can surely be said that Jesus fulfilled Moses' prophecy of the true intentional purpose and significance of the Passover lamb, because it is through His blood that we can be saved from sin, or passed over by death.

Another question also arises... One might wonder, how is it then that Jesus ate the Passover meal (the Last Supper) in the evening with His disciples when He was supposed to be crucified at the same time as all the other Passover sacrificial lambs in the land were? This is kind of a trick question, but it ties-into the greater significance of the 14th day. And this becomes very interesting because nowhere in the New Testament is there made mention of a lamb being eaten at the Last Supper with Jesus. It was not mentioned as being part of the meal. The only foods mentioned are bread and wine. Matthew 26:17 (The Last Supper) - On the first day of the Feast of Unleavened Bread, the disciples came to Jesus and asked, Where do you want us to make preparations for you to eat the Passover? Matthew 26:26-30 - While they were eating, **Jesus took bread,** gave thanks and broke it, and gave it to his disciples, saying, Take and eat; this is my body. Then he took the cup, gave thanks and offered it to them, saying, Drink from it, all of you. This is my blood of the covenant, which is poured out for many for the forgiveness of sins. I tell you, I will not drink of this **fruit of the vine** from now on until that day when I drink it anew with you in my Father's kingdom. When they had sung a hymn, they went out to the Mount of Olives.

So why was there no lamb eaten at this particular time? Well lets examine what happened next... After eating only bread and wine, they had gone out to the "Mount of Olives". In addition, from there they went to "Gethsemane" where Jesus was arrested. This all took place in the evening, and at sunrise early in the morning (the next day) Jesus was handed over to Pilate. There was evening and there was morning, the 14th day. Matthew 27:1-2 - Early in the morning, all the chief priests and the elders of the people came to the decision to put Jesus to death. They bound him, led him away and handed him over to Pilate, the governor. And it was in the evening on this day (the 14th - which was a Friday) that the disciple Matthew tells us that Jesus died at the ninth hour (which is three o'clock in the evening). He died at the exact moment when all the other lambs in the land were being slaughtered. Matthew 27:46 - About the ninth hour Jesus cried out in a loud voice, Eloi, Eloi, lama sabachthani? which means, My G-d, my G-d, why have you forsaken me? But remember Jesus was not actually saying that G-d had forsaken Him, He was merely quoting the prophesy about Himself from the Psalms (Psalms 22:1-31).

Now seeing that there was no lamb eaten at the particular time when Jesus had His last supper with the disciples... this is basically telling us that it must have been the day beforehand (Thursday the 13th that He had His last supper). This would be indicating that Jesus' meal represented a break (rather a new opening) in the old covenant law and tradition. He clarifies this for us when He tells us that He is the NEW SPIRITUAL COVENANT that is given in His blood. 1 Corinthians 11:25 - In the same way, after supper he took the cup, saying, **This cup is the "new covenant" in my blood;** do this, whenever you drink it, in remembrance of me. 1 Corinthians 5:7 - Get rid of the old yeast that you may be a new batch without yeast—as you really are. For Christ, our **Passover lamb**, has been sacrificed.

Such an observation is made more prominent that Jesus and His disciples celebrated the Passover meal a day in advance (on the 13th) before the start of the actual Passover on the evening of the 14th. We can detect and determine this when we read Luke's account in the New Testament. Luke 22:15-18 - And he said to them, **I have eagerly desired to eat this Passover with you before I suffer.** For **I tell you, I will not eat it again until it finds fulfillment in the kingdom of G-d.** After taking the cup, he gave thanks and said, Take this and divide it among you. **For I tell you I will not drink again of the fruit of the vine until the kingdom of G-d comes.** If you read the highlighted words here accurately, you can detect that "twice" Jesus emphasized to His disciples that He would not be sharing this meal with them again for a long time. We also read how He had eagerly wanted to eat the Passover with them before His impending suffering. The reason Jesus responded "twice" in this way was because the disciples must have been asking Him if they will be doing this whole Passover ritual all over again tomorrow (seeing as there was no lamb)? Therefore, as we can see the timing is in accordance to what Jesus knew (in His mind) what was going to happen next and immanently transpire upon the world stage for generations to come. There is one further observation that is also essential here… We can almost hear the disciples asking Jesus, just as if Isaac asked Abraham, "behold, the unleavened bread and the bitter herbs, but WHERE IS THE LAMB?" Genesis 22:7 - Isaac spoke up and said to his father Abraham, Father? Yes, my son? Abraham replied. The fire and wood are here, Isaac said, but **where is the lamb** for the burnt offering? And like Abraham on the same mountain (Mt. Moriah) several thousand years earlier, Jesus' answer to His disciples was, "G-d will provide for Himself the lamb for the sacrifice". Genesis 22:8 - Abraham answered, G-d himself will provide the lamb for the burnt offering, my son. And the two of them went on together.

There is even something of greater significance here… Lets again revisit when the sacrifice of Isaac had been called off, and Abraham noticed that behind him was the ram that was caught in the thicket by its horns. The capture of the ram in the thicket is significant for several reasons. Firstly, scripture states it was a ram. This makes perfect sense because a lamb doesn't have horns yet and therefore can't be caught in a thicket as a ram can. Secondly, if the ram had been trapped in the thicket any other way, it might have been blemished and therefore defective and imperfect as a sacrifice animal. Thirdly, since the ram had been held in place by its horns, which may suggest that, something more than the thicket was holding it in place since horns are the strong point of strength. Lastly, the ram was offered in the place of Isaac, who clearly illustrates (no matter which way the Jews choose to look at it), the substitutional nature of sacrifice between a human and an animal (lamb/ram).

However to fully understand the inner recesses in all of this, G-d did not want the sacrifice of a son but rather the surrender of a father. What G-d had desired was not Isaac's life, but Abraham's loyalty and faithfulness. It is a misconception to think that G-d wants anything from us when He asks for sacrifice; rather what He really-truly wants from us is our complete obedience. Sacrifice is nothing other than the surrender of the will. (Again, as aforementioned), this is the only way anything could conclude because of total self-sacrifice (Jesus giving His life for us on the cross as a pardon for sins).

Moreover, the Jews need to give this further considerable thought since G-d called for the sacrifice of Isaac on a mountain that He revealed to Abraham, which was called Mount Moriah. The name "Moriah" actually means "shown/vision of G-d". It was this actual mountain that was later the temple site for Solomon's temple. 2 Chronicles 3:1 - Then Solomon began to build **the house of the Lord at Jerusalem in mount Moriah**, where the Lord appeared unto David his father, in the place that David had prepared in the threshingfloor of Ornan the Jebusite. Additionally, after receiving the command from G-d to sacrifice Isaac, it took Abraham 3 days from his camp at Beersheba to arrive at Mount Mariah (the site G-d had shown to him). Genesis 22:2-5 - Then G-d said, Take your son, your only son, Isaac, whom you love, and go to the region of Moriah. Sacrifice him there as a burnt offering on one of the mountains I will tell you about. Early the next morning Abraham got up and saddled his donkey. He took with him two of his servants and his son Isaac. When he had cut enough wood for the burnt offering, he set out for the place G-d had told him about. **On the third day Abraham looked up and saw the place in the distance.** He said to his servants, Stay here with the donkey while I and the boy go over there. We will worship and then we will come back to you.

Without even realizing it, this 3 days is a symbolic reference to the "resurrection" of Jesus. In fact, much of this whole account has a huge representational significance to the offering of Jesus by G-d the Father on Mount Moriah (which is Calvary) later. Hebrews 11:17-19 - By faith Abraham, when he was tried, offered up Isaac: and he that had received the promises offered up his only begotten son, Of whom it was said, That in Isaac shall thy seed be called: Accounting that G-d was able to raise him up, even from the dead; from whence also he received him in a figure. Exodus 19:11 - and be ready by **the third day,** because on that day the Lord will come down on Mount Sinai in the sight of all the people. Calvary is a hill just outside the city walls of ancient Jerusalem where the crucifixion of Jesus took place according to the Bible. In Hebrew, this hill is called "Golgotha", and it was so called this from the shape of the hill (which means the place of the skull). It was on this very same hill that Abraham was looking forward to that distant day when G-d in human flesh would offer Himself as the ultimate burnt offering and sacrifice for sin.

There is something else of interest here even though it is taken out of context... This has to do with the time that Jesus was captured before He was brought to crucifixion. Jesus went to the Garden of "Gethsemane" mentioned in John 18:1 - When he had finished praying, Jesus left with his disciples and crossed the Kidron Valley. On the other side there was an olive grove, and he and his disciples went into it. Mark 14:32 - They went to a place called Gethsemane, and Jesus said to his disciples, Sit here while I pray. Matthew 26:36 - Then Jesus went with his disciples to a place called Gethsemane, and he said to them, Sit here while I go over there and pray. He took Peter and the two sons of Zebedee along with him, and he began to be sorrowful and troubled. Then he said to them, My soul is overwhelmed with sorrow to the point of death. Stay here and keep watch with me. The name "Gethsemane" is very typical, perhaps wholly why Jesus chooses this place... since it comes from the Hebrew word "Gat Shmanim" - meaning, "oil press". In addition, since oil is used in the Bible to symbolize the Holy Spirit, it follows that this garden is where "the Spirit of G-d was crushed". Because it was here where Jesus agonized in prayer over what was to occur next. Matthew 27:26 - Then released he Barabbas unto them: and when he had scourged Jesus, he delivered him to be crucified.

So lets recap in conclusion here… the events on Mount Moriah involving Isaac and the ram in his stead were a future figurative (almost allegorical) vision of the Gospel (the story of the life, death, and teachings of Jesus recorded in the New Testament). Allegorical in terms of a work in which the characters and event are to be understood as representing other things (acted out by human and animal) and symbolically expressing a deeper spiritual and moral meaning. It is also forthcoming to know that both Mount Moriah and Calvary are identified within the city of Jerusalem as the same exact place. It was also Abraham's son Isaac, like Jesus, that was the obedient son unto death. Philippians 2:8 - And being found in fashion as a man, he humbled himself, and became obedient unto death, even the death of the cross. As well as Abraham being characteristic of G-d the Father who did "not spare His own son, but delivered Him up for us all" Romans 8:32 - He that spared not his own Son, but delivered him up for us all, how shall he not with him also freely give us all things? And the ram, which was offered in Isaac's place is distinctive of Jesus offered in our place. Hebrews 10:5-10 - Therefore, when Christ came into the world, he said: Sacrifice and offering you did not desire, but a body you prepared for me; with burnt offerings and sin offerings you were not pleased. Then I said, Here I am—it is written about me in the scroll—I have come to do your will, O G-d. First he said, Sacrifices and offerings, burnt offerings and sin offerings you did not desire, nor were you pleased with them (although the law required them to be made). Then he said, Here I am, I have come to do your will. He sets aside the first to establish the second. And by that will, we have been made holy through the sacrifice of the body of Jesus Christ once for all. Finally, the continuation (preservation) of Isaac is viewed as equally typical of the resurrection in the New Testament. Hebrews 11:17-19 - By faith Abraham, when he was tried, offered up Isaac: and he that had received the promises offered up his only begotten son, Of whom it was said, That in Isaac shall thy seed be called: Accounting that G-d was able to raise him up, even from the dead; from whence also he received him in a figure. Some biblical scholars also view Abraham's statement in Genesis 22:14 which states, "On the mountain of the Lord it will be provided", as a prophecy that upon this holy spot (Mt. Moriah/Calvary) G-d would provide the substitutional sacrifice of Jesus (His son). Genesis 22:14 - So Abraham called that place The Lord Will Provide. And to this day it is said, **"On the mountain of the Lord it will be provided".** Several things here are a type of microcosm… Both G-d and Abraham offered their sons as an offering on Mt. Moriah, and both Isaac and Jesus are of supernatural birth.

Talking about a "resurrection" of sorts, (and perhaps deviating off subject somewhat)… Many people have no inkling about this one… that the prophesy of the "Holocaust" was foretold by the prophet Ezekiel CENTURIES BEFORE it even happened. Ezekiel was a priest in the Bible who prophesied for approximately 22 years sometime in the 6th century BC while exiled in Babylon. Traditionally, the book of Ezekiel is thought to have been written in 500(s) BC. Ezekiel 37:1-14 (The valley of dry bones) - The hand of the Lord was upon me, and he brought me out by the Spirit of the Lord and set me in the middle of a valley; it was full of bones. He led me back and forth among them, and **I saw a great many bones on the floor of the valley, bones that were very dry.** He asked me, Son of man, can these bones live? I said, O Sovereign Lord, you alone know. Then he said to me, Prophesy to these bones and say to them, Dry bones, hear the word of the Lord! This is what the Sovereign Lord says to these bones: **I will make breath enter you, and you will come to life. I will attach tendons to you and make flesh come upon you and cover you with skin; I will put breath in you, and you will come to life.** Then you will know that I am the Lord. So I prophesied as I was commanded. And as I was prophesying, there was a noise, a rattling sound, and the bones came together, bone to bone. I looked, and tendons and flesh appeared on them and skin covered them, but there was no breath in them. Then he said to me, Prophesy to the breath; prophesy, son of man, and say to it, This is what the Sovereign Lord says: **Come from the four winds,** O breath, and breathe into these slain, that they may live. So I prophesied as he commanded me, and breath entered them; they came to life and stood up on their feet—a vast army. Then he said to me: Son of man, **these bones are the whole house of Israel.** They say, Our bones are dried up and our hope is gone; we are cut off. Therefore prophesy and say to them: This is what the Sovereign Lord says: O my people, **I am going to open your graves and bring you up from them; I will bring you back to the land of Israel.** Then you, my people, will know that I am the Lord, when I open your graves and bring you up from them. I will put my Spirit in you and you will live, and I will settle you in your own land. Then you will know that I the Lord have spoken, and I have done it, declares the Lord. The dry bones are the 6 million dead bodies of Jewish people who died in the Holocaust. To dispose of the bodies, the Nazis literally dumped them down hills and into the gorges and valleys. Today these mysterious words (found in the prophesy of Ezekiel) foreshadow the "population of Israel" which is now approximately 6 million Jews (and rising). The Jews have literally been "resurrected" just as the prophecy states.

What is also missed by the Jews here is a twofold deliverance from Egypt that is in every way related to Jesus' ministry... Since Israel's deliverance from Egypt was by blood and by water - blood by the lamb, and water baptism when they crossed the sea opened by G-d / Moses. With the blood (and belief) of the sacrificial lamb upon one (Jesus), along with baptism by water - are the first steps of obedience and observance for a Judeo-Christian.

Moving forwards... I can assure you that the Jews do not stop here to demagnetize the legitimacy of Jesus as the Messiah. Aside from all this, there is something else that the Jews bring forth which they claim is essentially missing in determining Jesus as the prophesized Messiah. This claim beckons us to revisit Daniels prophecy. It seems that the Jews have conveniently excluded Daniels prophesy about the coming Messiah which was previously mentioned (and determined) earlier in this chapter that Jesus did in fact come at that precise time in prophesized biblical history. Daniel 9:25 - Know therefore and understand, that **from the going forth of the commandment to restore and to build Jerusalem unto the Messiah** the Prince shall be seven weeks, and threescore and two weeks: the street shall be built again, and the wall, even in troublous times. However, the Jews refute this as well, and exclude Daniels prophesy even though Daniel was a prominent Old Testament prophet. They surmise that Daniels prophesy is incorrect because the rebuilding of Jerusalem did not start with the Persian Emperor Artaxerxes who issued the edict to rebuild Jerusalem on the 1st of Nissan in the 20th year of his reign (March 5th 444BC), as this was determined from Nehemiah 2:1-8 which was quoted earlier in this chapter. Rather, the Jews had determined that the actual edict to rebuild Jerusalem was not given by Artaxerxes but was given by "Cyrus" in 537BC (who was given the credit in Isaiah 45:1 and Isaiah 45:3). Isaiah 45:13 - Thus saith the Lord to his anointed, to **"Cyrus",** whose right hand I have holden, to subdue nations before him; and I will loose the loins of kings, to open before him the two leaved gates; and the gates shall not be shut. Isaiah 45:13 - I have raised him up in righteousness, and I will direct all his ways: **he shall build my city,** and he shall let go my captives, not for price nor reward, saith the Lord of hosts.

However, this can also be disputed, as pointed out in the New Testament when Peter asks Jesus how many times should he forgive, and Jesus "prophetically" answers in Matthew 18:21-23 - Then came Peter to him, and said, Lord, how oft shall my brother sin against me, and I forgive him? till seven times? Jesus saith unto him, I say not unto thee, Until seven times: but, **Until seventy times seven.** Therefore is the kingdom of heaven likened unto a certain king, which would take account of his servants. This is a prophetic prophecy where Jesus was again referring to Himself as "Daniels message" from the angel Gabriel concerning the prophesized Messiah - 70x7=490. G-d had given the Jews only 490 years (483 from the decree to rebuild Jerusalem plus 7 more years to acknowledge, accept and receive their Messiah). There is only three places in the Old and New Testament where it makes mention to "seven times" and "seventy times seven" and this can be found firstly in Genesis 4:24 - If Cain is avenged **seven times, then Lamech seventy-seven times.** And secondly in Matthew 18:22 - Jesus answered, I tell you, not **seven times, but seventy-seven times.** And thirdly of course Daniel 9:24-26 - Seventy weeks are determined upon thy people and upon thy holy city, to finish the transgression, and to make an end of sins, and to make reconciliation for iniquity, and to bring in everlasting righteousness, and to seal up the vision and prophecy, and to anoint the most Holy. Know therefore and understand, that from the going forth of the commandment to restore and to build Jerusalem unto the Messiah the Prince shall be **seven weeks, and threescore and two weeks (70x7)**: the street shall be built again, and the wall, even in troublous times. And after threescore and two weeks shall Messiah be cut off, but not for himself: and the people of the prince that shall come shall destroy the city and the sanctuary; and the end thereof shall be with a flood, and unto the end of the war desolations are determined.

So what we do know from these passages above is that Jesus Himself was clarifying for us that Daniels prophesy was indeed the proper estimated time period that would usher in the Messiah (which is precisely what happened). However, if as the Jewish scholars claim the edict went out to rebuild Jerusalem in 537BC by Cyrus as apposed to Artaxerxes in 444 BC, then our Messiah would have arrived on the scene some 93 years earlier than previous calculated. So where was our Messiah? Obviously, He never came at that particular time, because He wasn't there yet.

Therefore, this brings us back to the question of why the Jews do not except (or even acknowledge) their Messiah, and why they cleave so stringently to the Old Testament. Well firstly, I quickly realized that it is just another part of G-d's purpose and plan concerning future prophesy. Just as G-d did to Pharaoh, He is hardening our hearts so that we might further believe. Exodus 9:14-16 - Or this time I will send the full force of my plagues against you and against your officials and your people, **so you may know that there is no one like me in all the earth.** For by now I could have stretched out my hand and struck you and your people with a plague that would have wiped you off the earth. **But I have raised you up for this very purpose, that I might show you my power and that my name might be proclaimed in all the earth.** Exodus 10:1-2 Then the Lord said to Moses, Go to Pharaoh, for **I have hardened his heart and the hearts of his officials so that I may perform these miraculous signs of mine among them that you may tell your children and grandchildren how I dealt harshly with the Egyptians and how I performed my signs among them, and that you may know that I am the Lord.**

Secondly, Israel is blind of Messiah until fulfillment. The Bible (notably the New Testament) is considered a "theanthropic" book to the Jews (meaning is has the nature of both G-d and human). It is both divine and human in nature or quality and in accordance with Jesus. Remnants of Jesus can be found IN ALL THE SCRIPTURES especially those of the Old Testament (including the book of Enoch), but the Jewish scholars have a blockage (their hearts are hardened) and for that reason are unable to discern (and interpret) it properly. And because of this nature the Jews that are spiritually blind will neither read the New Testament nor believe in Jesus. Romans 11:8 - According as it is written, **G-d hath given them the spirit of slumber, eyes that they should not see, and ears that they should not hear;** unto this day. Deuteronomy 29:4 - **But to this day the Lord has not given you a mind that understands or eyes that see or ears that hear.** Romans 11:25-27 - For I would not, brethren, that ye should be ignorant of this mystery, lest ye should be wise in your own conceits; that **blindness in part is happened to Israel, until the fulness of the Gentiles be come in. And so all Israel shall be saved: as it is written, There shall come out of Sion the Deliverer, and shall turn away ungodliness from Jacob: For this is my covenant unto them, when I shall take away their sins.**

However, this will eventually change! In the end of days the Jews (whose hearts have been hardened against Jesus their Messiah) will again witness the full resurgent power of G-d when they encounter plagues and great storms of rain and hailstones with fire, brimstone and pestilence like what happened in Egypt will happen again so that all may witness the full power of G-d again, and BELIEVE! Ezekiel 38:22-23 - And I will plead against him with pestilence and with blood; and I will rain upon him, and upon his bands, and upon the many people that are with him, an overflowing rain, and great hailstones, fire, and brimstone. Thus will I magnify myself, and sanctify myself; and I will be known in the eyes of many nations, and they shall know that I am the Lord.

The Jew cannot help it anymore than the Christian. The Jews cling unswervingly to the Old Testament (5 books of Moses) and strongly believe that believing in another human being is strictly a pagan theology, and the Christians excepting and steadfastly believe in Jesus as the prophesized Messiah. Both these religions without realizing it yet, lack the ability to see that "the later is prerequisite to the former" and thus not allowing them to merge as one. Because unbelievably, some Christians refuse to associate Christianity with Judaism. They segregate the two as something completely different from one another, and neglect to see that all of Christianity (as well as all other religions) stemmed from Judaism. There are even anti-Semitic people that hate the Jews, and not only just hate them for baseless, unfounded and nefarious reasons, but hate them for killing Jesus! These people need to be corrected in their thinking, because it doesn't matter who killed Jesus; it was preordained, inescapable, and necessary for G-d's plan to be fulfilled. Using the excuse to persecute the Jews is going directly against holy writ. It is just another use of twisting Holy Scripture for evil ends. This type of thinking is actually a meme virus, which is a type of virus that effects the way the mind reasons and thinks about things. I call it a virus since it corrupts with a poisonous effect, especially on people's minds. Viruses are contagious just like ideas. These types of viruses have been passed down from the polluted idiosyncrasies of other people and are extremely contagious, toxic, ill intentioned, and just plain evil. Throughout history, there has always been this viral animosity and segregation towards the Jew (surprisingly as I've mentioned even from Christians). However, this segregation is not what G-d truly wants.

G-D WANTS HIS PEOPLE TO YEARN FOR "REDEMPTION"! (Salvation, deliverance, rescue, liberation and emancipation). This is all part of G-d's plan of future prophesy, which is eminent concerning the second coming of the Messiah. It is G-d's will through prophesy that ALL PEOPLES come to Him. John 6:40-47 - **For my Father's will is that everyone who looks to the Son and believes in him shall have eternal life, and I will raise him up at the last day.** At this the Jews began to grumble about him because he said, I am the bread that came down from heaven. They said, Is this not Jesus, the son of Joseph, whose father and mother we know? How can he now say, I came down from heaven? Stop grumbling among yourselves, Jesus answered. **No one can come to me unless the Father who sent me draws him,** and I will raise him up at the last day. **It is written in the Prophets: They will all be taught by G-d. Everyone who listens to the Father and learns from him comes to me.** No one has seen the Father except the one who is from G-d; only he has seen the Father. I tell you the truth, **he who believes has everlasting life.**

As Jesus said, "it is written in the Prophets", they will all be taught by G-d. Everyone who listens to the Father and learns from Him comes to me. This can be backed up by several Old Testament scriptures. Deuteronomy 31:12 - Assemble the people—men, women and children, and the aliens living in your towns—so they can **listen and learn** to fear the Lord your G-d and follow carefully all the words of this law. Job 34:2 - Hear my words, you wise men; **listen to me, you men of learning.** Proverbs 1:5 - **let the wise listen and add to their learning,** and let the discerning get guidance. This means learning, reading, and speaking the holy language, conversing in prayer, and understanding the meaning and significance of the holidays which He had firstly established upon all of creation (the heavenly luminaries which are a microcosm of everything below). Jesus was merely teaching us how to be **COMPLETE JEWS.** Thus, we must still respect the holy language, engage in Torah tradition, and observe the holidays (in some manner) just as Jesus did - as a tool of learning the significance and setting an example. However, some holidays need to be celebrated differently now, such as the Passover. This is because Jesus already came for us as the sacrificial lamb. To find out more about how to celebrate the holidays or observe traditions from both the Old and New Testament, you can contact or join with a messianic church (which supports Jews for Jesus). However, we can look to the Bible to find out just what exactly Jesus Himself though about observances and replicating the traditions of the Old Testament. Some may wonder the reason why Jesus needed to be baptized. In addition, the answer to this is that He did it to institute a sacrament. He didn't personally need John's baptism, but in a like manner, He experienced it to consecrate the sacramental rite for future generations. Matt 3:13-15 - Then Jesus came from Galilee to the Jordan to be baptized by John. But John tried to deter him, saying, I need to be baptized by you, and do you come to me? Jesus replied, Let it be so now; it is proper for us to do this to fulfill all righteousness. Then John consented.

And as also revealed in an earlier chapter, we must put more emphasis into studying the significance of the shapes and forms of the Hebrew letters themselves, in realizing how all things tie in with our creator - as the Hebrew letters (and language) are the tools in which He used in creating the world. If the Jews did not keep fast and preserve the Torah in all its perfect ways, how then are we to later learn from it? Therefore, the Jews are a necessary part in fulfilling G-d's prophecy. This is why they say that the inner calling of a Jew is to "set an example" to the world. The Jews are only fulfilling their "role and part" to keep safe the law and all the commandments. In the end, prophesy tells us that their eyes will eventually be opened. These future prophecies concerning the Jews excepting Jesus as their Messiah can be found emphasizes in the "Old Testament" where G-d is putting the Jewish people through an ultimate period of tribulation where He will re-gather all the believing Jews (in Jesus) in the world to Israel where they will then be established as the leading nation in the world during the second coming of Jesus. Deuteronomy 4:30 - When thou art in tribulation, and all these things are come upon thee, even in the latter days, if thou turn to the Lord thy G-d, and shalt be obedient unto his voice. Zechariah 12:10 - And I will pour upon the house of David, and upon the inhabitants of Jerusalem, the spirit of grace and of supplications: **and they shall look upon me whom they have pierced, and they shall mourn for him, as one mourneth for his only son,** and shall be in bitterness for him, as one that is in bitterness for his firstborn. Zechariah 13:8-9 - And it shall come to pass, that in all the land, saith the Lord **two parts therein shall be cut off and die; but the third shall be left therein. And I will bring the third part through the fire, and will refine them as silver is refined, and will try them as gold is tried:** they shall call on my name, and I will hear them: I will say, It is my people: and they shall say, The Lord is my G-d. Samuel 7:9-10 - And I was with thee whithersoever thou wentest, and have cut off all thine enemies out of thy sight, and have made thee a great name, like unto the name of the great men that are in the earth. **Moreover I will appoint a place for my people Israel, and will plant them, that they may dwell in a place of their own, and move no more; neither shall the children of wickedness afflict them any more, as beforetime.** Micah 4:1-7 - **But in the last days it shall come to pass,** that the mountain of the house of the Lord shall be established in the top of the mountains, and it shall be exalted above the hills; and people shall flow unto it. **And many nations shall come, and say, Come, and let us go up to the mountain of the Lord, and to the house of the G-d of Jacob; and he will teach us of his ways, and we will walk in his paths:** for the law shall go forth of Zion, and the word of the Lord from Jerusalem. And he shall judge among many people, and rebuke strong nations afar off; and they shall beat their swords into plowshares, and their spears into pruninghooks: **nation shall not lift up a sword against nation, neither shall they learn war any more.** But they shall sit every man under his vine and under his fig tree; and none shall make them afraid: for the mouth of the Lord of hosts hath spoken it. For all people will walk every one in the name of his god, and we will walk in the name of the Lord our G-d for ever and ever. In that day, saith the Lord, will I assemble her that halteth, and I will gather her that is driven out, and her that I have afflicted; And I will make her that halted a remnant, and her that was cast far off a strong nation: and the Lord shall reign over them in mount Zion from henceforth, even for ever. Jeremiah 30:7-11 (the time of Jacobs trouble when the Jews will realize their Messiah) - Alas! for that day is great, so that none is like it: it is even the time of Jacob's trouble, but he shall be saved out of it. For it shall come to pass in that day, saith the Lord of hosts, that I will break his yoke from off thy neck, and will burst thy bonds, and strangers shall no more serve themselves of him: But **they shall serve the Lord their G-d, and David their king, whom I will raise up unto them.** Therefore fear thou not, O my servant Jacob, saith the Lord; neither be dismayed, O Israel: for, lo, I will save thee from afar, and thy seed from the land of their captivity; and Jacob shall return, and shall be in rest, and be quiet, and none shall make him afraid. **For I am with thee, saith the Lord, to save thee: though I make a full end of all nations whither I have scattered thee, yet I will not make a full end of thee: but I will correct thee in measure, and will not leave thee altogether unpunished.**

And if you believe that any of these biblical passages (cited above) are just pertaining to old and previous time prophesies - think again! These seemingly old and previous prophesies that were accurately fulfilled will pop-up again and unfold into a momentous culmination of end time events. As we have discovered, the Bible is written in parables, which gives it a special dittoing quality where the same thing applies elsewhere, especially when comparing concordances! You are gravely mistaken if you don't look past the fact that they are in part (comparatively) a small version of something larger yet to come - A "MICROCOSM" IN THE WORKINGS.

Over and above everything that religion has to teach us, G-d wants us ALL to discover that His pervading presents is always there even when we cannot see it, we should inherently know that the world is not just defined by our subjective perception - hence FAITH! John 20:29 - Jesus saith unto him, Thomas, because thou hast seen me, thou hast believed: blessed are they that have not seen, and yet have believed.

Last but not lease the most important question left to ask after reading this entire chapter? How do we know that the Bible was divinely inspired (written) by G-d? The answer is, FULFILLED PROPHESY! Isaiah 46:10 - Declaring the end from the beginning, and from ancient times the things that are not yet done. Who but G-d can proclaim the future let alone fulfill it - Jesus did just that!

Again, no other religion in the world has repeated and unfailing fulfillment of predictions that were foretold many years in advance, and all events to which the forecaster had no control over. Such as the many prophesies that were out of Jesus' control like His ancestry from King David, or His place and time of birth, along with His time of death. The odds of these fulfilled prophesies are like mixing all of them up in a whirlwind and having a blind person catch the marked one at random the very first time. There is no evidence recorded anywhere in the arsenals of writing that Jesus was a manipulator or a deceiver. In fact, He made believers out of His disciples whom He practically lived with on a daily basis, including many others in the land, even His enemies. In order to does this Jesus would have needed supernatural powers, and if He had such powers it's highly likely that He was indeed who He claimed to be - which was truly the Messiah!

If one still thinks these odds were predictable they ought to reconsider how the Old Testament prophet Micah prophesied in 700BC that the birthplace of the Messiah would take place in Bethlehem of Judea. How is it that Micah choose Bethlehem to be the birthplace of the Messiah (approximately 700 years beforehand) out of hundreds of cities (or nations) in existence of the world?

If you think that's not anything to marvel at, one should consider the fact that the prophesy made in 1012BC by King David himself (in the Psalms) that the Messiah would have His hands and feet pierced. This is a clear reference to death by crucifixion. This prophecy was made approximately 800 years before the Romans even instituted crucifixion as a form of punishment! Yes, impalement can be traced all the way to ancient Babylon and even Egypt, Genesis 40:19 - Within three days Pharaoh will lift off your head and hang you on a tree. And the birds will eat away your flesh. But this does not explain how the practice that was meant specifically for the wicked (or criminals) could bring about peace, healing or resolution. Psalm 22:16 - Dogs have surrounded me; a band of evil men has encircled me, **they have pierced my hands and my feet.** Isaiah 53:5 - But **he was pierced for our transgressions, he was crushed for our iniquities; the punishment that brought us peace**

was upon him, and by his wounds we are healed. Zechariah 12:10 - And I will pour out on the house of David and the inhabitants of Jerusalem a spirit of grace and supplication. **They will look on me, the one they have pierced, and they will mourn for him as one mourns for an only child,** and grieve bitterly for him as one grieves for a firstborn son. Revelation 1:7 - Look, he is coming with the clouds, and every eye will see him, **even those who pierced him;** and all the peoples of the earth will mourn because of him. So shall it be! Amen.

20 - REVEALING ANTICHRIST...

Some people do not believe that the forces of evil truly exist. Alternatively, they believe that evil is just an essential part of the workings of nature (keeping balance between light and darkness, good and bad, life and death). These also include acts of nature beyond the control of humankind - such as earthquakes, floods, tornadoes, disease, birth defects. In addition, it involves immoral evil acts committed by humans. Both the natural and immoral aspect of evil creates a problem for us in how we think about G-d. However, before evil can be accurately defined we must first start by examining "once again" what biblical scripture teaches us about it.

Before continuing, I would like to reaffirm for all that read that the red captioned segments highlighted in red are quoted straight out of the Bible so you know I'm not just making this STUFF up myself. Some may also ask what actual truth or authority does the Bible have? Well we can see that it's been right on track thus far, though some still stubbornly refuse to submit themselves. Therefore, my suggestion is to keep reading.

The first biblical notation of evil that creeps into "the history of humankind" starts in Genesis 2:9 - And the Lord G-d made all kinds of trees grow out of the ground - trees that were pleasing to the eye and good for food. In the middle of the garden were the tree of life and the tree of the knowledge of **good and evil.** Notice how it clearly states both "good and evil". Genesis 2:16-17 - And the Lord G-d commanded the man, You are free to eat from any tree in the garden; but you must not eat from the tree of the knowledge of good and evil, for when you eat of it **you will surely die.** Genesis 3:1 - Now **the serpent** was more crafty than any of the wild animals the Lord G-d had made. He said to the woman, Did G-d really say, You must not eat from any tree in the garden?

So as we can read and interpret evil started with "humankind" in the Garden. However, in the Bible, certain segments suggest that evil started before the creation of humankind in Heaven. One of these segments can be found in the Old Testament from the book of Job, which describes the angels worshipping G-d as He was creating the world. Job 38:4-7 - Where wast thou when I laid the foundations of the earth? declare, if thou hast understanding. Who hath laid the measures thereof, if thou knowest? or who hath stretched the line upon it? **Whereupon are the foundations thereof fastened? or who laid the corner stone thereof; When the morning stars sang together, and all the sons of G-d shouted for joy?**

This informs us that G-d had created the angels before He created the physical Earth and we know that Satan was once an angel that dwelled in G-d's great abode in Heaven. "Exactly when" G-d created the angels is hard to pin down. This may have been when He created the heavenly hosts in Genesis sometime between the 3rd and 4th day. Genesis 1:16 - And G-d made two great lights; the greater light to rule the day, and the lesser light to rule the night: he made the stars also. Genesis 2:1-4 - Thus the heavens and the earth were finished, **and all the**

host of them. And on the seventh day G-d ended his work which he had made; and he rested on the seventh day from all his work which he had made. And G-d blessed the seventh day, and sanctified it: because that in it he had rested from all his work which G-d created and made. These are the generations of the heavens and of the earth when they were created, in the day that the Lord G-d made the earth and the heavens. Psalm 148:2 - Praise him, **all his angels, praise him, all his heavenly hosts.** On the other hand, it may have even been sometime before the Genesis account. We also find in Revelation (the last book of the New Testament) that there was a war in Heaven that is perceived to be before the creation of the Earth, where Michael a good angel fights against Satan (the dragon) and wins. Which amounts to the dragon (Satan) and his hosts (1/3 of the other angels) being cast out of Heaven and into the Earth? Revelation 12:4-9 - And his tail drew the third part of the stars of heaven, and did cast them to the earth: and the dragon stood before the woman which was ready to be delivered, for to devour her child as soon as it was born. And she brought forth a man child, who was to rule all nations with a rod of iron: and her child was caught up unto G-d, and to his throne. And the woman fled into the wilderness, where she hath a place prepared of G-d, that they should feed her there a thousand two hundred and threescore days. **And there was war in heaven: Michael and his angels fought against the dragon; and the dragon fought and his angels, And prevailed not; neither was their place found any more in heaven. And the great dragon was cast out, that old serpent, called the Devil, and Satan, which deceiveth the whole world: he was cast out into the earth, and his angels were cast out with him.**

We also find recorded in the book of Ezekiel that Satan was in the Garden of Eden and it tells us that he was perfect (compared to every precious stone) until "iniquity" (evil, sin and injustice) was found in him. He corrupted himself because of his pride and beauty. This can become very confusing since we know that Satan was cast out of Heaven and down to the Earth during the war between G-d, Michael and Him, as well as the other angels. What this is clueing us into is that G-d's human creation was very good until Satan defiled it in the garden. This is also suggesting that "iniquity" was not put there by G-d. But that Satan himself had created it upon his own free will, however his evil nature was already present upon his original fall from Heaven. And this is when Satan had interrupted the infinite stream of G-d's creation by interfering with humans, and thus created upon himself the persona that he is noted for. Ezekiel 28:13-15 - Thou hast been in Eden the garden of G-d; every precious stone was thy covering, the sardius, topaz, and the diamond, the beryl, the onyx, and the jasper, the sapphire, the emerald, and the carbuncle, and gold: the workmanship of thy tabrets and of thy pipes was prepared in thee in the day that thou wast created. Thou art the anointed cherub that covereth; and I have set thee so: thou wast upon the holy mountain of G-d; thou hast walked up and down in the midst of the stones of fire. **Thou wast perfect in thy ways from the day that thou wast created, till iniquity was found in thee.** Ezekiel 28:17 - Thine heart was lifted up because of thy beauty, thou hast corrupted thy wisdom by reason of thy brightness: I will cast thee to the ground, I will lay thee before kings, that they may behold thee. In addition, this is where Satan was foolish enough to think that he was equal or greater than G-d. Isaiah 14:13 - For thou hast said in thine heart, I will ascend into heaven, I will exalt my throne above the stars of G-d: I will sit also upon the mount of the congregation, in the sides of the north.

Now all of the above captioned paragraphs specified from the Bible suggest several things… that G-d had indeed created evil contingent to free will and choice. However, the

Bible is ever so clear that G-d is not the dramatist of evil, and insists that He is incapable of doing so. However again, its makes sense that He incorporated evil in lieu of rebellion. If your not warmed up to the concept yet… I will repeat it again. G-d created it as a precautionary "contingent to free will" in the event that angels or human beings should become sinful, rebelling and defiant (which of course they/we did). You see, in order for "free will" to exist, evil is always a precursory no matter which way G-d looked at it. You see, even though evil is an unpleasant and undesirable necessity; it is needed to achieve a desired result. There could not even be "partial free will" since free will is free will. Free will included allowing us the choice of loving or rejecting G-d. Yes, G-d could have eliminated all evil in the design of His universe had He wanted to. However, He did not want beings that were perpetually frozen without free will or choice, as this would not make Him a fair G-d.

Some religious thinkers complain that G-d set Adam and Eve up to fail by placing the "Tree of knowledge" of good and evil in the garden. Nevertheless, what these religious thinkers fail to realize is that G-d had to give Adam and Eve a choice because without free will (to choose) they would literally amount to being puppets. This is not something that G-d had wanted, but He realized that true love always requires choice, and the only way to give this choice was for G-d to command something that was not allowed (i.e. TREE of Knowledge). But make no mistake about G-d's intentions here… it was not G-d's plan for Adam and Eve to experience evil or know that such a thing even existed. This may clearly explain why we find this biblical verse in Genesis. Genesis 3:5 - For G-d doth know that in the day ye eat thereof, then your eyes shall be opened, and ye shall be as gods, knowing good and evil. In other words, we would be like G-d knowing about evil. Again, this is not something that G-d had intentionally wanted.

So G-d had to be unwavering in this respect (regarding free will and choice) and had determined to permit evil even though He knew the nature of it, and the influence it would have on His creation, and what He would do to save us from it. He had determined that evil (sin) was to be overruled by good, making salvation available to us through His only Son (Jesus). In addition, for those of us who choose to be evil-free and choose to love G-d instead of rejecting Him, such a place exists for us called Heaven. 1 John 3:8 - He that committeth sin is of the devil; **for the devil sinneth from the beginning. For this purpose the Son of G-d was manifested, that he might destroy the works of the devil.** Acts 17:31 - Because he hath appointed a day, in the which he will judge the world in righteousness by that man whom he hath ordained; **whereof he hath given assurance unto all men, in that he hath raised him from the dead.**

Now lets take a closer look at what qualities are noted about angels. The Bible tells us that man was created a little lower than the angels. Psalm 8:5 - For thou hast made him a little lower than the angels, and hast crowned him with glory and honour. The angels were created to continually serve G-d and His will, while man was given free choice to serve G-d. Here is what the Bible tells us about angels… **They praise G-d.** Psalm 148:1-2 - Praise ye the Lord. Praise ye the Lord from the heavens: praise him in the heights. Praise ye him, all his angels: praise ye him, all his hosts. Isaiah 6:3 - And one cried unto another, and said, Holy, holy, holy, is the Lord of hosts: the whole earth is full of his glory. **They worship G-d.** Hebrews 1:6 - And again, when he bringeth in the firstbegotten into the world, he saith, And let all the angels of G-d worship him. Revelation 5:11-13 - And I beheld, and I heard the voice of many angels round about the throne and the beasts and the elders: and the number of them was ten thousand times ten thousand, and thousands of thousands; Saying with a loud voice, Worthy is the Lamb that was slain to receive power, and riches, and wisdom, and strength, and honour, and glory, and blessing. And every creature which is in heaven, and on the earth, and under the earth, and such as are in the sea, and all that are in them, heard I saying, Blessing, and honour, and glory, and power, be unto him that sitteth upon the throne, and unto the Lamb for ever and ever. **They serve G-d.** Psalm 103:20 - Bless the Lord, ye his angels, that excel in strength, that do his commandments, hearkening unto the voice of his word. Revelation 22:8-9 - And I John saw these things, and heard them. And when I had heard and seen, I fell down to worship before the feet of the angel which shewed me these things. Then saith he unto me, See thou do it not: for I am thy fellowservant, and of thy brethren the prophets, and of them which keep the sayings of this book: worship G-d. **They appear before G-d.** Job 1:6 - Now there was a day when the sons of G-d came to present themselves before the Lord, and Satan came also among them. **They rejoice in what G-d does.** Job 38:6-7 - Whereupon are the foundations thereof fastened? or who laid the corner stone thereof; When the morning stars sang together, and all the sons of G-d shouted for joy? **They are instruments of G-d's judgments.** Revelation 7:1 - And after these things I saw four angels standing on the four corners of the earth, holding the four winds of the earth, that the wind should not blow on the earth, nor on the sea, nor on any tree. Revelation 8:2 - And I saw the seven angels which stood before G-d; and to them were given seven trumpets. **They bring answers to prayer.** Acts 12:5-10 - Peter therefore was kept in prison: but prayer was made without ceasing of the church unto G-d for him. And when Herod would have brought him forth, the same night Peter was sleeping between two soldiers, bound with two chains: and the keepers before the door kept the prison. And, behold, the angel of the Lord came upon him, and a light shined in the prison: and he smote Peter on the side, and raised him up, saying, Arise up quickly. And his chains fell off from his hands. And the angel said unto him, Gird thyself, and bind on thy sandals. And so he did. And he saith unto him, Cast thy garment about thee, and follow me. And he went out, and followed him; and wist not that it was true which was done by the angel; but thought he saw a vision. When they were past the first and the second ward, they came unto the iron gate that leadeth unto the city; which opened to them of his own accord: and they went out, and passed on through one street; and forthwith the angel departed from him. **They care for the righteous at the time of death.** Luke 16:22 - And it came to pass, that the beggar died, and was carried by the angels into Abraham's bosom: the rich man also died, and was buried. **They do not marry or reproduce like humans.** Matthew 22:30 - For in the resurrection they neither marry, nor are given in marriage, but are as the angels of G-d in heaven. **All angels were created good, however they did not remain all good! Some angels opposed G-d under the leadership of Satan.** Matthew 25:41 - Then shall he say also unto them on the left hand, Depart from me, ye cursed, into everlasting fire, prepared for the devil and his angels. 2 Peter 2:4 - For if G-d spared not the angels that sinned, but cast them down to hell, and delivered them into chains of darkness, to be reserved unto judgment. Jude 1:6 - And the angels which kept not their first estate, but left their own habitation, he hath reserved in everlasting chains under darkness unto the judgment of the great day. Ephesians 6:12 - For we wrestle not against flesh and blood, but against principalities, against powers, against the rulers of the darkness of this world, against spiritual wickedness in high places.

Moreover, there is not one instance in the Jewish "Old Testament" Bible that mentions any angel, (Satan included), opposing G-d's will. One of the great books of the Old Testament Bible that specifically specifies this for us can be found in the book of Job (G-d's faithful servant, in which G-d permits Satan to put the virtue of Job to the test). In this story G-d permits Satan to test Job. G-d gives Satan power over Jobs property. He takes away all of Jobs riches, his livestock, his house, his servants, and his children. G-d also allows Satan to smite him with terrible boils. However, G-d had forbidden Satan to take Job's life. Both Job's wife and friends cursed G-d, but Job does not. He replies even after suffering and enduring these calamities, Job 1:20-21 - At this, Job got up and tore his robe and shaved his head. Then he fell to the ground in worship and said: Naked I came from my mother's womb, and naked I will depart. The Lord gave and the Lord has taken away; **may the name of the Lord be praised.** In all this, Job did not sin by charging G-d with wrongdoing. He replies again, Job 2:10 - He replied, **shall we accept good from G-d, and not trouble?** In all this, Job did not sin in what he said.

Inadvertent to this, Satan not having the authority to do as he pleases without G-d allowing him to do so, does not appear to be the case when we delve into the New Testament where Satan seems to roam free. Yes, Satan is characterized as that member of the divine council who watches over human activity with the main purpose of searching out peoples sins and appearing as their accuser. He is a celestial prosecutor who sees only iniquity. While comparable to angels, man was made to serve G-d also, but only through the act of total free will. G-d did not want mandatory subservience from all the beings He created, as this would not make Him a fair G-d. Therefore, in a strange humanistic way of understanding, Satan is a blessing for without the evil influence, goodness, desirable quality, righteousness, integrity, honesty and morality, would not be possible. Our faith could not be tested and our obedient relationship with G-d could not exist. We would all be glorified walking and talking machines. How then could one truly love G-d unless one had the opportunity to reject a relationship with Him first? This opportunity could not exist without Satan. In a sort of sardonic way, the creation of Satan was "good", for through the creation of this angel, love and piety became possible. It would seem that G-d allows the assault of Satan and demons onto us and turns it into a human opportunity and benefit. G-d had all the intention of bestowing every goodness onto us, FOR HUMANKINDS GOAL IN LIFE IS A SPIRITUAL AND INTELLECTUAL PERFECTION. Here we find in scripture that G-d created evil in order to provide His creation with free will. Deuteronomy 30:15 - See, I set before you today life and prosperity, death and destruction. Isaiah 45:7 - I form the light and create darkness, I bring prosperity and create disaster; I, the Lord, do all these things. But be careful in properly interpreting this scripture, because it does not say that G-d is the cause of "moral evil". It is merely saying that G-d is the one who brings catastrophe, disaster and devastation, which includes judgment, chastening, discipline and shaking us into repentance. Still we may wonder, if indeed G-d wanted us to be perfect, why did He create us in such a manner that allows us the capability to sin. Because what He wants of us is our very humanness. Sometimes we do well and sometimes we fail. However, we keep on struggling and eventually make some real change in order to create an almost unimaginable perfect world. On this path of struggling and making change, the older we get the wiser we get; we must also be humbling to those who are younger than us and have not matured into this arena yet. What separates us from the less matured is that we can see beyond our own needs and desires and learn to share our material and spiritual aspect with others. We make room for other people, stooping to their emotional level, recognizing their ideas and thoughts. The greater the span of concern we have, the further is our reach. This is also an experience of liberation and freedom. The more mature we grow the freer we are towards giving to the world at large. Bondage mentality is also a form of evil that we must contend with in order to free our mind of restricting thoughts that prevent us from becoming totally liberated. Thoughts such as thinking: I'm a down to earth kind of person… I'm trying to manage real life… I can't get into this spirituality stuff... I know who I am and I don't plan on changing… Look what I have achieved… Lets just get through another day… I am free already because I live in a land of freedom and who needs more freedom than that… Been there, done that. These are all thoughts of bondage! Now the real liberated thoughts should be: I can master my world by staying above it... I am religious to get the strength I need from G-d… I am lacking much… I have only started to grow... I don't belong to my habits... My challenges in life help me to discover the strength of my soul... I am a conductor and the world is my orchestra to play a symphony for its Creator. And also remember from two chapter ago (on expanding the quality of your life) that food can also be an inhibitor to good health and thus total liberation of both body and spirit when one is a slave to food (with the mentality that tells us that we must live to eat), when we should be thinking that we are a food liberator and only eating to live. Therefore, we must continually evaluate our way of thinking as Satan uses reverse psychology to pervert our thoughts against G-d's original design for us.

Now, G-d also allows evil to exist as an essential part of nature when we begin to realize that the physical laws of cause and effect can also be contrasting and complimentary to the cycle of life. We witness this in the operation of the Earths weather system that creates tornadoes and thunderstorms that in turn cleans, waters, and nourishes the Earth. And also in a likewise manner nature can also bring about blazing hellish fires that purposely dry out the Earth to rejuvenate forests so that seeds may germinate. The same is also true of certain kinds of bacteria that occasionally make us sick while they also equally yield chemical substances to the human body that bring about healing. In a likewise sense G-d uses Satan and his followers (1/3rd of the fallen angels) as pruning and weeding tools of His kingdom, and the morally demon influenced world as the fertilizer of His vineyard. Properly put, Satan is G-d's assistant. We have not even begun to scratch the surface of understanding such mechanisms of nature that work together for our long term benefit, let alone understand the complex and intricate interrelationships of the forces set in place by G-d on our planet, so how are we to truly define what is inherently "good or evil" when it is these very forces "together" that allow G-d to work His purpose. Romans 8:28 - And we know that all things work together for good to them that love G-d, to them who are the called according to his purpose.

All the same, evil is more or less viewed as an essential part of nature. However, G-d's true intentions were clearly indicated in the beginning of Genesis when it tells us that G-d saw that everything He created was "very good". Genesis 1:31 - And G-d saw every thing that he had made, and, behold, it was "very good". And the evening and the morning were the sixth day. This clearly tells us that "evil" was not the initial or the intentional natural balance of G-d. Evil crept in as a consequence that humankind had chosen, for it tells us in the Bible that G-d created us "very good" in the beginning, and that there was no evil, death or destruction.

Many of us have no clue as to how a world that is structured without sin/evil would be. It raises the logical question, that if there was sin/evil in the beginning how could everything be "very good", when it's evident today that death, disease and extinctions are prevalent which are all part of the law of entropy and thermodynamics (as discussed earlier). This does not equate to "very good"! So how could everything have been very good? The answer to this query is that G-d had, in effect, SUSTAINED Adam, Eve, and all living things. And we can later witness this phenomenon during the 40 years that the Jews wandered in the desert wilderness on their way to the holy land of Israel. Deuteronomy 8:4 - Thy raiment waxed not old upon thee, neither did thy foot swell, these forty years. Deuteronomy 29:5 - And I have led you forty years in the wilderness: your clothes are not waxen old upon you, and thy shoe is not waxen old upon thy foot. Nehemiah 9:21 - Yea, forty years didst thou **sustain** them in the wilderness, so that they lacked nothing; their clothes waxed not old, and their feet swelled not. If G-d had indeed sustained us while we were in the Garden of Eden, He also had sustained the animals as well. This should give us a good idea about the existence of dinosaurs and their enormous size. Fact in point, reptiles do not stop growing. They have no definite instruction in their DNA that indicates when to stop growing (which is also dependent upon their environment). For instance, snakes, fish, lizards and alligators keep growing and growing, and this even includes some mammals like Kangaroos. They seem to keep growing until disease, old age or another predator takes them down. Another contingent that inhibits growth is the body organs - the skeleton may continue to grow, but the organs remain the same size, so the size of the body is constrained by the ability of the organs to support it (dinosaurs had a brain the size of a nut). Reptiles are also cold blooded - so dependant on their environment in order to regulate their body temperature (if its cold) they may need to hibernate - in which hibernation stops growth. Thus in a warm environment cold-blooded animals grow faster, but in a cold environment they stop growing in order to conserve energy. If indeed G-d had sustained everything, dinosaurs (which are reptiles) would not have died and they would have continued to grow bigger and bigger. Job was a witness to the existence of living dinosaurs. Job 40:15-24 - Behold now **behemoth,** which I made with thee; he eateth grass as an ox. Lo now, his strength is in his loins, and his force is in the navel of his belly. He moveth his **tail like a cedar**: the sinews of his stones are wrapped together. His bones are as strong pieces of brass; his bones are like bars of iron. He is the chief of the ways of G-d: he that made him can make his sword to approach unto him. Surely the mountains bring him forth food, where all the beasts of the field play. He lieth under the shady trees, in the covert of the reed, and fens. The shady trees cover him with their shadow; the willows of the brook compass him about. Behold, he drinketh up a river, and hasteth not: he trusteth that he can draw up Jordan into his mouth. He taketh it with his eyes: his nose pierceth through snares. It is of my opinion that huge dinosaurs did not die because of an asteroid, but rather Noah's great flood. Ignorance is bliss concerning this subject. Nobody can know the real answer for why the dinosaurs went extinct if they don't read the Bible. If they read the Bible, they would implicitly know that before the flood the environment was different. There was no rain. Even humans lived longer lives. However, after the flood, life spans began to decrease. If the change in environment affected the lifespan of humans, which are warm-blooded creatures, then it also affected the lifespan of reptiles especially in ideal weather conditions. We know from the Bible that the state of the Earth before the flood was different consisting of a canopy of water (in some form) that encircled the Earth. It tells us this in Genesis 1:6-7 - And G-d said, Let there be a firmament in the midst of the waters, and let it divide the waters from the waters. In addition, G-d made the firmament, and divided the waters, which were under the firmament from the waters, which were above the firmament: and it was so. The firmament, as the Bible states, is the sky. Therefore, the Bible is saying that there was water above the sky as well as below it. The only thing that can successfully and almost completely block the Sun's harmful rays is water. The Earth was a sort of hyperbaric chamber (the use of oxygen operating at pressures higher than normal atmospheric pressure) allowing the beneficial rays of the Sun to shine through these high concentrations of oxygen in the tissues of living things may have been beneficial in keeping it alive longer, warding off diseases and even starving cancer since cancer cannot grow in an oxygen rich environment (as mentioned in an earlier chapter). These same conditions have been simulated successfully in a lab with stunning results. Plants that have been grown under hyperbaric conditions, after several years have increased doubling its size as well as harvest. If we still had these same conditions on Earth, both the atmospheric pressure and oxygen levels, living things would be healthier, live longer and grow to enormous sizes. So, the scenario probably went something like this... Noah most likely had taken smaller reptiles on board the ark to save space. Afterwards (upon generations of breeding) it appears they were killed for food. Psalm 74:14 - Thou brakest the heads of leviathan in pieces, and gavest him to be meat to the people inhabiting the wilderness.

Getting back on track here... So, what else does the Bible tell us about Satan/evil/sin? Well as mentioned, it starts by telling us in the Old Testament after Satan was found with iniquity, we then read that he was cursed by G-d in Genesis 3:14 - And the Lord G-d said unto the serpent, Because thou hast done this, thou art cursed above all cattle, and above every beast of the field; upon thy belly shalt thou go, and dust shalt thou eat all the days of thy life. This is G-d's way of indicating that the serpent would forever be despised as a despicable creature and an object of hatred. Therefore, by G-d cursing the serpent, it is a "future illustration" that the serpent will one day be the fate of Satan himself. Revelation 12:9 - And the great dragon was cast out, that old serpent, called the Devil, and Satan, which deceiveth the whole world: he was cast out into the earth, and his angels were cast out with him. Revelation 20:2 - And he laid hold on the dragon, that old serpent, which is the Devil, and Satan, and bound him a thousand years. Revelation 20:10 - And the devil that deceived them was cast into the lake of fire and brimstone, where the beast and the false prophet are, and shall be tormented day and night for ever and ever.

However, why would G-d want to literally destroy Satan when He had created him? What G-d is hell bent after DESTROYING is the "temptation" to sin through the act of free will? Without choice (including the will to sin) goodness could not be possible in the world.

Lets turn aside here and examine something else of great interest. As discussed in the earlier chapter regarding some of the differences of opinion the Jews have concerning Jesus, being the Messiah - for some strange reason they leave out, and rather not consider several of the Old Testament lessons that are specified regarding "SIN". In their rejection of Jesus, they neglect to recognize the symbolic resemblance between the "Fiery Serpent" mentioned in the Old Testament book of "Numbers" (made of brass) and set upon a pole, which G-d had instructed Moses to build in the wilderness for the removal of sins. What the Jews neglect to understand is that the same symbolic meaning of the "Fiery Serpent" is no different than looking to/upon Jesus for the cure of sins since He too was set upon a pole (crucified) much in the same way that the fiery serpent was set upon a pole for removal of sins. For just like the lesson of the Fiery Serpent that was used for the removal of sins for all who beheld it - later in history Jesus is the antidote. Numbers 21:5-9 - And the people spake against G-d, and against Moses, Wherefore have ye brought us up out of Egypt to die in the wilderness? for there is no bread, neither is there any water; and our soul loatheth this light bread. And the Lord sent fiery serpents among the people, and they bit the people; and much people of Israel died. Therefore the people came to Moses, and said, We have sinned, for we have spoken against the Lord, and against thee; pray unto the Lord, that he take away the serpents from us. And Moses prayed for the people. And the Lord said unto Moses, Make thee a fiery serpent, and set it upon a pole: and it shall come to pass, that every one that is bitten, when he looketh upon it, shall live. And Moses made a serpent of brass, and put it upon a pole, and it came to pass, that if a serpent had bitten any man, when he beheld the serpent of brass, he lived.

So as one can see the meaning of the brass serpent that Moses was instructed to set upon a pole is significant in understanding G-d's logic. We can also apply this concept to the holistic field of Homeopathy (as was discussed in an earlier chapter) where medicines were/are used to treat, like with like - fighting poison with poison. (Remembering in Exodus when Moses was taught by G-d how to cure bitter water with a bitter tree - we learned that, as like as cures like). In addition, it is in this same manner that G-d heals with the very same thing with which He smites. In addition, as was also mentioned earlier, this is the very same method and approach that modern medicine uses today to produce vaccines that immunize the body against deadly diseases and viruses, as illnesses could be healed by taking tiny doses of the natural substance that is actually causing the suffering. This tiny dose (to the point of being non-detectable) is mixed in with water. It is non-detectible in terms of only a "vibration" remaining of it, or in other terms, only the characteristics (or a minute amount) of the matter remains. In essence, what's actually occurring is that the "information" is copied to the water (since water retains memory) and is being used to cancel out the information of the symptoms. So as one can perceive, the serpent of brass set upon a pole by Moses was merely an antidote for sin, treating, like with like, and fighting poison with poison, (which was later fulfilled with the crucifixion of Jesus). What this literally means is that Jesus took on the sins of the world as though they were His own sins. Even though He Himself was sinless, yet by the very act of taking on our sins He appeared to G-d the Father as a sinful serpent worthy of death. This is an extremely hard concept to grip (especially for Jews), it is almost incomprehensible. Yes, looking upon another human being as though they were G-d is certainly paganistic. However, in this case (understanding G-d's logic) there appears to be an overriding clause, which G-d has allowed as the healing antidote for sin. John in the New Testament clarifies this connection for us John 3:14-18 - **And as Moses lifted up the serpent in the wilderness, even so must the Son of man be lifted up:** That whosoever believeth in him should not perish, but have eternal life. For G-d so loved the world, that he gave his only begotten Son, that whosoever believeth in him should not perish, but have everlasting life. For G-d sent not, his Son into the world to condemn the world; but that the world through him might be saved. He that believeth on him is not condemned: but he that believeth not is condemned already, because he hath not believed in the name of the only begotten Son of G-d. Peter in the New Testament tells us in 1 Peter 2:24 - Who his own self bare our sins in his own body on the tree, that we, being dead to sins, should live unto righteousness: by whose stripes ye were healed. Jesus also tells us in Luke 10:19 - **Behold, I give unto you power to tread on serpents and scorpions,** and over all the power of the enemy: and nothing shall by any means hurt you.

The question now arises, how did evil/sin officially get its name. It was given the name "Satan", a Hebrew word that was passed down to Greek and thence into English. Also commonly, know as the "Devil" (a word which is derived from the Latin and Greek word for "diabolical"). However, when we revert to thoroughly examine what the Hebrew Bible (Old Testament) has to say about evil. It speaks of a character known as Ha-Satan (the accuser, adversary or challenger). As earlier read, Job in the literature of the Hebrew Bible refers to him as "Satan" (pronounced Shaataan) (the Hebrew letters: Shin, Tet and Nun). Job 1:6 - Now there was a day when the sons of G-d came to present themselves before the Lord, and **Satan** came also among them. As discussed earlier, according to the story of Job, Satan acts like a prosecuting attorney, or a district attorney, in G-d's court - for G-d and not against Him. According to the Old Testament, Satan has no power or authority in and of himself; rather he must get permission from the Judge, G-d, in order to do anything. He is one of G-d's perpetual servants and a member of the divine council whose function is to test the faith of humans. However, the concept of Satan is different for Christians. In the New Testament, Satan has power and authority in and of himself. While the Jews hold strong that any entity that competes with G-d, and having power and authority of his own is in opposition to G-d, and this violates the basic idea of "one G-d". Both Testaments will be examined in detail as you continue to read.

Why is he called Satan in Hebrew? Well when we examine Genesis 3:14 it tells us - So the Lord G-d said to the serpent, Because you have done this, Cursed are you above all the livestock and all the wild animals! You will crawl on your belly and you will eat dust all the days of your life. This suggests that the serpent was transformed into something we as humans recognize in the animal kingdom as a SNAKE. This can be further evaluated by the spelling of the actual Hebrew word itself starting with the Hebrew letter "Shin" (shown below). Shin literally means fire or the fiery spirit. It is also known as the cosmic snake. The shape of the Hebrew letter Shin refers to a shape composed of three Hebrew letter "Vuvs" placed next to each other, giving the gamatria number of 666. It is said that when Shin is not in the middle of a word it turns to anger and resentment. The form of this letter also refers to the three roots of the giant molar teeth, because the tooth refers to the eating or taking of food, as Eve bit into the forbidden fruit of the Tree of Knowledge in the Garden of Eden.

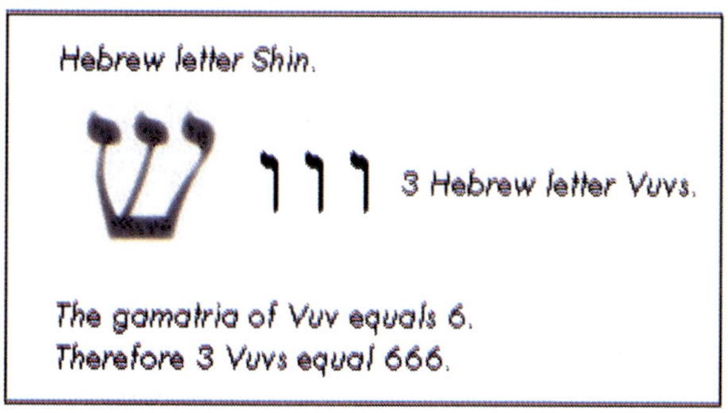

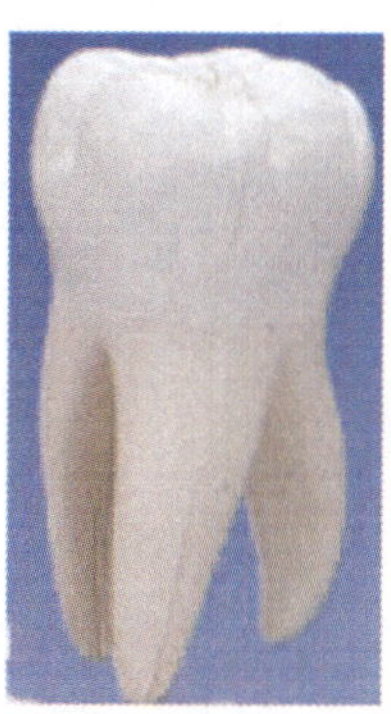
3 root molar tooth.

"Nun" depicted a snake, and the letter Tet has also been noted to represent a snake.

In several way, both its image, its numeric value of the first letter and it sound (Shhh -the sound that a snake makes) all coalesce together to symbolize the curse.

Additionally, 666 is also the gamatria equivalent to the name Emperor Nero (pronounced in Hebrew Neron keser). When it is spelled in Hebrew (along with its gamatria equivalent) it reads:

It literally stands for a resurrected reincarnation (or counterfeit) of satan. On the other hand, anyone that bears his same characteristics, likeness or demeanor. In addition, these characters are: Misleading, deceiving, corruptive and destroying even though the motives may seem good. 2 Corinthians 11:14–15 - And no wonder, for Satan himself masquerades as an angel of light. It is not surprising, then, if his servants masquerade as servants of righteousness. Their end will be what their actions deserve.

However, the parallels don't end here, as aforementioned in the previous chapter on "Global Consciousness" I discussed what I thought true evil was. This is what I said: If there be such a thing as true evil, I would describe it as an opposite and opposing force. And I'm estimating that it would "also" fall within the extremes of Phi/Pi, Fibonacci sequence, and Golden Mean Spiral. In which its frequency (low thoughts) describes defiance and deformity with a degree of irregularity against the designated growth sequence of Phi/Pi, Fibonacci sequence and Golden Mean Spiral. Where Phi/Pi represents growth and life, negative Phi/Pi (anti Phi/Pi) represents decay, darkness and death (entropy and thermodynamics). Therefore, I came to the understanding that the beast described in the Bible numbering 666 is the exact opposite process of Phi/Pi. Interesting enough, when dividing the 360 degrees of a circle by 54 degrees you get 6.66, the other side of a 54 degree angle in a right angle is 36 degrees, and 36 divided by 54 is also .666. Revelations 13:18 - this calls for wisdom: let anyone with understanding calculate the number of the beast, for it is the number of a person. Its number is six hundred sixty six. That little revelation in itself is quite interesting, as we know that Phi/Pi appears throughout the entire natural world and that includes every physical proportion of the human body - as was shown in the chapter on "Earths natural law". In that sense, perhaps it is the number of "mankind" in turning our thoughts against G-d's nature. We should also not forget (from the earlier chapter on the "Primordial Language") that each Hebrew letter is also formed by natures law, in shape, form and spiral - a mathematical process that is instructed to grow only in one specific way. The Hebrew letter "Shin" falls into that category. SHIN IS THE ONLY HEBREW LETTER THAT IS SEEN IN THE MOST OPEN AND DIFFERING WAY! (With three prongs facing upwards). But I also found that there is more to the number 666… the number 6 seems to be the chosen number used to describe the antichrist which in an evil sense is measured as the "perfected man of malice". Now when we combine the number 6 in a cluster of the same number that occurs "three times", this symbolizes that it has the most powerful of influences. For instance as we have learned in a previous chapter that the number 3 symbolizes a distinct perfection of power. Such as holy, holy, holy… Isaiah 6:3 - And they were calling to one another: Holy, holy, holy is the Lord Almighty; the whole earth is full of his glory. The Father, the Son and Holy Spirit… Matthew 28:19 - Therefore go and make disciples of all nations, baptizing them in the name of the Father and of the Son and of the Holy Spirit. Therefore, the number of the antichrist will be the number of the "thrice perfected man", and his number will be "666". What's more, similarly to the Father, the Son and the Holy Spirit, Satan himself too uses the trinity (3 in one nature) as a matter of organizing himself. Satan does something very similar in a deceiving way. He uses Lucifer, the False Prophet, and the Antichrist.

Satan is described in the Hebrew Bible (Old Testament) only in a few places, and in each place, Satan asks G-d for permission to conduct his business. As briefly discussed earlier in the book of "Job", we notice that it is G-d who is doing the talking. Let's examine it again. Job 2:3-6 - And the Lord said unto Satan, Hast thou considered my servant Job, that there is none like him in the earth, a perfect and an upright man, one that feareth G-d, and escheweth evil? and still he holdeth fast his integrity, although thou movedst me against him, to destroy him without cause. And Satan answered the Lord, and said, Skin for skin, yea, all that a man hath will he give for his life. But put forth thine hand now, and touch his bone and his flesh, and he will curse thee to thy face. And the Lord said unto Satan, Behold, he is in thine hand; but save his life. Therefore, we can see that Satan must get permission from G-d in order to go about his business, and when he does, he obediently follows his Creator's instructions. We later read in Zechariah 3:1-2 - And he shewed me Joshua the high priest standing before the angel of the Lord, and Satan standing at his right hand to resist him. And the Lord said unto Satan, The Lord rebuke thee, O Satan; even the Lord that hath chosen Jerusalem rebuke thee: is not this a brand plucked out of the fire? We also read something to this same effect again in Psalms 109:6-7 - Set thou a wicked man over him: and let Satan stand at his right hand. When he shall be judged, let him be condemned: and let his prayer become sin.

However, it should not surprise us that we can find reference to Satan in even older literature, and one can only Surmise where? In Enoch again! Yes, good ole Enoch describes and makes mention of this diabolical character in 2 Enoch 18:3 - And I threw him out from the height with his angels, and he was flying in the air continuously above the bottomless. 2 Enoch 29:4 - The devil is the evil spirit of the lower places, as a fugitive he made Sotona from the heavens as his name was "Satanail", thus he became different from the angels, but his nature did not change his intelligence as far as his understanding of righteous and sinful things. Then in 2 Enoch 31:4 it contains references to a Watcher called "Satanail". The text describes Satanail as being the prince of the Watchers who was cast out of heaven, and an evil spirit who knew the difference between what was righteous and sinful. Followers of his regime are known as demons or rebellious fallen angels.

In Enoch, these Watchers are fallen angels apparently lingering around in the Inverness of Earth lusting for things of the flesh - namely human women that are appealing to them. Which in effect, they produce children that are called "Nephilim" (giants who corrupt the earth with malevolent technology) (which was the discussion of an earlier chapter). Some may ask how did they genetically produce "giants". And the answer and reason these infants grew into giants is because they were in-between half light beings and human flesh - thus the human flesh could not contain the embodiment or the over generous magnitude of effervescence of the light being part, therefore producing "giants". This is against G-d's creation of interbreeding angel with human. In effect, G-d sends the Great Flood to rid the Earth of this evil. He spares only Noah and his family and bounds Satan and the Watchers (fallen angels) in the valley of the Earth until Judgment Day. Jude 1:6 - And the angels which kept not their first estate, but left their own habitation, he hath reserved in everlasting chains under darkness unto the judgment of the great day. The Watchers story in Enoch is later verified and determined from Genesis, which describes the origin of the "Nephilims" and mention the "Sons of G-d" who beget them. Genesis 6:1-4 - And it came to pass, when men began to multiply on the face of the earth, and daughters were born unto them, That the sons of G-d saw the daughters of men that they were fair; and they took them wives of all which they chose. And the Lord said, My spirit shall not always strive with man, for that he also is flesh: yet his days shall be an hundred and twenty years. There were giants in the earth in those days; and also after that, when the sons of G-d came in unto the daughters of men, and they bare children to them, the same became mighty men which were of old, men of renown.

Isaiah also writes something to this effect. Isaiah 14:12-15 - How you have fallen from heaven, O morning star, **son of the dawn!** You have been cast down to the earth, you who once laid low the nations! You said in your heart, I will ascend to heaven; I will raise my throne above the stars of G-d; I will sit enthroned on the mount of assembly, on the utmost heights of the sacred mountain. I will ascend above the tops of the clouds; I will make myself like the Most High. But you are brought down to the grave, to the depths of the pit. However some suggest that Isaiah isn't talking about Satan, because he refers to him as "Lucifer". And as we read in the paragraph above, it doesn't actually read (both in English or the original Hebrew) the name of Lucifer, rather it's specified as "son of the dawn". However its correct name and pronunciation in the Hebrew Bible reads (Heylel) הילל. So, since the word is not actually "Lucifer" which is a Latin term that essentially means "light bearing", the Hebrew word "Heylel" means something else. The word "Heylel" is used only once in the entire Bible, and as mentioned it is only found is Isaiah 14:12. So to narrow down the actual meaning of this word "Heylel" we must look for another conceptual word in the Hebrew Bible, and we find that this word is derived from the primal root word "Halal" and is used 165 times in the Old Testament, and translated to mean: Mad, foolish, rage, and TO BOAST. So when we read Isaiah 14:13-14 where "boasting" is very obvious and is noted of Satan's character, that is obviously why G-d calls him "Heylel" - because he BOASTED.

As a side not, when we start pulling apart elements of the Bible and suggesting other meanings, we are also overlooking what's called "double prophecy". In Isaiah the writer was describing the King of Babylon (who was an adversary) as a type of Satan, (not *the* Satan), and was attributing that same likeness and demeanor of evil from the description of the earlier writings of Enoch. Therefore, we must look beyond the Babylonian King to that of Satan. We know that Lucifer (who is also known as Satan, but is actually "Heylel") is spoken of in this verse, because he will go to the pit at the second coming. Isaiah 14:12-15 - How art thou fallen from heaven, **O Lucifer, (Heylel)** son of the morning! how art thou cut down to the ground, which didst weaken the nations! For thou hast said in thine heart, I will ascend into heaven, I will exalt my throne above the stars of G-d: I will sit also upon the mount of the congregation, in the sides of the north: I will ascend above the heights of the clouds; I will be like the most High. But you are brought down to the grave, to the depths of the pit.

So, what transpired between the Old Testament and the New Testament that allowed Satan to roam free about his business? Firstly as aforementioned, the Jews (according to the Old Testament) do not believe he is free upon his own authority to go about his business, while the Christians (according to the New Testament) say that he has free reign to do his dealings without restraint from G-d. When we look to the book of Jude, it tells us that G-d has "reserved" Satan and his angels in a place of darkness and chains. Jude 1:6 - And the angels which kept not their first estate, but left their own habitation, he hath reserved in everlasting chains under darkness unto the judgment of the great day. The word "reserved" means set aside for a future time of prophetic fulfillment. 2 Peter 2:4 - For if G-d spared not the angels that sinned, but cast them down to hell, and delivered them into chains of darkness, to be reserved unto judgment. Jude 1:6 - And the angels which kept not their first estate, but left their own habitation, he hath reserved in everlasting chains under darkness unto the judgment of the great day. However there are also others fallen angels that have populated the Earth. Ephesians 6:11-12 - Put on the whole armour of G-d, that ye may be able to stand against the wiles of the devil. For we wrestle not against flesh and blood, but against principalities, against powers, against the rulers of the darkness of this world, against spiritual wickedness in high places. Either way it is evident that evil is still abounding in the world, which suggests that Satan does in fact have free reign. He may have free reign, but his business does not escape G-d, as G-d sees all. And this business of evil will be "reserved" unto the judgment of the great day. Just to be certain, the Bible additionally states that he still roams Heaven and Earth. As aforementioned in Job 1:6 where it states that Satan appeared with other angels before the Lord. Presumably in heaven. Job 1:6 - Now there was a day when the sons of G-d came to present themselves before the Lord, and Satan came also among them. When G-d asked Satan where he had been, Satan replied, "From roaming through the earth and going back and forth in it". Satan does in fact have free reign to go about his business and as some people may think that he is in Hell, the evidence suggest that he is not. Peter in the New Testament declares that he walks about seeking whom he might devour. 1 Peter 5:8 - Be sober, be vigilant; because your adversary the devil, as a roaring lion, walketh about, seeking whom he may devour. Revelation 20:1 (future tense) - And I saw an angel come down from heaven, having the key of the bottomless pit and a great chain in his hand. We also know from other verses of the Bible that Satan is still intensely lurking around and more so to an unusual extent after his confrontation (war of words) with Jesus. Matthew 4:1 - Then Jesus was led up by the Spirit into the wilderness to be tempted by the devil (Satan).

Again, it is clear from passages such as these that Satan and his followers are not bound in Hell and probably spend most of their time on Earth wandering around seeking to destroy the lives of human beings and to keep them separated from G-d. However, as aforementioned the Jews do not believe that Satan is bad; they believe that he is only the tempter and tester - that which allows us to learn the difference between good and bad so that we might desire G-d. While in the New Testament we do not fully understand (plus the Bible does not really say) why certain angels are free to roam the Earth and others are reserved in chains unto judgment. However what we do read is that demons are afraid of Jesus and fearful of being tormented before their time. Perhaps the angels that have been bound are only pertaining to the fallen angels known as the "Watchers" specified in Enoch, who interbred with Earth women. These angels were the most sinful fallen angels that were so immorally malevolent in their covetousness and lust for the flesh of womankind that it brought the judgment of G-d upon the earth, which was the flood whereupon G-d judged the earth (as this was the second assault on the woman). The point here is that sin will condemn even angels to an eternity in hell. 2 Peter 2:4 - For if G-d spared not the angels that sinned, but cast them down to hell, and delivered them into chains of darkness, to be reserved unto judgment.

According to the Bible, we are heading towards what's called the "tribulation period". This is where Satan and the Fallen Angels (who are not one of those bound below) will wreak havoc on the Earth for 7 years (cited from Daniel 9:24-27). And during this 7 year period there will be distress and suffering upon the Earth as there has never been before. But before this happens, it is also mentioned in the Bible that all the believers in Jesus (who are still alive on the earth) will be "raptured" (a transliteration of our human body into the ethereal realm) and reunited with - Jesus before (or at the start) of this 7 year period (this is considered the "seconded coming" of Jesus). These believers will not die a physical death but will be caught up to meet Jesus in the sky along with all the other believers that died before them. They will have "new glorified bodies". 1Theolosians 4:15-17 - For this we say to you by the word of the Lord, that we who are alive and remain until the coming of the Lord will by no means precede those who are asleep. For the Lord Himself will descend from heaven with a shout, with the voice of an archangel, and with the trumpet of G-d. And the dead in Christ will rise first. Then we who are alive and remain shall be caught up together with them in the clouds to meet the Lord in the air. And thus we shall always be with the Lord. After the 7 years of tribulation, Satan will be cast into the bottomless pit for a time of 1,000 years (a millennium). And it is during the tribulation period (1,000 years) that many will be martyred (put to death) for their faith. There will also be people that made it through the great tribulation that did not worship the beast (antichrist) and they will reign with Jesus and the believers that went up to heaven for a thousand years "here on the earth". Revelation 20:4 - And I saw thrones, and they sat on them, and judgment was committed to them. Then I saw the souls of those who had been beheaded for their witness to Jesus and for the word of G-d, who had not worshiped the beast or his image, and had not, received his mark on their foreheads or on their hands. And they lived and reigned with Christ for a thousand years.

By the way, this 1,000-year reign may relate to the extra 1,000 years that was calculated from the previous chapter concerning the end of days. Cited from the previous chapter: And there may be something even more spectacular to the number 7 yet... and this has to do with recorded history (in other words, "measured time"). It is my estimation that there will be exactly only 7 thousand years of measured time - as this is G-d's great culminating plan.

We are now nearing the 7 thousand years of measured time - which is the Grand Jubilee?

Adam - 930
Lamech - 874-1651
Noah - 1056-2006
Terah - 1879-2084
Abraham - 2009-2184
Flood - 1650

So if we estimate the timeline today from Jesus was approximately 2,000 years, and from Jesus to Abraham was 2,000 years, and from Abraham back to Adam is about 2,000 years, which gives us an approximate age of 6,000 years of recorded (and measured) time.

(Now from 6,000 years), if we estimate that Jesus died in 33AD, and this is now the year 2009 AD... that means that 33 AD was exactly 1,976 (one thousand nine hundred and seventy six) years ago. So if we subtract 1,976 from the 2,000 year mark, we are roughly 24 years short of the 6,000th year which would bring us to the year 2033. After these 24 years, we have another 1,000 years left to make the Grand Jubilee of 7,000 years! We may very well be in (or nearing) the 1,000 years specified in Revelation 20:1-11. However what is not known is weather or not this extra 1,000 years takes place before the rapture of the church (which is the body of believers) or takes place afterwards as the remaining 1,000 years after the rapture when Jesus dwells with those left on the earth (which I believe will be the Jews when they finally realize their Messiah). I also am personally in favor of believing that the extra 1,000 years will take place before the rapture.

(Trying not to make things to complicated)... Let me also quickly add that the Jewish calendar system for the end of days according to some Rabbinical Rabbis is for the year 5777 which is eerily very close to our present day. The number 777 is G-d's number (which you will read and learn about in the coming chapter). And with this in mind, there is strong speculation that the number 777 also suggests that it is the final hour, day and year. Now, since it is currently the Hebrew year 5770, (which equates to the modern calendar year of 2009-2010) + 7 years is 2017 - thus the Hebrew year 5777. Therefore according to some religious thinkers who study scripture (put in terms that we can understand according to the modern year calendar), the year 2017 is the end of days for those that are left unbelieving! This they correlate with the "rapture" (the second coming of Jesus).

Then after the thousand year reign of Jesus, in which He will rule with enforced righteousness on the Earth, Satan will be loosed and try to deceive the nations once more but just for a short time, and even though people lived in the millennial reign of Jesus, there will still be some people that are rebellious (free will) and rather be a fan and enthusiast of Satan and follow his evil and corrupt ways again. It will mainly be the facilitators of Government again that will utterly accompany this spirit in. This is what will usher in the final battle between

G-d and Satan, also called the battle of Armageddon. At which time Jesus will judge and cast all evil (everybody that worshiped the beast) along with Satan, the Antichrist and the False Prophet into the place of torment (lake of fire - hell) for eternity. Revelation 19:20 - Then the beast (antichrist) was captured, and with him the false prophet who worked signs in his presence, by which he deceived those who received the mark of the beast and those who worshiped his image. These two were cast alive into the lake of fire burning with brimstone. Revelation 20:10 - The devil, who deceived them, was cast into the lake of fire and brimstone where the beast and the false prophet are. And they will be tormented day and night forever and ever. Then the final judgment will come when all unbelievers will stand before the great white throne of G-d. Revelation 20:11 - And I saw a great white throne, and him that sat on it, from whose face the earth and the heaven fled away; and there was found no place for them. And it is at this last and final draw that unfortunately all the people that have rejected Jesus as the Messiah will be judged for what they have done in their lives, and will be cast into the lake of fire and for evermore be eternally separated from G-d. Revelation 20:12-15 - And I saw the dead, small and great, standing before G-d, and books were opened. And another book was opened, which is the Book of Life. And the dead were judged according to their works, by the things which were written in the books. The sea gave up the dead who were in it, and Death and Hades delivered up the dead who were in them. And they were judged, each one according to his works. Then Death and Hades were cast into the lake of fire. This is the second death. And anyone not found written in the Book of Life was cast into the lake of fire.

Quickly, while was on the subject of the apostle John who wrote the book of Revelations… we also find another gem amongst the rocks. A prophetic prophesies that has to do with a massive army rising against Israel in the last days (possibly the battle of Armageddon). Revelation 9:14-16 - It said to the sixth angel who had the trumpet, release the four angels who are bound at the great river Euphrates. And the four angels who had been kept ready for this very hour and day and month and year were released to kill a third of mankind. **The number of the mounted troops was two hundred million. I heard their number.** The question is, how did John arrive at the immense number of "200 million people". Because John did not have a framework to work with those numbers (such vast sums) especially some 2000 years ago when the Earth wasn't even that populated. This number was unheard of back then. This huge number is surely something to marvel at, and John makes this message clear when he states: **I heard their number.**

Now on the subject of hell… Hell is a place of ruin and waste, it is a literal dump! There is no pleasure, contentment or cheerfulness there despite what some evil and disobedient people may hypothesize and make a laughing joke about. In hell, there is perpetual grief and terror, darkness, remorse, crying, anguish, misery, agony, guilt, shame, weeping and gnashing of teeth, along with intense pain. Matthew 25:30 - And throw that worthless servant outside, into the darkness, where there will be weeping and gnashing of teeth.

Once one is in hell there is also no forgiveness! There is also no chance for change. Hell is place that was designed by G-d for those who no longer serve the purpose for which they were designed, which is to serve, work for, and glorify G-d. Hell is an ultimate, irreversible permanent "separation" away from G-d. This place is for people who don't/didn't want anything to do with G-d during their life and rejected His offer of salvation provided though Jesus. 2 Thessalonians 1:7-9 - And to you who are troubled rest with us, when the Lord Jesus shall be revealed from heaven with his mighty angels, In flaming fire taking vengeance on them that know not G-d, and that obey not the gospel of our Lord Jesus Christ: Who shall be punished with everlasting destruction from the presence of the Lord, and from the glory of his power.

Those that died before hearing Jesus' message are less accountable than those who heard His message and did not openly receive it. G-d will not send people to hell who through no fault of their own have never had a chance to hear, receive and believe in Jesus. HOWEVER, ONCE YOU HEAR HIS MESSAGE YOU ARE FOREVERMORE WITHOUT EXCUSE AND ACCOUNTABLE! According to the Bible G-d desires all people to come to the knowledge of Him and be saved. 1 Timothy 2:1-7 - I urge, then, first of all, that requests, prayers, intercession and thanksgiving be made for everyone— for kings and all those in authority, that we may live peaceful and quiet lives in all G-dliness and holiness. **This is good, and pleases G-d our Savior, who wants all men to be saved and to come to a knowledge of the truth.** For there is one G-d and one mediator between G-d and men, the man Christ Jesus, who gave himself as a ransom for all men—the testimony given in its proper time. And for this purpose I was appointed a herald and an apostle—I am telling the truth, I am not lying—and a teacher of the true faith to the Gentiles. 2 Peter 3:9 - The Lord is not slow in keeping his promise, as some understand slowness. He is patient with you, **not wanting anyone to perish, but everyone to come to repentance.**

Then there are those who think that they have enough time left in the grave awaiting to find out if Jesus was truly the Messiah. They think that when they die they will then find out the truth, but alas, they really will, and it will be to late. We know this again from what scripture tells us. The Bible implicitly tells us that the unbelieving dead who have already died and were judged and sent to the pit of torment do not have a second chance to enter the kingdom of Heaven, not even during Jesus' second coming. We know this from what Jesus tells us in Luke 16:19-31 - There was a certain rich man, which was clothed in purple and fine linen, and fared sumptuously every day: And there was a certain beggar named Lazarus, which was laid at his gate, full of sores, And desiring to be fed with the crumbs which fell from the rich man's table: moreover the dogs came and licked his sores. And it came to pass, that the beggar died, and was carried by the angels into Abraham's bosom: the rich man also died, and was buried; And in hell he lift up his eyes, being in torments, and seeth Abraham afar off, and Lazarus in his bosom. And he cried and said, Father Abraham, have mercy on me, and send Lazarus, that he may dip the tip of his finger in water, and cool my tongue; for I am tormented in this flame. But Abraham said, Son, remember that thou in thy lifetime receivedst thy good things, and likewise Lazarus evil things: but now he is comforted, and thou art tormented. And beside all this, **between us and you there is a great gulf fixed: so that they which would pass from hence to you cannot; neither can they pass to us,** that would come from thence. Then he said, I pray thee therefore, father, that thou wouldest send him to my father's house: For I have five brethren; that he may testify unto them, lest they also come into this place of torment. Abraham saith unto him, They have Moses and the prophets; let them hear them. And he said, Nay, father Abraham: but if one went unto them from the dead, they will repent. And he said unto him, If they hear not Moses and the prophets, neither will they be persuaded, though one rose from the dead.

Now despite what some may think, this is not a parable, Jesus it telling us of an actual account. We know that its an actual account because "parables never mention names" and Jesus tells us of a beggar named "Lazarus" who suffered greatly during his life on Earth and ended up going to Heaven (Abraham's bosom at the time). Instead, the rich man that had died had opened his eyes and found himself in hell. In hell, it tells us that he could "see" across the chasm of the bottomless pit, but that he could not "get across" it because there is a great gulf fixed separating them. By the way this tell us that even after our physical material body dies our souls are still fully conscious along with our senses being in operation whether or not we are in hell or paradise. We can also detect from this scripture that along with being fully conscious we also have memory, emotions, feelings, knowledge, affection, desires and will power. Additionally, the dead both in hell and in Heaven are completely conscious and able to watch everything that is going on down here on Earth.

And in effect, if you do not believe Moses and the prophets (the 5 books of Moses) as Abraham said to the rich man in hell, you wont believe either the word of G-d. Abraham was literally saying that there is no hope for those who do not believe on Earth as well as in hell. To implicitly understand this: There is no plan of salvation available anywhere in the entire Universe outside the word of G-d - the Bible!

Now when Jesus died He went down to Paradise (Abraham's bosom) and took with him (out of Paradise) all the souls of those who have been watching and waiting for 4,000 years for His coming, and then took them out up on high and established a second Paradise. And it is in this new Paradise (Heaven) that all the souls that are presently there and even those that are newly arriving, are now awaiting Jesus' 2nd coming, and helping in preparing the new kingdom for this coming event (we are never dead in G-d's service). Revelation 21:1-7 - And I saw a new heaven and a new earth: for the first heaven and the first earth were passed away; and there was no more sea. And I John saw the holy city, new Jerusalem, coming down from G-d out of heaven, prepared as a bride adorned for her husband. And I heard a great voice out of heaven saying, Behold, the tabernacle of G-d is with men, and he will dwell with them, and they shall be his people, and G-d himself shall be with them, and be their G-d. And G-d shall wipe away all tears from their eyes; and there shall be no more death, neither sorrow, nor crying, neither shall there be any more pain: for the former things are passed away. And he that sat upon the throne said, Behold, I make all things new. And he said unto me, Write: for these words are true and faithful. And he said unto me, It is done. I am Alpha and Omega, the beginning and the end. I will give unto him that is athirst of the fountain of the water of life freely. He that overcometh shall inherit all things; and I will be his G-d, and he shall be my son.

Here in the picture below is a rough illustration of what the great separating gulf may look like:

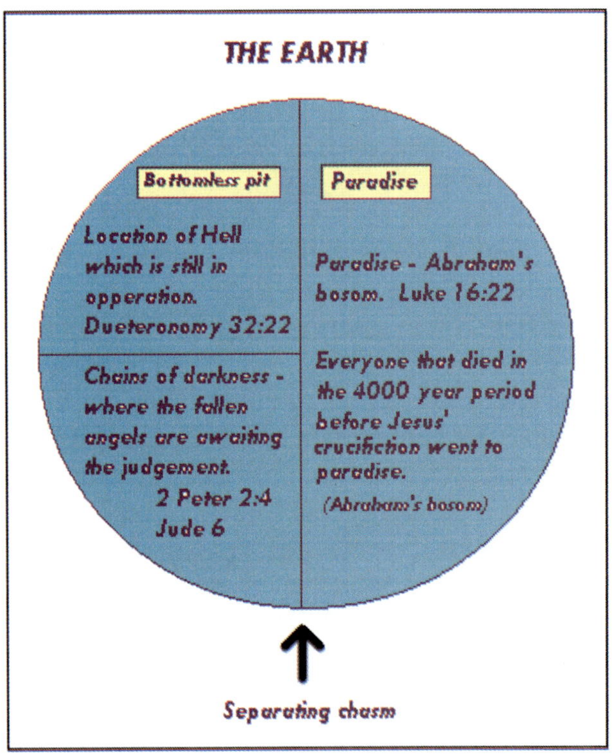

The Bible also tells us exactly what happens when we die. Again, it is not true that one looses their consciousness... you don't go into a grave and lay permanently asleep (this only pertains to the physical material body). Ones wakeful soul becomes fully conscious after death finding them either with G-d, or in hell. Firstly, when a believer in Jesus dies the spirit departs from the body and the angels of G-d usher us into the presents of G-d. Luke 16:22 - And it came to pass, that the beggar died, and was carried by the angels into Abraham's bosom. Secondly, it is appointed unto us to die only once and than to be judged. We are not roaming around on the Earth as ghosts. Hebrews 9:27 - And as it is appointed unto men once to die, but after this the judgment. Thirdly, G-d calls us by name. John 10:3 - The watchman opens the gate for him, and the sheep listen to his voice. He calls his own sheep by name and leads them out. Fourthly, all those that believe in Jesus are pardoned by G-d. 2 Timothy 2:19 - Nevertheless the foundation of G-d standeth sure, having this seal, The Lord knoweth them that are his. And, let every one that nameth the name of Christ depart from iniquity. In addition, we know that those that are pardoned by G-d will live with Him in Heaven. Revelation 1:17-18 - And when I saw him, I fell at his feet as dead. And he laid his right hand upon me, saying unto me, Fear not; I am the first and the last: I am he that liveth, and was dead; and, behold, I am alive for evermore, Amen; and have the keys of hell and of death. Here is more scripture that confirms for us that when we die we are not dead. Luke 8:49-55 - While Jesus was still speaking, someone came from the house of Jairus, the synagogue ruler. Your daughter is dead, he said. Don't bother the teacher any more. Hearing this, Jesus said to Jairus, Don't be afraid; just believe, and she will be healed. When he arrived at the house of Jairus, he did not let anyone go in with him except Peter, John and James, and the child's father and mother. Meanwhile, all the people were wailing and mourning for her. Stop wailing, Jesus said. She is not dead but asleep. They laughed at him, knowing that she was dead. But he took her by the hand and said, My child, get up! Her spirit returned, and at once she stood up. Then Jesus told them to give her something to eat.

Now to pick up where we left off on the subject of Jesus' 2nd coming... G-d has also given us a timetable in the Bible of when all this will transpire. For an exact date, we will never know until the actual appointed time occurs. However, there are certain things of surety that we can count on because of their symbolic significance to G-d. For instance, we may not know the actual date, but we can get an adequate estimation (if we can recall from the previous chapter) that the Jewish holiday "Rosh Hashanah" corresponds with, as well as represents the "day of judgment". "Rosh Hashanah" is the Jewish New Year characterized by blowing the Shofar to awaken the listener to coming judgment, followed by 10 days of repentance just before "Yom Kippur" which is the "day of atonement". In effect, G-d gave us this holy event to teach us that we will be judged on Rosh Hashanah and sealed unto the closing of the gates on Yom Kippur. This is where the righteous are separated and gathered up to be with G-d. This holiday is mentioned in Leviticus 23:24 - Speak unto the children of Israel, saying, in the seventh month, in the first day of the month, shall ye have a Sabbath, a memorial of blowing of trumpets, an holy convocation. It also mentions something to this fact in Isaiah 26:2 - The gates of Heaven are opened on Rosh Hashanah so the righteous nation may enter. It also makes mention of this in Psalms 118:19-20 - Open to me the gates of righteousness, I will go into them, and I will praise the Lord. This gate of the Lord, into which the righteous shall enter. Because the gates of Heaven are understood to be open on Rosh Hashanah, this is further evidence that the "rapture of the believers in the Messiah" will take place between Rosh Hashanah and Yom Kippur. Additionally, as discussed in the previous chapter, this will be on the Jubilee (Leviticus 25:9-19).

Along with this analysis we also have Daniels prophecy that foretells of the Antichrist with remarkable prophetic foresight (corresponding to this day and age), thus serving as another piece of flawless and undeniable evidence (as you will read and see below). Firstly please read Daniels prophesies concerning the antichrist. Daniel 2:31-41 - You looked, O king, and there before you stood a large statue—an enormous, dazzling statue, awesome in appearance. **The head of the statue was made of pure gold, its chest and arms of silver, its belly and thighs of bronze, its legs of iron, its feet partly of iron and partly of baked clay.** While you were watching, a rock was cut out, but not by human hands. **It struck the statue on its feet of iron and clay and smashed them. Then the iron, the clay, the bronze, the silver and the gold were broken to pieces at the same time** and became like chaff on a threshing floor in the summer. The wind swept them away without leaving a trace. But the rock that struck the statue became a huge mountain and filled the whole earth. This was the dream, and now we will interpret it to the king. You, O king, are the king of kings. The G-d of heaven has given you dominion and power and might and glory; in your hands he has placed mankind and the beasts of the field and the birds of the air. Wherever they live, he has made you ruler over them all. **You are that head of gold. After you, another kingdom will rise, inferior to yours. Next, a third kingdom, one of bronze, will rule over the whole earth. Finally, there will be a fourth kingdom, strong as iron—for iron breaks and smashes everything—and as iron breaks things to pieces, so it will crush and break all the others.** Just as you saw that the **feet and toes** were partly of baked clay and partly of iron, so this will be a divided kingdom; yet it will have some of the strength of iron in it, even as you saw iron mixed with clay.

Now in the text above Daniel is telling us that in the future of the worlds history there will be 5 nations coming out of this statue. Starting at the head of gold, which represents Nebuchadnezzar and Babylon (this era lasted for about 100 years from 626-539BC). The arms and chest of silver represents Persia (the Medes and the Persians which sacked Babylon and took it over). This era lasted for about 200 years from 539-334BC. (This is also now modern day Iran). Then the midsection (the waist of brass) represents Greece that broke up into 4 parts: Greece (the Roman Empire), Turkey, Syria and Egypt. This era lasted about 200 years from 334-150BC, which history confirms again resulting from Alexander the Great. Now keep in mind that Daniel saw these things centuries before they happened (approximately 2,500 years ago). And when we get down to the statues legs of iron, it is these legs that correspond precisely to what history tells us occurred with the forming of the Roman Empire that broke into "two parts", which defined a Western part and an Eastern part. This can be found in Roman history to have occurred in 133BC. The Roman Empire era lasted to about 476AD. Now in the scripture above it describes two feet and toes, and as we know that there are 5 toes on each foot and 10 toes altogether.

Here in the picture below we see the 10 toes that are said to be made of partly iron and baked clay that correspond to the 10 horns mentioned in Daniel 7:7 - After that, in my vision at night I looked, and there before me was a fourth beast—terrifying and frightening and very powerful. It had large iron teeth; it crushed and devoured its victims and trampled underfoot whatever was left. It was different from all the former beasts, and it had **ten horns**. Daniel 7:20 - I also wanted to know about the **ten horns** on its head and about the other horn that came up, before which three of them fell—the horn that looked more imposing than the others and that had eyes and a mouth that spoke boastfully. Daniel 7:24 - **The ten horns are ten kings** who will come from this kingdom. **After them another king will arise, different from the earlier ones;** he will subdue three kings. These 10 horns which we read about are also in essence the 10 kings - hence the 10 toes… and it is these 10 horns, 10 kings, and 10 toes that are actually the 10 "nations" shown below. The 11th king is the Antichrist (not shown).

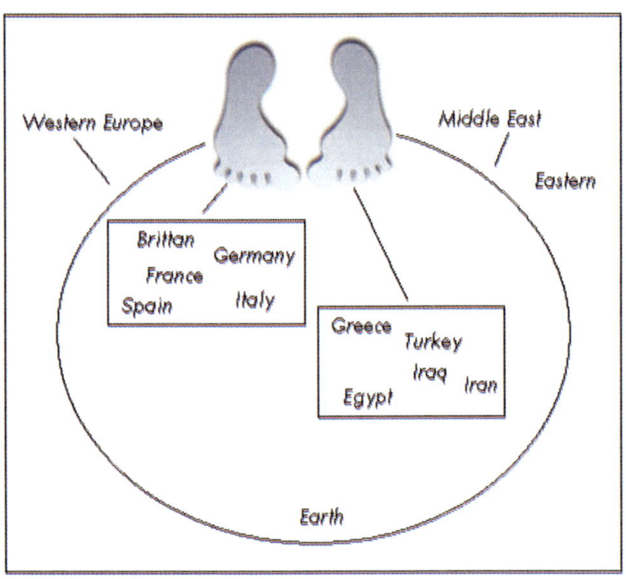

He is head of the 10 kings and nations. The antichrist will rise from the Eastern leg.

What does Daniels prophetic foresight convey to us? That it is forthcoming to know that these turn of events prophesized in the Bible regarding the end of days are imminent and will INDEED transpire, since they are right on track with prophesy (as you can see in the picture above that indeed there are 10 individual nations that are presently there today). Those that say the end of days scenario is all fictional fanatical religious nonsense do not truly know their history, since history has revealed otherwise to those that truly pay attention to what's happening in the world. These historical facts are DOCUMENTED and VERIFIABLE! What really irks me is that history buffs (completely blinded by these facts) cannot make the connection to things mentioned in the Bible simply because they are blinded by a godless worldview.

Since these are reliable facts corresponding to this day and age, we can surely be rest assured on what the Bible tells us will happen next in history. This is in the bag, IT IS FAIL-SAFE! The 5th kingdom is the literal earthy kingdom of Jesus when He comes back to reign on the Earth for 1,000 years at His 2nd coming, and than into eternity (time without end). Daniel 2:43-45 (G-d will intervene and create a New Kingdom, and it will stand forever) - And whereas thou sawest iron mixed with miry clay, they shall mingle themselves with the seed of men: but they shall not cleave one to another, even as iron is not mixed with clay. **And in the days of these kings shall the G-d of heaven set up a kingdom, which shall never be destroyed: and the kingdom shall not be left to other people, but it shall break in pieces and consume all these kingdoms, and it shall stand for ever.** Forasmuch as thou sawest that the stone was cut out of the mountain without hands, and that it brake in pieces the iron, the brass, the clay, the silver, and the gold; **the great G-d hath made known to the king what shall come to pass hereafter: and the dream is certain, and the interpretation thereof sure.**

There is also something else I will disclose here that was not previously mentioned before. This has to do with lunar and solar eclipses which also play a strong role in "prophetic prophesy" as they correspond to Jewish holidays as well as significant events that correlate with Israel's history in the making. For instance, in the 20th century two events of lunar eclipses transpired. Once in 1949-1950, following the year Israel became a state, and the second time lunar eclipses occurred was on the holiday of Passover and the Feast of Tabernacles, which occurred in 1949-1950 in addition to 1967-1968. This was the well-known year Israel liberated Jerusalem. Indecently prior to this in history, these same eclipses happened in 1493, which was the year that Spain expelled all the Jews, which put an end to the largest most distinguished Jewish settlement in all of Europe. This was one of the relatively modern tumultuous times in the history of Judaism. Now, these same eclipses corresponding to the Jewish holidays (both lunar and solar) are estimated to occur again in the coming years of 2014-2015. These eclipses will be noticeable by darkening of the sky and Stars, and the Moon turning blood red in color which happens when the Earth (which is in-between the Sun and Moon) blocks the Suns light from hitting the Moon. But some of the Suns light curves (bends and refracts) around the Earth and reaches the Moon, thus illuminating it and giving it its orange-red color. The Bible links these times of darkness's or interruptions of the Sun and Moon with judgments from G-d. Matthew 24:29-30 - Immediately after the tribulation of those days shall **the sun be darkened, and the moon shall not give her light, and the stars shall fall from heaven,** and the powers of the heavens shall be shaken: And then shall appear the sign of the Son of man in heaven: and then shall all the tribes of the earth mourn, and they shall see the Son of man coming in the clouds of heaven with power and great glory.

In addition, this is where the rubber meets the road in relation to these eclipses. The theory goes that if these eclipses marked a transition period in the past they will do so again in the future (to occur in 2014-2015). This period of great darkening was prophesied in the Bible (by several prophets) as the invasion of an army coming against Israel - otherwise known as the battle of Armageddon, the rise of the Antichrist and the great tribulation. Joel 2:1-2 - Blow ye the trumpet in Zion, and sound an alarm in my holy mountain: let all the inhabitants of the land tremble: for the day of the Lord cometh, for it is nigh at hand; **A day of darkness and of gloominess, a day of clouds and of thick darkness,** as the morning spread upon the mountains: a great people and a strong; there hath not been ever the like, neither shall be any more after it, even to the years of many generations. Joel 2:10-11 - **The earth shall quake before them; the heavens shall tremble: the sun and the moon shall be dark, and the stars shall withdraw their shining:** And the Lord shall utter his voice before his army: for his camp is very great: for he is strong that executeth his word: for the day of the Lord is great and very terrible; and who can abide it? The Bible also mentions something that sounds like it could imply a nuclear bomb. Zechariah 14:12 - And this shall be the plague wherewith the Lord will smite all the people that have fought against Jerusalem; **Their flesh shall consume away while they stand upon their feet, and their eyes shall consume away in their holes, and their tongue shall consume away in their mouth.** Now some people think that mankind is going to blow the entire Earth up with nuclear weapons and they relate it to the battle of Armageddon. But I will tell you that G-d is not going to allow mankind the ability to destroy the entire planet by nuclear war no matter how many hundreds of nuclear weapons each nation has. We will never ignite a chain reaction that will destroy everything in its path simply because G-d has reserved (saved) this day for Himself, when He Himself is going to completely and thoroughly destroy the whole Earth, which will be even more devastating than we could ever possibly imagine. 2 Peter 3:10 - But the day of the Lord will come as a thief in the night; in the which the heavens shall pass away with a great noise, and the elements shall melt with fervent heat, the earth also and the works that are therein shall be burned up.

Moving right along here… there is something else concerning Antichrist. The Bible also mentions something called the "Seat of Satan", which may also play a strong role in the emergence of the antichrist. This seat is mentioned in Revelation 2:12-13 - And to the angel of the church in **Pergamos** write; These things saith he which hath the sharp sword with two edges; **I know thy works, and where thou dwellest, even where Satan's seat is:** and thou holdest fast my name, and hast not denied my faith, even in those days wherein Antipas was my faithful martyr, who was slain among you, **where Satan dwelleth.** This is referring to King Pergomus (spelled, Pergamos, Pergamum or Pergamon) who built an alter (seat) to the Greek god Zeus.

Latter on in history (in the 1880's), German archeologists went to Pergomus and dug up the alter of Zeus and took it back to Berlin Germany, where it now rests in the Berlin museum. Now on this alter there are statues that depict the battle of the giants where Zeus is fighting against his father Chronos who headed up the Titans. Nevertheless, this battle has its roots in a long chain of history. In fact this battle goes all the way back to Genesis 6 where we read about the sons of G-d (AKA Watchers/Fallen Angels) who looked upon the daughters of men. This also parallels the Greek god Zeus with the main fallen angel "Semjaze". Semjaze according to the book of Enoch is portrayed as the leader of the band of Fallen

Angels. Enoch 6:3-5 - **And Semjaza who was their leader,** said unto them: I fear ye will not indeed agree to do this deed, and I alone shall have to pay the penalty of a great sin. And they all answered him and said: Let us all swear an oath, and all bind ourselves by mutual imprecations not to abandon this plan but to do this thing. Then sware they all together and bound themselves by mutual imprecations upon it. Enoch 10:11-14 - And the Lord said unto Michael: Go, bind **Semjaza and his associates** who have united themselves with women so as to have defiled themselves with them in all their uncleanness. And when their sons have slain one another, and they have seen the destruction of their beloved ones, bind them fast for seventy generations in the valleys of the earth, till the day of their judgement and of their consummation, till the judgement that is for ever and ever is consummated. In those days they shall be led off to the abyss of fire: to the torment and the prison in which they shall be confined for ever. And whosoever shall be condemned and destroyed will from thenceforth be bound together with them to the end of all generations.

Now when Hitler rose to power in the 1930's he had a replica of this alter built where he stood giving his famous Nuremberg speeches blaming the Jews for everything, which led to the Holocaust - the murder of 6 million Jews. So Hitler stood right before the alter of Zeus - the Seat of Satan himself as the personification of evil. Incidentally, there were also other people in history (relatively recent history) that also used this alter as a backdrop. One such person happened to be President Barrack Obama. He first encountered this alter when he visited Berlin Germany to make his speech on foreign policy standing at the Brandenburg gate while he gave his speech approximately to one hundred thousand plus people. Then again when Obama gave his presidential exception speech in Denver Colorado he used this same backdrop of the alter of Zeus. When Obama was questioned about it, he commented that it represented the U.S. capital and the White House West Wing combined. Keep in mind that there are many religious objects surrounding Washington (as mentioned earlier in this book). However, this indeed may be a dead ringer of things to come. Why Obama would pick the alter of Zuis (the Seat of Satan) for his political coming out speech just as Hitler did many years ago is certainly bizarre! What are we supposed to make of it? This may sound like I am suggesting that President Obama is the Antichrist, but this happenstance may just be an insight for another to come. What we need to do is continue to watch all prestigious people that come to power. This Antichrist will be a man of all nations and culture. All people of the world will be looking to him to solve all the problems (which includes racial indifferences).

Even peace issues between Israel and Palestine. This person will bring false peace - then war! This will be one of the indicators to look for in the Antichrists leadership and term. The Antichrist wants to rule over the entire world (seeking world government utopia) - therefore he will settle for nothing less than being leader of the worldwide government which may indeed be Washington D.C. since it is the most powerful focal point in the world today. We can assess that the Antichrist wants worldwide power from the book of Isaiah, which tells us he will exalt his throne above the stars of G-d and expect worldwide worship when he sits upon the mount of the congregation. Isaiah 14:13-14 - For thou hast said in thine heart, I will ascend into heaven, I will exalt my throne above the stars of G-d: I will sit also upon the mount of the congregation, in the sides of the north: I will ascend above the heights of the clouds; I will be like the most High. Throughout history, he strives through false religions and governments. He will be a "counterfeit" such as to appear as something genuine. However, keep in mind that it is an imitation. Such as to deceive, mislead and defraud. The

Antichrists religion would appear to be a form of godliness but denying the power thereof. 2 Timothy 3:5 - Having a form of godliness, but denying the power thereof: from such turn away. He will mimic all the aspects of the true faith in order to deceive humanity. He will speak marvelous things. Daniel 7:8 - I considered the horns, and, behold, there came up among them another little horn, before whom there were three of the first horns plucked up by the roots: and, behold, in this horn were eyes like the eyes of man, and **a mouth speaking great things.** He will be given power over all kindred's, tongues and nations. Revelation 13:7 - And it was given unto him to make war with the saints, and to overcome them: **and power was given him over all kindreds, and tongues, and nations.** He will have great military power that will stand behind him to enforce his laws. Daniel 11:31 (This may be the U.N. - United Nations) - His armed forces will rise up to desecrate the temple fortress and will abolish the daily sacrifice. Then they will set up the abomination that causes desolation. The Antichrists mother will even be a counterfeit of Jesus' mother Mary (the virgin Mary). She will be an anti-Mary. In addition, have a religion that accepts people of all religions, as well as passing onto her son the same persona of the acceptance of all races and religions melding them into one pot. The anti-Mary is actually the counterfeit of the "magna mater" (mother) of ancient Babel who is the prelude to the anti-Mary. She began with Nimrod at Babel. He was the man that Satan used to motivate the people to build the tower. Genesis 10:8 - And Cush begat Nimrod: he began to be a mighty one in the earth. He was a mighty hunter before the Lord: wherefore it is said, Even as Nimrod the mighty hunter before the Lord. In addition, the beginning of his kingdom was Babel, and Erech, and Accad, and Calneh, in the land of Shinar. Nimrods wife was a harlot named Semiramis. Her name Semiramis was a prename of Shinar (which is lower Mesopotamia from which can be derived Samaria). This is symbolized by the region of Babylon, which was destroyed by G-d to never be rebuilt. At Babel, she initiated a religion that was united and joined all nations and religions combined with world government. Legend has it that she was carried off by a god described as a bull (woman on a beast) and had a son named Tammuz (mentioned in Ezekiel 8:14). Ezekiel 8:14 - Then he brought me to the door of the gate of the Lord's house, which was toward the north; and, behold, there sat women weeping for **Tammuz.** Throughout history, Satan has tried to produce the same counterfeit. Revelation 17:5 - And upon her forehead was a name written, MYSTERY, BABYLON THE GREAT, THE MOTHER OF HARLOTS AND ABOMINATIONS OF THE EARTH. This anti-Mary will unite herself (as well as her son) with all the major religions of the world (both race, ethnic and religion) with cultural connections allowing each religion to continue their individualized worship and traditions. This may sound appealing to people of our time because of all the racial tensions. We hear the constant reminder that we shouldn't be prejudice, and that we should marry our own kind. No, we should not be prejudice, and yes, we should respect other peoples culture, race and religion, but what most people don't understand is that this very ideology leads directly to what G-d considers "false gods and religion". Hence, history will repeat itself like at Babel - and G-d will again destroy it.

Now for those of us who still don't understand the nature of Satan, lets take another look at his attributes. He is both tempter and tester, he even does the will of G-d, and he still roams the Earth and brings with him "ill will" as that is his nature. He carries a tumor of rotten principles. He can be detected in anything that is "against" or in opposition to G-d's word. This may include disobedience to G-d as well as sins also mentioned both in the Old and New testament including such things like vanity, adultery, pride, lies, stealing, hate, killing, swearing, bad temper, gossiping, revenge, grudge, greed, materialism, impatience, defiant, rebellion, resentment, grumblers and faultfinders, the love of money, bad habits, bad impulses, or evil desires, and all the works of the flesh (this also includes bad habits such as drinking, smoking, drugs, overeating, etc… remember, if we continue with these sorts of habits we become enslaved (or in bondage) to them - hence we are not liberated (have no freedom). Galatians 5:19-21 - Now the works of the flesh are manifest, which are these; Adultery, fornication, uncleanness, lasciviousness, idolatry, witchcraft, hatred, variance, emulations, wrath, strife, seditions, heresies, Envyings, murders, drunkenness, revellings, and such like: of the which I tell you before, as I have also told you in time past, that they which do such things shall not inherit the kingdom of G-d. Ephesians 4:25-32 - Wherefore putting away lying, speak every man truth with his neighbour: for we are members one of another. Be ye angry, and sin not: let not the sun go down upon your wrath: Neither give place to the devil. Let him that stole steal no more: but rather let him labour, working with his hands the thing which is good, that he may have to give to him that needeth. Let no corrupt communication proceed out of your mouth, but that which is good to the use of edifying, that it may minister grace unto the hearers. And grieve not the holy Spirit of G-d, whereby ye are sealed unto the day of redemption. Let all bitterness, and wrath, and anger, and clamour, and evil speaking, be put away from you, with all malice: And be ye kind one to another, tenderhearted, forgiving one another, even as G-d for Christ's sake hath forgiven you. 2 Timothy 3:1-5 - This know also, that in the last days perilous times shall come. For men shall be lovers of their own selves, covetous, boasters, proud, blasphemers, disobedient to parents, unthankful, unholy, Without natural affection, trucebreakers, false accusers, incontinent, fierce, despisers of those that are good, Traitors, heady, highminded, lovers of pleasures more than lovers of G-d; Having a form of godliness, but denying the power thereof: from such turn away. Revelation 21:8 - But the fearful, and unbelieving, and the abominable, and murderers, and whoremongers, and sorcerers, and idolaters, and all liars, shall have their part in the lake which burneth with fire and brimstone: which is the second death. Jude 1:14-16 - Enoch, the seventh from Adam, prophesied about these men: See, the Lord is coming with thousands upon thousands of his holy ones to judge everyone, and to convict all the ungodly of all the ungodly acts they have done in the ungodly way, and of all the harsh words ungodly sinners have spoken against him. These men are grumblers and faultfinders; they follow their own evil desires; they boast about themselves and flatter others for their own advantage.

Now for any of us who may be plagued with evil, or inflicted by any of these attributes (mention above) in some way, there is help available. And this is where I am going to give everyone a linguistic self-defense course against Satan and his followers (demons and evil spirits). Essentially anything that is in opposition of G-d must be “rebuked“ in G-d’s name - and ONLY IN HIS NAME, as there is NO ONE that is higher in authority than G-d, not even in the name of biblical patriarchs, saints, the Virgin Mary, or good angels. What’s more, the evil spirits explicitly know this! James 2:19 - Thou believest that there is one G-d; thou doest well: the devils also believe, and tremble. Even the good angels know this as well. Jude 1:9 - But even the archangel Michael, when he was disputing with the devil about the body of Moses, did not dare to bring a slanderous accusation against him, but said, “The Lord rebuke you“! Furthermore, to denote the divine authority given to Jesus by G-d… all we have to do is examine New Testament scripture in which we find that all the “driving out of evil spirits” performed by Jesus were done in an INSTANT! All He had to do was to tell Satan to depart, and the evil spirit left the victim at His very words. Luke 10:17 - The seventy-two returned with joy and said, Lord, even the demons submit to us in your name. Matthew 8:31 - The demons begged Jesus, If you drive us out, send us into the herd of pigs. Mark 5:12 - The demons begged Jesus, send us among the pigs; allow us to go into them. Mark 1:34 - And Jesus healed many who had various diseases. He also drove out many demons, but he would not let the demons speak because they knew who he was. Luke 4:41 - Moreover, demons came out of many people, shouting, You are the Son of G-d! But he rebuked them and would not allow them to speak, because they knew he was the Christ. Luke 8:31 - Legion, he replied, because many demons had gone into him. And they begged him repeatedly not to order them to go into the Abyss. Luke 8:32-33 - A large herd of pigs was feeding there on the hillside. The demons begged Jesus to let them go into them, and he gave them permission. When the demons came out of the man, they went into the pigs, and the herd rushed down the steep bank into the lake and was drowned. Mark 9:14-32 - When they came to the other disciples, they saw a large crowd around them and the teachers of the law arguing with them. As soon as all the people saw Jesus, they were overwhelmed with wonder and ran to greet him. What are you arguing with them about? he asked. A man in the crowd answered, Teacher, I brought you my son, who is possessed by a spirit that has robbed him of speech. Whenever it seizes him, it throws him to the ground. He foams at the mouth, gnashes his teeth and becomes rigid. I asked your disciples to drive out the spirit, but they could not. O unbelieving generation, Jesus replied, how long shall I stay with you? How long shall I put up with you? Bring the boy to me. So they brought him. When the spirit saw Jesus, it immediately threw the boy into a convulsion. He fell to the ground and rolled around, foaming at the mouth. Jesus asked the boy's father, how long has he been like this? From childhood, he answered. It has often thrown him into fire or water to kill him. But if you can do anything, take pity on us and help us. If you can? said Jesus. Everything is possible for him who believes. Immediately the boy's father exclaimed, I do believe; help me overcome my unbelief! When Jesus saw that a crowd was running to the scene, he rebuked the evil spirit. You deaf and mute spirit, he said, I command you, come out of him and never enter him again. The spirit shrieked, convulsed him violently and came out. The boy looked so much like a corpse that many said, he's dead. But Jesus took him by the hand and lifted him to his feet, and he stood up. After Jesus had gone indoors, his disciples asked him privately, why couldn't we drive it out? He replied, this kind can come out only by prayer. They left that place and passed through Galilee. Jesus did not want anyone to know where they were, because he was teaching his disciples. He said to them, the Son of Man is going to be betrayed into the hands of men. They will kill him, and after three days he will rise. But they did not understand what he meant and were afraid to ask him about it. Mark 9:38-41 - Teacher, said John, we saw a man driving out demons in your name and we told him to stop, because he was not one of us. Do not stop him, Jesus said. No one who does a miracle in my name can in the next moment say anything bad about me, for whoever is not against us is for us. I tell you the truth, anyone who gives you a cup of water in my name because you belong to Christ will certainly not lose his reward. Matthew 12:43-45 - When an evil spirit comes out of a man, it goes through arid places seeking rest and does not find it. Then it says, I will return to the house I left. When it arrives, it finds the house unoccupied, swept clean and put in order. Then it goes and takes with it seven other spirits more wicked than itself, and they go in and live there. And the final condition of that man is worse than the first. That is how it will be with this wicked generation.

We find this in both the Old and New Testaments, when we read about the "evil spirit" (destroying angels) and the "rebuking" of them. Psalm 78:49 - He unleashed against them his hot anger, his wrath, indignation and hostility a band of destroying angels. Mark 8:33 - But when Jesus turned and looked at his disciples, he rebuked Peter. Get behind me, Satan! He said. You do not have in mind the things of G-d, but the things of men. Luke 4:41 - Moreover, demons came out of many people, shouting, You are the Son of G-d! But he rebuked them and would not allow them to speak, because they knew he was the Christ. The chief means of expelling these evil spirits (or anything that was in opposition to G-d) in the Old Testament was done directly in the name of the Lord G-d Himself. The New Testament enhances this for us and it becomes obvious that one of the MAIN REASONS for the New Testament is its power to drive out evil spirits in the name of Jesus! It is for this reason that we have so much evil and the uprising of demonic forces in the world today. The deluge of possessions, which began with Jesus coming into the world, has set an all-time high pattern for the fulfillment of impending prophecy. This can be witnessed in the New Testament which records the activity, which indicates that the number of possessed persons have increased dramatically. In addition to this, the "hostility" of these evil spirits has also increased. Perhaps the reason for this stems from when Jesus was tempted three times by Satan, and each time Jesus quoted appropriate and exact scripture (specified below). If there is one characteristic of the evil spirit, it is its hatred of Jesus and of the woman, (Eve, the first women who brought about the messianic seed) because of Satan's smug slyness in tempting Eve to eat of the forbidden fruit and the aftermath of G-d's curse upon him. As a result, Satan and his agents are even more angered! As a District Attorney of sorts, he does not want to loose his case. His attacks have become relentless with the power to provoke, tempt, enter into (possessing and occupying both human and animal bodies, even in adamant objects), inflict disease, vex and transform himself. He and his agents engage in opposing G-d's work through deception. He is a cunning trap setter, and his business with us is to test humankind. And he will continue to get nastier and nastier as judgment day (his trial court date) approaches.

Things started to get nasty right around the time that Satan tempted Jesus at the start of His ministry. Jesus was very hungry after 40 days of fasting, and the tempter (Satan) came to Him. Matthew 4:3 - And when the tempter came to him, he said, If thou be the Son of G-d, command that these stones be made bread. Matthew 4:4 - But he (Jesus) answered and said, **It is written,** Man shall not live by bread alone, but by every word that proceedeth out of the mouth of G-d. Then he brought Jesus to the highest point of the temple. Matthew 4:5 - Then the devil taketh him up into the holy city, and setteth him on a pinnacle of the temple. Matthew 4:6 - And saith unto him, If thou be the Son of G-d, cast thyself down: for it is written, He shall give his angels charge concerning thee: and in their hands they shall bear thee up, lest at any time thou dash thy foot against a stone. Satan might have been implying Jesus' impending crucifixion. And in Matthew 4:7 Jesus replies - Jesus said unto him, **It is written again,** Thou shalt not tempt the Lord thy G-d. Matthew 4:8-9 - Again, the devil taketh him up into an exceeding high mountain, and sheweth him all the kingdoms of the world, and the glory of them; And saith unto him, All these things will I give thee, if thou wilt fall down and worship me. Matthew 4:10 - Then saith Jesus unto him, Get thee hence, Satan: **for it is written,** Thou shalt worship the Lord thy G-d, and him only shalt thou serve. Matthew 4:11 - Then the devil leaveth him, and, behold, angels came and ministered unto him. Therefore, as we can read, even as hungry or as weak as Jesus was in fasting, Satan did not win over His flesh. This probably inflamed Satan to an even reddened state! From this point on, the grudge and resentment brought on my Satan towards Jesus was established and was never to be forgotten or absolved. And as one can learn, quoting the Bible "word-for-word" is a good example of how important written scripture is - especially to Jesus when He quotes the biblical passages and replies to Satan's temptation to deny and pervert the Fathers word, He said "IT IS WRITTEN" three times, and accurately quoted appropriate passages from the Old Testament, (remember something repeated 3 times is a perfection of power). This is also a prime example of why it is important for us to attain and understand biblical knowledge, and to be able to apply it with the use of properly quoted scripture in order to eliminate any evil or sinful attacks that we may encounter in our lives. If a person has no clue that such real forces as these exists or are really playing out in the lives of all of us, he/she will continue to be plagued by malevolent forces throughout their life. No ghost busters, no new age spiritualist, no Wicca or incenses of any kind can help here either. These types of things only serve to make matters worse. The only real (and lasting) way to get rid of such plaguing forces is by firstly understanding the word of G-d by reading His Bible, and secondly by implementing into your life the special ordinance that He put in place to rid of this evil/sin which is only by/through Jesus the Messiah.

Lets for a moment examine other instances in the Bible that give an account of evil spirits. Such as the spirits that we read about in the Bible that afflicted King Solomon. 1 Kings 11:4-9 - For it came to pass, when Solomon was old, that his wives turned away his heart after other gods: and his heart was not perfect with the Lord his G-d, as was the heart of David his father. For Solomon went after Ashtoreth the goddess of the Zidonians, and after Milcom the abomination of the Ammonites. And Solomon did evil in the sight of the Lord, and went not fully after the Lord, as did David his father. Then did Solomon build an high place for Chemosh, the abomination of Moab, in the hill that is before Jerusalem, and for Molech, the abomination of the children of Ammon. And likewise did he for all his strange wives, which burnt incense and sacrificed unto their gods. And the Lord was angry with Solomon, because his heart was turned from the Lord G-d of Israel, which had appeared unto him twice.

In the above passage, King Solomon was noted for being inflicted with demons, but he was also noted as one of the wisest of men and also a master of handling demons. 1 Kings 4:30 - Solomon's wisdom was greater than the wisdom of all the men of the East, and greater than all the wisdom of Egypt. Solomon could handle demons with no problem, it was with women that he had his greatest problem, as we read in the above passage with one of Solomon's wife's who is a pagan and he makes a token sacrifice to her gods to appease her. He listens to his pagan wife who is worshiping foreign gods, in doing so, his heart is drawn away from the one true G-d, and he looses the very wisdom that made him so great. He then begins to speak gibberish. Even though Solomon was able to control demons, he tells us in his testament (the "Testament of Solomon" which was excluded from the Hebrew Bible) that he was given a ring by the archangel Gabriel. In the Testament of Solomon - With it thou shalt lock up demons of the earth, and with their help thou shalt build up Jerusalem. One ring of power, one ring to rule them all, one ring to find them all and in the darkness bind them. Using this ring, Solomon summons Beelzebub himself in order to enslave a host of demons. Solomon was able to tap into the good side of the forces of evil spirits and use it to subdue the dark side and cause the dark side to work for good. There are many questions as to what this ring of Solomon was? It is obvious that it was the Star of David ring that he most likely inherited from his father King David. I have also heard the association of the "Star of David" to be that of a pagan symbol because Solomon used it to summon up demons. Some associate this "Star of David" pattern from Acts 7:42-43 - Then G-d turned, and gave them up to worship the host of heaven; as it is written in the book of the prophets, O ye house of Israel, have ye offered to me slain beasts and sacrifices by the space of forty years in the wilderness? Yea, ye took up the tabernacle of Moloch, and the **star** of your god Remphan, figures which ye made to worship them: and I will carry you away beyond Babylon. Amos 5:26-27 - But ye have borne the tabernacle of your Moloch and Chiun your images, the **star** of your god, which ye made to yourselves. Therefore will I cause you to go into captivity beyond Damascus, saith the Lord, whose name is The G-d of hosts. However Just a short clarification. I don't believe this star to be the 6 pointed Star of David, but rather that of the Egyptian 5 pointed star since we had discussed in an earlier chapter that the 5 pointed star symbolized the revolution of planetary stars in the sky in which the Egyptians denoted their false gods after. However in the case of the 6 pointed star originating from King David, and King David who was later the father of King Solomon (and therefore predates Solomon), and as aforementioned in an earlier chapter that King David got this original star design from the "pomegranate" which he later adorned in the design of his temple. This does not make the 6 pointed Star of David a pagan symbol. However, if one prays to such a symbol it then becomes something of a pagan nature? It also seems that much in the same way a crucifix is noted for repelling demons, so was the Star of David, probably even more so at that time considering that it was later to represent the 613 commandments. But one must also not forget that this design which was originally produced from the Pomegranate (as discussed in an earlier chapter) was also what caused sin to enter the world in the first place when Eve took of it from the Tree of knowledge of good and evil and bit into it. This is why Solomon was able to use it to summon up demons, subdue them, and cause them to work for good.

In fact Jesus Himself makes mention of King Solomon after they brought him a demon-possessed man who was blind and mute, and Jesus healed him, so that he could both talk and see. Jesus tells them, someone greater than Solomon is here (referring to Himself). Matthew 12:42 - The Queen of the South will rise at the judgment with this generation and condemn it; for she came from the ends of the earth to listen to Solomon's wisdom, **and now one greater than Solomon is here.**

Ironically, in the "Testament of Solomon" which was written centuries before Jesus, cites a passage that makes mention of the coming crucifixion of Jesus. The demons themselves foretell of it. In the Testament of Solomon - (And the demon said) we shall lead astray the inhabited world for a long season, until the Son of G-d is stretched up on a cross. Some say this was tainted by Christian scholars after the fact. However, newly discovered older texts verify this same account at an even older date.

King Solomon was a prime example of a person who dealt with spiritual demons. Be careful what you read here… these were specifically "spiritual demons". However many people to this day are convinced that there are lingering ghosts of relatives or un-restful spirits of once living people that perhaps had their life cut short by death in some way and are now lurking around in our dimension. However, the Bible does not agree with this, it plainly tells us that when a person dies they are judged by G-d and pass on to another place. And once they die and are judged they can no longer interact with the living. Hebrews 9:27 - And as it is appointed unto men once to die, but after this the judgment.

However, the "biblically uneducated" continually press the issue that these spirits (both good and bad) have not passed on or are at rest. They attempt to intercede and talk with the dead to perhaps assist them in passing on. But the Bible is clear on this point too, telling us that such practices are detestable. Deuteronomy 18:9-13 (concerning detestable practices) - When you enter the land the Lord your G-d is giving you, do not learn to imitate the detestable ways of the nations there. Let no one be found among you who sacrifices his son or daughter in the fire, who practices divination or sorcery, interprets omens, engages in witchcraft, **or casts spells, or who is a medium or spiritist or who consults the dead.** Anyone who does these things is detestable to the Lord, and because of these detestable practices the Lord your G-d will drive out those nations before you. You must be blameless before the Lord your G-d. In effect, these ghosts are actually demons; they are not the ghosts of past relatives or people that have not passed on. They are deceiving demons spoofing around with us. Their main objective is to want us to doubt the existence of G-d. And I can say this with full confidence that they are nothing more than malevolent spirits (fallen angels) masquerading around as our past relatives or acquaintances. This also includes the astral bodies (Merkabah vehicles) of people who are still alive (as discussed in an earlier chapter).

So how do these malevolent spirits accomplish this deception? Well we know according to the Bible that evil spirits have existed for centuries since G-d created the world and the Garden of Eden with Adam and Eve. In addition, the SAME EVIL SPIRITS that are present today were here when all you're past relatives were alive. In addition, since they have been lingering around all this time they know all about our past relatives and us, even retaining knowledge of secret things that they personally knew or did. They have watched us throughout the centuries of family generations. Additionally this works in the same way for people who believe they have been reincarnated, which feel that they've lived another life centuries ago. These accounts appear to be very real to the individual persons involved, they can even accurately account for peoples names and places. These memories can only account for evil spirits impressing thoughts in the minds of people. Reincarnation is nothing more than a transference of energy, which contains information and memory. However, we need to ask the question of why this information is being transferred. In most if not all cases, there are negative forces involved, which reek of malevolent spirits. For instance, most people that claim reincarnation remember a cheated or violet death of some sort. They may have also been shorted in life somehow, or they may have even been somebody of notability and power. However, to this same account, the Bible speaks against reincarnation and clearly specifies that death is final, that we live only once and die only once. Hebrews 9:27 - And as it is appointed unto men once to die, but after this the judgment. Therefore, in the process of elimination, what we are dealing with is evil spirits.

Lets further this process of elimination… Some ghosts have even appeared to people claiming that they want to help them. 2 Corinthians 11:14-15 - And no marvel; for Satan himself is transformed into an angel of light. Therefore it is no great thing if his ministers also be transformed as the ministers of righteousness; whose end shall be according to their works. While others plainly are out to cause harm. We hear about this all the time concerning haunted houses or grounds. Demons are lying spirits, and we must not forget that demons do have the ability to manifest themselves, and manipulate the matter around them. Satan can vex, inflict, posses and impersonate by taking the form of anything he pleases, just as he took the form of the serpent in the Garden of Eden, as well as inflicting the patriarch Job. There is also another way to tell that these spirits are of an evil nature. All one has to do is simply observe (as well as innately feel) the surrounding nature of these spirits. Usually these apparitions occur in places that have a "negative" frequency resonating from them. You hear reports of people committing suicide and hanging themselves there. A professional comes to investigate the scene and tells us that this person hung himself or herself because they were depressed. However, one must ask the question of what caused the depression in the first place. It was most likely the negative energy in the place to begin with. Some of these places (like old hotels) now claim that they see these peoples ghosts there. Look folks, all these grounds had negative energy there to begin with, that's why these people later killed themselves or committed heinous crimes there. It may also be that somebody that attracted evil spirits at one time came to visit or live in these places. Ones best option when it comes to ghosts (demons) is to ignore them since what they are seeking is attention and are fueled by our reactions to them. If they know they are not accomplishing anything, they will eventually leave. If they continue to linger around then you must subject them to an intolerable atmosphere that is opposite to their nature. This is by filling yourself with the holy spirit of the written word, and fashioning your house (or environment) in the same manner. However, if you are intrigued by and go seeking after such things, they will adhere to you like glue and plague you everywhere you go until the day you die!

Jesus never taught that ghosts were real or that they were the spirits of our past relatives or friends returning to visit us. On the contrary, he said that when we die we either are with G-d in heaven, or will be waiting final judgment in hell. To be sure that ghosts do not exist we find Jesus accounting for His state of transfiguration. Luke 24:39 - Behold my hands and my feet, that it is I myself: handle me, and see; for a spirit hath not flesh and bones, as ye see me have. Do not misread this… This is not Jesus telling us that ghosts are real. It is him telling us that he is very real, and not that of a ghost or astral body (as discussed in an earlier chapter). In effect, this is what our "resurrected body" will be like. However, Jesus did tell us that many "evil spirits" roam the Earth.

Do not become confused here, evil spirits and ghosts are not separate entities, they are both demons. They are not the ghosts of our past relatives, they are demons in disguise. There are simply no exceptions! These ghostly apparitions are ALL of an evil negative inclinational nature even though they may appear good or even beautiful. Even the person who is projecting their astral body. This is because it is still umbilical-corded and bonded to the sinful flesh unless otherwise separated by death.

However, to ones satisfaction there is a scriptural way to test the spirits. Whether you are dealing with so called ghosts, aliens, psychics or astral bodies, the only way to know if you are being deceived is to know the word of G-d (scripture) and know it well. Test the spirits to see if they are of G-d, or if you are being deceived by a lying spirit. 1 John 4:1-6 - Beloved, believe not every spirit, but try (test) the spirits whether they are of G-d: because many false prophets are gone out into the world. Hereby know ye the Spirit of G-d: Every spirit that confesseth that Jesus Christ is come in the flesh is of G-d: And every spirit that confesseth not that Jesus Christ is come in the flesh is not of G-d: and this is that spirit of antichrist, whereof ye have heard that it should come; and even now already is it in the world. Ye are of G-d, little children, and have overcome them: because greater is he that is in you, than he that is in the world. They are of the world: therefore speak they of the world, and the world heareth them. We are of G-d: he that knoweth G-d heareth us; he that is not of G-d heareth not us. Hereby know we the spirit of truth, and the spirit of error.

When you do test the spirits and find that they are "of course" evil… an "exorcism" is in order. An "exorcism" is the act of driving out or warding off evil spirits from persons, places, or things. It is a science in itself. Mark 16:17 - These signs shall attend those who believe: in my name they shall cast out devils. In order to keep these evil spirits out PERMANENTLY one must dedicate their life to "proper" biblical knowledge and continue to uphold their spirituality for the rest of their lives. Once a person is inflicted with evil spirits they will continue to be subjected and prone to them. This then becomes like a habitual habit, a lifetime battle. If one does not cast them out "properly and permanently" the first time, they will gain a notch (a foothold). No other means will suffice… not Wicca or any other religion; it must be cast out in Jesus name that is under the authority of the "one G-d" for this very purpose. The reason why one must continue to uphold their spirituality for the rest of their lives is because Jesus additionally tells us in Matthew 12:43-45 - When an evil spirit comes out of a man, it goes through arid places seeking rest and does not find it. Then it says, I will return to the house I left. When it arrives, it finds the house unoccupied, swept clean and put in order. **Then it goes and takes with it seven other spirits more wicked than itself, and they go in and live there.** And the final condition of that man is worse than the first. That is how it will be with this wicked generation. Matthew 6:9-15 - This, then, is how you should pray: Our Father in heaven, hallowed be your name, your kingdom come, your will be done on earth as it is in heaven. Give us today our daily bread. Forgive us our debts, as we also have forgiven our debtors. **And lead us not into temptation, but deliver us from the evil one.** For if you forgive men when they sin against you, your heavenly Father will also forgive you. But if you do not forgive men their sins, your Father will not forgive your sins. To enforce this, we must also uphold the commandment of Deuteronomy, which includes not seeking after any sort of divination that might invite evil spirits. Deuteronomy 18:9-13 - When you enter the land the Lord your G-d is giving you, do not learn to imitate the detestable ways of the nations there. Let no one be found among you who sacrifices his son or daughter in the fire, who practices divination or sorcery, interprets omens, engages in witchcraft, or casts spells, or who is a medium or spiritist or who consults the dead. Anyone who does these things is detestable to the Lord, and because of these detestable practices the Lord your G-d will drive out those nations before you. You must be blameless before the Lord your G-d.

Biblical scripture works best to eliminate evil spirits, but you can even make your own prayer up as long as it testifies of divinely written scripture. Such as praying the prayer: I cast you out, unclean spirit and all your fallen companions in the name of Jesus. Be gone and stay far from this person of G-d. For it is He who commands you, He who hurled you from the heights of heaven into the depths of hell. He commands you. He who once stilled the sea, the wind, and the storm. You cannot resist, knowing as you do that Jesus brings your plans to nothing? It is He who clothed Adam and Eve and in Isaac was offered in sacrifice, slain as the sacrificial lamb, crucified as man, yet triumphed over the powers of hell. Be gone, in the name of the Father of Enoch, Abraham, Isaac, Jacob, Moses and Jesus - there is none higher than He is. James 2:19 - Thou believest that there is one G-d; thou doest well: the devils also believe, and tremble.

Additionally, along with scripture and prayer there are certain tools that one can use to facilitate exorcism - such as "baptism with water" in which the ancient Jewish ritual of the "Mikvah" and its purpose will soon be revealed. The ancient ritual of the "Mikvah" (which is basically a baptism) involves the immersion of the human body in water. It is a ritual bath of repentance usually followed by a woman's menstruation cycle and is used in marital law and sexual discipline in marriage (generally referred to as Taharat HaMishpach). The Mikvah is a repentance ritual of ensuing purity. Which is followed by a Hebrew blessing that goes like this: Baruch atah Adonai Elohanu melech ha-olam, asher kid'shanu b'mitzvotav v'tzivanu al ha-t'vila. Translation: Blessed are You, Lord, our G-d, King of the universe, who has sanctified us with His, commandments and commanded us concerning immersion (in/of water). One might ask the question of why "WATER" is so important, and was later used for baptism/mikvah?

Firstly, lets remember (from a past chapter) that the Hebrew word for water is "Mayim" and it is found within the Hebrew word "Shamayim" which means "Heaven". Therefore, it can be said that water is the closest thing to Heaven on earth. What this is telling us is that when water falls from the sky (from Heaven) upon things it brings about a cleansing (from whence was impurity), as well as bringing about rebirth for seeds. It also replenishes all living things and keeps them alive and growing. John 3:3-5 - Jesus answered and said unto him, Verily, verily, I say unto thee, **Except a man be born again, he cannot see the kingdom of G-d.** Nicodemus saith unto him, How can a man be born when he is old? can he enter the second time into his mother's womb, and be born? Jesus answered, Verily, verily, I say unto thee, **Except a man be born of water and of the Spirit, he cannot enter into the kingdom of G-d.**

With that in mind, when Jesus answered**:** Except a man be born of water and of the spirit, he cannot enter into the kingdom of G-d… figuratively speaking, "water" in itself is another microcosm of sorts, but sadly without the spiritual aspect applied to it since many people take water for granted.

Now to fully answer the question of the importance of water, we must again examine the Hebrew word for water, which is "Mayim". It is the primary element that was created on the second and third day of creation when G-d separated the upper from the lower waters, and when G-d gathered them into one place, calling them "Yamim" (Seas). Water is used literally and symbolically in the Bible as well as throughout all of biblical history and prophecy. Genesis 1:1-2 - In the beginning G-d created the heaven and the earth. And the earth was without form, and void; and darkness was upon the face of the deep. And the Spirit of G-d moved upon the face of the waters. Genesis 1:9-10 - And G-d said, Let the waters under the heaven be gathered together unto one place, and let the dry land appear: and it was so. And G-d called the dry land Earth; and the gathering together of the waters called he Seas: and G-d saw that it was good.

Now lets get back to the Hebrew word "Mikvah" (mentioned above) which literally means a pool of "mayim chayim" which is living (non-stagnant) waters. This is usually rain, clean ocean, river or flowing lake water, even frozen ice, snow, hail frost and salt (since it is still likely to again become living water after melting). Immersion in a Mikvah is that of renewal and rebirth, a form of transcendence. A return to ones inner, pure self, and a revitalizing restoration of ones spiritual status. Of course, it does more than just "spiritually" cleaning us; it also "physically" cleans us. Water symbolized anew. It is no coincidence that a human being grows within a bag of water before birth. Just as well, drinking water is one of the primary needs of life that we cannot live without, and not just us, but other living things as well. Aside from the familiar daily chore of drinking and bathing in it, there are many other scientific laws that apply to it. We can literally say that water blesses every cell of our body. Water is the most abundant liquid on the planet, covering approximately 70% of the earth's surface. Coincidently, the human body is also mostly water. Pure water is tasteless, odorless and colorless, although it does have a blue color in large volumes. The familiar chemical symbol H_2O means that water molecules are composed of 2 hydrogen atoms and 1 oxygen atom. And as you are about to read backed up by biblical scripture, water is also used as a method of becoming spiritually clean after having become spiritually unclean - like by contact with a human corpse mentioned in the laws of Leviticus (as aforementioned in an earlier chapter). Additionally, Jewish women do the immersion of a Mikvah after their menstrual cycle once a month and a bride on the eve of her wedding day. The Mikvah is exclusive for Jewish women because woman, not man, is the creator of new life. The woman conceives and gives birth forging a new soul. The Mikvah is also used by converts to Judaism and by religious Jewish men every morning. It even applies with pots, pans, dishes, and utensils before they are used in a kosher kitchen, which is considered a spiritual place where food nourishes the body not just physically but spiritually, as well as complying to scientific sanitary laws of keeping kosher as discussed in an earlier chapter.

We must also remember (as previously discussed) that water is vibratory in nature and retains memory. The Old Testament states, "In the beginning there was the word". John 1:1 - In the beginning was the Word, and the Word was with G-d, and the Word was G-d. The word was vibratory in nature. Genesis 1:3 - **And G-d said**, Let there be light, and there was light. Genesis 1:6 - **And G-d said**, Let there be an expanse between the waters to separate water from water. Genesis 1:9 - **And G-d said,** Let the water under the sky be gathered to one place, and let dry ground appear. And it was so. Genesis 1:11 - **Then G-d said,** Let the land produce vegetation: seed-bearing plants and trees on the land that bear fruit with seed in it, according to their various kinds. And it was so. Genesis 1:14 - **And G-d said,** Let there be lights in the expanse of the sky to separate the day from the night, and let them serve as signs to mark seasons and days and years. Genesis 1:20 - **And G-d said,** Let the water teem with living creatures, and let birds fly above the earth across the expanse of the sky. Genesis 1:24 - **And G-d said,** Let the land produce living creatures according to their kinds: livestock, creatures that move along the ground, and wild animals, each according to its kind. And it was so. Genesis 1:26 - **Then G-d said,** Let us make man in our image, in our likeness, and let them rule over the fish of the sea and the birds of the air, over the livestock, over all the earth, and over all the creatures that move along the ground. Genesis 1:28 - **G-d blessed them and said to them,** Be fruitful and increase in number; fill the earth and subdue it. Rule over the fish of the sea and the birds of the air and over every living creature that moves on the ground. Genesis 1:29 - **Then G-d said,** I give you every seed-bearing plant on the face of the whole earth and every tree that has fruit with seed in it. They will be yours for food. Ten times, "He said" and it was so! The fact that everything is in a state of vibration also means that everything is creating sound. If we delve even deeper we realize that everything consist of atoms, and each atom having a nucleus with electrons rotating around it gives each atom a particular set of vibration frequencies - there is only a nucleus surrounded by an endlessly rotating wave. Therefore, existence as we know it is vibration! Even the science of Quantum Mechanics (as aforementioned in an earlier chapter) generally acknowledges that substance is nothing more than vibration, and we find this out when we try and separate something into smaller parts we continually discover that all that exist is particles and waves.

So the point that I'm attempting to make concerning the special purpose of water is that most all life contingent to our planet consists of it - and the fact that water stores memory suggest that it can also contain negative vibrations that are malevolent (evil) in nature. And they can influence us if the vibrations vibrate at us and begin to resonate within us. We as

Human beings are vibrating, and each of us vibrates at a unique frequency. A person who is full of hate or sadness will emit a negative frequency, and someone who is happy and full of love will emit a positive frequency. By the way, others can intuitively sense these vibrating feelings as well. Not to be fooled though, these same principles also apply to physical objects and locations, such as the walls or the woodwork of a house, furniture, even jewelry like crystal gemstones. The reason for this is because they all contain or trap water in some form. A person or family who has lived in a house for some time generally implants their vibrating frequencies, which can be either positive or negative. Some rooms or areas of the house may have absorbed negative emotions that were transferred into vibrations which are most likely embedded and absorbed in places and things that consists of water like woodwork, walls and furniture because wood is primarily made of water (as it needed it to grow) and also many man made substances. This is also why some houses appear to be haunted, or upon entering or living in them make one sick or possessed. This also applies to specific locations on the Earth that have been known to bring or cause tragedy. The main reason for all this is that the vibration and resonance is still there retained within the water molecule of the Earths dirt/ground itself. This may sound like nonsense but I've just pointed out that "vibration" is indeed a fact of our physical and material nature. Now, to counteract this negative menace (which is a form of evil) one must "override" these negative vibrations/frequency with a vibration that is opposite to the negative emotion. By combining two opposite waves, the negative vibration disappears. This again involves the same principles of homeopathy, which dilutes a substance in water. As it turns out, WATER ITSELF that is programmed with positive vibration/frequency can be used to override negative vibrations/frequency. If the water is prayed over with the scriptural words of G-d, it will indeed retain that ultra positive memory and can be used to counteract and override any negative or evil influence. This is the main reason why water is used during a baptism/Mikvah. Holy water (which is water that has been prayed over by a priest/pastor/minister or by any persons correctly quoting biblical scripture) can be used for purification, and to rid evil. Surprisingly, this can also be done by vocalizing, singing and reciting the "Ahhh" sound which is vocalized in many Hebrew songs and prayers (remember, this is the restructuring sound). It also forms the "Star of David".

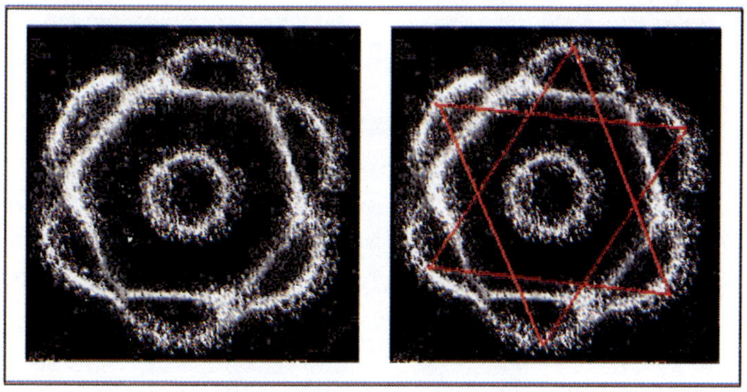

Here are some biblical passages that specify the use of water as a means of purification... The waters of Noah's great flood were used as a sort of baptism that brought an end to sin, and a new beginning of goodness. 2 Peter 3:6 - By these waters also the world of that time was deluged and destroyed. 1 Peter 3:20-21 - Which sometime were disobedient, when once the longsuffering of G-d waited in the days of Noah, while the ark was a preparing, wherein few, that is, eight souls were saved by water. The like figure whereunto even baptism doth also now save us (not the putting away of the filth of the flesh, but the answer of a good conscience toward G-d,) by the resurrection of Jesus Christ. Psalm 136:6 - To him that stretched out the earth above the waters: for his mercy endureth for ever. Moses and the crossing of the Red sea. Exodus 14:21 - Then Moses stretched out his hand over the sea, and all that night the Lord drove the sea back with a strong east wind and turned it into dry land. The waters were divided. In the waters of the Jordan, Jesus was baptized by John and anointed with the Spirit. Matthew 3:13 - Then Jesus came from Galilee to the Jordan to be baptized by John. Water and blood flowed from Jesus' side as He hung upon the cross. John 19:34 - Instead, one of the soldiers pierced Jesus' side with a spear, bringing a sudden flow of blood and water. 1 John 5:8 - The Spirit, the water and the blood; and the three are in agreement. After His resurrection, He told His disciples: Go out and teach all nations, baptizing them in the name of the Father and of the Son and of the Holy Spirit. Matthew 28:19 - Therefore go and make disciples of all nations, baptizing them in the name of the Father and of the Son and of the Holy Spirit. Numbers 5:17 - Then he shall take some holy water in a clay jar and put some dust from the tabernacle floor into the water. 2 Kings 2:19-22 - And the men of the city said unto Elisha, Behold, I pray thee, the situation of this city is pleasant, as my lord seeth: but the water is naught, and the ground barren. And he said, Bring me a new cruse, and put salt therein. And they brought it to him. And he went forth unto the spring of the waters, and cast the salt in there, and said, Thus saith the Lord, I have healed these waters; there shall not be from thence any more death or barren land. So the waters were healed unto this day, according to the saying of Elisha which he spake. Exodus 30:35 - And make a fragrant blend of incense, the work of a perfumer. It is to be salted and pure and sacred.

Therefore, water (including salt, since all salt comes from evaporated sea salt). Salt is also a disinfectant as well as a preservative and may be used to dispel demons and rid some sicknesses. This also includes any harmful thing including pestilent spirit or hostile power, corrupting atmosphere or schemes of a hidden enemy or foe. There is also a Hebrew blessing for "surviving" illness or danger… this is called in Hebrew the Birkhat Ha-Gomel. It reads like this: Baruch atah Adonai Elohanu melech ha-olam, ha-gomel lahayavim tovot sheg'malani kol tov. Translation: Blessed are You, Lord, our G-d, King of the Universe, who bestows good things on the unworthy, and has bestowed on me every goodness.

Ones work is never ending! We must not forget that our world is accompanied by and dependent upon the everyday experiences of the material realm and this includes the temptations of evil and sin. However, as I mentioned earlier in this book, that the spiritual should not be divorced from the material, but sees the material as a way to attain the spiritual. Additionally as I also mentioned earlier at the beginning of this chapter: FOR HUMANKINDS GOAL IN LIFE IS A SPIRITUAL AND INTELLECTUAL PERFECTION. This means by first interpreting how to relate to the fleshly impulses of our nature, we then learn how to relate to G-d. In doing so, one soon realizes that what G-d is in theory, we must realize in devote religious and spiritual practice - for these spiritual practices in themselves are in fact also science. Ephesians 6:11-12 - Put on the whole armour of G-d, that ye may be able to stand against the wiles of the devil. For we wrestle not against flesh and blood, but against principalities, against powers, against the rulers of the darkness of this world, against spiritual wickedness in high places.

In addition to this, I'd like to briefly clarify a few things before we move on. Besides exorcism there is no G-dly sign miracles happening in the world today. The only obligation left for the Bible to pursue is the fulfillment of the written prophesies. There are also no other books to be written and included in the Bible. The apostle John in the last book of the Bible (Revelation) clarified this for us when he said: Revelation 22:18 - I warn everyone who hears the words of the prophecy of this book: If anyone adds anything to them, G-d will add to him the plagues described in this book. There are also no more apostles or prophets, fortunetellers or psychics, (not even weather predictions are 100% accurate). This is because there is no more revelation and thus no more signs miracles to substantiate any more revelation because everything has already been incorporated into the Bible! What remains after Jesus and the apostles is spiritual healings such as the holly spirit coming in as well as exercises to rid evil. There is also personal prayer for healing. Additionally, do not even believe in any miraculous healings (like Jesus performed) going on in the world such as on the news stations or on Christian television programs. In addition to this, there are no speaking in tongues. The purpose of speaking in tongues was specifically used only for one reason and time in history and this was when 3,000 people (each in their own individual language) heard the message about Jesus being the Messiah (and excepted Him) and were saved. Acts 2:41 - Those who accepted his message were baptized, and about three thousand were added to their number that day.

The reason Jesus performed sign miracles during His time on Earth was specifically to confirm, support and demonstrate that He was indeed the true Messiah and the one and only savior that had the power to forgive sin. John 20:30-31- Jesus did many other miraculous signs in the presence of his disciples, which are not recorded in this book. But these are written that you may believe that Jesus is the Christ, the Son of G-d, and that by believing you may have life in his name. Acts 2:22 - Men of Israel, listen to this: Jesus of Nazareth was a man accredited by G-d to you by miracles, wonders and signs, which G-d did among you through him, as you yourselves know. Here we find Jesus fulfilling what He (G-d) promised (said) to Israel hundreds of years earlier. Deuteronomy 18:20-22 - But a prophet who presumes to speak in my name anything I have not commanded him to say, or a prophet who speaks in the name of other gods, must be put to death. You may say to yourselves, How can we know when a message has not been spoken by the Lord. If what a prophet proclaims in the name of the Lord does not take place or come true, that is a message the Lord has not spoken. That prophet has spoken presumptuously. Do not be afraid of him.

Along these same lines, Jesus told the disciples that they would perform sign miracles in His name, but with greater works. John 14:12 - I tell you the truth, anyone who has faith in me will do what I have been doing. He will do even greater things than these, because I am going to the Father. Now one may wonder just what kind of miracles was Jesus referring to? Well, it was witnessed and later recorded in the New Testament that when Peter preached to thousands of Jews and in a single day 3000 people where saved! This is astounding because saving thousands of people in a single day never happened during Jesus' ministry on Earth! This is because it is a misconception (especially to the Jews) that Jesus came to convert the world at that particular time in history. Rather He came to die in order that there would be a message and an imbursement power available to renovate the entire world through the "Church".

In fact so astounding was the saving of 3,000 individuals by the disciples compared to Jesus' temporary healing miracles (since everyone that He healed died anyways). It is only when we focus on the "finished message" of Jesus that we see great achievement and triumph, since His pervading message has made its way throughout the ends of the Earth and has served to permanently change, modify, and heal forever!

Getting back on the subject of Satan and exorcism… So, the question now arises in this chapter of how Satan (the lawless one) will attempt to work his ways up until judgment day. First off, remember as a District Attorney of sorts, he does not want to loose his case even if the verdict is not in his favor and looks grim, he will continue to fight it all the way to the end. His attacks have become relentless with the power to provoke, tempt, enter into (possessing and occupying both human and animal bodies, even in adamant objects), inflict disease, vex and transform himself. He and his agents engage in opposing G-d's work through deception. He is a cunning trap setter, and his business with us is to test humankind. And he will continue to get nastier and nastier as judgment day (his trial court date) approaches. 2 Thessalonians 2:1-12 - Concerning the coming of our Lord Jesus Christ and our being gathered to him, we ask you, brothers, not to become easily unsettled or alarmed by some prophecy, report or letter supposed to have come from us, saying that the day of the Lord has already come. Don't let anyone deceive you in any way, for (that day will not come) until the rebellion occurs and **the man of lawlessness** is revealed, the man doomed to destruction. He will oppose and will exalt himself over everything that is called G-d or is worshiped, so that he sets himself up in G-d's temple, proclaiming himself to be G-d. Don't you remember that when I was with you I used to tell you these things? And now you know what is holding him back, so that he may be revealed at the proper time. **For the secret power of lawlessness is already at work;** but the one who now holds it back will continue to do so till he is taken out of the way. **And then the lawless one will be revealed,** whom the Lord Jesus will overthrow with the breath of his mouth and destroy by the splendor of his coming. The coming of **the lawless one will be in accordance with the work of Satan** displayed in all kinds of counterfeit miracles, signs and wonders, and in every sort of evil that deceives those who are perishing. They perish because they refused to love the truth and so be saved. For this reason G-d sends them a powerful delusion so that they will believe the lie and so that all will be condemned who have not believed the truth but have delighted in wickedness.

There is one thing that Satan cannot dispute or argue with, and that is the word of G-d, which was recorded in the Bible by His divinely inspired servants - the prophets. We must also not forget that the Old Testament translated to Moses by G-d is written in His holy script. And we witness this by Jesus (the judge) quoting scripture (law) out of the Old Testament Bible back at Satan. However, Satan can pervert, twist and partially bootleg the religious scriptures! Remember its part of his job in winning his opposing case - he tends to twist the law and look for loopholes. We witness this with attorneys today in the courtroom winning the cases of guilty people. Therefore, Satan's best option is to plagiarize and pervert the original scripture of G-d into the form of other religious documents and religions of the world (that "he" most likely fashioned and influenced). Which ones do you think he has influenced? This is a very conflict-ridden question, however herein lies the answer. As it happens, throughout recent history there has shown to be a central link between the Roman Empire (Rome and the Pope) and Islam - two of the fastest growing and influential religions dominating and prevaricating in the present day. One may ask what are the twisted particulars regarding both these religions and why focus on only these. Well, clearly it is undeniable that throughout history the Jews were persecuted more than any other people were. This will never change! (And the reason for this will soon reveal itself). We focus on these two religions because the linkage that they share is that they both have a long history of persecuting the Jews! In terms of twisting scripture, both these religions include revelations from extra biblical and non-biblical sources in such ways that the Pope says they worship the same G-d… They believe that salvation alone can be obtained without Jesus by good works... Both seek world dominion... The Virgin Mary is highly revered - the only woman mentioned in the Koran 34 times (as a model of submission). The Roman Church and the Judeo Christian evangelical (Lutheran) are also divided. The Roman Church believes in baptism, faith and works. They also believe that one is purified of sin only through the fires of purgatory. Additionally they believe that a priest, the Virgin Mary or the Pope can mediate between G-d and man. They believe the pope is the Vicar/embodiment of Jesus here on earth. This is a blatant outrage to the truth! Nobody, (and I do mean NOBODY) can be the embodiment unless the person was Jesus Himself). While the only ones who have it right are, the Judeo Christians believing that one can only be saved solely through the faith of the One G-d/Jesus. This is the only scriptural correct aspect!

Can one see what has happened here, how scripture was twisted? For instance to facilitate this understanding lets examine the Koran (Qur'an). There is much in the Koran (Qur'an) that is consistent with the Old Testament Bible. One can see similarities, but overall the Koran seems to have twisted and reversed the Old and New Testament teachings. However, the Qur'an referrers to a number of additional sayings of Muhammad that are derivatives in some form or way from the Old and New Testaments. The gospel of Jesus is also studied and considered revelations, although it is believed by followers of the Koran to have been corrupted through time and thus today Islam does not recognize the role of Jesus and Israel in salvation the way the scriptures unfold it. The Islam people see Muhammad, as if Christians see Jesus and feel that since they have Muhammad there is no need or reason to convert to Christianity. This is the confusing catch-22 about it that reveals otherwise - trickery by Antichrist (which will soon be recognized).

Islam (which means "submission") is one of the fastest growing religions around the world. The word "Muslim" means Follower of Muhammad - "the one who submitted or submits". Islam believes in one G-d (Allah), but their prophet Muhammad (unlike Jesus) was not prophetic. Now the word Koran means to "recite" what the angel Gabriel told Muhammad. But was it really the angel Gabriel? We cannot say for sure whether this was truly divinely inspired, but we can ultimately distinguish and witness that Islam/Muslim took a turn of events that corrupted and twisted its initial foundation. The question remains, why would the angel Gabriel declare to Muhammad that a woman's value is half that of a mans? This will become clear when the nature of antichrist is soon revealed. The "submission" of Islam does not mean salvation. Submission means compliance, surrender, giving in, proposal, suggestion, and you will soon find out to which this surrendering is for.

Lets first look at the history of Muhammad. He was born in Mecca in the Arabian Peninsula in the year 570 CE. He would go with his uncle on long journeys through the desert and in time he became a caravan manager and gained great admiration and esteem for his trustworthiness. When he was 25 years old, he went to work for a very wealthy widow whom he later married. In 610 CE Muhammad was meditating in a cave near Mecca when the supposed angel Gabriel appeared to him and revealed to him the Koran, eventually in the duration of time these words were written down and they became the basis of Islam/Muslim doctrine. In 622, CE Muhammad began an organized community and military with himself as the head of it. During this time, the Muslims and Jews lived side-by-side in peace and even prayed in the direction of Jerusalem until Muhammad's reign took hold over Mecca. Mecca is now the direction in which Muslims pray. During Muhammad's reign, he gave the order to besiege the Jews that lived in the surrounding area. However there are no real specifics that exist as to the real reason for why Muhammad (who became a tyrant) gave the order that warranted the total genocide of the Jews? Some suggest that the genocide of the Jews was all about harboring defiance against Islam and rejecting Muhammad as a prophet of G-d. Alongside of the Jews rejection of Muhammad, Muhammad most likely feared that they harbored a large arsenal of weapons (what Muhammad called "treachery" harbored against Islam). Sarcastically after the genocide of the Jews, these weapons were taken into the hands of the Muslims.

It is inconceivable that a messenger of G-d would kill hundreds of Jews and banish thousands more during his reign to continue with the ethnic cleansing that he had initiated. The man that is called a prophet was full of hate, killing, death and vengeance. He was certainly, not, what is claimed of him to be a "mercy of G-d to mankind". We cannot rightfully call him a messenger of G-d. He was also a thief who enriched himself with the wealth and weapons of his victims - the Jews.

Today the greatest threat of Islam lies with the radical violent Jihad (consisting of one sector of the Islam world). Within this sector, there is a conviction that all governments should be destroyed and replaced by a religious caliphate (territory or reign). They will battle any form of democracy because to them equality is blasphemous because it says the citizens not G-d shape the law, they find the idea of human equality to be equally offensive. Overall, they hate everything a democracy believes about freedom.

Contrary to Muslims including Jesus in the Koran, they believe Jesus never died on the cross. Jesus was never G-d and that G-d never became flesh. However, they do believe that Jesus was a prophet. They believe the New Testament was corrupted by the Christians and the Old Testament was corrupted by the Jews. The whole reason for Islam coming into being is to correct the Judeo Christian religion that was supposedly (in the minds of the Muslims) corrupted. And we can begin to get a glimpse (almost a sense) of Antichrist orchestrating himself throughout the woodwork of religious aspects, and adamantly seeking to spin religious aspects. The apostle John in the New Testament tells us that whoever denies Christ is antichrist. 1 John 2:22 - Who is a liar but he that denieth that Jesus is the Christ? He is antichrist, that denieth the Father and the Son.

Now if you're interested in understanding the spirit of Antichrist you must search in the Bible for Antichrist and how Antichrist thinks (along with his demeanor and motives). He is Antichrist, which means against Christ. He is not going to come like Jesus and say here I am, he will deny the attributes of Christ and cause "confusion". The Antichrist is a master of confusion and this can be traced back to Genesis in the garden when he created confusion with Eve. Genesis 3:1-7 - Now the serpent was more crafty than any of the wild animals the Lord G-d had made. He said to the woman, Did G-d really say, You must not eat from any tree in the garden? The woman said to the serpent, We may eat fruit from the trees in the garden, but G-d did say, You must not eat fruit from the tree that is in the middle of the garden, and you must not touch it, or you will die. You will not surely die, the serpent said to the woman. For G-d knows that when you eat of it your eyes will be opened, and you will be like G-d, knowing good and evil. When the woman saw that the fruit of the tree was good for food and pleasing to the eye, and also desirable for gaining wisdom, she took some and ate it. She also gave some to her husband, who was with her, and he ate it. Then the eyes of both of them were opened, and they realized they were naked; so they sewed fig leaves together and made coverings for themselves. The question is why did antichrist pick on the woman instead of the man. Is it because she was more influential or easily swayed? No! The reason antichrist picked on the woman is because he has a special hatred toward the woman. Why? Because it is her seed that brought about the redemptive Messiah. In fact, this is the whole reason for the gospel of Jesus in a nutshell! Antichrist also adamantly hates Israel because Israel is the sowing grounds that brought about the seed of the messiah and eventually will bring about the 2nd messianic fulfillment (the 2nd coming of Jesus).

Jesus tells us something to this fact mentioned in Matthew 10:16 - I am sending you out like sheep among wolves. Therefore be as shrewd as snakes (antichrist) and as innocent as doves. Therefore, by understanding the serpentine wisdom is to understand Antichrist! Without the Old Testament Bible we can never understand why Arabs hate Jews, Israel, America or any other country or persons that represents or backs those countries, or why there is hostility, hate, antagonism, or bad feelings between the races. This is because Antichrist is involved.

Muslims in their confused minds think that they love G-d. Nevertheless, they have the spirit of Antichrist. Remember, the Antichrist detests and despises the woman! In fact according to the Koran, Muslim woman are considered second-class citizens. Muhammad himself said that the woman has half the value of a man! Antichrist doesn't honor the desire of woman - he has no respect for women. Daniel 11:37 - Neither shall he regard the G-d of his fathers, **nor the desire of women,** nor regard any god: for he shall magnify himself above all. He is also constantly about change. He wants to also change times and laws - as Antichrist wanted to supersede G-d and change the laws (making himself equal with G-d). We can see that Islam constantly wants to change the whole world to one world government under "Sharia law" (one world religion that is administered under an Islamic state that only acknowledges Allah and Mohamed as its prophet). Daniel 7:25 - And he shall speak great words against the most High, and shall wear out the saints of the most High, **and think to change times and laws:** and they shall be given into his hand until a time and times and the dividing of time. The true definition of Islam is not really a religion. The government enforces religion on its entire people for the mere purpose of converting them to seek salvation only by the submission of death (as aforementioned, the word "Islam" means submission). Islam is not a religion; it is actually a holy declaration of war! Additionally the only way a Muslim can reach salvation is by them dying for the cause of Allah. They become in a way the Messiah/G-d incarnate. In other words, they claim to be G-d (or an incarnation of G-d). 2 Thessalonians 2:4 - He will oppose and will exalt himself over everything that is called G-d or is worshiped, so that he sets himself up in G-d's temple, proclaiming himself to be G-d. However, for the Judeo Christian, remember, Jesus wanted us to offer our body as a "living sacrifice", not as a dead sacrifice. Romans 12:1-2 - I beseech you therefore, brethren, by the mercies of G-d, that you present your bodies a **LIVING SACRIFICE**, holy, acceptable to G-d, which is your reasonable service. And do not be conformed to this world, but be transformed by the renewing of your mind, that you may prove what is that good and acceptable and perfect will of G-d. 1 Peter 2:5 - you also, like living stones, are being built into a spiritual house to be a holy priesthood, offering spiritual sacrifices acceptable to G-d through Jesus Christ. Matthew 12:7 - If you had known what these words mean, I desire mercy, not sacrifice, you would not have condemned the innocent. This is trickery all the way around… as one can see Satan did a first-rate number on a people at attempting to spin this truth. Let us be able to clearly see that the one that causes death is the Antichrist (even from the beginning in the garden).

If you want to know what business, the Antichrist is involved in now. Follow where all the confusion and corruption is happening in the world, both throughout history and at the present moment (this will follow through to the future). The Antichrist is called the King of Babylon mentioned in Isaiah 14:12-23 (which is ancient Iraq). This also includes all the Arab and Muslim countries. They claim Islam is a peaceful religion but overlooking the fact that Muhammad turned out to be a prophet of doom. This peculiar mind-set should have us all questioning, because no leader in today's world can literally come out and say that Islam is a violent religion, not even the Pope without uproar and threats of demise mainly from the Arab nations as well as other nations were Arabs reside. There is a reason for this uprising in attitude for those who talk down on Islam … because the prophesized passage about the Antichrist from the Old Testament tells us that by peace the Antichrist will deceive many, most likely by saying that he is about peace, but in reality he is really about war. Matthew 24:4-5 - And Jesus answered and said unto them, Take heed that no man deceive you. **For many shall come in my name, saying, I am Christ; and shall deceive many.** Daniel 8:25 - And through his policy also he shall cause craft to prosper in his hand; and he shall magnify himself in his heart, **and by peace shall destroy many**: he shall also stand up against the Prince of princes; but he shall be broken without hand.

I am not saying that the Muslims or their religion is the actual Antichrist in disguise, but I do think that they play hand in hand and exist together for malevolent and uncanny purposes. Remember, like the Father, the Son and the Holy Spirit, Satan too uses the trinity (3 in one nature) as a matter of organizing himself. And I believe that somehow these 3 elements will play out together and finally culminate as one, revealing the Antichrist. Revelation 19:20 - Then the **beast (antichrist)** was captured, and with him the **false prophet** who worked signs in his presence, by which he deceived those who received the mark of the beast and those who worshiped his image. **These two** were cast alive into the lake of fire burning with brimstone. Revelation 20:10 - **The devil,** who deceived them, was cast into the lake of fire and brimstone where the **beast and the false prophet** are. And they will be tormented day and night forever and ever.

Now, according to this passage (adjoining below in red) all the Arab/Muslim nations seek to destroy Israel from the face of the Earth! In addition, since America is a democracy that was originally born from the roots of Israel, they are now considered by the Muslim nations to be part of Israel. This also goes for any other country who supports Israel and to the fanatical Muslims these countries and their people must be destroyed! It has always been this outward "fanatical" agent that we witness from the ideologies of the Muslim people that involve destruction and corruption. Psalm 83:1-18 - O G-d, do not keep silent; be not quiet, O G-d, be not still. See how your enemies are astir, how your foes rear their heads. With cunning they conspire against your people; they plot against those you cherish. **Come, they say, let us destroy them as a nation, that the name of Israel be remembered no more.** With one mind they plot together; they form an alliance against you- the tents of Edom and the Ishmaelites, of Moab and the Hagrites, Gebal, Ammon and Amalek, Philistia, with the people of Tyre. Even Assyria has joined them to lend strength to the descendants of Lot. Selah Do to them as you did to Midian, as you did to Sisera and Jabin at the river Kishon, who perished at

Endor and became like refuse on the ground. Make their nobles like Oreb and Zeeb, all their princes like Zebah and Zalmunna, who said, Let us take possession of the pasturelands of G-d. Make them like tumbleweed, O my G-d, like chaff before the wind. As fire consumes the forest or a flame sets the mountains ablaze, so pursue them with your tempest and terrify them with your storm. Cover their faces with shame so that men will seek your name, O Lord. May they ever be ashamed and dismayed; may they perish in disgrace. Let them know that you, whose name is the Lord—that you alone are the Most High over all the earth.

Muhammad himself said in Volume 4, Book 52, Number 176 and 177 of the Koran: The hour and day of judgment shall not come to pass until the tribes of Islam defeat the tribes of Israel in Jerusalem and the surrounding nations and than the stones will cry out, there is a Jew hiding behind me, come O Muslim, (slave of Allah), come and kill him till not one Jew is left. The only way the Muslims know that the end is coming is when they destroy all the Jews in the surrounding nations. In other words, what the Antichrist did was brilliant! He was crafty all right! He caused confusion; he took biblical prophecies and corrupted, spun and tweaked them to make them for Islam (he did this by reversing the scriptures). The Muslims according to the Koran see the Jews as the bad guys, while on the other hand it is the Jews that are told through biblical scripture that the Muslims are the bad guys that will eventually be destroyed by G-d, as witnessed in Zachariah in the Old Testament it tells us (in future tense) that the surrounding Arab Muslim nations will come to defile and kill the Jews and G-d will put an end to them. Zachariah 14:1-20 - A day of the Lord is coming when your plunder will be divided among you. I will gather all the nations to Jerusalem to fight against it; the city will be captured, the houses ransacked, and the women raped. Half of the city will go into exile, but the rest of the people will not be taken from the city. Then the Lord will go out and fight against those nations, as he fights in the day of battle. On that day his feet will stand on the Mount of Olives, east of Jerusalem, and the Mount of Olives will be split in two from east to west, forming a great valley, with half of the mountain moving north and half moving south. You will flee by my mountain valley, for it will extend to Azel. You will flee as you fled from the earthquake in the days of Uzziah king of Judah. Then the Lord my G-d will come, and all the holy ones with him. On that day there will be no light, no cold or frost. It will be a unique day, without daytime or nighttime, a day known to the Lord. When evening comes, there will be light. On that day living water will flow out from Jerusalem, half to the eastern sea and half to the western sea, in summer and in winter. The Lord will be king over the whole earth. On that day there will be one Lord, and his name the only name. The whole land, from Geba to Rimmon, south of Jerusalem, will become like the Arabah. But Jerusalem will be raised up and remain in its place, from the Benjamin Gate to the site of the First Gate, to the Corner Gate, and from the Tower of Hananel to the royal winepresses. It will be inhabited; never again will it be destroyed. Jerusalem will be secure. **THIS IS THE PLAGUE WITH WHICH THE LORD WILL STRIKE ALL THE NATIONS THAT FOUGHT AGAINST JERUSALEM:** Their flesh will rot while they are still standing on their feet, their eyes will rot in their sockets, and their tongues will rot in their mouths. On that day men will be stricken by the Lord with great panic. Each man will seize the hand of another, and they will attack each other. Judah too will fight at Jerusalem. The wealth of all the surrounding nations will be collected, great quantities of gold and silver and clothing. A similar plague will strike the horses and mules, the camels and donkeys, and all the animals in those camps. Then the survivors from all the nations that have attacked Jerusalem will go up year after year to worship the King, the Lord Almighty, and to celebrate the Feast of Tabernacles. If any of the peoples of the earth do not go up to Jerusalem to worship the King, the Lord Almighty, they will have no rain. If the Egyptian people do not go up and take part, they will have no rain. The Lord will bring on them the plague he inflicts on the nations that do not go up to celebrate the Feast of Tabernacles. This will be the punishment of Egypt and the punishment of all the nations that do not go up to celebrate the Feast of Tabernacles. On that day **HOLY TO THE LORD** will be inscribed on the bells of the horses, and the cooking pots in the Lord's house will be like the sacred bowls in front of the altar. Every pot in Jerusalem and Judah will be holy to the Lord Almighty, and all who come to sacrifice will take some of the pots and cook in them. And on that day there will no longer be a Canaanite in the house of the Lord Almighty.

So, as we can clearly see, it is the Muslims (Antichrist) who are the corrupters of the Bible. The corrupters are calling the innocent corrupters, and everything is turned around and in reverse! Let us not be confused here in knowing which religion is the correct religion. The correct religion is obviously the one that came first, which is the Old Testament followed by the New Testament that prophetically feeds off the Old! The correct religion is the one religion that tells us that we are of a living spirit of salvation (Jesus), rather than the submission spirit of death (Antichrist).

We can also witness just what Antichrist has done with the Muslim religion by reversing, turning, and twisting religious aspects around. This is exactly what happened when Adam and Eve sinned in the garden, which brought about entropy. Satan said to Eve, you should not surely die. Genesis 3:4 - You will not surely die, the serpent said to the woman. We witness this with Mahmoud Ahmadinejad the 6th president of Iran. He wants to completely annihilate the Jewish nation of Israel. He believes that once he destroys the Jewish nation, the holy day that Mohamed spoke about will have come. This is the Muslim version of the Messiah coming. Without the Old Testament Bible we can never understand the nature of Antichrist or understand why Arabs hate Jews, Israel, America or any other country or persons that represents or backs those countries, or why there is hostility, hate, antagonism, or bad feelings between the races. This is because Antichrist is involved. I cannot stress this enough!

This same aspect also follows through with the land of Israel since the Arabs claim that it is Palestine because they occupied that land first. Therefore, lets set the record straight on this as well. The truth is, there never was a Palestinian people, nation, religion, language, or even culture (there is no differentiating them from the surrounding Arab milieu). The claim of the Palestinian people who claim to have lived long ago in the land of "Palestine" is a deliberate lie! (Again fashioned by Satan). This is because the land was "originally called Canaan" and it was inherited and inhabited by Canaanites (which were ancient Jews). And it was the land of Canaan that later became the land of Israel. Palestinians by birth are Arabs, the name "Palestine" comes from the "Philistines" who were ancient Israel's chief enemies. The switchover from the land of Canaan to Palestine reluctantly transpired in 130AD when the Romans took over Jerusalem and rebuilt a pagan city with an idol temple to Jupiter in the place where the Jewish temple once stood. In the interim of this, the Romans renamed the land of Israel "Syria Palestina" (after the Philistines), and that is where the name Palestine comes from. From that time on Jews, living there became known as Palestinians. I am a Palestinian, but don't like the name. Palestine is a name the Romans gave to Israel with the express purpose of infuriating defeated Jews… Why should we use a spiteful name meant to humiliate us? - Golda Meir However Israel's claim to the land goes back 4,000 years to Abraham's purchase of the cave of "Machpella". There Abraham buries his wife Sarah, and later Isaac, Rebecca, Jacob, Leah are also buried there, and eventually himself. This site is not related to Arabs or Islam.

Furthermore, concerning the Islamic people one of their holiest sites sits in Jerusalem in place where the Jewish Temple once stood. This site is called the Dome of the Rock mosque and within it is said to be the topmost part of Mount MORIAH where Abraham was directed by G-d to offer his son Isaac as a sacrifice. It was also this same place (directed by G-d) that King David latter purchased to build the first Jewish Temple. Mohamed came later in history and melded the two monotheistic religions of the day together, which are Judaism and Christianity. He rewrote and adjusted them to his own narrative, which can now be found in the Koran (known as Islam). Although here is the kicker in all this… according to the Muslim/Islamic tradition Mohamed took a ride on his divine horse (or angel) and flew at night from the Dome of the Rock mosque and ascended up to heaven. But it was originally (and ultimately) the Jewish tradition that told about the temple mount being the gates of heaven that enabled the Muslims centuries later to develop the tradition about Mohamed going up to heaven from that very spot. Nevertheless, in truth Mohamed never set a foot in the holy land or saw with his eyes the temple mount.

And this is where Satan sneaks his way into the social and political world stage, feeding off the conflict (that he created) between the Jews and the surrounding Arab and Muslim (Islamic) nations. As it looks now, he will work his way through the UN (United Nations).

The United Nations officially came into existence on the 24th of October 1945. It is a group of people from several leading countries that come together for several purposes. These purposes are to bring all nations of the world together to work for peace and development as well as to maintain (and keep) international peace and security throughout the world and to develop friendly relations among all nations, as well as to cooperate in solving international, economic, social, and cultural issues. Their principles are supposed to be based on justice and human dignity with the motto for the well-being of all people. This includes helping people live better, as well as eliminating poverty and disease, which includes stopping environmental destruction. There are currently just fewer than 200 Members of the United Nations. Their main headquarters is in New York City. They are a world parliament all to themselves, and make decisions that become resolutions. In addition, you can bet that the UN had their part in the establishment of the state of Israel, which took place in November 1947. The resolution that they had called for (number 181) was for both a Palestinian state as well as Jewish state. It also called for a free capital of Israel. However, the city of "Jerusalem" is what is considered a district with Israel on one side, and a Palestinian state on the other - shared by both. The term for a "district" in this respect is considered an area such as a town or country that has been divided into geopolitical (and cultural) areas, especially one with a particular distinguishing national aspect like Jerusalem that is holy to three religions (Judaism, Islam and Christianity among others). Therefore, Jerusalem became a "district" under the leadership of the UN (United nations). The resolution (number 181) stated: Independent Arab and Jewish states, and the "special international regime" for the city of Jerusalem shall come into existence and Palestine. The UN petitioned Jerusalem and Palestine to both share Jerusalem, giving less of it to the Jews and more to the Arabs, but with the stipulation that the city of Jerusalem is to continue to remain under Jewish control. So what does all this mean? This means that the UN has within its power the authority and control to take over the city of Jerusalem, especially during a time of great turmoil, instability, and disorder between Jews and Arabs. Zechariah 12:2-3 - Behold, I will make Jerusalem a cup of trembling unto all the people round about, when they shall be in the siege both against Judah and against Jerusalem. And in that day will I make Jerusalem a burdensome stone for all people: all that burden themselves with it shall be cut in pieces, though all the people of the earth be gathered together against it. And it is at this point in time that the UN will opt for Jerusalem to become an "international capital". The conflict between the Jews and Arabs has long held many (if not most) of the UN deliberations (discussions). For that reason, it is eventually in the plans for the UN to take over Jerusalem (both securing and acquiring it) in which they will become the ruling power over both states, and it will be at this point in time that both states will have lost their sovereign rights. I have a gut feeling that this is exactly what will happen in the middle of the tribulation period that will allow the Antichrist to come in and take charge. He will tell the people that they are no longer ruling separately… that Israel and Palestine now fall under the jurisdictions of the UN. Therefore, the head honcho of the UN council (whomever that might be at the time) is going to rule over both of them. This head honcho will be the leader as the antichrist. Whoever that might be at the time, we cannot say? But since it was the United States that had originally helped to found the UN and continues to retain a vital stake in advancing it, it is a good estimation (and certainly possible) that the President of the United States might step into this place to play this part. Incidentally, President Obama is the first U.S. leader to head a United Nations Security Council meeting (which was on the reduction of nuclear weapons). President Obama led the meeting because the United States holds the revolving presidency of the Security Council in September 2009. Now as I stated earlier in this chapter, I am not saying that President Obama is the Antichrist, but we must certainly keep our eyes open to the matter, as it just may be an insight for another to come. In addition, as I also said previously, what we need to do is continue to watch all prestigious people that come to power; this might also include the Russian president.

However, the UN as some may think is a very good organization moving in favor of the people however, history has shown us differently, that the UN council has not always been so fair. For some unwarranted reason, they have always been anti-Israel. This outlook arose simply because Israel went to war against its Arabs neighbors without the UN's approval. The reason Israel did this was because there were rumors that the Arabs were planning a sneak attack on them. Because of this, the retaliation for Israel defending herself against the Arabs resulted in 61 nations breaking diplomatic relations with her. However, none of these nations, including the UN had reprimanded the Arabs for their sneak attack on Israel. In fact, the UN has condemned Israel over 300 times for defending herself, but has never once condemned the terrorists and aggressors who attack her. What's more, the UN even supported and sheltered PLO terrorists in Lebanon and gave them military equipment and training as well as assisted terrorists squads on there way to attack Israel. Therefore, not everything is so cut and dry with the UN as we can see, and it should be no surprise to us that this behavior will be repeated again.

Amidst all this, the Jews eventually intend to build a 3rd temple in Jerusalem, and along with the temple they fully intend to reestablish the sacrificial system. This includes every devout Jewish household sacrificing a lamb. The slaughter in the land would be so horrendous that the world organizations against animal cruelty would be all over this… there is no way the world is going to put up with it! But the Jews fully intend to follow through with it anyways! And we shouldn't be surprised about this either because this is exactly what the Bible says will happen in the end. Daniel 9:27 - And he shall confirm the covenant with many for one week: **and in the midst of the week he shall cause the sacrifice and the oblation to cease,** and for the overspreading of abominations he shall make it desolate, even until the consummation, and that determined shall be poured upon the desolate. Moreover, the Jewish Messiah "who has already come", would never support the reestablishment of the sacrificial system since He had already came as the last sacrificial lamb for all of sin. Rather what the Bible tells us is that the person who resides in the 3rd temple will call an end to the sacrifices and cause the oblation to cease calling himself the sacrificial lamb and the long awaited Messiah. However, this individual will actually be the Antichrist. 2 Thessalonians 2:3-4 - Let no man deceive you by any means: for that day shall not come, except there come a falling away first, and that man of sin be revealed, the son of perdition; Who opposeth and exalteth himself above all that is called G-d, or that is worshipped; so that he as G-d sitteth in the temple of G-d, shewing himself that he is G-d. Revelation 13:15 - And he had power to give life unto the image of the beast, that the image of the beast should both speak, and cause that as many as would not worship the image of the beast should be killed.

Therefore, the Jews, who are waiting for their Messiah, will actually be welcoming the antichrist. 2 Thessalonians 2:9-12 - The coming of the lawless one will be in accordance with the work of Satan displayed in all kinds of counterfeit miracles, signs and wonders, and in every sort of evil that deceives those who are perishing. They perish because they refused to love the truth and so be saved. **For this reason G-d sends them a powerful delusion so that they will believe the lie** and so that all will be condemned who have not believed the truth but have delighted in wickedness. Daniel 9:27 - And he shall confirm the covenant with many for one week: and in the midst of the week he shall cause the sacrifice and the oblation to cease, and for the overspreading of abominations he shall make it desolate, even until the consummation, and that determined shall be poured upon the desolate.

However, not to be confused here since this seven years of duration (when Israel flees) is also noted as the time "after the rapture" (which is the second coming of Jesus when all of the true believers will rise up to join Him in Heaven). However (in future tense) the time after the "rapture" is the time that the Antichrist will come onto the scene, which is specified in the Bible as the "great tribulation" (which is a time of great suffering, difficulty, affliction, and distress).

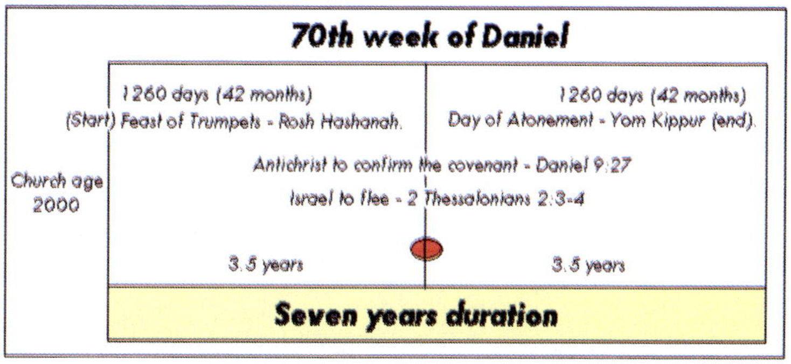

Let me also cite what I wrote in a previous chapter concerning the appointed time of the 2nd coming of Jesus (this will be the Rapture which will occur before the "great tribulation"): A Jewish holiday that corresponds to this is "Rosh Hashanah" which represents the "day of judgment" (it is the Jewish New Year characterized by blowing the Shofar to awaken the listener to coming judgment) followed by 10 days of repentance just before "Yom Kippur" which is the "day of atonement". If effect, G-d gave us this holy event to teach us that we will be judged on Rosh Hashanah and sealed unto the closing of the gates on Yom Kippur. This is where the righteous are separated and rapture up to be with G-d. This holiday is mentioned in Leviticus 23:24 - Speak unto the children of Israel, saying, in the seventh month, in the first day of the month, shall ye have a Sabbath, a memorial of blowing of trumpets, an holy convocation. It also mentions something to this fact in Isaiah 26:2 - The gates of Heaven are opened on Rosh Hashanah so the righteous nation may enter. It also makes mention of this in Psalms 118:19-20 - Open to me the gates of righteousness, I will go into them, and I will praise the Lord. This gate of the Lord, into which the righteous shall enter. Because the gates of Heaven are understood to be open on Rosh Hashanah, this is further evidence that the "rapture of the believers in the Messiah" will take place between Rosh Hashanah and Yom Kippur.

This is also why I believe that the 3rd "authentic" temple is not a temple that is being built on Earth, but rather in Heaven as the Church (the Church is the body of believers). The 3rd temple being built on Earth (in Jerusalem) will eventually be defiled just as the 1st and 2nd temples were (and we should not be surprised about this either). Moreover, in order for the UN to make peace in the area, without destroying the Muslim Mosque, it will decide to build the 3rd temple in the same area just adjacent (bordering) to the dome of the rock Muslim Mosque (which is the slab of stone that was at the top of Mt. Moriah where Abraham was to sacrifice Isaac). It is said that the Ark of the Covenant sat upon this rock in the 1st and 2nd temples. However, due to the peacekeeping problem the third temple will not be built to this same spec's (where the ark of the covenant needs to truly sit). Therefore, technically it will not be a proper (or accurate) temple to begin with. Oh and by the way, some think that the Ark of the Covenant will miraculously come out of hiding and be given back to the Jews to place in the 3rd holy temple. However, according to scripture the Ark is not needed for the earthly 3rd Millennium Temple. According to the prophet Ezekiel, he describes the prefigure Temple in detail in Ezekiel chapters 40-42. And not once does it make mention of the Ark of the Covenant being present in it, rather there is a Holy of Holies, but it is empty - there is also no separating veil. This is because the Bible tells us that there is no more need for the Ark of the Covenant on the Earth. Jeremiah 3:16 - And it shall come to pass, when ye be multiplied and increased in the land, in those days, saith the Lord, they shall say no more, The ark of the covenant of the Lord: neither shall it come to mind: neither shall they remember it; neither shall they visit it; neither shall that be done any more. In short, the glory of G-d has left the Ark of the Covenant on Earth long ago. Even if we find the Ark and open it up, nothing will happen because G-d had left the box centuries ago, and the original prototype of the Ark of the Covenant (at the present moment) rests in Heaven. This is another reason why I believe the 3rd "authentic" temple is a Heavenly structure that will rise out of the sky. Revelation 11:19 - And the temple of G-d was opened in heaven, and there was seen in his temple the ark of his testament: and there were lightnings, and voices, and thunderings, and an earthquake, and great hail. Additionally, if the 3rd temple is not to be built next to the Muslim Mosque (aka, the Dome of the Rock) the Bible tells us that sometime in the future there will be a great earthquake, and all the walls of the city of Jerusalem in the land of Israel will fall down. This will transpire in the beginning of the Great Tribulation period. If this great earthquake occurs, the Dome of the Rock will be destroyed and the new Jewish Temple can than be built in its place. Ezekiel 38:18-20 - And it shall come to pass at the same time when Gog shall come against the land of Israel, saith the Lord G-D, that my fury shall come up in my face. For in my jealousy and in the fire of my wrath have I spoken, Surely in that day **there shall be a great shaking in the land of Israel**; So that the fishes of the sea, and the fowls of the heaven, and the beasts of the field, and all creeping things that creep upon the earth, and all the men that are upon the face of the earth, shall shake at my presence, **and the mountains shall be thrown down, and the steep places shall fall, and every wall shall fall to the ground.**

Now before I end with this chapter I'd like to confer on a lingering subject that has sporadically been mentioned in several other chapters, and this again deals with the scientific notion of "Time Travel". The thought of "Time Travel" has been the talk of many scientists for many years already. In addition, as I had said in several places within this book that "Time Travel" is strictly forbidden by G-d since it would destroy all prophesy. We know that "Time Travel" has not occurred yet on the mere ground that prophesies are being fulfilled (and right on target), which was aptly proven in the previous chapter. "Time Travel" as it stands, is the influence of Antichrist. If such a thing were invented, Antichrist would want it all for himself! Remember, Satan himself appears as an angel of light and rides on the premise of good (things that are helpful and beneficial) to entice his victim. 2 Corinthians 11:14-15 - And no marvel; for Satan himself is transformed into an angel of light. Therefore it is no great thing if his ministers also be transformed as the ministers of righteousness; whose end shall be according to their works. Antichrist compromises his integrity by only telling half-truths and using confusion to make it sound good. "Time Travel" is not good if it is used to change the past, the present, especially if it goes against G-d's prophesy, or His will. Remember we already have that ability to have some bearing on the future, and its called freedom of choice (free will). So how do we absolutely know that the G-d that wrote the Jewish Old Testament Bible is not a Time Traveler Himself? Well, as discussed in the earlier chapter on the Bible Code, G-d gave the gift of Torah (the Bible), which facilitates choice - which not even Satan can change no matter how hard he tries. And we know that G-d is ALREADY omniscient (all-knowing), omnipresent (all-pervading), and omnipotent (all-powerful). He is therefore "BEYOND" the workings of Time Travel, in which He lives outside of time and knows past, present and future alike. Time is in G-d's control; He owns it, while TIME TRAVEL FOR US "MORTALS" IS NOT POSSIBLE. If it were possible, things would literally be chaos, which would mean changing prophecy. Again, G-d would never allow such a thing since everything alluded to and organized in the Bible uses a chronological time scale that engulfs all of history and prophesy. If Time Travel were to exist, it would make all those chronological time durations null and void. Furthermore (as I also said) if Time Travel were possible, it would give Satan a foot in the door to change things to his desires. Thank G-d this is not the case! However, in one full swoop, fallen angels (being deceitful in nature) CAN even take on the form and likeness of our loved ones or someone you recognize, EVEN YOURSELF, and make them appear onto you as Time Travelers. Do not be deceived here!

What becomes so confusing in all of this is that G-d gives us prophecy, but He also gives us "choice". Choices that we can choose from in life. Choice in a way is a kind of "Time Travel". Even dreams can alert us to a future prophetic or probable event. Additionally in a sense, the quantum mechanics of cell phones (particle physics) and satellite technology is also a form of "Time Travel" especially the new science of "particle entanglement" which may also send particles of information at the speed of light across vast distances. However, not to be confused here again… these things are whatever the scientists want to claim they are. It is not in any way MORTAL Time Travel. Incidentally science is ushering in something else to this effect at the (LHC) Large Hadron Collider at CERN (European Organization for Nuclear Research), which is a large cylinder that catches atomic particles and causes them to collide. Here, scientists study these collisions in the hopes of simulating the same conditions of the Big Bang (again the Big Bang is only in theory). In the interim, I've also heard some speculation that smashing these particles together and observing their hidden paths could uncover hidden dimensions and possibly lead to "Time Travel" (though not in the way that we understand it). Scientist seems to be looking for what they call the Higgs Boson, or the G-d particle. A theoretical particle that creates mass. CERN also claims to have created "tiny black holes" that are harmless and evaporate. Again, all these things revolve around Quantum Mechanics. Plainly put, Quantum Mechanics is the law! It engulfs everything from particle physics to the migration of butterflies.

Actually, to ones amazement, what CERN is looking for is in fact right before their very eyes! All they need to focus on is the "path of the particles" since as was shown in an earlier chapter that the life and death of a particle is nonetheless but the "SPIRAL ITSELF". Science still hasn't adequately answered the mystery of where the spiral shape itself comes from, or whose order it is? And as we've since learned it is this "spiral" that creates all the Hebrew letters which G-d used to create all things (therefore, the spiral is G-d's order). The evidence is fleeting right before the scientists eyes at CERN and additionally at Fermi national laboratory, yet they are still entranced and beguiled and led astray by the even smaller elements in the equation.

However, we are not quite yet finished here in regards to "Time Travel". The fallen angels (who are light beings, NOT MORTALS) have access to hidden dimensions to which they might try to abuse in some hostile way towards mortals... However to clarify the matter, these dimensions are not set apart in terms of time dilations... remember they are just filtrations of light that physically (in the flesh) separate us from G-d. Therefore, time remains the same for us in our dimension as it does for the angels in their dimensions. (To learn more about these dimensions, read the book of 2 Enoch).

What the angels CAN DO from their dimension (point of observation) is to see with better peripheral vision since their sight can differentiate through the spectrums of light (since they can easily translate and decipher the Hebrew language of G-d). They have a better marginal view from a higher status so to speak. They can almost look down (as well as up) into the very microcosms of all things, and since they have this higher understanding, they can pretty much estimate or predict outcomes. In a sense, they are almost looking between things and can literally see the peanut butter and jelly between the two pieces of bread. They can therefore estimate the future to a certain degree, but it is not any sort of definite (fixed) future. Nevertheless, their predictions are in vain since there is other ingredients at play here. The Bible Code comes into play as well as the beauty of free will (choice), which is all part of Quantum Mechanics. This was the well-designed system set up by G-d.

These predictions on their part are NOT THE PATH OF RIGHTEOUSNESS since no righteous or G-dly angel would use these means without the authorization of G-d. But yes, the biblical facts remain clear that G-D'S APPOINTED ANGELS do indeed hand over and deliver prophecy - the Bible shows us this in the book of Daniel with the angel Gabriel when he gave Daniel the vision of the Messiah as well as the Antichrist. Likewise, Satan (and his minions) additionally study prophecy to try and figure out what is going to happen and then attempt to launch efforts to try and thwart it - but they can't fully do it.

We clearly witness in several places in the scriptures Satan's attempts to try to manipulate, revert and thwart G-d's plan. We witness this when Satan tried to wipe out the lineage of Shem (the Jews). G-d in essence preserved Noah and his family, which brought forth Noah's son Shem, and it was from the lineage of Shem that King David came about and thus Jesus the Messiah. On top of this, Satan having diligently studied prophesy knew that the Messiah would come into the world as G-d in the flesh, but Satan as we can see is not all-knowing about G-d's plan. And we know this because at the time that Jesus was born, the 3 Wiseman came from the east and asked King Herod about the King of the Jews, in effect upon hearing and fearing this, Herod had all the male children less than 2 years of age killed, but again Satan's plans were thwarted by G-d.

So, as one can see, Satan has been very active on the Earth.

Satan's plans were even thwarted in the book of Esther when Satan attempted to use Haman to kill all the Jews, which G-d by divine intervention also prevented. And it's remarkably spine-chilling that a certain prophetic Bible Code exists concerning this very happenstance. This code even foretold of Hitler's reign and the Holocaust. It comes from the book of Esther and the story of Purim regarding evil Haman who wanted to annihilate the Jewish people. When Haman was killed and the Jews were allowed to defend themselves, Queen Esther was asked by her husband the king what else he could do for her, and she said tomorrow hang the 10 sons of Hamen. Esther 9:6-14 - And in Shushan the palace the Jews slew and destroyed five hundred men. And **Parshandatha,** and **Dalphon**, and **Aspatha**, And **Poratha**, and **Adalia**, and **Aridatha**, And **Parmashta,** and **Arisai,** and **Aridai,** and **Vajezatha**, **The ten sons of Haman** the son of Hammedatha, the enemy of the Jews, slew they; but on the spoil laid they not their hand. On that day the number of those that were slain in Shushan the palace was brought before the king. And the king said unto Esther the queen, The Jews have slain and destroyed five hundred men in Shushan the palace, and the ten sons of Haman; what have they done in the rest of the king's provinces? now what is thy petition? and it shall be granted thee: or what is thy request further? and it shall be done. Then said Esther, If it please the king, let it be granted to the Jews which are in Shushan to do to morrow also according unto this day's decree, and **let Haman's ten sons be hanged upon the gallows.** And the king commanded it so to be done: and the decree was given at Shushan; and they hanged Haman's ten sons.

Now if one opens a "Hebrew Bible" (the Torah/Tenach) to the book of Esther you will notice that the 10 names of Haman's sons can be found in the Hebrew texts "one on top of the other" just like they were hung (seen in the picture below). In this ancient text, you can also find little mini letter versions sandwiched in-between these 10 names, which were found to symbolize 10 "FUTURE" sons of Haman!

How is this so? Well in 1946, the Nuremberg trial put 11 Nazis on trial which ended in conviction and death sentences. One of them named "Hermann Goering" cheated the gallows and committed suicide before he could be hung. Instead 10 "MODERN DAY" sons of Haman who wanted to annihilate the Jewish people where hung for their crimes. The last one of these 10 was a man by the name of "Julius Streicher"; he was the main man of the Nazi regime! The news week magazine of October 28, 1946 reports foreign affair page 45 (also found within the New York Times on October 16th 1946). At the end of this article is one paragraph that reads**:** Only Julius Striecher went without dignity. He had to be pushed across the floor with wild eyes screaming hail Hitler, and mounting the steps he shouted out AND NOW I GO TO G-D. He then stared at the witnesses facing the gallows, and shouted **PURIM FEST 1946!**

"Purim fest" is the Jewish holiday that is celebrated today to recount the story of Esther. So when this man Julius Striecher recited "Purim Fest 1946" as his last words, he must have (in some way) either consciously or subconsciously recognized that his own death was the culmination of the story of Purim where the 10 sons of Haman were put to death.

Now when we analyze the number of the year of "1946" in Hebrew, we find that it coincides and equals up (using gamatria) with the ten smaller names that are sandwiched in-between the present names of the 10 sons of Haman back in the biblical day of Esther meaning, the date in which the 10 sons of Haman were hung as well as the future Holocaust murderers that were put to death by hanging. Like I said very spine-chilling! To witness that the past had coincided with the future - hence, another microcosm of sorts.

Let us now examine the text itself found in the book of Esther (chapter 9:6-9), which contains the names of the 10 sons of Haman.

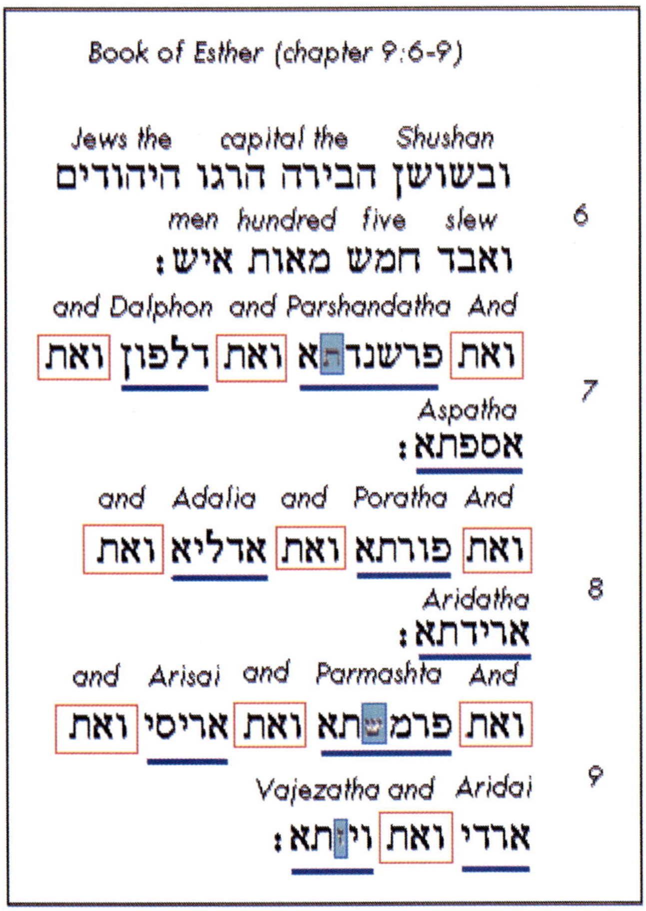

The 10 sons of Haman are underlined in blue. And boxed in red we have the Hebrew word V'at ואת (Vuv, Aleph and Tov), which appears 10 times. This word means "and". This word is used to denote **"replication"** which leads us to deduce that another 10 (future) people were to be hung in addition to Haman's 10 sons. So how do we know exactly who these other (future) 10 people are? Well, when we begin to examine the text in more detail we see that there are also 3 Hebrew letters that are minimized (written smaller) in the text ז ש ת (shown above boxed and highlighted in light blue).

Now these three letters together תשז form the "Jewish year" **5707** (which equates to the year 1946CE in the modern calendar). This is the year that the 10 Nazi criminals were put to death (by hanging in the gallows). The gamatria equivalent to these three letters are Tov =400, Shin=300 and Zayin=7, added together 400+300+7=**707.** These numbers are what are called in gamatria "the absolute values". But the "ordinal" values that we are more familiar with are Tov=22, Shin=21 and Zayin=7. Now if we add the ordinal numbers up together, 2+2+2+1+7=14, and then add the 1+4=**5.**

Now when we put the "ordinal value" in front of the "absolute value" we get **5707,** which is the Jewish year that equates to the modern day calendar of 1946. Even stranger when we add the 5+7+0+7=19, and then the 1+9, it equates out to 10 again (the number of people hung). As if to verify to us that our arithmetic is accurate!

As you can see, the names appear one above the other in the text. This is very remarkable to say the least, since the true prophetic meaning was unbeknownst to the men that actually penned this text at that time, but were somehow led by the Holy Spirit to encode it. What is more, this code is letting us know that it is G-d who is in charge of "TIME" and future events, and not Satan.

It also appears in the Hebrew Torah/Tenach like this:

Esther 9:7

ואת
פרשנדתא ואת
דלפון ואת
אספתא: 8 ואת
פורתא ואת
אדליא ואת
ארידתא: 9 ואת
פרמשתא ואת
אריסי ואת
ארדי ואת
ויזתא:

Yet, if this still were not enough proof to truly hit home for some, additionally found within this very text are other words that also pertain to the Holocaust and Hitler's reign.

Additionally, found within this very text (shown bellow) we can also find the word "hanged" (Hebrew letters: Tov, Lamed, Vuv, Yud), as well as the notorious word "Nazi" (Hebrews letters: Nun, Aleph, Tsadi, yud) can be found at equidistant letter skips.

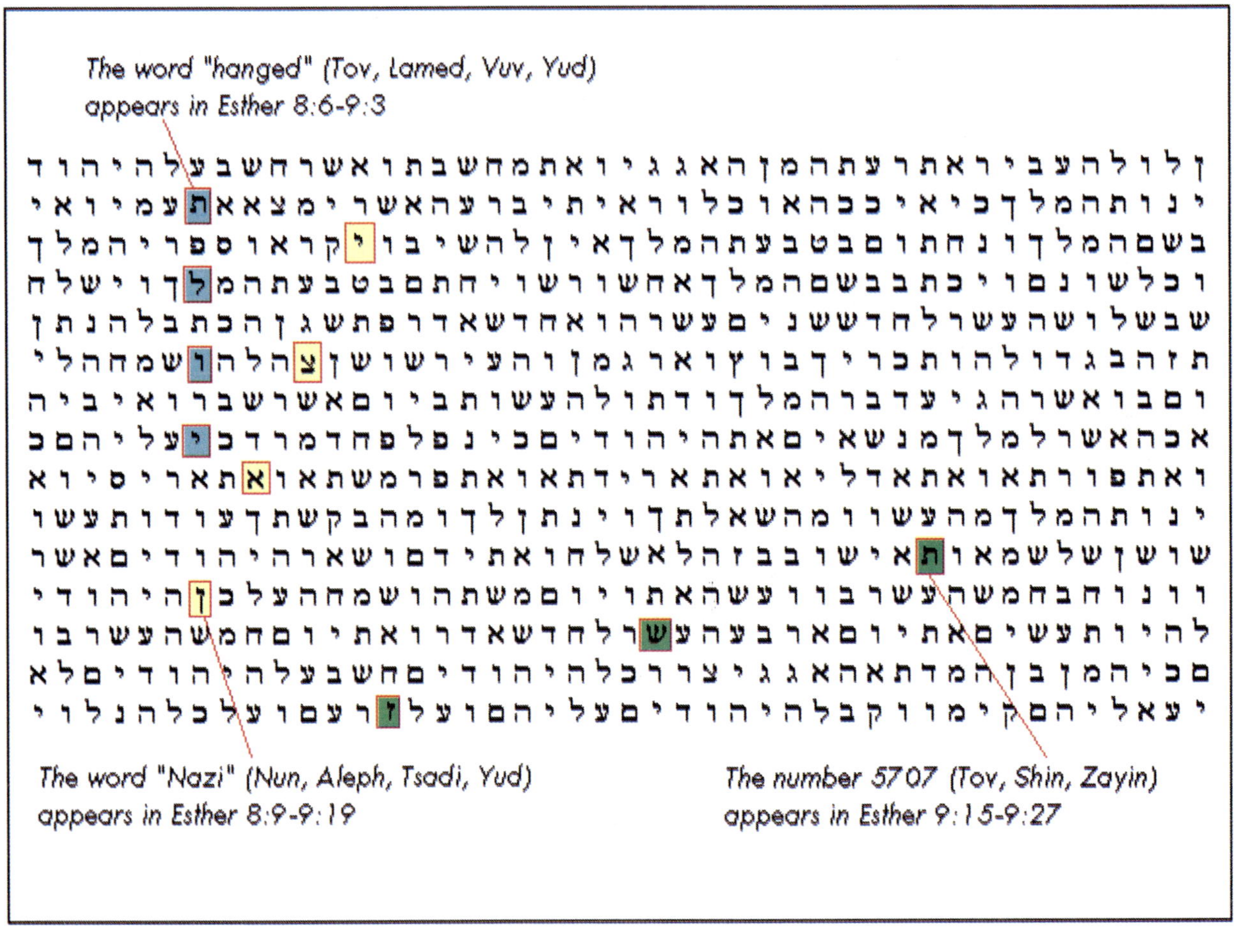

Here we find the Hebrew word “hanged” in very close proximity to the notorious word “Nazi”. The word “Nazi” is known in all other languages as the word Nazi. But during Esther’s time, there was no such word. The odds of finding just the word Nazi alone in this very text is beyond needing any more proof!

There is even more words in this very text that pertain to the Holocaust, Hitler’s reign, and the 10 sons of Haman. Research this for yourself, it is truly amazing!

As I said, this code is letting us know that it is G-d who is in charge of “TIME” and future events, and not Satan. The looking glass that Satan may have is elite. It may even allow him many windows of opportunity to really mess things up. However, no matter how hard Satan tries to defeat, or get one-up on G-d, his plans are always thwarted, this is because nothing can travel faster than the speed of light. Not even the angels who are light beings (Satan included). G-d is supreme, and resides externally (outside) of time and light, since He alone created it. What’s more, “time” is a contingency to the “spiral”, since the structuring of the spiral is due to gravity guiding it to grow in units, (hence units of “time“). Satan meticulously reads this language (and scripture) to try to skew it by setting into reverse motion anti Pi/Phi. But he has yet (hitherto) to learn that report of his demise travels faster! Nothing travels faster than light, with the possible exception of bad news, which follows its own rules - Douglas Adam. In the end, Satan is ultimately thwarted!

The number 7 is G-d's identification (ID) number, which is regarded as the divine number. While the number 6 is the number of man since man was created on the sixth day. The number 7 is also G-d's number since He rested on the 7th day (thus a completion). If there were a limitless number that G-d would choose to reveal His triune nature, 777(s) would be it. Satan is not, nor can he ever be a number 7. He will forever be a number 6, which is man's number, which is one less than a number 7. Satan is a counterfeit; he is literally absorbed in a "Role Playing Game" (RPL). 777 is G-d's number... it represents the trinity (3 part nature) of G-d the Father, the Son and the Holy Spirit. While the numbers 666 is a number that reveals the trinity of Satan (whom we can consider is the anti-Father), the Beast (anti-Christ), and the False Prophet.

When Satan started his Role Playing Game, I think he bargained for a little something more than he thought he would get. There is hidden somewhere in the inner recesses of the Bible Code a predestined coded message within G-d's prophesy forlorn to the date of Satan's termination and demise, (not once, not twice, but TRICE! When this code is finally revealed, it will be a haunting indication to Satan that all his future endeavors are futile (pointless). We can relentlessly look for such a code, but will be unsuccessful. Nobody, not even Satan could guess it even if he lived forever. Nobody knows what it will actually be, only G-d Himself knows, but when He does release this future coded date, it will be verifiably consistent with the dates that are linked to Satan's demise in the past. And at that time, will be eerily recognized.

However to just get a taste that such a code for Satan's demise actually does exist. We can roughly estimate what such a code might be.

We know that the number 7 is intimately tied to G-d as was previously revealed. And we have learned in a past chapter that it is a number that is numerously linked to in the Bible that denotes completion and finalization (such as the Jubilee). 7 IS G-D'S NUMBER! (It is the final say-so).

We can witness the number 7 tip-off and cue straight from the very beginning. Found within the very first documented words from G-d Himself given to Moses starting in Genesis 1:1. Within these very multi-layered words, we can find G-d's all-time enduring signpost, exclaiming: **Hello, I'm G-d!** Genesis 1:1 is not just merely saying**:** In the beginning G-d created the Heaven and the Earth. It is saying much more!

Take a look**:**

> *There are 7 words and 28 letters. The number 28 is also a multiple of 7 since 7x4=28. So as we can see the number 7 is evident from the very beginning of G-d's creation. This is evidently His coded number indicator.*
>
> **בראשית ברא אלהים את השמים ואת הארץ**
>
> *In the beginning G-d created the heaven and the earth.*
> *Barashith bara Elohim et hashamayim ve'et ha'arets.*

An even greater component can be witness "first-hand" which relates directly to the trinity of G-d's 3 part nature (777) when we analyze it in a singular (yet plural) way, by segmenting the very first word of Genesis 1:1.

Take notice that the first 3 letters in Genesis 1:1 (from the box above) are: (Hebrew letters**: B**et, **R**esh, **A**leph).

Now it is these 3 Hebrew letters that each begin with the 3 names of the trinity… G-d the Father (**A**bba), the Son (**B**en), and the Holy Spirit (**R**uach). Incidentally (for a Jew at heart) this little tidbit goes beyond proving the 3-in-one nature of G-d as Jesus the Messiah (G-d in the flesh).

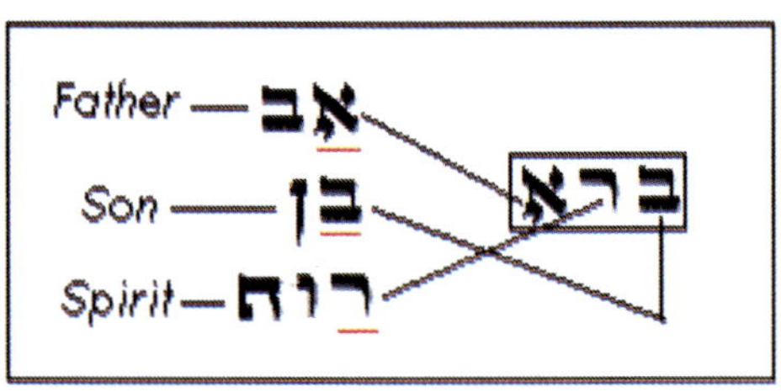

In addition, just when you though you've heard it all already, WAIT, there is even more! What becomes even more interesting in this calculation is that the gamatria value of the individual Hebrew words that read**:** "Father, Son and Spirit" also miraculously equate to the numbers 777. G-d in all His splendorous mathematical ways breaks it down for us very efficiently. And we can begin to see another dexterous triad system at work here (shown below), for there truly is no set way (or real rules) in gamatria to calculate as long as there remains an intelligence and consistency to it. This consistency should be evenly balanced or matched. For instance lets examine the Hebrew letter gematria system in deeper depth**:**

The most common gematria system that is used with the Hebrew letters is the "standard method", but there are several variations that can be used to produce a completely different string of numbers.

This is the "standard method":

100	ק	10	י	1	א
200	ר	20	כ	2	ב
300	ש	30	ל	3	ג
400	ת	40	מ	4	ד
		50	נ	5	ה
		60	ס	6	ו
		70	ע	7	ז
		80	פ	8	ח
		90	צ	9	ט

This is the "ordinal value method":

19	ק	10	י	1	א
20	ר	11	כ	2	ב
21	ש	12	ל	3	ג
22	ת	13	מ	4	ד
23	ך	14	נ	5	ה
24	ם	15	ס	6	ו
25	ן	16	ע	7	ז
26	ף	17	פ	8	ח
27	ץ	18	צ	9	ט

This is the "at bash method": This is where each letter is exchanged with its opposite letter, by substituting the first letter (Aleph) with the last letter (Tov).

4	ק	40	י	400	א
3	ר	30	כ	300	ב
2	ש	20	ל	200	ג
1	ת	10	מ	100	ד
		9	נ	90	ה
		8	ס	80	ו
		7	ע	70	ז
		6	פ	60	ח
		5	צ	50	ט

Moreover, here we can see in the boxed picture below that G-d's seal also pervades creation as though it were woven into the very fabric of nature. It used both the gamatria "standard method" for the words Son and Spirit, and the "at bash method" for the word Father by substituting the values in a reverse manner. This is emphasizing the absolute engulfing and all pervading measures of the Father.

Abba	Father	אב	400+300=700
Ben	Son	בן	2+50=52 and 5+2=7
Ruach	Spirit	רוח	20+6+8=34 and 3+4=7

However, another one takes the cake! In past chapters I've mentioned that the most important verse (prayer) in the Hebrew Old Testament Bible (the Torah) is called the

"Shema". It is found in Deuteronomy 6:4 - Hear, O Israel: The Lord our G-d is one Lord. The passage reads in Hebrew: "Shema Yisroel Adonai Elohenu Adonai Echad." For the Jew this verse solely declares that G-d is a single entity. Here are some other verses that the Jews cling to that declare that G-d is not a multiple G-d, but is a solely a single G-d. Exodus 20:3 - Thou shalt have no other gods before me. Isaiah 44:6 - Thus saith the Lord the King of Israel, and his redeemer the Lord of hosts; I am the first, and I am the last; and beside me there is no G-d. Isaiah 45:5 - I am the Lord, and there is none else, there is no G-d beside me: I girded thee, though thou hast not known me. Isaiah 46:9 - Remember the former things of old: for I am G-d, and there is none else; I am G-d, and there is none like me.

Jewish people reject the Trinity because of Deuteronomy 6:4 as well as the other verses cited above (which is found in the Old Testament). Nevertheless, contrary to these verses by thoroughly examining the Shema in depth, we unravel a whole other expression of its true meaning, which is actually that of a 3-part nature. Thus emphasizing the Trinity (G-d's 3-part nature).

To fully understand this 3-part nature we need to dissect and explore the context of each word used in the Shema. For instance, the Hebrew word "echad" means "one", but there is another word that's found in the Hebrew Bible that also denotes "one" and this Hebrew word is "yachid". The Hebrew word (echad) is plural (which implies more than one) and the other Hebrew word (yachid) is singular (which implies absolutely only one). Lets examine some verses in the Bible that demonstrate the difference between the usage of these two words for us: The first example for the Hebrew word "echad" can be found in Genesis 1:5 - And G-d called the light Day, and the darkness he called Night. And the evening and the morning were the first **(echad)** day. The word "echad" was specifically used here to emphasize that this day consisted of both light and darkness (two parts, but still "one" day). Again, this same precept can be found in Genesis 2:24 - Therefore, shall a man leave his father and his mother, and shall cleave unto his wife: and they shall be one **(echad)** flesh. This is where Adam and Eve (male and female) came together (joined) as "one". Numbers 13:23 - When they reached the Valley of Eshcol, they cut off a branch bearing a single **(echad)** cluster of grapes. Two of them carried it on a pole between them, along with some pomegranates and figs. Here we find the use of the word "echad" that there is "one" cluster, but many grapes. Ezra 2:64 - The whole congregation together **(echad)** was forty and two thousand three hundred and threescore. Here again, we find the plurality of the whole congregation being derived from "one". Jeremiah 32:38-39 - And they shall be my people, and I will be their G-d. And I will give them one **(echad)** heart, and one **(echad)** way, that they may fear me for ever, for the good of them, and of their children after them. Yet again, many people are seen as "one".

Now when we examine how the Hebrew word "yachid" is used in the Hebrew Bible we find that is indicates exactly ONLY "one": Genesis 22:2 - And he said, Take now thy son, thine only **(yachid)** son Isaac, whom thou lovest, and get thee into the land of Moriah; and offer him there for a burnt offering upon one of the mountains which I will tell thee of. Genesis 22:12 - And he said, Lay not thine hand upon the lad, neither do thou any thing unto him: for now I know that thou fearest G-d, seeing thou hast not withheld thy son, thine only **(yachid)** son from me. Genesis 22:16 - And said, By myself have I sworn, saith the Lord, for because thou hast done this thing, and hast not withheld thy son, thine only **(yachid)** son. Proverbs 4:3 - For I was my father's son, tender and only **(yachid)** beloved in the sight of my mother. Psalm 22:20 - Deliver my soul from the sword; my darling **(yachid)** from the power of the dog. Psalm 68:6 - G-d setteth the solitary **(yachid)** in families: he bringeth out those which are bound with chains: but the rebellious dwell in a dry land. Judges 11:34 - And Jephthah came to Mizpeh unto his house, and, behold, his daughter came out to meet him with timbrels and with dances: and she was his only child; **(yachid)** beside her he had neither son nor daughter. Jeremiah 6:26 - O daughter of my people, gird thee with sackcloth, and wallow thyself in ashes: make thee mourning, as for an only son, **(yachid)** most bitter lamentation: for the spoiler shall suddenly come upon us. Amos 8:10 - And I will turn your feasts into mourning, and all your songs into lamentation; and I will bring up sackcloth upon all loins, and baldness upon every head; and I will make it as the mourning of an only **(yachid)** son, and the end thereof as a bitter day. Zechariah 12:10 - And I will pour upon the house of David, and upon the inhabitants of Jerusalem, the spirit of grace and of supplications: and they shall look upon me whom they have pierced, and they shall mourn for him, as one mourneth for his only **(yachid)** son, and shall be in bitterness for him, as one that is in bitterness for his firstborn.

So when G-d instructed Moses to write Deuteronomy 6:4, it is obvious that He intended to use the plural word for "one" (echad) to demonstrate His plurality. (Plural meaning more than one).

All right, so now, we can see that there is plurality in the Hebrew word "echad" which is used in the Shema, but you may also be surprised to find out that there is additionally plurality in the NAME OF G-D (Elohenu) that is used in the Shema (Elohenu is a form of Elohim). Remember from the chapter on the Bible Code how it was shown that the name of G-d (Elohim) was plural. It was found 5 different times in the same verse with the name of G-d (Yahovah) sandwiched in-between. And it was determined that the particular name for G-d that is used as "Elohim" is the name that is used in creating all things by way of a filtering system of light. Therefore, the name of G-d that is used as "Elohim" shows the multiplicity of the Trinity.

So, the question remains, why did G-d choose to have Moses write the name Elohenu instead of Elohim. What is the slight difference of significance here? Well if one takes away the letter M (mem) from Eloh**im** and instead adds an N and a U to it, (nun and vuv shown below) another significant meaning takes hold. If you can also remember, we had learned from a previous chapter that the Hebrew letter "nun" has a mystical representation to it. It not only represents the number 14, that equals King David's name who is the progenitor of the Messiah (Jesus), but the Hebrew letter "nun" is also represented by its shape and form as it resembles a bent over vessel (or a bent over servant). It is this bent over servant that is the receiving vessel - the Messiah. Moreover, something much more significant is being displayed here by the very LAST LETTER of the Hebrew word "Eloha**nu**", which is the Hebrew letter "vuv" pronounced in this case U.

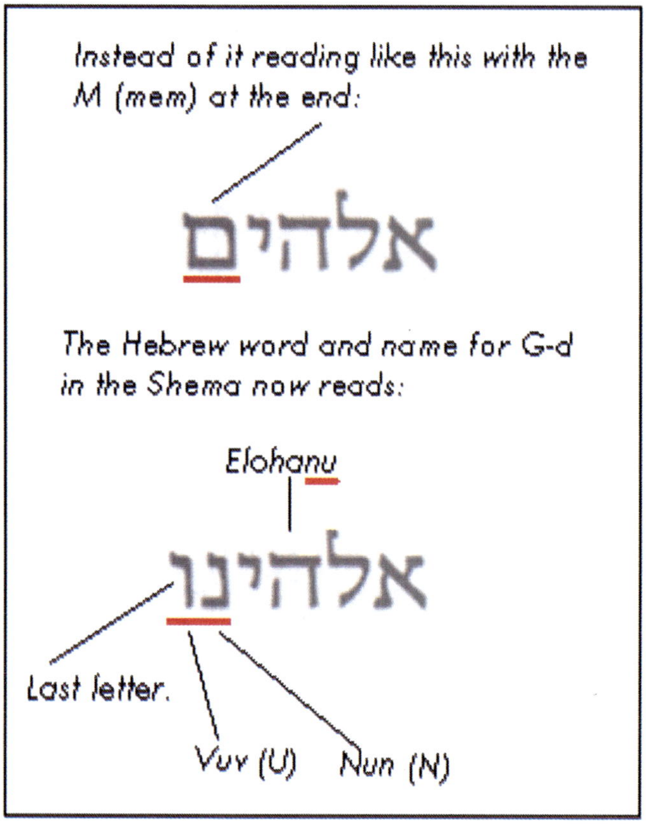

I cannot begin to tell you how significant this is! That the last letter in the name of G-d used in the Shema is the Hebrew letter "vuv". The letter vuv is symbolically known as the "hook"… it is a slightly curved top with a straight line down. And it is by this straight line that light fills all reality - G-d (figuratively speaking) drew down this line of light. Though this line is "singular" in appearance, it embraces two dimensions (an external and an internal), both of which play a part in the process of creation. The external part of the line is the dynamism to separate (set apart) the various aspects of reality, thereby establishing hierarchical order. While the internal line, is the dynamism to reveal the inner inclusion of the various aspects of reality, one in the other, thereby joining them together as a whole? And it is this special quality of the letter vuv that can be referred to as the "go-between". We see this in the very first "vuv" that is used in the Torah. Genesis 1:1 - In the beginning G-d created the heavens and (vuv) the earth. As we can see it is the "vuv" that joins spirit and matter (Heaven and the Earth). Additionally this "vuv" which first appears at the beginning of the 6th word of the Torah also happens to be the 22nd letter of the verse. This alludes to the connective force that interrelates all of the 22 individual letters of the Hebrew alphabet from Aleph to Tov.

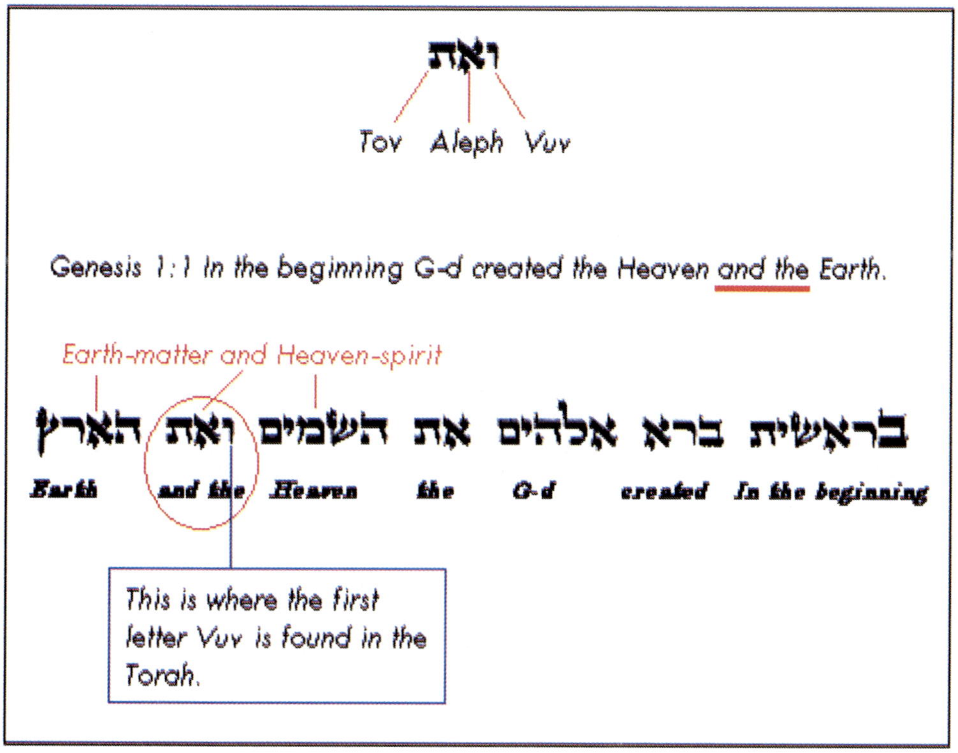

The Hebrew letter "vuv" is not only represented as a hook, but the complete stature of a man standing upright. This man can be known as the Messiah (Jesus), which is the connecting hook of the Tabernacle (representing the place where the presence of G-d is). This man is the connecting link between G-d and man - the Holy Spirit - the Messiah!

Now the question can be appropriately asked. Why would G-d instruct Moses to write Elohanu (the plural form) for His name, and in addition to this sandwiching it in-between two Adonai's? (G-d's other name). Likewise, why didn't G-d instruct Moses to use the singular form of the Hebrew word "yachid" instead of "echad" (which is plural) to establish that He alone is a singular unity? Don't you think that if G-d explicitly wanted to imply that He was a singular G-d He would have straightforwardly done so? It's obvious that G-d's intent here was to show His fused unity (3-part nature) as all being part of His oneness (a unity in plurality). Additionally, within this 3-part nature, the Father, the Son, and the Holy Spirit, they are always operating in complete agreement with each other. Here is a passage in the Old Testament that identifies all three persons of the G-dhead acting together as one**:** Isaiah 11:1-2 - **And there shall come forth a rod out of the stem of Jesse, and a Branch shall grow out of his roots**: And **the spirit of the Lord shall rest upon him**, the spirit of wisdom and understanding, the spirit of counsel and might, the spirit of knowledge and of the fear of the Lord. The Lord… the Branch of Jesse (as the Messiah)… and the Holy Spirit. So in essence, to say that G-d cannot take on a human body (such as Jesus that operated in complete agreement with G-d) is saying that G-d is not transcendent (beyond limits). However, we wholeheartedly know that for G-d to be all-powerful, all knowing, and every at once, He has to be unequivocally without a doubt "TRANSCENDENT"!

I'd also like to concur here that it is also an undeniable fact of physics that nothing on this Earth can exists without 3 constituents (parts) to it! Proverbs 30:18 - There are **three things** that are too amazing for me, four that I do not understand. How is this so? Well for starters, look at any object. For instance lets use the example of a light switch on the wall. Firstly, it is a light switch… secondly, it is attached to the wall... and thirdly, and it has a purpose. Another example would be a car. Firstly, it is a car… secondly, it rolls on the street… thirdly, and it has a purpose. Another example would be a plant. Firstly, it is a plant… secondly, it grows in the ground… and thirdly, it has a purpose. How about the calendar hanging on the wall. Firstly, it is a calendar… secondly, it is hanging (attached somehow) to the wall… and thirdly, it serves a purpose. Another example would be a camera. Firstly, it is a camera… secondly; it is made to be held in the hand or on a tripod or even set on a counter when not in use… and thirdly, it serves a purpose. How about a simple pen. Firstly, it is a pen… secondly, it sits in a cup holder, or a drawer, or perhaps just lays on a desk… and thirdly, it serves a purpose. How about a house. Firstly, it is a house… secondly, it rests upon a lot of land... and thirdly, it serves a purpose. How about a lake of water. Firstly, it is a lake… secondly, it rests upon or within the ground… and thirdly it serves a purpose. Anyone can do this observant experiment. Even an atom can uphold to this principle, since it is an atom (we know they exist)… it is in a given space (though it be invisible to the human eye)… and it serves a purpose. The 3 principle parts to an atom are the proton, neutron, and electron.

Amazingly the pine tree drops pine cones onto the ground, which in turn has a 3-part nature to it, and to which releases seeds which implant into the ground and grow another pine tree, thus the whole process starts over again. Seeds may also serve as food for another living thing.

Thus, the fact of G-d existing as a "3-in one nature" (G-d the Father, the Son, and the Holy Spirit) is an undeniable fact. Remember, everything is a MICROCOSM from above. But not to mistake this "microcosm" as actually being G-d Himself, since He is a separate individual entity that WORKS His ways through the workings of nature (His designed program, which is a filtering system "using the plural name of Elohim" as was discussed in a previous chapter). But G-d in His highest name is "El Elyon" (Lord of Lords), which is described as G-d's greatness over all things, while the name of G-d used as "Adonai" (Yahovah) is beyond nature. Job 31:26-28 - If I have regarded the sun in its radiance or the moon moving in splendor, so that my heart was secretly enticed and my hand offered them a kiss of homage, then these also would be sins to be judged, for I would have been unfaithful to G-d on high. And (as I briefly mentioned in a previous chapter) this can become confusing to some people who are inclined to think in terms of the "new age/new spirituality movement" that has been sweeping the era that serves to reverse our thinking and put into our minds that "G-d is in everything one sees", and in everyone and everything, which eventually leads directly to wrongly thinking that the recognition of G-d is the recognition of oneself, and this is Satan's reverse psychology of deception at work in the minds of humans. Do not get confused here… G-D IS NOT MAN! For this "new age spirituality" teaches man that he is a god and therefore does not sin or do wrong and thus has no need of salvation. The Bible is clear that we are not god(s), nor will we ever be.

One final point is in order here before I end this chapter on revealing Antichrist… Throughout this entire chapter, evil should not be confused with G-d… remember, G-d is not the dramatist of evil. Although, G-d does take the position of being both protector as well as the punisher. However to clarify this, G-d does not cause the punishment, rather G-d just allows the punishment, which is a natural consequence of distance or detachment from G-d or the rejection of Him. We can witness G-d as both the protector and punisher within the Bible in cycles of prophesy, sin and then destruction (prophesy as the protector, sin as the detachment, and destruction as the punisher) (a system of 3 again). Along with this, we should continually bear in mind that in Genesis when G-d created everything, that it was "very good". Genesis 1:31 - G-d saw all that he had made, **and it was very good.** In addition, there was evening, and there was morning—the sixth day. The Bible motto directly tells us that there was no place in G-d's plan for evil in the beginning, and there will be no tolerating it thereof. The sooner we realize this we can begin to strive for something much more divine and significant in life. Romans 8:28 - And we know that all things work together for good to them that love G-d, to them who are the called according to his purpose.

21 - IN CLOSING...

568 pages and almost close to being complete! Time to weigh in here…

The knowledge that one has obtained from reading this book can never be perfectly reiterated to another person the way it was presented in this book. Meaning, that to share this knowledge with somebody outside the bounds of this book, and for them to be able to relate to it all is highly unlikely. One might find that they are scoffed at when they try to share sparse pieces of information from this book in the attempts to try to explain it to another person (especially an atheist or unbeliever) who does not have the privilege of knowing the whole. Therefore, I ask all readers who wish to pass on this knowledge of everything that they have learned in this book, by asking another person to READ THIS BOOK! Pass it on, or give it as a gift to another person. Keeping this knowledge for oneself is a selfish act. I myself give freely of my ideas allowing and encouraging all to share from this book. I am a free and liberated person and I impart my freedom to all. If you will, the creation of this book is in part an illustration of this spirit.

I would give a synopsis of all the chapters written in this book including what was learned from each of them. However, I have chosen not to do so since I must now leave that intuitive impression up to you. I know that there will be many commentators that will pick apart the pieces of this book and throw much darkness on these subjects. However in the end we must realize that we are all just searching for truth, **that which is greater than ourselves... Namely G-d!**

Can one see why I've chosen to name this book the "Primordial Language"?

Can it be said that G-d is certainly provable and believable?

Can it be determined that the Bible is a credible and authentic piece of evidence concerning G-d?

It seems to be the norm these days that in order to get through life it just means being compassionate and doing everything you can to help others. This is the only kind of religion most people think they need. However, this in itself is a woeful testimony that something is missing. Most people don't get involved in discussions of religion or G-d because they feel that there are no valid answers to those questions, the attitude is - so why fool with them. Alternatively, they may feel that they are not knowledgeable enough on the subject. However, after reading this book (in full) hopefully you have discovered that aside from the disarray of this thinking, THIS IS NOT TRUE! Despite what we do not know, there are answers to every single question in the Bible if of course one truly wants to know, or rather if one has the true determination, motivation, desire and willpower to search. The answers to every single question can be found within the Bible (both Old and New Testament). There are no contradictions in the Bible, just supplements to enhance our perspectives and viewpoints - every concordance counts. To find an answer all we need is a single key word to go on. The next step is using an approach of comparisons found within the concordance. 1 Corinthians 2:13 - Which things also we speak, not in the words which man's wisdom teacheth, but which the Holy Ghost teacheth; **comparing spiritual things with spiritual.**

Yet (as we've discovered) despite the Bibles unequalled position as an educational document, the Bible is ignored, sidelined, boycotted and scoffed at. There are still a great many people who fail to realize the effect and full impact it has had on the history of civilization, along with its linkage to science, spirituality and religion. I think its about time we start standing up for it! May this book be a helpful and handy tool in facilitating the Bible.

Even after reading this book, whether or not each of us chooses to believe in a G-d or not is becoming a bit of a dated question. However whether or not you believe, changes are occurring in the world, and in ourselves as we progress from our material state towards our path to spiritual perfection! Those changes are going to lead to a major shift in our thinking, and how we see the world. So whether or not you believe in G-d, science is already ushering in a new era where you will have to come to grips with what lies beyond. One will have to eventually come to grips with the ultimate power in our universe, JUST AS HAS BEEN PROPHESIED - we will once again be reunited with G-d. This will culminate when all the people of the Earth have heard Jesus' message. Matthew 24:14 - And this gospel of the kingdom shall be preached in the whole world for a witness to all the nations, and then the end shall come. 1 Thessalonians 4:16-17 (We will be reunited and caught up with G-d in the clouds) - For the Lord himself shall descend from heaven with a shout, with the voice of the archangel, and with the trump of G-d: and the dead in Christ shall rise first: Then we which are alive and remain shall be caught up together with them in the clouds, to meet the Lord in the air: and so shall we ever be with the Lord. 1 Corinthians 15:51-52 - Behold, I shew you a mystery; We shall not all sleep, but we shall all be changed, In a moment, in the twinkling of an eye, at the last trump: for the trumpet shall sound, and the dead shall be raised incorruptible, and we shall be changed.

Then there is the final draw of Jesus that follows, mentioned in Revelation 22:7-21 - Behold, I am coming soon! Blessed is he who keeps the words of the prophecy in this book. I, John, am the one who heard and saw these things. And when I had heard and seen them, I fell down to worship at the feet of the angel who had been showing them to me. But he said to me, Do not do it! I am a fellow servant with you and with your brothers the prophets and of all who keep the words of this book. Worship G-d! Then he told me, Do not seal up the words of the prophecy of this book, because the time is near. Let him who does wrong continue to do wrong; let him who is vile continue to be vile; let him who does right continue to do right; and let him who is holy continue to be holy. Behold, I am coming soon! My reward is with me, and I will give to everyone according to what he has done. I am the Alpha and the Omega, the First and the Last, the Beginning and the End. Blessed are those who wash their robes, that they may have the right to the tree of life and may go through the gates into the city. Outside are the dogs, those who practice magic arts, the sexually immoral, the murderers, the idolaters and everyone who loves and practices falsehood. I, Jesus, have sent my angel to give you this testimony for the churches. I am the Root and the Offspring of David, and the bright Morning Star. The Spirit and the bride say, Come! And let him who hears say, Come! Whoever is thirsty, let him come; and whoever wishes, let him take the free gift of the water of life. I warn everyone who hears the words of the prophecy of this book: If anyone adds anything to them, G-d will add to him the plagues described in this book. And if anyone takes words away from this book of prophecy, G-d will take away from him his share in the tree of life and in the holy city, which are described in this book. He who testifies to these things says, Yes, I am coming soon. Amen. Come, Lord Jesus. The grace of the Lord Jesus be with G-d's people. Amen. In addition, it is here at this appointed time that the dead are judged. Revelation 20:11-15 (The dead are judged) - Then I saw a great white throne and him who was seated on it. Earth and sky fled from his presence, and there was no place for them. And I saw the dead, great and small, standing before the throne, and books were opened. Another book was opened, which is the book of life. The dead were judged according to what they had done as recorded in the books. The sea gave up the dead that were in it, and death and Hades gave up the dead that were in them, and each person was judged according to what he had done. Then death and Hades were thrown into the lake of fire. The lake of fire is the second death. If anyone's name was not found written in the book of life, he was thrown into the lake of fire.

And it is at this time (the complete and final end) that ALL BELIEVERS and NONBELIEVERS in Jesus (this includes the Jews) WILL ACKNOWLEDGE, CONFESS AND BOW BEFORE JESUS THAT HE INDEED IS LORD. Yes, there will come a time whether on Earth or in hell that the "unbeliever" will acknowledge and finally concede, that Jesus is Lord even though they did not believe. Philippians 2:10-11 - That at the name of Jesus every knee should bow, of things in heaven, and things in earth, **and things under the earth**; And that every tongue should confess that Jesus Christ is Lord, to the glory of G-d the Father. John 12:32 - And I, if I be lifted up from the earth, will draw all men unto me. Ephesians 1:10 - That in the dispensation of the fullness of times he might gather together in one all things in Christ, both which are in heaven, and which are on earth; even in him. Colossians 2:2 - My purpose is that they may be encouraged in heart and united in love, **so that they may have the full riches of complete understanding, in order that they may know the mystery of G-d, namely, Christ.** 1 Corinthians 15:28 - And when all things shall be subdued unto him, then shall the Son also himself be subject unto him that put all things under him, **that G-d may be all in all.**

Following this there is the complete and utter destruction and doing away with all those that are in hell (Revelation 20:11-15). This includes the evil doers and the ultimate unbelievers. John 12:48 - There is a judge for the one who rejects me and does not accept my words; that very word which I spoke will condemn him at the last day. Remember, once one is in hell there is no forgiveness! Hell is place that was designed by G-d for those who no longer serve the purpose for which they were designed, which is to serve, work for, and glorify G-d. Hell is an ultimate, irreversible permanent "separation" away from G-d. This is for people who don't/didn't want anything to do with G-d during their life and rejected His offer of salvation provided though Jesus. Even though they may have done good deeds in their life and innocently denied G-d, they had nevertheless compromised with a godless worldview. 2 Thessalonians 1:7-9 - And to you who are troubled rest with us, when the Lord Jesus shall be revealed from heaven with his mighty angels, In flaming fire taking vengeance on them that know not G-d, and that obey not the gospel of our Lord Jesus Christ: Who shall be punished with everlasting destruction from the presence of the Lord, and from the glory of his power. Matthew 10:28 - And fear not them, which kill the body, but are not able to kill the soul: but rather fear him, which is able to destroy both soul and body in hell.

Going back to the very beginning of the Bible it tells us that after the curse occurred in Genesis G-d had set into action His plan of redemption for fallen mankind. This is the way it had to be, which was to end up being the result of total self-sacrifice with regard to Jesus. This was the only way anything could conclude. G-d would not have us to be ignorant of His divine plan of rescue. This in some sense is really the holy grail of all scripture... the concept and understanding of "redemption" (rescue, liberation, deliverance, salvation, emancipation and recovery) is the most important focal point found in the word of G-d (the Bible) for lost mankind. In fact so inextinguishable is the Bible concerning this message that its duplicate texts have survived all the wars of the world and all the Caesars (thanks to the Jews for preserving it). Matthew 5:18 - I tell you the truth, until heaven and earth disappear, not the smallest letter, not the least stroke of a pen, will by any means disappear from the Law until everything is accomplished. The Bible has also miraculously survived other opposing religions. All of these things have not the staying power of the Bible (G-d's word). And the fact remains undeniable that the Bible continues to be the best selling book of all time. Millions of people are still buying them. One need not wonder why, as it has the power of G-d behind it.

As it happens, there is no redemptive plan available anywhere else in the entire Universe except for on Earth! In addition, there are no aliens anywhere else in the entire Universe either (not even microbes); this is because there is no meaning, sense, or importance for life to exist on other planets. Everything that means anything is right here on Earth because this is the place that G-d had decided to place humans in environmental harmony (synergizing) with all living thing. A place that was specially designed and engineered by G-d to have a perfect recycling energy source that sustain and renews all things. This can be witnessed by the chemical engineering of carbon, oxygen and nitrogen cycles facilitated by plant, land, soil, animal, insect, water and ocean life in combination with the Earths hydraulic system. So important is this particular place to G-d that He sent His one and only Son to die for the releases of sin. We are it, the only ones on this little speck in space! We are all that matters! Therefore, let us be in service (for which we were created) as a workman for G-d in some way. There is no other job or career in the entire Universe as important as this one! It is the only occupation that is worth doing, or that truly counts for anything. Therefore, let us start by doing one act of random kindness at a time.

Some may also inquisitively ask the question for why we need "redemption". Well, to make a long antidote short, redemption is a spiritual law, a scientific law, a physical law, as well as a biological law (all work the same way). However, redemption becomes illuminated by the main ingredient mentioned in the Bible, which is by the biologically genetic blood of Jesus. This is because "biologically" blood is necessary to make things "perfect" in nature by cleansing (sin), since our blood is a life force, which contains DNA that determines everything about us. DNA is the substance that carries genetic information found in all living things, which is responsible for passing along hereditary characteristics from one generation to the next. So in essence, if we are to be made "perfect", it will be by the blood - the perfect blood of Jesus! That's why the Bible continually proclaims to us that it is by the blood (DNA) of Jesus (a perfect sinless human being) that we are made complete again. Ephesians 1:7 - In him we have redemption through his blood, the forgiveness of sins, in accordance with the riches of G-d's grace. 1 Corinthians 13:10 - But when perfection comes, the imperfect disappears. So in short, "redemption" is a biological necessity. It is the law whether or not certain godless people have a different worldview or think otherwise! If I might also add here... some people may wonder (from prophesy) how on earth is G-d going to work the magical reconstruction of resurrecting us from the dead, or rapture us up to be with Him? Well there is a valid scientific answer to this... as it happens, DNA is very significant to G-d in this matter since no two DNA is the same - each persons DNA is a genetic code. Although identical twins have the same DNA (genotype), they have different "phenotypes", meaning that the same DNA is expressed in different ways. Their fingerprints are also different after birth, and the Rh factor (positive or negative) may not be the same even before birth. Additionally identical twins do not always share the same right or left-handedness. Additionally, there is a huge impact the environment plays on the genes, and this includes "spirituality". If one can recall from chapter 17 (Worldwide consciousness) it talks about how ones thoughts can change their DNA. There was the simple rice experiment that showed how good thoughts preserved rice from rotting. However, new studies now show that DNA can be influenced and reprogrammed by words and frequencies much in the same way the rice experiment proved. Our DNA is not only responsible for the formation of our body but it also serves as information storage and for communication. A persons thoughts are an important key to changing their DNA structures. How does the body and brain accomplish this? Well, it stems from the neurotransmitters in our brain that then sends signals to the cells of our body. But before any of this can happen information needs to be fed into the mind by 2 of the 5 senses... sight by seeing the world around us, and hearing such as in the word of G-d. In addition, for the blind by feeling the world around them and by feeling/reading brail about G-d. This might finally explain why biblical doctrine, music, praying, and hypnosis have such strong effects on our minds and bodies.

Again, words are frequency, but when its completely broken-down (deduced), words and frequencies are nothing more than energy and light particles. 1 John 1:5 - This is the message we have heard from him and declare to you: **G-d is light**; in him there is no darkness at all. Can one imagine how easily it would be for G-d to be able to fix and repair our DNA with light? Remember G-d used a filtering process of light when He created all things (as we've learned from chapter 13 on the Bible Codes). Therefore, when it comes time for G-d to resurrect the believers, and glorify (making new) our dead bodies from the grave, G-d could in essence repair genetic defects in the DNA code. For instance, such genetic defects like Downs Syndrome and many others.

Amazingly consistent ole Enoch was privy to how this resurrection (the rapture) of the believers bodies will be done. He describes something that looks to him like long cords. And this is most interesting when we stop to consider that today when scientists look under a high-powered microscope they see that DNA looks like a coiled cord. What is also interesting about DNA is that it is organized in fine long threads in the nucleus and during mitosis the DNA thickens and begins to coils into its distinguishing form. Additionally, DNA's natural color state is translucent. But Enoch saw the vision of these cords (just like the DNA strand) that were given to angels to measure the righteous (all the believers). These "cords" had the special ability to record and bring back living things that were once destroyed. Enoch book of parables 61:1-5 - And I saw in those days how long cords were given to those angels, and they took to themselves wings and flew, and they went towards the north. And I asked the angel, saying unto him: Why have those (angels) taken these cords and gone off? In addition, he said unto me: They have gone to measure. And the angel who went with me said unto me: These shall bring the measures of the righteous, And the ropes of the righteous to the righteous, That they may stay themselves on the name of the Lord of Spirits for ever and ever. The elect shall begin to dwell with the elect, And those are the measures which shall be given to faith And which shall strengthen righteousness. And these measures shall reveal all the secrets of the depths of the earth, And those who have been destroyed by the desert, And those who have been devoured by the beasts, And those who have been devoured by the fish of the sea, That they may return and stay themselves On the day of the Elect One; For none shall be destroyed before the Lord of Spirits, And none can be destroyed. In other words, if one loses a body part such as a limb, or perhaps was an organ donor, or ones body is scattered in various places G-d can reconstruct your body simply by ones DNA code, which if one is a believer their DNA code is recorded in the lambs book of life.

Now to sum it all up... We should be confident to know that the victory will be G-d's (to destroy sin) because He will eventually destroy hell in the end. Then, for the first time, it will be true that there is no DEATH, not because Satan told the truth when he said in Genesis 3:4 - You will not surely die, the serpent said to the woman, but because G-d will literally destroy DEATH! (That was brought about by sin). Then, as we read in the very last book of the Bible (Revelation) that there will be no more death, neither shall there be any more pain, for the former things are passed away. Revelation 21:4 - And G-d shall wipe away all tears from their eyes; and there shall be no more death, neither sorrow, nor crying, neither shall there be any more pain: for the former things are passed away. Then for the first time, there will be NO MORE CURSE. The terrible curse of sin and death (entropy and thermodynamics) that has emerged from the waywardness of sin and rested upon humanity, which has robbed the happiness and peace of all mankind for centuries. But G-d so loved the race of lost and dying sinners that He provided a Savior, the seed of promise who as a precious unblemished lamb, willingly coming to die, and in the most horrific, agonizing and painful way known to mankind by crucifixion, giving His life in sacrifice for the price of redemption (the removal of sin). 1 Corinthians 15:22 - For as in Adam all die, even so **in Christ shall all be made alive.** So pervasive is this promise that the apostle Paul assures us that G-d will destroy death once and for all. 1 Corinthians 15:55 - O death, where is thy sting**? O grave, where is thy victory?**

Now scripture tells us that heaven will be a place that will have an entirely different set of physical laws that are void of thermodynamics and entropy. This is because it is not only void of sin but that there will be no more need of the Sun or the Moon. Revelation 21:23 - And the city has no need of the sun or of the moon to shine upon it, for the glory of G-d has illumined it, and its lamp is the Lamb. In addition to this, it will not have a sea either. Revelation 21:1 - And I saw a new heaven and a new earth; for the first heaven and the first earth passed away, and there is no longer any sea. There will also be NO MORE death, suffering or pain. Revelation 7:17 - For the Lamb in the center of the throne shall be their shepherd, and shall guide them to springs of the water of life; and G-d shall wipe every tear from their eyes. Revelation 21:4 - And He shall wipe away every tear from their eyes; and there shall no longer be any death; there shall no longer be any mourning, or crying, or pain; the first things have passed away. Scripture also tells us that the believers will receive new bodies. 1 Corinthians 15:42-44 - So also is the resurrection of the dead. It is sown a perishable body, it is raised an imperishable body; it is sown in dishonor, it is raised in glory; it is sown in weakness, it is raised in power; it is sown a natural body, it is raised a spiritual body. If there is a natural body, there is also a spiritual body.

This new Heaven implies several things… since there is no more need of the Sun or the Moon this tells us that the laws of physics will be vastly different. This is because without the Sun, Moon or Sea there would be no gravity, and without gravity there would not be the elements of heat and light (this is because G-d Himself will provide the illumination which is a different kind of light). Additionally, no sea implies that there is no more need of a water cycle. That is because the water cycle cannot function without gravity either since it is the Moon and the Sun that tug and pull these forces. Logically speaking, without the Sun, there is no heat and therefore the second law of thermodynamics and entropy is done away with. Without the Sun, stars cannot shine, no living thing can eat food to produce energy and live, and no chemical reaction can occur. Heaven is a place that will be continually sustained by G-d (as He will permanently reside there). Revelation 21:1-7 - And I saw a new heaven and a new earth: for the first heaven and the first earth were passed away; and there was no more sea. And I John saw the holy city, new Jerusalem, coming down from G-d out of heaven, prepared as a bride adorned for her husband. And I heard a great voice out of heaven saying, Behold, the tabernacle of G-d is with men, and he will dwell with them, and they shall be his people, **and G-d himself shall be with them,** and be their G-d. And G-d shall wipe away all tears from their eyes; and there shall be no more death, neither sorrow, nor crying, neither shall there be any more pain: for the former things are passed away. And he that sat upon the throne said, Behold, I make all things new. And he said unto me, Write: for these words are true and faithful. And he said unto me, It is done. I am Alpha and Omega, the beginning and the end. I will give unto him that is athirst of the fountain of the water of life freely. He that overcometh shall inherit all things; and I will be his G-d, and he shall be my son.

Get the message, (can one truly grasp what is going on here?) What John says in Revelation IS TRUE, it can be substantiated by physics today? Now riddle me this… how is it that John knew these scientific and physical laws some 2000 years ago without it really-truly being privy from G-d? The answer is, as you would expect the pieces of the puzzle fit perfectly, and it all makes perfect and logical sense! The Bible remains consistent and scientifically sound. To be more precise, it is scripture not science that is the ultimate test of truth!

The Bible also gives us somewhat of a sequence description of how all these end-time events will transpire. I have already gone through them, but I am doing so again so that the sequence is correctly understood. It goes something like this**:** <u>When we die (the believers) will be with G-d/Jesus in Heaven. We will remain there until the rapture occurs. At the rapture, we are resurrected,</u> (this is where our body is put back together with our spirit and soul). <u>Those that are still alive are translated (instantaneously with their glorified body) to meet G-d in Heaven.</u> The Rapture will occur BEFORE the Tribulation. As a sigh of relief, the Church (body of believers) WILL NOT have to go through the Tribulation period. 1 Thessalonians 4:13-18 - But I would not have you to be ignorant, brethren, concerning them which are asleep, that ye sorrow not, even as others which have no hope. For if we believe that Jesus died and rose again, even so them also which sleep in Jesus will G-d bring with him. For this we say unto you by the word of the Lord, that we which are alive and remain unto the coming of the Lord shall not prevent them which are asleep. For the Lord himself shall descend from heaven with a shout, with the voice of the archangel, and with the trump of G-d: and the dead in Christ shall rise first: Then we which are alive and remain shall be **caught up** together with them in the clouds, to meet the Lord in the air: and so shall we ever be with the Lord. Wherefore comfort one another with these words. John 14:1-4 - Let not your heart be troubled: ye believe in G-d, believe also in me. **In my Father's house are many mansions: if it were not so, I would have told you. I go to prepare a place for you. And if I go and prepare a place for you, I will come again, and receive you unto myself; that where I am, there ye may be also.** And whither I go ye know, and the way ye know. While there in heaven, each one of us is recognized for our accomplishments, and rewards are handed out. 2 Corinthians 5:10 - For we must all appear before the judgment seat of Christ, **that each one may receive what is due him for the things done while in the body,** whether good or bad. 1 Corinthians 3:10-15 - By the grace G-d has given me, I laid a foundation as an expert builder, and someone else is building on it. But each one should be careful how he builds. For no one can lay any foundation other than the one already laid, which is Jesus Christ. If any man builds on this foundation using gold, silver, costly stones, wood, hay or straw, his work will be shown for what it is, because the Day will bring it to light. It will be revealed with fire, and the fire will test the quality of each man's work. **If what he has built survives, he will receive his reward.** If it is burned up, he will suffer loss; he himself will be saved, but only as one escaping through the flames. 1 Corinthians 9:24-27 - Do you not know that in a race all the runners run, but only one gets the **prize?** Run in such a way as to get the prize. Everyone who competes in the games goes into strict training. They do it to get a crown that will not last; but we do it to **get a crown** that will last forever. Therefore I do not run like a man running aimlessly; I do not fight like a man beating the air. No, I beat my body and make it my slave so that after I have preached to others, I myself will not be disqualified for the prize. Then we sit down in heaven with Jesus to celebrate the marriage feast of the lamb. Revelation 19:6-8 - Then I heard what sounded like a great multitude, like the roar of rushing waters and like loud peals of thunder, shouting: Hallelujah! For our Lord G-d Almighty reigns. Let us rejoice and be glad and give him glory! For **the wedding of the Lamb has come, and his bride has made herself ready.** Fine linen, bright and clean, was given her to wear. (Fine linen stands for the righteous acts of the saints). And when the wedding feast is all over Jesus says, it's time, ok lets go! He breaks from Heaven on a great white horse, and we follow behind Him all the way down to the Mount of Olives. Revelation 19:14 - **The armies of heaven were following him, riding on white horses** and dressed in fine linen, white and clean. Zechariah 14:4-5 - **On that day his feet will stand on the Mount of Olives, east of Jerusalem**, and the Mount of Olives will be split in two from east to west, forming a great valley, with half of the mountain moving north and half moving south. You will flee by my mountain valley, for it will extend to Azel. You will flee as you fled from the earthquake in the days of Uzziah king of Judah. Then the Lord my G-d will come, and all the holy ones with him. Jude 1:14-15 - **Enoch, the seventh from Adam, prophesied about these men: See, the Lord is coming with thousands upon thousands of his holy ones to judge everyone,** and to convict all the ungodly of all the ungodly acts they have done in the ungodly way, and of all the harsh words ungodly sinners have spoken against him. When Jesus speaks here at the Mount of Olives, the Antichrist and his evil world empire will be destroyed. After this, we will then reign with Jesus here on the Earth for 1,000 years, which will be flourishing with peace, righteousness, and justice. However, there will still remain a significant population of people on the Earth that will be deceived by Satan and still want to follow evil ways rather than chose to follow Jesus. Revelation 20:7-10 (Satan's Doom) - **When the thousand years are over, Satan will be released from his prison and will go out to deceive the nations in the four corners of the earth—Gog and Magog—to gather them for battle.** In number they are like the sand on the seashore. They marched across the breadth of the earth and surrounded the camp of G-d's people, the city he loves. But fire came down from heaven and devoured them. And the devil, who deceived them, was thrown into the lake of burning sulfur, where the beast and the false prophet had been thrown. They will be tormented day and night for ever and ever. Revelation 20:11-15 (The Dead Are Judged) - Then I saw a great white throne and him who was seated on it. Earth and sky fled from his presence, and there was no place for them. And I saw the dead, great and small, standing before the throne, and books were opened. Another book was opened, which is the book of life. **The dead were judged according to what they had done as recorded in the books.** The sea gave up the dead that were in it, and death and Hades gave up the dead that were in them, and each person was judged according to what he had done. **Then death and Hades were thrown into the lake of fire. The lake of fire is the second death.** If anyone's name was not found written in the book of life, he was thrown into the lake of fire. When this is occurring G-d/Jesus will take us off this Earth and put us in that New Jerusalem for a time being, and from this vantage point we will see the greatest fireworks show of all time - when the Earth is refined with fire. 2 Peter 3:10 - But the day of the Lord will come like a thief. The heavens will disappear with a roar; **the elements will be destroyed by fire, and the earth and everything in it will be laid bare.** And out of this comes the New Heaven and the New Earth, and then in our glorified bodies (that are inside the New Jerusalem) we are lowered down to the New Earth and it is here that we will LIVE FOREVER. 2 Peter 3:13 - But in keeping with his promise we are looking forward to a new heaven and a new earth, the home of righteousness. Isaiah 65:17 - Behold, I will create new heavens and a new earth. The former things will not be remembered, nor will they come to mind. Isaiah 65:20 - Never again will there be in it an infant who lives but a few days, or an old man who does not live out his years; **he who dies at a hundred will be thought a mere youth**; he who fails to reach a hundred will be considered accursed.

Now, for those that are avidly awaiting the "Rapture", this will proceed after "Elul". "Elul" is the Hebrew name for the month which proceeds (comes before) the Jewish New Year Rosh Hashanah. "Elul" takes place 29/30 days right before Rosh Hashanah and is noted as the time when Moses went back up Mount Sinai (for a 3rd time) to petition and plead with G-d to give the Jews another chance after wrongly worshiping the Golden Calf. "Elul" is a very special time... it is a time when G-d is out there "campaigning" (in a manner of speaking) for us to nominate Him as the supreme King. "Elul" is also a month of introspection where we self-examine our feelings, thoughts, and motives. "Elul" is also a "call to action" since it is during this time that we relate individually to G-d. This call is to prepare us to be ready to go out and meet Him in order to go to the "wedding feast", which is the "marriage supper of the Lamb" mentioned in the New Testament book of Matthew where the foolish virgins were unprepared and missed their calling. Matthew 25:1-13 - Then shall the kingdom of heaven be likened unto ten virgins, which took their lamps, and went forth to meet the bridegroom? And five of them were wise, and five were foolish. They that were foolish took their lamps, and took no oil with them: But the wise took oil in their vessels with their lamps. While the bridegroom tarried, they all slumbered and slept. And at midnight there was a cry made, Behold, the bridegroom cometh; go ye out to meet him. Then all those virgins arose, and trimmed their lamps. And the foolish said unto the wise, Give us of your oil; for our lamps are gone out. But the wise answered, saying, Not so; lest there be not enough for us and you: but go ye rather to them that sell, and buy for yourselves. And while they went to buy, the bridegroom came; and they that were ready went in with him to the marriage: and the door was shut. Afterward came also the other virgins, saying, Lord, Lord, open to us. But he answered and said, Verily I say unto you, I know you not. **Watch therefore, for ye know neither the day nor the hour wherein the Son of man cometh.**

In addition to this, it is common practice for a Jewish wedding to recite this Hebrew prayer: Ani ledodi vidodi lee, which means: I am to my beloved (G-d) and my beloved is to me (song of songs 6:3). Amazingly, the significance here is that the "first letters of each word" in this prayer spells "Elul" in Hebrew (Ani (aleph) (E), ledodi (lamed) (L), vidodi (vuv) (U), lee (lamed) (L). This signifies and consecrates not only our relationship with our significant other, but also (in the foremost and highest degree) our relationship with G-d.

We can also look at "Elul" as the preparation time before the Jewish New Year of "Rosh Hashanah". This is when (as we've discussed before in earlier chapters) many believe the Messiah will return to gather and judge, followed by 10 days latter on Yom Kippur where we (the believers in Jesus the Messiah) are forgiven. Starting with "Elul", "Rosh Hashanah" is a time of "repentance", and "Yom Kippur" is the time of "forgiveness". As such, the High Holidays will also inaugurate the final consummation at the end of the age when the Groom (Jesus) returns for His Bride (the Church/body of believers), and ushers in the Messianic Age. To thoroughly understand this, the 30 days of "Elul" are preparation days for the Bride, while (Rosh Hashanah) will be the Rapture (or the catching away) of the Bride.

There is also something else very elusive about the month of "Elul" which is followed by the Jewish holiday of "Rosh Hashanah". "Rosh Hashanah" is the beginning of the New Year, "Rosh" in Hebrew means "Chief or Head", "Ha" means "the", and "Shanah" means "Year". So there we have it, Rosh HaShanah is the Head of the Year. For the record, "Rosh Hashanah" is also known as the BIRTHDAY of the world since after the month of "Elul", it then turns into the month of "Tishrei" (which starts exactly on the first day of "Rosh Hashanah"). Interestingly as it turns out (and not by accident) the Hebrew word "Tishrei" is the very first word of the Bible "Breshith" (Genesis 1:1) SPELLED BACKWARDS. And as so many times before in this book we have already come to know the Hebrew word "Breshith" to mean: "in the beginning". Interestingly enough when you turn the letters around in the Hebrew word "Breshith", this then reads: T(tuv), i (yud), sh (shin), rei (resh), which reads (and means) the first of "Tishrei". So there, we have "Rosh Hashanah" as the beginning of the world - thus creation. This is extremely significant since when 7,000 years rolls around which will be the "Grand Jubilee" (the gathering of the first fruits) as well as "Rosh Hashanah" (the day of judgment). To correlate with this "Rosh Hashanah" which falls in the month of "Tishrei" also happens to be on the 7th month of the Jewish calendar as well, which signifies it as a day of perfection and completion, this will correspond precisely to the "end of the world" followed by "Yom Kippur" (the day of forgiveness). This is the day that G-d makes His decisions, and when "Yom Kippur" is over, all the books are closed, judgment is finished, and everything is sealed. The Hebrew word "Kippur" actually means "a covering", otherwise known as the "ransom by means of a substitute". This can be explained further in the Old Testament. Leviticus 16:8-10 - He is to cast lots for the two goats—one lot for the Lord and the other for the scapegoat. Aaron shall bring the goat whose lot falls to the Lord and sacrifice it for a sin offering. But the goat chosen by lot as the scapegoat shall be presented alive before the Lord to be used for making atonement by sending it into the desert as a scapegoat. Leviticus 16:21-22 - He is to lay both hands on the head of the live goat and confess over it all the wickedness and rebellion of the Israelites—all their sins—and put them on the goat's head. He shall send the goat away into the desert in the care of a man appointed for the task. The goat will carry on itself all their sins to a solitary place; and the man shall release it in the desert. What these biblical verses are saying is… this scapegoat (Jesus) was the goat that all the sins of Israel were placed upon, and when it was lead out of into the desert, it made"atonement" for all Israel (Jesus as the sacrificial lamb). "Yom Kippur" in truth is really the holiest day of all since it is actually on this day that we are united in our heart, mind, and soul with G-d. For it was on this very day that G-d revealed the depth of His relationship with us. What is now essentially required of us is to return to Him, to relate to Him, and to fulfill His wishes, which He related to us in the Bible/Torah. This is a time that we need to recognize our sins (wrongdoings) and ask for pardon (forgiveness), and to come to grips with completely understanding the full purpose of "repentance through the blood (covering/Kippur) of Jesus the Messiah" and to determinedly and resolutely become loyal to G-d, to heed His will, and study His Bible/Torah. The Hebrew letters are bent (and spiral) for a specific and special reason… this reason is to teach us "humility" in which to "bend" our hearts and minds before G-d our creator (as He also did for us). Do you wish to rise? Begin by descending. You plan a tower that will pierce the clouds? Lay first the foundation of humility - Saint Augustine

So, in preparation for this great event… firstly, we need keep our eyes and hearts open for "Elul". For it is that love of G-d for us and us for G-d that is the full retrospective picture of "Elul" (incorporated with the "high holidays"). "Elul" is our preparation time to meet with G-d, which is for the coming of the Messiah, and for the preparation Marriage Supper of the Lamb. Revelation 19:7-9 - Let us rejoice and be glad and give him glory! For **the wedding of the Lamb has come, and his bride has made herself ready.** Fine linen, bright and clean, was given her to wear. Then the angel said to me, Write: **Blessed are those who are invited to the wedding supper of the Lamb!** And he added, These are the true words of G-d. I hope we're all ready for the month of "Elul" and all that it brings. Greater than anything else, the message is clear. It is a love story between our creator and us. Understanding the end conclusion of this "High Holiday" is of vital importance to G-d, that is why the Jews and well as Jesus continually celebrate(d) these holidays because they are a rehearsal (as well as continual reminder and indication) to teach us and prepare us for the end time prophesy. The glorious Day of Judgment and forgiveness for all those that are steadfast in believing.

There is also something else about "Elul" (that is also no coincidence). This has to do with another special season known as "Teshuvah" (which was also briefly discussed in a previous chapter) and to which happens to fall and begin on the first day of the month of "Elul". Now, the Hebrew word "Teshuvah" means to "return or repent" - hence the full meaning of the compressed holiday (Elul, Rosh Hashanah and Yom Kippur). And it is 30 days into "Teshuvah" (on the 1st of Tishrei) that starts Rosh Hashanah. It is also no coincidence that Rosh Hashanah falls on the New Moon (precisely 29.5 days after the last new moon). This means that the New Moon might occur on either the 29th or the 30th, nobody knows for sure the day, the hour, or even really the year (even though it was surmised about in this book), but we can say with confidence that we know the season. Matthew 24:32-36 - Now learn this lesson from the fig tree: **As soon as its twigs get tender and its leaves come out, you know that summer is near.** Even so, when you see all these things, you know that it is near, right at the door. I tell you the truth, this generation will certainly not pass away until all these things have happened. Heaven and earth will pass away, but my words will never pass away. **No one knows about that day or hour,** not even the angels in heaven, nor the Son, but only the Father.

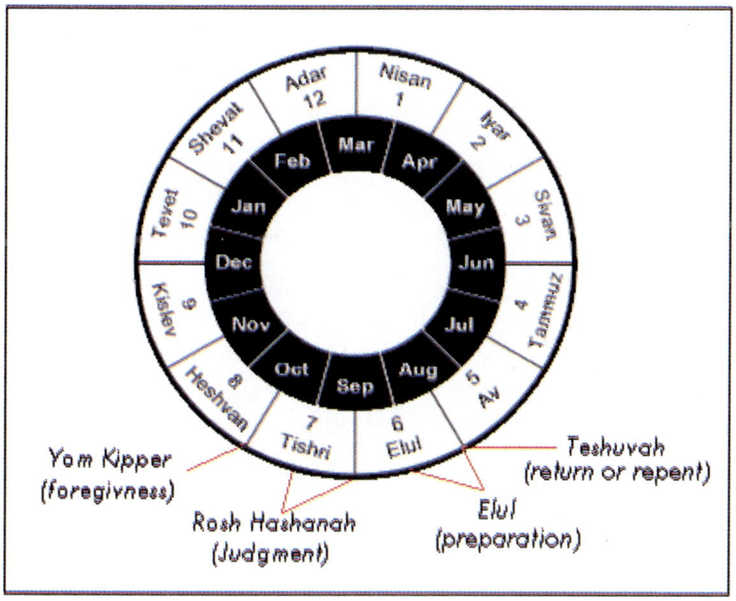

Now some may wonder when the Rapture (catching away) occurs, how this transfiguration will actually happen. I guess the big question is "how" (scientifically) can human flesh be transmuted. This is a very good question, and even I don't know all the answers. I suspect as I previously mentioned earlier that this would be done with the DNA blood of Jesus. However, I have somewhat of another idea of just how this might interrelate to work.

Since there has been a lot of knowledge presented in this book, we (now hopefully the believers) have been elevated to a HIGHER STATE of understanding because we can NOW see the whole picture. Therefore, we can literally say that we are in the process of "raising our consciousness" (synthesizing) about our entire human experience. This higher state of thinking, belief, philosophy, view, and understanding has actually changed (corrected) our DNA. Put into plain language, not only is our place in the Universe becoming clear, we are finally becoming clear about the big picture of our "human story" on this planet, of who we are and from where we originally came from. In addition, we can now hope to see the full spectrum and grand scheme of G-d's creation. More importantly, we are finally gaining back the knowledge of what our true purpose and intent for being alive on this planet is for. Along with this we are beginning to understand just what a state of "lowered consciousness" is, and how we can raise above it by being in a perpetual state of "service" (use or benefit to G-d in some way). In other words, when we have finally figured out our higher-life purpose, we then have to learn to meld this higher-level consciousness into our human-level mind in order to gain the full understanding and purpose of things, which includes the complete understanding of G-d's divine plan of "repentance". What's actually happening here is that we are allowing more light into our bodies (literally emitting a higher electromagnetic vibration). Thus, we are becoming "enlightened" and ready for "Rapture" (the catching away). I'd like to make clear that there's no mystery to the "Rapture" reserved for the chosen few. Either you have this innate frequency in your heart, mind and soul to ascend, or you don't. Those that have been diligently watching and waiting for the day and the hour for Jesus' return will instinctively know with irresistible love, affection, pledge, and devotion (as a Bride awaits for her Groom).

Thus, the transforming of our "understanding" will then transform our outer-shell (DNA) of reality into a "glorified body". This will be the first time people will be able to "ascend with their physical body intact". This means we will shift (raise our consciousness) by reconnecting to our higher-self while still in our physical body without needing to die a physical death. Jesus possessed this higher-level consciousness. This is because He clearly understood who He was, His mission, and His full purpose. This is why He said, "the kingdom of heaven is within". But not to be confused here... Jesus was the one who paved the way for us, and it is only through Him that we might ascend. In addition, I must make clear that this state of mind and affairs will only be opened to those who truly want to go to "Heaven".

Now there is also a huge misconception of what some seem to think Heaven will be like. Some think that this is a place where we will be flapping our wings and living in frivolous royalty. Yes, it will be magnificent and majestic, but it is truly a place of rest and respite (merely a place to take a breather). Some people think that even though they were genuinely good in all respects of life which may include serving in the military, never committing a crime, recycling to help save the Earths resources, being exceptionally good to family members, giving money away, or helping those in need or trouble can buy them a free ticket to Heaven. Yes, these things are good, however, I will flat-out tell you that these type of things WILL NOT buy your way into Heaven. Heaven isn't really, what most of us perceive it to be, even for the believer. Firstly, in Heaven there will be all the believers of all the ages of time. This will include all the G-dly people who have done what G-d required and expected of them. These people are those that love G-d, serve G-d, respect and revere G-d, and those who were absolutely committed to doing what G-d wanted them to do during their disposition of time (life) on Earth. Heaven is explicitly reserved for those who are under the blood of Jesus (accredited to as righteousness)... for only with this are they imputed in righteousness... THIS IS THE ONLY "RIGHTOUSNESS" THAT G-D WILL ACCEPT!

Now for the believers... firstly, we are going to Heaven to be judged, to give an accounting to G-d for everything that we've done in this body and life. This will be a joyous time, but also a time of tears! (Tears of joy and glory that is), also regretful and apologetic tears for all the times we have failed G-d during our life. Heaven for the believer is a place were one goes to serve G-d, and unless you are prepared to take up into Heaven with you the same religious disposition in the continual service of G-d (and mind you, like doing it), which includes serving, and more serving (not only serving G-d, but along with this helping others in some way, if you are not this type of serving person then this place (heaven) is not for you. And it is only these types of people that will eventually inherit and end up in the new heavenly city of Jerusalem, which will be on the New Earth. Why would G-d give the New Earth to people that don't care about anything else except for themselves? He will only give the New Earth to people that are truly saved under the blood of Jesus. These people will have compassion for others, they will be in a perpetual state of service to others, they will care and nurture the land, not seeking to build, destroy or pollute, these people will by all means be farmers and carpenters. They will also be kind to animals (not seeking to hurt or kill them, but rather to preserve them in their natural habitats. These people are those that when an animal or even an insect gets in their house they seek to catch and release, instead of kill. What will the houses be like? They will be gloriously made from stone and mud mortar (plentiful things of the earth), but very little wood will be used as it means killing trees. There will be no manufacturing companies anywhere to be found. And we will make due with natural resources just as the people of old did. But don't let this be discouraging because the people that will inherit the New Earth will be like biblical scientists in their own right and they will seek the natural ways simply because they understand the chemical construct of the earth, oh and they will like, enjoy (and prefer) this lifestyle (and take benefit from it). You must be the change you wish to see in the world - Mahatma Gandhi.

Some people also have the biblical misconception that they are going to go to Heaven and live forever. Nevertheless, this is not what the Bible says. Psalm 115:16 - The heaven, even the heavens, are the Lord's: but the earth hath he given to the children of men. We are not destined (or intended) to go to heaven and live forever. This was not G-d's original intent for us when He had originally created Adam on theEarth. The heavens are the Lord's, He has reserved this place for Himself, but the Earth hath He given to the children of men and we are coming back with Jesus to reign on the Earth (the New Jerusalem - the holy city of G-d) for ever and ever where there will be no more death, and it will be reestablished like the day of Adam and Eve before they had sinned. We've come full-circle around.

Finally in closing… In writing this book, it has left me with an inspirational sense of hope. It is a message of such immense importance worth leaving behind, (not only for oneself), but for those who come after us. Beyond all else, I feel something more powerful is taking place... It is a story of the future, of something trying to happen, a transformation of consciousness, biblical knowledge, culture, and society, a new perspective of G-d attached to the understanding of the world around us... A mystical experience in which the spirit is exalted to knowledge of divine things. And most importantly, LIBERATION! That, complete liberated freedom is when you can turn-around and liberate all the elements of your world from their earthly material state, and make them transcendent as well. This has been another one of my goals in writing this book. Not just for me though! G-d has used me to work His way, and for that I am deeply honored, grateful, thankful, gratified and even indebted that little ole me could be of some use in G-d's grandiose plan. It is also no accident that you've picked up this book to read at this particular time in your life. I would like to emphasize that everything a person experiences in life is for a divine purpose, (remember coincidence is not a kosher word). This also includes everything a person sees - they are meant to see and everything that a person hears - they are meant to hear (this also takes into account evil as a way of learning right from wrong). All knowledge is important in putting the pieces together. Whenever something happens that makes a particularly strong impression on a person, that person needs to be aware that this experience was custom created by G-d specifically for him/her in order to give direction and insight in fulfilling the divine mission. The fact that I'm writing this together with all who read is surely significant. Again, nothing is ever an accident in life. If I had told a different story, my life would not have taken the purposeful driven path that it had, and you would not be reading these words today. In addition, mind you, there were many MANY deceiving and evil influences surrounding me throughout my life and even in the making of this book. Several times had this book file on my computer become corrupt, and many times did my computer become infected with viruses, and twice it had completely gone kaput. Additionally my electrical power in my house went out along with my internet connection on several occasions for freaky reasons. All of these mishaps stealing valuable time as a deterrent to misdirect me. In addition, it was now when things in my life took a heart falling toll on me. It almost seemed that literally everything, in in every which way, was out to get me. Yet G-d willingly, I persevered and prevailed to present before you this book!

My message to all who read**:** Only he who sees takes off his shoes! Exodus 3:5 - Do not come any closer, G-d said. Take off your sandals, for the place where you are standing is holy ground. It is more important than ever that you notice where "YOU" are going and to watch for those signposts and significant messages along the path of your life, and to sometimes permit those messages to change your life - heed the whispers, LEARN THE LANGUAGE! - Anita Meyer

Tam ve'nishlam - It is finished - the gift is given.

I wish to pass this book onto my family, my children, and their children's children, and to the future generations of humankind - with G-d speed!

Authors biography:

Anita Meyer was born in Augsburg Bavaria Germany on June 29 1967. She was born on an American Army base where her father was serving in the US army. She graduated from Maine East High School in Park Ridge Illinois where many other prominent people like the one Hillary Clinton attended, and whose insights and ideologies were shaped by the many excellent and worldly-minded teachers. Anita's younger years can be characterized by the desire to study her Jewish heritage, which resulted in additionally attending Hebrew school and latter living in Israel for a year. While in Israel she learned all about Christianity, which strongly continues with her throughout her later years. She also attended college for two years in the field of Criminology. Though she did not follow through in this field, within her still remained the cunning knack and burning desire to investigate and thoroughly examine evidences of a different kind - namely for G-d's existence. Call her a religious procurement specialist if you will. She currently resides in Kenosha Wisconsin, is married and has two children.

Dear Readers:

Anita Meyer, welcomes your comments and she will communicate with you personally about any questions pertaining to topics within this book.

For those that, wish to communicate with Anita Meyer, please feel free to contact her at the following email:
anitameyer1@hotmail.com

In Search Of The Universal Truth Publisher
Book Ordering Form

The Primordial Language
Confirmation of the Divine Creator
by
Anita Meyer

ISBN 13: 978-1-61500-002-9
LCCN: 2009939397
584 pages 7 ¾ x 10 ¾ Trade Paperback

Genre: Religion / Spiritual / Science
Ages: 14 years and up

Release Date: December 20, 2009

The Primordial Language Confirmation Of The Divine Creator	Retail Price	10% off	Qty.	Total
	$38.95 USD	$35.05 USD		$
Shipping Price $6.00 USD per Book			Shipping Total	$
			Grand Total	$

Make payable to: In Search OF The Universal Truth Publisher
Send order with Check or Money Order to:

In Search Of The Universal Truth Publisher
8824 E. Florida Ave. Unit G-13
Denver, CO. 80247

Shipping Address:
Name____

Address____

City____

State / Province ______________________________ Zip / Postal Code _______________

E-Mail (optional)______________________________________

Books will be shipped out same day as we receive your orders.

You may also order this book through our online website bookstore:
www.insearchoftheuniversaltruthpublisher.com/bookstore/html.